*AIR DISPERSION MODELING*

# *AIR DISPERSION MODELING*

Foundations and Applications

**ALEX DE VISSCHER**
*University of Calgary*

Published by John Wiley & Sons, Inc., Hoboken, New Jersey.
Published simultaneously in Canada.



*Library of Congress Cataloging-in-Publication Data*

De Visscher, Alex, 1970–
Air dispersion modeling : foundations and applications / Alex De Visscher, Canada Research Chair in Air Quality and Pollution Control Engineering, Department of Chemical and Petroleum Engineering, and Centre for Environmental Engineering Research and Education (CEERE), Schulich School of Engineering, University of Calgary.
pages cm
Includes bibliographical references and index.
ISBN 978-1-118-07859-4 (hardback)
1. Air–Pollution–Simulation methods. 2. Atmospheric diffusion–Simulation methods. I. Title.
TD883.1.D45 2013
628.5′3011–dc23

2013009404

10 9 8 7 6 5 4 3 2 1

*To Julie*

# CONTENTS

# PREFACE

Air dispersion modeling has become one of the main tools in the academic study of air quality and in environmental engineering practice. It is a key element in most environmental impact assessments. Despite its importance, the field has not seen any new comprehensive textbooks in more than a decade. During the last decade, the air dispersion modeling regulations of the U.S. Environmental Protection Agency (EPA) have undergone some drastic changes. Gone is the stability-class-based approach of ISC3, which is the main subject of most air dispersion modeling textbooks. The new EPA standard in Gaussian dispersion modeling is the similarity-theory-based model AERMOD. With these changes, the level of expertise that is expected of air quality professionals today is dramatically different from the expectations just a decade ago.

Historically, air dispersion modeling has always been a field that was mainly populated by meteorologists, and most air dispersion models used today were developed by meteorologists. However, the users of these models are often engineers with no training in meteorology, who speak a scientific language very different from the language of meteorologists, and this has proven to be a barrier that often hampers the development of the field.

The purpose of this book is twofold. First, it is meant as a textbook for graduate students in environmental engineering and meteorology who wish to specialize in air dispersion modeling. Second, it is meant as a reference book for professionals in air quality. A complaint I have heard from many professional air dispersion modelers, even with a meteorology background, is that they feel they have an insufficient understanding of the models they are using. Enlightening those people was a major motivation for writing this book. Boundary layer meteorology is usually the stumbling block. There are some excellent textbooks in this area, but these are not entry-level courses. They are hard to understand without a substantial amount of background knowledge.

To serve all the audiences envisioned for this book, I have tried to strike a balance between mathematical rigor and intuitively clear explanations. This had some repercussions on the choice of notation. Because of the varied background of the audience, I chose to avoid vector notation and Einstein's index notation. This choice makes many of the equations longer than strictly necessary, but easier to understand.

The structure of this book was also chosen to accommodate a varied audience. After the Introduction (Chapter 1), the reader is offered a "primer" (Chapter 2) containing a simple but functional air dispersion model. This gives the reader an early sense of achievement but also a sense of what is to follow in the rest of the book. Readers who are already familiar with Gaussian dispersion modeling may prefer to skip this chapter.

Air dispersion modelers should have some understanding as to what pollutants to look for, and why, and this background is given in Chapter 3. Chapter 4 touches upon the subject of regulation.

The meteorology that bridges the gap between the engineers and the meteorologists is given in Chapter 5. It is the backbone of the book, and I recommend any professional in the field not trained in meteorology to treat this chapter as required reading. The six chapters that follow discuss various aspects of air dispersion modeling, including Gaussian, (stochastic) Lagrangian, and Eulerian approaches. Most readers will not be interested to read all of them, but rather select those chapters that align best with their modeling needs. I tried to make each of these chapters stand on its own, but they all assume Chapter 5 to be read and digested. Throughout these chapters numerous examples are included to show how the theory is applied in practice.

Chapter 12 offers some practical information that did not find its way in any of the other chapters. Chapters 13–16 provide detailed descriptions of the models ISC3, AERMOD, and CALPUFF and a brief introduction to the model CMAQ. These chapters refer extensively to Chapters 5–11, so readers will be able to put these models in perspective.

For graduate courses in air dispersion modeling, I would recommend teaching Chapters 1–6, selected topics of Chapter 7, and a selection of the remaining chapters, based on time availability, research needs, and professional needs of the region. If the audience is not well-versed in meteorology, instructors are encouraged to spend at least one third of the course on Chapter 5.

Many of the calculations outlined in this book are included in files that can be found online. These are Excel and Matlab files, and they are referred to in the book as they are encountered and summarized at the end of each chapter. Readers are free to use these files as they please and are encouraged to experiment with them. However, acknowledgment of the source is required if the files are distributed in any way, including the incorporation into other software, or if results from these files are reported or published. The files can be found by visiting http://booksupport.wiley.com and entering the ISBN.

Writing this book has been an intense but enjoyable experience, and I have many people to thank for their support. I would like to thank the University of Calgary for granting me a sabbatical leave, which gave me the time to write this book. I am very indebted to Prof. Wolfgang Voigt and Prof. Martin Bertau, who made it possible for me to spend my sabbatical at their institution, the TU Bergakademie Freiberg, Germany. This book grew out of the lecture notes on air dispersion modeling I accumulated over the years, and I would like to thank all the students who took this elective course, as they were the testing ground for the presentation of the material. In particular, Zoe Pfeiffer offered many practical comments, Ishpinder Kailey tirelessly pointed out typos and errors in my handouts, Scott Fraser taught me many things about geospatial data handling, and Guiqin Li, one of the brightest graduate students I have ever dealt with, is now more of an advisor to me than the other way around. Many other professionals in "downtown" Calgary have offered me useful advice over the years, and I thank them all for their efforts. Of those I would like to thank Christian Reuten, whose monthly gatherings have offered me opportunities to solicit feedback on my book idea, Michael Zelensky, who shared some useful flare modeling ideas with me, and modeling veteran Mervyn Davies, who offered me custody over an immensely valuable air dispersion data set that unfortunately did not make it into this book due to time constraints, but that will undoubtedly be the subject of future research and model refinements. I'd like to thank my colleagues Danielle Marceau, Ann-Lise Norman, Jalal Abedi, and Sheldon Roth for collaborations and useful discussions on the topic of air dispersion modeling. Discussions with colleagues at other universities, like John D. Wilson and John Feddes (University of Alberta, Canada) and Matthew Johnson (Carleton University, Canada) have influenced the way I think about air dispersion modeling. I took my first steps as an air dispersion modeler in the mid-1990s under the supervision of Herman Van Langenhove (Ghent University, Belgium), and solidified my knowledge in a CALPUFF course by Joe Scire. I would not have been able to write this book without the lessons learned from these people. Last but not least, my beloved wife and amazing artist Julie Hunter Denoncourt, who believed in my book writing abilities from the very beginning, has always been there with support and advice, and put up with a very absent-minded husband at times. Many, many thanks for everything.

# LIST OF SYMBOLS

| | |
|---|---|
| $a$ | acceleration (m s$^{-2}$) |
| | activity (–) |
| $A$ | droplet surface area per unit air volume (m$^2$ m$^{-3}$ = m$^{-1}$) |
| | frequency factor [s$^{-1}$ (first order); cm$^3$ molecule$^{-1}$ s$^{-1}$ (second order)] |
| | absorbance (–) |
| $A_f$ | frontal area (m$^2$) |
| $A_p$ | plan area (m$^2$) |
| $A_T$ | total area (m$^2$) |
| $A_s$ | stack cross-sectional area (m$^2$) |
| $B$ | Bowen ratio (–) |
| $b_a$ | empirical constant (m$^2$ μg$^{-1}$) |
| $c$ | speed of light in vacuum (m s$^{-1}$) |
| | amount concentration or mass concentration (mol m$^{-3}$ or g m$^{-3}$) |
| $C$ | Cunningham slip correction factor (–) |
| | number concentration (molecules cm$^{-3}$) |
| $c_0$ | gas-phase concentration in equilibrium with the surface (μg m$^{-3}$) |
| $C_0$ | Lagrangian structure function constant (–) |
| $c_{air}$ | air-phase pollutant concentration (μg m$^{-3}$) |
| $C_D$ | drag coefficient (–) |
| $C_G$ | net radiation fraction going into the ground (–) |
| $c_{i,air}$ | air-phase concentration of pollutant at interface (μg m$^{-3}$) |
| $c_{i,water}$ | water-phase concentration of pollutant at interface (μg m$^{-3}$) |
| $c_m$ | mass concentration (g m$^{-3}$) |
| $c_{odor}$ | odor concentration (ou m$^{-3}$) |
| $c_p$ | specific heat (J kg$^{-1}$ K$^{-1}$) |
| $c_{p,air}$ | specific heat of air (J kg$^{-1}$ K$^{-1}$) |
| $C_{vis}$ | visual contrast (–) |
| $C_{vis,x=0}$ | visual contrast at zero distance (–) |
| $c_w$ | droplet-phase pollutant concentration (μg m$^{-3}$) |
| $c_{water}$ | water-phase pollutant concentration (μg m$^{-3}$) |
| $C_x$ | $yz$ plane integrated plume concentration (g m$^{-1}$) |
| cov() | covariance (product of units of correlating functions) |
| $d$ | displacement height (m) |
| | day of the year (d) |
| $d_p$ | particle diameter (m) |
| $D$ | molecular diffusivity (gas phase) (m$^2$ s$^{-1}$) |
| $D_{ij}$ | binary diffusion coefficient of compound $i$ in compound $j$ (m$^2$ s$^{-1}$) |
| $D_w$ | molecular diffusivity (water phase) (m$^2$ s$^{-1}$) |
| $\overline{e}$ | turbulent kinetic energy (m$^2$ s$^{-2}$) |
| $E$ | activation energy (J mol$^{-1}$) |
| | irradiance (W m$^{-2}$ or W cm$^{-2}$) |
| $e_p$ | photon energy (J photon$^{-1}$ or simply J) |

$E_p$ photon irradiance (photons $cm^{-2}$ $s^{-1}$ or simply $cm^{-2}$ $s^{-1}$)
$E_{p,\lambda}$ spectral photon irradiance (photons $cm^{-2}$ $s^{-1}$ $nm^{-1}$ or simply $cm^{-2}$ $s^{-1}$ $nm^{-1}$)
$E_\lambda$ spectral irradiance (W $m^{-2}$ $nm^{-1}$ or W $cm^{-2}$ $nm^{-1}$)
$f$ Coriolis parameter ($s^{-1}$)
$F$ deposition flux (μg $m^{-2}$ $s^{-1}$)
broadening factor in falloff kinetics (–)
$f_0$ reactivity factor (–)
$F_b$ buoyancy flux parameter ($m^4$ $s^{-3}$)
$F_c$ "complementary" (stochastic) force (N)
falloff parameter (–)
$F_d$ pollutant flux at droplet surface (μg $m^{-2}$ $s^{-1}$)
drag force (N)
$F_m$ momentum flux parameter ($m^4$ $s^{-2}$)
$f_y$ dispersion factor in the $y$ direction (–)
$f_z$ dispersion factor in the $z$ direction (–)
$Fr_H$ Froude number for hill height (–)
$Fr_W$ Froude number for hill width (–)
$g$ acceleration of gravity (m $s^{-2}$)
$h$ Planck constant (J s)
effective source height (m)
hour angle (rad or °)
vertical thickness scale (m)
denser-than-air plume height (m)
$H$ enthalpy (J)
height of a hill (m)
height of a building (m)
dimensionless Henry constant (–)
$h_a$ average plume height (m)
$H_c$ height of the cavity (depends on location) (m)
$H_{cp}$ Henry constant on a concentration/pressure basis (mol $m^{-3}$ $Pa^{-1}$)
$h_d$ dividing streamline height (m)
$h_{fh}$ horizontal flame extent (m)
$h_{fv}$ vertical flame height (m)
$h_{mix}$ mixing layer height (m)
$H_r$ height of obstacles defining roughness (m)
$H_R$ height of the cavity at its maximum (m)
$h_s$ source height (m)
$h_t$ altitude of terrain (m)
$H_W$ height of the far wake (m)
$h_*$ vertical height scale (m)
$I$ actinic flux (photons $cm^{-2}$ $s^{-1}$ $nm^{-1}$ or simply $cm^{-2}$ $s^{-1}$ $nm^{-1}$)
ionic strength (mol $kg^{-1}$)
$i_{background}$ light intensity of background
$iF$ intake fraction (–)
$i_{object}$ light intensity of object
$i_u$ turbulent intensity in the $x$ direction (m)
$i_v$ turbulent intensity in the $y$ direction (m)
$i_w$ turbulent intensity in the $z$ direction (m)
$J$ mass flux (moving frame of reference) (g $m^{-2}$ $s^{-1}$)

| | |
|---|---|
| $j_i$ | photochemcial rate constant of reaction $i$ ($s^{-1}$) |
| $k$ | von Kármán constant (–) |
| | Boltzmann constant (J $molecule^{-1}$ $K^{-1}$ or J $K^{-1}$) |
| $K$ | equilibrium constant (–) |
| $k_0$ | apparent second-order rate constant in low-pressure limit ($cm^3$ $molecule^{-1}$ $s^{-1}$) |
| $k_G$ | air-phase mass transfer coefficient (m $s^{-1}$) |
| $K_h$ | turbulent heat diffusivity ($m^2$ $s^{-1}$) |
| $k_i$ | rate constant of reaction $i$ [($s^{-1}$ (first order); $cm^3$ $molecule^{-1}$ $s^{-1}$ (second order)] |
| $k_L$ | water-phase mass transfer coefficient (m $s^{-1}$) |
| $K_m$ | turbulent momentum diffusivity ($m^2$ $s^{-1}$) |
| $K_s$ | solubility constant (–) |
| $K_x$ | turbulent mass diffusivity in the $x$ direction ($m^2$ $s^{-1}$) |
| $K_y$ | turbulent mass diffusivity in the $y$ direction ($m^2$ $s^{-1}$) |
| $K_z$ | turbulent mass diffusivity in the $z$ direction ($m^2$ $s^{-1}$) |
| $k_\infty$ | rate constant at high pressure limit ($cm^3$ $molecule^{-1}$ $s^{-1}$) |
| $L$ | Obukhov length (m) |
| | length of a building, projected along the wind direction (m) |
| $l$ | path length (m or cm) |
| $L_f$ | flame length (m) |
| $L_I$ | integral length scale (m) |
| $L_R$ | length of the cavity (m) |
| LAI | leaf area index (–) |
| $m$ | mass (kg) |
| $m_a$ | mass of ambient air (kg) |
| $m_w$ | mass of water vapor (kg) |
| $M$ | molar mass (g $mol^{-1}$ or kg $mol^{-1}$) |
| $\mathbf{M}$ | variance–covariance matrix of velocity fluctuation matrix ($m^2$ $s^{-2}$) |
| | material balance matrix ($s^{-1}$) |
| $M_{air}$ | molar mass of air (g $mol^{-1}$ or kg $mol^{-1}$) |
| $M_c$ | pollutant molar mass (kg $mol^{-1}$) |
| $M_u$ | mass exchange rate constant between layers ($s^{-1}$) |
| $n$ | amount of substance (i.e., number of moles) (mol) |
| | fractional cloud cover (–) |
| | integrated stochastic perturbation function (m $s^{-1}$) |
| $N$ | mass flux (fixed frame of reference) (g $m^{-2}$ $s^{-1}$) |
| $N_A$ | Avogadro constant (molecules $mol^{-1}$ or $mol^{-1}$) |
| $N_{BV}$ | Brunt–Väisälä frequency ($s^{-1}$) |
| $N_i$ | flux of $i$ in a fixed frame of reference (mol $m^{-2}$ $s^{-1}$) |
| $n_{iA}$ | order of reaction $i$ with respect to compound A (–) |
| ou | odor unit |
| $p$ | pressure (Pa = kg $m^{-1}$ $s^{-2}$) |
| | probability density function of location and velocity (s $m^{-2}$) |
| $P$ | probability density function (inverse of units of the distributed variable) |
| $p_0$ | pressure at the surface (Pa) |
| $p_c$ | critical pressure (Pa) |
| $p^*$ | nondimensionalized pressure (–) |
| $PM_{2.5}$ | mass concentration of particulate matter smaller than 2.5 μm diameter (μg $m^{-3}$) |
| $q$ | heat (J = kg $m^2$ $s^{-2}$) |
| | sensible heat flux (J $m^{-2}$ $s^{-1}$) |

| | |
|---|---|
| | water/air mass ratio (–) |
| | pollutant mass fraction (–) |
| $Q$ | emission rate (g s$^{-1}$) |
| $q_A$ | anthropogenic heat flux (J m$^{-2}$ s$^{-1}$) |
| $Q_c$ | heating value (kW) |
| $q_G$ | heat flux into the ground (J m$^{-2}$ s$^{-1}$) |
| $q_i$ | mass fraction of compound $i$ (–) |
| $q_L$ | latent heat flux (J m$^{-2}$ s$^{-1}$) |
| $Q_{mdi}$ | downward mass flux between two layers (kg m$^{-2}$ s$^{-1}$) |
| $Q_{mi}$ | exchanged mass flux between two layers (kg m$^{-2}$ s$^{-1}$) |
| $Q_{mui}$ | upward mass flux between two layers (kg m$^{-2}$ s$^{-1}$) |
| $Q_s$ | sensible heat emission (kW) |
| $r$ | albedo (–) |
| $r_a$ | aerodynamic resistance (s m$^{-1}$) |
| $r_{ac}$ | surface canopy resistance (s m$^{-1}$) |
| $r_b$ | quasi-laminar resistance (s m$^{-1}$) |
| $r_c$ | surface resistance (s m$^{-1}$) |
| $r_{cf}$ | foliar resistance (s m$^{-1}$) |
| $r_{cg}$ | ground resistance (s m$^{-1}$) |
| $r_{cl}$ | resistance of leaves, twigs, bark, … (s m$^{-1}$) |
| $r_{cut}$ | cuticle resistance (s m$^{-1}$) |
| $r_{cut,d}$ | dry cuticle resistance (s m$^{-1}$) |
| $r_{cut,w}$ | wet cuticle resistance (s m$^{-1}$) |
| $r_{cw}$ | water resistance (s m$^{-1}$) |
| $r_{dc}$ | canopy resistance |
| $r_{fg}$ | correlation coefficient between $f$ and $g$ (–) |
| $r_{gd}$ | dry ground resistance (s m$^{-1}$) |
| $r_{gw}$ | wet ground resistance (s m$^{-1}$) |
| $r_{gs}$ | ground resistance below surface canopy (s m$^{-1}$) |
| $r_i$ | resistance parameter (s m$^{-1}$) |
| $r_{lu}$ | cuticle resistance parameter (s m$^{-1}$) |
| $r_m$ | mesophyll resistance (s m$^{-1}$) |
| $r_{st}$ | stomatal resistance (s m$^{-1}$) |
| $R$ | ideal gas constant (J mol$^{-1}$ K$^{-1}$) |
| | solar radiation energy flux (J m$^{-2}$ s$^{-1}$) |
| | spread parameter in bi-Gaussian convective cycling model (–) |
| | building length scale (m) |
| | precipitation rate (mm h$^{-1}$) |
| | atmospheric resistance (s m$^{-1}$) |
| | reflection probability (–) |
| $R_A$ | reaction rate of compound A (molecules cm$^{-3}$ s$^{-1}$) |
| $R_i$ | reaction rate of reaction $i$ (molecules cm$^{-3}$ s$^{-1}$) |
| $R_L$ | Lagrangian autocorrelation coefficient (–) |
| $R_N$ | net radiation energy flux (J m$^{-2}$ s$^{-1}$) |
| $r_s$ | stack radius (m) |
| $r_t$ | total resistance to deposition (s m$^{-1}$) |
| $R_{vv}$ | autocorrelation coefficient (–) |
| $R'_{vv}$ | autocorrelation function (m$^2$ s$^{-2}$) |
| Re | Reynolds number ($=du\rho/\mu = du/\nu$) (–) |

| | |
|---|---|
| $Re_*$ | Reynolds number based on friction velocity ($=z_0 u_* \rho/\mu = z_0 u_*/\nu$) (–) |
| RH | relative humidity (%) |
| $s$ | stability parameter ($s^{-2}$) |
| | distance traveled by plume center (m) |
| $S$ | skewness of vertical wind speed distribution ($m^3\ s^{-3}$) |
| $S^\circ$ | standard entropy ($J\ mol^{-1}\ K^{-1}$) |
| $S_r$ | spacing of obstacles defining roughness (m) |
| $S'_{uu}$ | spectrum function of wind speed variance ($m^2\ s^{-1}$ if function of $\nu$) |
| $S_{uu}$ | normalized spectrum function of wind speed variance (s if function of $\nu$ or $\omega$; m if function of $\kappa$) |
| $S_w$ | saturation (–) |
| Sc | Schmidt number ($=\nu/D$) |
| $t$ | time (s or h) |
| $t_0$ | time of the solar noon (h) |
| $t_{1/2}$ | half life (s) |
| $T$ | temperature (K) |
| $t_a$ | apparent time (s) |
| $T_a$ | ambient temperature (K) |
| $T_c$ | critical temperature (K) |
| $T_i$ | integral time scale (s) |
| $T_{i,L}$ | Lagrangian integral time scale (s) |
| $T_s$ | stack gas temperature (K) |
| | surface air temperature (K) |
| $T_v$ | virtual temperature (K) |
| $u$ | wind speed in the $x$ direction ($m\ s^{-1}$) |
| $U$ | wind speed in the $X$ direction, Lagrangian frame of reference ($m\ s^{-1}$) |
| | internal energy (J) |
| $\overline{u}$ | average wind speed ($m\ s^{-1}$) |
| $u'$ | wind speed fluctuation ($m\ s^{-1}$) |
| $u_0$ | wind speed variable in nocturnal atmosphere stability calculations ($m\ s^{-1}$) |
| $u_1$ | wind speed in the $x_1$ (i.e., $x$) direction ($m\ s^{-1}$) |
| $u_2$ | wind speed in the $x_2$ (i.e., $y$) direction ($m\ s^{-1}$) |
| $u_3$ | wind speed in the $x_3$ (i.e., $z$) direction ($m\ s^{-1}$) |
| $U_a$ | ambient wind speed ($m\ s^{-1}$) |
| $U_{atm}$ | wind speed ($m\ s^{-1}$) |
| $u_c$ | characteristic velocity ($m\ s^{-1}$) |
| $U_c$ | wind speed in a cavity ($m\ s^{-1}$) |
| $u_{H_r}$ | wind speed at height $H_r$ ($m\ s^{-1}$) |
| $U_{sc}$ | velocity of plume centerline ($m\ s^{-1}$) |
| $U_W$ | wind speed in a far wake ($m\ s^{-1}$) |
| $u_*$ | friction velocity ($m\ s^{-1}$) |
| $u_{*w}$ | water-side friction velocity ($m\ s^{-1}$) |
| $u_\tau$ | local friction velocity (between obstacles) ($m\ s^{-1}$) |
| $v$ | wind speed in the $y$ direction ($m\ s^{-1}$) |
| $V$ | wind speed in the $Y$ direction, Lagrangian frame of reference ($m\ s^{-1}$) |
| $\mathbf{V}$ | velocity fluctuation matrix ($m\ s^{-1}$) |
| $\overline{V}$ | total wind speed ($m\ s^{-1}$) |
| $v_d$ | deposition velocity ($m\ s^{-1}$) |
| $v_s$ | settling velocity ($m\ s^{-1}$) |

| | |
|---|---|
| $v_t$ | transfer velocity (m $s^{-1}$) |
| $w$ | wind speed in the $z$ direction (m $s^{-1}$) |
| | work (J) |
| $W$ | wind speed in the $Z$ direction, Lagrangian frame of reference (m $s^{-1}$) |
| | width of a hill (m) |
| | width of a building, projected across the wind flow (m) |
| | Wiener process |
| $w_*$ | convective velocity (m $s^{-1}$) |
| $\overline{w_1}$ | average vertical wind speed of updrafts (m $s^{-1}$) |
| $\overline{w_2}$ | average vertical wind speed of downdrafts (m $s^{-1}$) |
| $w_b$ | emission velocity at the surface (m $s^{-1}$) |
| $W_c$ | half-width of the cavity (m) |
| | wet fraction of the cuticle (–) |
| $w_e$ | entrainment velocity (m $s^{-1}$) |
| $W_g$ | wet fraction of the ground (–) |
| $w_s$ | stack gas velocity in the vertical direction (m $s^{-1}$) |
| $W_{st}$ | wet fraction of the stomata (–) |
| $w_w$ | water weight fraction of air (–) |
| $W_W$ | width of the far wake (m) |
| $w'_w$ | water–air weight ratio (–) |
| $x$ | downwind distance from the source (m) |
| | coordinate in the west–east direction (Eulerian models) (m) |
| | distance from the upwind side of a building (m) |
| $X$ | downwind distance coordinate, Lagrangian frame of reference (m) |
| $x_1$ | coordinate in the west–east direction (Eulerian models) (m) |
| $x_2$ | coordinate in the south–north direction (m) |
| $x_3$ | vertical distance (m) |
| $x_f$ | downwind distance to the source of final plume rise (m) |
| $x_v$ | distance to virtual source (m) |
| $x_{vis}$ | visibility (m) |
| $y$ | crosswind distance from the plume axis (m) |
| | coordinate in the south–north direction (m) |
| $Y$ | crosswind distance coordinate, Lagrangian frame of reference (m) |
| $y_i$ | mole fraction of $i$ in the gas phase (–) |
| $z$ | vertical distance from the surface (terrain following coordiates) (m) |
| | vertical distance from the surface at the source (Cartesian coordinates) (m) |
| $Z$ | vertical distance coordinate, Lagrangian frame of reference (m) |
| | vertical distance coordinate, terrain following coordinates (m) |
| $z_0$ | roughness length (m) |
| $z_j$ | junction height (m) |

## Greek letters

| | |
|---|---|
| $\alpha$ | along-plume entrainment rate parameter (–) |
| | quasi-laminar layer thickness parameter (–) |
| | mass transfer enhancement factor (–) |
| $\alpha_a$ | absorption coefficient ($m^{-1}$) |
| $\alpha_c$ | sea roughness parameter (–) |
| $\alpha_{ext}$ | extinction coefficient ($m^{-1}$) |

$\alpha_k$ Kolmogorov constant (–)
$\alpha_s$ scattering coefficient ($m^{-1}$)
$\alpha_{s,air}$ scattering coefficient for air molecules ($m^{-1}$)
$\alpha_{s,PM}$ scattering coefficient for PM ($m^{-1}$)
$\beta$ ratio of Lagrangian to Eulerian integral time scales (–)
across-plume entrainment rate parameter (–)
parameter describing height dependence of density in denser-than-air plumes (–)
$\gamma$ vertical slope of potential temperature above the planetary boundary layer (K $m^{-1}$)
wavelength (m)
activity coefficient
$\Gamma$ dry adiabatic lapse rate (K $m^{-1}$)
$\Gamma_s$ moist adiabatic lapse rate (K $m^{-1}$)
$\delta$ solar declination (rad or °)
$\Delta_f H°$ standard enthalpy of formation (kJ $mol^{-1}$)
$\Delta h$ plume rise (final) (m)
$\Delta h_{sd}$ stack downwash (m)
$\Delta_r G°$ standard Gibbs free energy of reaction (kJ $mol^{-1}$)
$\Delta_r H°$ standard enthalpy of reaction (kJ $mol^{-1}$)
$\Delta_r S°$ standard entropy of reaction (J $mol^{-1}$ $K^{-1}$)
$\Delta U$ velocity deficit (m $s^{-1}$)
$\Delta_{vap} H$ enthalpy of vaporization (J $kg^{-1}$)
$\Delta z$ plume rise (transitional) (m)
$\Delta\theta$ temperature jump at capping inversion (K)
$\varepsilon$ height of obstacles defining surface roughness (m)
energy dissipation rate ($m^2$ $s^{-3}$)
emissivity (–)
molar decadic absorption coefficient ($cm^2$ $molecule^{-1}$ or simply $cm^{-2}$)
$\zeta$ dimensionless height variable in Monin–Obukhov theory (–)
$\eta$ Kolmogorov length microscale (m)
$\theta$ potential temperature (K)
dimensionless temperature variable (–)
terrain slope (rad)
$\theta_C$ temperature (°C)
$\theta_F$ temperature (°F)
$\theta_m$ vertical average potential temperature in mixing layer (K)
$\theta_v$ virtual potential temperature (K)
$\theta_*$ friction temperature (K)
$\kappa$ thermal diffusivity ($m^2$ $s^{-1}$)
wave number (rad $m^{-1}$ or simply $m^{-1}$)
$\lambda$ height scale for Coriolis effects (m)
thermal conductivity (J $m^{-1}$ $K^{-1}$)
latitude (rad or °)
molecule mean free path (m)
scavenging coefficient ($s^{-1}$)
wavelength (m or nm)
$\lambda_1$ area fraction of updraft (–)
$\lambda_2$ area fraction of downdraft (–)
$\lambda_f$ frontal area parameter (–)
$\lambda_l$ spatial Taylor microscale (m)

| | |
|---|---|
| $\Lambda_L$ | Lagrangian integral length scale (m) |
| $\lambda_p$ | plan area parameter (–) |
| $\Lambda$ | temperature lapse rate ($K\ m^{-1}$) |
| | scavenging ratio ($s^{-1}$) |
| $\mu$ | dynamic viscosity (Pa s) |
| $\nu$ | kinematic viscosity ($m^2\ s^{-1}$) |
| | frequency ($s^{-1}$) |
| $\rho$ | density ($kg\ m^{-3}$) |
| $\rho_a$ | ambient air density ($kg\ m^{-3}$) |
| $\rho_p$ | particle density ($kg\ m^{-3}$) |
| $\rho_s$ | stack gas density ($kg\ m^{-3}$) |
| $\sigma$ | Stefan–Boltzmann constant ($s^{-1}\ m^{-2}\ K^{-4}$) |
| | absorption cross-section ($cm^2$ molecule$^{-1}$ or simply $cm^2$) |
| $\sigma_n$ | standard deviation of perturbation function ($m\ s^{-1}$) |
| $\sigma_u$ | turbulent velocity in the $x$ direction ($m\ s^{-1}$) |
| $\sigma_v$ | turbulent velocity in the $y$ direction ($m\ s^{-1}$) |
| $\sigma_w$ | turbulent velocity in the $z$ direction ($m\ s^{-1}$) |
| $\sigma_{w1}$ | turbulent velocity in the $z$ direction of updraft ($m\ s^{-1}$) |
| $\sigma_{w2}$ | turbulent velocity in the $z$ direction of downdraft ($m\ s^{-1}$) |
| $\sigma_x$ | dispersion parameter along the horizontal wind direction (m) |
| $\sigma_y$ | dispersion parameter across the horizontal wind direction (m) |
| $\sigma_z$ | dispersion parameter in the vertical direction (m) |
| $\tau$ | shear stress (Pa) |
| | time constant (s) |
| $\tau_0$ | shear stress at the surface (Pa) |
| $\tau_s$ | Taylor microscale (s) |
| $\varphi$ | solar elevation (rad) |
| $\phi$ | latitude (rad) |
| | angle between plume path and the horizontal (rad) |
| | quantum yield (–) |
| $\phi_h$ | Monin–Obukhov similarity function for heat transfer (–) |
| $\phi_m$ | Monin–Obukhov similarity function for momentum transfer (–) |
| $\varphi_y$ | dilution factor due to lateral dispersion ($m^{-1}$) |
| $\varphi_z$ | dilution factor due to vertical dispersion ($m^{-1}$) |
| $\omega$ | angular frequency (rad $s^{-1}$ or simply $s^{-1}$) |
| $\Omega$ | angular velocity of Earth ($s^{-1}$) |

CHAPTER 1

# INTRODUCTION

## 1.1 INTRODUCTION

Almost every human activity and natural process leads to some form of air pollution. Fortunately, not every form of air pollution is problematic, but many forms are, and it is estimated that polluted outdoor air causes tens of thousands of premature deaths per year in North America alone and nearly a million premature deaths per year worldwide (Ezzati et al., 2002). Infants and elderly people are the groups most sensitive to air pollution.

The concentration at which an air pollutant reaches the population depends on the degree of **dispersion** of the pollutant in the air. Dispersion is the spreading of a compound in a fluid resulting from the random motions of the fluid and its molecules. In the physical literature (e.g., meteorology) this is often referred to as **diffusion**, or sometimes **turbulent diffusion** to distinguish from molecular diffusion, a related process in laminar fluids. In the engineering literature, the word *diffusion* is often limited to mass transfer in nonturbulent fluids, and the word *dispersion* is preferred when the process is driven by turbulence. This book will follow the latter practice, except in cases where the use of the word *dispersion* could be a source of ambiguity.

Air dispersion modeling is a powerful technique to evaluate if a source of air pollution creates a problem or not. An air pollution **problem** (as opposed to air pollution proper) can be loosely defined as an ambient concentration of an air pollutant or combination of air pollutants at any location that exceeds the acceptable level for that location. What is acceptable can be based on regulations, toxicology, ecotoxicology, or simply the well-being of the local population. There are several reasons why dispersion models are used, depending on the application:

- It is not possible to measure the air quality at every relevant location all of the time.
- If a new source of air pollution is planned, air dispersion models can predict the impact of the source on the air quality.
- When air pollution is found, dispersion models can help determine the source.

*Air Dispersion Modeling: Foundations and Applications*, First Edition. Alex De Visscher.

- When a source creates an air pollution problem, dispersion models can determine the emission reduction required to solve the problem.
- Air dispersion models can be used to plan the response to emergencies such as accidental leaks.

The purpose of this book is to teach you the different aspects of air pollution modeling: to select appropriate data and an appropriate model to evaluate the environmental impact of an air pollution source. This includes a thorough understanding of the science underlying different models, the strengths and weaknesses of these models, and their range of applicability. After reading this book you will understand all the major aspects of air dispersion modeling—you will know what you are doing if you enter air dispersion modeling practice.

This book is *not* a hands-on tutorial on how to use particular software. In other words, you will learn the inner workings of a model such as CALPUFF, but you will not learn which buttons to press in the commercially available CALPUFF View 2.3 or on which form to find a particular option in the original CALPUFF Modeling System. There are specialized courses for that purpose, organized by professional organizations such as the Air and Waste Management Association and distributors of air dispersion modeling software. Hands-on courses tend to cover the theory only superficially. This book is complementary to those courses.

Figure 1.1 shows the context of air dispersion models. An air pollution problem is never the result of an isolated event, but the result of a **chain of events**: from the formation of a pollutant in a process to the emission (with or without waste gas treatment), the dispersion and chemical transformation in the atmosphere, to the uptake by a receptor (e.g., a person breathing in polluted air, a plant absorbing a pollutant), to a health and/or wellness effect. A complete understanding of an air pollution problem requires knowledge of all steps in the process.

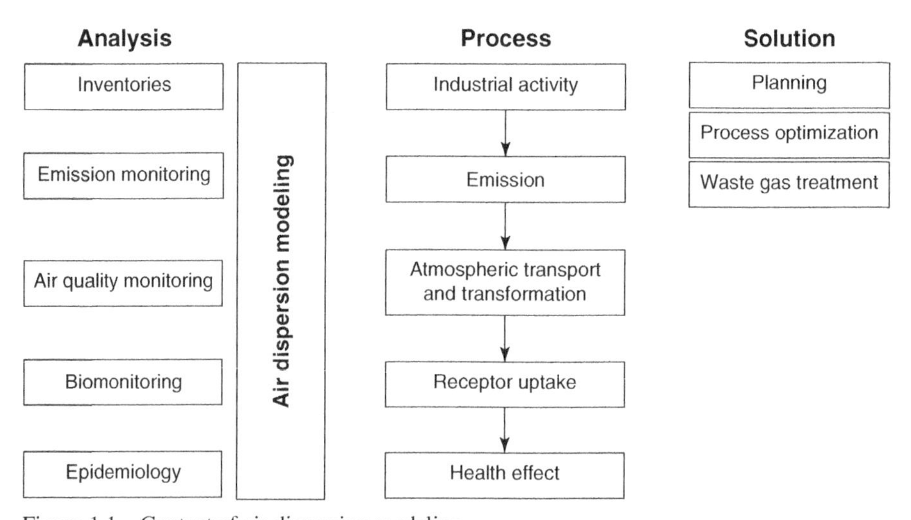


Figure 1.1 Context of air dispersion modeling.

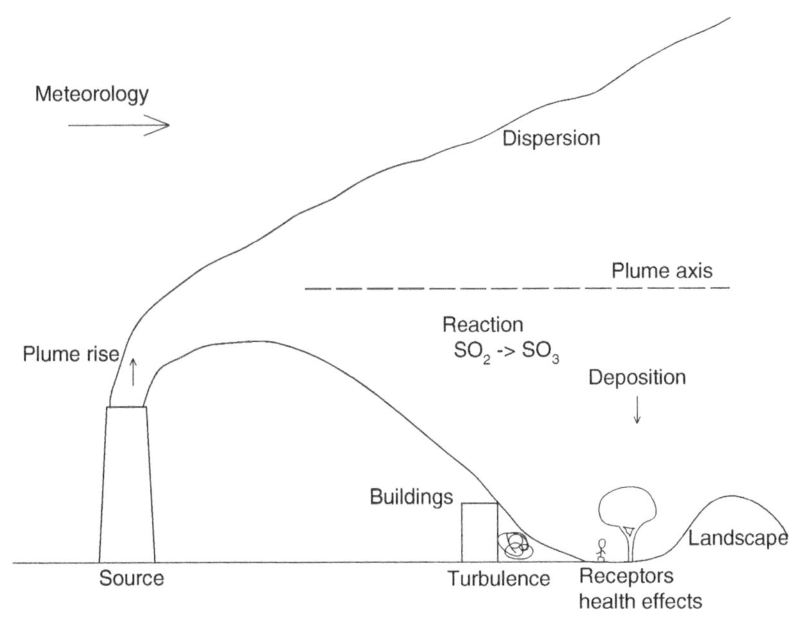


Figure 1.2 Events of an air pollution problem.

Figure 1.2 shows the chain of events leading to air pollution problems and the main factors governing these events. We can see that air pollution is influenced by the **weather** (wind speed, wind direction, temperature, rainfall, sunshine), the **terrain** (buildings, hills, surface water), the **source** (stack height and diameter, stack gas temperature and velocity, concentration of the pollutant), and the physical and chemical **properties** of the pollutant (reactivity, solubility). In a comprehensive model, all this information needs to be included. Simpler models use only the most relevant of these influencing factors as input.

It is important to note that the atmosphere is *turbulent* and *chaotic*. Consequently, pollutant concentrations are not constant even when the source and the weather are constant. Furthermore, not all the factors influencing air dispersion can be included in a model. Hence, the accuracy of air dispersion models can seem disappointing to the inexperienced user. We consider a model as *successful* when the following criteria are met:

- The predicted *hourly average* concentration is within a *factor of 2* of the actual concentration most of the time.
- Over a long time, the average concentration predicted by the model is close to the actual average concentration.
- When predicted concentrations at different locations and different times are ranked from lowest to highest, and the same is done with measured concentrations at the same locations and times, then a very similar distribution is obtained. Corresponding points in space and time do *not* necessarily have the same rank.

Simple models like the one outlined in the primer (Chapter 2) are moderately successful (within a factor of 2 half of the time) when the conditions are favorable (flat terrain, constant weather, and short distance). The models SCREEN3 and ISC3 are in this category. For less favorable conditions more sophisticated models are needed. AERMOD, and especially CALPUFF, are in this category.

## 1.2 TYPES OF AIR DISPERSION MODELS

There is a wide variety of air dispersion models, ranging from the conceptually and computationally simple **Gaussian plume models** to very demanding models based on **computational fluid dymamics**. In general, dispersion models are potentially more accurate as the models become more fundamental and hence computationally more intensive. However, the latter types of models have their own *pitfalls* and require more expertise to run. Hence, higher accuracy is not guaranteed. In practice, the highest accuracy will be obtained when the model is chosen that best suits the needs and resources of the modeler. In most cases the practically most feasible models are Gaussian plume models and Gaussian puff models.

### 1.2.1 Gaussian Plume Models

When it is assumed that wind speed and wind direction are constant in space and time, as well as the turbulent diffusivity (i.e., the propensity of the atmosphere to disperse pollutants), and the source emits a constant stream of pollution, then it can be proven mathematically that the resulting plume will have a Gaussian concentration profile in the lateral and vertical directions. These Gaussian profiles will fan out as the distance to the source increases. This plume shape is the main feature of Gaussian plume models.

*None of the above assumptions is even remotely accurate*. Not surprisingly, when we look at an *instantaneous* snapshot of a plume, the concentration profile is not Gaussian. However, we know from experience that the hourly average concentration is very nearly Gaussian in many cases, and as long as we use well-established empirical correlations to estimate the model parameters, Gaussian plume models can be very accurate. In practice, this class of models does not really assume that wind speed is independent of height. Instead, corrections are made to incorporate wind speed at the appropriate height into the model.

The advantage of Gaussian plume models is that they are simple and efficient (i.e., require little computational effort). The disadvantage is that they are less accurate and less generally applicable than more sophisticated models. Their accuracy tend to decrease quickly after 10–20 km. Examples of such models are SCREEN3 (EPA, 1995a; http://www.epa.gov/scram001/dispersion_screening.htm) and ISC3 (EPA, 1995b; http://www.epa.gov/scram001/dispersion_alt.htm). AERMOD (Cimorelli et al., 2004; http://www.epa.gov/scram001/dispersion_prefrec.htm) is a Gaussian plume model with a number of modifications to improve its accuracy and applicability.

### 1.2.2 Gaussian Puff Models

Gaussian puff models were designed to resolve the most glaring weaknesses of Gaussian plume models. They assume that the emission of a stack can be subdivided in a series of pollutant puffs that have a Gaussian concentration profile in all directions. Each puff follows the local wind speed and wind direction independently of other puffs. Hence, when the wind direction changes in space or time, the puff will follow a *curved path*.

Puff models are more accurate than plume models. They perform well for distances up to at least 50 km and are routinely used for distances up to 200 km. However, because each source emits hundreds of puffs that have to be tracked, they are computationally more intensive than plume models. An example of a puff model is CALPUFF (Scire et al., 2000; http://www.epa.gov/scram001/dispersion_prefrec.htm).

Puff models are **Lagrangian** models. This means that puff properties are characterized in a coordinate system that moves with the puff. In contrast, **Eulerian** models use a coordinate system that is fixed in space.

### 1.2.3 Stochastic Lagrangian Particle Models

Stochastic Lagrangian particle models assume that each source emits a large number of particles, and each particle follows a *random path* around the mean wind vector. This path is updated with every time step. Predictions of pollutant concentrations are obtained by counting the number of particles in a given volume of air. Stochastic Lagrangian particle models potentially follow the actual physics of air dispersion better than any other model. Hence, models of this type are very accurate and can be used up to long distances (thousands of kilometers). However, they are computationally very intensive because tens of thousands to millions of particles need to be considered for each source, making this model practical only when there is a small number of sources. An example is the German model AUSTAL2000 (Janicke, 2008; http://www.austal2000.de/en/home.html).

### 1.2.4 Eulerian Advection and Dispersion Models

Eulerian advection and dispersion models are based on a *grid* that is fixed in space. In each grid point the advection (i.e., the flow) and dispersion of pollutants is calculated based on local concentration gradients. Dispersion is calculated with a turbulent version of Fick's law of diffusion, which states that a diffusion flux is proportional to the concentration gradient. Models of this type are accurate at long distances, but they have low resolution, often 10-km cell size or more. They are computationally intensive, and the computational power needed increases very rapidly with increasing resolution. Furthermore, the physics underlying Eulerian dispersion models breaks down at scales smaller than a few kilometers. The main advantage of this type of model is that it is able to handle complex chemistry, unlike simpler dispersion models. This type of model is used for the prediction of photochemical smog. An example is CMAQ (Byun and Ching, 1999; http://www.cmaq-model.org/index.cfm).

### 1.2.5 Computational Fluid Dynamics

**Computational fluid dynamics (CFD)** is the practice of solving the **Navier–Stokes** equations on a fixed grid in order to explicitly calculate turbulent motion of air masses. The Navier–Stokes equations are **momentum balances**. They only provide bulk properties like the wind vector at every location; thus, for the calculation of plume dispersion, an Eulerian or Lagrangian advection and dispersion model is needed.

Computational fluid dynamics is an extremely computationally intensive technique and is usually feasible only at small scale. Even then, not all scales of turbulence can be covered because the smallest turbulent eddies in the atmosphere are on the millimeter scale, whereas the largest turbulence elements can be on the kilometer scale. For that reason, grid cells larger than the smallest turbulence elements are defined, and subgrid turbulence and dispersion are parameterized. This technique is called **large eddy simulation (LES)**. A further simplification is the practice of parameterizing *all* levels of time-dependent turbulence. This technique is known as **Reynolds averaged Navier–Stokes (RANS)**. RANS can be seen as the solution of an equivalent laminar flow problem with a high, location-dependent viscosity.

## 1.3 STANDARD CONDITIONS FOR TEMPERATURE AND PRESSURE

Stack gas emissions are often presented as volumetric flow rates at **standard** conditions of **temperature and pressure (STP)**. Because stack gas can reasonably be assumed to be an ideal gas, the volume at standard conditions is an unambiguous measure of the amount of gas. Unfortunately, different standard conditions circulate in the technical literature, so the practice of defining standard conditions probably creates more ambiguity than it solves. Within a technical field, the definition of the standard is usually well-known, but air dispersion modelers are often active in multiple fields and may not be aware that several standards are in use. For that reason it is recommended that the standard be defined whenever it is used. A number of commonly used standards are given below.

The **International Union of Pure and Applied Chemistry (IUPAC)** defines standard conditions as 0 °C temperature and 100,000 Pa pressure (1 bar). This convention will be used in this text. Under the IUPAC STP, 1 mol of ideal gas has a volume of 22.711 L. Many publications still use the pre-1982 standard pressure of 101,325 Pa (1 atm), which leads to a molar volume of 22.414 L (Mills et al., 1993).

In physical chemistry the standard state is the standard ambient temperature and pressure, also defined by IUPAC: 25 °C and 100,000 Pa. Here also, the old standard pressure of 101,325 Pa is still frequently used.

The Society of Petroleum Engineers (SPE) uses either 15 °C and 100,000 Pa or 60 °F and 14.696 psi. Note that the two are not exactly the same.

Other temperatures that are sometimes used are 20 °C and 70 °F. Other pressures are 14.504 psi (which is nearly identical to 1 bar) and 14.73 psi (which is slightly above 1 atm).

## 1.4 CONCENTRATION UNITS IN THE GAS PHASE

Two types of concentration units are commonly used in the gas phase. The air quality expert must be familiar with both and capable of converting between one scale and the other.

The first type of concentration unit is ppm (parts per million) and ppb (parts per billion). They are essentially *mole fraction* multiplied by $10^6$ and $10^9$, respectively, and are often referred to as *mixing ratio*. This should not be confused with the ppm and ppb scales in the liquid phase, where they usually denote mg $L^{-1}$ and μg $L^{-1}$, respectively. To avoid confusion, the scales are often denoted ppmv and ppbv, where ppmv denotes part per million by volume assuming ideal gas behavior.

The second type of concentration unit is mg $m^{-3}$ and μg $m^{-3}$, which represent *mass concentration*. Unlike ppm and ppb, they depend on temperature and pressure.

**Example 1.1.** Convert a mole fraction of $x$ ppm of a compound A with molar mass $M$ g $mol^{-1}$ to mass concentration in mg $m^{-3}$ at temperature $T$ and pressure $p$. Apply to 1 ppm ozone ($M = 48$ g $mol^{-1}$) at STP ($T = 273.15$ K, $p = 100{,}000$ Pa).

***Solution.*** When a mole fraction needs to be converted to mg $m^{-3}$, it is advisable to consider, for instance, one million moles of gas. This gas contains $x$ moles of A. With a molar mass of $M$ g $mol^{-1}$, the gas contains $M \cdot x$ g of A. The total volume of gas can be found from the ideal gas law:

$$pV = nRT \tag{1.1}$$

where $V$ is the volume ($m^3$), $n$ the amount of substance (mol), $R$ the ideal gas constant (8.314472 J $mol^{-1}$ $K^{-1}$), and $T$ the temperature (K). Hence:

$$V = \frac{1{,}000{,}000RT}{p}$$

and the mass concentration of A is

$$c = \frac{Mxp}{1{,}000{,}000RT} \quad \text{in g m}^{-3} \qquad \text{or} \qquad c = \frac{Mxp}{1{,}000RT} \quad \text{in mg m}^{-3}$$

For instance, 1 ppm of ozone ($M = 48$ g $mol^{-1}$) at STP ($T = 273.15$ K, $p = 100{,}000$ Pa) leads to a mass concentration of

$$48 \cdot 1 \cdot 100{,}000 / 1000 / 8.314472 / 273.15 = \mathbf{2.11\ mg\ m^{-3}}$$

We can infer a rule of thumb from the example: the concentration in mg $m^{-3}$ is roughly equal to the concentration in ppm for $M = 25$ g $mol^{-1}$ and is proportional with $M$. However, *do not use any rules of thumb when high accuracy is required.*

**Example 1.2.** Convert a mass concentration of the compound A (see Example 1) of $y$ mg $m^{-3}$ to ppm. Apply to ozone at 1 mg $m^{-1}$ at STP.

***Solution.*** Now we assume 1 $m^3$ of gas, which contains $y$ mg $= y/M$ mmol $= y/1000M$ mol of A. The total number of moles is calculated from the ideal gas law:

$$n = \frac{pV}{RT}$$

Hence the mole fraction is $yRT/1000MpV$, or

$$\frac{1{,}000{,}000\,yRT}{1000Mp}\,\text{ppm} = \frac{1000\,yRT}{Mp}\,\text{ppm}$$

For instance, 1 mg $m^{-3}$ ozone at STP equals

$$1000 \cdot 1 \cdot 8.314472 \cdot 273.15/\left(48 \times 100{,}000\right) = \mathbf{0.473\,ppm} = 473\,\text{ppb}$$

Barometric pressure depends on **altitude**. It is important to be aware of the fact that *barometric pressures given in weather reports are not actual pressures* but the pressure adjusted to sea level. In other words, the reported pressure is the pressure of an air parcel at sea level in hydrostatic equilibrium with the actual air. To convert back to actual pressure, the following rule of thumb is helpful:

$$p = p_{\text{sea}} \exp(-0.00012\,\text{alt}) \tag{1.2}$$

where $p$ is the actual pressure, $p_{\text{sea}}$ is the reported pressure, and alt is altitude (m). A derivation of eq. (1.2) can be found in Chapter 5.

**Example 1.3.** Convert a $p_{\text{sea}}$ of 101 kPa in Aspen, Colorado (2400 m altitude), to actual pressure.

***Solution.***

$$p = 101\,\text{kPa} \times \exp(-0.00012 \cdot 2400) = 75.7\,\text{kPa}$$

It is clear from Example 1.3 that pressure corrections can be significant. Hence, rules of thumb for concentration conversions that do not incorporate actual barometric pressures should be avoided at high altitudes. Some air quality textbooks provide such rules of thumb and present them as if they are strict mathematical relationships. It is prudent always to use the ideal gas law in concentration conversions.

A third type of concentration, often used in photochemistry and atmospheric chemistry, is the *number concentration* $C$, usually expressed in molecules $cm^{-3}$.

**Example 1.4.** Calculate the total concentration of air at 1 bar and 25 °C in molecules $cm^{-3}$.

***Solution.*** The total concentration of air is calculated from the ideal gas law:

$$c = \frac{n}{V} = \frac{p}{RT}$$

Hence, $c = 100{,}000$ Pa/(8.314473 J $mol^{-1}$ $K^{-1}$ · 298.15 K) = 40.3395 mol $m^{-3}$
Multiply by Avogadro's constant ($N_A = 6.0221415 \times 10^{23}$ molecules $mol^{-1}$) to obtain

$$\begin{aligned} C = c\,N_A &= 40.3395\,\text{mol}\,\text{m}^{-3} \times 6.0221415 \times 10^{23}\,\text{molecules}\,\text{mol}^{-1} \\ &= 2.429 \times 10^{25}\,\text{molecules}\,\text{m}^{-3} \end{aligned}$$

Divide by $10^6$ $cm^3$ $m^{-3}$ to obtain

$$\mathbf{C = 2.429 \times 10^{19}\,molecules\,cm^{-3}}$$

## 1.5 UNITS

This book will use SI (metric) units throughout. However, imperial units are still common practice in North American industry. Hence, it is important to be able to convert units. *Never* make assumptions when units are not specified. For instance, a U.S. consulting company once presented incorrect air pollution data to a Canadian client by plotting pollutant plumes from U.S. software (in miles) on a Canadian map (in kilometers), leading to grossly inaccurate results.

Below is a list of the main dimensions with their SI unit:

| | |
|---|---|
| Length | meter (m) |
| Time | second (s) |
| Mass | kilogram (kg) |
| Temperature | kelvin (K) |
| Amount of substance | mole (mol) |

Some *derived units* are the following:

| | | |
|---|---|---|
| Force | newton = kilogram meter second$^{-2}$ | (N = kg m s$^{-2}$) |
| Energy | joule = newton meter | (J = N m) |
| Pressure | pascal = newton meter$^{-2}$ | (Pa = N m$^{-2}$) |
| Power | watt = joule second$^{-1}$ | (W = J s$^{-1}$) |

The following prefixes multiply the units in steps of 1000:

| | | |
|---|---|---|
| nano | n | $10^{-9}$ |
| micro | μ | $10^{-6}$ |
| milli | m | $10^{-3}$ |
| kilo | k | $10^{3}$ |
| mega | M | $10^{6}$ |
| giga | G | $10^{9}$ |

Here are some common *alternative units*. They should be used with care because unit consistency is not guaranteed once alternative units are involved.

1 cm = 0.01 m

1 mile = 1609.344 m = 1.609344 km

1 g = 0.001 kg

1 atm = 101,325 Pa

1 bar = 100,000 Pa

760 Torr = 760 mm Hg = 101,325 Pa

1 cal = 4.184 J (thermochemistry)

1 cal = 4.1868 J (international steam table)

1″ = 0.0254 m (inch)

1′ = 0.3048 m (foot)

1 lb = 0.45359237 kg (pound) (rule of thumb: 1 kg = 2.2 lb with 0.2% error)

1 tonne = 1000 kg (metric ton)

1 ton (U.S.) = 907.18474 kg (short ton)

1 ton (UK) = 1016.046909 kg (long ton)

1 psi = 6894.75729... Pa (pound per square inch)

A special derived unit is the *degree Celsius* (°C), which is derived from the kelvin unit as follows:

$$\theta_C/°C = T/K - 273.15$$

The kelvin unit is defined as the temperature of the triple point of water divided by 273.16. It follows that the triple-point temperature of water is 0.01 °C. At 1 atm the melting point of water is at 0.003 °C and the boiling point is at 99.974 °C.

The Fahrenheit scale is defined as follows:

$$\theta_F/°F = 1.8\theta_C/°C + 32$$

## 1.6 CONSTANTS AND APPROXIMATELY CONSTANT VARIABLES

A list of constants and variables that can be assumed constant are given in Table 1.1. A useful source of internationally accepted values for fundamental constants is Mohr and Taylor (2005). The molar mass of air was taken from Dubin et al. (1976). At the time of writing, IUPAC is considering a redefinition of the kilogram, which will affect some of the numerical values in Table 1.1. However, the changes will be negligible.

**TABLE 1.1 Constants and Approximately Constant Variables**

| Constant | Value | Meaning |
|---|---|---|
| $c$ | 299,792,458 m $s^{-1}$ | Speed of light in vacuum |
| $c_{p,air}$ | 1006 J $kg^{-1}$ $K^{-1}$ | Specific heat of dry air |
| $g$ | 9.80665 m $s^{-2}$ | Acceleration of gravity |
| $h$ | $6.6260693 \times 10^{-34}$ J s | Planck constant |
| $k$ | 0.40 | von Karman constant |
| $k$ | $1.3806505 \times 10^{-23}$ J $K^{-1}$ | Boltzmann constant |
| $M_{air}$ | 28.964 g $mol^{-1}$ = 0.028964 kg $mol^{-1}$ | Molar mass of dry air |
| $N_A$ | $6.0221415 \times 10^{23}$ molecules $mol^{-1}$ | Avogadro constant |
| $R$ | 8.314472 J $mol^{-1}$ $K^{-1}$ | Ideal gas constant |
| $\Omega$ | $7.292 \times 10^{-5}$ $s^{-1}$ | Angular velocity of the earth |

## 1.7 FREQUENTLY USED GREEK SYMBOLS

Here are the names of some Greek letters commonly used in air dispersion modeling:

| | | |
|---|---|---|
| α | Α | alpha |
| β | Β | beta |
| γ | Γ | gamma |
| δ | Δ | delta |
| ε | Ε | epsilon |
| ζ | Ζ | zeta |
| η | Η | eta |
| θ | Θ | theta |
| κ | Κ | kappa |
| λ | Λ | lambda |
| μ | Μ | mu |
| ν | Ν | nu |
| ξ | Ξ | xi |
| π | Π | pi |
| ρ | Ρ | rho |
| σ | Σ | sigma |
| τ | Τ | tau |
| ϕ, φ | Φ | phi |
| ψ | Ψ | psi |
| ω | Ω | omega |

## PROBLEMS

1. Calculate the volume of 1 mol ideal gas using the two standards of the Society of Petroleum Engineers.
2. The barometric pressure in Reno, Nevada (1300 m altitude), is 87.7 kPa. Calculate the approximate barometric pressure corrected to sea level.
3. Calculate the speed of light (299,792,458 m $s^{-1}$) in miles per leap year.
4. The 24-h sulfur dioxide standard in the United States is 0.14 ppm. Calculate the standard in μg $m^{-3}$ on a cold winter day (–20 °C) and on a hot summer day (+40 °C), both at 1 bar pressure.
5. Your client reports an $H_2S$ emission as 20 L $min^{-1}$ under the SPE standard of 60 °F and 14.696 psi. Your colleague needs the data for an air dispersion calculation, but he prefers the data presented as $m^3$ $h^{-1}$ under the current IUPAC STP. Make the conversion.

## REFERENCES

Byun D.W. and Ching J.K.S. (1999). *Science Algorithms of the EPA Models-3 Community Multiscale Air Quality (CMAQ) Modeling System*. Report EPA/600/R-99/030, US-EPA, Research Triangle Park, NC.

Cimorelli A.J., Perry S.G., Venkatram A., Weil J.C., Paine R.J., Wilson R.B., Lee R.F., Peters W.D., Brode R.W., and Paumier J.O. (2004). *AERMOD: Description of Model Formulation.* Report EPA-454/R-03-004, US-EPA, Research Triangle Park, NC.

Dubin M., Hull A.R., and Champion K.S.W. (eds.) (1976). *US Standard Atmosphere.* Report NOAA-S/T 76-1562, NOAA, NASA, USAF, Washington DC.

EPA (1995a). *SCREEN3 Model User's Guide.* Report EPA-454/B-95-004, US-EPA, Research Triangle Park, NC.

EPA (1995b). *User's Guide for the Industrial Source Complex (ISC3) Dispersion Models. Volume II — Description of Model Algorithms.* Report EPA-454/B-95-003b, US-EPA, Research Triangle Park, NC.

Ezzati M., Lopez A.D., Vander Hoorn S., and Murray C.J.L. (2002). Selected major risk factors and global and regional burden of disease. *Lancet* **360**, 1347–1360.

Janicke (2008). *AUSTAL2000. Program Documentation of Version 2.4.4.4.* Janicke Consulting, Dunum, Germany.

Mills I., Cvitaš T., Homann K., Kallay N., and Kuchitsu K. (1993). *Quantities, Units and Symbols in Physical Chemistry ("Green Book")*, 2nd ed. IUPAC and Blackwell Science, Oxford, UK.

Mohr P.J., and Taylor B.N. (2005). CODATA recommended values of the fundamental physical constants: 2002. *Rev. Mod. Phys.* **77**, 1–107.

Scire J.S., Strimaitis D.G., and Yamartino R.J. (2000). *A User's Guide for the CALPUFF Dispersion Model.* Earth Tech, Concord, MA.

CHAPTER 2

# AN AIR DISPERSION MODELING PRIMER

## 2.1 INTRODUCTION

There are many different types of dispersion models, ranging from Gaussian plume models based on atmospheric stability classes to models based on computational fluid dynamics. The air dispersion modeler should be familiar with several of those, with their strengths and weaknesses. The background knowledge needed to build up to that level of expertise takes up a major portion of this book. However, simple air dispersion models can be developed with only a minimum of background knowledge, and we can learn a lot from such models.

When we look at the atmosphere, we see that there are *many factors that influence the behavior of a pollutant plume*: wind speed, atmospheric stability, the occurrence of temperature inversion, plume temperature, plume exit speed, landscape, and obstacles near the pollution source. Except for the last two, all these effects are incorporated in even the most basic dispersion model, the **Gaussian plume model**. This has some important consequences:

- Unless the terrain surrounding the source is complex, Gaussian plume dispersion models can accurately predict concentrations around a source (within a factor of 2).
- Quick calculations with Gaussian plume dispersion models can give you a feel for factors influencing atmospheric dispersion.

For these reasons very simple Gaussian plume models are routinely used for *screening* purposes: When a source of air pollution has been identified, a quick calculation can establish if it is worthwhile to conduct more detailed calculations to estimate the impact of the source. Regulators recommend this approach in their guidelines (e.g., EPA, 2005; Idriss and Spurrell, 2009).

To understand air dispersion, a good knowledge of many meteorological principles is needed. These principles will be discussed in much detail in subsequent chapters. The link between these principles and air dispersion is not always obvious. This primer will help you see that link.

---

*Air Dispersion Modeling: Foundations and Applications*, First Edition. Alex De Visscher.

At the end of this primer you will be able to:

- Conduct dispersion calculations for screening a potential source of air pollution
- Understand the effect of the main factors influencing air dispersion
- Understand the main weaknesses of simple models so we can improve on them.

The model outlined in this chapter is similar to ISC3, but with a number of simplifications. The material presented here is similar to introductory texts on air dispersion modeling in environmental engineering textbooks such as Cooper and Alley (2011). Hence, readers familiar with the basics of Gaussian plume modeling may prefer to skip the main part of this chapter and go straight to Section 2.5. If not, the primer will provide a quick sense of accomplishment, as you will be conducting your own air dispersion calculations by the end of this chapter.

## 2.2 BASIC CONCEPTS OF AIR DISPERSION

When we watch a plume emitted by a stack, we usually observe the following:

- The plume rises and stabilizes at a certain height.
- The plume fans out horizontally and vertically.
- The plume shape fluctuates randomly.

The random fluctuations of a plume cannot be captured by a deterministic model and will not be considered in this primer. However, the *average* pollutant concentration over a given period of time (e.g., 1 h) can be captured. The concepts are shown in Figure 2.1.

In Figure 2.1 the following definitions apply:

$h_s$ = source height (m)

$\Delta h$ = plume rise (m)

$h$ = effective source height (m)

Both the fanning out of the plume and the random fluctuations are the result of **turbulence** in the wind flow (eddies). Turbulence can be caused by heat released in the atmosphere (thermal turbulence) and by air passing obstacles and roughness of the surface (mechanical turbulence). It follows that *atmospheric dispersion can only be quantified if something is known about the atmosphere* and *the surface*.

Plume rise can be caused by the **momentum** of the plume as it leaves the stack or by the **buoyancy** as hot plumes are lighter than ambient air. Usually, buoyancy is the main driving mechanism for plume rise.

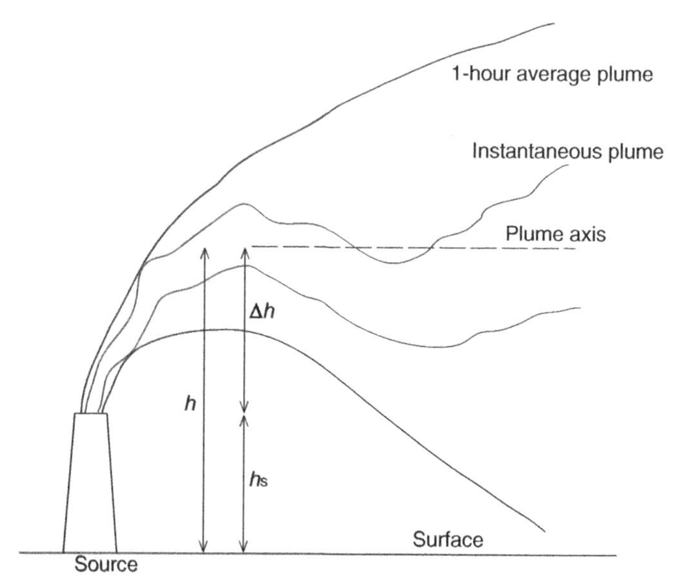


Figure 2.1 Concepts of plume dispersion.

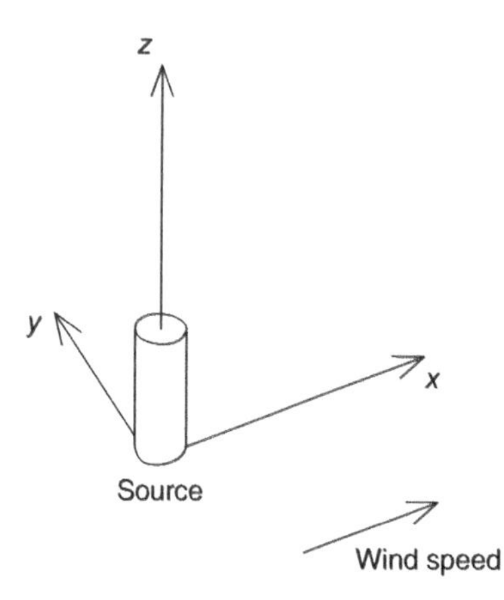


Figure 2.2 Coordinate system in simple Gaussian dispersion models.

A *convention* in Gaussian plume dispersion modeling is the definition of the coordinate axes (see Fig. 2.2):

$x$ = direction of the wind ($x=0$ at the source; $x>0$ downwind).

$y$ = horizontal direction perpendicular to the wind ($y=0$ at the center of the plume; positive on your left when you look downwind).

$z$ = vertical direction ($z=0$ at the surface and positive above the surface).

In Eulerian air quality modeling we will use a different convention: we define the $x$ coordinate from west to east and the $y$ coordinate from south to north.

The angle of the wind direction itself is also subject to a convention. Typically 0° is associated with a northern wind (i.e., moving south), 90° is associated with an eastern wind, and so forth.

## 2.3 GAUSSIAN DISPERSION MODEL

### 2.3.1 Assumptions Underlying the Gaussian Plume Concept

If the fluctuations in plume shape in the $y$ and $z$ directions are completely random, then the instantaneous concentration profile of a pollutant in a plume will be irregular, while the time-averaged concentration will be a Gaussian distribution in the $y$ and $z$ directions. This is illustrated in Figure 2.3. Figure 2.4 shows top and side projection of a Gaussian pollutant plume. As a rule of thumb, the plume edge is 2.15 standard deviations from the plume center (Gifford, 1961).

Strictly speaking, the Gaussian plume model is only valid under certain simplifying conditions:

- The plume starts from a mathematical point referred to as a **point source**.
- The source of pollution is constant.
- Wind direction and wind speed are constant in space and time.
- Atmospheric turbulence is constant in space and time.

This means that the Gaussian plume model can only be an approximation, as none of these conditions is ever satisfied. It can be expected that the model is *approximately valid* (within a factor of 2) when the above conditions are approximately valid, and the model becomes less accurate as the real conditions deviate more from these idealized assumptions. Accounting for hills and valleys is especially challenging.

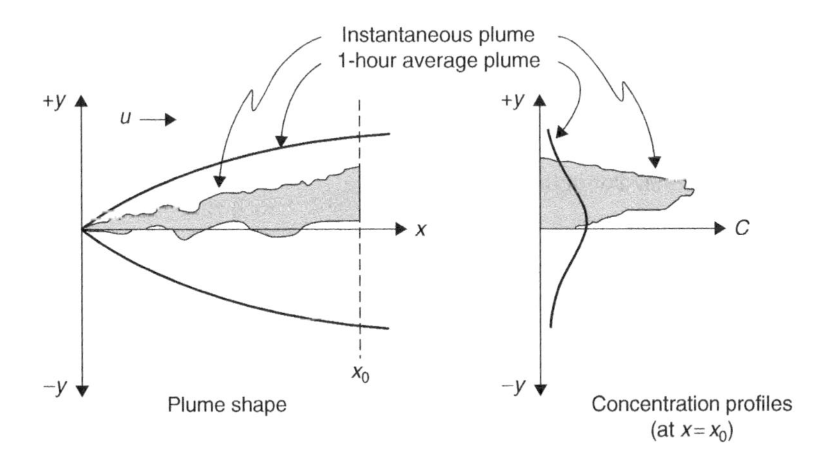


Figure 2.3 Top view of an instantaneous plume and a 1-h average plume and their corresponding concentration profiles (from Cooper and Alley, 2011).

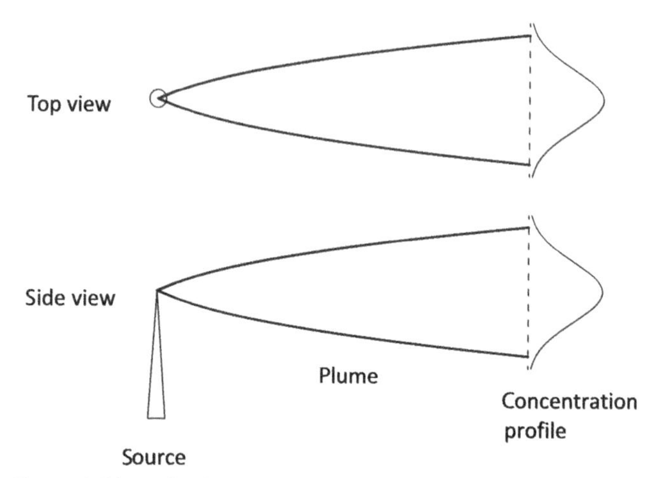


Figure 2.4 Top and side projection of a Gaussian plume, with concentration profiles.

How to handle conditions that deviate from the idealized conditions given above will be discussed in the next chapters.

### 2.3.2 Quantitative Description

***2.3.2.1 Gaussian Plume Equation*** In the *absence of boundaries*, the equation for pollutant concentrations in Gaussian plumes is as follows:

$$c = \frac{Q}{2\pi u \sigma_y \sigma_z} \exp\left(-\frac{1}{2}\frac{y^2}{\sigma_y^2}\right) \exp\left[-\frac{1}{2}\frac{(z-h)^2}{\sigma_z^2}\right] \tag{2.1}$$

where

$c$ = concentration at a given point (g m$^{-3}$)

$Q$ = emission rate (g s$^{-1}$)

$u$ = wind speed (m s$^{-1}$)

$\sigma_y$ = dispersion parameter in the horizontal (lateral) direction (m)

$\sigma_z$ = dispersion parameter in the vertical direction (m)

and $x$, $y$, and $z$ have been defined before, as well as the effective source height $h$. They all have meters as units. Further justification and interpretation of eq. (2.1) is provided in Chapter 6.

Some remarks:

- Because wind speed depends on height, the wind speed *at the effective source height h* should be used for $u$.
- The spread parameters depend on the distance from the source and on weather conditions.

In practice, *boundaries* usually play a role in air dispersion, in particular the *surface* (i.e., the ground or a water surface). To account for the surface, another assumption is needed. Most pollutants deposit to the surface only slowly, so the conservative assumption that there is no deposition at all is usually made. Hence, the plume behaves as if it *reflects* on the surface. To calculate this effect, an *imaginary* source is defined as shown in Figure 2.5. Equation (2.1) is extended to:

$$C = \frac{Q}{2\pi u \sigma_y \sigma_z} \exp\left(-\frac{1}{2}\frac{y^2}{\sigma_y^2}\right)\left\{\exp\left[-\frac{1}{2}\frac{(z-h)^2}{\sigma_z^2}\right] + \exp\left[-\frac{1}{2}\frac{(z+h)^2}{\sigma_z^2}\right]\right\} \quad (2.2)$$

Figure 2.6 shows a vertical concentration profile of the direct plume, the reflected plume, and the total plume. The figure clearly demonstrates that it is crucial to take

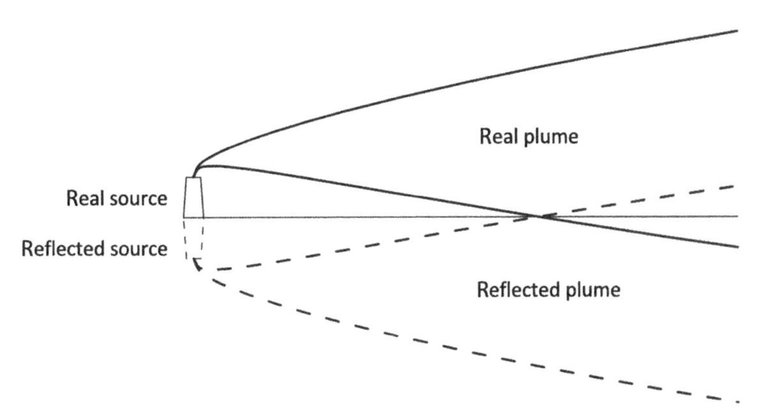


Figure 2.5 Plume reflection on the ground (after Cooper and Alley, 2011).

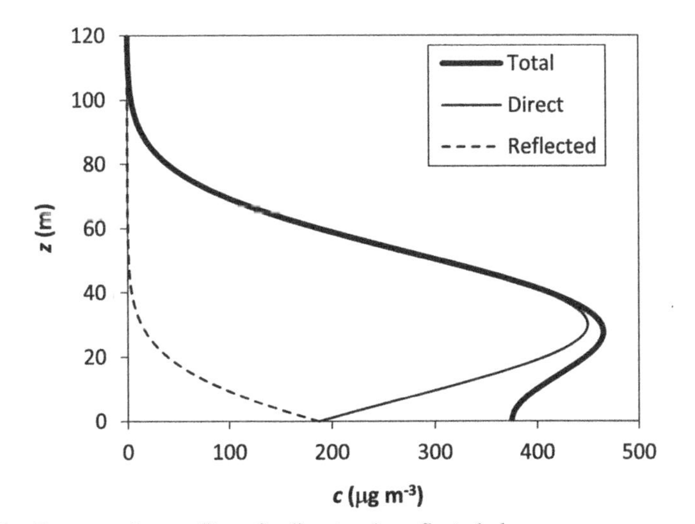


Figure 2.6 Concentration profiles of a direct and a reflected plume.

plume reflection into account. Without the reflection, the ground-level concentration would be underestimated by a factor of 2.

From looking at eq. (2.2), a number of observations can be made directly:

- The pollutant concentration is proportional to the pollutant emission.
- The pollutant concentration is inversely proportional to the wind speed: High pollutant concentrations can be expected at low wind speeds.
- If emission and wind speed affect the plume rise, then the above two statements are only approximations.

Further inferences will be made based on calculated concentration profiles. First, we need to find a way to calculate $\sigma_y$ and $\sigma_z$.

***2.3.2.2 Dispersion Parameters—Stability Classes*** In order to calculate $\sigma_y$ and $\sigma_z$, we need to know something about the weather conditions at the site of the emission. Until recently, most air dispersion calculations were made based on **stability classes**. Recently, more sophisticated approaches have found their way into the regulatory models. These will be discussed in Chapter 6. At this point, stability classes will suffice. The most commonly used classification of atmospheric stability was developed by Pasquill and Gifford (Pasquill, 1961; Gifford, 1961). They defined six classes, named A through F, with A the most unstable class, D neutral atmosphere, and F the most stable class:

A = Very unstable

B = Moderately unstable

C = Slightly unstable

D = Neutral

E = Slightly stable

F = Stable

Criteria for each class are given in Table 2.1. The criteria for solar radiation are not sharply defined for all possible conditions and require some interpretation by the user. Hence, stability classes are subjective to some degree. As the evaluation of the incoming solar radiation depends on the position of the sun above the horizon, the stability class will depend on the latitude. The key observations to remember from Table 2.1 are the following:

- The atmosphere is unstable on sunny days, neutral on overcast days and nights, and stable on clear nights.
- Increasing the wind speed leads to more neutral conditions.

In the original formulation of the Gaussian dispersion model the dispersion parameters were read from graphs (e.g., Turner, 1970). Several studies have put forward empirical equations to describe these graphs. The most succesful ones are by Briggs (1973), who suggested equations that reflect the autocorrelated stochastic nature of the dispersion process (see Chapters 5 and 6). The equations were calibrated using an extensive data set. The result is shown in Table 2.2 for open (rural)

**TABLE 2.1 Criteria for Pasquill–Gifford Stability Classes**

| | Day | | | Night | |
|---|---|---|---|---|---|
| | Incoming solar radiation | | | Cloudiness | |
| $u$ (m s$^{-1}$)[a] | Strong[b] | Moderate[c] | Slight | Cloudy($\geq 4/8$)[d] | Clear ($\leq 3/8$)[e] |
| <2 | A | A–B[f] | B | E | F |
| 2–3 | A–B | B | C | E | F |
| 3–5 | B | B–C | C | D | E |
| 5–6 | C | C–D | D | D | D |
| >6 | C | D | D | D | D |

Source: After Turner and Schulze (2007).
[a] Measured at 10-m height.
[b] Clear summer day with sun higher than 60° above the horizon.
[c] Summer day with a few broken clouds or a clear day with the sun 35–60° above the horizon.
[d] Fall afternoon or cloudy summer day with the sun 15–35° above the horizon.
[e] Fractional cloud cover.
[f] Take average dispersion values of two classes.
Note: *Always* use class D for overcast conditions.

**TABLE 2.2 Briggs (1973) Equations for Dispersion Parameters in Rural Terrain (Pasquill–Gifford Parameters)[a]**

| Stability Class | $\sigma_y$ (m) | $\sigma_z$ (m) |
|---|---|---|
| A | $0.22x(1+0.0001x)^{-0.5}$ | $0.2x$ |
| B | $0.16x(1+0.0001x)^{-0.5}$ | $0.12x$ |
| C | $0.11x(1+0.0001x)^{-0.5}$ | $0.08x(1+0.0002x)^{-0.5}$ |
| D | $0.08x(1+0.0001x)^{-0.5}$ | $0.06x(1+0.0015x)^{-0.5}$ |
| E | $0.06x(1+0.0001x)^{-0.5}$ | $0.03x(1+0.0003x)^{-1}$ |
| F | $0.04x(1+0.0001x)^{-0.5}$ | $0.016x(1+0.0003x)^{-1}$ |

[a] $x$ is the distance to the source in meters.

**TABLE 2.3 Briggs (1973) Equations for Dispersion Parameters in Urban Terrain (McElroy–Pooler Parameters)[a]**

| Stability Class | $\sigma_y$ (m) | $\sigma_z$ (m) |
|---|---|---|
| A–B | $0.32x(1+0.0004x)^{-0.5}$ | $0.24x(1+0.0001x)^{0.5}$ |
| C | $0.22x(1+0.0004x)^{-0.5}$ | $0.2x$ |
| D | $0.16x(1+0.0004x)^{-0.5}$ | $0.14x(1+0.0003x)^{-0.5}$ |
| E–F | $0.11x(1+0.0004x)^{-0.5}$ | $0.08x(1+0.0015x)^{-0.5}$ |

[a] $x$ is the distance to the source in meters.

terrain and in Table 2.3 for urban or industrial terrain. The latter were based on the experimental work of McElroy and Pooler (1968).

The equations apply up to a 10-km distance and become increasingly unreliable at larger distances. They are *not recommended* beyond a 30-km distance.

It is generally accepted that these and other calculation schemes of the dispersion parameters represent *10-min. averages*. However, the U.S. Environmental

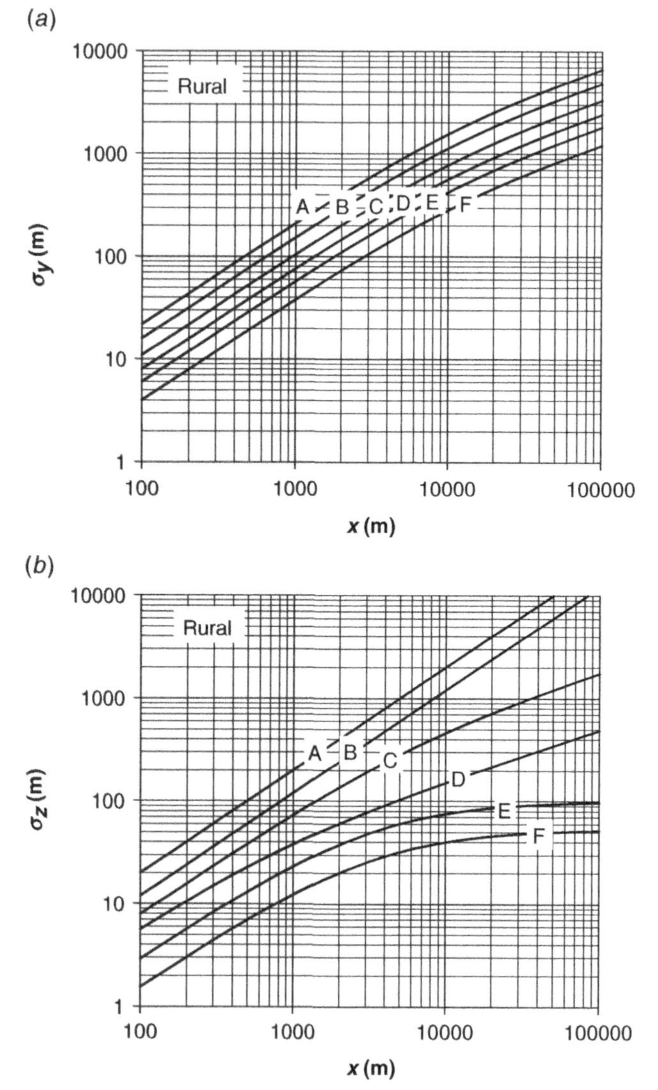


Figure 2.7 (*a*) Horizontal and (*b*) vertical dispersion parameters for rural terrain based on Briggs (1973).

Protection Agency (EPA) treats them as if they are 1-h averages (Beychok, 2005). Consequently, many models, including CALPUFF, use the above equations as if they are hourly averages (Scire et al., 2000a). The significance of this will be discussed further. Note that the Briggs equations, while still the default setting at the time of writing, are by no means the only option in CALPUFF. It is anticipated that the default setting in future releases of CALPUFF will be based on the approaches discussed in Chapters 5 and 6 (J.S. Scire, personal communication).

For quick reference, the Briggs equations are plotted in Figure 2.7 (rural) and Figure 2.8 (urban). The data is included in the enclosed CD (file "Briggs Dispersion Parameters.xlsx").

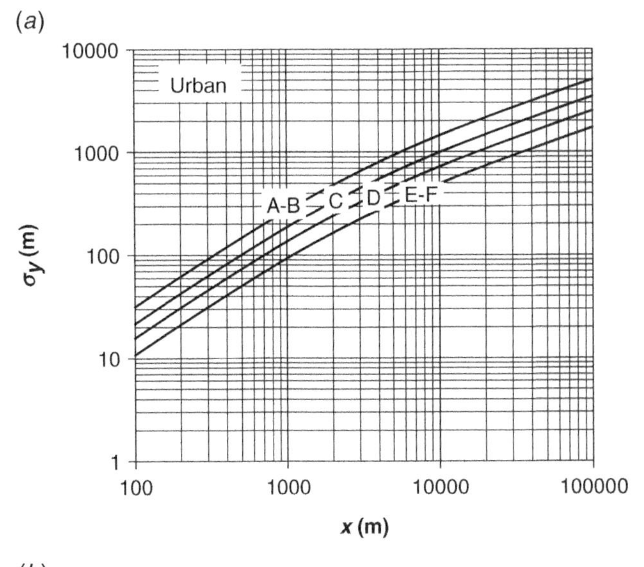


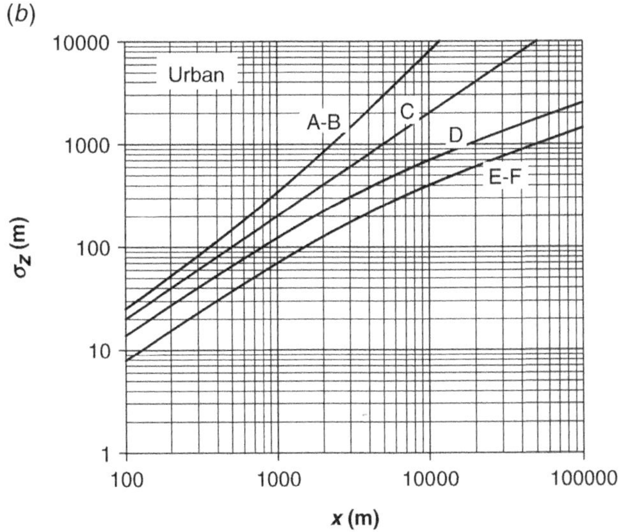


Figure 2.8 (*a*) Horizontal and (*b*) vertical dispersion parameters for urban terrain based on Briggs (1973).

With this introduction, we are ready to attempt our first air dispersion calculation.

**Example 2.1.** A coal-fired power plant in rural Pennsylvania emits 100 g s$^{-1}$ $SO_2$ from a stack with height 75 m. Plume rise is 15 m. Wind speed at the effective emission height is 7 m s$^{-1}$. The weather is overcast. Calculate the $SO_2$ concentration at ground level 1.5 km downwind from the source at the plume centerline (i.e., directly downwind) and at 100 m lateral distance from the plume centerline. Compare the values with the 24-h ambient air quality standard of 0.14 ppm (assume a temperature of 15 °C, a sea-level corrected barometric pressure of 1 atm, and an altitude of 200 m for the concentration conversion). Do the calculations warrant a further investigation with a more sophisticated model?

***Solution.*** In overcast conditions the stability class is D.

At 1500 m from the source the dispersion parameters are calculated by the Briggs equations for rural conditions:

$$\sigma_y = 0.08 \cdot 1500 \cdot (1 + 0.0001 \cdot 1500)^{-0.5} = 111.9\,\text{m}$$

$$\sigma_z = 0.06 \cdot 1500 \cdot (1 + 0.0015 \cdot 1500)^{-0.5} = 49.9\,\text{m}$$

At the plume centerline ($y=0$), at ground level ($z=0$) the concentration is given by

$$\begin{aligned} C &= \frac{100}{2\pi \cdot 7 \cdot 111.9 \cdot 49.9} \exp\left(-\frac{1}{2}\frac{0^2}{\sigma_y^2}\right)\left\{\exp\left[-\frac{1}{2}\frac{(0-90)^2}{49.9^2}\right] + \exp\left[-\frac{1}{2}\frac{(0+90)^2}{49.9^2}\right]\right\} \\ &= 160.3 \times 10^{-6}\,\text{g m}^{-3} \\ &= \mathbf{160.3\,\mu g\,m^{-3}} \end{aligned}$$

At 100-m lateral distance from the plume centerline ($y = 100$ m) the concentration is given by

$$\begin{aligned} C &= \frac{100}{2\pi \cdot 7 \cdot 111.9 \cdot 49.9} \exp\left(-\frac{1}{2}\frac{100^2}{111.9^2}\right)\left\{\exp\left[-\frac{1}{2}\frac{(0-90)^2}{49.9^2}\right] + \exp\left[-\frac{1}{2}\frac{(0+90)^2}{49.9^2}\right]\right\} \\ &= 160.3 \exp\left(-\frac{1}{2}\frac{100^2}{111.9^2}\right) \\ &= \mathbf{107.5\,\mu g\,m^{-3}} \end{aligned}$$

The ambient air quality standard of 0.14 ppm $SO_2$ is converted to the same units as the air dispersion calculations:

Ideal gas law ($V = nRT/p$): volume of 1 million moles of gas
$= 1{,}000{,}000\,\text{mol} \times 8.314472\,\text{J mol}^{-1}\,\text{K}^{-1} \times (15 + 273.15)\,\text{K} / [101{,}325 \exp(-0.00012 \times 200)\,\text{Pa}]$
$= 24{,}219\,\text{m}^3$

One million moles of gas contains 0.14 mol $SO_2$ or

$$0.14\,\text{mol} \times 64\,\text{g mol}^{-1} = 8.96\,\text{g SO}_2$$

The concentration is

$$\frac{8.96\,\text{g}}{24{,}219\,\text{m}^3} = 0.000370\,\text{g}\,\text{m}^{-3} = 370\,\mu\text{g}\,\text{m}^{-3}$$

Without further calculations, we have no indication whether the calculated value of $160.3\,\mu g\,m^{-3}$ represents a worst case. Under the given meteorological conditions the highest concentration is found at about 1900m from the source, and amounts to about $170\,\mu g\,m^{-3}$. If other meteorological conditions can be discarded, then the worst case is more than a factor of 2 below the standard, and more sophisticated calculations are not required. But very likely more unstable conditions and lower wind speeds will lead to higher concentrations, which would prompt more sophisticated calculations.

Comparison of Figure 2.7 with Figure 2.8 shows that dispersion parameters for urban terrain are generally larger than the ones for rural terrain. This is due to the extra turbulence generated by heat (thermal turbulence) and by the shear exerted by the buildings on the air going past. The difference can be a factor of 2 or more for each of $\sigma_y$ and $\sigma_z$. This means that concentration predictions of the two types of terrain can differ by a factor of 4. In practice, all intermediate conditions between fully rural and fully urban exist. Clearly a better parameterization of terrain would be helpful. This problem will be tackled in further chapters.

***2.3.2.3 Example Concentration Profiles at Ground Level*** For a better understanding of what affects pollutant concentrations, it is useful to look at a number of calculation results.

Figure 2.9 shows typical concentration results versus distance (stability class D, rural terrain, $u=2\,m\,s^{-1}$, $Q=5\,g\,s^{-1}$, $h=30\,m$). The data is included in the enclosed CD ("Concentration vs distance.xlsx"). It is seen that, while the plume axis ($y=0$, $z=h$) concentration decreases with increasing distance, the ground-level concentration passes through a maximum. Why?

A more unexpected result is that the ground-level concentration exceeds the plume axis concentration at large distances. Why?

Figure 2.10 shows the ground-level pollutant concentration for each of the six stability classes. It can be observed that at sufficiently large distance, the concentrations under unstable conditions are lower than the concentrations at stable conditions. This is because the turbulence dilutes the pollutant. As a result, the impact zone of an air emission is always smaller under unstable conditions than under stable conditions. However, the peak concentration is highest under unstable conditions. This is because turbulence has a larger impact on the vertical dispersion than on the horizontal dispersion. Remember that unstable conditions correspond with sunny days, whereas stable conditions correspond with clear nights. During the day, high pollutant concentrations can occur but they dissipate quickly. At night, pollutants can travel much further than during daytime.

Under most circumstances the difference between concentrations occurring in two adjacent stability classes is less than a factor of 2, but close to the source and far (>5km) away from the source the difference can be much more. Likewise, away from the plume centerline ($y \neq 0$) the difference between consecutive stability classes

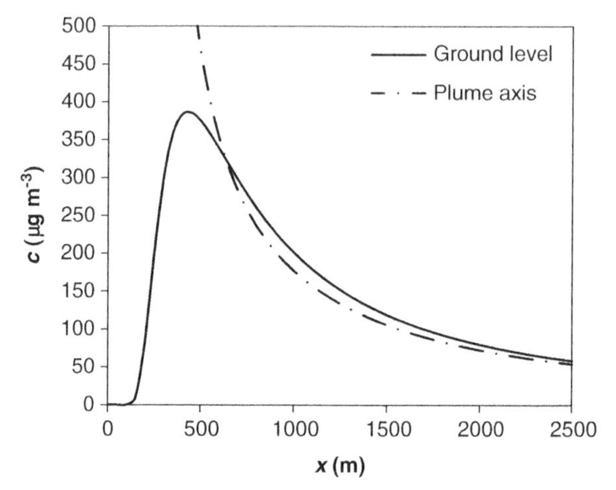


Figure 2.9 Pollutant concentration versus distance from the source.

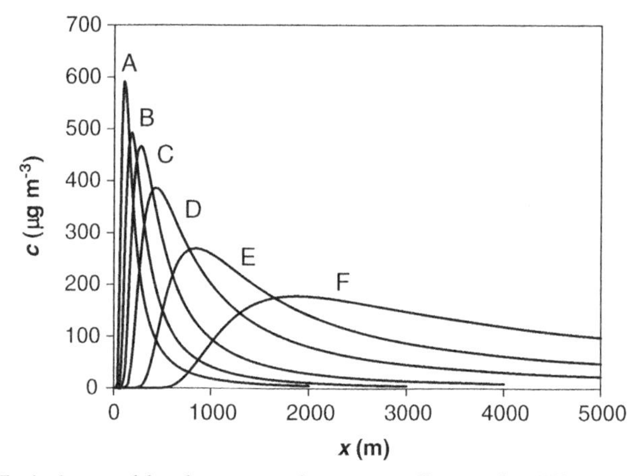


Figure 2.10 Typical ground-level concentrations versus distance for different stability classes.

can be substantially more than a factor of 2. To avoid this, a continuous measure of stability is needed. Such measures will be discussed in Chapters 5 and 6.

Figure 2.11 shows the influence of the effective source height $h$ on pollutant concentrations downwind from the source. The conditions are the same as for Figure 2.9. It is clear that increasing the source height decreases the ground-level concentrations. It also increases the distance of the peak concentration. At large distances, the influence of source height is very limited.

### 2.3.3 Refinements

***2.3.3.1 Height Dependence of Wind Speed*** So far the wind speed at the effective source height was treated as a given. In practice, wind speed is usually known at a different height (at best). Therefore, an equation is needed to relate wind

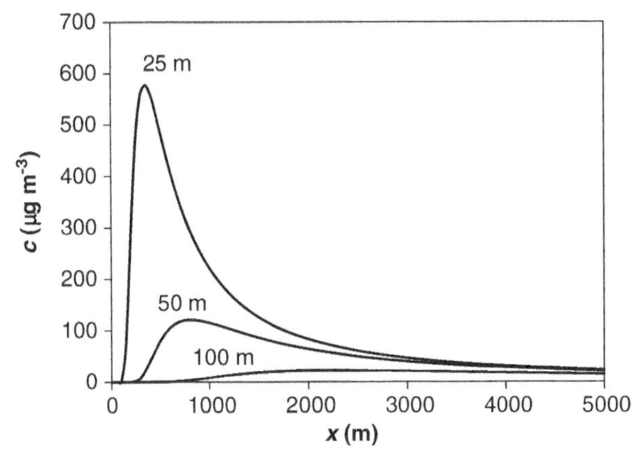


Figure 2.11 Ground-level concentrations versus distance downwind from the source, for three different effective source heights.

**TABLE 2.4 Values of *p* for Use in eq. (2.3) to Predict Wind Speed Profiles**

| Stability Class | *p* for Rural Terrain | *p* of Urban Terrain |
|---|---|---|
| A | 0.11 | 0.15 |
| B | 0.12 | 0.15 |
| C | 0.12 | 0.20 |
| D | 0.17 | 0.25 |
| E | 0.29 | 0.40 |
| F | 0.45 | 0.60 |

Source: Beychok (2005).

speeds at different heights. A sophisticated treatment of wind velocity profiles will be given in Chapter 5. For now an empirical equation will suffice. The following equation is commonly used:

$$u_2 = u_1 \left( \frac{z_2}{z_1} \right)^p \tag{2.3}$$

where $u_1$ and $u_2$ are wind speeds 1 and 2, and $z_1$ and $z_2$ are heights 1 and 2. There is no agreement in the literature concerning recommended values for $p$. Based on a comparison of Arya (1999), Scire et al. (2000b), Cooper and Alley (2011), Beychok (2005), on estimates based on Seinfeld and Pandis (2006), and on comparison with more sophisticated meteorological theory, the values given in Table 2.4 are put forward. They are values of Beychok, 2005, based on Touma (1977) and Petersen (1978), for rural and urban terrain, respectively.

Calculated wind speed profiles for urban terrain are shown in Figure 2.12 at constant wind speed at a 10-m height. It is clear that the most pronounced wind speed profiles are found in a stable atmosphere.

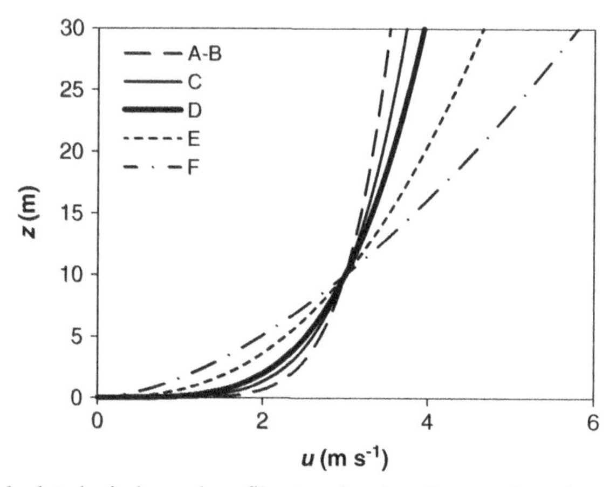


Figure 2.12 Calculated wind speed profiles in urban terrain. $u_{10} = 3\,\mathrm{m\,s^{-1}}$.

**Example 2.2.** Assume that the wind speed in Example 2.1 was measured at a 10-m height, not at the effective source height. How would that affect the calculation?

***Solution.*** For stability class D in rural terrain a value for $p$ of 0.17 is used. This leads to the following wind speed at a 90-m height:

$$
\begin{aligned}
u &= 7\,\mathrm{m\,s^{-1}} \cdot (90/10)^{0.17} \\
&= 10.17\,\mathrm{m\,s^{-1}}
\end{aligned}
$$

This reduces the ground-level centerline concentration at a 1.5-km distance from $160.3\,\mu\mathrm{g\,m^{-3}}$ to $110.3\,\mu\mathrm{g\,m^{-3}}$, and at a 100-m lateral distance from that point from $107.5\,\mu\mathrm{g\,m^{-3}}$ to $74.0\,\mu\mathrm{g\,m^{-3}}$. It follows that height corrections for wind speed can be substantial.

***2.3.3.2 Temperature Inversion Layer*** A temperature inversion layer is a layer in the atmosphere where the temperature increases with height. Such layers are extremely stable, and act as an effective barrier to further dispersion. Most daytime atmospheres are capped by a temperature inversion, confining dispersion to a mixed layer (or mixing layer) with a thickness ranging from a few hundreds of meters to a few kilometers. In the nighttime atmosphere the temperature inversion usually reaches the surface, and the mixing layer is defined by mechanical mixing effects.

When the atmosphere has a well-defined mixing layer, this needs to be accounted for in the calculation of air dispersion. This is possible by considering the inversion layer capping the mixing layer as another reflecting layer. Hence, the plume is represented as trapped between two reflecting layers, as shown in Figure 2.13.

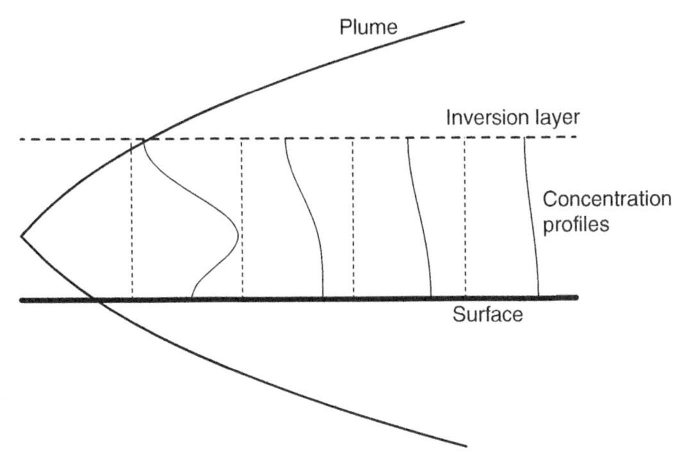


Figure 2.13 Plume dispersion between the Earth's surface and an elevated temperature inversion (after Cooper and Alley, 2011).

In this case, the concentration in the plume can be described by the following equation:

$$C = \frac{Q}{2\pi u \sigma_y \sigma_z} \exp\left(-\frac{1}{2}\frac{y^2}{\sigma_y^2}\right) \sum_{j=-\infty}^{+\infty} \left\{ \exp\left[-\frac{1}{2}\frac{(z-h+2jh_{\text{mix}})^2}{\sigma_z^2}\right] + \exp\left[-\frac{1}{2}\frac{(z+h+2jh_{\text{mix}})^2}{\sigma_z^2}\right] \right\} \tag{2.4}$$

where $h_{\text{mix}}$ (m) is the distance of the temperature inversion from the ground, also known as the height of the mixed layer, or the height of the mixing layer. After a few reflections the plume can be considered well mixed. In practice, eq. (2.4) can be limited to $j = -1, 0, +1$ to within 2% for values of $\sigma_z$ up to $h_{\text{mix}}$. Once $\sigma_z$ exceeds $h_{\text{mix}}$, the concentration can be approximated to within 1.5% by the following approximate equation, which assumes complete mixing in the vertical direction up to $h_{\text{mix}}$:

$$C = \frac{Q}{\sqrt{2\pi} u \sigma_y h_{\text{mix}}} \exp\left(-\frac{1}{2}\frac{y^2}{\sigma_y^2}\right) \tag{2.5}$$

CALPUFF uses this equation once $\sigma_z > 1.6\ h_{\text{mix}}$. From eq. (2.5) we can see that $C \sim 1/h_{\text{mix}}$ under these conditions. Hence, *good estimates of $h_{\text{mix}}$ are important when a plume is trapped under an inversion layer*. Ways to estimate $h_{\text{mix}}$ when no measured value is available are presented in Chapter 5.

## 2.4 PLUME RISE

### 2.4.1 Plume Rise Correlations

There are two possible reasons why a plume rises when it leaves the stack:

- Because of its buoyancy (in case of hot exhaust gases)
- Because of its momentum (in case of high velocity)

The momentum contribution to plume rise is usually small and will be ignored here.

Buoyancy is usually the result of a temperature difference, but it can also occur when the waste gas has a molar mass that deviates considerably from the molar mass of air. However, that would mean that the pollutants are highly concentrated. In practice, regulators will not allow emissions in such conditions, and waste gas treatment will be fairly simple. Hence, this case will not be considered here. It is usually reasonable to assume that the waste gas has the same molecular weight as air. Accounting for buoyancy due to molecular weight differences can be important in the case of accidental releases and is especially worrisome when the gas is heavier than air. The EPA model SLAB (Ermak, 1990; http://www.epa.gov/scram001/dispersion_alt.htm) is specifically designed for that purpose.

Several equations have been proposed to predict plume rise. Unfortunately, the predictions of the different models are more than a factor of 10 apart (Briggs, 1975). Of these, Briggs (1968) developed the most well-conceived equations. They are used in many regulatory models.

First, the buoyancy flux parameter $F_b$ is defined:

$$F_b = \left(1 - \frac{\rho_s}{\rho}\right) g r_s^2 w_s \tag{2.6}$$

with $\rho_s$ the density of the stack gas, $\rho$ the density of the surrounding air, $g$ the acceleration due to gravity (9.80665 m s$^{-2}$), $r_s$ (m) the stack radius, and $w_s$ (m s$^{-1}$) the stack gas velocity in the vertical direction.

The *transitional* plume rise is the local plume rise, before the plume has reached its maximum height. It is given by the following equation (Briggs, 1972):

$$\Delta h = \frac{1.6 F_b^{1/3} x^{2/3}}{u} \tag{2.7}$$

where $x$ (m) is the distance downwind from the source, and $u$ is the wind speed (m s$^{-1}$). However, plumes do not rise indefinitely but stabilize at a certain height, the *final* plume rise height. This height is achieved at a distance $x_f$ (m) from the source (Briggs, 1975):

$$x_f = 49 F_b^{5/8} \qquad \text{for} \quad F_b < 55\,\mathrm{m^4\,s^{-3}} \tag{2.8}$$

$$x_f = 119F_b^{2/5} \quad \text{for} \quad F_b > 55\,\text{m}^4\,\text{s}^{-3} \tag{2.9}$$

Note that eqs. (2.8) and (2.9) are dimensionally not homogeneous and are only valid when metric units are used. At distances greater than $x_f$, the plume rise is assumed constant and given by

$$\Delta h = \frac{1.6F_b^{1/3}x_f^{2/3}}{u} \tag{2.10}$$

**Example 2.3.** Calculate the plume rise at a 1000-m distance of a waste gas stream leaving a stack at 100 °C into an atmosphere at 25 °C and 100 kPa. The flow rate is $20\,\text{m}^3\,\text{s}^{-1}$; the stack diameter is 2 m. Wind speed at the stack exit is $3\,\text{m}\,\text{s}^{-1}$.

***Solution.*** First, we check if the plume rise is transitional or final at $x = 1000$ m. For that we need $F_b$. The variables of eq. (2.7) are

$$\rho = 1.170\,\text{kg}\,\text{m}^{-3} \quad (\text{see Problem 8})$$

$$\rho_s = \rho \times 298.15 / 373.15 = 0.935\,\text{kg}\,\text{m}^{-3}$$

$$g = 9.80665\,\text{m}\,\text{s}^{-2}$$

$$r_s = 1\,\text{m}\,(\text{diameter} / 2)$$

$$w_s = Q / A_s$$

where

$Q$ = volumetric flow rate ($20\,\text{m}^3\,\text{s}^{-1}$)

$A_s$ = stack cross-sectional area ($= \pi r_s^2 = 3.1416\,\text{m}^2$)

Hence $w_s = 6.366\,\text{m}\,\text{s}^{-1}$.

Substitution into eq. (2.7) yields

$$F_b = 12.55\,\text{m}^4\text{s}^{-3}$$

This is substituted into eq. (2.8) ($F_b < 55\,\text{m}^4\,\text{s}^{-3}$):

$$x_f = 238.1\,\text{m} < x$$

It follows that the final plume rise is reached. Equation (2.10) is used, with $u = 3\,\text{m}\,\text{s}^{-1}$:

$$\mathbf{\Delta h = 47.6\,m}$$

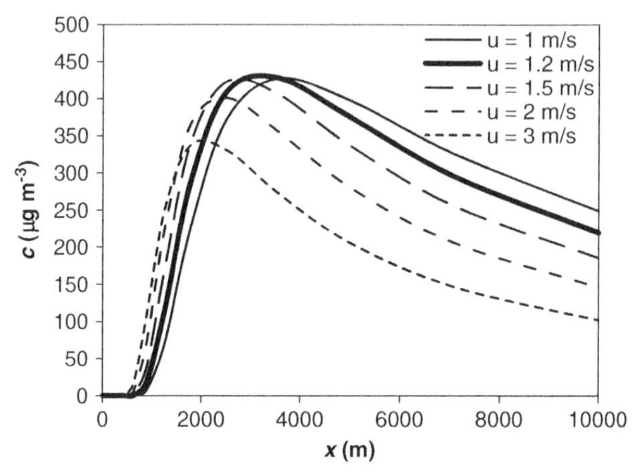


Figure 2.14 Concentration at ground level versus distance from the source at different wind speeds. Critical wind speed is about 1.2 m s$^{-1}$.

Plume rise is a complex process and depends on different factors that in turn depend on atmospheric conditions. Hence, one set of equations cannot cover all cases, and eqs. (2.7) and (2.10) should be used with care. Here are some issues:

- When the wind speed is low, eq. (2.7) predicts infinite plume rise. This is unrealistic.
- The equations are not designed for strongly unstable or strongly stable conditions.
- The equations do not account for momentum-dominated plume rise.

These issues will be considered in Chapter 7.

## 2.4.2 Critical Wind Speed

From eq. (2.2) it is clear that the concentration of a pollutant in a plume *decreases* with increasing wind speed. On the other hand, eq. (2.7) shows that the plume rise decreases with increasing wind speed, which leads to a *increase* of the concentration with increasing wind speed. The overall result of these counteracting effects is that the concentration passes through a maximum at a certain wind speed. This wind speed is known as the **critical wind speed**. It can be determined by trial and error. To illustrate the principle of critical wind speed, ground-level concentrations were calculated at different wind speeds for a source with considerable plume rise ($h_s$ = 75 m, $Q$ = 100 g s$^{-1}$, rural, stability class D, $F_b$ = 4 m$^4$ s$^{-3}$). The result is shown in Figure 2.14. Here the critical wind speed is around 1.2 m s$^{-1}$. The file that generated the figure and calculated the critical wind speed ("Figure 2.14. Critical wind speed. xlsx") is included on the enclosed CD. The critical wind speed was calculated with the Solver function (menu Data → Solver). In the file, the Solver is configured

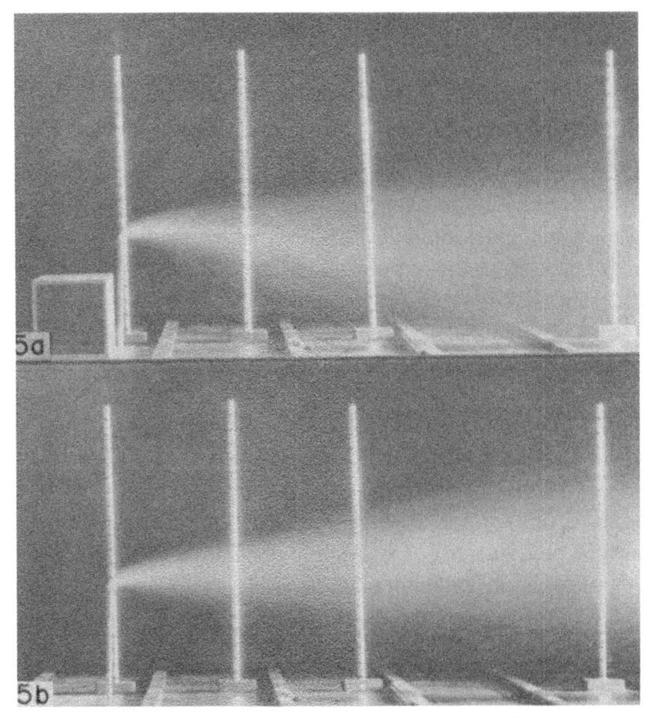


Figure 2.15 Plume dispersion in a wind tunnel experiment with downwash (top) and without downwash (bottom). [Reprinted from Huber and Snyder (1982) with permission from Elsevier.]

to maximize the cell I42 (concentration) by changing cells I4 (wind speed) and A42 (distance). A wind speed of $1.21096\,\mathrm{m\,s^{-1}}$ is obtained, with a maximum concentration of $432.5773\,\mu\mathrm{g\,m^{-3}}$ at a distance 3157.663 m from the source.

The critical wind speed is a crucial concept in screening calculations. It represents the worst-case scenario for the impact of an emission on ambient concentrations.

### 2.4.3 Rules of Thumb

The plume rise equations discussed above are only valid when there is no interference between the plume and surrounding buildings. A well-known interference is **downwash**: the plume gets trapped in the wake of a building or is dragged down by the air that follows the wake. An illustration of downwash from a wind tunnel experiment is shown in Figure 2.15 (Huber and Snyder, 1982).

Downwash decreases the effective source height and increases ground-level concentrations. To avoid downwash, the following rules of thumb are generally accepted:

- The stack should be at least 2.5 times as tall as the tallest of the nearest buildings.
- The exit velocity of the exhaust gas should be at least 1.5 times the maximum expected wind speed.

These criteria cannot always be met. For instance, the second criterion can lead to unrealistically high pressure drops in the stack. Sometimes this criterion can be met by narrowing the stack at the top only, thus limiting the friction-based pressure drop. The main reason to impose this criterion is to avoid that the plume gets trapped in the wake of the stack itself (**stack downwash**).

## 2.5 NEED FOR REFINEMENTS TO THE BASIC GAUSSIAN PLUME DISPERSION MODEL

From the above sections it will be clear that the basic Gaussian dispersion model is far from perfect. Here is a list of issues that need to be resolved in order to obtain a universally applicable dispersion model:

- Sources that are variable in time
- Changing wind speed and wind direction
- Incorporating landscapes (hills, valleys) in dispersion models
- Sources close to the ground, where the wind speed profile is pronounced
- Vertical turbulence profile
- Plume deposition
- Chemical reaction
- Avoiding the use of dispersion classes
- Terrain characterization that is more refined than rural–urban settings
- Better wind speed profiles
- Ways to estimate the height of the mixing layer
- More universal methodology to estimate plume rise

All these issues will be addressed in Chapters 5–11.

## PROBLEMS

1. Show that the units of eq. (2.1) are consistent. (*Hint*: exponentials are always dimensionless.)
2. Simplify eq. (2.2) for the case of ground-level concentrations. Simplify further for locations exactly downwind from the source (the plume centerline).
3. What is the highest possible elevation of the sun above the horizon in the area where you live?

4. What is the highest possible latitude for stability class A to occur?
5. Compare the peak concentrations in Figure 2.11. What ratio are they? What is the ratio you would expect?
6. Based on eq. (2.3) and Table 2.4, how does the wind speed profile compare with the velocity profile of a fluid in a pipe?
7. Show that the units of $F_b$ are $m^4\ s^{-3}$.
8. Calculate the density of air at standard ambient temperature and pressure (SATP) (25 °C, 100,000 Pa). The molar mass of air is $29\,g\,mol^{-1}$. Use the ideal gas law.
9. A stack with a height of 50 m emits $20\,g\,s^{-1}$ of $NO_2$. Plume rise is negligible. The wind speed is $2.5\,m\,s^{-1}$. Stability class is C. The terrain is rural. Make a plot of the downwind $NO_2$ concentration at ground level at the centerline as a function of distance from the stack. Determine the maximum concentration and the distance where this maximum is observed. Make a plot of the ground-level $NO_2$ concentration versus distance from the centerline on a transect through the concentration maximum (i.e., a lateral, or cross-wind concentration profile).
10. As a result of odor complaints, a "nasal ranger" is sent to the site of the odor for an investigation. The nasal ranger reports a distinct $H_2S$ smell downwind of a gas plant, up to about 5 km downwind of the plant. The investigation was done on a cloudy fall day at 10 °C, with a wind speed at a 10-m height of $7\,m\,s^{-1}$ and a barometric pressure of 88 kPa. Under lab conditions the nasal ranger could smell $H_2S$ concentrations down to 1 ppb. The gas plant is located in a rural area. All the waste gas is emitted through a 25-m-high stack with inside diameter at the top of 1 m. The plant emits $7\,m^3$ (STP) gas per second containing mainly air, at a temperature of 70 °C. Estimate the $H_2S$ emission based on this information.
11. In an experiment to characterize a particulate matter emission, an array of monitors is placed 200 m downwind from the source to form a cross-wind transect. The monitors are essentially at ground level. The terrain is rural. The effective source height, based on visual inspection of the plume, is determined to be 10 m. The wind speed at a 10-m height is $2\,m\,s^{-1}$. The average concentration measured at the monitors are the following:

| $y$ (m) | $c$ ($\mu g\,m^{-3}$) |
|---|---|
| −50 | 0 |
| −40 | 2 |
| −30 | 20 |
| −20 | 120 |
| −10 | 350 |
| 0 | 500 |
| 10 | 340 |
| 20 | 125 |
| 30 | 25 |
| 40 | 1.5 |
| 50 | 0 |

Determine the stability class that corresponds best with the data, as well as the emission.

## MATERIALS ONLINE

- "Briggs dispersion parameters.xlsx": data to Figures 2.7 and 2.8
- "Concentration vs distance.xlsx": data to Figure 2.9
- "Figure 2.14. Critical wind speed.xlsx": data to Figure 2.14, with calculation of the critical wind speed

## REFERENCES

Arya S.P. (1999). *Air Pollution Meteorology and Dispersion*. Oxford University Press, Oxford, UK.

Beychok M.R. (2005). *Fundamentals of Stack Gas Dispersion*, 4th ed. Beychok, Newport Beach, CA.

Briggs G.A. (1968). Momentum and buoyancy effects. In Slade D.H. (ed.) *Meteorology and Atomic Energy*. US Department of Energy, Technical Information Center, Oak Ridge, TN, pp. 189–202.

Briggs G.A. (1972). Chimney plumes in neutral and stable surroundings. *Atmos. Environ.* **6**, 507–510.

Briggs G.A. (1973). *Diffusion Estimation of Small Emissions*. Contribution No. 79, Atmospheric Turbulence and Diffusion Laboratory, Oak Ridge, TN.

Briggs G.A. (1975). Plume rise predictions. In *Lectures on Air Pollution and Environmental Impact Analyses*. American Meteorological Society, Boston, MA, pp. 59–111.

Cooper C.D. and Alley F.C. (2011). *Air Pollution Control*, 4th ed., Waveland Press, Long Grove, IL.

EPA (2005). Revision to the Guideline on Air Quality Models: Adoption of a Preferred General Purpose (Flat and Complex Terrain) Dispersion Model and Other Revisions. *Federal Register*, **70**, No. 216, 68218–68261. (http://www.epa.gov/scram001/guidance/guide/appw_05.pdf).

Ermak D.L. (1990). *User's Manual for SLAB: An Atmospheric Dispersion Model of Denser-Than-Air Releases*. Livermore, CA, UCRL-MA-105607.

Gifford F.A. (1961). Use of routine meteorological observations for estimating atmospheric dispersion. *Nuclear Safety* **2**, 47–51.

Huber A.H. and Snyder W.H. (1982). Wind tunnel investigation of the effects of a rectangular-shaped building on dispersion of effluents from short adjacent stacks. *Atmos. Environ.* **16**, 2837–2848.

Idriss A. and Spurrell F. (2009). *Air Quality Model Guideline*. Government of Alberta, Edmonton, Alberta, Canada. (http://environment.gov.ab.ca/info/library/8151.pdf).

McElroy J.L. and Pooler F. (1968). *The St. Louis Dispersion Study*. Report AP-53, US Public Health Service, National Air Pollution Control Administration, Arlington, VA.

Pasquill F. (1961). Estimation of the dispersion of windborne material. *Meteorol. Mag.* **90**, 33–49.

Petersen W.B. (1978). *User's Guide for PAL. A Gaussian-Plume Algorithm for Point, Area, and Line Sources*. Report EPA-600/4-78-013. US-EPA, Research Triangle Park, NC.

Scire J.S., Strimaitis D.G., and Yamartino R.J. (2000a). *A User's Guide for the CALPUFF Dispersion Model*. Earth Tech, Concord, MA.

Scire J.S., Robe F.R., Fernau M.E., and Yamartino R.J. (2000b). *A User's Guide for the CALMET Meteorological Model*. Earth Tech, Concord, MA.

Seinfeld J.H. and Pandis S.N. (2006). *Atmospheric Chemistry and Physics*, 2nd ed. Wiley, Hoboken, NJ.

Touma J.S. (1977). Dependence of the wind profile power law on stability for various locations. *J. Air Pollut. Control Assoc.* **27**, 863–866.

Turner D.B. (1970). *Workbook of Atmospheric Dispersion Estimates*. US EPA, Washington DC.

Turner D.B. and Schulze R.H. (2007). *Practical Guide to Atmospheric Dispersion Modeling*. Trinity Consultants, Inc., Dallas, TX, and Air and Waste Management Association, Pittsburgh.

CHAPTER 3

# AIR POLLUTANTS: AN OVERVIEW

## 3.1 INTRODUCTION

There is a nearly endless variety of pollutants in the atmosphere. Air pollutants differ in their physical properties (e.g., vapor pressure), chemical properties (e.g., reactivity), their state (gas, liquid, or solid), the way they enter the atmosphere (from a point source, a diffuse source, or from chemical reaction of a different air pollutant), the way they leave the atmosphere (chemical reaction, dry deposition, wet deposition), their atmospheric lifetime (from microseconds to thousands of years), to the reason they can be problematic (toxicity, visibility, reactivity, global warming potential). All these factors affect the proper choice of air dispersion model to describe them, as well as the scale and resolution needed for the implementation.

Hence, if we want to understand air pollution, it is necessary to have a feel for what air pollutants we need to look for, at what concentrations they are problematic, and why. The objective of this chapter is to provide that background.

## 3.2 TYPES OF AIR POLLUTION

There are several ways to categorize air pollutants. One way is based on the *properties* of the pollutants. This leads to the following classification:

- Particulate matter (PM)
- Gases: NO, $NO_2$, $SO_2$, $H_2S$
- Vapors: volatile organic compounds
- Secondary pollutants: $O_3$, $H_2SO_4$—pollutants that are not emitted directly but produced in the atmosphere from other pollutants
- Greenhouse gases
- Ozone depleting substances: chlorofluorocarbons (CFCs)

*Air Dispersion Modeling: Foundations and Applications*, First Edition. Alex De Visscher.

A different way to categorize air pollutants is according to the *chemistry* of the compounds:

- Sulfur compounds: $H_2S$, $SO_2$, dimethyl sulfide, and so forth
- Nitrogen compounds: $N_2O$, NO, $NO_2$, $NH_3$, peroxyacetyl nitrate, and so forth
- Hydrocarbons: methane, benzene, and so forth
- Halocarbons: CFCs and so forth
- Metals: Hg, Pb, and so forth

This chapter will largely follow the second classification.

### 3.2.1 Sulfur Compounds

The *background* sulfur mixing ratio (i.e., the mixing ratio in unpolluted air) is on the order of 1 ppb. About half of the background is carbonyl sulfide (COS). The rest is mainly dimethyl sulfide, $SO_2$, and sulfate. Polluted air can contain much higher concentrations of sulfur: tens to hundreds of parts per billion. The main sulfur components in *polluted air* are $SO_2$ and sulfate. The global sulfur emission is estimated at 100–120 Mton S/year (Seinfeld and Pandis, 2006).

Dimethyl sulfide (DMS), or $CH_3SCH_3$, is mainly produced by algae in the oceans. Anthropogenic sources of DMS are anaerobic processes such as composting. It is not toxic at the concentrations normally encountered in ambient air, but it is strongly odorous (threshold level about 1 ppb). DMS oxidizes in the atmosphere to $SO_2$ and traces of COS in less than one day and is the main source of background $SO_2$ and COS.

Carbonyl sulfide is mostly produced by oxidation of DMS. The production is small, but due to its long atmospheric lifetime [6 years; Seinfeld and Pandis (2006)], it accumulates to about 0.52 ppb (Lelieveld et al., 1997). Almost two-thirds of the COS in the atmosphere is of natural origin.

A related air pollutant is carbon disulfide ($CS_2$). Although there are anthropogenic sources of $CS_2$ (as a solvent or by-product of some chemical processes), it is mostly of natural origin (oceans). The $CS_2$ concentration in the atmosphere is usually below 0.1 ppb.

The most important sulfur compound in the atmosphere in terms of emission (not counting sulfate from breaking waves) is sulfur dioxide ($SO_2$). The global emission is about 80 Mton S year$^{-1}$ or 160 Mton $SO_2$ year$^{-1}$. Almost 90% of this emission is from the combustion of fossil fuels. Typical concentrations in polluted air are 0.1–10 ppb. Atmospheric $SO_2$ oxidizes to sulfate in a couple of hours to a couple of days (Seinfeld and Pandis, 2006) by the following reactions:

$$SO_2 + {}^{\cdot}OH + M \rightarrow HOS^{\cdot}O_2 \tag{3.1}$$

$$HOS^{\cdot}O_2 + O_2 \rightarrow HO_2^{\cdot} + SO_3 \tag{3.2}$$

$$SO_3 + H_2O + M \rightarrow H_2SO_4 + M \tag{3.3}$$

and by reaction with hydroxen peroxide and ozone in cloud droplets ($M = O_2$ or $N_2$). The atmospheric reactions of $SO_2$ are discussed in more detail in Chapter 11.

It follows that the majority of sulfate in the atmosphere is secondary pollution. Global sulfate ($SO_4^{2-}$) emissions are on the order of 10 Mton S year$^{-1}$ (Seinfeld and Pandis, 2006). A possible exception is sulfate originating from the ocean: breaking waves create a spray that evaporates to PM. The amount of sulfate emitted this way is very uncertain (40–320 Mton year$^{-1}$; Seinfeld and Pandis, 2006). As a large fraction of this particulate matter deposits immediately, estimating this emission is somewhat arbitrary. Because sulfate occurs in the atmosphere as particulate matter, it will be discussed in more detail in Section 3.2.6.

$SO_2$ causes respiratory and cardiovascular diseases (Bernstein et al., 2004), and leads to increased mortality (Pope et al., 2002). It is also corrosive to concrete and marble. It is the main contributor to acid precipitation.

Hydrogen sulfide ($H_2S$) is an important air pollutant from a safety perspective. The global emission is on the order of 1 Mton S year$^{-1}$ and is mainly from volcanic activity and biological processes (Seinfeld and Pandis, 2006). However, it is the concentrated sources that are the most harmful ones as $H_2S$ is highly toxic (more toxic than CO). For that reason it is the number one safety issue in the petroleum industry, especially natural gas processing. The lethal concentration of $H_2S$ is less than 1000 ppm.

### 3.2.2 Nitrogen Compounds

The main nitrogen air pollutants are $N_2O$, $NH_3$, NO, and $NO_2$. The last two are usually categorized collectively as $NO_x$.

Nitrous oxide ($N_2O$) is a greenhouse gas responsible for about 6% of the anthropogenic greenhouse effect. Its background concentration was about 319 ppb in 2005, up from about 270 ppb before the industrial revolution (IPPC, 2007). The global warming potential, defined as the capacity of a tonne of greenhouse gas to trap heat in the atmosphere, relative to one tonne of $CO_2$ over a certain time horizon, is 298 over a 100-year time horizon. It has an atmospheric lifetime of 114 years. The global background emission of $N_2O$ is about 9 Mton N year$^{-1}$ and is coming from soils. The anthropogenic emission is about 7 Mton N year$^{-1}$ (Mosier et al., 1998). The main anthropogenic source of $N_2O$ is agriculture. $N_2O$ is emitted in nitrification and denitrification processes, which are very pronounced in fertilized soils. A brief discussion of the greenhouse effect will follow in Section 3.2.9.

The global emission of ammonia ($NH_3$) is about 60 Mton N year$^{-1}$. About two-thirds of this emission is from agriculture. Ammonia stays in the atmosphere for about 10 days. Typical concentrations are 0.1–10 ppb. It is odorous and irritating at high concentrations. In combination with $SO_2$ it will form particulate matter.

The global emission of $NO_x$ ($= NO + NO_2$) is about 50 Mton N year$^{-1}$. About two-thirds of this emission is from combustion processes and traffic. Together with volatile organic compounds, it is the major source of tropospheric ozone (photochemical smog). The background $NO_x$ concentration is less than 0.1 ppb, but in polluted areas the concentration can be hundreds of parts per billion. These high concentrations are problematic because $NO_x$ is emitted by traffic, where the receptor of the pollution is right at the source. $NO_x$ pollution has been linked with cardiovascular and respiratory health effects (Qian et al., 2007). The increased risk of dying from a 10-μg-m$^{-3}$ increase in the $NO_x$ concentration has been estimated at 8% (Nafstad et al., 2004), which is almost as high as the exposure risk of PM2.5 (see Section 3.2.6).

$NO_2$ can also lead to acid rain, due to the following reaction in the atmosphere:

$$NO_2 + {}^{\bullet}OH + M \rightarrow HNO_3 + M \tag{3.4}$$

There is a myriad of secondary nitrogen pollutants, such as nitric acid, the nitrate radical, $HNO_2$, $HNO_4$, $N_2O_5$, and organic nitrates such as peroxyacetyl nitrate (PAN) ($CH_3CO$-O-O-$NO_2$). PAN causes eye irritation. The combination $NO_x$ + secondary nitrogen pollutants are sometimes categorized as $NO_y$.

### 3.2.3 Volatile Organic Compounds

Volatile organic compounds (VOCs) are compounds with a sufficiently high vapor pressure at ambient conditions to be a potential air pollutant (excluding particulate matter). For a practical classification, they should be divided into methane, nonmethane hydrocarbons, and halocarbons. Methane is a greenhouse gas. Nonmethane hydrocarbons are the most important contributors to tropospheric ozone (along with $NO_x$). Halocarbons are greenhouse gases and deplete stratospheric ozone.

Methane ($CH_4$) is the second most important contributor to the anthropogenic greenhouse effect (22%). Its global warming potential is 25 over a 100-year time horizon (IPCC, 2007). The global methane emission is almost 600 Mton year$^{-1}$. The main sources are the oil and gas industry, landfills, wetlands, and ruminants. Its concentration increased from about 700 ppb before the industrial revolution to 1775 ppb in 2005. The perturbation lifetime of methane in the atmosphere is about 12 years and depends on its background concentration. Because of this fairly short lifetime, methane has been a good target for greenhouse gas mitigation. Furthermore, reductions in methane emissions do not lead to reductions of particulate matter, which is a global dimmer, making it particularly effective in the battle against climate change. The atmospheric methane concentration has been fairly constant since 1999 (IPCC, 2007) with only slight (about 1%) increases in the last 10 years (Dlugokencky et al., 2009).

The majority of nonmethane hydrocarbons are emitted biogenically (trees), but the emission is very diffuse, making it a less relevant source than anthropogenic sources. The main anthropogenic sources are solvent use, oil production, and incomplete combustion, including traffic. The main hydrocarbons found in polluted air are BTX (benzene, toluene, and xylene) and alkanes. Some nonmethane hydrocarbons are carcinogenic (e.g., benzene) or possibly carcinogenic (e.g., styrene).

Nonmethane hydrocarbon concentrations in the atmosphere are sometimes expressed in ppbC, which means ppb carbon. This is the concentration in ppb that is obtained by pretending that each carbon atom in hydrocarbons is a separate molecule. The main reasons of this notation are because it is a more relevant unit than ppb and because the most widely used detector for hydrocarbons, the flame ionization detector, measures ppbC. The total nonmethane hydrocarbon concentration can run up to 500 ppbC.

The most problematic halogenated hydrocarbons are the chlorofluorocarbons (CFCs), which destroy ozone in the stratosphere (Molina and Rowland, 1974; note the typing error in the title of this Nobel Prize winning article) and are strong greenhouse gases. They have been phased out by the 1987 Montreal Protocol and replaced by hydrochlorofluorocarbons (HCFCs) for that reason. They were used as refrigerants, propellants, blowing agents, and as solvents. The concentration of most CFCs in the atmosphere has declined in recent years, and even for the most stable CFCs the concentrations have stabilized and show the first signs of a decline.

CFCs such as $CF_2Cl_2$ have only one degradation pathway: photolysis in the stratosphere.

$$CF_2Cl_2 + h\nu \rightarrow CF_2Cl^{\cdot} + Cl^{\cdot} \tag{3.5}$$

The chlorine radical destroys ozone through the following chain reaction:

$$Cl^{\cdot} + O_3 \rightarrow ClO + O_2 \tag{3.6}$$

$$ClO + O^{\cdot} \rightarrow Cl^{\cdot} + O_2 \tag{3.7}$$

Each chlorine radical can react with up to 100,000 ozone molecules. However, most of the chlorine in the stratosphere is in the form of HCl and $ClNO_3$.

The hole in the ozone layer above Antarctica is the result of a chemistry that is unique to the Antarctic stratosphere in winter. The key reaction is as follows:

$$HCl + ClNO_3 \rightarrow Cl_2 + HNO_3 \tag{3.8}$$

This reaction is catalyzed by stratospheric clouds, ice particles containing sulfuric acid, and nitric acid. The Antarctic stratosphere is the only location sufficiently cold to form such clouds. In spring, the accumulated chlorine molecules are photolyzed by sunlight:

$$Cl_2 + h\nu \rightarrow 2Cl^{\cdot} \tag{3.9}$$

This initiates the ozone destruction cycle.

### 3.2.4 Inorganic Carbon

This category consists of the toxic CO and the greenhouse gas $CO_2$.

Carbon monoxide (CO) is mainly produced by atmospheric oxidation of VOCs, biomass burning (e.g., forest fires), and combustion. Human exposure to CO originates mostly from vehicle emissions. Typical concentrations are 40–200 ppb. It is a highly toxic gas. What makes it more dangerous is the fact that it is odorless.

Carbon dioxide ($CO_2$) is the main anthropogenic greenhouse gas (60% of the anthropogenic greenhouse effect). It is emitted by all types of combustion. The 2005 $CO_2$ concentration was 380 ppm, up from 280 ppm before the industrial revolution. Its atmospheric lifetime distribution is a complicated function with different time

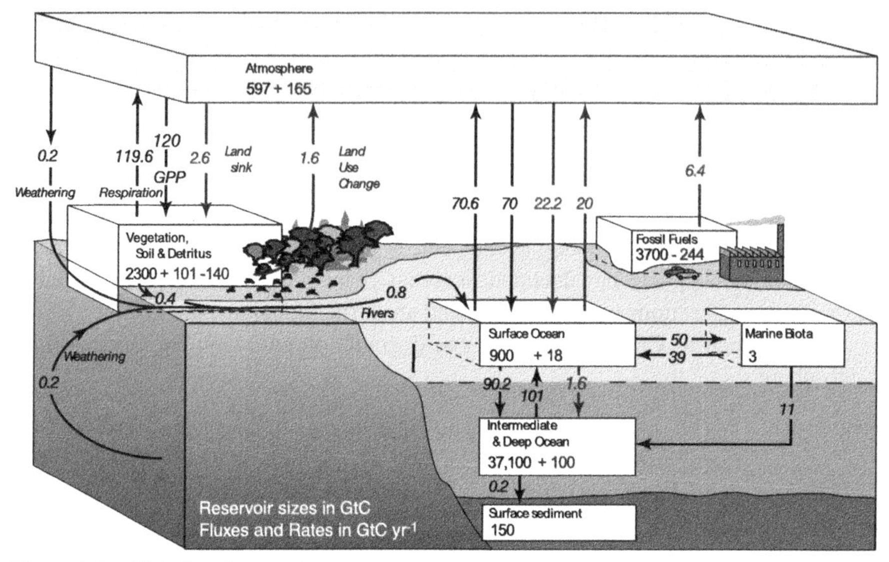


Figure 3.1 Global carbon cycle.
(*Source*: *Climate Change 2007: The Physical Science Basis*. Working Group I Contribution to the Fourth Assessment Report of the Intergovernmental Panel on Climate Change, Figure 7.3. Cambridge University Press, Cambridge, UK.)

scales, from a few years to several centuries. This is due to the complexity of the carbon cycle, which is shown in Figure 3.1 (IPCC, 2007).

## 3.2.5 Ozone

Ozone is a secondary pollutant produced by photochemical reactions between $NO_x$ and VOCs. The main reactions involved in ozone production are given below. In the absence of hydrocarbons, ozone is produced and destroyed in a cycle involving NO and $NO_2$:

$$NO_2 + h\nu \rightarrow NO + O(^3P) \tag{3.10}$$

$$O(^3P) + O_2 \rightarrow O_3 \tag{3.11}$$

$$NO + O_3 \rightarrow NO_2 + O_2 \tag{3.12}$$

In the above reactions, $O(^3P)$ denotes the oxygen atom in the ground state. Based on this mechanism, ozone is not expected to rise dramatically in the presence of $NO_x$ and the absence of nonmethane hydrocarbons. However, a number of additional reactions become relevant when nonmethane hydrocarbons are present:

$$O_3 + h\nu \rightarrow O_2 + O(^1D) \tag{3.13}$$

$$H_2O + O(^1D) \rightarrow 2HO^{\bullet} \tag{3.14}$$

$$HO^{\bullet} + RH \rightarrow H_2O + R^{\bullet} \tag{3.15}$$

$$R^{\bullet} + O_2 \rightarrow R-O-O^{\bullet} \tag{3.16}$$

$$R-O-O^{\bullet} + NO \rightarrow R-O^{\bullet} + NO_2 \tag{3.17}$$

In the above reactions, $O(^1D)$ denotes the oxygen atom in the excited state, and RH stands for a hydrocarbon. From these reactions it is clear that one ozone molecule can potentially lead to the production of two ozone molecules through the formation of two hydroxyl radicals. This is why it is the combination of nonmethane hydrocarbons and $NO_x$ that leads to excessive ozone formation. Methane is not very reactive to the hydroxyl radical and does not contribute substantially to ozone formation. The kinetics of ozone formation in the troposphere is discussed in detail in Chapter 11.

The background ozone concentration is about 10 ppb. Even in an unpolluted environment the ozone concentration can run up to 30–40 ppb because $NO_x$ is produced in soils and hydrocarbons are produced by trees. In very densely populated areas, ozone levels can occasionally peak above 100 ppb. Ironically, air pollution can sometimes lead to a reduction of ozone concentrations. When the emission is mostly NO, and there is no direct sunlight to produce the hydroxyl radical in large quantities, then reaction (3.12) will lead to a depletion of the ambient ozone concentration (ozone titration). This phenomenon is common in northern regions.

Ozone has a significant effect on the risk of death from respiratory causes (Jerrett et al., 2009). It attacks the lung lining, which in turn can lead to inflammation of the lungs. Ozone only attacks the surface of the lung liner, but ozonation products penetrate more deeply into the lungs (Cvitaš et al., 2005). Ozone has been linked with asthma as well (Cvitaš et al., 2005; Curtis et al., 2006).

For the sake of completeness the difference between tropospheric ozone and stratospheric ozone will be explained here. *Stratospheric ozone* is referred to as the ozone layer and is an accumulation of ozone at 15–30 km altitude. This ozone is what protects us from most of the ultraviolet light from the sun. This ozone is not harmful to humans because humans do not live in the stratosphere. CFCs, because of their long atmospheric lifetime, can penetrate the stratosphere and destroy ozone in photochemical chain reactions (Molina and Rowland, 1974). A concentration profile of stratospheric ozone is given in Figure 3.2 (Seinfeld and Pandis, 2006). *Tropospheric ozone*, on the other hand, is ozone at ground level, and can be breathed in by humans. It is produced photochemically in mixtures of $NO_x$ and VOCs.

Ozone in the stratosphere is formed by the following reactions:

$$O_2 + h\nu \rightarrow 2O^{\bullet} \tag{3.18}$$

$$O_2 + O^{\bullet} + M \rightarrow O_3 + M \tag{3.19}$$

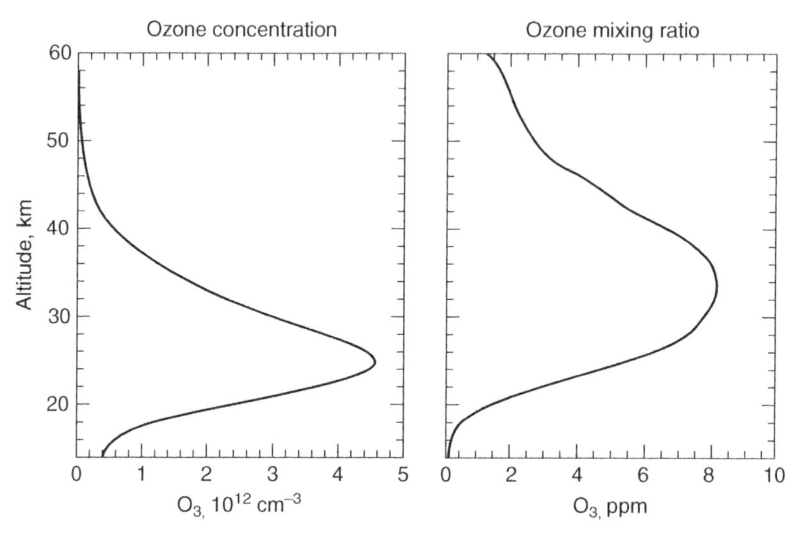


Figure 3.2 Ozone concentration profile in the stratosphere (Seinfeld and Pandis, 2006; reprinted with permission).

### 3.2.6 Particulate Matter

Particulate matter (PM), also described as an *aerosol*, is defined as a finely dispersed solid or liquid phase or phases in the atmosphere. Sometimes the words *aerosol* and *particulate matter* are used to indicate solids only.

In terms of health effects, particulate matter is probably the most harmful type of air pollution. It is produced by combustion processes, diesel motors, and in grinding and milling processes. Background concentrations are typically around $5\,\mu g\,m^{-3}$, except in coastal regions where sea salt can add to the PM load in the air. In urban centers the PM concentration is typically around $30\,\mu g\,m^{-3}$, with peaks of around $100\,\mu g\,m^{-3}$.

Particulate matter can be a primary pollutant (dust, soot, sea salt) or a secondary pollutant ($NH_4SO_4$, $NH_4NO_3$). Nonvolatile organic compounds such as polycyclic aromatic compounds (PAHs), polychlorinated biphenyls (PCBs), and dioxins usually occur either as PM or adsorbed on PM. Many nonvolatiles are carcinogenic.

Particulate matter sizes range from less than 10 nm to more than 10 μm in size. Very small and very large particles are removed from the atmosphere quickly, but particles of sizes between 0.1 and 1 μm have the longest atmospheric lifetime (a couple of days in the absence of rain) because there is no efficient natural removal mechanism for them. Unfortunately, these are among the most harmful particles, as toxicity increases with decreasing particle size. This is why total PM is not the only regulated pollutant, but also $PM_{2.5}$ (PM with diameter smaller than 2.5 μm).

Particulate matter causes lung cancer, respiratory and cardiovascular problems, and acute respiratory infection (Ezzati et al., 2002). Chronic $PM_{2.5}$ exposure has been linked to subclinical artheriosclerosis in epidemiological studies (Kunzli et al., 2005).

There are indications that PM does not increase mortality at concentrations below 7.5 $\mu g\,m^{-3}$ $PM_{2.5}$ (Cohen et al., 2005), but other studies have not confirmed this (Pope and Dockery, 2006). Studies have indicated that each 10 $\mu g\,m^{-3}$ of $PM_{2.5}$ concentration increase increases mortality from *chronic* exposure with 4% (Pope et al., 2002). More recent reanalyses of these and extended data sets have indicated that the relative mortality increase is much more pronounced. Current estimates range from 11 to 18%, depending on the techniques used to eliminate confounding factors (Jerrett et al., 2005; Laden et al., 2006). Results of epidemiological studies with sensitive groups such as infants and elderly people are more uncertain, but there are indications that the relative mortality increase is smaller in these groups (Pope and Dockery, 2006). This counterintuitive result is due to the fact that absolute mortality rates are much higher than average among sensitive groups. Among smokers, the relative risk of $PM_{2.5}$ pollution is larger than for nonsmokers (Pope et al., 2004). The *acute* relative mortality effect of a $PM_{2.5}$ concentration increase by 10 $\mu g\,m^{-3}$ is estimated at about 1%. The American Heart Association published an official statement confirming both acute and chronic effects of $PM_{2.5}$ pollution on cardiovascular mortality and morbidity (Brook et al., 2010).

### 3.2.7 Metals

The main metals found in polluted air are lead, mercury, cadmium, and arsenic (Curtis et al., 2006). The main sources of this pollution are coal-fired power plants, waste incineration, and metal mining and smelting. Volcanic emissions also contain metals. Metals occur in the atmosphere mainly as particulate matter.

Lead and mercury exposure can lead to neurological symptoms. Lead exposure in children has been linked with IQ decrease. Cadmium exposure can cause kidney damage. It has been linked with cancer, but the link is uncertain. Exposure to arsenic can cause lung cancer (Järup, 2003).

Currently, air pollution is not the main source of exposure to heavy metals in the developed world. Ingestion of food and drinking water are the main uptake pathways (Järup, 2003). Nevertheless, air pollution can be an indirect source of metals ingested, as metals can deposit on plants or in surface water. Hence, deposition is an important process to consider when modeling the dispersion of metals in the atmosphere.

### 3.2.8 Air Pollution and Health

It is clear from the previous sections that most air pollutants can cause health problems. Two types of health effects need to be borne in mind: acute toxicity and chronic toxicity. Air pollution can create acute toxic effects in the case of extreme air pollution episodes. For instance, the smog that hit London from December 5 to 12, 1952, killed an estimated 4000 people. This estimate is based on the crude epidemiological methods that were available at the time. It has been argued that the actual death toll could be three times higher (Bachmann, 2007). Accidental releases can also create acute toxic effects. The most serious incident was probably the release of methyl isocyanate in Bhopal, India, in 1984, killing 3800 people instantly and an estimated 20,000 in total. About 100,000 people have permanent injuries as a result of the incident.

Chronic or long-term effects can occur as the result of pollution levels commonly encountered in urban and industrial areas. They lead to lung cancer, respiratory and cardiovascular diseases, irritation of eyes, nose, and throat, and aggravation of asthma. Many of these have been linked with $PM_{2.5}$, $NO_x$, or generally with traffic air pollution (Hoek et al., 2002; Nafstad et al., 2004; Kunzi et al., 2005; Schikowski et al., 2005; Pope and Deckery, 2006; Sahsuvaroglu et al., 2009). Traffic pollution has also been linked with rheumatoid arthritis (Hart et al., 2009).

The worldwide premature death burden due to outdoor air pollution is estimated at 800,000 per year (Ezzati et al., 2002). This study estimates the death toll of air pollution in North America at 28,000. The majority of the casualties are older people with weakened health. Because these estimates only include particulate matter, and a relative mortality increase of 4% per 10 $\mu g\ m^{-3}$ of $PM_{2.5}$ was assumed, these numbers are probably underestimates.

### 3.2.9 Global Warming

From the previous sections it is clear that a significant amount of air pollution consists of greenhouse gases. These pollutants enhance the naturally occurring greenhouse effect. The gas with the strongest natural greenhouse effect is water vapor. The main gases responsible for enhancing the greenhouse effect are $CO_2$, $CH_4$, $N_2O$, CFCs, and ozone. Another form of air pollution, partuculate matter, counteracts the greenhouse effect.

The natural greenhouse effect is over 30 °C. Without water vapor and naturally occurring greenhouse gases, the average temperature on Earth would be almost –20 °C. These are theoretical numbers, as such a large temperature change would almost certainly change the albedo of Earth, changing Earth's heat balance further. While a major part of the natural greenhouse effect is due to water vapor, direct anthropogenic emissions of water vapor do not affect climate markedly because they are negligible in comparison with natural evaporation. Water vapor is not referred to as a greenhouse gas for that reason. However, human activity has an indirect effect on the water vapor concentration in the atmosphere. Warming induced by greenhouse gases emitted by human activity causes additional water to vaporize, adding to the water vapor concentration in the atmosphere. This is one of the *feedback* cycles that amplify the greenhouse effect. Other feedback cycles are melting of permafrost, creating new swamps that emit methane, decomposition of gas hydrates, leading to $CO_2$ and methane emissions, and melting of the ice in the Arctic, leading to increased absorption of sunlight in the ocean instead of ice reflecting sunlight back into space.

The above feedback cycles are crucial to the understanding of the ice ages. Figure 3.3 (IPCC, 2007) shows a history of the concentration of three major greenhouse gases in the last 20,000 years. It is clear that each of them saw a marked increase between 20,000 and 10,000 years ago. These are the result of the feedback cycles mentioned above. Without those the temperature difference between an ice age and an interglacial would be only a fraction of the 4–7 °C that is estimated from geological records (IPCC, 2007).

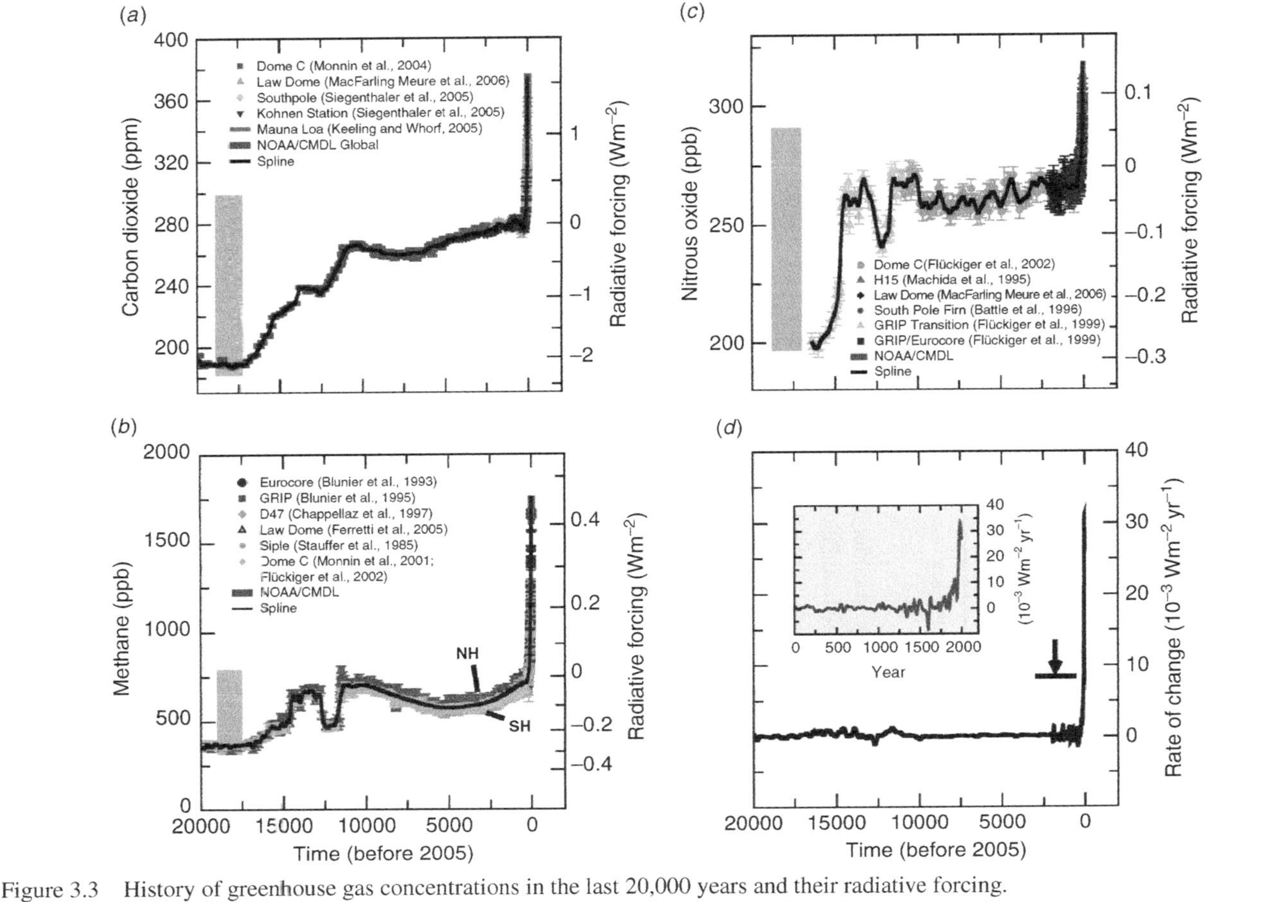


Figure 3.3 History of greenhouse gas concentrations in the last 20,000 years and their radiative forcing.
(*Source*: *Climate Change 2007: The Physical Science Basis*. Working Group I Contribution to the Fourth Assessment Report of the Intergovernmental Panel on Climate Change, Figure 6.4. Cambridge University Press, Cambridge, UK.)

**TABLE 3.1 Radiative Forcing of Main Climate Factors (W m$^{-2}$)[a]**

| Factor | IPCC estimate (1750-2005) | GISS estimate (1880-2003) |
|---|---|---|
| Greenhouse gases | 2.63 (2.45 – 2.85) | 2.75 |
| Ozone | 0.30 (0.10 – 0.70) | 0.22 |
| Stratospheric water vapor | 0.07 (0.02 – 0.12) | 0.06 |
| Solar irradiance | 0.12 (0.06 – 0.30) | 0.22 |
| Land use effect on albedo | –0.19 (–0.39 – 0.00) | –0.09 |
| Snow albedo (black carbon) | 0.10 (0.00 – 0.20) | 0.15 |
| Aerosols: direct effects | –0.50 (–0.90 - –0.10) | Black carbon: 0.43<br>Reflective aerosols: –1.05<br>Total: –0.62 |
| Aerosols: cloud effect | –0.70 (–1.80 - –0.30) | –0.77 |
| **Total** | **1.6 (0.6 – 2.4)** | **1.92** |

[a]as Estimated by IPCC (2007) and by GISS (http://data.giss.nasa.gov/modelforce/RadF.txt).

An important quantity in understanding climate change is **radiative forcing**. It expresses the *change* in the heat balance of the Earth that is caused by a particular change (e.g., a concentration increase of a greenhouse gas, a change of the solar energy flux, etc.). Radiative forcing is expressed in Wm$^{-2}$. See IPCC (2007) for a rigorous definition of radiative forcing. Table 3.1 summarizes the radiative forcings of the main climate factors, as estimated by IPCC (2007) and by NASA's Goddard Institute for Space Science (GISS) (http://data.giss.nasa.gov/modelforce/RadF.txt).

From Table 3.1 it is clear that the main drivers of climate change are the greenhouse gases. The main cooler in the climate system is PM, which offsets about half of the effect of the greenhouse gases. Increasing solar radiation is responsible for about 10% of global warming. The uncertainties estimated by IPCC are large, but these estimates do not incorporate our knowledge of **climate sensitivity** (i.e., the amount of global warming caused by a unit of radiative forcing), which nails down overall radiative forcing with much greater accuracy. The GISS estimate is consistent with our current understanding of climate sensitivity. The heat balance of the Earth is shown in Figure 3.4 (IPCC, 2007).

The ice ages can help us understand the current climate change. As mentioned before, 20,000 years ago it was 4–7 °C colder than it is now. From Figure 3.3 it is clear that the three major anthropogenic greenhouse gases created a radiative forcing of –2.8 W m$^{-2}$ at that time. All radiative forcings combined amounted to about –8 Wm$^{-2}$. Hence, we can estimate that 1.4–2.5 °C of cooling was caused by a depletion of greenhouse gases, while the rest was due to other forcings. Currently, global environmental changes have created a radiative forcing of 1.6–1.9 W m$^{-2}$ (Table 3.1). Based on the ice age experience, this forcing should create an increase of the temperature by about 0.8–1.7 °C. The temperature increase since 1880 is about 0.8 °C, but due to the large buffering effect of the oceans, it has been estimated that there is currently still about 0.6 °C of global warming "in the pipeline." Hence, we expect that the current atmosphere has a total potential for 1.4 °C of global warming, consistent with the estimate of 0.8–1.7 °C based on the ice age experience. Hence, our current understanding of global warming is consistent with our current understanding of the ice ages.

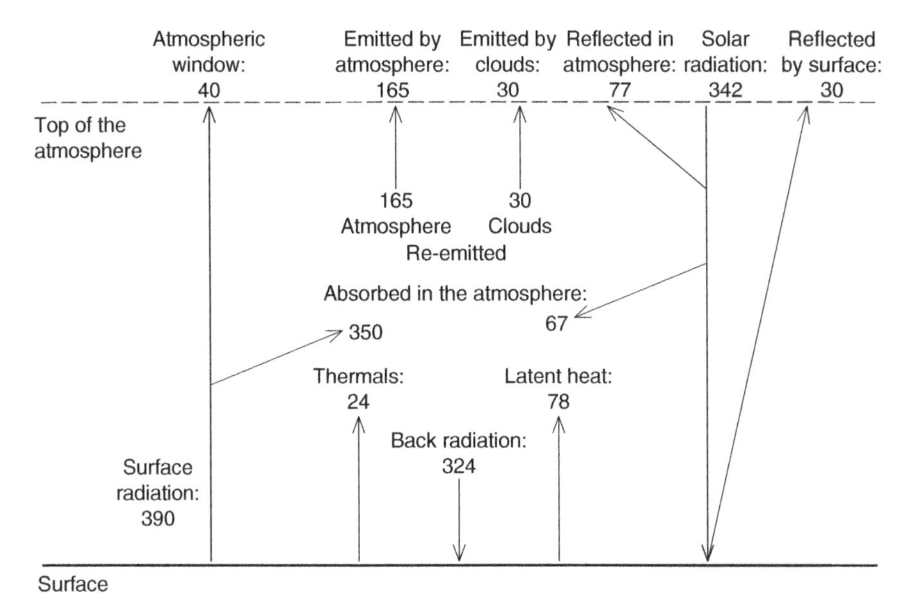


Figure 3.4 Heat balance of Earth.

## 3.2.10 Air Pollution and Visibility

Visibility is usually defined as the maximum distance at which a dark object can be observed against a bright background. Visibility is related to the visual contrast between the object and the background. Visual contrast is defined as

$$C_{vis} = \frac{i_{background} - i_{object}}{i_{background}} \tag{3.20}$$

where $i_{background}$ is the light intensity of the background, and $i_{object}$ is the light intensity of the object. It can be assumed that the visual contrast decreases exponentially with distance:

$$C_{vis} = C_{vis,x=0} \exp(-\alpha_{ext} x) \tag{3.21}$$

where $C_{vis,x=0}$ is the visual contrast at zero distance, $\alpha_{ext}$ is the extinction coefficient ($m^{-1}$), and $x$ (m) is the distance. For a sufficiently dark object, we can assume $C_{vis,x=0} = 1$, and eq. (3.21) reduces to

$$C_{vis} = \exp(-\alpha_{ext} x) \tag{3.22}$$

Most people can see objects down to a visual contrast of about 0.02 (the threshold contrast). Hence, the visibility $x_{vis}$ is found by solving eq. (3.22) for $x_{vis}$ (Koschmieder, 1925):

$$x_{vis} = -\frac{\ln(0.02)}{\alpha_{ext}} = \frac{3.91}{\alpha_{ext}} \tag{3.23}$$

Extinction of contrast can be caused by light absorption and by light scattering. Hence:

$$\alpha_{ext} = \alpha_a + \alpha_s \tag{3.24}$$

where $\alpha_a$ ($m^{-1}$) is the absorption coefficient, and $\alpha_s$ ($m^{-1}$) is the scattering coefficient.

The only air pollutant that absorbs light significantly is $NO_2$. It absorbs mainly blue light, moving colors in the orange direction. Because of the wavelength dependence of the absorption, the absorption coefficient depends on the light. A simple calculation providing a representative absorption for visible sunlight is as follows:

$$\alpha_a = b_a c_{m,NO_2} \tag{3.25}$$

where $b_a = 0.215 \times 10^{-6}\,m^2\,\mu g^{-1}$, and $c_{m,NO_2}$ ($\mu g\,m^{-3}$) is the mass concentration of $NO_2$.

Light can scatter on air molecules and on particles. Hence, the scattering coefficient can be calculated as

$$\alpha_s = \alpha_{s,air} + \alpha_{s,PM} \tag{3.26}$$

The value for air, $\alpha_{s,air}$, equals $13.2 \times 10^{-6}\,m^{-1}$ (Seinfeld and Pandis, 2006). For particulate matter, the following equation applies (Scire et al., 2000):

$$\alpha_{s,PM} = b_{s,PM} c_{m,PM} f_{RH} \tag{3.27}$$

where $b_{s,PM}$ equals $3 \times 10^{-6}\,m^2\,\mu g^{-1}$ for sulfates and nitrates, $4 \times 10^{-6}\,m^2\,\mu g^{-1}$ for organic aerosols, $1 \times 10^{-6}\,m^2\,\mu g^{-1}$ for soil particles, $0.6 \times 10^{-6}\,m^2\,\mu g^{-1}$ for coarse PM (>2.5 μm), and $10 \times 10^{-6}\,m^2\,\mu g^{-1}$ for elemental carbon; $f_{RH}$ is a correction factor to incorporate the effect of relative humidity (RH) and ranges from 1 for RH = 0 to 16 for RH = 98%.

The overall effect of absorption and scattering on visibility is a theoretical maximum visibility of 296 km for absolutely pure dry air, a visibility on the order of 100 km for normal clean to slightly polluted air, and a visibility on the order of 10 km for moderately to heavily polluted air.

### 3.2.11 Odor Nuisance

Odor nuisance is an important aspect of air pollution. Some air pollutants are detectable by the human olfactory system at concentrations as low as 1 ppb. Their major issue is therefore not their toxicity but the nuisance of their odor. Odors negatively affect the well-being of the receptor population. They cause psychological stress and even respiratory symptoms, nausea, and headache (Brattoli et al., 2011).

Because many pollutants contribute to odor, and the sensitivity of the human olfactory system to different odorous substances differs widely, it is not practical to quantify odor by the compounds that contribute to it. Instead, it is more practical to define "odor" as a separate compound, with its own concentration scale.

The quantification of odor is based on the **odor threshold**, the concentration that can be distinguised from pure air by 50% of the population. Odor thresholds can be determined by panels whose members have average olfactory. For the purpose of standardization, what constitutes "average" has been defined using *n*-butanol as a standard compound. To be allowed on an odor panel, a person's individual odor threshold for *n*-butanol should be between 20 and 80 ppb on average (CEN, 2003). There are indications that the average olfactory is actually more sensitive than this standard, but sensitivity to *n*-butanol correlates well with the sensitivity to other odors (Defoer and Van Langenhove, 2004).

The amount of odor required to contaminate 1 $m^3$ of air to the odor threshold is defined as one odor unit (ou). Hence, an odorous compound at its odor threshold has an odor concentration of 1 ou $m^{-3}$ by definition. Odor concentrations are measured by dilution: the dilution that can be distinguished from clean air by 50% of a panel with average olfactory is the odor concnetration $c_{odor}$:

$$\frac{c_{odor}}{1\,\mathrm{ou\,m^{-3}}} = \frac{V_{odor} + V_{dil}}{V_{odor}} \tag{3.28}$$

where 1 ou $m^{-3}$ is included to balance units, $V_{odor}$ is the volume of the odorous air, and $V_{dil}$ is the volume of the diluent air.

The disadvantage of the idea of an odor unit is that it assumes that odor is additive across different compounds, which is not always the case. However, the quantification of odor allows its incorporation into an air dispersion model. Unfortunately, the interpretation of air dispersion modeling results is not as straightforward as with other pollutants. This is because the response to odor is instantaneous, and the hourly average values predicted by models are not always representative of the actual odor sensation, which may be experienced as an intermittent nuisance even though the hourly average odor concentration is less than 1 ou $m^{-3}$. Resolving this issue would require a model capable of predicting statistics of instantaneous concentrations rather than averages. An attempt in this field is the stochastic puff model of Vanderschaege (1987).

The most odorous compounds are organic sulfides and mercaptans, $H_2S$, amines, aldehydes and ketones, phenols, and higher alcohols.

## PROBLEMS

1. If water pollution creates an algal bloom, releasing large quantities of DMS, would you consider that a natural source of air pollution or an anthropogenic source?
2. Should anthropogenically emitted $N_2$ be considered air pollution?
3. It is said that $NH_3$ leads to acidification. However, ammonia is a base, not an acid. How is that possible?
4. What is the global warming potential of $CO_2$?
5. Calculate the visibility of air polluted with 20 μg $m^{-3}$ $NO_2$, 35 μg $m^{-3}$ $PM_{2.5}$ (half sulfate and half elemental carbon), and 85 μg $m^{-3}$ $PM_{10}$. The humidity is such that $f_{RH} = 2$.

## REFERENCES

Bachmann J. (2007). Will the circle be unbroken: A history of the U.S. national ambient air quality standards. *J. Air Waste Manage. Assoc.* **57**, 652–697.

Bernstein J.A., Alexis N. Barnes C., Bernstein I.L., Bernstein J.A., Nel A., Peden D., Diaz-Sanchez D., Tarlo S.M., and Williams P.B. (2004). Health effects of air pollution. *J. Allergy Clin. Immunol.* **114**, 1116–1123.

Brattoli M., de Gennaro G., de Pinto V., Demarinis Loiotile A., Lovascio S., and Penza M. (2011). Odour detection methods: Olfactory and chemical sensors. *Sensors* **11**, 5290–5322.

Brook R.D., Rajagopalan S., Pope C.A., Brook J.R., Bhatnagar A., Diez-Roux A.V., Holguin F., Hong Y.L., Luepker R.V., Mittleman M.A., Peters A., Siscovick D., Smith S.C., Whitsel L., and Kaufman J.D. (2010). Particulate matter air pollution and cardiovascular disease. An update to the scientific statement from the American Heart Association. Circulation **121**, 2331–2378.

CEN (2003). *Air Quality—Determination of Odour Concentration by Dynamic Olfactometry*. Report EN13725, Committee for European Normalization, Brussels, Belgium.

Cohen A.J., Anderson H.R., Ostro B., Pandey K.D., Krzyzanowski M., Künzli N., Gutschmidt K., Pope A., Romieu I., Samet J.M., and Smith K. (2005). The global burden of disease due to outdoor air pollution. *J. Toxicol. Environ. Health A* **68**, 1301–1307.

Curtis L., Rea W., Smith-Willis P., Fenyves E., and Pan Y. (2006). Adverse health effects of outdoor air pollutants. *Environ. Int.* **32**, 815–830.

Cvitaš T., Klasinc L., Kezele N., McGlynn S.P., and Pryor W.A. (2005). New directions: How dangerous is ozone? *Atmos. Environ.* **39**, 4607–4608.

Defoer N. and Van Langenhove H. (2004). Variability and repeatability of olfactometric results of *n*-butanol, pig odour and a synthetic gas mixture. *Water Sci. Technol.* **50**(4), 65–73.

Dlugokencky E.J., Bruhwiler L., White J.W.C., Emmons L.K., Novelli P.C., Montzka S.A., Masarie K.A., Lang P.M., Crotwell A.M., Miller J.B., and Gatti L.V. (2009). Observational constraints on recent increases in the atmospheric $CH_4$ burden. *Geophys. Res. Lett.* **36**, L18803.

Ezzati M., Lopez A.D., Vander Hoorn S., and Murray C.J.L. (2002). Selected major risk factors and global and regional burden of disease. *Lancet* **360**, 1347–1360.

Hart J.E., Laden F., Pueff R.C., Costenbader K.H., and Karlson E.W. (2009). Exposure to traffic pollution and increased risk of rheumatoid arthritis. *Environ. Health Persp.* **117**, 1065–1069.

Hoek G., Brunekreef B., Goldbohm S., Fischer P., and van den Brandt P.A. (2002). Association between mortality and indicators of traffic-related air pollution in the Netherlands: A cohort study. *Lancet* **360**, 1203–1209.

IPCC (Intergovernmental Panel on Climate Change). (2007). *Climate Change 2007—The Physical Basis*. IPCC, Cambridge University Press, Cambridge, UK.

Järup, L. (2003). Hazards of heavy metal contamination. *Brit. Med. Bull.* **68**, 167–182.

Jerrett M., Burnett R.T., Ma R., Pope C.A. III, Krewski D., Newbold K.C., Thurston G., Shi Y., Finkelstein N., Calle E.E., and Thun M.J. (2005). Spatial analysis of air pollution and mortality in Los Angeles. *Epidemiology* **16**, 727–736.

Jerrett M., Burnett R.T., Pope C.A. III, Ito K., Thurston G., Krewski D., Shi Y., Calle E., and Thun M. (2009). Long-term ozone exposure and mortality. *N. Engl. J. Med.* **360**, 1085–1095.

Koschmieder H. (1925). Theorie der horizontalen Sichtweite. II. Kontrast und Sichtweite. *Beit. Phys. Freien Atmos.* **12**, 171–181.

Kunzli N., Jerrett M., Mack W.J., Beckerman B., LaBree L., Gilliland F., Thomas D., Peters J., and Hodis H.N. (2005). Ambient air pollution and atheriosclerosis in Los Angeles. *Environ. Health Persp.* **113**, 201–206.

Laden F., Schwartz J., Speizer F.E., and Dockery D.W. (2006). Reduction in fine particulate air pollution and mortality. Extended follow-up of the Harvard six cities study. *Am. J. Respir. Crit. Care Med.* **173**, 667–672.

Lelieveld J., Roelofs G.J., Ganzeveld L., Feichter J., and Rodhe H. (1997). Terrestrial sources and distribution of atmospheric sulphur. *Phil. Trans. R. Soc. London B* **352**, 149–158.

Molina M.J. and Rowland F.S. (1974). Stratospheric sink for chlorofluoromethanes: Chlorine atomc-atalysed destruction of ozone. *Nature* **249**, 810–812.

Mosier A, Kroeze C., Nevison C., Oenema O., Seitzinger S., and Van Cleemput O. (1998). Closing the global $N_2O$ budget: Nitrous oxide emissions through the agricultural nitrogen cycle. *Nutr. Cycl. Agroecosyst.* **52**, 225–248.

Nafstad P., Haheim L.L., Wisloff T., Gram F., Oftedal B., Holme I., Hjermann I., and Leren P. (2004). Urban air pollution and mortality in a cohort of Norwegian men. *Environ. Health Persp.* **112**, 610–615.

Pope C.A. III and Dockery D.W. (2006). Health effects of fine particulate air pollution: Lines that connect. *J. Air Waste Manage. Assoc.* **56**, 709–742.

Pope C.A. III, Burnett R.T., Thun M.J., Calle E.E., Krewski D., Ito K., and Thurston G.D. (2002). Lung cancer, cardiopulmonary mortality, and long-term exposure to fine particulate air pollution. *J. Am. Med. Assoc.* **287**, 1132–1141.

Pope C.A. III, Burnett R.T., Thurston G.D., Thun M.J., Calle E.E., Krewski D., and Godleski J.J. (2004). Cardiovascular mortality and long-term exposure to particulate air pollution. Epidemiological evidence of general pathophysiological pathways of disease. *Circulation* **109**, 71–77.

Qian Z., He Q., Lin H.M., Kong L., Liao D., Yang N., Bentley C.M., and Xu S. (2007). Short-term effects of gaseous pollutants on cause-specific mortality in Wuhan, China. *J. Air Waste Manage. Assoc.* **57**, 785–793.

Sahsuvaroglu T., Jerrett M., Sears M.R., McConnell R., Finkelstein N., Arain A., Newbold B., and Burnett R. (2009). Spatial analysis of air pollution and childhood asthma in Hamilton, Canada: Comparing exposure methods in sensitive subgroups. *Environ. Health* **8**, 14.

Schikowski T., Sugiri D., Ranft U., Gehring U., Heinrich J., Wichmann H.E., and Kramer U. (2005). Long-term air pollution exposure and living close to busy roads are associated with COPD in women. *Respir. Res.* **6**, 152.

Scire J.S., Strimaitis D.G., and Yamartino R.J. (2000). *A User's Guide for the CALPUFF Dispersion Model.* Earth Tech, Concord, MA.

Seinfeld J.H. and Pandis S.N. (2006). *Atmospheric Chemistry and Physics*, 2nd ed. Wiley, Hoboken, NJ.

Vanderschaege P. (1987). *Simulatiemodel voor Geurverspreiding in de Atmosfeer.* M.Sc. Thesis, Ghent University, Belgium.

CHAPTER 4

# REGULATION OF AIR QUALITY AND AIR QUALITY MODELING

## 4.1 INTRODUCTION

Typically, air dispersion models are run for regulatory purposes: to investigate whether a planned installation will lead to exceedances of the air quality objectives or not. Hence we need to know what is the regulatory status of air dispersion models: What models do regulators expect to be used? How should they be used? We also need to know what the air quality objectives are. This topic cannot be covered in full because each state/province/country has its own regulations and guidelines, and they change from time to time. Hence, the reader should always be aware of the current regulations and guidelines that apply when conducting an air dispersion simulation. However, we can get a good idea of what to expect by looking at sample regulations and guidelines. The objective of this chapter is to provide some basic regulatory background.

No attempt was made to provide a complete picture of the regulatory framework for air quality and air dispersion modeling. Because of the large mandate of the states (U.S.) and provinces (Canada), there is a wide variety of approaches, regulations, and guidelines. In this chapter, only the most common among these approaches are given. The reader should check local regulations applicable to the case at hand before attempting to model air dispersion for regulatory purposes.

## 4.2 AIR QUALITY REGULATION

A comprehensive review of the history of air quality regulation is given by Bachmann (2007). A brief overview will be given here.

The first major federal initiative in the United States to regulate air quality was the **Clean Air Act** of 1963. Before that time, air quality was mostly regulated on the municipal level. The federal regulatory framework was further developed with the 1970 and 1977 amendments to the Clean Air Act.

The Clean Air Act required the Environmental Protection Agency (EPA) to develop air quality standards. Air emissions are regulated by the states. They are

*Air Dispersion Modeling: Foundations and Applications*, First Edition. Alex De Visscher.

**TABLE 4.1 United States National Ambient Air Quality Standards**

| | Primary Standards | | Secondary Standards | |
|---|---|---|---|---|
| Pollutant | Level | Averaging Time | Level | Averaging Time |
| CO | 9 ppm (10 mg m$^{-3}$) | 8 h | None | |
| | 35 ppm (40 mg m$^{-3}$) | 1 h | None | |
| Pb | 0.15 μg m$^{-3}$ | Rolling 3-month average | Same as primary | |
| $NO_2$ | 53 ppb | Annual (arithmetic average) | Same as primary | |
| | 100 ppb | 1 h | None | |
| $PM_{10}$ | 150 μg m$^{-3}$ | 24 h | Same as primary | |
| $PM_{2.5}$ | 12 μg m$^{-3}$ | Annual (averaged over 3 years) | 15 μg m$^{-3}$ | |
| $PM_{2.5}$ | 35 μg m$^{-3}$ | 24 h | Same as primary | |
| Ozone | 75 ppb | 8 h | Same as primary | |
| $SO_2$ | 75 ppb | 1 h | 500 ppb | 3 h |

From http://www.epa.gov/air/criteria.html.

responsible to ensure that the air quality standards are met. Standards were made up for $SO_2$, PM, $NO_2$, CO, ozone, and lead.

The 1990 amendments are a drastic departure from the original philosophy underlying the Clean Air Act. The EPA now plays a more central role in controlling air pollution. First, EPA is now mandated to set emission standards for certain sources, instead of leaving all emission standards at the state level. Furthermore, the EPA is empowered to enforce the maximum achievable air emission control technology, rather than just regulating emissions. Table 4.1 contains the current U.S. national ambient air quality standards set by EPA. The most significant recent changes are the 24-h average $PM_{2.5}$ guideline, which changed from 65 to 35 μg m$^{-3}$, the introduction of a 1-h standard for $NO_2$ (100 ppb), and the introduction of a 1-h standard for $SO_2$ (75 ppb), which is stricter than the 24-h standard (140 ppb) set in 1971. Exceptions apply in some cases, and the interpretation of the standards varies from pollutant to pollutant. More details are given on the EPA website (http://www.epa.gov/air/criteria.html).

The 1990 amendments also included a list of 189 hazardous air pollutants. The EPA has the authority to make changes to the list. Table 4.2 contains all pollutants that have been on the list. Any source emitting more than 10 tons per year of any pollutant on the list, or 25 tons per year of any combination of pollutants on the list, must be regulated.

Other topics in the 1990 Clean Air Act include urban pollution, vehicle standards, the permitting process, penalties for noncompliance, acid deposition, industry performance standards, and the like.

In Canada many air quality regulations are based on the U.S. framework. In general, the provinces are responsible for adopting and maintaining air quality objectives, with some Canada-wide standards (CWSs) and national ambient air quality objectives (NAAQOs). The CWSs are considered priorities for Canada-wide

**TABLE 4.2 EPA List of Hazardous Air Pollutants**

Acetaldehyde
Acetamide
Acetonitrile
Acetophenone
2-Acetylaminofluorene
Acrolein
Acrylamide
Acrylic acid
Acrylonitrile
Allyl chloride
4-Aminobiphenyl
Aniline
*o*-Anisidine
Asbestos
Benzene (including benzene from gasoline)
Benzidine
Benzotrichloride
Benzyl chloride
Biphenyl
Bis(2-ethylhexyl)phthalate (DEHP)
Bis(chloromethyl)ether
Bromoform
1,3-Butadiene
Calcium Cyanamide
Caprolactam
Captan
Carbaryl
Carbon Disulfide
Carbon tetrachloride
Carbonyl sulfide
Catechol
Chloramben
Chlordane
Chlorine
Chloroacetic acid
2-Chloroacetophenone
Chlorobenzene
Chlorobenzilate
Chloroform
Chloromethyl methyl ether
Chloroprene
Cresols/Cresylic acid (isomers and mixture)
*o*-Cresol
*m*-Cresol
*p*-Cresol
Cumene
2,4,-D, salts and esters
DDE
Hydrazine
Hydrochloric acid
Hydrogen fluoride
Hydrogen sulfide
Hydroquinone
Isophorone
Lindane
Maleic anhydride
Methanol
Methoxychlor
Methyl bromide
Methyl chloride
Methyl chloroform
Methyl ethyl ketone
Methyl hydrazine
Methyl iodide
Methyl isobutyl ketone
Methyl isocyanate
Methyl methacrylate
Methyl *tert*-butyl ether
4,4-Methylene bis(2-chloroaniline)
Methylene chloride
Methylene diphenyl diisocyanate (MDI)
4,4-Methylenedianiline
Naphthalene
Nitrobenzene
4-Nitrobiphenyl
4-Nitrophenol
2-Nitropropane
*N*-Nitroso-*N*-methylurea
*N*-Nitrosodimethylamine
*N*-Nitrosomorpholine
Parathion
Pentachloronitrobenzene
Pentachlorophenol
Phenol
*p*-Phenylenediamine
Phosgene
Phosphine
Phosphorus
Phthalic anhydride
Polychlorinated biphenyls
1,3-Propane sultone
Beta-Propiolactone
Propionaldehyde
Propoxur (Baygon)
Propylene dichloride
Propylene oxide

**TABLE 4.2** (Continued)

| | |
|---|---|
| Diazomethane | 1,2-Propylenimine |
| Dibenzofuranes | Quinoline |
| 1,2-Dibrome-3-chloropropane | Quinone |
| Dibutylphthalate | Styrene |
| 1,4-Dichlorobenzene | Styrene oxide |
| 3,3-Dichlorobenzidene | 2,3,7,8-Tetrachlorodibenzo-*p*-dioxin |
| Dichloroethyl ether | 1,1,2,2-Tetrachloroethane |
| 1,3-Dichloropropene | Tetrachloroethylene |
| Dichlorvos | Titanium tetrachloride |
| Diethanolamine | Toluene |
| *N,N*-Diethyl aniline | 2,4-Toluene diamine |
| Diethyl sulfate | 2,4-Toluene diisocyanate |
| 3,3-Dimethoxybenzidine | *o*-Toluidine |
| Dimethyl aminoazobenzene | Toxaphene |
| 3,3′-Dimethyl benzidine | 1,2,4-Trichlorobenzene |
| Dimethyl carbamoyl chloride | 1,1,2-Trichloroethane |
| Dimethyl formamide | Trichloroethylene |
| 1,1-Dimethyl hydrazine | 2,4,5-Trichlorophenol |
| Dimethyl phthalate | 2,4,6-Trichlorophenol |
| Dimethyl sulfate | Triethylamine |
| 4,6-Dinitro-*o*-cresol, and salts | Trifluralin |
| 2,4-Dinitrophenol | 2,2,4-Trimethylpentane |
| 2,4-Dinitrotoluene | Vinyl acetate |
| 1,4-Dioxane | Vinyl bromide |
| 1,2-Diphenylhydrazine | Vinly chloride |
| Epichlorohydrin | Vinylidene chloride |
| 1,2-Epoxybutane | Xylenes (isomer mixtures) |
| Ethyl acrylate | *o*-Xylene |
| Ethyl benzene | *m*-Xylene |
| Ethyl carbamate | *p*-Xylene |
| Ethyl chloride | Antimony compounds |
| Ethylene dibromide | Arsenic compounds (inorganic including arsine) |
| Ethylene dichloride | Beryllium compounds |
| Ethylene glycol | Cadmium compounds |
| Ethylene imine | Chromium compounds |
| Ethylene oxide | Cobalt compounds |
| Ethylene thiourea | Coke oven emissions |
| Ethylidene dichloride | Cyanide compounds |
| Formaldehyde | Glycol ethers |
| Heptachlor | Lead compounds |
| Hexachlorobenzene | Manganese compounds |
| Hexachlorobutadiene | Mercury compounds |
| Hexachlorocyclopentadiene | Fine mineral fibers |
| Hexachloroethane | Nickel compounds |
| Hexamethylene-1,6-diisocyanate | |
| Hexamethylphosphoramide | Radionuclides (including radon) |
| Hexane | Selenium compounds |

Source: http://www.epa.gov/ttn/atw/orig189.html.

**TABLE 4.3 Canadian National Ambient Air Quality Objectives**[a]

| Pollutant | Maximum Acceptable Level | Averaging Time |
|---|---|---|
| $SO_2$ | 23 ppb | Annual |
| | 115 ppb | 24 h |
| | 334 ppb | 1 h |
| Total suspended particulates | 70 μg m$^{-3}$ | Annual |
| | 120 μg m$^{-3}$ | 24 h |
| CO | 13 ppm | 8 h |
| | 31 ppm | 1 h |
| $NO_2$ | 53 ppb | Annual |
| | 213 ppb | 1 h |
| $O_3$ | 82 ppb | 1 h |

[a]From http://www.ec.gc.ca/rnspa-naps/default.asp?lang=En&n=24441DC4-1.

**TABLE 4.4 California Ambient Air Quality Standards**[a]

| Pollutant | Concentration | Averaging Time |
|---|---|---|
| Ozone | 90 ppb (180 μg m$^{-3}$) | 1 h |
| | 70 ppb (137 μg m$^{-3}$) | 8 h |
| PM10 | 50 μg m$^{-3}$ | 24 h |
| | 20 μg m$^{-3}$ | Annual arithmetic mean |
| PM2.5 | 12 μg m$^{-3}$ | Annual arithmetic mean |
| CO | 20 ppm (23 mg m$^{-3}$) | 1 h |
| | 9 ppm (10 mg m$^{-3}$) | 8 h |
| | 6 ppm (7 mg m$^{-3}$) | 8 h (Lake Tahoe) |
| $NO_2$ | 180 ppb (339 μg m$^{-3}$) | 1 h |
| | 30 ppb (57 μg m$^{-3}$) | Annual arithmetic mean |
| $SO_2$ | 250 ppb (655 μg m$^{-3}$) | 1 h |
| | 40 ppb (105 μg m$^{-3}$) | 24 h |
| lead | 1.5 μg m$^{-3}$ | 30 day |
| Visibility reducing particles | 0.23 km$^{-1}$ | 8 h |
| | 0.07 km$^{-1}$ | 8 h (Lake Tahoe) |
| sulfates | 25 μg m$^{-3}$ | 24 h |
| $H_2S$ | 30 ppb (42 μg m$^{-3}$) | 1 h |
| Vinyl chloride | 10 ppb (26 μg m$^{-3}$) | 24 h |

[a]From http://www.arb.ca.gov/research/aaqs/aaqs2.pdf.

implementation of air quality standards. The CWSs are an 8-h standard for ozone (65 ppb) and a 24-h standard for $PM_{2.5}$ (30 μg m$^{-3}$). The NAAQOs are shown in Table 4.3.

Each state (U.S.) and province (Canada) has its own air quality standards, based on the local conditions, industry, and the like. As an example, the air quality standards of California are shown in Table 4.4.

## 4.3 AIR DISPERSION MODELING GUIDELINES

In the United States, air dispersion modeling for permit applications is regulated by the EPA (2005). The principle of the guideline is based on two levels of modeling: a screening level and a refined level. The purpose of the approach is to achieve the necessary level of sophistication without wasting resources in the process. Canada has similar guidelines.

A **screening technique** in air dispersion modeling is the use of a simple model such as SCREEN3 to calculate the worst-case scenario resulting from a proposed air pollution source. At all climatic conditions the critical wind speed is determined for a given source, and the maximum ground-level concentration at the critical wind speed is stored. The highest of these maximum ground-level concentrations is added to the background concentration of the pollutant, and the result is compared with the ambient air quality objective. When the objective is not exceeded, the proposed source of air emissions is not problematic, and further modeling is not required. When the terrain is complex, CTSCREEN should be used for screening instead of SCREEN3.

When screening techniques predict the ambient air quality objectives to be exceeded, a **refined modeling technique** must be used to evaluate the proposed pollution source. The model of choice is AERMOD. Other models can be accepted on a case-by-case basis, when adequate justification is given.

Some situations cannot be adequately modeled by AERMOD, and the EPA provides special regulations for those cases. For instance, for ozone, the use of CMAQ is encouraged. This model can also be used for $PM_{2.5}$ predictions when there is a large contribution of secondary formation of PM. For adequately modeling CO, a traffic model is required in conjunction with a dispersion model. When pollution over 50 km downwind of the source needs to be considered, CALPUFF should be used.

Some regulators provide representative meteorological data sets for air dispersion modeling. An example is the province of Alberta, Canada (Idriss and Spurrell, 2009), which provides a 5-year data set of the entire province based on the MM5 meteorological model (http://www.albertamm5data.com/Home.aspx). Alberta's modeling guideline also defines a third modeling level, "advanced" modeling, but does not put forward any recommended model in that category.

As indicated above, this is by no means a complete overview of the regulatory framework related to air emissions and air dispersion modeling. The reader should consult the websites of the regulators that have the mandate to regulate the situation at hand.

## REFERENCES

Bachmann J. (2007). Will the circle be unbroken: A history of the U.S. national ambient air quality standards. *J. Air Waste Manage. Assoc.* **57**, 652–697.

Environmental Protection Agency (EPA). (2005). 40 CFR Part 51. Revision to the guideline on air quality models: Adoption of a preferred general purpose (flat and complex terrain) dispersion model and other revisions; Final rule. *Federal Register* **70(216)**, 68218–68261 (http://www.epa.gov/scram001/guidance/guide/appw_05.pdf).

Idriss A. and Spurrell F. *Air Quality Model Guideline*. Government of Alberta, Edmonton, Alberta, Canada (2009). (http://environment.gov.ab.ca/info/library/8151.pdf).

CHAPTER 5

# METEOROLOGY FOR AIR DISPERSION MODELERS

## 5.1 INTRODUCTION

The primer (Chapter 2) demonstrated that changing weather conditions can change the ambient concentration of a pollutant by more than a factor of 10 even if the wind direction does not change. Clearly, it is crucial to account for meteorological conditions if air dispersion is to be modeled with any degree of success.

The field of meteorology that is most relevant for air dispersion modelers is **boundary layer meteorology**, the study of the lower part of the atmosphere. We are particularly interested in the properties of the atmosphere at the microscale, about 1 km. Meteorological processes on a larger scale are relevant to air dispersion modeling, but they mainly affect air pollution on a time scale longer than one hour. These processes do not concern us at this point.

The objective of this chapter is to learn the basic concepts of microscale meteorology. In particular, the calculation of the properties of the atmosphere and their vertical profiles will be covered. This chapter lays the foundation that will be built upon in Chapters 6–11. Every air dispersion model more sophisticated than the one discussed in Chapter 2 requires the concepts discussed in this chapter.

Here is a list of subjects that will be covered in this chapter:

- General introduction to the structure of the atmosphere.
- How does the barometric pressure decrease with increasing altitude?
- How does temperature decrease with increasing altitude in a neutral atmosphere?
- Definition of potential temperature
- How does the vertical temperature profile determine the stability of the atmosphere?
- How does wind speed change with height?
    - In a neutral atmosphere
    - In a nonneutral atmosphere
- The concept of turbulence and its estimation.
- Estimation methods of the height of the mixing layer.

*Air Dispersion Modeling: Foundations and Applications*, First Edition. Alex De Visscher.

One of the key concepts in this chapter is similarity theory. It will be applied throughout the chapter.

## 5.2 STRUCTURE OF THE ATMOSPHERE

Air dispersion occurs mostly in the lower layers of the atmosphere, which are affected by Earth's surface. However, for the sake of completeness, we begin with a brief outline of the structure of the entire atmosphere.

The structure of the atmosphere is shown in Figure 5.1. The atmosphere is subdivided into four layers, based on the temperature profile.

The first 10–15 km of the atmosphere is the **troposphere**. The decreasing temperature with increasing height is due to the pressure gradient. For most air pollution processes this is the relevant layer.

The next 30–40 km is the **stratosphere**. In this zone there is a temperature inversion due to the ultraviolet light absorbed by ozone and by ozone-forming reactions. Most of the ozone in the atmosphere is located in the stratosphere. This is a very stable layer.

The next 30–40 km is the **mesosphere**. This is where the coldest temperatures occur.

Above the mesosphere is the **thermosphere**, where the highest temperatures occur (>1000 °C). The high temperature is due to photodissociation of $O_2$ and $N_2$.

The lowest layers of the troposphere are the relevant ones for air pollution. Several slightly overlapping concepts can be defined here: the planetary boundary layer, the mixing layer, and the surface layer.

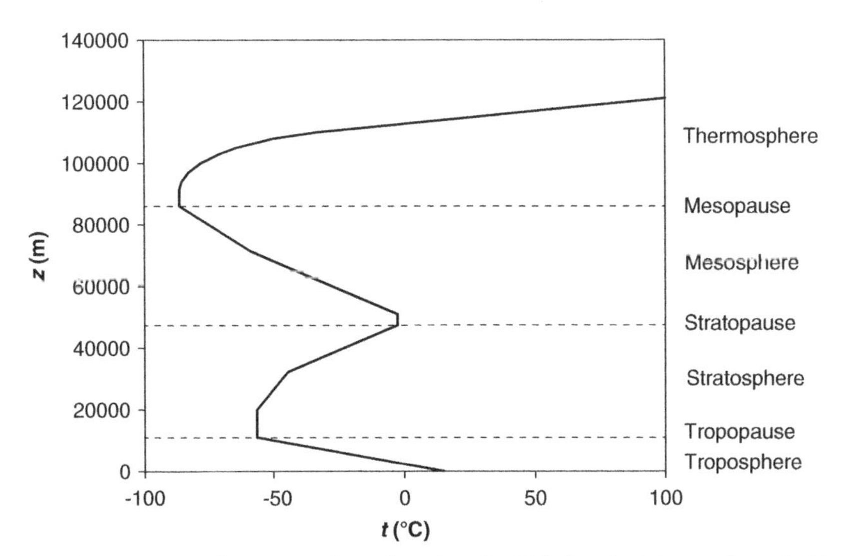


Figure 5.1 Structure of the atmosphere (based on the U.S. Standard Atmosphere; Dubin et al., 1976).

The **planetary boundary layer** (PBL) is the layer of the atmosphere that is influenced by Earth's surface. It is usually 1–4 km thick during the day and about 400 m thick at night (although there is a layer above 400 m at night that is still influenced by effects that occurred the day before). Boundary layer meteorology is the study of this layer.

The **mixing layer**, or mixed layer, is the layer that has enough turbulence to allow complete mixing. It usually takes up almost the entire PBL. The height of the mixing layer is the **mixing height**, $h_{mix}$.

The **surface layer** is the bottom part of the mixing layer. There are more precise definitions of the surface layer, but for now we will simply state that the surface layer has a wind speed that is mainly influenced by the surface. It is usually 100–400 m thick. Sometimes it is simply assumed that the surface layer is the first 10% of the mixed layer (Driedonks and Tennekes, 1984). During daytime this is a reasonable assumption. The part of the PBL above the surface layer has a constant wind speed. However, sometimes during nighttime a wind speed that decreases with height is observed above the surface layer. The wind speed peak that occurs in those cases is known as the **nocturnal jet**. In regions conducive to nocturnal jets, they occur in up to 30% of the nights (Stull, 1988).

In this chapter a similarity theory will be discussed to calculate the wind speed profile. This theory applies to the surface layer.

## 5.3 ALTITUDE DEPENDENCE OF BAROMETRIC PRESSURE

The first step in the understanding of the properties of the atmosphere is the altitude dependence of the barometric pressure. Its calculation is based on the **hydrostatic equation**, which is a force balance applied to a thin horizontal air layer of area 1 m$^2$ and thickness $dz$ (see Fig. 5.2):

$$\left(p+\frac{\partial p}{\partial z}dz\right)+\rho\cdot g dz = p \tag{5.1}$$

where $p$ is barometric pressure (Pa), $\rho$ is the density of the air (kg m$^{-3}$), $g$ is the acceleration of gravity (9.80665 m s$^{-2}$), and $z$ is the altitude (m). Because we consider only one dimension of the atmosphere in steady conditions, the partial derivative in eq. (5.1) is a total derivative. Canceling out $p$ and rewriting leads to the following:

$$\boxed{\frac{dp}{dz}=-\rho g} \tag{5.2}$$

The **ideal gas law** is used to calculate the density of the air:

$$pV = nRT = \frac{m}{M_{air}}RT \tag{5.3}$$

where $V$ is the volume of the layer of air (m$^3$), $n$ is the amount of substance (mol), $R$ is the ideal gas constant (8.314472 J mol$^{-1}$ K$^{-1}$), $T$ is the absolute temperature (K), $m$

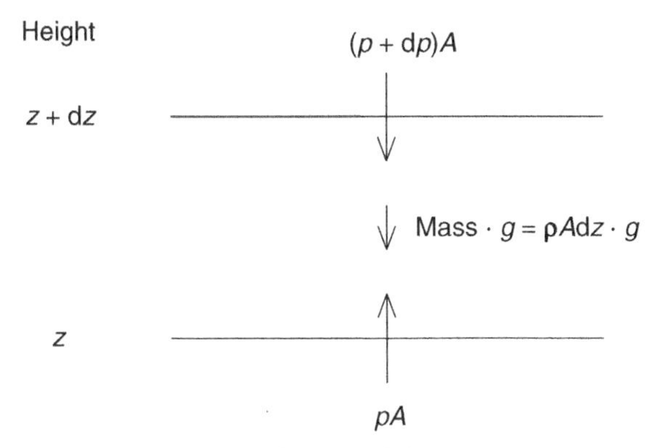


Figure 5.2 Force balance applied to a thin layer of air.

is the mass of the layer of air (kg), and $M_{air}$ is the molar mass of air (0.028964 kg $mol^{-1}$; Dubin et al., 1976). Thus,

$$\rho = \frac{m}{V} = \frac{pM_{air}}{RT} \tag{5.4}$$

Substitution into eq. (5.2) yields

$$\boxed{\frac{dp}{dz} = -\frac{pM_{air}\,g}{RT}} \tag{5.5}$$

Over short distances the temperature is sufficiently constant to have a negligible effect on the density of air. Hence, the above differential equation can be integrated easily:

$$\frac{dp}{p} = -\frac{M_{air}\,g}{RT}dz \tag{5.6}$$

Both sides of the equation are integrated with initial conditions $p = p_0$ at $z = 0$:

$$\ln\left(\frac{p}{p_0}\right) = -\frac{M_{air}\,g}{RT}z \tag{5.7}$$

or

$$\boxed{p = p_0 \exp\left(-\frac{M_{air}\,g}{RT}z\right)} \tag{5.8}$$

On average, the temperature in the atmosphere is 288 K. Hence, $pM_{air}/(RT) = 1.2 \times 10^{-4}\ m^{-1}$ with sufficient accuracy for our purpose [see eq. (1.2) in Chapter 1].

**Example 5.1.** A **rawinsonde** is a small meteorological station attached to a balloon. It contains a barometer and a thermometer and broadcasts the temperature and pressure

**TABLE 5.1 Analysis of Rawinsonde Data from the Prairie Grass Data Set (Barad, 1958), July 10, 1956**

| $p$ (mb) | $t$ (°C) | $\rho$ (kg m$^{-3}$) | $\Delta z$ | $z - z_0$ |
|---|---|---|---|---|
| 946 | 31.0 | 1.0835 | | 0 |
| 938 | 26.8 | 1.0894 | 75.1 | 75.1 |
| 904 | 24.5 | 1.0580 | 322.9 | 398.0 |
| 900 | 24.1 | 1.0547 | 38.6 | 436.6 |
| 850 | 19.5 | 1.0118 | 493.4 | 930.1 |
| 800 | 14.6 | 0.9685 | 514.9 | 1445.0 |
| 732 | 7.4 | 0.9089 | 738.7 | 2183.7 |
| 702 | 5.5 | 0.8776 | 342.5 | 2526.1 |
| 700 | 5.4 | 0.8754 | 23.3 | 2549.4 |
| 666 | 4.1 | 0.8368 | 405.0 | 2954.4 |
| 655 | 5.1 | 0.8200 | 135.4 | 3089.8 |
| 633 | 3.8 | 0.7962 | 277.6 | 3367.4 |
| 600 | 1.0 | 0.7624 | 431.8 | 3799.2 |

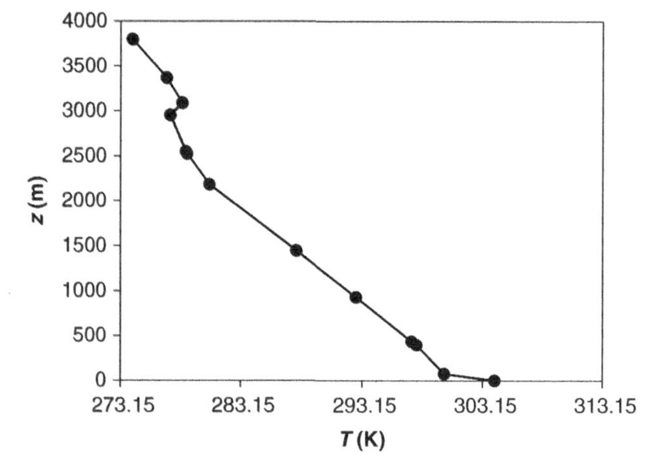


Figure 5.3 Temperature profile derived from rawinsonde data.

data to the main meteorological station. An example of such a data set is shown in Table 5.1. The data was taken from the Prairie Grass Data Set (Barad, 1958).

Draw a temperature profile of the atmosphere based on the rawinsonde data.

***Solution.*** The result is shown in Table 5.1 and in Figure 5.3. First, we rewrite eq. (5.2) as

$$dz = -\frac{dp}{\rho g}$$

Integration leads to

$$z - z_0 = -\int_{p_0}^{p_n} \frac{dp}{\rho g}$$

This integral can be evaluated with the **trapezoid rule** (see Appendix B4). Because the pressure decrements in the data set are not constant, we cannot use any of the Simpson rules. The trapezoid rule is

$$y = \int_{x_0}^{x_1} f(x)\,dx = (x_1 - x_0)\frac{f_0 + f_1}{2} = \frac{\Delta x}{2}(f_0 + f_1)$$

The density is calculated with eq. (5.4). First, pressure is converted to Pa by multiplying by 1000, and temperature is converted to K by adding 273.15. The trapezoid rule is used to calculate $\Delta z$, and the height increments are added to obtain $z - z_0$. The calculations are only approximate because water vapor, which is lighter than air, was not accounted for in the calculation. Hence, all densities are slightly lower than estimated in Table 5.1, and consequently all heights are slightly higher than estimated in the table.

As we will see in Section 5.5.3, this is a typical daytime temperature profile. Note the sharp gradient near the surface and the short inverse gradient at a 3000-m height.

The calculations of this example are included in the enclosed CD, in the file "Example 5.1 Analysis rawinsonde data.xlsx."

## 5.4 HEIGHT DEPENDENCE OF TEMPERATURE—ADIABATIC CASE

### 5.4.1 Adiabatic Lapse Rate

We will start by calculating the altitude dependence of the temperature of what we will call a *neutral* atmosphere, which coincides with an **adiabatic** atmosphere. What that means is that we will consider a parcel of air, move it up or down the atmosphere without exchanging any heat, causing it to experience a pressure change as described in the previous section, and determine the temperature change. This section will assume a basic understanding of thermodynamics.

We will start from the **energy conservation law**, and write the energy terms as a function of the temperature to the extent possible. The energy conservation law states that the change of the internal energy of an air parcel equals the heat added to it plus the work added to it:

$$dU = dq + dw \tag{5.9}$$

in which $U$ is the internal energy (J) of the air parcel, $q$ is the heat added to the air parcel (J), and $w$ is the work added to the air parcel (J). Because we assume that this is an adiabatic process, the heat exchanged between the air parcel and the environment equals zero, or $dq = 0$. Hence:

$$dU = dw \tag{5.10}$$

By definition, the internal energy is related to the **enthalpy**, $H$, as

$$U = H - pV \tag{5.11}$$

Differentiation leads to

$$dU = d(H - pV) = dH - p\ dV - V\ dp \tag{5.12}$$

The enthalpy change with temperature at constant pressure is the definition of the heat capacity at constant pressure:

$$dH = mc_{p,\text{air}}\ dT \tag{5.13}$$

where $m$ is the mass of the air parcel (kg), $c_{p,\text{air}}$ is the specific heat of air (J kg$^{-1}$ K$^{-1}$), and $T$ is the absolute temperature (K). As an air parcel moves up or down, its pressure equalizes with the pressure of the surrounding air. Hence, the surrounding air exerts **compression work** on the air parcel:

$$dw = -p\ dV \tag{5.14}$$

The minus sign indicates that the work exerted on the air parcel is positive when the air parcel volume decreases (compression) and negative when the volume increases (expansion). Substitution of eq. (5.14) into eq. (5.10) leads to

$$dU = -p\ dV \tag{5.15}$$

Further substitution of eqs. (5.13) and (5.15) into eq. (5.12) leads to

$$-p\ dV = mc_{p,\text{air}}\ dT - p\ dV - V\ dp \tag{5.16}$$

This reduces to

$$mc_{p,\text{air}}\ dT = V\,dp \tag{5.17}$$

Ideally, the equations should describe $dT$ in terms of $dz$. But we know the relationship between $dp$ and $dz$ from the previous section. Substituting eq. (5.5) from the previous section into eq. (5.17) leads to

$$mc_{p,\text{air}}\,dT = -V\frac{pM_{\text{air}}g}{RT}dz \tag{5.18}$$

From the ideal gas law [eq. (5.3)] we know that $pVM_{\text{air}}/(RT)$ equals the mass of the air parcel. Hence, mass can be canceled on both sides:

$$c_{p,\text{air}}\ dT = -g\ dz \tag{5.19}$$

After reorganizing, we obtain the desired equation:

$$\boxed{\frac{dT}{dz} = -\frac{g}{c_{p,\text{air}}}} \tag{5.20}$$

Note that this equation only applies to *dry air*. We know that $g$ equals 9.80665 m s$^{-2}$, and $c_{p,\text{air}} = 1006$ J kg$^{-1}$ K$^{-1}$. Hence:

$$\frac{dT}{dz} = -0.00975\,\mathrm{K\,m^{-1}} \tag{5.21}$$

Per hundred meter altitude change the temperature of dry air decreases by about 1 °C. This is called the **dry adiabatic lapse rate**, $\Gamma = 0.00975$ K m$^{-1}$. Note that $\Gamma$ is defined as a cooling rate, so it is positive. Hence, in a *neutral* atmosphere,

$$\frac{dT}{dz} = -\Gamma \tag{5.22}$$

Integration leads to

$$T = T_0 - \Gamma z \tag{5.23}$$

with $T_0$ the absolute temperature at $z = 0$.

In humid air without condensation the adiabatic lapse rate is slightly less than the dry adiabatic lapse rate because water vapor has a larger $c_p$ than air (about 1900 J kg$^{-1}$ K$^{-1}$). However, because the water content in air is never more than a few percent, the adiabatic lapse rate of humid air without condensation is never more than a few percent below the dry adiabatic lapse rate. Hence, this case can be treated as dry air without much loss of accuracy.

In moist air where there is cloud or fog formation or evaporation, a rise of an air parcel is associated with **condensation**, that is, a change of the mass of water vapor in the air. In this case, neglecting the heat capacity of the liquid water in the cloud, an approximate enthalpy equation can be defined:

$$dH = mc_{p,\mathrm{air}}\ dT + \Delta_{\mathrm{vap}}H\ dm_{\mathrm{w}} \tag{5.24}$$

where $\Delta_{\mathrm{vap}}H$ is the enthalpy of vaporization of water (J kg$^{-1}$), $m$ now is the mass of the air and the water vapor, $c_{p,\mathrm{air}}$ the specific heat of the moist air, and $m_{\mathrm{w}}$ is the mass of water vapor (kg). Calculations similar to the ones given above lead to the following equation:

$$\frac{dT}{dz} = -\frac{g}{c_{p,\mathrm{air}} + \Delta_{\mathrm{vap}}H\left(dw_{\mathrm{w}}/dT\right)} \tag{5.25}$$

where $w_{\mathrm{w}}$ is the water mass fraction in the gas phase (kg water per kg gas phase at saturation). This is the **moist adiabatic lapse rate**, $\Gamma_{\mathrm{s}}$. In the temperature range $\theta_{\mathrm{C}} = 0$–$40\,°\mathrm{C}$, the enthalpy of vaporization can be calculated as

$$\Delta_{\mathrm{vap}}H/(\mathrm{kJ\,kg^{-1}}) = 2501.4 - 2.365\theta_{\mathrm{C}}/(°\mathrm{C}) \tag{5.26}$$

where kJ kg$^{-1}$ and °C are included to balance units.

It can be shown that the temperature dependence of $w_{\mathrm{w}}$ is nonlinear, so $\Gamma_{\mathrm{s}}$ depends on the temperature as well; and $w_{\mathrm{w}}$ does not have a unique relationship with temperature but depends on pressure too. Hence, $\Gamma_{\mathrm{s}}$ depends on barometric pressure

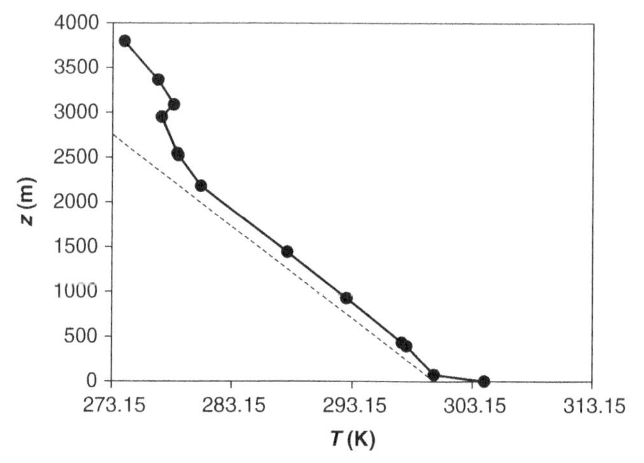


Figure 5.4 Temperature profile derived from rawinsonde data, compared with the dry adiabatic lapse rate.

as well and can be as small as 0.0014 K m$^{-1}$ at 40 °C and 1 bar. At 0 °C and 1 bar, $\Gamma_s$ is 0.0058 K m$^{-1}$. Temperature and pressure dependence of $\Gamma_s$ is demonstrated in Section A5.1 (Appendix A). The following approximate equation can be used to predict $\Gamma_s$:

$$\Gamma_s = \frac{a}{1 + b\ \exp\left(ct^d\right)} \tag{5.27}$$

where $a = 0.01028$ K m$^{-1}$, $b = 0.75586$, $c = 0.074363$ °C$^{-1}$, and $d = 0.915735$.

**Example 5.2.** Look at the temperature profile obtained from rawinsonde data in Example 5.1. Is there a zone where the temperature profile is adiabatic?

***Solution.*** An adiabatic temperature profile is shown in Figure 5.4 (dotted line) along with the actual temperature profile. The curve is nearly parallel to the adiabatic line in the range 400–2200 m. Based on linear regression of the data in this range, a temperature gradient of −0.00955 K m$^{-1}$ is obtained, slightly less negative than the dry adiabatic lapse rate. The difference is due to water vapor.

The calculation of the slope is included in the same file solving Example 5.1. on the enclosed CD file "Example Analysis rawinsonde data.xlsx."

## 5.4.2 Potential Temperature

The **potential temperature** $\theta$ of a parcel of air is the temperature the parcel would have if we moved it to a surface at pressure $p_0 = 100$ kPa in a dry adiabatic way.

The simplest way to calculate the relationship between $T$ and $\theta$ is in terms of altitude. By definition, a parcel of air heats up $\Gamma z$ degrees when it moves down $z$ meters. Hence, eq. (5.23) becomes

$$\theta = T + \Gamma z \tag{5.28}$$

where $z$ is defined as the height above a surface at 100 kPa. Sometimes the requirement $p_0 = 100$ kPa is dropped, and $z$ refers to the height above the actual surface (Stull, 1988).

Sometimes we do not know the altitude of an air parcel, but we do know its pressure. In that case, we need eq. (5.17) ($mc_{p,\text{air}}\, dT = V\, dp$) to derive an equation for $\theta$. Combination with the ideal gas law [eq. (5.3)] leads to

$$c_{p,\text{air}}\, dT = \frac{RT}{pM_{\text{air}}} dp \tag{5.29}$$

In order to integrate this equation, it is rewritten as

$$\frac{dT}{T} = \frac{R}{c_{p,\text{air}} M_{\text{air}}} \frac{dp}{p} \tag{5.30}$$

Both sides of the equation are integrated from $\theta$ to $T$ and from $p_0$ to $p$:

$$\ln\left(\frac{\theta}{T}\right) = \frac{R}{c_{p,\text{air}} M_{\text{air}}} \ln\left(\frac{p_0}{p}\right) = \ln\left(\frac{p_0}{p}\right)^{R/(c_{p,\text{air}} M_{\text{air}})} \tag{5.31}$$

The relationship between $\theta$ and $T$ is obtained by taking the exponential of both sides of eq. (5.31):

$$\boxed{\theta = T\left(\frac{p_0}{p}\right)^{R/(c_{p,\text{air}} M_{\text{air}})}} \tag{5.32}$$

With $R = 8.314472$ J mol$^{-1}$ K$^{-1}$, $c_{p,\text{air}} = 1006$ J kg$^{-1}$ K$^{-1}$, and $M_{\text{air}} = 0.028964$ kg mol$^{-1}$, we obtain a power law coefficient of 0.285.

**Example 5.3.** Draw a potential temperature profile based on the rawinsonde data of Example 5.1.

***Solution.*** The real temperature data are taken from Table 5.1 and repeated in Table 5.2. From these, the potential temperatures are calculated with eq. (5.28). The result is shown in Table 5.2 and Figure 5.5. As expected, the adiabatic zone (400–2200 m) shows a vertical potential temperature profile.

The calculations are included on the enclosed CD, in the same file as Example 5.1, file “Example 5.1 Analysis rawinsonde data.xlsx.”

**TABLE 5.2 Calculation of Potential Temperature Profile from Actual Temperature Profile.**[1]

| $z - z_0$ | $T$ (K) | $\theta$ (K) |
|---|---|---|
| 0 | 304.15 | 304.15 |
| 75.1 | 299.95 | 300.68 |
| 398.0 | 297.65 | 301.53 |
| 436.6 | 297.25 | 301.51 |
| 930.1 | 292.65 | 301.72 |
| 1445.0 | 287.75 | 301.84 |
| 2183.7 | 280.55 | 301.84 |
| 2526.1 | 278.65 | 303.28 |
| 2549.4 | 278.55 | 303.40 |
| 2954.4 | 277.25 | 306.05 |
| 3089.8 | 278.25 | 308.37 |
| 3367.4 | 276.95 | 309.78 |
| 3799.2 | 274.15 | 311.19 |

[1] Basis: Rawinsonde data from the Prairie Grass Data Set (Barad, 1958), July 10, 1956.

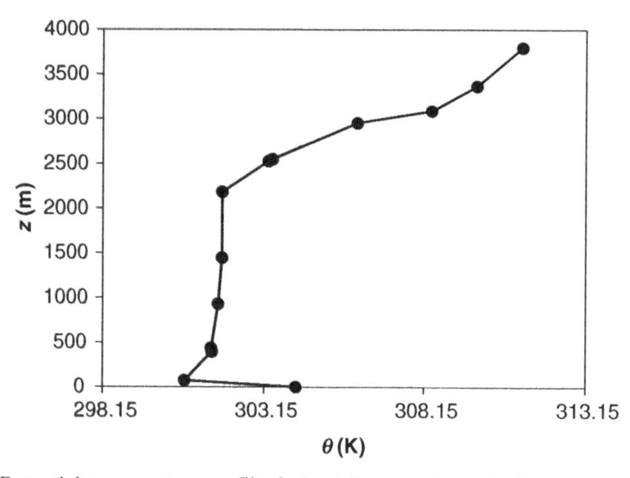


Figure 5.5 Potential temperature profile derived from rawinsonde data.

## 5.5 STABILITY

### 5.5.1 General Description of Stability

This section will outline what happens to an air parcel when it is *not adiabatic*. First, we define $\Lambda$ as the actual lapse rate (remember that $\Gamma$ is the adiabatic lapse rate).

**Case 1:** $\Lambda < \Gamma$ (subadiabatic) This case is shown in Figure 5.6*a*. When the air parcel moves, its temperature changes more strongly than the surrounding air because the parcel follows the adiabatic temperature profile. As a result, if the parcel rises, it becomes colder than the surrounding air and denser. It will have a tendency to sink back to its original location. If the parcel sinks, it will heat up more than the surrounding air. That makes the air parcel hotter and less dense than the surrounding air, giving it a tendency to rise back to its original location. This is called a *stable* atmosphere: Whenever an air parcel moves, the surrounding air will exert a force moving the parcel back to its original position. This type of atmosphere has relatively low turbulence, as turbulent motions are damped quickly by the thermal stratification.

We expect the dispersion of pollutants in the air to be very *slow* in a stable atmosphere.

**Case 2:** $\Lambda > \Gamma$ (superadiabatic). This case is shown in Figure 5.6*b*. In this case the temperature of a moving air parcel changes less than the surrounding air. When the parcel moves up, it becomes hotter and lighter than the surrounding air, causing it to prolong its upward movement. When an air parcel moves down, it becomes colder and denser than the surrounding air, causing it to prolong its downward movement. This is called an *unstable* atmosphrere. This type of atmosphere is characterized by large air movements convecting heat from the surface to higher air layers. It is therefore also known as a *convective* atmosphere.

We expect the dispersion of pollutants in the air to be very *fast* in an unstable atmosphere.

**Case 3:** $\Lambda = \Gamma$ (neutral). In this case, an air parcel at the same temperature as the surrounding air *stays* at the same temperature as the surrounding air when it moves and has no tendency to continue or reverse any movement. The atmosphere is neither stable nor unstable.

Some remarks:

- The overall (nonlocal) stability can differ from what a local stability analysis indicates. This can be seen in Figure 5.7. It appears that most of the atmosphere is neutral because $\Lambda = \Gamma$ (remember that $\Lambda$ and $\Gamma$ refer to the slope in the graph). However, it is clear that an air parcel at the ground becomes hotter than the surrounding air when it rises, so it will continue to rise. The overall stability is unstable. Here the use of the potential temperature can be handy. The unstable part of the atmosphere can be readily identified in Figure 5.8: It is the part of the atmosphere with a potential temperature higher than the ground-level potential temperature. For instance, in Example 5.3, the unstable part of the atmosphere is about 2700 m high (Fig. 5.5). The mixing layer can be slightly higher because rising air currents may overshoot into the stable layer aloft due to its momentum.

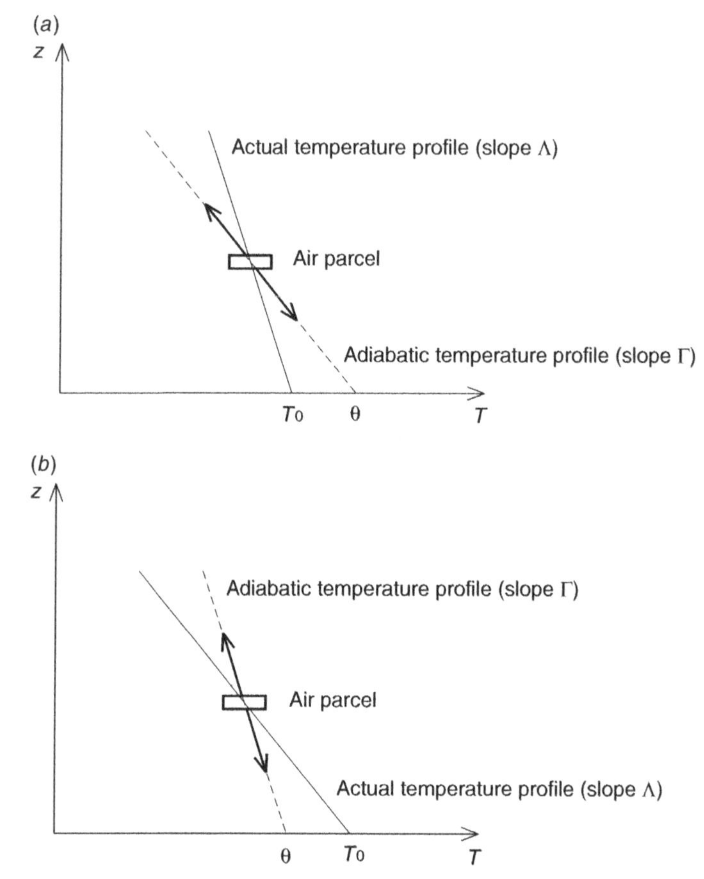


Figure 5.6 Movement of an air parcel in a (*a*) subadiabatic and (*b*) superadiabatic atmosphere.

- An extreme case of a stable atmosphere is a **temperature inversion**. In this case the temperature *increases* with increasing height. Temperature inversions can occur at ground level or on top of an unstable layer. An example is shown in Figure 5.3 in Example 5.1.

### 5.5.2 Stability Parameter

The **stability parameter** $s$ can be defined as the tendency of an air parcel to accelerate against a given vertical movement (i.e., to slow it down) per meter of that movement.

By tendency we mean that friction is neglected. The **acceleration** is the result of gravitation and buoyancy. By against we mean that if the parcel moves up and tends to accelerate down, then $s > 0$.

The acceleration of an air parcel in the absence of friction is the effect of **gravity**, **Archimedes' law** (describing the buoyant forces), and **Newton's law** (relating force to acceleration).

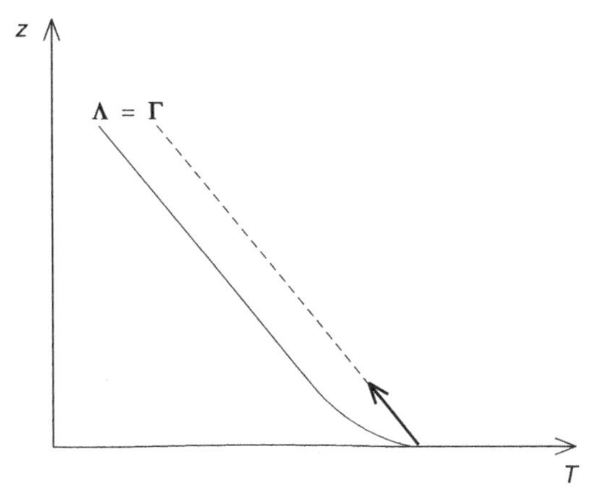


Figure 5.7 Atmosphere with overall unstable conditions in spite of the fact that the local stability is neutral.

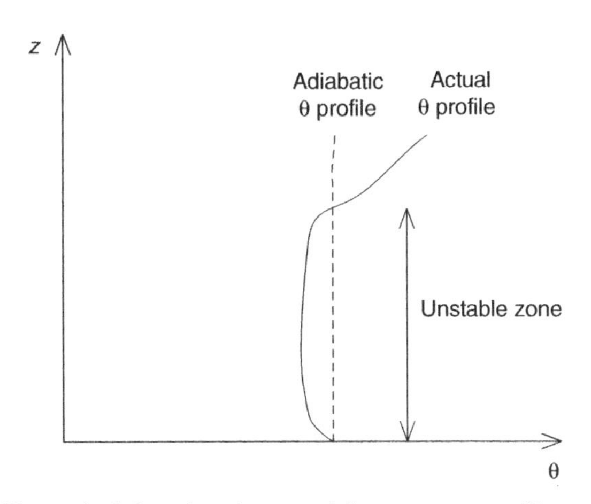


Figure 5.8 Stability analysis based on the potential temperature profile.

The total force acting on an air parcel is the sum of the force of gravity and the buoyant force. Since we will consider a downward acceleration as positive (contrary to our usual convention), we will consider a downward force as positive as well. Hence, the sum of gravitational and buoyant forces is

$$F = mg - m_a g = (m - m_a)g = (\rho - \rho_a)Vg \tag{5.33}$$

where $m$ is the mass of the air parcel (kg), $m_a$ is the mass of an equal volume of surrounding air (kg), $g$ is the acceleration due to gravity (9.80665 m s$^{-2}$), $\rho$ is the density of the air parcel (kg m$^{-3}$), $\rho_a$ is the density of the surrounding air (kg m$^{-3}$), and $V$ is the volume of the air parcel (m$^3$). Newton's law states

$$F = ma = \rho Va \tag{5.34}$$

where $a$ is the acceleration of the air parcel. By setting eq. (5.33) equal to eq. (5.34), one obtains

$$\rho Va = (\rho - \rho_a)Vg \tag{5.35}$$

Solving for $a$ yields

$$\boxed{a = \frac{\rho - \rho_a}{\rho} g} \tag{5.36}$$

The density can be eliminated from eq. (5.36) by using the ideal gas law, eq. (5.4):

$$\rho = \frac{m}{V} = \frac{pM_{air}}{RT} \tag{5.4}$$

Hence:

$$\rho_a = \frac{m}{V} = \frac{pM_{air}}{RT_a} \tag{5.37}$$

where $T_a$ is the temperature of the surrounding atmosphere. Substitution into eq. (5.36) leads to

$$a = \frac{pM_{air}/(RT) - pM_{air}/(RT_a)}{pM_{air}/(RT)} g = \frac{1/T - 1/T_a}{1/T} g \tag{5.38}$$

Multiplying the numerator and denominator by $T T_a$ leads to

$$\boxed{a = \frac{T_a - T}{T_a} g} \tag{5.39}$$

Remember that we are interested in the acceleration that would ensue if the parcel of air was lifted by 1 m. We assume that the air parcel is initially at equilibrium with the surrounding air at a height $z$:

$$T(z) = T_a(z) \tag{5.40}$$

where $(z)$ means a functional dependence on $z$. Now we push the parcel up by 1 m and calculate what happens:

Parcel:

$$T(z + 1\,\text{m}) = T(z) - \Gamma \tag{5.41}$$

Surroundings:

$$T_a(z + 1\,\text{m}) = T_a(z) - \Lambda = T(z) - \Lambda \tag{5.42}$$

Subtract eq. (5.41) from eq. (5.42):

$$T_a(z+1\,\text{m}) - T(z+1\,\text{m}) = T_a(z) - \Lambda - (T(z) - \Gamma) = \Gamma - \Lambda \tag{5.43}$$

Substitution into eq. (5.39) leads to the acceleration at $z+1$ m, which, by definition, is the stability parameter:

$$\boxed{a = \frac{\Gamma - \Lambda}{T_a} g = s} \tag{5.44}$$

We can also link $s$ with the potential temperature, $\theta$:

The relationship between the temperature of an air parcel, $T$, and its potential temperature, $\theta$, is given by eq. (5.28):

$$\theta = T + \Gamma z \tag{5.28}$$

Remember that $T$ is not a constant. Take the derivative of eq. (5.28):

$$\frac{d\theta}{dz} = \frac{dT}{dz} + \Gamma \tag{5.45}$$

By definition:

$$\frac{dT}{dz} = -\Lambda \tag{5.46}$$

Substitution of eq. (5.46) into eq. (5.45) leads to

$$\frac{d\theta}{dz} = \Gamma - \Lambda \tag{5.47}$$

Substitution of eq. (5.47) into eq. (5.44) leads to

$$\boxed{a = \frac{g}{T_a}\frac{d\theta}{dz} = s} \tag{5.48}$$

This is the stability parameter: $s$ is positive when the atmosphere is stable (potential temperature increases with height), and negative when the atmosphere is unstable.

Also $s$ is one of the variables that determine plume rise. Hence, we will use $s$ when we look at plume rise in detail (Chapter 7).

### 5.5.3 Diurnal Cycle of Stability

Figure 5.9 shows a typical example of the temperature profile in the atmosphere during daytime and nighttime. Based on this information, we can learn something about the diurnal cycle of the stability of the atmosphere.

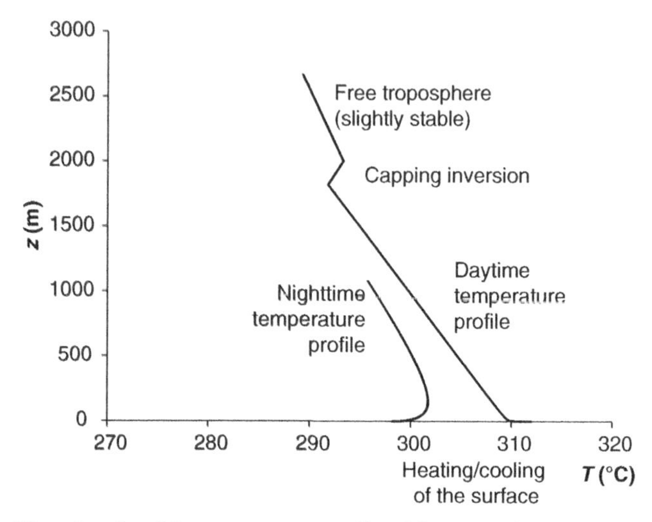


Figure 5.9 Diurnal cycle of the temperature profile of the atmosphere.

At high altitude, where the atmosphere is not influenced by Earth's surface, the temperature gradient is about –0.006 K $m^{-1}$ (see Example 5.1: at 3100–3800 m height, the temperature gradient is –0.0058 K $m^{-1}$). Hence, the temperature profile is slightly subadiabatic (stable). At the surface, there is pronounced heating during the day and pronounced cooling during the night. This affects the temperature of the atmosphere just above it, and this affected layer grows with time. During daytime, a superadiabatic layer is created, which is unstable. This layer grows as the day proceeds. Pronounced mixing occurs in this layer, hence its name as the mixing layer or mixed layer. The mixed layer is capped by a shallow inversion layer called the **capping inversion**. This inversion is the result of the convection below, which forces the temperature gradient to be larger than the temperature gradient of the overlying air. This causes the temperature of the upper part of the mixed layer to be slightly below the temperature at the bottom of the free atmosphere. An example is the layer at 2900–3100 m in Example 5.1. At nighttime, the air cools and creates a temperature inversion that increases in depth as the night proceeds. This layer is stable. In spite of the stability, there is still what can be considered a mixing layer, but this layer is much thinner and less sharply defined and only created by mechanical mixing.

We conclude that the atmospheric stability oscillates between unstable ($s < 0$; stability class A, B, or C) during daytime, and stable ($s > 0$; stability class E or F) during nighttime. On overcast days, the surface exchanges very little heat with the atmosphere, and the atmosphere is neutral ($s = 0$; stability class D).

## 5.6 HEAT BALANCE

We can predict the stability of the atmosphere by looking at a **heat balance** applied to the surface of Earth (Fig. 5.10):

$$R_N = q + q_L + q_G \tag{5.49}$$

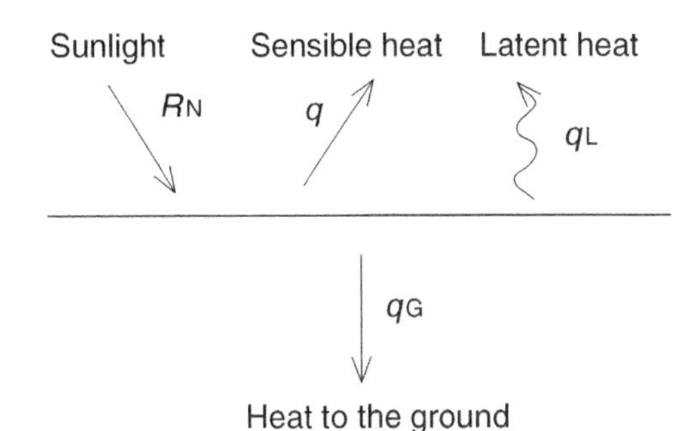


Figure 5.10 Heat balance of Earth's surface.

where $R_N$ is the net radiation energy flux received by the surface (W m$^{-2}$), $q$ is the sensible heat flux given off by the surface (W m$^{-2}$), $q_L$ is the latent heat flux given off by the surface (W m$^{-2}$), and $q_G$ is the heat flux going into the ground (W m$^{-2}$).

Sensible heat is heat transfer driven by a temperature gradient. Latent heat is heat carried by water vapor.

In urban areas anthropogenic heat flux can be added to the left-hand side of eq. (5.49). However, this term is usually small (Scire et al., 2000), and most of the urban heat island effect is included in other terms.

Sometimes an accumulation term is added to eq. (5.49), but that term is already covered by $q_G$. In order to partition between sensible heat flux and latent heat flux, the **Bowen ratio**, $B$, is introduced (Bowen, 1926):

$$B = \frac{q}{q_L} \tag{5.50}$$

The Bowen ratio $B$ is usually between 0 (open water) and 10 (deserts). Typical values are given in Table 5.3. Oke (1982) suggested the following values for $B$: 0.5 (range 0.1–1.5) for rural landscape, 1 (range 0.25–2.5) for suburban terrain, and 1.5 (range 0.5–4) for urban terrain.

The heat flux to the ground is often estimated with the following equation (Scire et al., 2000):

$$q_G = C_G R_N \tag{5.51}$$

A typical value of $C_G$ is 0.1, which is the value used in AERMOD (Cimorelli et al., 2004); $C_G$ ranges from 0.05 to 0.3 based on terrain. Typical values are 0.05–0.25 for rural areas, 0.25–0.30 for urban terrain, and 0.1 for grasslands (Scire et al., 2000).

Substitution of eqs. (5.50) and (5.51) into eq. (5.49) leads to

$$R_N = q + \frac{q}{B} + C_G R_N \tag{5.52}$$

**TABLE 5.3 Default Values of Albedo at 90° Solar Elevation and Bowen Ratio in CALPUFF**

| Area | Albedo, $r'$ | Bowen Ratio, $B$ |
|---|---|---|
| Urban | 0.18 | 1.5 |
| Rural | 0.15–0.25 | 1 |
| Irrigated | 0.15 | 0.5 |
| Water | 0.1 | 0 |
| Forest | 0.1 | 1 |

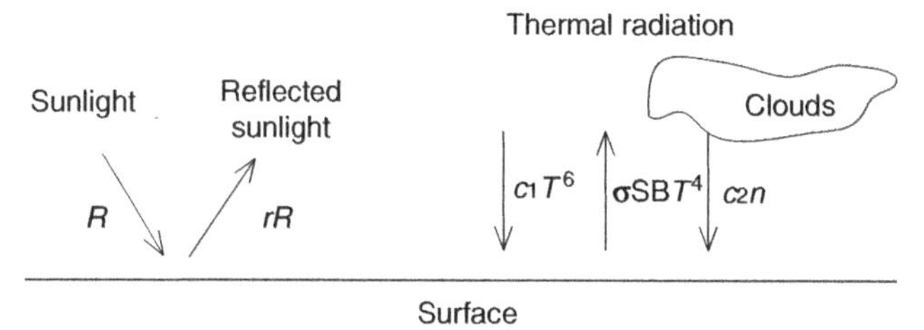


Figure 5.11 Radiation balance of Earth's surface.

The equation is solved for $q$, which yields

$$\boxed{q = \frac{(1 - C_G)R_N}{1 + 1/B}} \tag{5.53}$$

The net radiation is calculated from the solar radiation energy flux $R$ using a **radiation balance**. An equation for $R$ (in W m$^{-2}$) is the following (Kasten and Czeplak, 1980; Holtslag and van Ulden, 1983):

$$R = (990 \sin\varphi - 30)(1 - 0.75n^{3.4}) \tag{5.54}$$

where $\varphi$ is the solar elevation (angle of the sun above the horizon), and $n$ the fractional cloud cover (a number between 0 and 1, usually given in steps of one eighth). Equations for the solar elevation are given in Section A5.2 (Appendix A).

Next, we calculate net radiation from solar radiation with a radiation balance (Fig. 5.11). The following radiation flows are observed.

The main driver of the daytime heat balance is the incoming solar radiation. Part of this radiation is reflected back into space. The ratio between the reflected radiation and the incoming radiation is the **albedo**, $r$, of the surface. The albedo is typically 0.1–0.25 and depends on the solar elevation (See Section A5.3, Appendix A: The albedo is higher when the sun is near the horizon). Note that the albedo of

Earth as a whole, 0.3, is larger than the albedo of the surface because part of the reflection happens in the atmosphere as well.

Because of the temperature of Earth's surface, some infrared radiation will be emitted. This radiation is proportional to $T^4$, with $T$ the surface temperature in kelvins (the Stefan–Boltzmann law). The clouds block some of the sunlight, but they also emit some infrared radiation, some of which is absorbed by Earth's surface. Finally, some of the infrared radiation of Earth is absorbed by greenhouse gases and reemitted back toward Earth. This radiation is taken as proportional to $T^6$ in the Holtslag–van Ulden theory.

Semiempirical parameterization of each of these contributions to the radiation balance leads to the following equation:

$$R_N = \frac{R(1-r) + 60n - 5.67 \times 10^{-8} T^4 + 5.31 \times 10^{-13} T^6}{1.12} \tag{5.55}$$

The value of $R_N$ obtained with eq. (5.55) can be used to calculate $q$ in eq. (5.53). When $q$ is *positive*, there is net sensible heat deposition into the atmosphere. The atmosphere will be superadiabatic and therefore *unstable*. When $q$ is *negative*, there is net sensible heat removal from the atmosphere. The atmosphere is subadiabatic and therefore *stable*.

Values for $B$ and $r$ recommended as default in CALPUFF are given in Table 5.3.

**Example 5.4.** Calculate the sensible heat flux $q$ on a cloudless day when the solar elevation is 60° above the horizon. The terrain is rural, and an albedo $r$ of 0.2 can be assumed, a $C_G$ value of 0.1, and a Bowen ratio of 0.5. How does the sensible heat flux change during a severe draught, when $B$ rises to 1.5?

***Solution.*** Equation (5.54) is used to calculate the solar radiation $R$. With $n = 0$ and $\sin \varphi = 3^{0.5}/2$, a value of 827.4 W m$^2$ is obtained.
Next, eq. (5.55) is used to calculate the net radiation $R_N$. A value of 524.0 W m$^{-2}$ is obtained.

Equation (5.53) calculates $q$. With $B = 0.5$, a value of **157.2 W m$^{-2}$** is obtained. With $B = 1.5$, a value of **282.9 W m$^{-2}$** is obtained. Dry conditions increase the sensible heat flux because less heat is transported in water vapor.

Some of the subtleties of the **nocturnal** heat balance are *not* resolved by eqs. (5.53) and (5.55). For instance, the nighttime Bowen ratio can differ substantially from the daytime value and can even become negative. Hence, a different approach is needed. However, we need to understand temperature profiles in nonneutral conditions before we can approach the nocturnal heat balance. Nonneutral temperature profiles will be discussed in Section 5.8; the nocturnal heat balance is discussed in Section 5.9.

Figure 5.12 illustrates the influence of season, cloudiness, wetness of the ground, and albedo on the diurnal variation of $q$ as calculated with the above equations. Note that the nighttime conditions in this figure are only approximations for the reasons mentioned above.

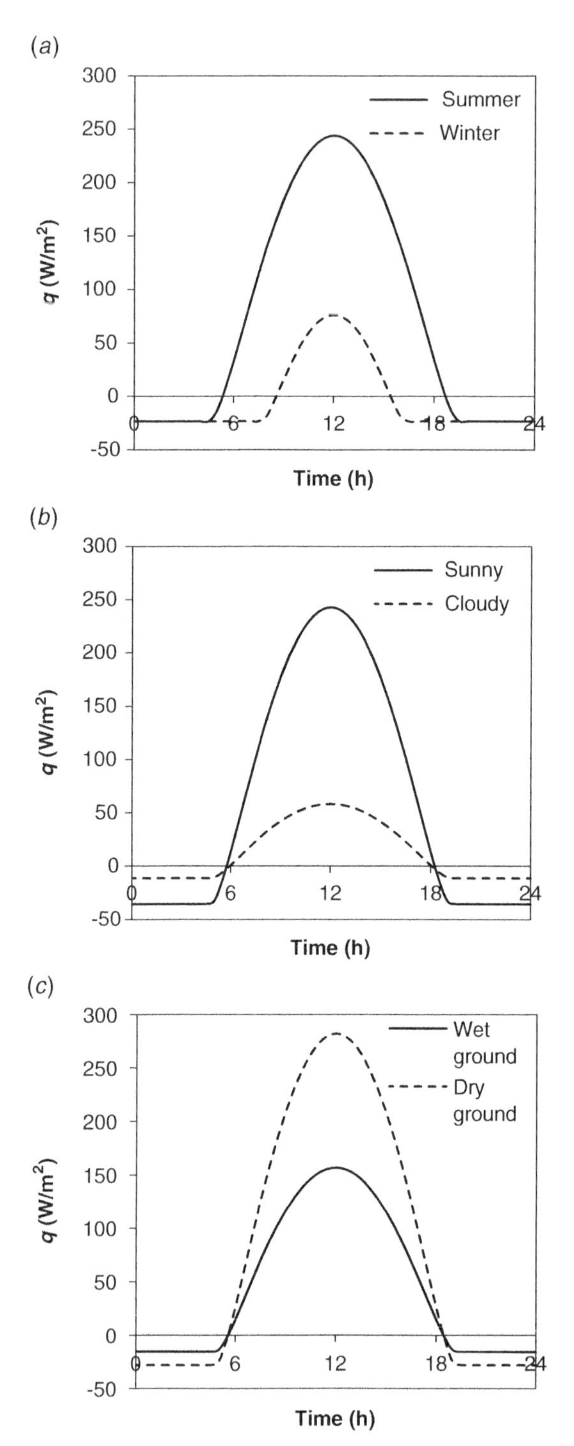


Figure 5.12 Typical calculated diurnal variation of $q$ (*a*) in summer and winter, (*b*) on sunny and cloudy days, (*c*) on dry and wet soils, and (*d*) for different albedos.

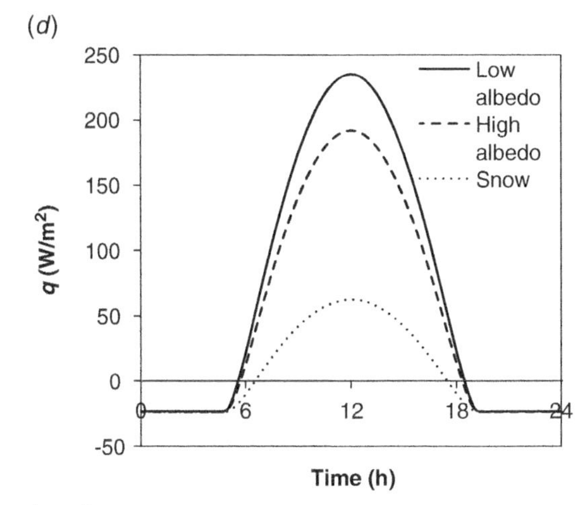


Figure 5.12 (continued)

## 5.7 WIND SPEED PROFILE

In this section we start the outline of the **similarity theory**. This theory assumes that all atmospheres behave essentially the same, but the scale of behavior is different. The method used to derive the similarity theory is known in engineering as **dimension analysis**. In dimension analysis we try to narrow down the number of possible relationships between variables by grouping them into **dimensionless numbers**.

### 5.7.1 Case 1: Smooth Surface, Adiabatic Conditions

We start with a calculation for surfaces with roughness elements too small to influence wind flow or surface friction. Such a surface is said to be *aerodynamically smooth*. This case rarely happens in the environment, but it is a good starting point for calculations.

We try to identify all the variables that describe the conditions of the wind flow. The following come to mind:

- $u$ wind speed (m $s^{-1}$)
- $\nu$ kinematic viscosity ($m^2$ $s^{-1}$)
- $\rho$ air density (kg $m^{-3}$)
- $z$ height (m)

We need a fifth variable: a measure of the friction caused by the flow. We could choose $\tau_0$, the shear stress at the surface. **Shear stress** is stress caused by a force parallel to the surface upon which the force acts. Numerically, it is the force divided by the area of the surface. It has the same units as pressure (Pa = N $m^{-2}$ = kg $m^{-1}$ $s^{-2}$).

When we look at the five variables, it is clear that only two of them, $\rho$ and $\tau_0$, have mass in the dimensions. Hence, to obtain dimensionless numbers, they can only occur as a ratio. For that reason, we define a new variable, the **friction velocity**, $u_*$:

$$u_* = \sqrt{\frac{|\tau_0|}{\rho_0}} \tag{5.56}$$

where $\rho_0$ is the air density at the surface. The friction velocity should not be interpreted as a real velocity but rather as a measure of the friction that happens to have the same units (m $s^{-1}$) as a velocity. Hence, we have four variables ($u$, $v$, $z$, and $u_*$) that incorporate two dimensions (length and time). According to the Buckingham $\pi$ theorem, there are two (= 4−2) dimensionless groups of variables that can be defined at the same time without one being a combination of the others. We will call those variables $\pi_1$ and $\pi_2$. If the four variables we defined have any relationship with each other at all, then there has to be a functional relationship between the two variables:

$$\pi_2 = G(\pi_1) \tag{5.57}$$

The theorem does not tell us what the function is, but the fact that we narrowed down a function of the form $u=f(v, z, u_*)$ to eq. (5.57) is helpful.

In meteorology we are lucky in a sense because the wind flow is fully turbulent. That means that the effect of $v$ is much smaller than the effect of turbulent eddies. This is demonstrated with a numerical example in Section A 5.4 (Appendix A).

If we can neglect the effect of viscosity, then there are only three variables ($u$, $z$, and $u_*$) containing two dimensions. Hence, there is only one dimensionless number, and that number has to be a constant.

We see immediately that we cannot use the three variables without some modification because two of them have the same units, making it impossible to create a dimensionless number that incorporates the three variables. We solve this by replacing variable $u$ by variable $du/dz$. Hence, our three variables are $du/dz$, $z$, and $u_*$.

We observe that $z\, du/dz$ has the same units as $u_*$. Hence, the choice of dimensionless number is obvious:

$$\pi = \frac{z}{u_*}\frac{du}{dz} \tag{5.58}$$

Hence:

$$\frac{z}{u_*}\frac{du}{dz} = \text{const.} = \frac{1}{k} \tag{5.59}$$

where $k$ is the von Kármán constant and is about 0.4. We can write

$$du = \frac{u_*}{k}\frac{dz}{z} \tag{5.60}$$

Integration leads to

$$u = \frac{u_*}{k}(\ln z + \text{const.}) \tag{5.61}$$

This equation is known as the **law of the wall**. The constant is $\ln(9u_*/\nu)$, but in most real situations related to meteorology the surface is too rough for this value to be applicable.

Recent research indicates that $k$ might be closer to 0.39 (Andreas et al., 2006; Andreas, 2009). However, there is no consensus about this yet. If true, the effect on meteorological calculations will be small.

### 5.7.2 Case 2: Rough Surface, Adiabatic Conditions

When the surface is not smooth, then a new variable needs to be introduced, the height of the obstacles determining the roughness of the surface, $\varepsilon$. Rather than repeating the dimension analysis, we will simply assume that the wind speed depends on $z/\varepsilon$, not on $z$. Therefore, eq. (5.60) is rewritten as

$$du = \frac{u_*}{k}\frac{d(z/\varepsilon)}{z/\varepsilon} \tag{5.62}$$

Define $z_0$ as the (extrapolated) height where $u=0$. Integration of eq. (5.62) leads to

$$u = \frac{u_*}{k}\left[\ln\left(\frac{z}{\varepsilon}\right) - \ln\left(\frac{z_0}{\varepsilon}\right)\right] \tag{5.63}$$

which simplifies to

$$\boxed{u = \frac{u_*}{k}\ln\frac{z}{z_0}} \tag{5.64}$$

and $z_0$ is called the **roughness length** (or roughness height). It typically varies between $\varepsilon/30$ and $\varepsilon/3$, with an average value of $\varepsilon/10$. Table 5.4 gives typical values of $z_0$, based on the CALPUFF default values.

Note that $z_0$ is not a physical height, unlike $\varepsilon$. When we talk about **roughness** we do not refer to the size of any particular object.

By definition, the wind speed prediction is 0 when $z=z_0$. In reality, the wind speed does not normally become zero above the surface. The equation is *inaccurate* when $z < 2\varepsilon$. Equations for wind speed close to the surface are discussed in Chapter 8.

The disadvantage of eq. (5.64) is that the friction velocity $u_*$ is not a routinely measured variable. Fortunately, it can be calculated based on a wind speed measurement $u_m$ at a known height $z_m$. To that effect, eq. (5.64) is solved for $u_*$:

$$u_* = \frac{ku_m}{\ln(z_m/z_0)} \tag{5.65}$$

The value of $u_*$ obtained is substituted in eq. (5.64) to obtain a wind speed profile.

**TABLE 5.4 Recommended Values of Roughness Length, $z_0$**

| Area | Roughness Length, $z_0$ (m) (CALPUFF default in bold) | Comments |
|---|---|---|
| Mountains | 5–70 | Rocky Mountains: 50–70 |
| Urban | 0.5–1 (suburbs); 2–3 (city centres); **1** | |
| Rural | **0.05–0.25**; average 0.1 | |
| Farmland | 0.03–0.1 | |
| Prairie | 0.005–0.03 | cut grass: 0.005–0.01 |
| Irrigated | **0.25** | |
| Water, snow | 0.00001–0.0001; **0.001** | open sea: 0.0001–0.001 |
| Forest | 0.5–**1** | lower when the canopy is very dense (skimming flow) |
| Ice and mud flats | 0.00001 | |
| Lawn | 0.01 | |

*Sources*: Adapted from Stull (1988), Scire et al. (200), and Hanna and Britter (2002).

**Example 5.5.** In the primer (Chapter 2), a wind speed of 7 m s$^{-1}$ measured under neutral conditions at a 10-m height was extrapolated to 10.17 m s$^{-1}$ at a 90-m height with an empirical equation. Calculate the extrapolation with the similarity theory, assuming average $z_0$ for rural terrain (0.1 m). Compare the result with the result from the primer.

***Solution.*** The friction velocity is estimated with eq. (5.65):

$$u_* = 0.4 \cdot 7\,\text{m}\,\text{s}^{-1} / \ln(90\,\text{m} / 0.1\,\text{m}) = 0.608\,\text{m}\,\text{s}^{-1}$$

The friction velocity is substituted in eq. (5.64):

$$u_{90} = (0.608 / 0.4) \times \ln(90 / 0.1)$$

$$u_{90} = \mathbf{10.34\,m\,s^{-1}}$$

The extrapolation agrees very well with the value from the primer.

Figure 5.13 shows neutral wind speed profiles for different roughness lengths, assuming constant wind speed at 200 m (i.e., above the surface layer). The wind speeds at a 10-m height and the friction velocities are shown in Table 5.5. It is clear that the roughness length has a pronounced effect on the wind speed profile. At a 10-m height the wind speed ranges from 4.35 to 7.55 m s$^{-1}$. It is also interesting to note the change of the friction velocity as surface roughness increases: from 0.328 to 0.755 m s$^{-1}$. This indicates that there is more friction, more shear stress on a rougher surface.

Returning to the aerodynamically smooth case, it follows from eq. (5.61) that the roughness length of a smooth surface is $\nu/(9u_*)$, where $\nu$ is the kinematic viscosity. Assuming $\nu = 1.5 \times 10^{-5}$ m$^2$ s$^{-1}$ and $u_*$ ranging from 0.01 to 1 m s$^{-1}$ leads to a roughness length ranging from $1.67 \times 10^{-4}$ to $1.67 \times 10^{-6}$ m. This is the minimum possible surface roughness of a rigid surface. A further discussion of the microscopic behavior of flow along a surface is given in Appendix C.

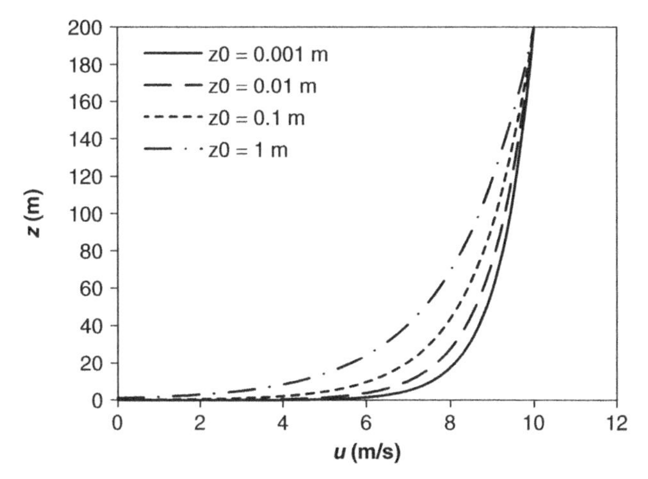


Figure 5.13 Neutral wind speed profiles at four different roughness lengths, assuming a wind speed of $10\,\mathrm{m\,s^{-1}}$ at a 200-m height.

**TABLE 5.5 Wind Speeds at 10-m Height and Friction Velocities Associated with Four Curves in Figure 5.13**

| $z_0$ (m) | $u_{10}$ (m s$^{-1}$) | $u_*$ (m s$^{-1}$) |
|---|---|---|
| 0.001 | 7.55 | 0.328 |
| 0.01 | 6.98 | 0.404 |
| 0.1 | 6.06 | 0.526 |
| 1 | 4.35 | 0.755 |

For a **water surface**, the surface roughness is also dependent on the friction velocity. The following equation has been proposed (Charnock, 1955):

$$z_0 = \frac{\alpha_c u_*^2}{g} \tag{5.66}$$

Hence, the wind speed profile above a water surface is given by

$$u = \frac{u_*}{k} \ln\left(\frac{zg}{\alpha_c u_*^2}\right) \tag{5.67}$$

Charnock (1955) used a value of $\alpha_c = 0.0067$. Peña and Gryning (2008) updated the coefficient $\alpha_c$ to 0.012 for open sea. This leads to typical roughness lengths of $10^{-5}$ to $10^{-4}$ m, somewhat less than recommended in CALPUFF. Equation (5.67) shows an opposite trend as found at a smooth surface. For friction velocities lower than 0.111 m s$^{-1}$ a roughness length *less* than a smooth surface is predicted. This is possible because a water surface moves with the wind, lowering the wind speed gradient.

Sometimes, when the surface is densely covered with obstacles, eq. (5.64) is not accurate. In that case, the following equation can be used instead:

$$u = \frac{u_*}{k} \ln \frac{z - d}{z_0} \tag{5.68}$$

where $d$ is known as the **displacement height**, typically around $0.5\varepsilon$ (Hanna and Britter, 2002). This equation can be useful in urban conditions, when wind speeds close to the roughness elements are needed. This situation is discussed in more detail in Chapter 8. Equation (5.68) is rarely used in dispersion modeling.

### 5.7.3 Case 3: Rough Surface, Nonneutral Conditions

The theory outlined in this section is known as the **Monin–Obukhov similarity theory**. It is one of the most important theories in air dispersion modeling.

Wind speed profiles strongly depend on the stability of the atmosphere. As discussed in Section 5.6, the stability of an atmosphere can be evaluated by considering the heat balance. The key variable is the sensible heat flux at the surface, $q$.

As with the neutral condition, wind speed profiles can be obtained by considering dimensionless variables. A variable that has been put forward is $\zeta$:

$$\zeta = -\frac{kgzq}{\rho_0 c_p T_0 u_*^3} \tag{5.69}$$

where $\zeta$ is a stability indicator that depends on height $z$, which is inconvenient for some purposes. Therefore, we define the **Obukhov length** $L$ (Monin, 1959; Obukhov, 1971):

$$\zeta = \frac{z}{L} \tag{5.70}$$

Hence:

$$\boxed{L = -\frac{\rho_0 c_p T_0 u_*^3}{kgq}} \tag{5.71}$$

We can consider three cases:

1. When $q > 0$, sensible heat is released from the ground to the atmosphere. This leads to *unstable* conditions. In that case, $\boldsymbol{\zeta < 0}$ and $\boldsymbol{L < 0}$.
2. When $\boldsymbol{q < 0}$, sensible heat is absorbed from the atmosphere to the ground. This leads to *stable* conditions. In that case, $\boldsymbol{\zeta > 0}$ and $\boldsymbol{L > 0}$.
3. When $\boldsymbol{q = 0}$, no sensible heat is released or absorbed. This leads to *neutral* conditions. In that case, $\boldsymbol{\zeta = 0}$ and $\boldsymbol{L = \infty}$.

It follows that the larger $|L|$, the more neutral the atmosphere. Small $|L|$ means extreme conditions. Because of this counterintuitive behavior of $L$, the inverse $(1/L)$

is sometimes used as a variable instead: $1/L$ is near zero under neutral conditions, it has a large positive value (e.g., 0.1 m$^{-1}$) under strongly stable conditions, and a large negative value (e.g., −0.1 m$^{-1}$) under strongly unstable conditions.

Like $z_0$, $L$ is not a physical length, but rather a measure of stability that has the units of length.

The second dimensionless group is the one we used before. Based on eq. (5.59) we find under neutral conditions:

$$\pi = \frac{kz}{u_*}\frac{\partial u}{\partial z} = 1 \tag{5.72}$$

Under nonneutral conditions $\pi$ is no longer constant. Through the Buckingham $\pi$ theorem we state

$$\frac{kz}{u_*}\frac{\partial u}{\partial z} = \phi_m(\zeta) \tag{5.73}$$

with $\phi_m$ a function known as the Monin–Obukhov similarity function for **momentum** transfer. Rearrangement leads to

$$\boxed{\frac{\partial u}{\partial z} = \frac{u_*}{kz}\phi_m(\zeta)} \tag{5.74}$$

From eq. (5.72) we know the value of $\phi_m$ for $\zeta=0$:

$$\phi_m(0) = 1 \tag{5.75}$$

The function $\phi_m(\zeta)$ has been determined experimentally for other values of $z$, and the following equations (Dyer, 1974) have found widespread use:

$\zeta > 0$ (stable):

$$\phi_m(\zeta) = 1 + 5\zeta \tag{5.76}$$

$\zeta < 0$ (unstable):

$$\phi_m(\zeta) = (1 - 16\zeta)^{-1/4} \tag{5.77}$$

These are the equations used in AERMOD. Over the years a wide range of equations has been proposed. Even though wind speed profile measurements have improved tremendously over time, the determination of $\phi_m(\zeta)$ is still prone to substantial experimental error, especially under stable conditions. Section A 5.5 (Appendix A) gives an overview of some equations found in the literature. A promising recent alternative to eq. (5.76) is the following equation of Cheng and Brutsaert (2005):

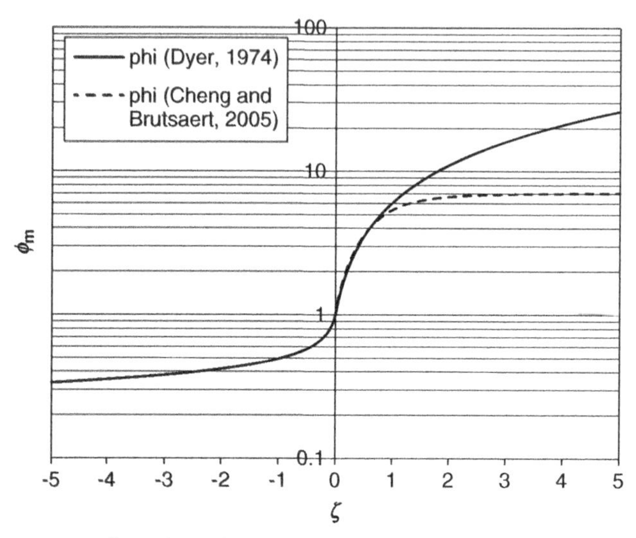


Figure 5.14 $\phi_m$ versus $\zeta$, semilogarithmic scale.

$\zeta > 0$ (stable):

$$\phi_m = 1 + a\left[\frac{\zeta + \zeta^b\left(1+\zeta^b\right)^{(1-b)/b}}{\zeta + \left(1+\zeta^b\right)^{1/b}}\right] \tag{5.78}$$

where $a = 6.1$ and $b = 2.5$. Equations (5.76) and (5.77) are plotted in semilog scale in Figure 5.14, together with eq. (5.78).

Equation (5.74) needs to be integrated. It is rewritten as follows:

$$du = \frac{u_*}{k}\phi_m(\zeta)\frac{dz}{z} \tag{5.79}$$

Since $z = L\zeta$ and $dz = L\,d\zeta$:

$$du = \frac{u_*}{k}\frac{\phi_m(\zeta)}{\zeta}d\zeta \tag{5.80}$$

When $\zeta > 0$ (stable atmosphere), eq. (5.80) can be integrated analytically. The solution with $\phi_m(\zeta)$ given by eq. (5.76) is

$$\boxed{u = \frac{u_*}{k}\left(\ln\frac{z}{z_0} + 5\frac{z - z_0}{L}\right)} \tag{5.81}$$

For $\zeta < 0$ (unstable atmosphere), there is no exact solution. Benoit (1977) found the following approximation:

$$u = \frac{u_*}{k}\left\{\ln\frac{z}{z_0} + \ln\left[\frac{(n_0^2+1)(n_0+1)^2}{(n^2+1)(n+1)^2}\right] + 2\left[\arctan(n) - \arctan(n_0)\right]\right\} \tag{5.82}$$

with

$$n_0 = \left(1 - 16\frac{z_0}{L}\right)^{1/4} \qquad n = \left(1 - 16\frac{z}{L}\right)^{1/4} \tag{5.83}$$

In eq. (5.82), arctan is a function that has radians as units, not degrees. Values are tabulated in Section B2 (Appendix B).

The general notation of eqs. (5.81) and (5.82) is

$$u = \frac{u_*}{k}\left[\ln\frac{z}{z_0} + \psi_\text{m}(\zeta_0) - \psi_\text{m}(\zeta)\right] \tag{5.84}$$

where

$$\psi_\text{m} = \int_0^\zeta \frac{1-\phi_\text{m}(x)}{x}\,dx \tag{5.85}$$

When wind speed profiles are calculated with eq. (5.78), then the integration leads to the following:

$$\psi_\text{m}(\zeta) = -a\ln\left[\zeta + \left(1+\zeta^b\right)^{1/b}\right] \tag{5.86}$$

It is easy to produce wind speed profiles when $u_*$ and $q$ are known. In practice, $u_*$ is usually unknown but $u$ is known at a certain height (typically 10 m), and $q$ can be estimated from the heat balance. The problem then arises that $u_*$ is needed to calculate $L$ [with eq. (5.69)], and $L$ is needed to calculate $u_*$ [with eq. (5.81) or (5.82) or eqs. (5.84) and (5.86)].

This problem can be solved iteratively. For instance, a first estimate of $u_*$ is calculated with the equation for neutral atmosphere, eq. (5.65). Length $L$ is then calculated from its definition [eq. (5.71)]. Equation (5.81) or (5.82) or eqs. (5.84) and (5.86) is/are then used to get a new estimate of $u_*$, and the procedure is repeated with the calculation of $L$ and the like until convergence. This usually does not take more than three or four iterations.

**Example 5.6.** Calculate the wind speed profile when $u_{10} = 5$ m s$^{-1}$, $q_0 = 200$ W m$^{-2}$, $z_0 = 0.5$ m, $T_0 = 300$ K, and $p_0 = 101$ kPa.

***Solution.*** With eq. (5.65), we obtain 0.668 m s$^{-1}$ as a first guess for $u_*$. With eq. (5.71), this leads to $L=-134.4$ m. We solve eq. (5.82) for $u_*$ to obtain a second guess for $u_*$:

$$u_* = \frac{ku}{\ln\frac{z}{z_0} + \ln\left[\frac{(n_0^2+1)(n_0+1)^2}{(n^2+1)(n+1)^2}\right] + 2\left[\arctan(n) - \arctan(n_0)\right]}$$

A value of 0.718 m s$^{-1}$ is obtained.

With eq. (5.71), this leads to a second guess of $L=-167.4$ m.

Our third guess is $u_*=0.710$ m s$^{-1}$ and $L=-161.5$ m. This is close enough to the exact solution of $u_*=0.711$ m s$^{-1}$ and $L=-162.3$ m.

From these data, an entire velocity profile can easily be constructed from eq. (5.82). The profile is shown in the Excel file "Wind speed profiles for problem solving.xlsx" on the enclosed CD.

When detailed meteorological conditions are not known, but the Pasquill stability class is known, then a rough estimate of $L$ can be made using the following equation (Seinfeld and Pandis, 2006):

$$\frac{1}{L} = a + b\log z_0 \tag{5.87}$$

where $z_0$ is roughness length in meter, and log is the logarithm with base 10. The coefficients $a$ and $b$ are given in Table 5.6.

Figure 5.15 shows the regions of the Pasquill stability classes in a $z_0$ vs. $1/L$ diagram (Golder, 1972).

Wind speed profiles of different stabilities are plotted in dimensionless form in Figures 5.16 and 5.17. They were generated with eqs. (5.81) and (5.82). These figures allow easy recognition of the conditions that generated a measured wind speed profile by comparison with the figures. However, inferring the surface sensible heat flux $q$ from a wind speed profile is far less accurate than estimating $q$ with a heat balance.

**TABLE 5.6 Coefficients *a* and *b* for Use in eq. (5.87) for Different Pasquill Stability Classes**

| Pasquill Stability Class | *a* | *b* |
|---|---|---|
| A | –0.096 | 0.029 |
| B | –0.037 | 0.029 |
| C | –0.002 | 0.018 |
| D | 0 | 0 |
| E | 0.004 | –0.018 |
| F | 0.035 | –0.036 |

*Source*: From Seinfeld and Pandis (2006).

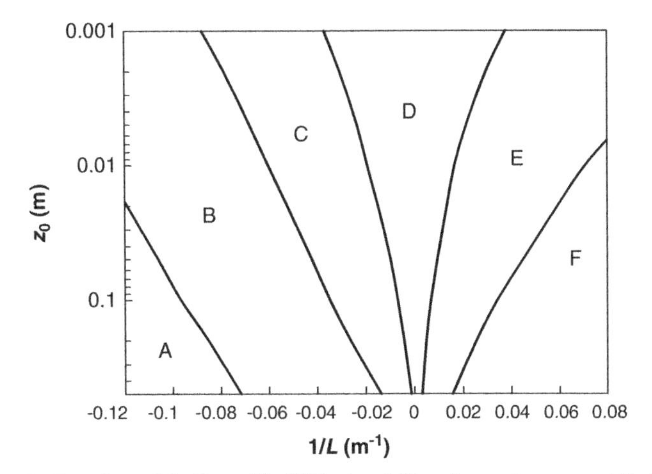


Figure 5.15 Delineation of the Pasquill–Gifford stability classes in a $z_0$ vs. $1/L$ diagram (after Golder, 1972).

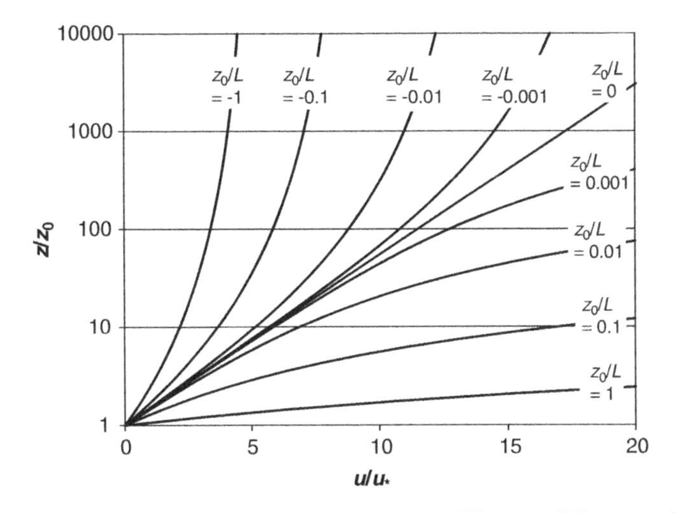


Figure 5.16 Wind speed profiles in dimensionless form at different stabilities (semilog scale).

Under stable conditions, eq. (5.81) may be out of date. In Figure 5.18, the traditional wind speed profiles based on Dyer (1974) are compared with profiles generated with the more recent equations of Cheng and Brutsaert (2005). The two give nearly identical results for $z/L$ values up to about 1.3 and diverge above that. It follows that the difference is only impotant in strongly stable conditions.

Not only the wind speed depends on height but wind **direction** as well. At high altitudes the wind is not influenced by the surface. This is known as the **geostrophic wind**. In the Northern Hemisphere, geostrophic winds move counterclockwise around low-pressure regions and clockwise around high-pressure regions. This is

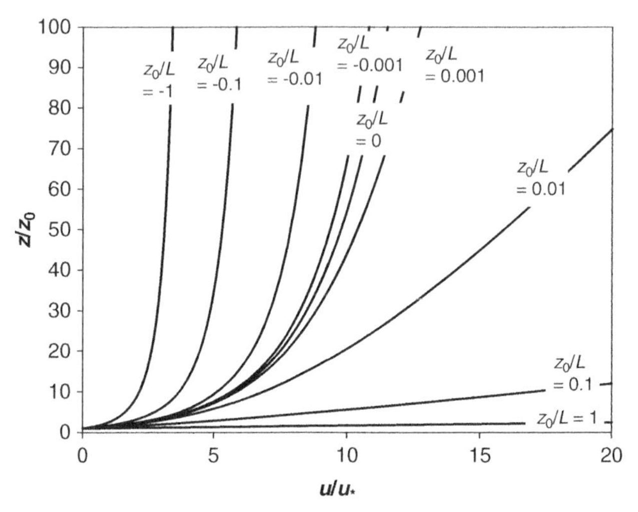


Figure 5.17 Wind speed profiles in dimensionless form at different stabilities (linear scale).

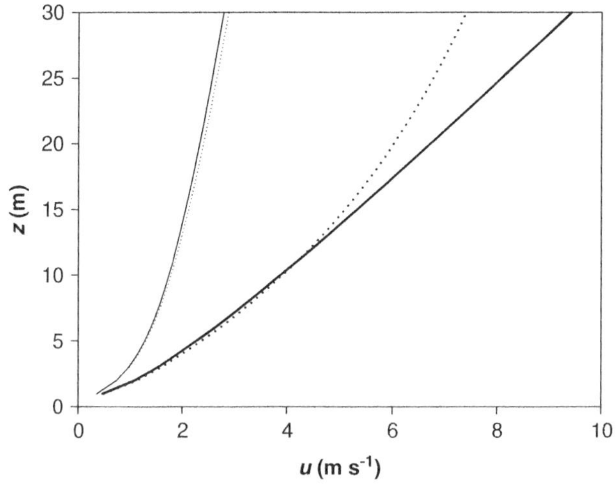


Figure 5.18 Wind speed profiles under moderately stable conditions ($L = 100$ m: thin lines) and strongly stable conditions ($L = 10$ m: thick lines) generated with the model of Dyer (1974) (solid lines) and with the model of Cheng and Brutsaert (2005) (dotted lines).

shown mathematically in Section A 5.6 (Appendix A). This is the result of the Coriolis force. Closer to the ground, the Coriolis force is largely undone by friction, so the wind tends to flow from high to low pressure. Hence, the wind direction at high altitude can be expected to be 90° clockwise from the wind speed at ground level. In practice, high-altitude winds are not entirely separated from ground-level winds, so the actual angle is less than 90°.

## 5.8 TEMPERATURE PROFILE REVISITED: NONNEUTRAL CONDITIONS

It has been discussed before that in a neutral atmosphere the potential temperature $\theta$ is independent of height, and the temperature $T$ decreases with a lapse rate of $-\Gamma=-0.01$ K m$^{-1}$ approximately. Estimates of $\theta$ profiles in the surface layer can also be made in nonneutral conditions using similarity theory.

First, the **friction temperature** $\theta_*$ is defined:

$$\boxed{\theta_* = -\frac{q}{\rho_0 c_p u_*}} \tag{5.88}$$

Just as the friction velocity is not a physical velocity, the friction temperature is not a physical temperature but a measure of the impact of the sensible heat flux on the temperature profile, which happens to have Kelvin as the unit.

The Obukhov length [eq. (5.71)] can be written in terms of $\theta_*$:

$$L = \frac{u_*^2 T_0}{kg\theta_*} \tag{5.89}$$

where $\theta_*$ has the same sign as $L$. Hence, the atmosphere is stable when $\theta_* > 0$.

Using dimension analysis, we can develop an equation very similar to eq. (5.74):

$$\boxed{\frac{\partial \theta}{\partial z} = \frac{\theta_*}{kz}\phi_h(\zeta)} \tag{5.90}$$

where $\phi_h$ is the Monin–Obukhov similarity function for **heat transfer**. The following expressions have been suggested (Dyer, 1974):

$\zeta > 0$ (stable):

$$\phi_h(\zeta) = \phi_m(\zeta) = 1 + 5\zeta \tag{5.91}$$

$\zeta < 0$ (unstable):

$$\phi_h(\zeta) = \phi_m^2(\zeta) = (1 - 16\zeta)^{-1/2} \tag{5.92}$$

As with wind speed profiles, many equations have been put forward to calculate $\phi_h$. Some examples are summarized in Section A 5.6 (Appendix A). Again, a promising development is the similarity function put forward by Cheng and Brutsaert (2005) for stable conditions:

$$\phi_h = 1 + c\left[\frac{\zeta + \zeta^d\left(1+\zeta^d\right)^{(1-d)/d}}{\zeta + \left(1+\zeta^d\right)^{1/d}}\right] \tag{5.93}$$

where $c=5.3$, and $d=1.1$.

Equation (5.90) is integrated with boundary condition $\theta=\theta_1$ for $z=z_1$. Note that, unlike wind speed calculations, where the boundary condition is at an extrapolated zero wind speed, there is no obvious interpretation of $z_1$, so the choice of $z_1$ is often made arbitrarily. The integration leads to the following equations. When the atmosphere is *stable* ($\zeta > 0$), the potential temperature is given by

$$\boxed{\theta = \theta_1 + \frac{\theta_*}{k}\left(\ln\frac{z}{z_1} + 5\frac{z - z_1}{L}\right)} \tag{5.94}$$

When the atmosphere is *unstable* ($\zeta < 0$), the potential temperature is given by

$$\boxed{\theta = \theta_1 + \frac{\theta_*}{k}\left\{\ln\frac{z}{z_1} + \ln\left[\frac{(1+x_1)^2}{(1+x)^2}\right]\right\}} \tag{5.95}$$

with

$$x_1 = \left(1 - 16\frac{z_1}{L}\right)^{1/2} \qquad x = \left(1 - 16\frac{z}{L}\right)^{1/2} \tag{5.96}$$

A general form of eqs. (5.94) and (5.95) is

$$\theta = \theta_1 + \frac{\theta_* \alpha}{k}\left[\ln\frac{z}{z_1} + \psi_h(\zeta_1) - \psi_h(\zeta)\right] \tag{5.97}$$

where $\alpha$ accounts for the fact that momentum and heat transfer coefficients are not necessarily the same under neutral conditions. Mass transfer coefficients and their link with similarity theory will be discussed in a later section. However, most observations find that $\alpha$ is 1 or slightly below. $\zeta_1$ is the value of $\zeta$ at height $z_1$. $\psi_h$ is given by

$$\psi_h = \int_0^{\zeta} \frac{1 - \phi_h(x)}{x} dx \tag{5.98}$$

When the equation of Cheng and Brutsaert (2005) [eq. (5.93)] is used, the integration leads to the following:

$$\psi_h(\zeta) = -c \ln\left[\zeta + \left(1 + \zeta^d\right)^{1/d}\right] \tag{5.99}$$

where $c=5.3$ and $d=1.1$. They found $\alpha=1$.

Potential temperature profiles of different stabilities are plotted in dimensionless form in Figures 5.19 and 5.20. The equations of Dyer (1974) [eqs. (5.94) and (5.95)] were used in the calculation. The charts are deceptive because the profile is never vertical, whereas the potential temperature profile is vertical in the case of a neutral atmosphere. For that reason, profiles of $T$ and $\theta$ versus height are given in

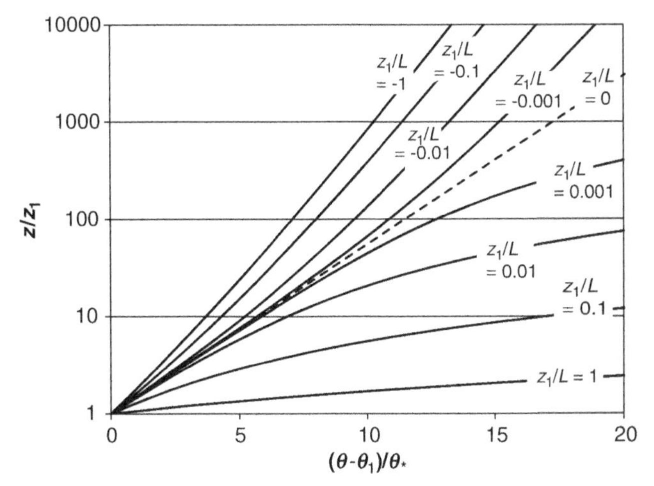


Figure 5.19 Potential temperature profiles in dimensionless form at different stabilities (semilog scale).

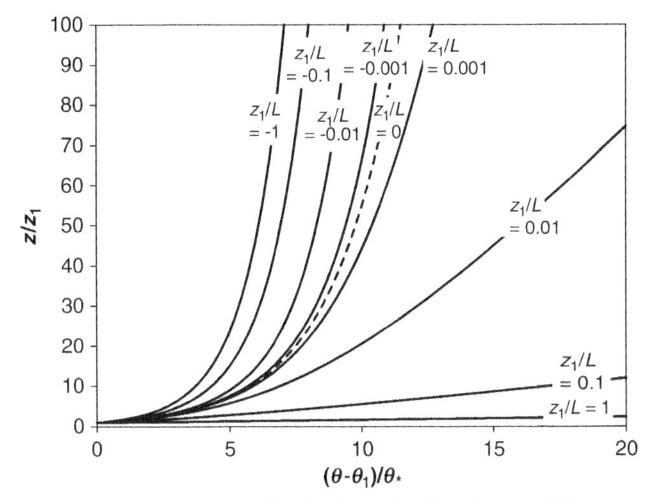


Figure 5.20 Potential temperature profiles in dimensionless form at different stabilities (linear scale).

Figure 5.21. The model of Cheng and Brutsaert (2005) was used for the stable cases. The qualitative temperature profiles of Figure 5.9 are reproduced by the calculations in Figure 5.21.

Section A 5.8 (Appendix A) briefly outlines how wind speed and temperature theories are interrelated and introduces two measures of stability, the **gradient Richardson number** and the **flux Richardson number**. Section A 5.9 (Appendix A) outlines two practical approaches to calculate $u_*$ and $L$ (and hence a complete temperature and wind speed profile) from knowledge of the temperature at two different heights. These approaches were taken from AERMOD (Cimorelli et al., 2004).

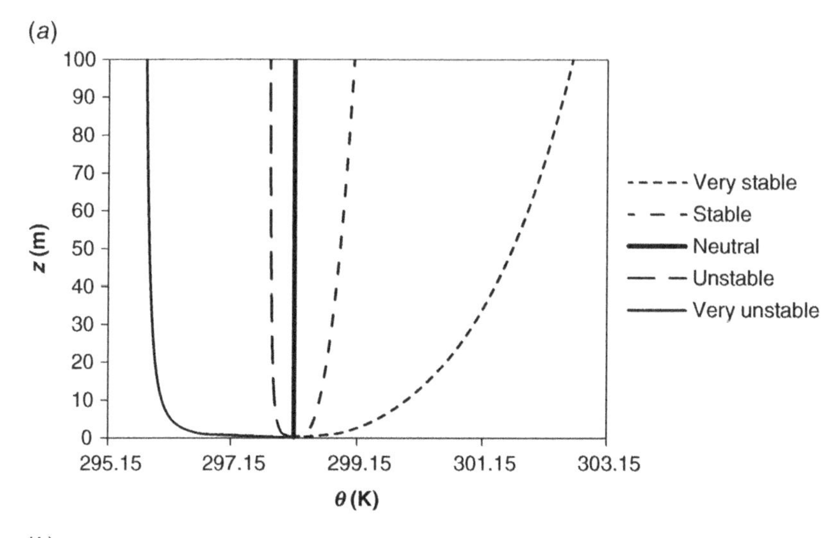


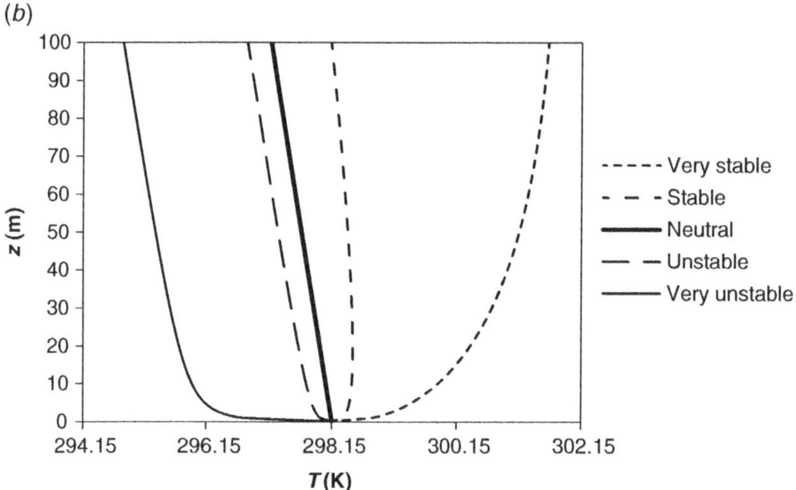


Figure 5.21 Vertical profile of (*a*) potential temperature and (*b*) actual temperature in very stable ($L=30\,\text{m}$), stable ($L=100\,\text{m}$), neutral, stable ($L=-100\,\text{m}$), and very unstable ($L=-10\,\text{m}$) conditions. $u_*=0.2\,\text{m s}^{-1}$, $z_1=0.1\,\text{m}$.

**Example 5.7.** In Example 5.6 we found that $u_*=0.711$ m s$^{-1}$ and $L=-162.3$ m when $u_{10}=5$ m s$^{-1}$, $q_0=200$ W m$^{-2}$, $z_0=0.5$ m, $T_0=300$ K, and $p_0=101$ kPa. Use these data to calculate $\theta_*$ and a temperature profile.

***Solution.*** Based on eq. (5.89), we find

$$\theta_* = \frac{u_*^2 T_0}{kgL}$$
$$= (0.711 \text{ m s}^{-1})^2 \cdot 300 \text{ K} / [0.4 \cdot 9.80665 \text{ m s}^{-2}(-162.3 \text{ m})] = -0.2382\,\text{K}$$

Based on this value, a potential temperature profile can be obtained by assuming $\theta_1 = 300$ K at $z_1 = 10$ m in eq. (5.95). For instance, a value of $\theta = 299.32$ K is obtained at $z = 100$ m. Because of the way we defined potential temperature here (at 10 m height), the real temperature is given by

$$T = \theta - \Gamma(z - z_1)$$

where $\Gamma = 0.00975$ K m$^{-1}$. The temperature at a 100-m height is 298.44 K.

The entire temperature profile, as well as the potential temperature profile, is given in the file "Temperature profiles for problem solving.xlsx."

## 5.9 HEAT BALANCE REVISITED: STABLE CONDITIONS

As indicated in Section 5.6, the heat balance of Holstslag and van Ulden (1983) [eq. (5.55)] is inadequate for the nocturnal atmosphere. However, accurate estimates of the surface heat flux $q$ at night can be made based on observations of the friction temperature at night. The friction temperature was defined as

$$\theta_* = -\frac{q}{\rho_0 c_p u_*} \tag{5.88}$$

At sufficiently high wind speed the following empirical relationship has been found (van Ulden and Holtslag, 1985):

$$\theta_* = 0.09(1 - 0.5n^2) \tag{5.100}$$

with $n$ the fractional cloud cover. Thanks to eq. (5.89):

$$L = \frac{u_*^2 T_0}{kg\theta_*} \tag{5.89}$$

we can calculate $L$ without the need for a value of $q$ using eq. (5.100) to substitute $\theta_*$ in eq. (5.89):

$$L = \frac{u_*^2 T_0}{0.09(1 - 0.5n^2)kg} \tag{5.101}$$

However, $u_*$ is still needed and can be obtained by substituting (5.89) into (5.81):

$$u = \frac{u_*}{k}\left( \ln\frac{z}{z_0} + 5kg\theta_* \frac{z - z_0}{T_0 u_*^2} \right) \tag{5.102}$$

Rearrangement and multiplication by $ku_*/\ln(z/z_0)$ leads to

$$u_*^2 - u_* \frac{ku}{\ln(z/z_0)} + \frac{5kg\theta_*(z-z_0)}{T_0 \ln(z/z_0)} = 0 \tag{5.103}$$

Define the **drag coefficient** $C_D$ as

$$C_D = \frac{k}{\ln(z/z_0)} \tag{5.104}$$

and define $u_0$ such that

$$u_0^2 = \frac{5g\theta_*(z-z_0)}{T_0} \tag{5.105}$$

so eq. (5.103) can be written as

$$u_*^2 - C_D u u_* + C_D u_0^2 = 0 \tag{5.106}$$

This quadratic equation can be solved for $u_*$:

$$u_* = \frac{C_D u}{2}\left[1 + \sqrt{1 - \left(\frac{2u_0}{\sqrt{C_D}u}\right)^2}\right] \tag{5.107}$$

The other solution for the quadratic equation is physically unrealistic because it would predict a decreasing friction velocity with increasing wind speed. With $u_*$ and $\theta_*$ known [eqs. (5.100) and (5.107)], $L$ can be calculated with eq. (5.89), and the wind speed profile can be calculated.

With eq. (5.107) it is clear that the method will not work when $u$ is less than a minimum value of

$$u_{\min} = \frac{2u_0}{\sqrt{C_D}} \tag{5.108}$$

This is because eq. (5.100) is no longer valid. At the minimum wind speed the friction velocity is given by eq. (5.107):

$$u_{*\min} = \frac{C_D u_{\min}}{2} = \sqrt{C_D}u_0 \tag{5.109}$$

When the wind speed is less than the minimum wind speed, Cimorelli et al. (2004) suggest the following:

$$u_* = u_{*\min}\frac{u}{u_{\min}} \tag{5.110}$$

$$\theta_* = 0.09\left(1 - 0.5n^2\right)\frac{u}{u_{\text{crit}}} \tag{5.111}$$

In that case, $L$ can be calculated directly from eq. (5.89).

The procedure outlined in this section is used in AERMOD to calculate nocturnal stability, wind speed profile, and potential temperature profile.

**Example 5.8.** Calculate $u_*$, $L$, and $q$ in nighttime conditions when the temperature is 10 °C, the pressure is 1 atm, the wind speed at a 10-m height is 5 m s$^{-1}$, the roughness height is 10 cm, and the fractional cloud cover is $\frac{4}{8}$. Compare with the heat balance approach of Section 5.6.

***Solution.*** Based on eq. (5.100) the friction temperature is calculated: $\theta_* = 0.0689$ K.

Based on eqs. (5.104) and (5.105) we obtain $C_D = 0.08686$; $u_0^2 = 0.1181$ m$^2$ s$^{-2}$.

We check the minimum wind speed: $u_{\text{min}} = 2.332$ m s$^{-1}$.

Hence, we can use eq. (5.107) for $u_*$: $u_* = 0.4092$ m s$^{-1}$.

Based on eq. (5.89) we can calculate $L$: $L = 175.4$ m.

Solving eq. (5.88) leads to a value of $q$: $q = -35.41$ W m$^{-2}$.

Based on the heat balance method, the following values are obtained. We assume the following data: $C_G = 0.1$; $B = 1$. Assuming $R = 0$ (nighttime), $R_N$ is obtained with eq. (5.55):

$$R_N = -54.30\ \text{W m}^{-2}$$

Substitution into eq. (5.53) leads to $q$: $q = -24.43$ W m$^{-2}$.

After trial and error, the following values are obtained for $u_*$ and $L$:

$$u_* = 0.4177\ \text{m s}^{-1}$$
$$L = 270.3\ \text{m}$$

where $u_*$ is almost the same for both methods, but the two other variables are substantially different.

## 5.10 MIXING LAYER HEIGHT

As we saw in the primer, the mixing layer height, or mixing height, $h_{\text{mix}}$, is an important parameter in air dispersion modeling. It is the thickness of the layer in the atmosphere that has sufficient turbulence to achieve well-mixed conditions given sufficient time. In convective conditions it is equal to the height of the capping inversion and is sometimes denoted $z_i$ for that reason. Ideally, $h_{\text{mix}}$ should be measured, for instance, with a **radiosonde** or a **rawinsonde**. The principle of a radiosonde has already been shown in Section 5.3, in Example 5.1. In practice, a radiosonde also contains a hygrometer, which allows for a more accurate calculation of the air density, which is influenced by the water vapor mole fraction of the air. The difference between a radiosonde and a rawinsonde is that a rawinsonde is tracked from the ground to determine the wind speed and wind direction as a function of height.

Unfortunately, $h_{mix}$ is not a variable that is measured on a routine basis. Even meteorological stations that have radiosondes use them only every 12 h, which is too infrequent to be used for direct input in air dispersion models. Furthermore, radiosondes can only determine daytime mixing heights easily. Hence, it is necessary to estimate $h_{max}$. Several approaches have been put forward. For neutral and stable conditions, similarity theory can provide fairly accurate estimates, whereas in unstable conditions it is better to use a heat balance approach.

Similarity theory provides three fundamental length scales that are relevant for the mixing height: $h_{mix}$, $L$, and $\lambda = ku_*/|f|$ (Zilitinkevich, 1972), where $f$ is the Coriolis parameter [$= 2\Omega \sin(\phi)$, with $\Omega$ the Earth's angular velocity ($7.292 \times 10^{-5}$/s) and $\phi$ the latitude].

Under *neutral* conditions, when $|L|$ is much larger than $h_{mix}$, only $h_{mix}$ and $\lambda$ remain as variables, so only one dimensionless number can be defined: $h_{mix}/\lambda$. This dimensionless number must, therefore, be constant. Hence, based on observations, the following equation has been proposed (Arya, 1999):

$$\boxed{h_{mix} = \frac{0.3u_*}{|f|}} \tag{5.112}$$

The coefficient 0.3 is uncertain. Values from 0.2 to 0.4 have been reported (Garratt, 1994). This equation does not work near the equator, where the Coriolis force does not have a dominant influence on $h_{mix}$.

For the *stable* atmosphere, two dimensionless variables can be defined: $h_{mix}/\lambda$ and $\mu = \lambda/L$. Based on the work of Zilitinkevich (1972), who argued that the stable boundary layer height must be inversely proportional to the square root of the Coriolis parameter, we can put forward the following equation:

$$h_{mix} = C\sqrt{\frac{u_* L}{|f|}} \tag{5.113}$$

where $C$ is usually taken to be equal to 0.4 (e.g., Scire et al., 2000). However, Venkatram (1980) established two empirical correlations that enable us to derive $C$ independently. These equations are

$$L = Au_*^2 \tag{5.114}$$

$$h_{mix} = Bu_*^{3/2} \tag{5.115}$$

where $A = 1100$ s$^2$ m$^{-1}$, and $B = 2400$ s$^{1.5}$ m$^{0.5}$ (AERMOD uses $B = 2300$ s$^{1.5}$ m$^{0.5}$). Combining these two equations, we obtain

$$h_{mix} = B\sqrt{\frac{u_* L}{A}} \tag{5.116}$$

Comparing eq. (5.116) with eq. (5.113), we obtain

$$C = B\sqrt{\frac{|f|}{A}} \tag{5.117}$$

Given the location of the experiments (Minnesota), we obtain an estimate of $C=0.741$, showing that this variable is prone to considerable uncertainty. Nevertheless, it establishes eq. (5.113) as a more universal equation than eq. (5.115). A calculation procedure for $u_*$ and $L$ under stable conditions is given in Section 5.9. Eventually, the following equations are obtained:

$$\boxed{h_{\text{mix}} = Cu_*^{3/2}\sqrt{\frac{T_0}{0.09(1-0.5n^2)kg|f|}}} \quad (u > u_{\text{min}}) \tag{5.118}$$

$$\boxed{h_{\text{mix}} = Cu_*^{1/2}u_{*\text{crit}}\sqrt{\frac{T_0 u}{0.09(1-0.5n^2)kg|f|u_{\text{crit}}}}} \quad (u < u_{\text{min}}) \tag{5.119}$$

The above equations only consider mechanical mixing. In an *unstable* atmosphere it is usually not mechanical turbulence that determines $h_{\text{mix}}$ but thermal turbulence. Such calculations are based on an estimation of the amount of heat needed to create an adiabatic air layer out of a stable layer. This is illustrated in Figure 5.22. The early morning potential temperature versus height is the function $\theta(z)$. Later in the day, when an adiabatic layer has been established, the potential temperature is constant and equal to the value at height $h_{\text{mix}}$, that is, $\theta(h_{\text{mix}})$.

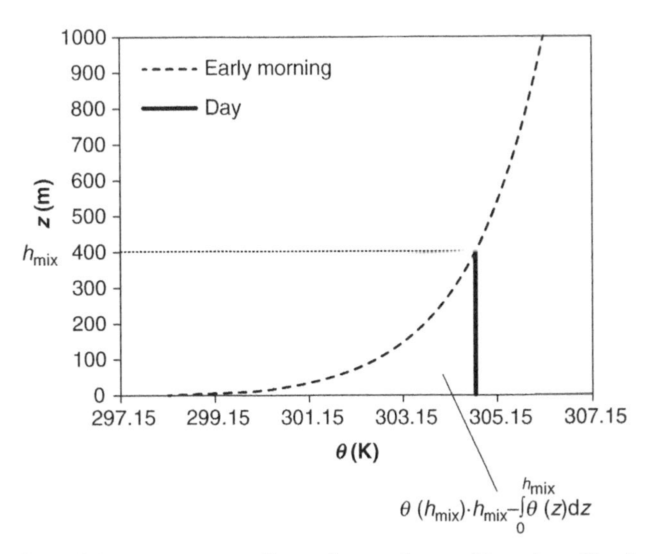


Figure 5.22 Potential temperature profile: early morning profile and profile after formation of a thermal mixed layer.

To heat up 1 kg of air from $\theta(z)$ to $\theta(h_{mix})$, the heat required is $c_p\,[\theta(h_{mix})-\theta(z)]$. The mass of air contained within 1 m$^2$ between height $z$ and height $z+dz$ is $\rho\,dz$. Hence, the heat needed to heat up 1 m$^2$ of air between $z$ and $dz$ is $\rho\,c_p[\theta(h_{mix})-\theta(z)]\,dz$. The total heat required to heat up a mixing layer is

$$\text{Heat added} = \int_0^{h_{mix}} \rho c_p\left[\theta(h_{mix})-\theta(z)\right]dz = \rho c_p\left[\theta(h_{mix})h_{mix} - \int_0^{h_{mix}}\theta(z)dz\right] \quad (5.120)$$

The heat deposited into the atmosphere from the soil per unit time is $q$. Hence, the heat added since sunrise is

$$\text{Surface heat added} = \int_{t_{sunrise}}^{t} q(t)\,dt \quad (5.121)$$

However, the sensible heat leaving the surface by convection is not the only heat source for the mixing layer. There is also heat transferred from above as a result of the temperature gradient in the capping inversion. This contribution is assumed to be about 0.4 times the contribution from the surface (see also Section 5.13.2). Hence, the heat balance of the mixing layer is

$$1.4\int_{t_{sunrise}}^{t} q(t)\,dt = \rho c_p \int_0^{h_{mix}}\left[\theta(h_{mix})-\theta(z)\right]dz \quad (5.122)$$

This is only an approximation because the density is actually height dependent. Equation (5.122) needs to be solved for $h_{mix}$. To illustrate this, the special case of a constant potential temperature gradient will be solved here. The potential temperature gradient is $\Gamma-\Lambda$. Hence:

$$\theta(h_{mix})-\theta(z) = (\Gamma-\Lambda)(h_{mix}-z) \quad (5.123)$$

Substitution into eq. (5.122) and solving the integral leads to

$$1.4\int_{t_{sunrise}}^{t} q(t)\,dt = \rho c_p(\Gamma-\Lambda)\left(h_{mix}^2 - \frac{h_{mix}^2}{2}\right) \quad (5.124)$$

Solving for $h_{mix}$ leads to

$$\boxed{h_{mix} = \sqrt{\frac{2.8\int_{t_{sunrise}}^{t} q(t)dt}{\rho c_p(\Gamma-\Lambda)}}} \quad (5.125)$$

When the atmosphere is only slightly unstable, it is possible that mechanical effects rather than thermal effects determine $h_{mix}$. There is no adequate equation for that case, although eq. (5.115) is sometimes used in that occasion (e.g., AERMOD). Based on considerations made by Zilitinkevich (1972), the following trend can be expected:

$$h_{\text{mix}} = D\sqrt{\frac{u_*^3}{|f|^3 L}} \tag{5.126}$$

where $D$ is a yet to be determined parameter. Qualitatively, this equation seems to agree with eq. (5.115).

## 5.11 CONCEPT OF TURBULENCE

### 5.11.1 Basic Properties of Turbulence

As mentioned before, the flow of the wind is *turbulent*: It is irregular and random and varies randomly with time. This randomness can be considered as the generation and disappearance of **eddies**: rotating air masses that transfer momentum, heat, and mass between adjacent air parcels.

It is customary in turbulence theory to subdivide any instantaneous wind speed component $u$ into an *average* wind speed $\overline{u}$ and a wind speed **fluctuation** $\boldsymbol{u'}$. The latter is the *random* or *stochastic* component of the wind speed. From these concepts it is clear that

$$u = \overline{u} + u' \tag{5.127}$$

This is called **Reynolds decomposition**. The concepts are illustrated in Figure 5.23.

From Figure 5.23 we observe that there is a certain degree of **regularity** in the random component $u'$; and $u'$ is not as random as white noise, where even the adjacent values are not correlated to each other. We see that $u'(t)$ is highly correlated with

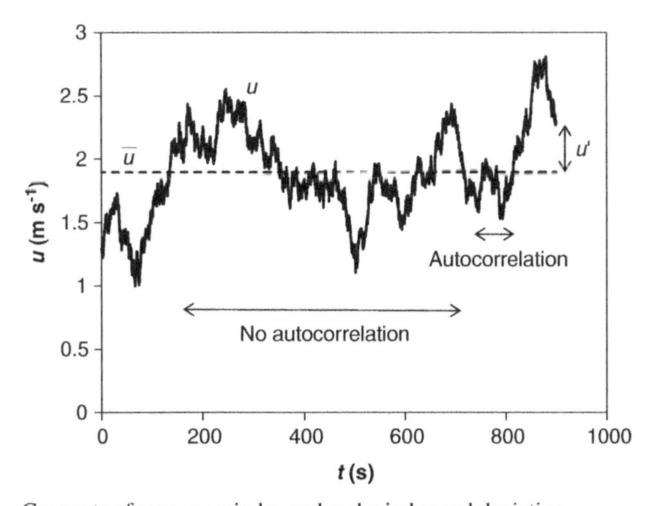


Figure 5.23 Concepts of average wind speed and wind speed deviation.

$u'(t+\Delta t)$ when $\Delta t$ is very small (seconds). On the other hand, $u'(t)$ is not correlated with $u'(t+\Delta t)$ when $\Delta t$ is large (> 1000 s). Therefore, when we average a wind speed, we need to take an averaging time large enough so $\overline{u}$ does not fluctuate due to the autocorrelation, but shorter than the time scale of diurnal trends or weather changes. Averaging times between 15 min and 1 h are reasonable.

Thus $\overline{u}$ is calculated as

$$\overline{u} = \frac{1}{\Delta t} \int_{t-\Delta t/2}^{t+\Delta t/2} u \, dt \tag{5.128}$$

By definition, for any function $f$:

$$\overline{f'} = 0 \tag{5.129}$$

When we work with random variables, we often need to average sums, products, and the like. These are calculated with the **Reynolds averaging rules**. The following rules apply:

$$\overline{f+g} = \overline{f} + \overline{g} \tag{5.130}$$

with $f$ and $g$ random variables;

$$\overline{cf} = c\overline{f} \tag{5.131}$$

with $c$ a constant. Since an average is a constant, it follows that

$$\overline{\overline{f}g} = \overline{f}\,\overline{g} \tag{5.132}$$

However, it is important to be aware that

$$\overline{fg} \neq \overline{f}\,\overline{g} \tag{5.133}$$

and as a result,

$$\overline{f^2} \neq \overline{f}^2 \tag{5.134}$$

We will discuss this in more detail when we discuss correlation. Two more averaging rules apply:

$$\overline{\left(\frac{\partial f}{\partial s}\right)} = \frac{\partial \overline{f}}{\partial s} \tag{5.135}$$

$$\overline{\int f \, ds} = \int \overline{f} \, ds \tag{5.136}$$

with $f$ a function of variable $s$.

### 5.11.2 Measures of Turbulence

There are several ways to quantify turbulence. One of the most fundamental ones is the **variance** of the wind speed:

$$\sigma_u^2 = \overline{u'^2} \qquad \sigma_v^2 = \overline{v'^2} \qquad \sigma_w^2 = \overline{w'^2} \tag{5.137}$$

Remember eq. (5.134), which means that $\overline{u'^2} \neq \overline{u'}^2$.

Likewise, we can define the variance of other variables, like pressure ($\sigma_p^2$), potential temperature ($\sigma_\theta^2$), or wind direction ($\sigma_\phi^2$).

The **turbulent kinetic energy** (or turbulence kinetic energy) is given by

$$\bar{e} = \tfrac{1}{2}\left(\overline{u'^2} + \overline{v'^2} + \overline{w'^2}\right) \tag{5.138}$$

This is the kinetic energy contained in the turbulence of one kilogram of air.

The **turbulent velocity** is defined as the standard deviation of the wind speed components, that is, the square root of the wind speed variance:

$$\sigma_u = \sqrt{\overline{u'^2}} \tag{5.139}$$

$$\sigma_v = \sqrt{\overline{v'^2}} \tag{5.140}$$

$$\sigma_w = \sqrt{\overline{w'^2}} \tag{5.141}$$

Note that these are *not* the same as the dispersion parameters $\sigma_x$, $\sigma_y$, and $\sigma_z$. However, there is a relationship between turbulent velocities and dispersion parameters that will be discussed in Chapter 6.

The **turbulent intensity** is a relative measure of the turbulence:

$$i_u = \frac{\sigma_u}{\overline{V}} \qquad i_v = \frac{\sigma_v}{\overline{V}} \qquad i_w = \frac{\sigma_w}{\overline{V}} \tag{5.142}$$

with $\overline{V}$ the average total wind speed:

$$\overline{V} = \sqrt{\overline{u}^2 + \overline{v}^2 + \overline{w}^2} \tag{5.143}$$

When the $x$ axis is chosen parallel with the wind speed, and the wind is horizontal, then $\overline{V} = \overline{u}$.

A common instrument to measure wind speed is the **cup anemometer**. However, cup anemometers respond rather slowly to changing wind speeds and may miss some of the wind speed variation. **Sonic anemometers** are better suited for that purpose. A sonic anemometer is a set of sensors that measures the time needed for sound pulses to travel back and forth across a gap. The difference between the forward travel time and the back travel time can be used to calculate the wind speed in

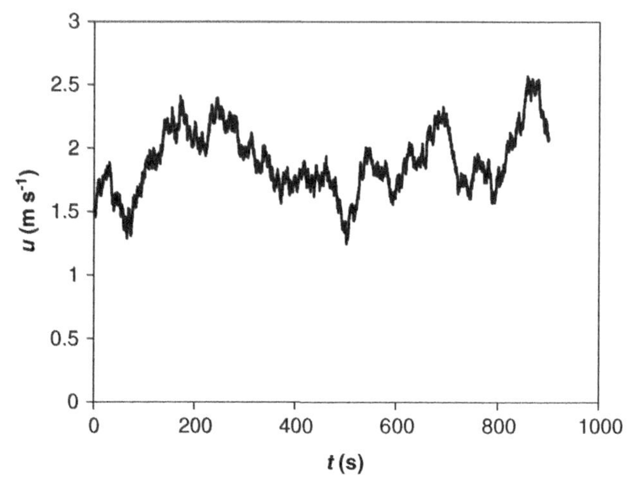


Figure 5.24 10-Hz wind speed data.

that direction. With three perpendicular paths, the three-dimensional wind speed vector can be calculated. A typical measurement frequency of a sonic anemometer is 10 Hz, that is, 10 measurements per second. As we will see, this is sufficient to cover all levels of turbulence.

**Example 5.9.** Figure 5.24 shows 15 min of 10-Hz wind speed data ($u$). The full data set is given in the Excel file "Measures of turbulence from 10 Hz data.xlsx" on the enclosed CD. Calculate the main measures of turbulence for this data set. Assume that the data was collected in the direction of the mean wind speed.

***Solution.*** The full calculation is included in the Excel file. The first second of data will be shown here for illustration (Table 5.7).

First the average wind speed is calculated, either with the Excel function AVERAGE(), or with the equation

$$\overline{u} = \frac{1}{9001}\sum_{i=0}^{9000} u_i$$

The result is $\overline{u} = 1.90139 \text{ m s}^{-1}$.

The wind speed variance can be calculated directly with the Excel funtion VAR() or based on $\overline{u}$ with the equation:

$$\sigma_u^2 = \frac{1}{9000}\sum_{i=0}^{9000}\left(u_i - \overline{u}\right)^2$$

The result is $\sigma_u^2 = 0.06394 \text{ m}^2 \text{ s}^{-2}$.

**TABLE 5.7 10-Hz Data of Wind Speed, Wind Speed Deviation, and Square of Wind Speed Deviation: First Second of Data**

| $t$ (s) | $u$ (m s$^{-1}$) | $u'$ (m s$^{-1}$) | $u'^2$ (m$^2$ s$^{-2}$) |
|---|---|---|---|
| 0.0 | 1.5164 | −0.38499 | 0.14822 |
| 0.1 | 1.5011 | −0.40029 | 0.16023 |
| 0.2 | 1.4886 | −0.41279 | 0.17040 |
| 0.3 | 1.5004 | −0.40099 | 0.16079 |
| 0.4 | 1.4991 | −0.40229 | 0.16184 |
| 0.5 | 1.5156 | −0.38579 | 0.14883 |
| 0.6 | 1.5001 | −0.40129 | 0.16103 |
| 0.7 | 1.5166 | −0.38479 | 0.14806 |
| 0.8 | 1.5265 | −0.37489 | 0.14054 |
| 0.9 | 1.5503 | −0.35109 | 0.12326 |
| 1.0 | 1.5370 | −0.36439 | 0.13278 |

The turbulent velocity is the square root of the wind speed variance:

$$\sigma_u = 0.2529 \, \text{m s}^{-1}$$

Because it is assumed that the measurements were taken along the mean wind direction, the turbulent intensity is the turbulent velocity divided by the average wind speed:

$$i_u = \frac{0.2529 \text{ m s}^{-1}}{1.90139 \text{ m s}^{-1}} = 0.1330$$

The full calculation is given in the Excel file "Measures of turbulence from 10 Hz data.xlsx" on the enclosed CD.

### 5.11.3 Similarity Theory and Turbulence

To relate turbulent velocities to other atmospheric variables, dimension analysis can be used. The main variables that come to mind in such an analysis are $\sigma_u$ (or $\sigma_v$ or $\sigma_w$), $u_*$, $L$, $h_{mix}$, and $z$. These five variables include two dimensions (length and time), so three dimensionless numbers can be defined. In neutral conditions, $L$ does not play a role, and only two dimensionless variables can be defined, for example, $\sigma_u/u_*$ and $z/h_{mix}$. In nonneutral conditions we can include either $z/L$ or $h_{mix}/L$ as a third dimensionless variable.

The following equations are recommended under *neutral* and *stable* conditions (Arya, 1999):

$$\frac{\sigma_u}{u_*} = 2.5\left(1 - \frac{z}{h_{mix}}\right)^a \tag{5.144}$$

$$\frac{\sigma_v}{u_*} = 1.9\left(1 - \frac{z}{h_{\text{mix}}}\right)^a \tag{5.145}$$

$$\frac{\sigma_w}{u_*} = 1.3\left(1 - \frac{z}{h_{\text{mix}}}\right)^a \tag{5.146}$$

where $a$ has a value between 0.5 and 1. Nieuwstadt (1984) recommends $a=0.75$ but used a coefficient of 1.4 instead of 1.3 in the $z$ direction. Lenschow et al. (1988) suggested $a=0.875$ but used a coefficient $4.5^{1/2}$ in the $x$ and $y$ directions and $3.1^{1/2}$ in the $z$ direction. AERMOD uses $a=0.5$ for $\sigma_w$. The overall result is fairly insensitive to the choice of $a$, so choosing the midpoint value of $a=0.75$ provides results that are fairly consistent with all the relevant data.

In the surface layer ($z < 0.1\ h_{\text{mix}}$), the above equations can be simplified to:

$$\frac{\sigma_u}{u_*} = 2.5 \tag{5.147}$$

$$\frac{\sigma_v}{u_*} = 1.9 \tag{5.148}$$

$$\frac{\sigma_w}{u_*} = 1.3 \tag{5.149}$$

However, between the obstacles that define the surface roughness, the numbers need to be adjusted (Hanna and Britter, 2002). This will be discussed in Chapter 8.

Under *unstable* conditions, the above equations are still valid, but they only represent the mechanical portion of the turbulent velocity. Under unstable conditions there is also thermally induced turbulence, which needs to be included. A new velocity scale is introduced, the **convective velocity** (Deardorff, 1970a,b):

$$\boxed{w_* = \left(\frac{gqh_{\text{mix}}}{T_0 \rho c_p}\right)^{1/3}} \tag{5.150}$$

The convective velocity is a grouping of three previously introduced variables:

$$w_* = u_*\left(\frac{h_{\text{mix}}}{Lk}\right)^{1/3} \tag{5.151}$$

The convective velocity is a variable associated with the degree of convective cycling in the boundary layer. Grouping three variables into one reduces the total number of variables from five to three (e.g., $\sigma_u$, $w_*$, and $z$). With only three variables, only one dimensionless number can be defined. However, because two of the three variables are velocities, the third ($z$) cannot be included, and any height variation cannot be included in a similarity theory. Hence, $h_{\text{mix}}$ is added to the variables, and the dimensionless numbers are $\sigma_u/w_*$ and $z/h_{\text{mix}}$.

Caughey and Palmer (1979) found that in strongly convective (i.e., unstable) conditions, $\sigma_u$, $\sigma_v$, and $\sigma_w$ are largely independent of height, and the following equations can be used (Arya, 1999):

$$\frac{\sigma_u}{w_*} = \frac{\sigma_v}{w_*} = \frac{\sigma_w}{w_*} = 0.6 \tag{5.152}$$

For $\sigma_u$ and $\sigma_v$ this applies to the entire mixing layer, with a gradual decline for $z > h_{\text{mix}}$. The data of Caughey and Palmer (1979) suggest an exponential decline, which can be described roughly as follows for $z > h_{\text{mix}}$:

$$\frac{\sigma_u}{w_*} = \frac{\sigma_v}{w_*} = 0.6 \, \exp\left(-2\frac{z - h_{\text{mix}}}{h_{\text{mix}}}\right) \tag{5.153}$$

For $\sigma_w$ the height dependence is more pronounced, with a pronounced decline for $z < 0.2h_{\text{mix}}$ and $z > 0.8h_{\text{mix}}$. Several formulations have been suggested for $\sigma_w$ in the surface layer (Wyngaard et al., 1971; Irwin, 1979; Sorbjan, 1989). The measured data of Caughey and Palmer (1979) suggest the following for $z < 0.2h_{\text{mix}}$:

$$\sigma_w = \frac{0.6}{0.2^{0.4}}\left(\frac{z}{h_{\text{mix}}}\right)^{0.4} \tag{5.154}$$

and the following for $z > 0.8h_{\text{mix}}$:

$$\frac{\sigma_w}{w_*} = 0.6 \, \exp(-0.6)\exp\left(-3\frac{z - h_{\text{mix}}}{h_{\text{mix}}}\right) \tag{5.155}$$

The total turbulent velocity in an unstable boundary layer is obtained by adding up the wind speed variances for mechanical and thermal turbulence:

$$\sigma_u = \sqrt{\sigma^2_{u,\text{mech}} + \sigma^2_{u,\text{therm}}} \tag{5.156}$$

where $\sigma^2_{u,\text{mech}}$ is calculated with eq. (5.144) and $\sigma^2_{u,\text{therm}}$ is calculated with eq. (5.152) or (5.153). The same reasoning applies to $\sigma_v$ and $\sigma_w$.

Several authors have proposed equations that combine both the thermal and mechanical contributions to the turbulent velocity (Panofsky et al., 1977; Irwin, 1979). However, these attempts do not always extrapolate well to neutral conditions.

### 5.11.4 Covariance and Turbulence

**Covariance** of two variables is a measure of the simultaneous occurrence of fluctuations of these variables from average conditions. The covariance of variables $f$ and $g$ is defined as

$$\mathrm{cov}(f,g) = \overline{f'g'} \tag{5.157}$$

Imagine two variables $f$ and $g$ such that $f' > 0$ whenever $g' > 0$, and $f' < 0$ whenever $g' < 0$. In that case, $f'g'$ is always positive, and $\overline{f'g'}$ will be positive as well. The properties $u$ and $v$ are *positively correlated*. Similarly, a negative correlation means that there is a tendency for positive values of $f'$ to coincide with negative values of $g'$, and vice versa. This is illustrated in Figure 5.25.

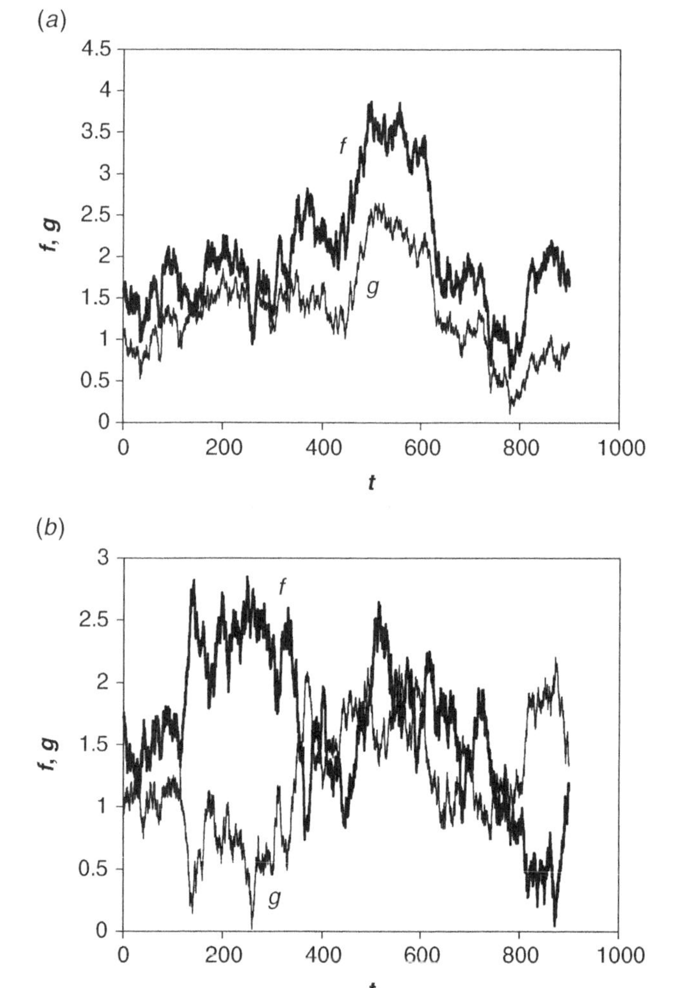


Figure 5.25 (*a*) Positively and (*b*) negatively correlated functions.

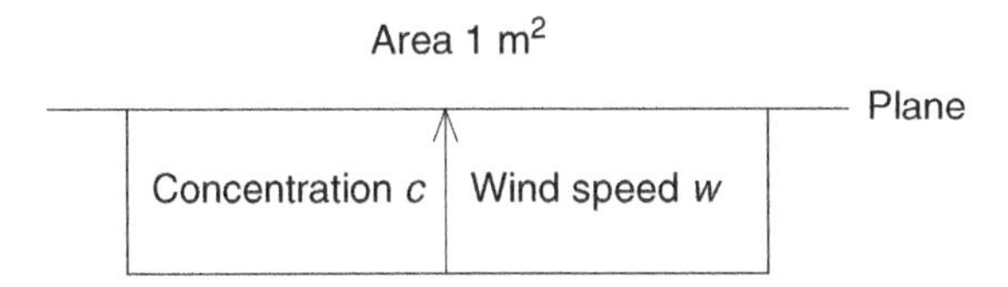


Figure 5.26 Vertical mass flux of a component with concentration $c$.

Covariance can be defined between any two fluctuating variables: $\overline{u'v'}$, $\overline{u'w'}$, $\overline{u'\theta'}$, and so forth. The **linear correlation coefficient** is the covariance normalized in such a way that the value is always between −1 and +1:

$$r_{fg} = \frac{\overline{f'g'}}{\sigma_f \sigma_g} \tag{5.158}$$

Many covariances indicate **fluxes** relevant for air dispersion modeling. This can be seen by decomposing the relevant variables in average and random components. For instance, Figure 5.26 illustrates the **vertical mass flux** of a pollutant through a fixed horizontal plane.

The vertical mass flux $N$ is given by

$$N = cw \tag{5.159}$$

where $c$ is the concentration of the pollutant, and $w$ is the vertical wind speed. When the units of $c$ are mg m$^{-3}$, and the units of $w$ are m s$^{-1}$, then the units of $N$ are mg m$^{-2}$ s$^{-1}$.

Equation (5.159) does not include the effect of molecular diffusion, which is usually negligible in comparison with turbulent diffusion.

The average flux is given by

$$\overline{N} = \overline{cw} = \overline{(\overline{c} + c')(\overline{w} + w')} \tag{5.160}$$

The product is decomposed into its four terms:

$$\overline{N} = \overline{\overline{c}\,\overline{w}} + \overline{\overline{c}w'} + \overline{c'\overline{w}} + \overline{c'w'} \tag{5.161}$$

Applying eq. (5.132) leads to

$$\overline{N} = \overline{c}\,\overline{w} + \overline{c}\,\overline{w'} + \overline{c'}\,\overline{w} + \overline{c'w'} \tag{5.162}$$

Since $\overline{f'} = 0$, for any variable $f$, the two middle terms on the right-hand side can be ignored. Equation (5.162) reduces to

$$\overline{N} = \overline{c}\,\overline{w} + \overline{c'w'} \tag{5.163}$$

In eq. (5.163) the first term corresponds with the flux associated with the regular motion of air masses, in other words, the **advective mass flux**. The second term corresponds with the mass transfer associated with random air motions, in other words the **turbulent diffusion** (or **dispersion) mass flux**.

In many cases the average vertical wind speed equals zero close to the surface. In that case, eq. (5.163) can be simplified as

$$\bar{N} = \overline{c'w'} \qquad (\bar{w} = 0) \tag{5.164}$$

When the concentration of a pollutant is positively correlated with the vertical wind speed, then there is a net upward flux of the pollutant. Remember that positive values of $w$ correspond with an upward motion.

Similarly, a heat flux by turbulence can be written as

$$\bar{q} = \rho c_p \overline{\theta' w'} \tag{5.165}$$

It can also be shown that the following property is the shear stress on a horizontal plane in the $x$ direction:

$$\tau_{xz} = -\rho \overline{u'w'} \tag{5.166}$$

This is shown intuitively in Section A 5.10 (Appendix A). The shear stress in eq. (5.166) is known as **Reynolds stress** (Stull, 1988). A similar Reynolds stress exists in the $y$ direction.

Based on the definition of $u_*$, we obtain

$$u_* = \sqrt{\frac{\tau_{xz}}{\rho}} = \sqrt{-\overline{u'w'}} \tag{5.167}$$

In this equation it is assumed that wind speed is along the $x$ direction. If this is not the case, then a second term needs to be added, representing $\tau_{yz}$:

$$u_* = \sqrt{\frac{\sqrt{\tau_{xz}^2 + \tau_{yz}^2}}{\rho}} = \sqrt[4]{\overline{u'w'}^2 + \overline{v'w'}^2} \tag{5.168}$$

This is a variable that can be measured directly with high-speed anemometers. Hence, $u_*$ can be measured without knowledge of the surface roughness, the temperature gradient, or the heat balance.

**Example 5.10.** Pollutant flux measurements based on covariances are known as the **eddy correlation technique**. An example is the $N_2O$ emission from fertilized soils. A tuneable diode laser (TDL) can measure $N_2O$ with 10 Hz frequency, sufficiently fast for flux measurements, provided they are sufficiently large.

The Excel file "Example 5.10. Flux measurement with eddy correlation technique.xlsx" contains vertical wind speed and $N_2O$ concentration data at 0.1-s intervals. Use this data to estimate the diffusive vertical $N_2O$ flux.

**TABLE 5.8 Calculations for Flux Measurements with Eddy Correlation Technique**

| $t$ (s) | $w$ (m s$^{-1}$) | $c$ ($\mu$g m$^{-3}$) | $c'$ ($\mu$g m$^{-3}$) | $w'c'$ ($\mu$g m$^{-2}$ s$^{-1}$) |
|---|---|---|---|---|
| 0.0 | −0.02533 | 366.05 | 0.9765 | −0.02473 |
| 0.1 | −0.02406 | 365.68 | 0.6065 | −0.01459 |
| 0.2 | −0.02957 | 365.92 | 0.8465 | −0.02503 |
| 0.3 | −0.03231 | 366.59 | 1.5165 | −0.04900 |
| 0.4 | −0.02226 | 366.38 | 1.3065 | −0.02908 |
| 0.5 | −0.02935 | 366.10 | 1.0265 | −0.03013 |
| 0.6 | −0.02558 | 365.94 | 0.8665 | −0.02217 |
| 0.7 | −0.02396 | 365.56 | 0.4865 | −0.01166 |
| 0.8 | −0.02270 | 365.19 | 0.1165 | −0.00264 |
| 0.9 | −0.02775 | 364.93 | −0.1435 | 0.003982 |
| 1.0 | −0.03608 | 365.00 | −0.03608 | 0.002652 |

***Solution.*** The average vertical wind speed and concentration are calculated with the AVERAGE() function. The result is $\overline{w} = 1.69 \times 10^{-7}$ m s$^{-1}$ and $\overline{c} = 365.0735$ $\mu$g m$^{-3}$.

Because the average wind speed is less than the measurement precision, we can approximate $w = w'$ in Table 5.8. However, for the calculations $\overline{w}$ was subtracted from $w$ to obtain $w'$; and $c'$ is calculated as $c - \overline{c}$.

The advective flux is $\overline{w} \cdot \overline{c}$ = **6.17 × 10$^{-5}$ μg m$^{-2}$ s$^{-1}$**.

The diffusive flux is calculated with eq. (5.164), applying the function COVAR() on the data of $w$ and $c$, or taking the average of $w' \cdot c'$. The result is **3.241 × 10$^{-3}$ μg m$^{-2}$ s$^{-1}$**.

The total flux is calculated either as the average of $w \cdot c$ or as the sum of the adjective and diffusive fluxes. Both lead to the same result, **3.302 × 10$^{-3}$ μg m$^{-2}$ s$^{-1}$**.

### 5.11.5 Introduction to Eddy Diffusivity and Gradient Transport Theory

Eulerian models like CMAQ are based on **gradient transport theory**, so it is important to have a basic understanding of this theory. Newton's law of viscosity, Fourier's law of conduction, and Fick's law of diffusion, each describing a macroscopic effect of molecular movements, are mathematically equivalent when they are written in appropriate terms:

Newton:

$$\tau = -\rho\nu \frac{\partial u}{\partial z} \Rightarrow \frac{\tau}{\rho} = -\nu \frac{\partial u}{\partial z} \tag{5.169}$$

Fourier:

$$q = -\rho c_p \kappa \frac{\partial \theta}{\partial z} \Rightarrow \frac{q}{\rho c_p} = -\kappa \frac{\partial \theta}{\partial z} \tag{5.170}$$

Fick:

$$J = -D\frac{\partial c}{\partial z} \Rightarrow J = -D\frac{\partial c}{\partial z} \tag{5.171}$$

In the above equations $\tau$ is the shear stress, $q$ is the sensible heat flux, $J$ is the mass flux, $\kappa$ is the thermal diffusivity, $\nu$ is the kinematic viscosity, and $D$ is the molecular diffusion coefficient. All three have units $m^2\ s^{-1}$. The following relations apply:

$$\rho\nu = \mu \qquad \text{(dynamic viscosity, Pa s)} \tag{5.172}$$

$$\rho c_p \kappa = \lambda \qquad \text{(thermal conductivity, J m}^{-1}\text{K}^{-1}\text{)} \tag{5.173}$$

Note that the mass flux in Fick's law is written as $J$, not $N$. The letter $N$ is used for a flux with respect to a plane that is fixed in space, whereas $J$ is used for a flux with respect to a plane that moves with the average wind speed. When the average wind speed is zero (e.g., in the vertical direction in many cases), then $J=N$.

Boussinesq (1877) proposed that turbulence simply increases the values of $\nu$, $\kappa$, and $D$. This can be written as follows:

$$\frac{\tau}{\rho} = \overline{u'w'} = -K_{\mathrm{m}}\frac{\partial \overline{u}}{\partial z} \tag{5.174}$$

$$\frac{q}{\rho c_p} = \overline{\theta' w'} = -K_{\mathrm{h}}\frac{\partial \overline{\theta}}{\partial z} \tag{5.175}$$

$$J = \overline{c'w'} = -K_z\frac{\partial \overline{c}}{\partial z} \tag{5.176}$$

where $\nu$, $\kappa$, and $D$ are replaced by $K_{\mathrm{m}}$, $K_{\mathrm{h}}$, and $K_z$, the turbulent momentum diffusivity (or turbulent kinematic diffusivity), the turbulent heat diffusivity, and the turbulent mass diffusivity in the $z$ direction, respectively. With respect to mass transfer, the following equations apply to the other directions:

$$\overline{c'u'} = -K_x\frac{\partial \overline{c}}{\partial x} \tag{5.177}$$

$$\overline{c'v'} = -K_y\frac{\partial \overline{c}}{\partial y} \tag{5.178}$$

Note that $K_x$, $K_y$, and $K_z$ are not necessarily the same.

Caution is required with eq. (5.174), which is only approximate, and which should never be used with a gradient of the average vertical wind speed, which is assumed zero in the application of this equation. If $\overline{w} \neq 0$, eq. (5.174) must be modified as discussed in Section 10.3.4.

Prantl (1925) suggested a mechanism to explain eqs. (5.171)–(5.175) in terms of eddies exchanging air parcels between adjacent locations at a distance $l$ (see Arya, 1999). The mechanism underlying Prantl's theory is incorrect, but the application is useful because it establishes a link between the $K$ values and the turbulent velocity. Prantl's theory leads to the following equations (Arya, 1999):

$$K_{\mathrm{m}} = c_{\mathrm{m}} l_{\mathrm{m}} \sigma_w \tag{5.179}$$

$$K_{\mathrm{h}} = c_{\mathrm{h}} l_{\mathrm{h}} \sigma_w \tag{5.180}$$

$$K_x = c l_x \sigma_w \tag{5.181}$$

$$K_y = c l_y \sigma_w \tag{5.182}$$

$$K_z = c l_z \sigma_w \tag{5.183}$$

Equations (5.179)–(5.183) introduce eight parameters to establish the link between the $K$ values and the turbulent velocities. The $l$ values are characteristic lengths, whereas the $c$ values are empirical constants. Both the $l$ values and the $c$ values need to be determined experimentally and have limited validity. A modernized version of this approach will be discussed in Chapters 10 and 11.

For now a more useful approach is to compare the Boussinesq equations with their respective similarity theories. For **momentum transfer** we compare the following two equations:

$$\frac{\tau}{\rho} = -K_{\mathrm{m}} \frac{\partial \bar{u}}{\partial z} \tag{5.174}$$

$$\frac{\partial u}{\partial z} = \frac{u_*}{kz} \phi_{\mathrm{m}}(\zeta) \tag{5.74}$$

Note that the shear stress in (5.174) is not necessarily the value at the surface. Hence, we cannot simply eliminate $\tau$ using the definition of $u_*$. Under *stable* conditions the following equation has been derived from measurements (Nieuwstadt, 1984):

$$\frac{\tau}{\rho} = -u_*^2 \left(1 - \frac{z}{h_{\mathrm{mix}}}\right)^{1.5} \tag{5.184}$$

Combining eqs. (5.174), (5.74), and (5.184), we can derive the following equation for $K_{\mathrm{m}}$:

$$K_{\mathrm{m}} = \frac{k u_* z (1 - z / h_{\mathrm{mix}})^{1.5}}{\phi_{\mathrm{m}}(\zeta)} \tag{5.185}$$

In the special case of *neutral* conditions this equation simplifies to

$$K_{\mathrm{m}} = ku_* z\left(1 - \frac{z}{h_{\mathrm{mix}}}\right)^{1.5} \tag{5.186}$$

Close to the surface, this equation can be further simplified to

$$K_{\mathrm{m}} = ku_* z \tag{5.187}$$

This equation is not valid within millimeters of the surface. In this layer viscosity affects turbulence, resulting in a quasi-laminar surface layer. Calculations probing the properties of this layer are given in Appendix C.

A similar approach is possible for **heat transfer** in the $z$ direction. Compare the following two equations:

$$\frac{q}{\rho c_p} = -K_{\mathrm{h}} \frac{\partial \overline{\theta}}{\partial z} \tag{5.175}$$

$$\frac{\partial \theta}{\partial z} = \frac{\theta_*}{kz} \phi_{\mathrm{h}}(\zeta) \tag{5.90}$$

Along with the definition of $\theta_*$:

$$\theta_* = -\frac{q}{\rho_0 c_p u_*} \tag{5.88}$$

Again, we cannot use this equation directly because $q$ in eq. (5.88) refers to the convective heat transfer at $z=0$. However, Lenschow et al. (1988) found the following relationship under *stable* conditions:

$$q(z) = q(z=0)\left(1 - \frac{z}{h_{\mathrm{mix}}}\right)^{1.5} \tag{5.188}$$

Nieuwstadt (1984) suggested a power law coefficient of 1 instead of 1.5, but 1.5 is probably more realistic. Combining the four equations leads to the following equation:

$$K_{\mathrm{h}} = \frac{ku_* z(1 - z/h_{\mathrm{mix}})^{1.5}}{\phi_{\mathrm{h}}(\zeta)} \tag{5.189}$$

Again, under *neutral* conditions, this equation simplifies to

$$K_{\mathrm{h}} = ku_* z\left(1 - \frac{z}{h_{\mathrm{mix}}}\right)^{1.5} \tag{5.190}$$

and close to the surface:

$$K_{\mathrm{h}} = ku_* z \tag{5.191}$$

It is generally assumed that $K_z = K_{\mathrm{h}}$ (e.g., Dyer, 1967, 1974). Hence, under neutral and stable conditions:

$$K_z = \frac{ku_* z(1 - z / h_{\mathrm{mix}})^{1.5}}{\phi_{\mathrm{h}}(\zeta)} \tag{5.192}$$

with the same simplifications as for heat transfer. Stull (1988) points out that there is evidence indicating that $K_{\mathrm{h}}$ and $K_z$ can be as high as 1.35 times $K_{\mathrm{m}}$ at neutral conditions. It is not clear what would cause this discrepancy, but dispersion calculations of Wilson et al. (1981) indicate that eq. (5.191) does underestimate turbulent diffusivity. See also Chapter 9.

There is no equivalent theory that enables us to derive values for $K_x$ and $K_y$. So far, fewer efforts have been made to develop such a theory because the main applications for such a theory, Eulerian grid models, tend to overestimate horizontal dispersion even with zero values of $K_x$ and $K_y$, due to a process called numerical dispersion (Byun and Ching, 1999). This will be discussed in more detail in Chapter 11. Furthermore, $K_x$ and $K_y$ are affected by mesoscale effects that cannot be resolved with a simple similarity approach (Wilson et al., 1983). As a conservative estimate, $K_x = K_y = K_z$ can be used. Some other options will be discussed further down.

All the above equations are limited to stable and neutral conditions. Because they account for momentum flux changes and heat flux changes with height, it can be assumed that they are valid in the entire mixing layer (with the exception of the simplifications).

Under unstable conditions Byun and Ching (1999) suggest the following equation for $K_{\mathrm{h}}$:

$$K_{\mathrm{h}} = kw_* z\left(1 - \frac{z}{h_{\mathrm{mix}}}\right) \tag{5.193}$$

with $w_*$ the convective velocity scale:

$$w_* = \left(\frac{gqh_{\mathrm{mix}}}{T_0 \rho c_p}\right)^{1/3} \tag{5.150}$$

In a narrow range of slightly unstable conditions, eq. (5.193) predicts a $K_{\mathrm{h}}$ (or $K_z$) value smaller than its neutral value. To avoid this, the following equation can be used in neutral and unstable conditions:

$$K_{\mathrm{h}} = K_z = ku_* z\left(1 - \frac{z}{h_{\mathrm{mix}}}\right)\left[\left(1 - \frac{z}{h_{\mathrm{mix}}}\right)^{1.5} - \frac{h_{\mathrm{mix}}}{kL}\right]^{1/3} \tag{5.194}$$

In the horizontal direction, Seinfeld and Pandis (2006) suggest the following:

$$K_x = K_y = 0.1w_* h_{\text{mix}} \tag{5.195}$$

This leads to typical values of 50–100 m$^2$ s$^{-1}$.

When we consider lateral dispersion over sufficiently long distances, the following equation holds (Arya, 1999):

$$\sigma_y = (2K_y t)^{1/2} = \left(2K_y \frac{x}{u}\right)^{1/2} \tag{5.196}$$

Under neutral conditions, Briggs suggested the following equation (see Chapter 2):

$$\sigma_y = 0.08x(1 + 0.0001x)^{-1/2} \tag{5.197}$$

Over long distances this leads to

$$\sigma_y = 8x^{1/2} \tag{5.198}$$

Combination of eqs. (5.196) and (5.198) leads to

$$K_y = C_y u \tag{5.199}$$

with $C_y = 32$ m. This is in general agreement with the Seinfeld and Pandis (2006) estimate. Similar equations can be derived for the other stability classes, provided that their asymptotic behavior is consistent with diffusion theory. As indicated above, $K_z$ offers a conservative estimate of $K_x$ and $K_y$. Hence, if any of these methods predict a $K_x$ or $K_y$ value that is less than $K_z$, the value of $K_z$ should be used instead.

In a recent report Park et al. (2009) found that $\phi_z$ for water vapor deviates from $\phi_h$. That would mean that $\phi_h$ in the mass transfer equations should be replaced by $\phi_z$. The equations for $\phi_z$ are (see also Section A 5.7, Appendix A):

$\zeta > 0$ (stable):

$$\phi_z(\zeta) = 1.21(1 - 60.4\zeta)^{1/3} \tag{5.200}$$

$\zeta < 0$ (unstable):

$$\phi_z(\zeta) = 1.21(1 - 13.1\zeta)^{-1/2} \tag{5.201}$$

So far there is no independent confirmation of these trends, so the use of eqs. (5.200) and (5.201) cannot yet be recommended.

In the presence of a displacement height [see eq. (5.68)], a modification of $K_m$ calculations near the surface may be needed. For instance, eq. (5.187) becomes

$$K_m = ku_*(z - d) \tag{5.202}$$

where $d$ is the displacement height.

## 5.12 SPECIAL TOPICS IN METEOROLOGY

This section and Section 5.13 contain topics that are relevant to air pollution meteorology but that do not fit anywhere else in this chapter. This section contains mainly essential information of a qualitative nature that all air dispersion modelers should be aware of. Section 5.13, and a similar section in Chapter 6, contains mainly quantitative information that may be optional for some students and users.

### 5.12.1 Convective Cycles: Qualitative Description

We have discussed convective cycles several times, and even defined a velocity scale ($w_*$) as a measure of the turbulence associated with convective cycles. So what are they? In unstable conditions an air parcel that rises continues to rise, and an air parcel that drops continues to drop. As a consequence, air in an unstable atmosphere moves in circles called **convective cycles**. They are illustrated in Figure 5.27.

As shown in Figure 5.27, the rising air currents (updrafts) occur on relatively small areas (roughly a third of the total area), whereas the sinking air currents (downdrafts) occur on relatively large areas. Because the two air flows must be the same, the updrafts are faster than the downdrafts. The updrafts are spaced about 1.5 times the mixing height and are typically about one degree warmer than the downdrafts (Turner and Schulze, 2007). The properties of convective cycles will be calculated in Section 5.13.1.

When the surface is homogeneous, the cycles will occur at random in an unpredictable manner, but this is often not the case. When the surface conditions are not homogeneous, the convective cycle is fixed in space.

An example of a convective cycle that is fixed in space is a **sea breeze**. This is illustrated in Figure 5.28. During daytime the land is usually warmer than the water. As a result, the air will rise above the land and sink above the sea. To complete the cycle, the wind will be toward the land close to the surface and away from the land at higher altitudes. During the nighttime, the sea is usually warmer than the land, and the atmosphere cycles in the opposite direction.

Another example is a **mountain valley breeze**. The side of a valley facing the sun is usually warmer than the other side. Hence a cycle is generated with the wind near the surface blowing toward the sun-facing side of the valley. This is also illustrated in Figure 5.28.

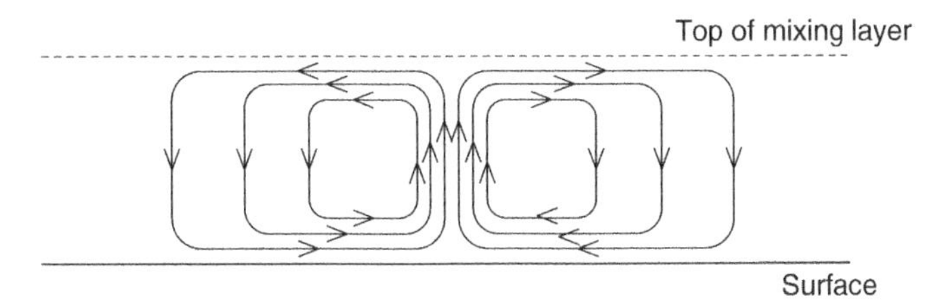


Figure 5.27 Convective cycles.

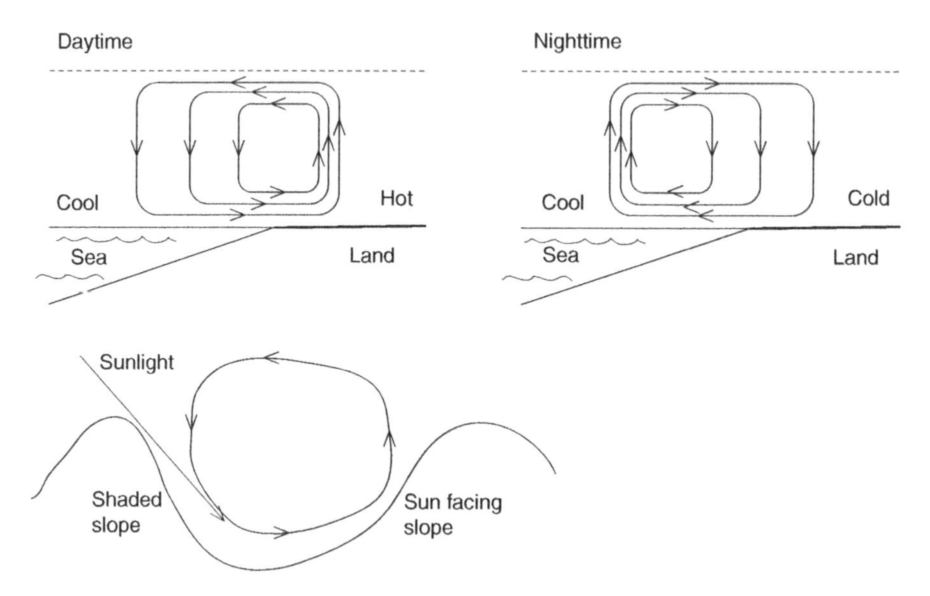


Figure 5.28 Sea breeze (*top*); mountain valley breeze (*bottom*).

## 5.12.2 Internal Boundary Layer: Qualitative Description

In this chapter we have discussed how the properties of a surface (roughness, Bowen ratio, etc.) affect the properties of the atmophere. No assumptions were made on the horizontal variability of these properties. Hence, it was assumed implicitly that there is no horizontal variance of the surface properties. The similarity theory as discussed in this chapter is a one-dimensional theory. This is not always an acceptable assumption.

When the conditions of the surface change drastically, this is not reflected in the atmospheric properties immediately. Instead, a new boundary layer is built up from the surface, within the planetary boundary layer. This new boundary layer is called the **internal boundary layer** (IBL). The source of the IBL can be *thermal*, when the stability of the atmosphere suddenly changes because of changing surface conditions, or *mechanical*, when the surface roughness suddenly changes. A sea breeze is a good example of a thermal IBL. This is shown in Figure 5.29.

Above the sea the atmosphere is usually near neutral. The stability above the sea depends on the temperature difference between the air and the water. Hence, stable conditions can occur above the sea when the atmosphere is unstable above the land. In that case, the atmosphere will be stable in the section of the atmosphere above the dotted line in Figure 5.29, irrespective of the location above sea or land. The atmosphere is only unstable below the dotted line.

The slope of the IBL depends on wind speed and heat flux, and can range from about 3% to about 30%. The average value is about 10%. More details are given in Section 5.13.3.

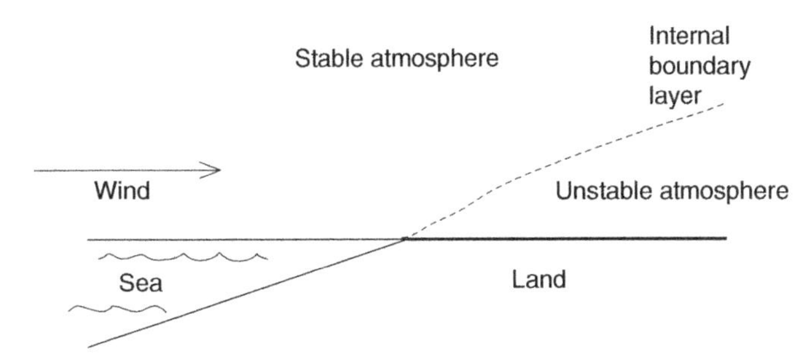


Figure 5.29 Internal boundary layer.

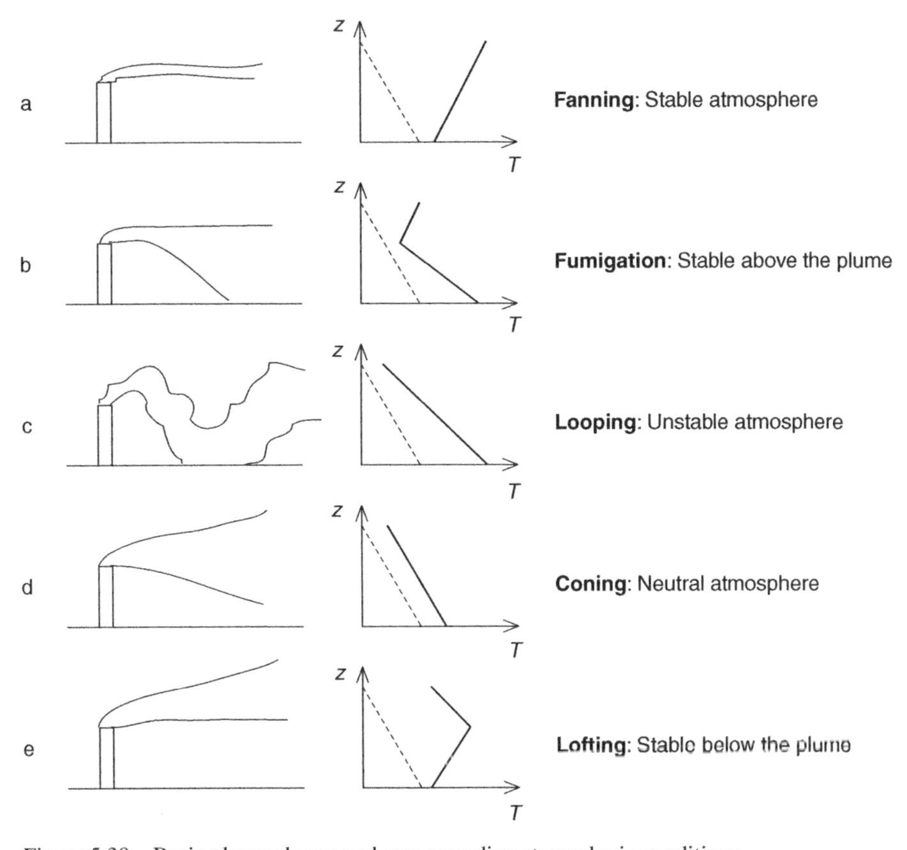


Figure 5.30 Basic plume shapes and corresponding atmospheric conditions.

### 5.12.3 Plume Shapes

We can learn a lot about the properties of the atmosphere by simply looking at the shape of a plume. A number of examples are given in Figure 5.30.

A temperature inversion above the plume is not the only cause of fumigation. Another cause is an IBL. This is illustrated in Figure 5.31.

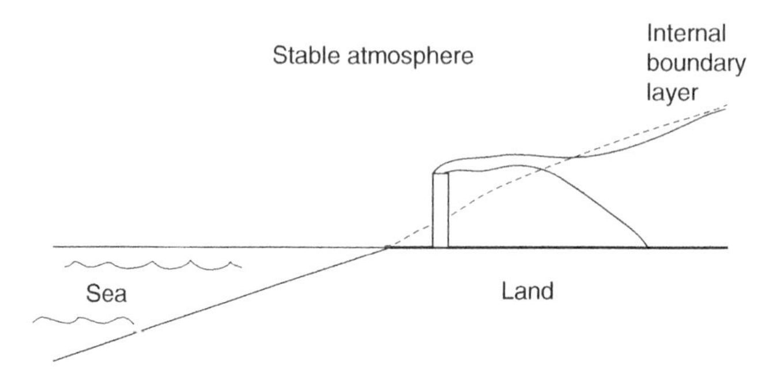


Figure 5.31 Fumigation at an internal boundary layer.

### 5.12.4 Virtual Temperature

Water vapor ($M_{H_2O} = 18$ g mol$^{-1}$) is lighter than air ($M_{air} = 29$ g mol$^{-1}$), and as a result moist air will have a tendency to rise when it is suspended in dry air, even when the temperature of the two is the same.

In other words, in terms of buoyancy, moist air behaves as if it is slightly hotter than it really is. For that reason, the **virtual temperature**, $T_v$, of a humid parcel of air is defined as the temperature of dry air with the same density as the humid air at equal pressure. It is approximated as

$$T_v = T(1 + 0.61q) \tag{5.203}$$

in which $q$ is the water/air mass ratio (i.e., kg water per kg dry air). When we use temperature to indicate buoyancy effects, we should use $T_v$ instead of $T$, especially when there are large humidity gradients. This means we should also define the **virtual potential temperature $\theta_v$**:

$$\theta_v = T_v + \Gamma z \tag{5.204}$$

The difference between $T$ and $T_v$ is the main mechanism that keeps clouds suspended in the air: air flows up through a cloud, compensating the settling velocity of cloud droplets. This process is also responsible for washout, one of the wet deposition mechanisms (see Chapter 7).

Many equations in this chapter refer to virtual temperatures rather than actual temperatures. However, this is usually not of concern to air dispersion modelers, unless they wish to develop their own models from first principles.

## 5.13 ADVANCED TOPICS IN METEOROLOGY

As with the previous section, this section consists of topics relevant to air dispersion modeling that do not fit anywhere else in this chapter. In particular, the quantification of convective cycling, internal boundary layers, and other landscape scale effects will be discussed. Not all advanced topics in meteorology can be discussed in this chapter because we have not covered all the basics yet. In particular, we have not considered the question of how long do wind speed deviations persist in the atmosphere. This will be covered in Chapter 6. Hence, Chapter 6 will be concluded with an advanced topics section as well.

### 5.13.1 Convective Cycles: Quantitative Description

As discussed in Section 5.12.1, the convective boundary layer (CBL) consists of updrafts and downdrafts forming a cycle. As a consequence, the distribution of $w'$, the vertical velocity fluctuation, is not Gaussian but has a negative mode and a positive skew. This has important repercussions for air dispersion: The vertical concentration profile of a plume in a CBL is also non-Gaussian. To properly account for non-Gaussian effects in dispersion modeling, we need to quantify the properties of the updrafts and the downdrafts in the CBL. This is the subject of this section.

The theory discussed here was first developed by Baerentsen and Berkowicz (1984) and incorporated in a stochastic Lagrangian particle model. It was adapted by Luhar and Britter (1989) and by Weil (1990) and incorporated in a bi-Gaussian dispersion model by Weil et al. (1997). The version of Weil et al. (1997) was incorporated in AERMOD (Cimorelli et al., 2004).

First, assume that the area fraction of the updraft is $\lambda_1$, and the area fraction of the downdraft is $\lambda_2$. The values of $\lambda_1$ and $\lambda_2$ are unknown for now, but their sum is immediately obvious:

$$\lambda_1 + \lambda_2 = 1 \tag{5.205}$$

The average vertical wind speed in the updrafts is $\overline{w_1}$; the average wind speed in the downdrafts is $\overline{w_2}$. If the terrain is homogeneous, then there is no net flow through any vertical surface. Hence, we can write

$$\lambda_1 \overline{w_1} + \lambda_2 \overline{w_2} = 0 \tag{5.206}$$

It is assumed that updrafts and downdrafts can be considered as separate entities with their own statistics. Hence, the vertical wind speed in an updraft fluctuates around $\overline{w_1}$ with standard deviation $\sigma_{w1}$. For the downdraft the "local" standard deviation is $\sigma_{w2}$. The overal turbulent velocity can be written in terms of its local values in updrafts and downdrafts:

$$\sigma_w^2 = \lambda_1 \overline{w_1^2} + \lambda_2 \overline{w_2^2} = \lambda_1 \overline{\left(\overline{w_1} + w'_1\right)^2} + \lambda_2 \overline{\left(\overline{w_2} + w'_2\right)^2} \tag{5.207}$$

Applying the Reynolds averaging rules leads to

$$\begin{aligned} \sigma_w^2 &= \lambda_1 \left(\overline{w_1}^2 + 2\overline{w_1} \cdot \overline{w'_1} + \overline{w_1'^2}\right) + \lambda_2 \left(\overline{w_2}^2 + 2\overline{w_2} \cdot \overline{w'_2} + \overline{w_2'^2}\right) \\ &= \lambda_1 \left(\overline{w_1}^2 + \overline{w_1'^2}\right) + \lambda_2 \left(\overline{w_2}^2 + \overline{w_2'^2}\right) \end{aligned} \tag{5.208}$$

Hence:

$$\sigma_w^2 = \lambda_1\left(\overline{w_1}^2 + \sigma_{w1}^2\right) + \lambda_2\left(\overline{w_2}^2 + \sigma_{w2}^2\right) \tag{5.209}$$

Define the skewness $S$ of the vertical wind speed distribution as

$$S = \frac{\overline{w'^3}}{\sigma_w^3} = \frac{\overline{w^3}}{\sigma_w^3} \tag{5.210}$$

Several empirical relationships have been proposed to link $S$ or a derived property to known properties of the atmosphere. Baerentsen and Berkowicz (1984) proposed:

$$\frac{\overline{w^3}}{w_*^3} = \frac{0.8(z/h_{\text{mix}})(1 - z/h_{\text{mix}})}{1 + 0.667(z/h_{\text{mix}})} \tag{5.211}$$

Luhar and Britter (1989) proposed:

$$\frac{\overline{w^3}}{w_*^3} = 0.8\left(\frac{\sigma_w^2}{w_*^2}\right)^{3/2} \tag{5.212}$$

Weil (1990) proposed:

$$\frac{\overline{w^3}}{w_*^3} = 0.84\frac{z}{h_{\text{mix}}}\left(1 - \frac{z}{h_{\text{mix}}}\right) \tag{5.213}$$

Weil et al. (1997) proposed:

$$S = 0.105\frac{w_*^3}{\sigma_w^3} \tag{5.214}$$

Other correlations proposed in the literature are summarized by Anfossi and Physick (2005).

The skewness is written in terms of the updraft and downdraft properties:

$$S\sigma_w^3 = \overline{w^3} = \lambda_1\overline{w_1^3} + \lambda_2\overline{w_2^3} = \lambda_1\overline{\left(\overline{w_1} + w'_1\right)^3} + \lambda_2\overline{\left(\overline{w_2} + w'_2\right)^3} \tag{5.215}$$

Applying the Reynolds averaging rules leads to

$$\begin{aligned} S\sigma_w^3 &= \lambda_1\overline{\left(\overline{w_1}^3 + 3\overline{w_1}^2 w'_1 + 3\overline{w_1}\,w'^2_1 + w'^3_1\right)} + \lambda_2\overline{\left(\overline{w_2}^3 + 3\overline{w_2}^2 w'_2 + 3\overline{w_2}\,w'^2_2 + w'^3_2\right)} \\ &= \lambda_1\overline{\left(\overline{w_1}^3 + 3\overline{w_1}\,\overline{w'^2_1}\right)} + \lambda_2\overline{\left(\overline{w_2}^3 + 3\overline{w_2}\,\overline{w'^2_2}\right)} \end{aligned} \tag{5.216}$$

Hence:

$$S\sigma_w^3 = \lambda_1\overline{\left(\overline{w_1}^3 + 3\overline{w_1}\sigma_{w1}^2\right)} + \lambda_2\overline{\left(\overline{w_2}^3 + 3\overline{w_2}\sigma_{w2}^2\right)} \tag{5.217}$$

We have developed four equations [eqs. (5.205), (5.206), (5.209), and (5.217)] with six unknowns, $\lambda_1$, $\lambda_2$, $\overline{w_1}$, $\overline{w_2}$, $\sigma_{w1}$, and $\sigma_{w2}$. Hence, two additional equations are needed to close the model. The following have been proposed:

$$\sigma_{w1} = R\overline{w_1} \tag{5.218}$$

$$\sigma_{w2} = -R\overline{w_2} \tag{5.219}$$

Several values for $R$ have been proposed. Baerentsen and Berkowicz (1984) and Luhar and Britter (1989) proposed $R=1$, Weil (1990) proposed $R=1.5$, whereas Weil et al. (1997) proposed $R=2$. Fortunately, the resulting air dispersion model is not very sensitive to this parameter.

Hence, six equations with six unknowns are obtained. A brief outline of the solution is given in Section A5.11 (Appendix A). The result is

$$\lambda_1 = \frac{1}{2}\left[1 - \sqrt{\frac{1}{1+4/\left(\gamma_1^2\gamma_2 S^2\right)}}\right] \tag{5.220}$$

$$\lambda_2 = \frac{1}{2}\left[1 + \sqrt{\frac{1}{1+4/\left(\gamma_1^2\gamma_2 S^2\right)}}\right] \tag{5.221}$$

$$\overline{w_1} = \frac{S\sigma_w\gamma_1}{2}\left(1 + \sqrt{1 + \frac{4}{\gamma_1^2\gamma_2 S^2}}\right) \tag{5.222}$$

$$\overline{w_2} = \frac{S\sigma_w\gamma_1}{2}\left(1 - \sqrt{1 + \frac{4}{\gamma_1^2\gamma_2 S^2}}\right) \tag{5.223}$$

$$\sigma_{w1} = \frac{RS\sigma_w\gamma_1}{2}\left(1 + \sqrt{1 + \frac{4}{\gamma_1^2\gamma_2 S^2}}\right) \tag{5.224}$$

$$\sigma_{w1} = \frac{RS\sigma_w\gamma_1}{2}\left(1 - \sqrt{1 + \frac{4}{\gamma_1^2\gamma_2 S^2}}\right) \tag{5.225}$$

where

$$\gamma_1 = \frac{1+R^2}{1+3R^2} \tag{5.226}$$

$$\gamma_2 = 1 + R^2 \tag{5.227}$$

As expected, we see that most of the area is taken by the downdrafts, and a smaller area is taken by the updrafts. Naturally, it follows that the updraft velocity is larger in absolute value than the downdraft velocity.

When we apply the equations to $S=0$ (neutral conditions), we find that the model still predicts updrafts and downdrafts, but now they are of equal size and velocity. Hence, the model predicts symmetrical but non-Gaussian concentration profiles. This is a weakness of the model. Luhar et al. (1996) proposed the following equation for $R$ to solve that problem:

$$R = \frac{2}{3} S^{1/3} \tag{5.228}$$

## 5.13.2 Simple Convective Boundary Layer Model

We have seen in Section 5.5.3 that a convective (unstable) boundary layer is characterized by an approximately constant potential temperature (except near the surface) and a capping inversion: a thin layer of temperature inversion. We have seen this temperature inversion in Example 5.1 (Section 5.3), where we found that the temperature increase in the capping inversion is on the order of 1°C. The purpose of this section is to provide a better understanding of this temperature inversion by means of a simple boundary layer model. The main concepts for this model are from Garratt (1994).

As simplifying assumptions for the model, we will assume that the convective boundary layer is well-mixed, so its potential temperature is constant and equal to the mean potential temperature $\theta_m$. The height $h$ of the mixed layer increases with time. The capping inversion is approximated by a temperature jump $\Delta\theta$ at height $h$. Above the capping inversion the potential temperature profile $\theta(z)$ has a slope $\gamma$. For convenience, the potential temperature at height $h$ but above the jump, $\theta(h)$ is simply denoted $\theta$. The assumptions and variables are summarized in Figure 5.32.

The boundary layer grows by mixing of air at the top. This mixing also supplies sensible heat $q_1$ to the mixing layer. Another source of heat to the mixing layer is

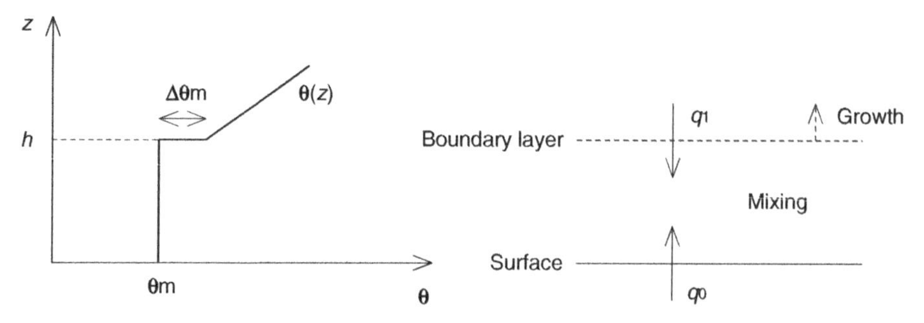


Figure 5.32 Assumptions and variables of a simplified convective boundary layer model.

sensible heat from the surface, $q_0$. Based on this information a heat balance of the boundary layer is made:

$$\text{Heat accumulated} = q_0 + q_1 \tag{5.229}$$

Expressing the heat balance per meter squared of surface, the heat accumulated is given by

$$\text{Heat accumulated} = \rho c_p \frac{\partial (h\theta_\mathrm{m})}{\partial t} = \rho c_p \left( h\frac{\partial \theta_\mathrm{m}}{\partial t} + \theta_\mathrm{m}\frac{\partial h}{\partial t} \right) \tag{5.230}$$

The heat flow $q_1$ is the heat contained in the air mixed into the mixing layer from above. Hence:

$$q_1 = \rho c_p \theta \frac{\partial h}{\partial t} \tag{5.231}$$

Substituting eqs. (5.230) and (5.231) into eq. (5.229) leads to

$$\rho c_p \left( h\frac{\partial \theta_\mathrm{m}}{\partial t} + \theta_\mathrm{m}\frac{\partial h}{\partial t} \right) = q_0 + \rho c_p \theta \frac{\partial h}{\partial t} \tag{5.232}$$

Rearranging leads to

$$\rho c_p h \frac{\partial \theta_\mathrm{m}}{\partial t} = q_0 + \rho c_p (\theta - \theta_\mathrm{m}) \frac{\partial h}{\partial t} = q_0 + \rho c_p \,\Delta\theta \frac{\partial h}{\partial t} \tag{5.233}$$

Divide by $\rho c_p$:

$$h\frac{\partial \theta_\mathrm{m}}{\partial t} = \frac{q_0}{\rho c_p} + \Delta\theta \frac{\partial h}{\partial t} \tag{5.234}$$

The temperature jump $\Delta\theta$ is not constant but tends to increase with time. Hence, we need a differential equation for $\Delta\theta$ as well. It is based on the definition of $\Delta\theta$:

$$\frac{\partial \Delta\theta}{\partial t} = \frac{\partial \theta}{\partial t} - \frac{\partial \theta_\mathrm{m}}{\partial t} = \frac{\partial \theta}{\partial h}\frac{\partial h}{\partial t} - \frac{\partial \theta_\mathrm{m}}{\partial t} = \gamma \frac{\partial h}{\partial t} - \frac{\partial \theta_\mathrm{m}}{\partial t} \tag{5.235}$$

We have two differential equations [eqs. (5.234) and (5.235)] with three dependent variables ($h$, $\Delta\theta$, and $\theta_\mathrm{m}$). Hence, we need a third differential equation to close the model. Here an assumption needs to be made. It is assumed that the heat effect of the air mixed into the mixed layer is proportional to the sensible heat flux from the surface:

$$\rho c_p \Delta\theta \frac{\partial h}{\partial t} = \beta q_0 \tag{5.236}$$

or

$$\frac{\partial h}{\partial t} = \frac{\beta q_0}{\rho c_p \Delta\theta} \tag{5.237}$$

where $\beta$ is a proportionality factor. Based on measurements, a value of 0.2 has been found for $\beta$.

To obtain a differential equation in $\Delta\theta$, first eq. (5.234) is substituted into eq. (5.235) to eliminate $\theta_m$:

$$\frac{\partial \Delta\theta}{\partial t} = \gamma \frac{\partial h}{\partial t} - \frac{q_0}{h\rho c_p} - \frac{\Delta\theta}{h}\frac{\partial h}{\partial t} \tag{5.238}$$

What we are really interested in at this point is the relationship between $\Delta\theta$ and $h$, rather than their relationship with time. Hence, we expand the left-hand side of eq. (5.238) as follows:

$$\frac{\partial \Delta\theta}{\partial h}\frac{\partial h}{\partial t} = \gamma \frac{\partial h}{\partial t} - \frac{q_0}{h\rho c_p} - \frac{\Delta\theta}{h}\frac{\partial h}{\partial t} \tag{5.239}$$

Next, eq. (5.237) is used to eliminate the time derivative of the height $h$:

$$\frac{\partial \Delta\theta}{\partial h}\frac{\beta q_0}{\rho c_p \Delta\theta} = \gamma \frac{\beta q_0}{\rho c_p \Delta\theta} - \frac{q_0}{h\rho c_p} - \Delta\theta \frac{\beta q_0}{h\rho c_p \Delta\theta} \tag{5.240}$$

which reduces to

$$\frac{\partial \Delta\theta}{\partial h} = \gamma - \frac{\Delta\theta}{h\beta} - \frac{\Delta\theta}{h} = \gamma - \frac{1+\beta}{\beta}\frac{\Delta\theta}{h} \tag{5.241}$$

To solve this differential equation, it is more convenient to write it in terms of $\Delta\theta/h$:

$$\frac{\partial(\Delta\theta/h)}{\partial h} = \frac{1}{h}\frac{\partial \Delta\theta}{\partial h} - \frac{\Delta\theta}{h^2} = \frac{1}{h}\left(\gamma - \frac{1+\beta}{\beta}\frac{\Delta\theta}{h}\right) - \frac{\Delta\theta}{h^2} = \frac{1}{h}\left(\gamma - \frac{1+2\beta}{\beta}\frac{\Delta\theta}{h}\right) \tag{5.242}$$

Simulations indicate that this equation reaches a steady-state value of $\Delta\theta/h$ fairly early on in the development of the boundary layer. Hence, we can set the left-hand side of eq. (5.242) equal to 0:

$$\frac{1}{h}\left(\gamma - \frac{1+2\beta}{\beta}\frac{\Delta\theta}{h}\right) = 0 \tag{5.243}$$

which leads to

$$\Delta\theta = \frac{\beta\gamma h}{1+2\beta} \tag{5.244}$$

As we can see from Example 5.3 as well as from the structure of the atmosphere, $\gamma$ is about 0.004 K m$^{-1}$. Hence, for a mixing layer height of 3000 m (Example 5.3), we find a $\Delta\theta$ value of about 1.7 K, which is indeed the correct order of magnitude. The main reason for the difference between the results of the examples (1 K) and the value found here is the lack of resolution of the rawinsonde data analyzed. Hence, this simple model describes the atmospheric boundary layer fairly well.

### 5.13.3 Internal Boundary Layer: Quantitative Description

As discussed in Section 5.12.2, a drastic change of the surface condition (roughness, Bowen ratio, etc.) does not translate into an immediate change of the wind speed profile and the other atmospheric properties. Instead, a new boundary layer builds up from the surface inside the planetary boundary layer. This new boundary layer is known as an internal boundary layer (IBL). A particularly important example of an internal boundary layer is the thermal internal boundary layer (TIBL) at the coast during a sea breeze because of the impact of a TIBL on air pollution from coastal industrial activities (see Fig. 5.31). During daytime the air above the sea is relatively stable, so a pollution plume emitted near the coast stays relatively compact until it enters the TIBL. Once inside the TIBL the air below the plume is unstable, and the air above is stable. The result is pronounced dispersion to the surface: fumigation. Hence, it is important to be able to predict internal boundary layers, especially coastline TIBLs. We will focus on this case.

The basis of internal boundary layer calculations is the model discussed in the previous section, which is applicable to both the planetary boundary layer and to internal boundary layers. Substitution of eq. (5.244) into eq. (5.237) leads to

$$\frac{\partial h}{\partial t} = \frac{q_0(1+2\beta)}{\rho c_p \gamma h} \tag{5.245}$$

The internal boundary layer height can be calculated in terms of downwind distance by expanding the left-hand side:

$$\frac{\partial h}{\partial t} = \frac{\partial h}{\partial x}\frac{\partial x}{\partial t} = \frac{\partial h}{\partial x}u = \frac{q_0(1+2\beta)}{\rho c_p \gamma h} \tag{5.246}$$

which leads to the following equation for the slope of the TIBL:

$$\frac{\partial h}{\partial x} = \frac{q_0(1+2\beta)}{\rho c_p \gamma h u} \tag{5.247}$$

For example, considering typical values of $q_0 = 100$ W m$^{-2}$, $1+2\beta = 1.4$, $\rho = 1.2$ kg m$^{-3}$, $c_p = 1006$ J kg$^{-1}$ K$^{-1}$, $\gamma = 0.004$ K m$^{-1}$, $h = 100$ m, and $u = 3$ m s$^{-1}$, the obtained slope is 0.097, within the expected 3–30% (see Section 5.12.2). When all variables except

$h$ and $x$ can be assumed approximately constant over a range of downwind distances from $x_0$ to $x$, eq. (5.247) can be integrated. First, separate variables:

$$h\partial h = \frac{q_0(1+2\beta)}{\rho c_p \gamma u}\partial x \tag{5.248}$$

Integration leads to

$$\frac{h^2}{2} - \frac{h_0^2}{2} = \frac{q_0(1+2\beta)}{\rho c_p \gamma u}(x - x_0) \tag{5.249}$$

Hence, the height of the TIBL at distance $x$ is calculated as

$$h = \sqrt{h_0^2 + \frac{2q_0(1+2\beta)}{\rho c_p \gamma u}(x - x_0)} \tag{5.250}$$

This equation is used to calculate thermal internal boundary layers for coastal regions in CALPUFF. More sophisticated models, as well as other types of internal boundary layers, are reviewed by Garratt (1990).

### 5.13.4 Effect of Complex Terrain in Meteorology

So far in this book we assumed that we always know the wind direction and wind speed needed for dispersion calculations. This is usually the case, at least approximately, because we can use data from a nearby weather station. When the terrain is flat, weather stations are representative to relatively large areas, but this is not always the case when the terrain is not flat. This section will discuss a number of ways in which a local wind can deviate from the measured wind at a nearby station due to local geographic effects.

The first type of terrain affected wind flow is the **drainage wind**, also known as gravity flow, slope flow, and as the katabatic wind (Stull, 1988). It refers to the tendency of cold air to flow down a slope by virtue of its larger density.

A simple approximate method to estimate the wind speed that can be generated by a slope is by application of the Bernouilli equation. The Bernouilli equation is the foundation of classical hydrodynamics in the absence of friction (viscous) forces and is as follows:

$$\frac{dp}{\rho} + \frac{du^2}{2} + gdz = 0 \tag{5.251}$$

The integrated form of eq. (5.251) is as follows and applies to two cross sections of the same streamline, labeled 1 and 2 (Fig. 5.33):

$$\frac{p_2 - p_1}{\rho} + \frac{u_2^2 - u_1^2}{2} + g(z_2 - z_1) = 0 \tag{5.252}$$

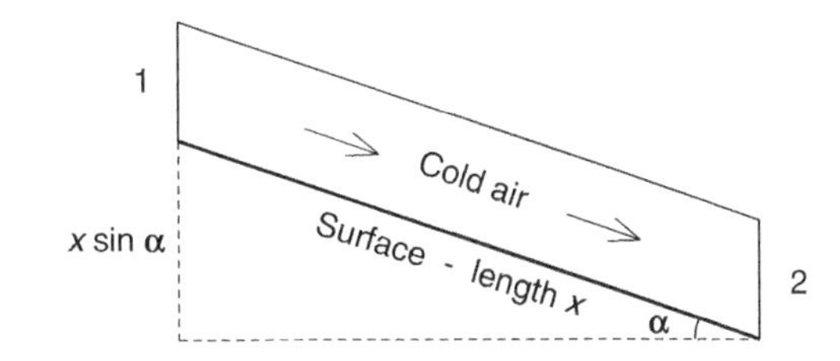


Figure 5.33 System with drainage wind.

The subscripts in eq. (5.252) refer to the cross sections. Assuming that the air is initially stationary at the top of the slope, we can state that $u_1=0$ and we solve the equation for $u_2$:

$$u_2 = \sqrt{2\left[-\frac{p_2 - p_1}{\rho} - g\left(z_2 - z_1\right)\right]} \tag{5.253}$$

The pressure difference $p_2-p_1$ is the result of the surrounding air and follows from the hydrostatic equation [eq. (5.2)]. Assuming an approximately constant density of the surrounding air, the pressure difference becomes

$$p_2 - p_1 = -\rho_a g(z_2 - z_1) \tag{5.254}$$

where $\rho_a$ is the density of the surrounding air. Substitution into eq. (5.253) leads to

$$u_2 = \sqrt{2\left[-\frac{\rho - \rho_a}{\rho} g(z_2 - z_1)\right]} = \sqrt{2\frac{\rho - \rho_a}{\rho} g(z_1 - z_2)} \tag{5.255}$$

The ratio $(\rho-\rho_a)/\rho$ is calculated in terms of temperature in a manner similar to Section 5.5.2. The result is $(T_a-T)/T_a$. Since we are considering a thin air layer near the surface, we can identify absolute temperature with potential temperature. The height difference $z_1-z_2$ can be written in terms of distance along the surface $x$, and the slope angle $\alpha$. Hence, eq. (5.255) becomes

$$u_2 = \sqrt{2\frac{\theta_a - \theta}{\theta_a} g x \sin\alpha} = \sqrt{2\frac{\Delta\theta}{\theta_a} g x \sin\alpha} \tag{5.256}$$

In practice, this is an upper limit because entrainment will slow down the process. Nighttime temperature inversions are typically one to a few kelvins. Hence, with a slope ($\sin\alpha$) of 0.01, we need a length $x$ of about a kilometer to find a drainage wind speed on the order of 1 m s$^{-1}$, and a length of about 10 km to find a value of 3 m s$^{-1}$. For a slope of 0.001 the required lengths are 100 and 1000 km. This is the case in the Great Plains in southeastern United States.

When the slopes are contained in a valley, the opposite slope flow happens during the early daytime: Warm air flows up along the slopes of the valley, whereas the cold air above it moves down toward the valley bottom to replace the warm outflowing air.

A question of importance to air dispersion modelers when an air flow approaches a hill is: Will the air flow over the hill or around it? The rest of this section is devoted to that question.

In unstable or neutral conditions, the answer is straightforward: There are no buoyancy effects restricting the vertical movement of the air, so the air can be expected to move over the hill. In stable conditions, buoyancy effects restrict the vertical movement of air parcels, and the air parcel will only move over the hill if it carries enough momentum to drive it over the hill.

To understand the interaction of a stable air flow and the underlying terrain, we return to the stability parameter $s$ (Section 5.5.2) and its square root, the Brunt–Väisälä frequency $N_{\mathrm{BV}}$:

$$N_{\mathrm{BV}} = \sqrt{\frac{g}{\theta}\frac{d\theta}{dz}} \tag{5.257}$$

As earlier in this section, the temperature has been replaced with the potential temperature. As an example, an atmosphere at the verge of a temperature inversion has a Brunt–Väisälä frequency of about 0.018 $\mathrm{s}^{-1}$.

In Section 5.5.2, the stability parameter was defined as the tendency of an air parcel to accelerate against any vertical motion, per meter of motion. Hence, if $z$ is defined as the vertical distance from the equilibrium position of the air parcel, the dynamics of the air parcel is described by the following differential equation:

$$\frac{d^2 z}{dt^2} = -N_{\mathrm{BV}}^2 z \tag{5.258}$$

Assuming that the air parcel is moving through its equilibrium position at $t=0$, the solution of eq. (5.258) is

$$z = a\sin(N_{\mathrm{BV}} t) \tag{5.259}$$

The parcel moves in an oscillatory fashion, with amplitude $a$ and frequency $N_{\mathrm{BV}}/(2\pi)$. Considering a wind speed $u$, the air parcel traces a sine wave with wavelength $\lambda$:

$$\lambda = \frac{2\pi u}{N_{\mathrm{BV}}} \tag{5.260}$$

For instance, at a wind speed of 3 m $\mathrm{s}^{-1}$, the wavelength in an atmosphere on the verge of a temperature inversion is about 1000 m.

Two properties of a hill on the path of an air parcel will determine the interaction between the air parcel and the hill: the **width** of the hill along the wind direction and the **height** of the hill above the parcel as it approaches the hill. When we consider the width of the hill, the following dimensionless number best describes the interaction between the air parcel and the hill (Stull, 1988):

$$\mathrm{Fr}_W = \frac{\lambda}{2W} = \frac{\pi u}{N_{\mathrm{BV}} W} \tag{5.261}$$

$Fr_W$ is the **Froude number** for the hill width. When $Fr_W = 1$, the width of the obstacle equals one-half of the natural wavelength of the air oscillations. An air parcel will follow the obstacle boundary at its resonance frequency and experience high-amplitude oscillations behind the obstacle.

When $Fr_W < 1$, the natural wavelength of the air oscillations is too short to follow the boundary of the hill easily, and the air will have a tendency to detach from the hill boundary on the downwind side, to oscillate at its natural frequency. When $Fr_W > 1$, the natural wavelength is too long to follow the hill boundary, and the air will leave room for a wake on the downwind side of the hill.

The above discussion does not indicate whether the air parcel follows the hill vertically (over the hill) or horizontally (around the hill). For that we need a different Froude number, based on the **height** of the hill. To arrive at the correct formulation of such a Froude number, consider the momentum of an air parcel. If the air parcel uses all its momentum to move vertically, then its initial vertical velocity is $u$. Take the time derivative of eq. (5.259):

$$\frac{dz}{dt} = aN_{BV}\cos(N_{BV}t) \tag{5.262}$$

The initial velocity is $aN_{BV}$. Hence, the maximum amplitude of vertical air movement is obtained by equating $aN_{BV}$ to $u$:

$$a_{max} = \frac{u}{N_{BV}} \tag{5.263}$$

The new Froude number is the ratio of this maximum amplitude to the height of the hill:

$$Fr_H = \frac{a_{max}}{H} = \frac{u}{N_{BV}H} \tag{5.264}$$

Note that, unlike $Fr_W$, $Fr_H$ does not have a factor $\pi$. When $Fr_H > 1$, any air parcel can travel over the hill. When $Fr_H < 1$, only the air that has to travel a fraction $Fr_H$ of the hill or less can travel over the hill. Hence, the **dividing streamline height** $h_d$ is defined as the height above the surface downwind of the hill where the air separates between parcels moving over the hill and parcels moving around the hill. It is given by

$$h_d = H(1 - Fr_H) = H - \frac{u}{N_{BV}} \tag{5.265}$$

Air parcels with an initial height greater than $h_d$ will predominantly travel over the hill, whereas air parcels with an initial height less than $h_d$ will predominantly travel around the hill.

## 5.14 SUMMARY OF MAIN EQUATIONS

The vertical pressure gradient is given by the barometric equation:

$$\frac{dp}{dz} = -\rho g \tag{5.2}$$

In integrated form, after substituting the ideal gas law, the barometric equation becomes

$$p = p_0 \exp\left(-\frac{M_{\text{air}} g}{RT} z\right) \tag{5.8}$$

The dry adiabatic lapse rate is given by

$$\frac{dT}{dz} = -\frac{g}{c_{p,\text{air}}} \tag{5.20}$$

The potential temperature is calculated from the actual temperature as

$$\theta = T + \Gamma z \tag{5.28}$$

$$\theta = T\left(\frac{p_0}{p}\right)^{R/\left(c_{p,\text{air}} M_{\text{air}}\right)} \tag{5.32}$$

The stability parameter is calculated as

$$a = \frac{g}{T_{\text{a}}} \frac{d\theta}{dz} = s \tag{5.48}$$

The sensible heat from the surface is calculated from the following heat balance:

$$q = \frac{(1 - C_{\text{G}}) R_N}{1 + 1/B} \tag{5.53}$$

The wind speed profile in a neutral atmosphere is given by

$$u = \frac{u_*}{k} \ln \frac{z}{z_0} \tag{5.64}$$

The Obukhov length is defined as

$$L = -\frac{\rho_0 c_p T_0 u_*^3}{kgq} \tag{5.71}$$

The wind speed profile in a stable and unstable atmosphere can be calculated with

$$u = \frac{u_*}{k}\left( \ln\frac{z}{z_0} + 5\frac{z - z_0}{L} \right) \quad \text{(stable)} \tag{5.81}$$

$$u = \frac{u_*}{k}\left\{ \ln\frac{z}{z_0} + \ln\left[ \frac{(n_0^2+1)(n_0+1)^2}{(n^2+1)(n+1)^2} \right] + 2\left[ \arctan(n) - \arctan(n_0) \right] \right\} \quad \text{(unstable)} \tag{5.82}$$

with

$$n_0 = \left( 1 - 16\frac{z_0}{L} \right)^{1/4} \quad n = \left( 1 - 16\frac{z}{L} \right)^{1/4} \tag{5.83}$$

The definition of the friction temperature is

$$\theta_* = -\frac{q}{\rho_0 c_p u_*} \tag{5.88}$$

The potential temperature in stable and unstable conditions can be calculated with

$$\theta = \theta_1 + \frac{\theta_*}{k}\left( \ln\frac{z}{z_1} + 5\frac{z - z_1}{L} \right) \quad \text{(stable)} \tag{5.94}$$

$$\theta = \theta_1 + \frac{\theta_*}{k}\left\{ \ln\frac{z}{z_1} + \ln\left[ \frac{(1+x_1)^2}{(1+x)^2} \right] \right\} \quad \text{(unstable)} \tag{5.95}$$

with

$$x_1 = \left( 1 - 16\frac{z_1}{L} \right)^{1/2} \quad x = \left( 1 - 16\frac{z}{L} \right)^{1/2} \tag{5.96}$$

In stable and neutral conditions the turbulent velocity is calculated with

$$\frac{\sigma_u}{u_*} = 2.5\left( 1 - \frac{z}{h_{\text{mix}}} \right)^a \tag{5.144}$$

$$\frac{\sigma_v}{u_*} = 1.9\left( 1 - \frac{z}{h_{\text{mix}}} \right)^a \tag{5.145}$$

$$\frac{\sigma_w}{u_*} = 1.3\left( 1 - \frac{z}{h_{\text{mix}}} \right)^a \tag{5.146}$$

In unstable atmospheres the convective velocity scale is defined as

$$w_* = \left( \frac{gqh_{\text{mix}}}{T_0 \rho c_p} \right)^{1/3} \tag{5.150}$$

In a convective atmosphere where convective cycling dominates air dispersion, the turbulent velocity is calculated as

$$\frac{\sigma_u}{w_*} = \frac{\sigma_v}{w_*} = \frac{\sigma_w}{w_*} = 0.6 \tag{5.152}$$

When both mechanical turbulence and convective cycling affect air dispersion, the turbulent velocity is calculated with

$$\sigma_u = \sqrt{\sigma_{u,\text{mech}}^2 + \sigma_{u,\text{therm}}^2} \tag{5.156}$$

In the gradient transport theory (*K* theory) the following fluxes can be calculated based on the relevant gradient:

$$\frac{\tau}{\rho} = \overline{u'w'} = -K_{\text{m}} \frac{\partial \bar{u}}{\partial z} \tag{5.174}$$

$$\frac{q}{\rho c_p} = \overline{\theta' w'} = -K_{\text{h}} \frac{\partial \bar{\theta}}{\partial z} \tag{5.175}$$

$$J = \overline{c'w'} = -K_z \frac{\partial \bar{c}}{\partial z} \tag{5.176}$$

The turbulent mass and heat diffusivity in the vertical direction can be calculated with the following equation:

$$K_{\text{h}} = ku_* z \left( 1 - \frac{z}{h_{\text{mix}}} \right) \left[ \left( 1 - \frac{z}{h_{\text{mix}}} \right)^{1.5} - \frac{h_{\text{mix}}}{kL} \right]^{1/3} \tag{5.194}$$

The virtual temperature is given by

$$T_{\text{v}} = T(1 + 0.61q) \tag{5.203}$$

## PROBLEMS

1. Prove that eq. (5.2) has consistent units.

2. At which altitude is the barometric pressure half of the value at zero altitude?

3. Derive an equation for barometric pressure versus height for an atmosphere with temperature profile $T = T_0 - \Gamma z$ where $\Gamma$ is a constant.

4. Based on its definition in Section 5.2, what units do you expect the stability parameter $s$ to have ? Check the units of eq. (5.44). Are these the units you expected?

5. Wind speed is measured on a site under overcast conditions. At a 10-m height the wind speed is 5 m s$^{-1}$; at 20-m height the wind speed is 6 m s$^{-1}$. Calculate $u_*$ and $z_0$.

6. Read the article of Zhang et al. (2008). The authors claim that there is a huge dependence of $k$ on the stability, although this has never been reported in the literature before. There seems to be something seriously wrong here. Can you find out what? What are the assumptions underlying the analysis of the authors? Are those reasonable assumptions? Can you come up with a better set of assumptions? (*Hint*: Under what conditions did we define $k$? Is it a coincidence that the authors find $k = 0.4$ under these conditions?)

7. Prove that $\theta = \theta_1$ when $z = z_1$. Also prove that $\theta = \theta_1$ under neutral conditions, at any value of $z$. Why is the latter not clear from Figures 5.19 and 5.20?

8. Calculate the height of the mixed layer after 6 h of sunshine with an average $q$ value of 200 W m$^{-2}$ and an early morning temperature gradient ($\Lambda$) of 2°C/km. Assume $T = 14$ °C at the surface, and $p = 97$ kPa.

9. Determine the units of each of the turbulence indicators introduced in Section 5.11.2.

10. Herndon (2010) claims that convective cycles in Earth's mantle are physically impossible because the density of Earth's mantle increases with increasing depth. This is due to the enormous pressure in the lower part of Earth's mantle, causing a compression that more than compensates for the thermal expansion. Hence, Herndon claims, Earth's mantle is in a state of stable stratification with the densest part at the bottom. Use your knowledge of the atmosphere to point out the error in this argument. (*Hint*: Earth's mantle is heated from below.)

11. In a farmland with surface roughness 0.1 m a wind speed at 10-m height of 2.5 m s$^{-1}$ is measured. Simultaneous high-speed measurements of wind speed fluctuations lead to an estimated friction velocity of 0.25 m s$^{-1}$. During the measurements the temperature near the surface is 20 °C, and the barometric pressure is 99 kPa.

    a. Is the atmosphere stable, neutral, or unstable?

    b. Calculate the Obukhov length.

    c. Calculate the sensible heat flux.

12. Careful heat balance measurements in a parkland at night indicate a sensible heat flux of −19.5 W m$^{-2}$. The local surface roughness is 0.25 m. The wind speed at 10 m height is 2.75 m s$^{-1}$. The temperature 1 m above the surface is 15 °C. The barometric pressure is 100 kPa.

    a. Calculate the friction velocity and the Obukhov length.

    b. Calculate the friction temperature.

**c.** Estimate the cloudiness.

**d.** Calculate the temperature and the potential temperature at 100-m height above the surface.

**13.** Calculate the virtual temperature of water saturated air at 1 bar pressure and 25 °C temperature.

**14.** In a peatland with a roughness length of 5 cm, an albedo of 0.3, and a Bowen ratio of 0.1, a temperature of 12 °C and a wind speed of 2.2 m $s^{-1}$ are measured at 2-m height. The barometric pressure corrected for sea level is 102 kPa, and the altitude is 300 m. The solar elevation is 68°, and the fractional cloud cover is $\frac{3}{8}$. Assume a typical value for $C_G$.

**a.** Calculate the sensible heat flux and the latent heat flux.

**b.** Calculate the Obukhov length, the friction velocity, and the friction temperature.

**c.** Calculate a wind speed and temperature profile up to 100-m height.

**15.** Calculate average wind speed, wind speed variance, turbulent velocity, and turbulent intensity assuming that the 10 wind speed measurements below are representative of the wind speed variability. Make reasonable estimates of the friction velocity, as well as the turbulent velocity in the cross wind and vertical directions, assuming neutral conditions and proximity to the surface.

| Measurement | $u$ (m $s^{-1}$) |
|---|---|
| 1 | 1.5248 |
| 2 | 2.2478 |
| 3 | 1.9245 |
| 4 | 1.2874 |
| 5 | 2.0215 |
| 6 | 1.6392 |
| 7 | 1.8989 |
| 8 | 2.4201 |
| 9 | 1.7675 |
| 10 | 1.9842 |

**16.** In a fertilized grassland, ammonia emissions are estimated based on the concentration gradient above the field. At a 0.6-m height a concentration of 100 ppb $NH_3$ is measured. At a 1-m height a concentration of 85 ppb is measured. The atmospheric stability is neutral, the wind speed at 10 m height is 3 m $s^{-1}$, the roughness length is 5 cm, the temperature is 25 °C, and the barometric pressure is 100 kPa. Use a representative turbulent diffusivity ($K_z$) and the concentration gradient to calculate the vertical ammonia mass flux above the field. Is the flux toward the surface (deposition) or away from the surface (volatilization)?

## MATERIALS ONLINE

- "Example 5.1. Analysis rawinsonde data.xlsx": Derivation of temperature profile from rawinsonde data.
- "Appendix A5.1. T and p dependence of GAMMAs.xlsx": Calculation of the temperature and pressure dependence of the wet adiabatic lapse rate $\Gamma_g$.
- "Wind speed profiles for problem solving.xlsx": Solution of Example 5.6.
- "Temperature profiles for problem solving.xlsx": Solution of Example 5.7.

- "Example 5.9. Measures of turbulence from 10 Hz data.xlsx": Solution of Example 5.9.
- "Example 5.10. Flux measurement with eddy correlation technique.xlsx": Solution of Example 5.10.

## REFERENCES


Andreas E.L. (2009). A new value of the von Kármán constant. Implications and implementation. *J. Appl. Meteorol. Climatol.* **48**, 923–944.

Andreas E.L., Claffey K.J., Jordan R.E., Fairall C.W., Guest P.S., Persson P.O.G., and Grachev A.A. (2006). Evaluations of the von Kármán constant in the atmospheric surface layer. *J. Fluid Mech.* **559**, 117–149.

Anfossi D. and Physick W. (2005). Lagrangian particle models. In Zannetti P. (ed.) *Air Quality Modeling. Volume II–Advanced Topics*. Enviro Comp Institute, Fremont, CA, and Air and Waste Management Association, Pittsburgh, pp. 93–161.

Arya S.P. (1999). *Air Pollution Meteorology and Dispersion*. Oxford University Press, Oxford, UK.

Baerentsen J.H. and Berkowicz R. (1984). Monte Carlo simulation of plume dispersion in the convective boundary layer. *Atmos. Environ.* **18**, 701–712.

Barad M.L. (1958). *Project Prairie Grass, a Field Program in Diffusion*. Geophysical Research Paper 59, Atmospheric Analysis Laboratory, Air Force Cambridge Research Center, US Air Force, Bedford, MA.

Benoit R. (1977). On the integral of the surface layer profile-gradient functions. *J. Appl. Meteorol.* **16**, 859–860.

Boussinesq J. (1877). Essai sur la theorie des eaux courantes. *Mem. Pres. Div. Sav. Acad. Sci. Paris* **23**, 1–680.

Bowen, I.S. (1926). The ratio of heat losses by conduction and by evaporation from any water surface. *Phys. Rev.* **27**, 779–787.

Byun D.W. and Ching J.K.S. (1999). *Science Algorithms of the EPA Models-3 Community Multiscale Air Quality (CMAQ) Modeling System*. Report EPA/600/R-99/030, US-EPA, Research Triangle Park, NC.

Caughey S.J. and Palmer S.G. (1979). Some aspects of turbulence structure through the depth of the convective boundary layer. *Quart. J. Roy. Meteorol. Soc.* **105**, 811–827.

Charnock H. (1955). Wind stress on a water surface. *Quart. J. Roy. Meteorol. Soc.* **81**, 639–640.

Cheng Y. and Brutsaert W. (2005). Flux-profile relationships for wind speed and temperature in the stable atmospheric boundary layer. *Bound. Layer Meteorol.* **114**, 519–538.

Cimorelli A.J., Perry S.G., Venkatram A., Weil J.C., Paine R.J., Wilson R.B., Lee R.F., Peters W.D., Brode R.W., and Paumier J.O. (2004). *AERMOD: Description of Model Formulation*. Report EPA-454/R-03-004, US-EPA, Research Triangle Park, NC.

Deardorff J.W. (1970a). Preliminary results from numerical integrations of the unstable planetary boundary layer. *J. Atmos. Sci.* **27**, 1209–1211.

Deardorff J.W. (1970b). Convective velocity and temperature scales for the unstable planetary boundary layer and for Rayleigh convection. *J. Atmos. Sci.* **27**, 1211–1213.

Driedonks A.G.M. and Tennekes H. (1984). Entrainment effects in the well-mixed atmospheric boundary layer. *Bound. Layer Meteorol.* **30**, 75–105.

Dubin M., Hull A.R., and Champion K.S.W. (eds.) (1976). *US Standard Atmosphere*. Report NOAA-S/T 76–1562, NOAA, NASA, USAF, Washington DC.

Dyer A.J. (1967). The turbulent transport of heat and water vapour in an unstable atmosphere. *Quart. J. Roy. Meteorol. Soc.* **93**, 501–508.

Dyer A.J. (1974). A review of flux-profile relationships. *Bound. Layer Meteorol.* **7**, 363–372.

Garratt J.R. (1990). The internal boundary layer—A review. *Bound. Layer Meteorol.* **50**, 171–203.

Garratt J.R. (1994). *The Atmospheric Boundary Layer*. Cambridge University Press, Cambridge, UK.

Golder D. (1972). Relations among stability parameters in the surface layer. *Bound. Layer Meteorol.* **3**, 47–58.

Hanna S.R. and Britter R.E. (2002). *Wind Flow and Vapor Cloud Dispersion at Industrial and Urban Sites*. AIChE–CPPS, New York.

Herndon J.M. (2010). Impact of recent discoveries on petroleum and natural gas exploration: Emphasis on India. *Curr. Sci.* **98**, 772–779.

Holtslag A.A.M. and van Ulden A.P. (1983). A simple scheme for daytime estimates of the surface fluxes from routine weather data. *J. Clim. Appl. Meteorol.* **22**, 517–529.

Irwin J.S. (1979). *Scheme for Estimating Dispersion Parameters as a Function of Release Height.* Report EPA-600/4-79-062, US-EPA, Research Triangle Park, NC.
Kasten F. and Czeplak G. (1980). Solar and terrestrial radiation dependent on the amount and type of cloud. *Solar Energy* **24**, 177–189.
Lenschow D.H., Li X.S., Zhu C.J., and Stankov B.B. (1988) The stably stratified boundary layer over the Great Plains. I. Mean and turbulence structure. *Bound. Layer Meteorol.* **42**, 95–121.
Luhar A. and Britter R.E. (1989). A random walk model for dispersion in inhomogeneous turbulence in a convective boundary layer. *Atmos. Environ.* **23**, 1911–1924.
Luhar A., Hibberd M., and Hurley P. (1996). Comparison of closure schemes used to specify the velocity PDF in Lagrangian stochastic dispersion models for convective conditions. *Atmos. Environ.* **30**, 1407–1418.
Monin A.S. (1959). On the similarity of turbulence in the presence of a mean vertical temperature gradient. *J. Geophys. Res.* **64**, 2196–2170.
Nieuwstadt F.T.M. (1984). The turbulent structure of the stable, nocturnal boundary layer. *J. Atmos. Chem.* **41**, 2202–2216.
Obukhov A.M. (1971). Turbulence in an atmosphere with a non-uniform temperature. *Bound. Layer Meteorol.* **2**, 7–29.
Oke T.R. (1982). The energetic basis of the urban heat island. *Quart. J. Roy. Meteorol. Soc.* **108**, 1–24.
Panofsky H.A., Tennekes H., Lenschow D.H., and Wyngaard J.C. (1977). The characteristics of turbulent velocity components in the surface layer under convective conditions. *Bound. Layer Meteorol.* **11**, 355–361.
Park S.J., Park S.U., Ho C.H., and Mahrt L. (2009). Flux-gradient relationship of water vapor in the surface layer obtained from CASES-99 experiment. *J. Geophys. Res.* **114**, article D08115.
Peña A. and Gryning S.E. (2008). Charnock's roughness length model and non-dimensional wind profiles over the sea. *Bound. Layer Meteorol.* **128**, 191–203.
Prantl L. (1925). Bericht uber Unterzuchungen zur ausgebildeten Turbulenz. *Z. Angew. Math. Mech.* **5**, 136–139.
Scire J.S., Robe F.R., Fernau M.E., and Yamartino R.J. (2000). *A User's Guide for the CALMET Meteorological Model.* Earth Tech, Concord, MA.
Seinfeld J.H. and Pandis S.N. (2006). *Atmospheric Chemistry and Physics*, 2nd ed. Wiley, Hoboken, NJ.
Sorbjan Z. (1989). *Structure of the Atmospheric Boundary Layer.* Prentice Hall, Englewood Cliffs, NJ.
Stull R.B. (1988). *An Introduction to Boundary Layer Meteorology.* Kluwer Academic, Dordrecht, The Netherlands.
Turner D.B. and Schulze R.H. (2007). *Practical Guide to Atmospheric Dispersion Modeling.* Trinity Consultants, Inc., Dallas, TX, and Air and Waste Management Association, Pittsburgh.
van Ulden A.P. and Holtslag A.A.M. (1985). Estimation of atmospheric boundary layer parameters for diffusion applications. *J. Climate Appl. Meteorol.* **24**, 1196–1207.
Venkatram A. (1980). Estimating the Monin–Obukhov length in the stable boundary layer for dispersion calculations. *Bound. Layer Meteorol.* **19**, 481–485.
Weil J.C. (1990). A diagnosis of the asymmetry in top-down and bottom-up diffusion using a Lagrangian stochastic model. *J Atmos. Sci.* **47**, 501–515.
Weil J.C., Corio L.A., and Brower R.B. (1997). A PDF dispersion model for buoyant plumes in the convective boundary layer. *J. Appl. Meteorol.* **36**, 982–1003.
Wilson J.D., Thurtell G.W., and Kidd G.E. (1981). Numerical simulation of particle trajectories in inhomogeneous turbulence, III: Comparison of predictions with experimental data for the atmospheric surface layer. *Bound. Layer Meteorol.* **21**, 443–463.
Wilson J.D., Legg B.J., and Thomson D.J. (1983). Calculation of particle trajectories in the presence of a gradient in turbulent-velocity variance. *Bound. Layer Meteorol.* **27**, 163–169.
Wyngaard J.C., Coté O.R., and Izumi Y. (1971). Local free convection, similarity, and the budgets of shear stress and heat flux. *J. Atmos. Sci.* **28**, 1171–1182.
Zhang Y., Ma J., and Cao Z. (2008). The von Kármán constant retrieved from CASES-97 dataset using a variational method. *Atmos. Chem. Phys.* **8**, 7045–7053.
Zilitinkevich S.S. (1972). On the determination of the height of the Ekman boundary layer. *Bound. Layer Meteorol.* **3**, 141–145.

CHAPTER 6

# GAUSSIAN DISPERSION MODELING: AN IN-DEPTH STUDY

## 6.1 INTRODUCTION

Gaussian dispersion models are the most commonly used models in regulatory air dispersion modeling. The basics of Gaussian dispersion models were presented in the primer (Chapter 2). The primer provided only limited theoretical background, and many of the details of contemporary Gaussian dispersion modeling were left out. In this chapter, Gaussian dispersion modeling will be discussed in more detail. While this chapter will start from first principles, the pace of this chapter will be substantially higher than the primer, so readers unfamiliar with the subject should not attempt to read this chapter without reading Chapter 2 first.

It was clear from the primer that the most basic Gaussian dispersion models have a number of weaknesses that need to be addressed. An example is the use of stability classes (A–F) and terrain types (rural–urban) that are too crude for accurate predictions. The main objective of this chapter is to develop more reliable calculation schemes, building on our expanded knowledge of air dispersion meteorology.

This chapter introduces some important concepts in meteorology such as autocorrelation of wind speed, the integral time scale, Taylor's hypothesis, and the Lagrangian frame of reference. These concepts are important to readers interested in advanced air dispersion models, even if they are not interested in Gaussian models.

There are two types of Gaussian dispersion models: plume models and puff models. The Primer only covered plume models. This chapter will cover both plume models and puff models.

In this chapter, much attention will be paid to parameterization schemes (i.e., ways to calculate $\sigma_y$ and $\sigma_z$). The next chapter will cover issues that are not exclusive to Gaussian dispersion models such as plume rise and deposition.

In particular, this chapter will discuss the following topics:

- An introduction to plume models
- An overview of parameterizations based on stability classes

*Air Dispersion Modeling: Foundations and Applications*, First Edition. Alex De Visscher.

- General principles of continuous parameterization schemes
- The link between turbulence and dispersion: autocorrelation and integral time scales
- Practical schemes of continuous parameterization
- Gaussian puff models

After completing this chapter you will be able to calculate dispersion parameters from meteorological input data without relying on stability classes.

## 6.2 GAUSSIAN PLUME MODELS

Within the context of the **Gaussian plume model**, it is assumed that a plume released in the atmosphere is diluted by the following mechanisms:

- In the $x$ direction (downwind) the plume is diluted by the wind.
- In the $y$ direction (crosswind) the plume is diluted by random motions of air parcels (eddies). The spread of the plume in the $y$ direction is unbounded.
- In the $z$ direction (vertical) the plume is diluted by random motions of air parcels (eddies). The spread of the plume is bounded by the ground and possibly by an elevated temperature inversion.

While the dilution in the $x$ direction is assumed to be the result of a nonrandom process (wind flow), the dilution in the $y$ and $z$ directions is the result of a random process, and the result will be a randomly fluctuating concentration. All we can hope to calculate with a Gaussian plume model is the Reynolds average concentration of the pollutant, $\overline{c}$. The calculation of instantaneous concentrations is not the objective of a Gaussian plume model.

Considering the three dilution mechanisms outlined above, the Gaussian plume equation can be subdivided into three parts, each representing a dilution mechanism:

$$\overline{c} = C_x \varphi_y \varphi_z \tag{6.1}$$

As a plume is emitted, it is diluted in the *downwind* direction by mixing with ambient air. The pollutant emitted in a time increment $dt$ ($s^{-1}$) is $Q\,dt$, where $Q$ (mg $s^{-1}$) is the mass emission rate. This pollution is spread across a distance $\overline{u}\,dt$, where $\overline{u}$ is the average wind speed (m $s^{-1}$). It follows that the pollution mass per meter of plume length, $C_x$, is given by

$$C_x = \frac{Q}{\overline{u}} \tag{6.2}$$

As a plume moves downwind, it is diluted in the *crosswind* direction by turbulent eddies. If the dilution is the result of a large number of motions each having a negligible effect on the plume concentration profile, then the crosswind concentration profile follows a Gaussian probability density function:

$$\varphi_y = \frac{1}{\sqrt{2\pi}\sigma_y} \exp\left(-\frac{1}{2}\frac{y^2}{\sigma_y^2}\right) \tag{6.3}$$

where $y$ (m) is the lateral (crosswind) distance from the plume axis (positive is left when looking downwind), and $\sigma_y$ (m) is the spread parameter in the horizontal (lateral) direction.

The plume is also diluted in the *vertical* direction. The dilution mechanism is the same, but the extent of the dilution depends on the circumstances. If the dilution is unbounded, the vertical concentration profile also follows a Gaussian probability density function:

$$\varphi_z = \frac{1}{\sqrt{2\pi}\sigma_z} \exp\left[-\frac{1}{2}\frac{(z-h)^2}{\sigma_z^2}\right] \tag{6.4}$$

where $z$ (m) is the height above the ground, $\sigma_z$ (m) is the spread parameter in the vertical direction, and $h$ (m) is the effective source height, which is given by

$$h = h_s + \Delta h \tag{6.5}$$

where $h_s$ is the stack height, and $\Delta h$ is the plume rise.

In practice, the plume dispersion is always bounded by the surface (i.e., the ground or a water surface). If there is no plume deposition to the ground, the plume is *reflected*, leading to the following probability density function:

$$\varphi_z = \frac{1}{\sqrt{2\pi}\sigma_z} \left\{ \exp\left[-\frac{1}{2}\frac{(z-h)^2}{\sigma_z^2}\right] + \exp\left[-\frac{1}{2}\frac{(z+h)^2}{\sigma_z^2}\right] \right\} \tag{6.6}$$

Combining eqs. (6.1)–(6.3) and (6.6), the classical Gaussian plume equation is obtained:

$$\bar{c} = \frac{Q}{2\pi\bar{u}\sigma_y\sigma_z} \exp\left(-\frac{1}{2}\frac{y^2}{\sigma_y^2}\right) \left\{ \exp\left[-\frac{1}{2}\frac{(z-h)^2}{\sigma_z^2}\right] + \exp\left[-\frac{1}{2}\frac{(z+h)^2}{\sigma_z^2}\right] \right\} \tag{6.7}$$

As discussed in the primer (Chapter 2), the Gaussian plume model is only valid under certain simplifying conditions:

- The source of pollution is constant.
- Wind direction and wind speed are constant in space and time.
- Atmospheric turbulence is constant in space and time.

There are other simplifying assumptions that were not mentioned before:

- *Sufficiently strong wind speed*, or sufficiently small turbulence, so the dispersion in the $x$ direction is negligible in comparison with the advection. Gaussian plume models become increasingly inaccurate when wind speeds are below 1 m $s^{-1}$. This limitation can be overcome by using a puff model instead of a plume model. Puff models are discussed in Section 6.10.

- *Conservation of mass* in the plume. This assumption is not met in the case of chemical reaction or deposition. The case of deposition will be discussed in Chapter 7. The case of chemical reaction is similar to deposition in the case of first-order kinetics. It is best treated in an Eulerian model if the chemistry is very complex. This situation is discussed in Chapter 11.

Some of the above assumptions, such as constant turbulence, do not need to be met for a Gaussian dispersion model to produce accurate predictions because the estimated values of $\sigma_y$ and $\sigma_z$ are based on model fits and validations to measured concentrations. Hence, the parameterization of $\sigma_y$ and $\sigma_z$ will compensate for any inaccuracy arising from the model itself.

When the plume is confined within a mixing layer, then an infinite number of plume reflections can occur. This is illustrated in Figure 6.1, where one real source and five imaginary sources are shown. Accounting for all possible plumes leads to the following probability density function:

$$\varphi_z = \sum_{j=-\infty}^{+\infty}\left(\exp\left\{-\frac{1}{2}\frac{\left[z-(h+2jh_{\mathrm{mix}})\right]^2}{\sigma_z^2}\right\}+\exp\left\{-\frac{1}{2}\frac{\left[z+(h+2jh_{\mathrm{mix}})\right]^2}{\sigma_z^2}\right\}\right) \tag{6.8}$$

where $h_{\mathrm{mix}}$ is the height of the mixing layer. Because of the symmetry in the summation, the innermost brackets in each of the exponentials can be omitted. In practice, eq. (6.8) can be limited to $j = -1, 0, +1$ to within 2% for values of $\sigma_z$ up to $h_{\mathrm{mix}}$. Adding $j = 2$ and $j = -2$ improves the accuracy to within 0.13% for values of $\sigma_z$ up to $1.2h_{\mathrm{mix}}$.

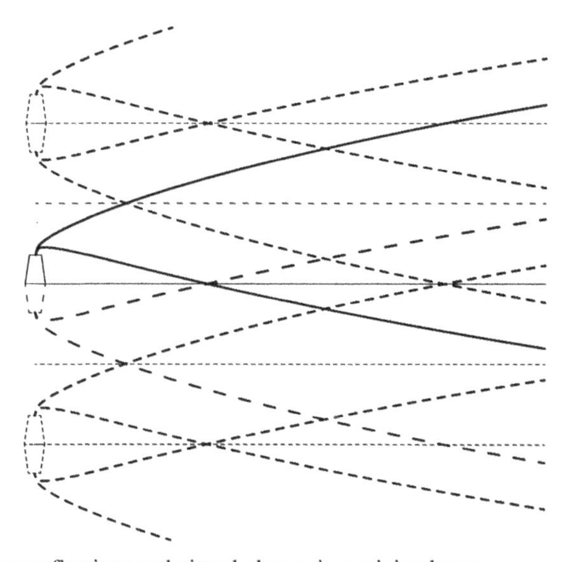

Figure 6.1 Plume reflections and virtual plumes in a mixing layer.

After a few reflections, the plume dispersion is sufficient to fill the entire mixing layer homogeneously, and the probablity density function in the $z$ direction can be simplified to a top hat function:

$$\varphi_z = \frac{1}{h_{\text{mix}}} \tag{6.9}$$

Equation (6.9) applies from $z = 0$ to $z = h_{\text{mix}}$. Outside this range, $\varphi_z = 0$.

Once $\sigma_z$ exceeds $h_{\text{mix}}$, eq. (6.9) is accurate to within 1.5%. When $\sigma_z$ exceeds $1.2h_{\text{mix}}$, eq. (6.9) is accurate to within 0.17%.

## 6.3 PARAMETERIZATIONS BASED ON STABILITY CLASSES

Although better characterizations of atmospheric stability based on the theories outlined in Chapter 5 exist and will be explained in later sections, the use of stability classes is still widespread in air dispersion modeling, especially for screening purposes. For that reason, air dispersion modelers should still be familiar with them. The most commonly used classification of atmospheric stability was developed by Pasquill (1961) and Gifford (1961). They defined six classes, named A through F, where A means very unstable, B means moderately unstable, C means slightly unstable, D means neutral, E means slightly stable, and F means stable. We repeat the criteria for each class in Table 6.1.

Originally, the dispersion parameters $\sigma_y$ and $\sigma_z$ were read from graphs (e.g., Turner, 1970). Many empirical equations have been put forward to describe

**TABLE 6.1 Criteria for Pasquill–Gifford Stability Classes**

| | Day | | | Night | |
|---|---|---|---|---|---|
| | Incoming solar radiation | | | Cloudiness | |
| $u$ (m s$^{-1}$)[a] | Strong[b] | Moderate[c] | Slight | Cloudy (≥4/8)[d] | Clear (≤3/8)[e] |
| <2 | A | A–B[f] | B | E | F |
| 2–3 | A–B | B | C | E | F |
| 3–5 | B | B–C | C | D | E |
| 5–6 | C | C–D | D | D | D |
| >6 | C | D | D | D | D |

Source: After Turner and Schulze (2007).

[a] Measured at 10-m height.

[b] Clear summer day with sun higher than 60° above the horizon.

[c] Summer day with a few broken clouds or a clear day with the sun 35–60° above the horizon.

[d] Fall afternoon or cloudy summer day with the sun 15–35° above the horizon.

[e] Fractional cloud cover.

[f] Take average dispersion values of two classes.

Note: *Always* use class D for overcast conditions.

**TABLE 6.2 Coefficients for the Briggs (1973) Equation [eq. (6.10)] for the Calculation of $\sigma_y$ and $\sigma_z$[a]**

Rural

| | $\sigma_y$ | | | $\sigma_z$ | | |
|---|---|---|---|---|---|---|
| Stability class | *a* | *b* | *c* | *d* | *e* | *f* |
| A | 0.22 | 0.0001 | 0.5 | 0.2 | 0 | — |
| B | 0.16 | 0.0001 | 0.5 | 0.12 | 0 | — |
| C | 0.11 | 0.0001 | 0.5 | 0.08 | 0.0002 | 0.5 |
| D | 0.08 | 0.0001 | 0.5 | 0.06 | 0.0015 | 0.5 |
| E | 0.06 | 0.0001 | 0.5 | 0.03 | 0.0003 | 1 |
| F | 0.04 | 0.0001 | 0.5 | 0.016 | 0.0003 | 1 |

Urban

| | $\sigma_y$ | | | $\sigma_z$ | | |
|---|---|---|---|---|---|---|
| Stability class | *a* | *b* | *c* | *d* | *e* | *f* |
| A–B | 0.32 | 0.0004 | 0.5 | 0.24 | 0.0001 | –0.5 |
| C | 0.22 | 0.0004 | 0.5 | 0.2 | 0 | — |
| D | 0.16 | 0.0004 | 0.5 | 0.14 | 0.0003 | 0.5 |
| E–F | 0.11 | 0.0004 | 0.5 | 0.08 | 0.0015 | 0.5 |

[a] All length scales are in meters.

these graphs. Of these, the parameterization of Briggs (1973) is recommended for reasons that will become clear in Section 6.6. It is based on the following equations:

$$\sigma_y = \frac{ax}{(1+bx)^c} \tag{6.10}$$

$$\sigma_z = \frac{dx}{(1+ex)^f} \tag{6.11}$$

where $a, b, \ldots, f$ are empirical constants. All length scales are in meters. Based on micrometeorological principles that will be outlined in Section 6.6, we expect $c$ and $f$ to be 0.5, but this is not always the case for $f$ when the atmosphere is extremely stable or extremely unstable. The empirical constants proposed by Briggs (1973) for rural and urban terrain are shown in Table 6.2. The data on urban terrain were based on the experimental work of McElroy and Pooler (1968). This parameterization is the default setting in CALPUFF at the time of writing. However, it is likely that the default setting will change to a parameterization based on similarity theory in future releases of CALPUFF.

The measurement data that formed the basis of the parameterization of Briggs (1973), and most other parameterizations, are based on 10-min average measurements (Beychok, 2005). However, EPA treats them as hourly averages. The significance of this will be discussed in Section 6.5.

This section will review other equations that have been proposed to calculate $\sigma_y$ and $\sigma_z$. The parameters are given in Section A6.1 (Appendix A). The earliest parameterizations of $\sigma_y$ and $\sigma_z$ were based on a simple power law (Klug, 1969; Turner, 1970):

$$\sigma_y = ax^b \tag{6.12}$$

$$\sigma_z = cx^d \tag{6.13}$$

Parameterizations for a 60-min averaging time based on eqs. (6.12) and (6.13) were presented in ASME (1973).

Martin (1976) used eq. (6.12) for $\sigma_y$ but extended eq. (6.13) for $\sigma_z$ to

$$\sigma_z = cx^d + f \tag{6.14}$$

Equations such as eqs. (6.12)–(6.14) are not very accurate across large distances. This is because in reality the coefficients $b$ and $d$ are not constants. This will become clear in Section 6.6. Somewhat more accurate is the following equation:

$$\sigma_y = \exp[a + b\ln x + c(\ln x)^2] \tag{6.15}$$

$$\sigma_z = \exp[d + e\ln x + f(\ln x)^2] \tag{6.16}$$

This equation was used by Turner (1970).

The Industrial Source Complex (ISC) model uses the following equation for $\sigma_y$ in rural terrain (EPA, 1995):

$$\sigma_y = 465.11628x\tan(\theta) \tag{6.17}$$

where

$$\theta = 0.017453293[c - d\ln(x)] \tag{6.18}$$

The values of $c$ and $d$ are given in Section A6.1 (Appendix A). For $\sigma_z$, ISC uses eq. (6.13) but with coefficients that depend on the distance from the source. The coefficients are given in Section A6.1 (Appendix A). This parameterization is available in CALPUFF as well. For urban terrain the Briggs parameterization (Table 6.2) is used in ISC.

CALPUFF also contains the MESOPUFF II parameterization (Scire et al., 1984a,b) as an option. This parameterization is based on a single power law and is not recommended.

To get an impression of the variance between parameterizations, the Briggs (1973) parameterization is compared with the Martin (1976) parameterization for rural terrain. The result is shown in Figure 6.2. In most cases the agreement is reasonable up to a 10-km distance, except for vertical dispersion in stability class A, where the disagreement becomes pronounced within the first kilometer.

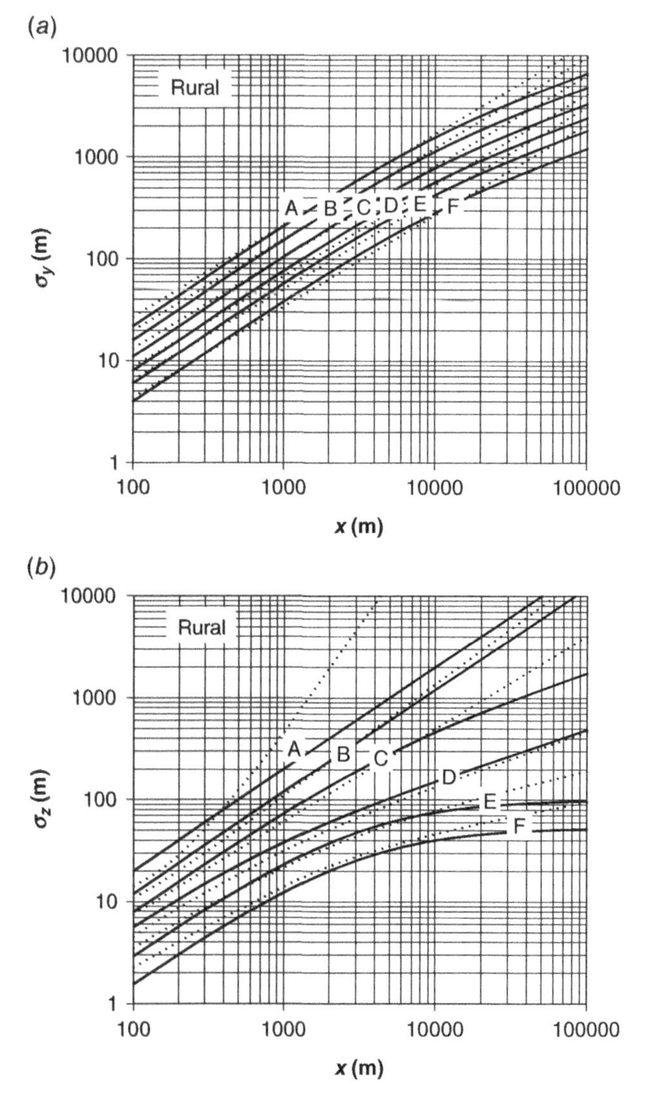


Figure 6.2 (*a*) Lateral and (*b*) vertical dispersion parameters calculated with the Briggs (1973) equation (solid lines) and with the Martin (1976) parameterization (dotted lines).

## 6.4 GAUSSIAN PLUME DISPERSION SHORT CUT

From eq. (6.7) it is clear that when $y = 0$ and $z = 0$, the variable $\bar{c}\bar{u}/Q$ is only dependent on $\sigma_y$, $\sigma_z$ (hence on stability class and $x$) and $h$. When we are only interested in the maximum concentration, $x$ becomes a dependent variable, and the variables $\bar{c}\bar{u}/Q$ and $x$ at the maximum concentration are uniquely defined by the stability class and the source height. Turner (1994) plotted the maximum $\bar{c}\bar{u}/Q$ value versus distance of maximum concentration in a variety of cases. The relationship is shown in Figure 6.3 for rural terrain and Figure 6.4 for urban terrain. These figures provide a

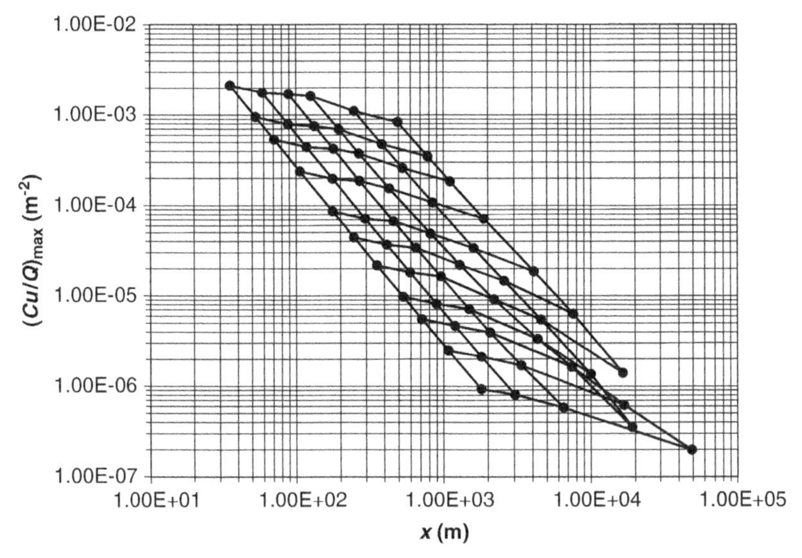


Figure 6.3 $\overline{c}\,\overline{u}/Q$ and $x$ at maximum concentration for different stability classes and effective source heights, rural terrain [after Turner (1994) but based on the Briggs (1973) parameterization]. Left to right data refer to stability class A, B, C, D, E, F. Top to bottom data refer to effective source height 10, 15, 20, 30, 50, 70, 100, 150, 200, 300, 500 m.

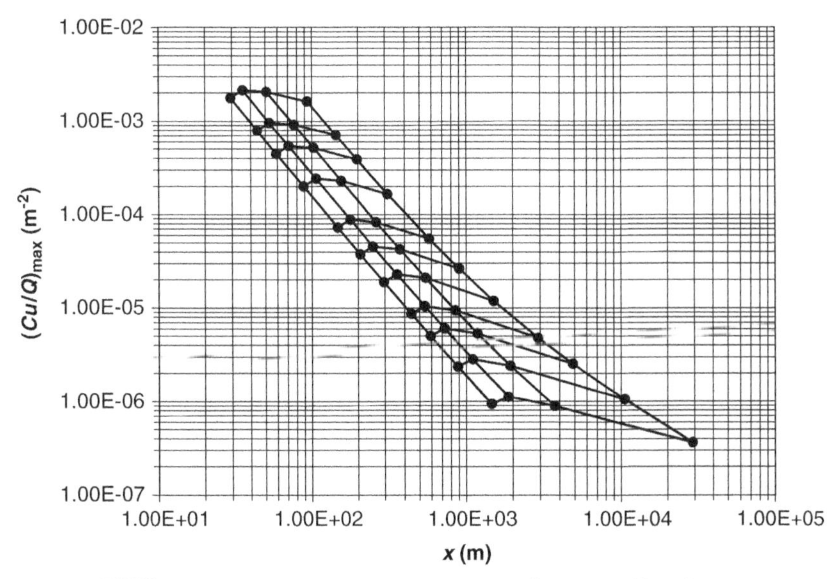


Figure 6.4 $\overline{c}\,\overline{u}/Q$ and $x$ at maximum concentration for different stability classes and effective source heights, urban terrain [after Turner (1994) but based on the Briggs (1973) parameterization]. Left to right data refer to stability class A-B, C, D, E-F. Top to bottom data refer to effective source height 10, 15, 20, 30, 50, 70, 100, 150, 200, 300, 500 m.

short cut estimation of worst-case scenarios in air dispersion based on Gaussian models. The data is based on the Briggs (1973) parameterizations. The result should be considered a rough estimate only.

## 6.5 PLUME DISPERSION MODIFIERS

After calculating the dispersion parameters $\sigma_y$ and $\sigma_z$, it is often necessary to apply corrections to suit a particular situation. Examples are corrections for averaging time, for surface roughness, for plume rise induced dispersion, and for wind direction shear.

The measurement data underlying the dispersion parameters of Briggs (1973) and most others are 10-min average values. **Dispersion parameters** increase with increasing averaging time because wind direction fluctuations have more time to reach a wider range. For the $y$ direction, a simple equation has been proposed to adjust dispersion parameters for a different averaging time:

$$\sigma_{y,2} = \sigma_{y,1}\left(\frac{t_2}{t_1}\right)^p \tag{6.19}$$

with $\sigma_{y,1}$ and $\sigma_{y,2}$ dispersion parameters obtained with averaging times of $t_1$ and $t_2$, respectively, and $p$ an empirical parameter of typically about 0.2, although higher values have been suggested as well (Beychok, 2005). A similar correction has not been proposed for the vertical dispersion parameter $\sigma_z$ because the time scale of vertical fluctuations does not extend past 10 min.

Although the averaging time of the measurements underlying most plume dispersion parameters is 10 min, EPA treats them as if they were hourly averages (Beychok, 2005). Presumably, this is to keep air dispersion models conservative. On the other hand, the way Briggs (1973) extrapolated measured dispersion parameters to longer distances suggests an averaging time far exceeding 10 min, so that assuming a 10-min averaging time at long distances will almost certainly overestimate the dispersion parameters. It is recommended to follow EPA in treating dispersion parameters as hourly averages in case the air dispersion modeling is conducted for regulatory purposes. However, when the purpose is scientific, it is recommended to use the Briggs dispersion parameters as lower limits and to use eq. (6.19) as a correction to obtain upper limits to the dispersion parameters.

The potential error introduced in $\sigma_y$ by not making this correction is a factor of $6^{0.2} = 1.43$. The actual hour-averaged plume widths will be 1.43 times the calculated values, and the actual plume centerline concentrations will be $1/1.43 = 0.70$ times the calculated plume centerline concentrations. This is one of the reasons models tend to overestimate pollutant concentrations.

It is not recommended to use eq. (6.19) past a one-hour averaging time, Instead, it is better to conduce a separate dispersion calculation for each hour and take the average of the results (Scire et al., 2000).

We can also adjust lateral plume dispersion for *roughness height*. The relationship is as follows:

$$\sigma_{y,2} = \sigma_{y,1}\left(\frac{z_{0,2}}{z_{0,1}}\right)^{q} \tag{6.20}$$

where $q = 0.2$, $\sigma_{y,2}$ is the lateral dispersion parameter at roughness height $z_{0,2}$, and $\sigma_{y,1}$ is the lateral dispersion at roughness height $z_{0,1}$. For the common parameterizations in rural terrain, a $z_{0,1}$ value of 0.03 m can be used (Scire et al., 2000). Note that this equation alone cannot explain the difference between dispersion in rural and urban terrain.

For tall stack emissions ($h > 100$ m), no surface roughness correction should be used (Scire et al., 2000). For short stack emissions, Scire et al. (2000) report a correction method of Smith (1972). However, it is only applicable to $\sigma_z$ formuations based on eq. (6.13) and will not be included here.

Plume dispersion can be *enhanced* by substantial **plume rise**. This is because plume rise generates turbulence of its own. To account for this effect, the following equations have been suggested (Pasquill, 1979):

$$\sigma_y = \sqrt{\sigma_{y,0}^2 + 0.1(\Delta h)^2} \tag{6.21}$$

$$\sigma_z = \sqrt{\sigma_{z,0}^2 + 0.1(\Delta h)^2} \tag{6.22}$$

where $\sigma_{y,0}$ and $\sigma_{z,0}$ are the plume dispersion parameters in the absence of plume rise, and $\Delta h$ is the plume rise. AERMOD uses the coefficient 0.08 instead of 0.1 in these equations. ISC and CALPUFF use 1/12.25.

**Wind direction shear**, the difference between wind directions at different heights, can also enhance lateral plume dispersion. When there is wind direction shear, eq. (6.21) can be extended to

$$\sigma_y = \sqrt{\sigma_{y,0}^2 + 0.1(\Delta h)^2 + (0.174x\,\Delta\overline{\theta})^2} \tag{6.23}$$

where $\Delta\overline{\theta}$ (in radians) is the difference between wind directions at the top and the bottom of the plume.

CALPUFF has a puff splitting algorithm that is sometimes invoked to account for plume shear. However, puff splitting in CALPUFF is limited to puffs that are well-mixed vertically.

**Example 6.1.** A coal-fired power plant in rural Pennsylvania emits 100 g s$^{-1}$ $SO_2$ from a stack with height of 75 m. Plume rise is 15 m. Wind speed at a 10-m height is 7 m s$^{-1}$. The weather is overcast. Calculate the hourly average $SO_2$ concentration at ground level 1.5 km downwind from the source at the plume centerline (i.e., directly downwind). The roughness length of the terrain is 10 cm. Incorporate all relevant corrections and modifiers.

***Solution.*** This example is the same as Example 2.1 but with a wind speed correction (Example 5.5), an averaging time correction, a roughness length correction, and a plume rise enhancement correction.

In overcast conditions the stability class is D. Use eqs. (6.10) and (6.11) to calculate $\sigma_y$ and $\sigma_z$ with the coefficients from Table 6.2. The result is

$$\sigma_y = 0.08 \cdot 1500 \cdot (1 + 0.0001 \cdot 1500)^{-0.5} = 111.9\,\text{m}$$

$$\sigma_z = 0.06 \cdot 1500 \cdot (1 + 0.0015 \cdot 1500)^{-0.5} = 49.9\,\text{m}$$

Use eq. (6.19) to correct $\sigma_y$ for averaging time from 10 to 60 min:

$$\sigma_y = 111.9 \times (60/10)^{0.2} = 160.1\,\text{m}$$

Use eq. (6.20) to correct $\sigma_y$ for surface roughness from 3 to 10 cm:

$$\sigma_y = 160.1 \times (10/3)^{0.2} = 203.7\,\text{m}$$

Use eqs. (6.21) and (6.22) to include plume-rise-enhanced dispersion:

$$\sigma_y = (203.7^2 + 0.1 \cdot 15^2) = 203.8\,\text{m}$$
$$\sigma_z = (49.9^2 + 0.1 \cdot 15^2) = 50.1\,\text{m}$$

Calculate the wind speed at a 90-m height. First, we calculate $u_*$ with eq. (5.65):

$$u_* = 0.4 \cdot 7\,\text{m}\,\text{s}^{-1} / \ln(10\,\text{m}/0.1\,\text{m}) = 0.608\,\text{m}\,\text{s}^{-1}$$

The friction velocity is substituted in eq. (5.64):

$$u_{90} = (0.608/0.4) \times \ln(90/0.1) = 10.34\,\text{m}\,\text{s}^{-1}$$

Enter all data in eq. (6.7), assuming $y = 0$ and $z = 0$:

$$c = \frac{100}{2\pi \cdot 10.34 \cdot 203.8 \cdot 51.1} \exp\left(-\frac{1}{2}\frac{0^2}{\sigma_y^2}\right)\left\{\exp\left[-\frac{1}{2}\frac{(0-90)^2}{51.1^2}\right] + \exp\left[-\frac{1}{2}\frac{(0+90)^2}{51.1^2}\right]\right\}$$
$$= 60.2 \times 10^{-6}\ \text{g}\,\text{m}^{-3} = \mathbf{60.2\,\mu g\,m^{-3}}$$

The concentration is 60.2 μg $\text{m}^{-3}$, markedly less than the value found in Example 2.1 without corrections (160.3 μg $\text{m}^{-3}$) or with only a correction of wind speed as found in Example 2.2 (110.3 μg $\text{m}^{-3}$). Older models, which do not include these corrections, and regulatory models, which use 10-min averages as if they were hourly averages, tend to overestimate actual pollutant concentrations at the plume centerline.

The calculation is included on the enclosed CD, in the Excel file "Example 6.1. Dispersion calculation stability classes.xlsx."

## 6.6 CONTINUOUS PARAMETERIZATION FOR GAUSSIAN DISPERSION MODELS

### 6.6.1 Introduction: From Turbulence to Dispersion

Plume dispersion in the lateral and vertical direction are caused almost exclusively by the wind speed fluctuations $v'$ and $w'$. Hence, it can be expected that there will be a relationship between $\sigma_v$ and $\sigma_y$ and between $\sigma_w$ and $\sigma_z$. Parameterizations for $\sigma_v$ and $\sigma_w$ were discussed in Chapter 5. The purpose of the next few sections is to establish the link between the turbulent velocities and the dispersion parameters.

Consider, for instance, an ensemble of pollutant particles that are emitted at $x = 0$ and $y = 0$, and we define $t = 0$ as the time of emission of each particle. Further, we assume that the average wind direction is along the $x$ axis, so $\overline{y} = 0$ and $y = y'$. If we neglect wind speed fluctuation in the $x$ direction (as we do in plume dispersion modeling), we find that all particles are at $x = \overline{u}t$, and we assume small $t$ (e.g., 1 s). Because the time is very short, we assume that $v'$ has remained constant during that time. However, in an ensemble of particles, $v'$ is different for each particle, and each particle has a different lateral position $y = v't$. We can calculate the standard deviation of the $y$ position of the particles:

$$\sigma_y = \sqrt{\overline{y'^2}} = \sqrt{\overline{y^2}} = \sqrt{\overline{(v't)^2}} = \sqrt{\overline{v'^2}}\,t = \sigma_v t \tag{6.24}$$

Note that eq. (6.24) is only valid for times sufficiently short to exclude any variations of $v$. Based on the stochastic nature of wind speed fluctuations, it can be proven that eq. (6.24) is the **theoretical maximum** value of $\sigma_y$. With increasing $t$, $\sigma_y$ will deviate increasingly from $\sigma_v t$. We conclude that the theoretical maximum values of $\sigma_y$ and $\sigma_z$ are given by

$$(\sigma_y)_{\max} = \sigma_v t = \sigma_v \frac{x}{\overline{u}} \tag{6.25}$$

$$(\sigma_z)_{\max} = \sigma_w t = \sigma_w \frac{x}{\overline{u}} \tag{6.26}$$

The *actual* values of $\sigma_y$ and $\sigma_z$ can be expressed as

$$\sigma_y = \sigma_v t f_y = \sigma_v \frac{x}{\overline{u}} f_y = i_v x f_y \tag{6.27}$$

$$\sigma_z = \sigma_w t f_z = \sigma_w \frac{x}{\overline{u}} f_z = i_w x f_z \tag{6.28}$$

where $f_y$ and $f_z$ are factors that are functions of $x$ (or $t$). They have a value of 1 at small values of $x$ and are less than 1 at large values of $x$. The $i_v$ and $i_w$ factors are turbulent intensities, as defined in Chapter 5. It is assumed here that $u$ represents the entire wind speed (i.e., zero average lateral and vertical wind speed).

Based on eqs. (6.27) and (6.28), it follows that all we need to derive $\sigma_y$ from $\sigma_v$ and $\sigma_z$ from $\sigma_w$ is an appropriate function for $f_y$ and for $f_z$. To achieve that, it is useful to revisit the concept of turbulence.

### 6.6.2 Autocorrelation of Wind Speed

From the preceding section it is clear that we need to get an idea of the likelihood that a wind speed going in a certain direction will continue to go in that direction. Put differently: We need a measure of the duration $v'$ keeps its value. This is captured by the concept of the **autocorrelation function**. We will concentrate on the autocorrelation function in the lateral direction, $R'_{vv}$, but the concept is the same for the other two directions:

$$R'_{vv}(\tau) = \overline{v'(t)v'(t+\tau)} \tag{6.29}$$

In words: The autocorrelation function is the **covariance** of a wind speed fluctuation at time $t$ and the same wind speed fluctuation at time $t + \tau$. By definition:

$$R'_{vv}(0) = \overline{v'^2} = \sigma_v^2 \tag{6.30}$$

The *normalized* autocorrelation function is the **autocorrelation coefficient** $R_{vv}$:

$$R_{vv}(\tau) = \frac{R'_{vv}(\tau)}{\sigma_v^2} = \frac{\overline{v'(t)v'(t+\tau)}}{\sigma_v^2} \tag{6.31}$$

Similar definitions apply for $R'_{uu}$, $R'_{ww}$, $R_{uu}$, and $R_{ww}$. The main properties of $R_{vv}$ (and $R_{uu}$ and $R_{ww}$) are

$$R_{vv}(0) = 1 \tag{6.32}$$

$$R_{vv}(\tau) < 1 \qquad \text{for all} \quad \tau > 0 \tag{6.33}$$

$$R_{vv}(\tau) \rightarrow 0 \qquad \text{for} \quad \tau \rightarrow \infty \tag{6.34}$$

The general behavior of $R_{vv}$ is shown in Figure 6.5. In this figure, $T_i$ is the **integral time scale**, which is defined as

$$T_i = \int_0^\infty R_{vv}(\tau)\, d\tau \tag{6.35}$$

In words: $T_i$ is the time at which the rectangle defined by the dotted lines $\tau = T_i$ and $R_{vv} = 1$ has an area equal to the area under the curve of $R_{vv}(\tau)$. $T_i$ is a *convenient measure of the average duration of autocorrelation*. $T_i$ has a different value in the $x$, $y$, and $z$ directions.

Coefficient $R_{vv}(\tau)$ has a couple of properties that are relevant for understanding energy dissipation but less so for understanding dispersion (Arya, 1999). They are included for completeness.

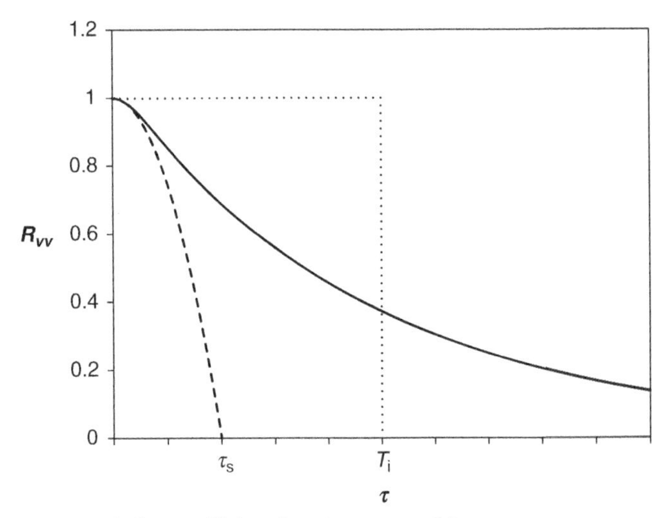


Figure 6.5 Autocorrelation coefficient ($\tau$ axis not to scale).

- $$\lim_{\tau \to 0} \frac{dR_{vv}}{d\tau} = 0 \tag{6.36}$$

- The parabola $R$ defined by

$$R(0) = 1 \qquad \lim_{\tau \to 0} \frac{dR}{d\tau} = 0 \qquad \lim_{\tau \to 0} \frac{d^2R}{d\tau^2} = \lim_{\tau \to 0} \frac{d^2R_{vv}}{d\tau^2}$$

intersects the $x$ axis at $\tau = \tau_s$, where $\tau_s$ is named the **Taylor microscale**: $\tau_s$ is on the order of 0.1 s and tells us something about the smallest turbulent eddies. Its value explains why sonic anemometers typically run at 10 Hz: It is on the order of the largest frequency found in fluctuations of the wind speed.

Time $T_i$ is on the order of 10–100 s and tells us something about the largest turbulent eddies. The largest turbulent eddies cause most of the dispersion, whereas the smallest turbulent eddies cause most of the friction. More about energy dissipation is given in Section 6.11.

### 6.6.3 Taylor's Hypothesis

In 1938, the British physicist Sir Geoffrey Taylor made the following hypothesis that has proven to be very useful in the study of turbulence: The changes of $u'$ at a fixed point over time are due to the passage of an unchanging pattern of turbulent motion over that point (Taylor, 1938).

This principle is illustrated in Figure 6.6. We observe

$$v'(x, t = 0) = v'(x + \bar{u}t, t) \tag{6.37}$$

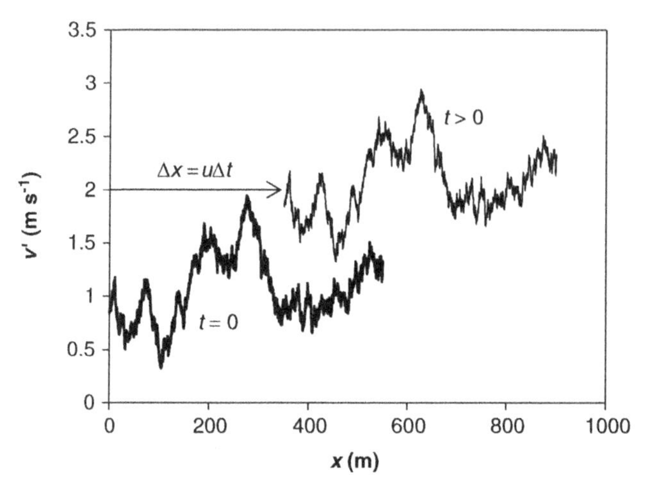


Figure 6.6 Taylor's hypothesis.

In eq. (6.37) we can switch between length scales and time scales by simply setting $x + \overline{u}t = 0$. Hence, we obtain

$$v'(x, t = 0) = v'\left(0, t = -\frac{x}{\overline{u}}\right) \tag{6.38}$$

In practice, this means that in any fluctuation caused by turbulence, $t$ can simply be replaced by $-x/\overline{u}$ to obtain a picture of the spatial structure of the fluctuation from knowledge of the temporal structure of the fluctuation. This leads to two definitions:

- The **spatial Taylor microscale**: $\lambda_l = \overline{u}\tau_s$
- The **integral length scale**: $L_l = \overline{u}T_i$

where $\lambda_l$ is a measure of the average size of the smallest eddies (0.1–1 m), whereas $L_l$ is a measure of the average size of the large eddies (100–1000 m). It can be expected that $L_l$ is less dependent on wind speed than $T_i$, which is useful to remember in dispersion studies.

## 6.6.4 Lagrangian Frame of Reference

So far our description of turbulence was based on a coordinate system that is fixed in space. This is called the **Eulerian** frame of reference. This frame of reference does not reveal the aspects of turbulence that are relevant for atmospheric dispersion. What really is needed is the turbulence a parcel of air undergoes as it moves in the atmosphere. A frame of reference that is fixed to an air parcel as it moves around is a **Lagrangian** frame of reference. Properties in a Lagrangian frame will be denoted with capital letters: $U$ for wind speed in the $X$ direction, for example.

As with the Eulerian view, we can decompose the velocity of the air parcel in a Lagrangian frame into two components (Reynolds decomposition). For instance, for the lateral wind speed $V$:

$$V = \bar{V} + V' \tag{6.39}$$

where $\bar{V}$ is the average velocity, and $V'$ is the velocity fluctuation. The **Lagrangian autocorrelation coefficient** is

$$R_{\mathrm{L}}(\tau) = \frac{R'_{\mathrm{L}}(\tau)}{\overline{V'^2}} = \frac{\overline{V'(t)V'(t+\tau)}}{\overline{V'^2}} \tag{6.40}$$

The **Lagrangian integral time** scale is

$$T_{\mathrm{i,L}} = \int_0^\infty R_{\mathrm{L}}(\tau)\, d\tau \tag{6.41}$$

As with $T_{\mathrm{i}}$, $T_{\mathrm{i,L}}$ is not the same in the three directions ($x$, $y$, and $z$). In this book, the direction is not included in the notation of $T_{\mathrm{i}}$ and $T_{\mathrm{i,L}}$ when discussing properties valid in all directions, or when the direction is obvious from the context. When the direction is relevant and not obvious from the context, it will be included as an additional subscript, for example, $T_{\mathrm{i,L},x}$.

If the flow is *stationary*, *homogeneous*, and *incompressible*, then the Eulerian and Lagrangian velocities have the same probability density functions and the same moments, that is, $\overline{U'^2} = \overline{u'^2}$. However, this is *not* true for the covariances and the autocorrelation properties. In particular, $T_{\mathrm{i,L}}$ is substantially larger than $T_{\mathrm{i}}$ (which justifies Taylor's hypothesis). Based on measurements and on some reasoning on eddie properties, the following relationship has been put forward to relate the two (Hanna, 1981; Arya, 1999):

$$\beta = \frac{T_{\mathrm{i,L}}}{T_{\mathrm{i}}} = \frac{\gamma}{i} \tag{6.42}$$

where $i = \sigma / \bar{u}$ the turbulence intensity in the appropriate direction, and $\gamma = 0.7$ (Hanna, 1981). Note that there is a fairly large uncertainty on the coefficient $\gamma$, which ranges from 0.4 to 0.8 (Degrazia and Anfossi, 1998). Typical values of $T_{\mathrm{i,L}}$ are 100–1000 s, which is about an order of magnitude larger than $T_{\mathrm{i}}$.

There are several formulations for $R_{\mathrm{L}}$. An adequate formulation for dispersion modeling purposes is the exponential function (Neumann, 1978; Tennekes, 1979):

$$R_{\mathrm{L}}(\tau) = \exp\left(-\frac{|\tau|}{T_{\mathrm{i,L}}}\right) \tag{6.43}$$

The absolute value $|\tau|$ is included to make sure that the equation provides the correct result for negative values of $\tau$ as well. However, we will not need this property.

A number of estimation methods have been developed for $T_{i,L}$. Under **convective** (unstable) conditions the following equation has been suggested for the entire boundary layer, in all directions (Hanna, 1981):

$$T_{i,L} = \frac{0.17 h_{mix}}{\sigma_{u,v,w}} \tag{6.44}$$

where $\sigma_{u,v,w}$ is $\sigma_u$, $\sigma_v$, or $\sigma_w$, depending of the direction. In the surface layer ($z \leq 0.1\ h_{mix}$) the following equation has been proposed in the $z$ direction:

$$T_{i,L} = 0.42 \frac{z\bar{u}}{u_* \sigma_w} \tag{6.45}$$

This equation replaces eq. (6.44) in this particular case. It should only be used if the result is less than the result of eq. (6.44).

Another equation that has been proposed for $T_{i,L}$ under *unstable conditions in the y direction* is

$$T_{i,L} = \frac{0.4 h_{mix}}{u_* \left[1 + 0.0013 \left(h_{mix} / -L\right)\right]^{1/3}} \tag{6.46}$$

Equation (6.46) leads to systematically higher $T_{i,L}$ estimates than eq. (6.44).

Under *stable conditions in the y direction* the following has been proposed (Hanna and Britter, 2002):

$$T_{i,L} = 1000\,\text{s} \tag{6.47}$$

Under *unstable conditions in the z direction* the suggested value is

$$T_{i,L} = 500\,\text{s} \tag{6.48}$$

Under *neutral and stable conditions in the z direction* the suggested value is

$$T_{i,L} = 100\,\text{s} \tag{6.49}$$

Further parameterizations require a better understanding of how $T_{i,L}$ affects dispersion and will be discussed at a later point (Section 6.6.7).

### 6.6.5 Practical Schemes for Continuous Parameterizations

As indicated before, continuous dispersion parameterization is based on the following equations:

$$\sigma_y = \sigma_v \frac{x}{\overline{u}} f_y \tag{6.50}$$

$$\sigma_z = \sigma_w \frac{x}{\overline{u}} f_z \tag{6.51}$$

Over short distances the wind speed fluctuations do not change very much, so we can assume that the dispersion is the maximum possible value: $f_y = f_z = 1$. This is the case for $t = x/\overline{u} << T_{i,L}$. Over long distances, when $t = x/\overline{u} >> T_{i,L}$, wind speed fluctuations change often between emission and receptor, and so do the paths of the pollution particles. Hence, we expect that $f_y < 1$ and $f_z < 1$. The air parcel can be assumed to follow a **random walk** in the lateral and vertical directions. A property of a random walk is that the *average* distance between the initial and the final locations is proportional to the square root of the number of steps taken (or, in our case, the square root of the time, or the square root of the distance in the $x$ direction). Therefore, we obtain

$$\sigma_{y,z} \propto \sqrt{x} \tag{6.52}$$

or

$$f_{y,z} \propto \frac{1}{\sqrt{x}} \tag{6.53}$$

A more detailed argument allows us to derive the following equation for large distances:

$$f_y = a\sqrt{\frac{L_{L,y}}{x}} \tag{6.54}$$

where $L_{L,y}$ is the Lagrangian integral length scale, given by

$$L_{L,y} = \frac{T_{i,L,y}}{\overline{u}} \tag{6.55}$$

This is shown in Section A6.2 (Appendix A). Based on the derivation in Section 6.6.6, we can argue that $a = 2^{1/2}$. Equation (6.54) applies when $x >> L_{L,y}$. Similar equations apply to the vertical ($z$) direction.

However, $L_{L,y}$ and $L_{L,z}$ are not the only length scales that can be defined in this context. Other scales are $T_{i,L,y}/\sigma_v$ and $T_{i,L,z}/\sigma_w$. These are important in stochastic Lagrangian particle modeling. They will be discussed in Chapter 9.

At this point it is clear why a single power law such as eqs. (6.12) and (6.13) cannot describe dispersion parameters accurately in a large interval of $x$. The power law index decreases from 1 at small distances to 0.5 at large distances.

Because of the above, we can assume that $f_y$ and $f_z$ are not functions of $x$ or $t$ but of $t/T_{i,L}$. Because $T_{i,L} = L_L / \bar{u}$ and $t = x / \bar{u}$, we can prove $t/T_{i,L} = x/L_L$, and $f_y$ and $f_z$ are functions of $x/L_L$ as well. Assuming that $L_L$ is a constant, $f_y$ and $f_z$ become dependent on $x$ only.

The following equations have been suggested for $f_y$ in the lateral direction:

$$f_y(x) = \frac{1}{1 + 0.0308x^{0.455}} \qquad x < 10{,}000\,\text{m} \tag{6.56}$$

$$f_y(x) = \frac{33.3}{x^{0.5}} \qquad x > 10{,}000\,\text{m} \tag{6.57}$$

This function has the desired properties for $x \to 0$ and $x \to \infty$. Comparing eq. (6.54) with eq. (6.57), we can infer a Lagrangian length scale of 555 m. A function was not proposed for the $z$ direction because $f_z$ is too dependent on atmospheric stability.

When we compare eqs. (6.27) and (6.28) with the Briggs equations for $\sigma_y$ and $\sigma_z$, then the theoretical foundation of the latter becomes immediately apparent. This is illustrated here for stability class D. The Briggs equations are

$$\sigma_y = \frac{0.08x}{\sqrt{1 + 0.0001x}} \tag{6.58}$$

$$\sigma_z = \frac{0.06x}{\sqrt{1 + 0.0015x}} \tag{6.59}$$

Hence, the following relationships are implied:

$$i_v = 0.08 \tag{6.60}$$

$$i_w = 0.06 \tag{6.61}$$

$$f_y(x) = \frac{1}{(1 + 0.0001x)^{0.5}} \tag{6.62}$$

$$f_z(x) = \frac{1}{(1 + 0.0015x)^{0.5}} \tag{6.63}$$

for $x \to 0$ we find that $f_y \to 1$ and $f_z \to 1$, as expected. At large values of $x$ we find that $f_y$ and $f_z$ are proportional with $x^{0.5}$, as expected. A value of $L_L$ can be obtained by comparing the above equations with eq. (6.54) for large values of $x$. Values of 5000 and 333 m are obtained for the $y$ and $z$ directions, respectively. Assuming an average wind speed of a few meters per second, this leads to a Lagrangian integral time of a few hundred to a few thousand seconds in the $y$ direction and a few hundred seconds in the $z$ direction. These numbers are within the range found earlier, or slightly above. Integral time scales up to a few thousand seconds are inconsistent with 10-min averaging times. Hence, while the measurement data underlying the Briggs dispersion parameters are 10-min averages, the Briggs parameters represent much longer averaging times, which may justify EPA's choice to treat them as hourly averages.

From Briggs' equations for the other stability classes, it is clear that $f_z$ does not follow the expected behavior for $x \to \infty$ in the case of stability classes A, B, E, and F. In the case of classes A and B (unstable) this is because of the convective cycling, causing $K_z$ to increase with height in the surface layer. In the case of classes E and F (stable), this is because the temperature inversion creates an effective barrier against strong dispersion. Diffusivity $K_z$ goes through a maximum close to the surface and decreases above that.

The following equations for $f_y$ and $f_z$ summarized by Seinfeld and Pandis (2006) explicitly account for $T_{i,L}$ and are potentially more accurate than the above equations (Draxler, 1976; Irwin, 1979, 1983).

For unstable atmosphere:

$$f_y = \frac{1}{1+\left[x/\left(\bar{u}\,T_{i,L}\right)\right]^{0.5}} \tag{6.64}$$

$$f_z = \frac{1}{1+0.9\left[\dfrac{x}{\left(\bar{u}\,T_{i,L}\right)}\right]^{0.5}} \tag{6.65}$$

For neutral and stable atmosphere:

$$f_y = \frac{1}{1+0.9\left[\dfrac{x}{\left(\bar{u}\,T_{i,L}\right)}\right]^{0.5}} \tag{6.66}$$

$$f_z = \frac{1}{1+0.945\left[\dfrac{x}{\left(\bar{u}\,T_{i,L}\right)}\right]^{0.806}} \tag{6.67}$$

None of these equations show the expected limiting behavior of eq. (6.54) at large distances. Hanna and Britter (2002) proposed a universal equation that has the expected limiting trend:

$$f_{x,y,z} = \frac{1}{\left[1 + x/(2\bar{u}T_{\mathrm{i,L}})\right]^{0.5}} = \frac{1}{\left[1 + x/(2L_{\mathrm{L}})\right]^{0.5}} \tag{6.68}$$

This equation is based on the work of Deardorff and Willis (1975). Equations (6.64)–(6.67) can be made to correspond with eq. (6.54) at large distances by changing the coefficient:

$$f_{x,y,z} = \frac{1}{1 + \left[\dfrac{x}{(2\bar{u}T_{\mathrm{i,L}})}\right]^{0.5}} = \frac{1}{1 + \left[\dfrac{x}{(2L_{\mathrm{L}})}\right]^{0.5}} \tag{6.69}$$

We will see in Section 6.6.6 that eq. (6.68) better corresponds with the behavior of plumes expected from autocorrelation theory than eq. (6.69). On the other hand, assuming a $T_{\mathrm{i,L}}$ value on the order of 1000 s, or using eq. (6.46) for $T_{\mathrm{i,L}}$, eq. (6.69) corresponds better with the Prairie Grass data set as reported by Hanna et al. (1990). On the other hand, these $T_{\mathrm{i,L}}$ values are much too large for the Prairie Grass data set, where the emission height was only 0.45 m. When eq. (6.44) is used to calculate $T_{\mathrm{i,L}}$, the Prairie Grass data set is between eq. (6.68) and (6.69), but even eq. (6.44) overestimates $T_{\mathrm{i,L}}$ near the surface. Better correlations, based on energy dissipation rates, will be discussed in Section 6.11.2. Degrazia et al. (2005) pointed out that eq. (6.68) corresponds well with both field data and large eddy simulation (LES) results. It follows that the optimal choice of equation for $f_y$ also depends on the choice of equation for other variables, but the best correlations of $T_{\mathrm{i,L}}$ work best with eq. (6.68).

These equations, combined with eqs. (6.44)–(6.49) for $T_{\mathrm{i,L}}$, eqs. (5.144)–(5.156) for $\sigma_v$ and $\sigma_w$, and eqs. (6.50) and (6.51) form a practical calculation scheme for the dispersion parameters $\sigma_y$ and $\sigma_z$. The dispersion parameters are determined at the *effective source height h*.

So far, the averaging time for $\sigma_y$ has not yet been considered. Averaging times can affect $\sigma_y$ through the values of $\sigma_v$ and of $T_{\mathrm{i,L},y}$. Turbulent velocity $\sigma_v$ is not very sensitive to the averaging time, even in the 2-min to 1-h range (Wilson, 2008). The effect of integration time on $T_{\mathrm{i,L},y}$ is more pronounced, but given the large values of $T_{\mathrm{i,L},y}$ sometimes found (>15 min), it is clear that the integration time is substantially longer than the 10 min applicable to the dispersion parameters based on stability classes, so it seems reasonable to assume that the dispersion parameters calculated in this section are hourly averages.

**Example 6.2.** Calculate $\sigma_y$ and $\sigma_z$ at 1500 m from the source, under neutal conditions, using the same data as Example 6.1 (emission height 90 m; $u_* = 0.608$ m s$^{-1}$; $u_{90} = 10.34$ m s$^{-1}$). Assume a latitude of 45°.

***Solution.*** The calculation of $\sigma_v$, $\sigma_w$, and $T_{\mathrm{i,L}}$ requires a mixing layer height. In neutral conditions eq. (5.112) can be used:

$$h_{\mathrm{mix}} = \frac{0.3u_*}{|f|} = \frac{0.3 \cdot 0.608}{7.292 \times 10^{-5} \sin 45^\circ} = 3538\,\mathrm{m}$$

The turbulent velocities are calculated with eqs. (5.145) and (5.146), assuming $a = 0.75$:

$$\sigma_v = 1.9 \times 0.608 \times (1 - 90/3538)^{0.75} = 1.133\,\mathrm{m\,s^{-1}}$$
$$\sigma_w = 1.3 \times 0.608 \times (1 - 90/3538)^{0.75} = 0.775\,\mathrm{m\,s^{-1}}$$

Because the emission is in the surface layer, eq. (6.45) can be used for $T_{\mathrm{i,L},z}$:

$$T_{\mathrm{i,L},z} = 0.42 \times 90 \times 10.34 / (0.608 \times 0.775) = 829\,\mathrm{s}$$

This is a rather large value for a surface layer Lagrangian integral time scale. In the $y$ direction, the calculation is more uncertain. Equation (6.46) could be used with $1/L = 0$, or eq. (6.44) could be used, but these equations may not be valid in the surface layer, which has lower integral time scales than the bulk of the boundary layer. Hence, eq. (6.44) is preferred here because it provides the lowest predictions. The following value is obtained:

$$T_{\mathrm{i,L},y} = 0.17 \times 3538 / 1.133 = 531\,\mathrm{s}$$

With eq. (6.46), a value of 2327 s is obtained, which is an extremely large value. The Lagrangian integral length scale is

$$L_{\mathrm{L},y} = 10.34 \times 531 = 5488\,\mathrm{m}$$
$$L_{\mathrm{L},z} = 10.34 \times 829 = 8573\,\mathrm{m}$$

The recommended equation for $f_y$ and $f_z$ is eq. (6.68). The calculated values are

$$f_y = 1/[1 + 1500/(2 \times 5488)]^{0.5} = 0.9380$$
$$f_z = 1/[1 + 1500/(2 \times 8573)]^{0.5} = 0.9589$$

Based on these values, the following values of $\sigma_y$ and $\sigma_z$ are obtained with eqs. (6.27) and (6.28):

$$\sigma_y = 1.133 \times (1500/10.34) \times 0.6437 = 154.2\,\mathrm{m}$$
$$\sigma_z = 0.775 \times (1500/10.34) \times 0.7246 = 107.9\,\mathrm{m}$$

The only plume dispersion modifier needed is plume-rise-enhanced dispersion. The result is as follows:

$$\sigma_y = (105.8^2 + 0.1\times 15^2) = 154.3\,\text{m}$$
$$\sigma_z = (81.5^2 + 0.1\times 15^2) = 108.0\,\text{m}$$

The value for $\sigma_y$ is greater than the uncorrected value calculated in Example 6.1 and less than the corrected values. The value of $\sigma_z$ is greater than the value from Example 6.1. The calculation will be repeated with better $T_{i,L}$ correlations based on energy dissipation rates in Section 6.11.2.

Substitution of the values obtained in the Gaussian equation [eq. (6.7)] leads to

$$c = \frac{100}{2\pi\cdot 10.34\cdot 154.3\cdot 108.0}\exp\left(-\frac{1}{2}\frac{0^2}{\sigma_y^2}\right)\left\{\exp\left[-\frac{1}{2}\frac{(0-90)^2}{108.0^2}\right]+\exp\left[-\frac{1}{2}\frac{(0+90)^2}{108.0^2}\right]\right\}$$
$$= 130.6\times 10^{-6}\,\text{g}\,\text{m}^{-3} = \mathbf{130.6\,\mu g\,m^{-3}}$$

The calculation is included on the enclosed CD, in the Excel file "Example 6.2. Dispersion calculation continuous parameterization.xlsx."

### 6.6.6 Dispersion Parameters Based on the Autocorrelation Function

Equations for $\sigma_y$ and $\sigma_z$ can be calculated based on a statistical argument using the autocorrelation function (Taylor, 1922). Parameter $\sigma_y$ will be used as an example here. By definition, $\sigma_y$ is the standard deviation of the $y$ coordinate of pollutant particles. We will derive an equation for the $y$ coordinate of a pollutant particle and use statistical theory to calculate its standard deviation.

During a time interval $\Delta t$, taken short enough so that $f_y = 1$, the change of the $y$ coordinate of a pollutant particle is given by

$$\Delta y = V\,\Delta t = V'\Delta t \tag{6.70}$$

In eq. (6.70), the wind direction is assumed to be along the $x$ axis. Assuming $y = 0$ at $t = 0$, the $y$ coordinate after $N$ time steps is

$$y = \sum_{i=1}^{N} V_i'\,\Delta t \tag{6.71}$$

In matrix form, this equation becomes

$$y = \mathbf{T}^{\mathrm{T}}\mathbf{V} \tag{6.72}$$

where

$$\mathbf{T} = \begin{pmatrix} \Delta t \\ \Delta t \\ \vdots \\ \Delta t \end{pmatrix} = \Delta t \begin{pmatrix} 1 \\ 1 \\ \vdots \\ 1 \end{pmatrix} \tag{6.73}$$

$$\mathbf{V} = \begin{pmatrix} V'_1 \\ V'_2 \\ \vdots \\ V'_N \end{pmatrix} \tag{6.74}$$

To calculate the standard deviation of eq. (6.72), we assume that **V** contains a set of correlated normally distributed variables. **T** is a constant vector. According to statistical theory, the variance of $y$ in eq. (6.72) is given by

$$\sigma_y^2 = \mathbf{T}^{\mathrm{T}}\mathbf{M}\mathbf{T} \tag{6.75}$$

where **M** is the variance–covariance matrix of **V**:

$$\mathbf{M} = \begin{pmatrix} \sigma_1^2 & \mathrm{cov}_{12} & \cdots & \mathrm{cov}_{1N} \\ \mathrm{cov}_{21} & \sigma_2^2 & \cdots & \mathrm{cov}_{2N} \\ \vdots & \vdots & \ddots & \vdots \\ \mathrm{cov}_{N1} & \mathrm{cov}_{N2} & \cdots & \sigma_N^2 \end{pmatrix} \tag{6.76}$$

Assuming that the autocorrelation function does not change in space and time, the variance–covariance matrix **M** has the following elements:

$$\sigma_1^2 = \sigma_2^2 = \cdots = \sigma_N^2 = \sigma_v^2 \tag{6.77}$$

$$\mathrm{cov}_{12} = \mathrm{cov}_{23} = \cdots = \mathrm{cov}_{(N-1)N} = R'_{\mathrm{L}}(\Delta t) \tag{6.78}$$

$$\mathrm{cov}_{13} = \mathrm{cov}_{24} = \cdots = \mathrm{cov}_{(N-2)N} = R'_{\mathrm{L}}(2\Delta t) \tag{6.79}$$

and so forth.

We can write

$$\mathbf{M} = \mathbf{N}\sigma_v^2 \tag{6.80}$$

where

$$\mathbf{N} = \begin{bmatrix} 1 & R_{\mathrm{L}}(\Delta t) & R_{\mathrm{L}}(2\Delta t) & \cdots \\ R_{\mathrm{L}}(\Delta t) & 1 & R_{\mathrm{L}}(\Delta t) & \cdots \\ R_{\mathrm{L}}(2\Delta t) & R_{\mathrm{L}}(\Delta t) & 1 & \cdots \\ \vdots & \vdots & \vdots & \ddots \end{bmatrix} = (N_{ij}) \tag{6.81}$$

Hence, eq. (6.75) can be written as

$$\sigma_y^2 = \mathbf{A}^{\mathrm{T}}\mathbf{N}\mathbf{A}(\Delta t)^2\,\sigma_v^2 = (\Delta t)^2\,\sigma_v^2 \sum_{i=1}^{N}\sum_{j=1}^{N} N_{ij} \tag{6.82}$$

where **A** is a unit column matrix. Solving the summations diagonal by diagonal, we obtain the following equation:

$$\sigma_y^2 = (\Delta t)^2\sigma_v^2\left\{n + 2(n-1)R_{\mathrm{L}}(\Delta t) + 2(n-2)R_{\mathrm{L}}(2\Delta t) + \cdots + 2R_{\mathrm{L}}[(n-1)\Delta t]\right\} \tag{6.83}$$

Substituting eq. (6.43) into eq. (6.83) yields

$$\sigma_y^2 = (\Delta t)^2\sigma_v^2\left\{\begin{array}{l} n + 2(n-1)\exp\left(-\dfrac{\Delta t}{T_{\mathrm{i,L}}}\right) + 2(n-2)\exp\left(-\dfrac{2\Delta t}{T_{\mathrm{i,L}}}\right) \\ + \cdots + 2R_{\mathrm{L}}\exp\left[-\dfrac{(n-1)\Delta t}{T_{\mathrm{i,L}}}\right] \end{array}\right\} \tag{6.84}$$

Bringing $\Delta t$ within brackets leads to

$$\sigma_y^2 = \Delta t\sigma_v^2\left\{\begin{array}{l} n\Delta t + 2(n-1)\Delta t\exp\left(-\dfrac{\Delta t}{T_{\mathrm{i,L}}}\right) + 2(n-2)\Delta t\exp\left(-\dfrac{2\Delta t}{T_{\mathrm{i,L}}}\right) + \cdots \\ +2\Delta t R_{\mathrm{L}}\exp\left[-\dfrac{(n-1)\Delta t}{T_{\mathrm{i,L}}}\right] \end{array}\right\} \tag{6.85}$$

In the limit to an infinite number of steps of infinitesimal size, the summation becomes an integral:

$$\sigma_y^2 = 2\sigma_v^2\int_0^t (t-\tau)\exp\left(-\frac{\tau}{T_{\mathrm{i,L}}}\right)d\tau \tag{6.86}$$

The integral can be written as two integrals:

$$\sigma_y^2 = 2\sigma_v^2 t\int_0^t \exp\left(-\frac{\tau}{T_{\mathrm{i,L}}}\right)d\tau - 2\sigma_v^2\int_0^t \tau\exp\left(-\frac{\tau}{T_{\mathrm{i,L}}}\right)d\tau \tag{6.87}$$

These are standard integrals as listed in Section B3 (Appendix B):

$$\int \exp(ax)\,dx = \frac{1}{a}\exp(ax) + C \tag{6.88}$$

$$\int x\exp(ax)\,dx = \left(\frac{x}{a} - \frac{1}{a^2}\right)\exp(ax) + C \tag{6.89}$$

Applying these integrals to eq. (6.87) yields

$$\sigma_y^2 = 2\sigma_v^2 t\left[-T_{\text{i,L}}\exp\left(-\frac{t}{T_{\text{i,L}}}\right)+T_{\text{i,L}}\right]-2\sigma_v^2\left[\left(-T_{\text{i,L}}t-T_{\text{i,L}}^2\right)\exp\left(-\frac{t}{T_{\text{i,L}}}\right)-\left(-T_{\text{i,L}}^2\right)\right] \tag{6.90}$$

Rearranging the equation and canceling two terms leads to

$$\sigma_y^2 = 2\sigma_v^2 T_{\text{i,L}}\left\{t-T_{\text{i,L}}\left[1-\exp\left(-\frac{t}{T_{\text{i,L}}}\right)\right]\right\} \tag{6.91}$$

To get an equation in terms of the downwind distance, $t$ is eliminated using the average wind speed:

$$t=\frac{x}{\bar{u}} \tag{6.92}$$

We obtain

$$\sigma_y^2 = 2\sigma_v^2 T_{\text{i,L}}\left\{\frac{x}{\bar{u}}-T_{\text{i,L}}\left[1-\exp\left(-\frac{x}{\bar{u}\,T_{\text{i,L}}}\right)\right]\right\} \tag{6.93}$$

Rearrangement, and considering that $T_{\text{i,L}} = L_{\text{L}}/\bar{u}$, leads to

$$\sigma_y^2 = 2\frac{\sigma_v^2}{\bar{u}^2}L_{\text{L}}\left\{x-L_{\text{L}}\left[1-\exp\left(-\frac{x}{L_{\text{L}}}\right)\right]\right\} \tag{6.94}$$

Take the square root to obtain:

$$\boxed{\sigma_y = \sqrt{2}\frac{\sigma_v}{\bar{u}}L_{\text{L}}^{1/2}\left\{x-L_{\text{L}}\left[1-\exp\left(-\frac{x}{L_{\text{L}}}\right)\right]\right\}^{1/2}} \tag{6.95}$$

This result was first obtained by Taylor (1922). We will now check the limiting behavior at small $x$ and at large $x$. At small $x$ we can approximate the exponential as

$$\exp\left(-\frac{x}{L_{\text{L}}}\right)=1-\frac{x}{L_{\text{L}}}+\frac{1}{2}\left(\frac{x}{L_{\text{L}}}\right)^2-\cdots \tag{6.96}$$

Substituting the second-order truncation of the exponential into eq. (6.95) leads to

$$\sigma_y = \sqrt{2}\frac{\sigma_v}{\bar{u}}L_L^{1/2}\left\{x - L_L\left[1 - 1 + \frac{x}{L_L} - \frac{1}{2}\left(\frac{x}{L_L}\right)^2\right]\right\}^{1/2} = \frac{\sigma_v}{\bar{u}}x = \sigma_v t \tag{6.97}$$

which is consistent with theory. At large $x$ the first term between brackets in eq. (6.95) becomes dominant. We obtain

$$\sigma_y = \sqrt{2L_L x}\frac{\sigma_v}{\bar{u}} \tag{6.98}$$

Combining with eq. (6.50) leads to eq. (6.54) with $a = 2^{1/2}$, which justifies the use of this value in Section 6.6.5. Combining eq. (6.95) with eq. (6.50) leads to

$$\boxed{f_y = \frac{\sqrt{2}L_L^{1/2}}{x}\left\{x - L_L\left[1 - \exp\left(-\frac{x}{L_L}\right)\right]\right\}^{1/2}} \tag{6.99}$$

In principle, the same equation applies to $f_z$ as well. However, under strongly stable or strongly unstable conditions, $\sigma_w$ is not constant but changes with height, so eq. (6.99) may not be applicable to $f_z$ under these conditions. Furthermore, $T_{i,L,z}$, and hence $L_L$ in the vertical direction, is not constant but increases with height.

Figure 6.7 compares eq. (6.99) with eqs. (6.68) and (6.69). Equations (6.99) and (6.68) are very similar, but eq. (6.69) shows much less dispersion close to the source. This comparison confirms that eq. (6.68) is a more appropriate equation than eq. (6.69).

Degrazia et al. (2005) inverted the procedure outlined here and derived an autocorrelation function based on the empirical $f_y$ function, eq. (6.68). The result is as follows:

$$R_L(\tau) = \frac{1}{1 + |\tau|/2T_{i,L}} - \frac{|\tau|}{T_{i,L}\left(1 + |\tau|/2T_{i,L}\right)^2} + \frac{\tau^2}{4T_{i,L}^2\left(1 + |\tau|/2T_{i,L}\right)^3} \tag{6.100}$$

### 6.6.7 More $T_{i,L}$ Relationships

In Section 6.6.4, a number of relationships were presented that can be useful for the estimation of the Lagrangian integral time scale $T_{i,L}$. The overview was incomplete because some important relationships needed to understand $T_{i,L}$ were not covered until Sections 6.6.5 and 6.6.6. The purpose of this section is to complete this review of $T_{i,L}$ relationships.

We start this section with the derivation of a fundamental relationship between $K_x$, $\sigma_u$, and $T_{i,L,x}$ and equivalent relationships in the $y$ and $z$ directions. The derivation refers to the $z$ direction but is equally valid in the $x$ and $y$ directions.

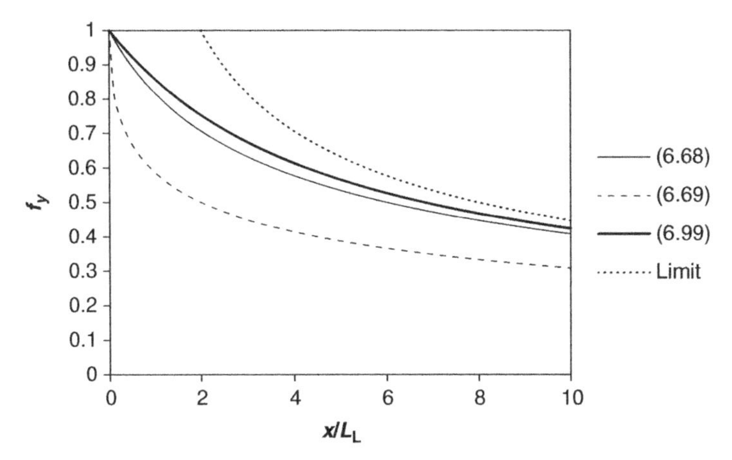


Figure 6.7 Predicted values of $f_y$ based on eq. (6.99) and on some equations from Section 6.6.5.

Consider eq. (5.196), which is valid for large distances, and apply it to the $z$ direction:

$$\sigma_z = \sqrt{2K_z t} \tag{6.101}$$

On the other hand, consider eq. (6.98), also valid for large distances, and apply it to the vertical direction:

$$\sigma_z = \sqrt{2L_L x}\frac{\sigma_w}{u} \tag{6.102}$$

Considering that $L_L = u\,T_{i,L}$ and $x = ut$, eq. (6.102) can be written as

$$\sigma_z = \sqrt{2T_{i,L}t}\,\sigma_w \tag{6.103}$$

Equating eq. (6.101) to (6.103):

$$\sigma_z = \sqrt{2K_z t} = \sqrt{2T_{i,L}t}\,\sigma_w \tag{6.104}$$

Rearranging leads to (Pasquill, 1974)

$$\boxed{T_{i,L} = \frac{K_z}{\sigma_w^2}} \tag{6.105}$$

Equation (6.105) is no longer limited to long distances. It connects the three main characteristics of turbulence. The correlations for $\sigma_z$ and for $K_z$ reviewed in Chapter 5 can be used to provide additional correlations for $T_{i,L,z}$. For instance, in the simplest case of neutral conditions near the surface, the following correlations were found:

$$\frac{\sigma_w}{u_*} = 1.3 \tag{5.149}$$

$$K_h = k u_* z = K_z \tag{5.191}$$

Hence, the Lagrangian integral time scale is estimated as

$$T_{i,L} = \frac{kz}{1.69 u_*} \tag{6.106}$$

What this derivation reveals is that the integral time scale is height dependent. The formulation of Weber et al. (1982) and Hanna (1982) has this property. In this formulation, the following equations are proposed in *stable* conditions:

$$T_{i,L,x} = \frac{0.15 z^{0.5} h_{mix}^{0.5}}{\sigma_u} \tag{6.107}$$

$$T_{i,L,y} = \frac{0.07 z^{0.5} h_{mix}^{0.5}}{\sigma_v} \tag{6.108}$$

$$T_{i,L,z} = \frac{0.1 z^{0.8} h_{mix}^{0.2}}{\sigma_w} \tag{6.109}$$

In *neutral* conditions, the following equation is proposed for all directions:

$$T_{i,L,x,y,z} = \frac{0.5z}{\sigma_w [1 - 15(fz)/u_*]} \tag{6.110}$$

where $f$ is the Coriolis parameter.

In *unstable* conditions the following equation applies to the $x$ and $y$ directions:

$$T_{i,L,x,y} = \frac{0.15 h_{mix}}{\sigma_{u,v}} \tag{6.111}$$

In the $z$ direction, three cases are distinguished.

For $z/h_{mix} < 0.1$ and $-(z - z_0)/L < 1$:

$$T_{i,L,z} = \frac{0.1z}{\sigma_w \{0.55 + 0.38[(z - z_0)/L]\}} \tag{6.112}$$

For $z/h_{mix} < 0.1$ and $-(z - z_0)/L > 1$:

$$T_{i,L,z} = \frac{0.059z}{\sigma_w} \tag{6.113}$$

For $z/h_{mix} > 0.1$:

$$T_{i,L,z} = \frac{0.15h_{mix}\left[1-\exp\left(-5z/h_{mix}\right)\right]}{\sigma_w} \tag{6.114}$$

Further parameterizations are given by Anfossi and Physick (2005) and in Byun and Ching (1999).

**Example 6.3.** Recalculate $\sigma_y$ and $\sigma_z$ at 1500 m from the source, under neutal conditions, using the same data as Example 6.1 (emission height 90 m; $u_* = 0.608$ m s$^{-1}$; $u_{90} = 10.34$ m s$^{-1}$), but with the newly found relationships for the Lagrangian integral time scale. Assume a latitude of 45°.

***Solution.*** In neutral conditions, no mixing layer height is needed in the calculation.

The turbulent velocities, calculated with eqs. (5.145) and (5.146), are $\sigma_v$ = 1.133 m s$^{-1}$ and $\sigma_w = 0.775$ m s$^{-1}$.

Equation (6.110) can be used for $T_{i,L}$:

$$\begin{aligned} T_{i,L,x,y,z} &= \frac{0.5z}{\sigma_w[1-15(fz)/u_*]} \\ &= 0.5\times 90/(0.775\times(1-15\times 7.292\times 10^{-5}\cos 45°\times 90/0.608)) = 65.5\,\text{s} \end{aligned}$$

The Lagrangian integral length scale is

$$L_{L,y} = L_{L,z} = 10.34\times 65.5 = 677.8\,\text{m}$$

The recommended equation for $f_y$ and $f_z$ is eq. (6.68). The calculated values are

$$f_y = f_z = 1/[1+1500/(2\times 677.8)]^{0.5} = 0.6890$$

Based on these values, the following values of $\sigma_y$ and $\sigma_z$ are obtained with eqs. (6.27) and (6.28):

$$\sigma_y = 1.133\times(1500/10.34)\times 0.6890 = 113.3\,\text{m}$$
$$\sigma_z = 0.775\times(1500/10.34)\times 0.6890 = 77.5\,\text{m}$$

With plume-rise-enhanced dispersion, the dispersion parameters become

$$\sigma_y = 113.3^2 + 0.1\times 15^2 = 113.4\,\text{m}$$
$$\sigma_z = 77.5^2 + 0.1\times 15^2 = 77.6\,\text{m}$$

The value for $\sigma_y$ is similar to the uncorrected value calculated in Example 6.1. The value of $\sigma_z$ is greater than the value from Example 6.1. The concentration at a 1500-m distance from the source is

$$c = \frac{100}{2\pi\cdot 10.34\cdot 133.4\cdot 77.6}\exp\left(-\frac{1}{2}\frac{0^2}{\sigma_y^2}\right)\left\{\exp\left[-\frac{1}{2}\frac{(0-90)^2}{77.6^2}\right]+\exp\left[-\frac{1}{2}\frac{(0+90)^2}{77.6^2}\right]\right\}$$
$$= 178.7\times 10^{-6}\,\text{g}\,\text{m}^{-3} = \mathbf{178.7\,\mu g\,m^{-3}}$$

The calculation is included on the enclosed CD, in the Excel file "Example 6.3. Dispersion Calculation Continuous parameterization.xlsx."

## 6.7 GAUSSIAN PLUME MODELS FOR NONPOINT SOURCES

**Nonpoint sources** are sources of air pollution that are not negligible in size, so they cannot be described as a mathematical point. They are subdivided into line sources, area sources, and volume sources. There are four ways to calculate plumes of non-point sources that will be described briefly below: analytical solutions, numerical solutions, dispersion parameter modification, and the virtual source concept. Virtual sources have a broader application than just nonpoint sources and will be discussed in Section 6.8.

**Analytical solutions** are typical of older models because they are computationally the least demanding, but they have a more limited applicability. They are no longer recommended, with the possible exception of preliminary calculations. Consider, for instance, a **line source**, a source that is a mathematical line. If we take simplifying assumptions that the line is perpendicular to the wind direction, and the line is so long that edge effects are negligible, then the analytical solution becomes very simple: Equation (6.1) ($\overline{c} = C_x\varphi_y\varphi_z$) remains valid, with

$$\varphi_y = \left(\frac{Q}{L}\right) \tag{6.115}$$

where $Q$ is the total emission mass flow rate (mg s$^{-1}$), and $L$ is the length of the line source (m). Hence, assuming no temperature inversion layer, the Gaussian equation for this case becomes

$$\overline{c} = \frac{Q}{\sqrt{2\pi}\overline{u}L\sigma_z}\left\{\exp\left[-\frac{1}{2}\frac{(z-h)^2}{\sigma_z^2}\right]+\exp\left[-\frac{1}{2}\frac{(z+h)^2}{\sigma_z^2}\right]\right\} \tag{6.116}$$

When the line source is too short, or the receptor is too close to one of the ends of the line, eq. (6.116) is no longer valid. In that case, we need to define the line source as an infinite number of point sources and integrate the resulting concentrations. The calculations are given in Section A6.3 (Appendix A). The result is

$$\overline{c} = \frac{C_x \varphi_z}{2L}\left[\text{erf}\left(\frac{y_2}{\sqrt{2}\sigma_y}\right)-\text{erf}\left(\frac{y_1}{\sqrt{2}\sigma_y}\right)\right] \tag{6.117}$$

where erf($x$) is defined as

$$\text{erf}(x) = \frac{2}{\sqrt{\pi}}\int_0^x \exp(-t^2)\,dt \tag{6.118}$$

A table of erf($x$) is given in Appendix B.

The full expression of the concentration is

$$\overline{c} = \frac{Q}{2\overline{u}L\sqrt{2\pi}\sigma_z}\left[\text{erf}\left(\frac{y_2}{\sqrt{2}\sigma_y}\right)-\text{erf}\left(\frac{y_1}{\sqrt{2}\sigma_y}\right)\right]$$
$$\left\{\exp\left[-\frac{1}{2}\frac{(z-h)^2}{\sigma_z^2}\right]+\exp\left[-\frac{1}{2}\frac{(z+h)^2}{\sigma_z^2}\right]\right\} \tag{6.119}$$

For an infinitely long line source, one obtains

$$\overline{c} = \frac{Q}{2\overline{u}L\sqrt{2\pi}\sigma_z}\left[1-(-1)\right]\left\{\exp\left[-\frac{1}{2}\frac{(z-h)^2}{\sigma_z^2}\right]+\exp\left[-\frac{1}{2}\frac{(z+h)^2}{\sigma_z^2}\right]\right\} \tag{6.120}$$

which reduces to eq. (6.116).

When the line source is not perpendicular to the wind direction, the integration involves both $\varphi_y$ and $\varphi_z$. Consequently, there is not usually an exact solution. An approximation is possible by considering that the emission of a line source with length $L$ is now concentrated within a crosswind section of length $L \sin(\theta)$, where $\theta$ is the angle between the wind direction and the line source. Hence, $L \sin(\theta)$ becomes the effective length of the source.

In such cases, a **numerical solution** provides higher accuracy, albeit at the expense of computational efficiency. The simplest way to calculate the concentration downwind of a line source by integration is by defining a series of point sources sufficiently closely spaced that the plumes overlap completely. Hence, line sources simply become a convenient way to define a large number of sources simultaneously.

Similarly, **area sources**, which are sources defined as objects on a mathematical plane, can be modeled by defining a sufficiently large number of point sources within the area source.

In principle, **volume sources** could be treated the same way. However, it is doubtful if the result would justify the computational effort. CALPUFF uses the virtual source concept (see Section 6.8) to model volume sources (Scire et al., 2000).

An alternative way to model nonpoint sources is by dispersion parameter modification: adding an initial plume dispersion $\sigma_{y,0}$ to the dispersion calculations $\sigma_{y,calc}$ as a sum of variances:

$$\sigma^2_{y,total} = \sigma^2_{y,calc} + \sigma^2_{y,0} \tag{6.121}$$

This procedure is applied to plume-rise-induced dispersion.

## 6.8 VIRTUAL SOURCE CONCEPT

The **virtual source concept** is a method to calculate dispersion of nonpoint sources and plumes moving in variable landscape by changing the location of the source in the calculations. This will be illustrated with an example.

**Example 6.4.** A plume moves over urban terrain for 1000 m under neutral conditions and then moves over rural terrain. Derive an equation for the plume parameters over the rural terrain.

***Solution.*** For simplicity, we will use the Briggs equations [eqs. (6.10) and (6.11)]. After 1000 m over urban terrain, the dispersion parameters are

$$\sigma_y = \frac{0.16 \times 1000\text{m}}{(1 + 0.0004 \times 1000)^{0.5}} = 135.22\text{ m} \tag{6.122}$$

$$\sigma_y = \frac{0.14 \times 1000\text{m}}{(1 + 0.0003 \times 1000)^{0.5}} = 122.79\text{ m} \tag{6.123}$$

We cannot use the rural plume parameters directly at the 1000-m distance because the Briggs equations predict $\sigma_y = 76.28$ m and $\sigma_z = 37.95$ m at this distance. Any realistic calculation scheme should yield smooth transitions from urban to rural. This is where the virtual source concept comes into play. Starting with the lateral direction, we require that the Briggs equation yields a $\sigma_y$ of 135.22 m:

$$\sigma_y = \frac{0.08 x_v}{(1 + 0.0001 x_v)^{0.5}} = 135.22\text{m} \tag{6.124}$$

Rearranging leads to the following equation:

$$0.08x_v^2 - 0.0001 \cdot 135.22^2\, x_v - 135.22^2 = 0 \tag{6.125}$$

This equation has one positive root:

$$x_v = \frac{0.0001 \cdot 135.22^2 + \sqrt{0.0001^2 \cdot 135.22^4 + 4 \cdot 0.08 \cdot 135.22^2}}{2 \cdot 0.08} = 1839.19\,\text{m} \tag{6.126}$$

We can create a smooth transition between urban and rural conditions by adding 839.19 m to $x$ once the plume reaches rural terrain. We say that the virtual source is located 1839.19 m upwind from the urban–rural boundary, 839.19 m upwind from the real source.

Similarly, we can solve

$$\sigma_z = \frac{0.06x_v}{(1+0.0015x_v)^{0.5}} = 122.79\,\text{m} \tag{6.127}$$

The solution is $x_v = 6889.90\,\text{m}$. In the vertical direction the virtual source is located 5889.90 m upwind from the real source. Hence, the equations for the dispersion parameters in the rural terrain are

$$\sigma_y = \frac{0.08 \times (x + 839.19\,\text{m})}{[1 + 0.0001 \times (x + 839.19\,\text{m})]^{0.5}}$$
$$= \frac{0.06 \times (x + 5{,}889.90\,\text{m})}{[1 + 0.0015 \times (x + 5{,}889.90\,\text{m})]^{0.5}}$$

Similarly, we can use the virtual source concept to model plumes with nonzero initial values of $\sigma_y$ and $\sigma_z$. Volume sources in CALPUFF are modeled this way (Scire et al., 2000).

## 6.9 SPECIAL ISSUES

### 6.9.1 Probability Density Functions for Plumes in Convective Boundary Layers

Plumes released in a convective boundary layer do not grow symmetrically but tend to have a descending mode due to the distribution of updrafts and downdrafts. This was predicted in numerical models by Lamb (1978) and confirmed experimentally by Willis and Deardorff (1978, 1981). Weil et al. (1997) developed a methodology for modeling plume dispersion in convective boundary layers that accounts explicitly for updrafts and downdrafts. The method was implemented in AERMOD (Cimorelli et al., 2004).

Updrafts and downdrafts were described in Sections 5.12.1 and 5.13.1. A fraction $\lambda_1$ of the atmosphere is in an updraft and moves up with average velocity $\overline{w_1}$. The rest of the atmosphere, fraction $\lambda_2$, moves down: It has a negative vertical velocity component $\overline{w_2}$. For that reason the vertical component of the Gaussian plume equation, eq. (6.4) (direct plume), is split onto two parts: one moving up and one moving down. The result is

$$\varphi_z = \frac{\lambda_1}{\sqrt{2\pi}\sigma_{z1}} \exp\left[-\frac{1}{2}\frac{(z-\Psi_1)^2}{\sigma_{z1}^2}\right] + \frac{\lambda_2}{\sqrt{2\pi}\sigma_{z2}} \exp\left[-\frac{1}{2}\frac{(z-\Psi_2)^2}{\sigma_{z2}^2}\right] \quad (6.128)$$

The dispersion parameters of the updraft, $\sigma_{z1}$, and of the downdraft, $\sigma_{z2}$, are given by the equation

$$\sigma_{zj} = \frac{\sigma_{wj} x}{u} \quad (6.129)$$

where $j$ is 1 or 2, $u$ is the wind speed, and $x$ is the downwind distance. In other words, it is assumed that $f_z = 1$ throughout. The turbulent velocities of the updraft, $\sigma_{w1}$, and of the downdraft, $\sigma_{w2}$, are calculated with eqs. (5.218) and (5.219):

$$\sigma_{wj} = R\left|\overline{w_j}\right| \quad (6.130)$$

Weil et al. (1997) suggested $R = 2$ based on field measurements. Alternative values of $R$ are given in Section 5.13.1. The effective source heights of the updraft, $\Psi_1$, and of the downdraft, $\Psi_2$, are given by

$$\Psi_j = h_s + \Delta h + \frac{\overline{w_j} x}{u} \quad (6.131)$$

More calculation details are given in Section 5.13.1.

At the surface, an ordinary reflection is assumed. At the temperature inversion, $z = h_{mix}$, Weil et al. (1997) proposed to introduce an additional plume rise $\Delta h_i$ to account for the fact that plumes reaching the top of the mixing layer tend to loft before getting caught in a downdraft. The calculation of $\Delta h_i$ is discussed in Chapter 14.

### 6.9.2 Emission from a Ground-Level Source

The behavior of a plume emitted from a ground-level source can deviate substantially from a plume emitted from an elevated source. At ground level, the wind speed is 0, which cannot be used in the Gaussian equation. However, the height of the average particle in a ground-level plume increases with height even in the absence of plume rise, due to reflection on the surface. This solves the issue of zero wind speed but introduces the problem of defining a representative plume height. An additional issue arises when the plume is located between the obstacles that define surface roughness. This special case is discussed in Chapter 8. In this section we will discuss vertical plume growth near the surface, assuming that deviations from similarity theory are negligible.

One of the first studies to tackle plume growth near the surface is by Briggs and McDonald (1978). As reported by Briggs (1982), they defined a vertical thickness scale as follows:

$$h = \frac{Q}{\bar{u}\int_{-\infty}^{+\infty} c\,dy} \qquad (6.132)$$

Where $c$ is ground-level concentration. When the plume is Gaussian, with a plume axis at ground level, the vertical thickness scale is related to $\sigma_z$ as follows:

$$h = \sqrt{\frac{2}{\pi}}\sigma_z \approx 0.8\sigma_z \qquad (6.133)$$

The vertical thickness scale will be discussed further in Chapter 8. Briggs and McDonald (1978) found the following relations for $h$:

For stable atmosphere:

$$\frac{h}{L} = \frac{X}{1+\sqrt{X}} \qquad (6.134)$$

where

$$X = \frac{u_* x}{\bar{u}L} \qquad (6.135)$$

where $\bar{u}$ was evaluated at a fixed height, 8 m. For unstable atmosphere:

$$\frac{h}{L} = \frac{X}{\sqrt{1+5X^2}} \qquad (6.136)$$

The relevant wind speed for plume rise would be an average wind speed experienced by the plume, not the value at the 8-m height. The height dependence of wind speed depends on the surface roughness, so eqs. (6.134) and (6.136) are only valid for a surface roughness close to 0.01 m, the value of the Prairie Grass data set (Barad, 1958), which was used to derive these equations. At neutral conditions, the equations reduce to

$$h = \frac{u_*}{\bar{u}}x \qquad (6.137)$$

To remove the inherent surface roughness dependence of the correlations, Briggs (1982) introduced the following height scaling:

$$h_* = \frac{Q}{u_*\int_{-\infty}^{+\infty} c\,dy} = \frac{\bar{u}}{u_*}h \qquad (6.138)$$

Equation (6.137), translated to the new height scale, is as follows:

$$h^* = x \tag{6.139}$$

We do not expect this relationship to hold exactly in a new formulation. However, eq. (6.139) shows the potential for $h^*$ to simplify relationships. Briggs (1982) suggested the following:

For stable atmosphere:

$$h^* = \frac{0.8x}{[1 + 0.13\ (x/L)]^{1/3}} \tag{6.140}$$

For unstable atmosphere:

$$h^* = 0.8x\sqrt{1 - 0.19\frac{x}{L} - 0.00014\left(\frac{x}{L}\right)^3} \tag{6.141}$$

The neutral case,

$$h^* = 0.8x \tag{6.142}$$

is close to Briggs and McDonald's (1978) original formulation. Venkatram (1992) proposed a formulation for the unstable atmosphere consistent with eq. (6.139):

$$h^* = x\sqrt{1 + 0.006\left(\frac{x}{L}\right)^2} \tag{6.143}$$

In the unstable region, the limiting behavior of $-h^*/L$ versus $-x/L$ for large values of $-x/L$ are a 2.5 power law for eq. (6.141) and a 2 power law for eq. (6.143). Lagrangian stochastic dispersion modeling results of Venkatram and Du (1997) indicate that the actual limiting power may be somewhat less than 2. However, the difference is small for $-x/L$ values less than 100.

To predict values of $h$ (and therefore $\sigma_z$) from $h^*$, the wind speed at a representative height is needed. The most logical choice is to use height $h$. This means the calculation of $h$ from $h^*$ is an interative process. The procedure is illustrated in Example 6.5.

**Example 6.5.** Calculate $\sigma_z$ at 1500 m from the source of a plume emitted at ground level on an overcast day with wind speed 7 m $s^{-1}$ (measured at a 10-m height). The surface roughness is 10 cm.

***Solution.*** The meteorological conditions are the same as in Example 6.3. The friction velocity was found there to be 0.608 m $s^{-1}$.

Based on Brigg's equations applied to neutral conditions [eq. (6.142)] the following value of $h^*$ is found:

$$h^* = 0.8 \times 1500\,\text{m} = 1200\,\text{m}$$

Applying the definition of $h^*$ [eq. (6.138)]:

$$h = \frac{h^* u^*}{u}$$

where $u$ is the wind speed at the as-yet unknown hight $h$. Use $u = 7\,\mathrm{m\ s^{-1}}$ as a first estimate:

$$h = 1200 \times 0.608 / 7 = 104.2\,\mathrm{m}$$

At a height of 104.2 m, the wind speed is $(0.608/0.4) \times \ln(104.2/10) = 10.56\,\mathrm{m\ s^{-1}}$.

Based on this wind speed a second height is estimated: $h = 1200 \times 0.608/10.56 = 69.1\,\mathrm{m}$.

In the next iteration a wind speed of $9.94\,\mathrm{m\ s^{-1}}$ is obtained, leading to a new height of 73.4 m.

After a few more iterations the calculation converges to a wind speed of $10.02\,\mathrm{m\ s^{-1}}$ and a height $h$ of 72.8 m.

Equation (6.133) allows us to estimate $\sigma_z$ from $h$:

$$\sigma_z = (\pi / 2)^{1/2} h = \mathbf{91.3\,m}$$

The dispersion near the surface is more than the dispersion from an elevated source. This is somewhat surprising because the turbulent diffusivity increases with height. However, the equations discussed in this section were not developed for such long distances, and linear growth of the plume is probably not a realistic assumption at 1500 m from the source.

In wind tunnel studies, it is more common to represent plume growth from a surface source by its entrainment velocity $w_e$, which can be defined as (Briggs et al., 2001; Britter et al., 2003)

$$w_e = \frac{d}{dx}\left( \frac{Q}{\int_{-\infty}^{+\infty} c\,dy} \right) \tag{6.144}$$

In neutral conditions, where plume growth is expected to be linear, Britter et al. (2003) suggested the following equation for $w_e$:

$$w_e = \alpha u_* \tag{6.145}$$

where $\alpha = 0.65$. Combining eq. (6.144) with eq. (6.138), it is clear that

$$\frac{dh_*}{dx} = \frac{w_e}{u_*} \tag{6.146}$$

Hence, combining eq. (5.145) with eq. (6.146), we find

$$h_* = 0.65x \tag{6.147}$$

Less effort has been made to determine lateral dispersion of ground-level sources, although the issues associated with vertical dispersion apply to lateral dispersion as well.

## 6.10 GAUSSIAN PUFF MODELING

### 6.10.1 Introduction

Plume models capture many of the essentials in air dispersion, but they lack the flexibility to deal with some situations such as wind direction changes. The needed flexibility is introduced in the Gaussian puff model.

For a description of the **Gaussian puff model**, let us first consider a single puff of smoke emitted by a stack (Fig. 6.8). In the figure, a new coordinate system ($X$, $Y$, $Z$) is defined, different from the coordinates ($x$, $y$, $z$) defined in the Gaussian plume model. Coordinates ($X$, $Y$, $Z$) make up the set of distances from the plume center in the wind direction, the lateral direction, and the vertical direction. In other words: ($X$, $Y$, $Z$) is the coordinate system in a **Lagrangian frame of reference**: a frame of reference that moves with the puff.

The average concentration defined by a Gaussian puff is not the average in time or in space but the **ensemble average**. Assume that an infinite number of puffs were released in identical conditions and each time the concentration would be measured at the same receptor point, at the same time after puff release. The resulting average concentration is the ensemble average. To get the average concentration at a point in

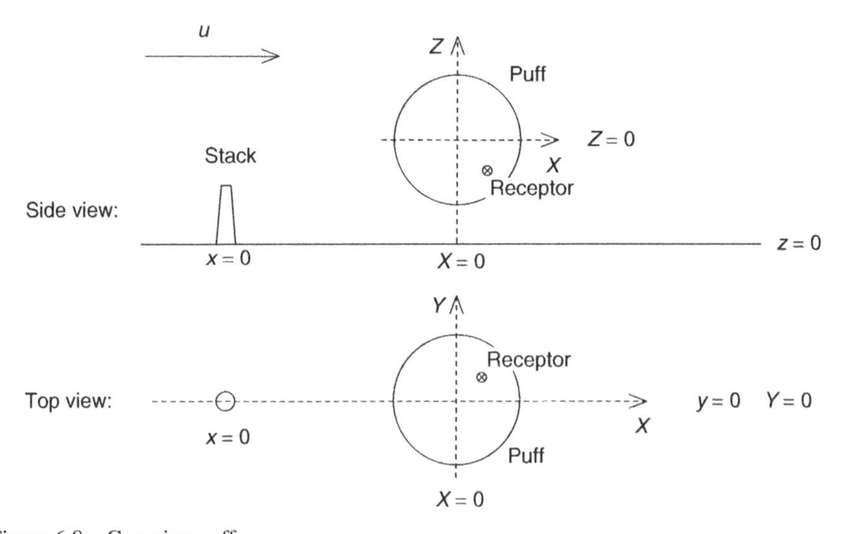


Figure 6.8 Gaussian puff.

space, we need to consider all the puffs overlapping with that point, and add up the ensemble average concentrations of all those puffs in that point.

Puff models have a number of advantages over plume models:

- Puff models, unlike plume models, can be made accurate at low wind speeds. They occasionally predict pollution upwind when the dispersion at small distances is faster than the wind speed.
- Puff models can handle situations when wind speed and/or wind direction changes in space or in time. Variation in space can happen in a valley, for instance, as illustrated in Figure 6.9.

Figure 6.9 shows how a puff model can predict effects of wind direction changes, whereas plume models cannot describe this situation accurately. This is the case for wind direction changes with time as well. A typical example is a sea breeze. A puff model can predict nonzero ground-level concentration at the source when the source is a point source and the height is not zero. With a plume model this is not possible.

### 6.10.2 Puff Models

In the mathematical description of Gaussian puffs, there is one fundamental difference with Gaussian plumes. For each puff individually, it is assumed that the dilution in the $x$ direction is purely random, due to turbulence. Wind speed does not dilute

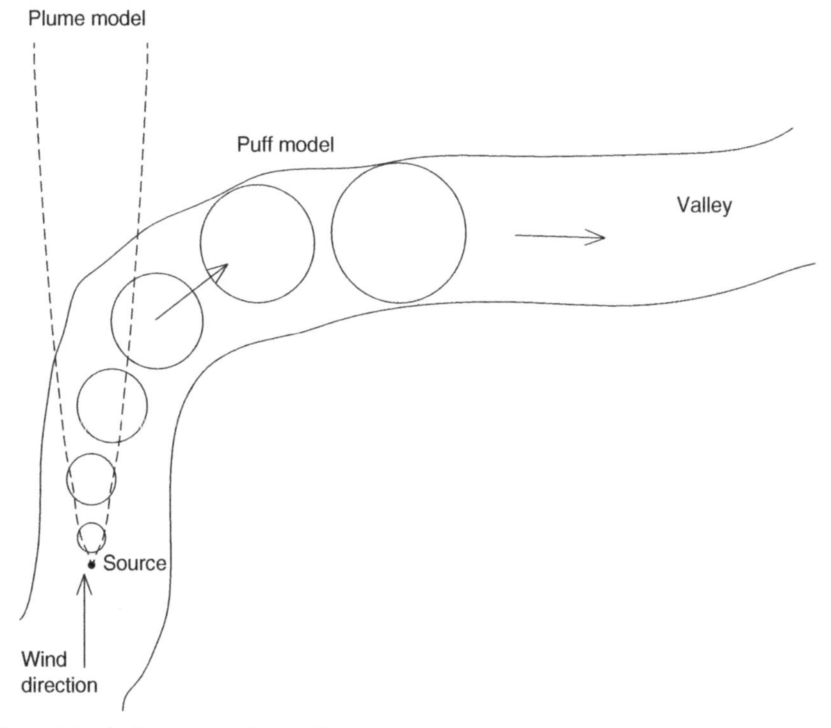


Figure 6.9 Puff movement in a valley.

individual plumes but increases the spacing between puffs in the $x$ direction, so a receptor experiences fewer puffs at any given time. Hence, the mathematical description of a puff is based on the following equation:

$$\bar{c} = Q_{\mathrm{p}} \varphi_x \varphi_y \varphi_z \tag{6.148}$$

where $Q_{\mathrm{p}}$ is the emission contained in the puff (mg); and $\varphi_x$ and $\varphi_y$ are defined in Lagrangian coordinates and have an identical form:

$$\varphi_x = \frac{1}{\sqrt{2\pi}\sigma_x} \exp\left(-\frac{1}{2}\frac{X^2}{\sigma_x^2}\right) \tag{6.149}$$

$$\varphi_y = \frac{1}{\sqrt{2\pi}\sigma_y} \exp\left(-\frac{1}{2}\frac{Y^2}{\sigma_y^2}\right) \tag{6.150}$$

and $\varphi_z$ is defined in Eulerian coordinates. Assuming puff reflection on the surface only, $\varphi_z$ is given by

$$\varphi_z = \frac{1}{\sqrt{2\pi}\sigma_z} \left\{ \exp\left[-\frac{1}{2}\frac{(z-h)^2}{\sigma_z^2}\right] + \exp\left[-\frac{1}{2}\frac{(z+h)^2}{\sigma_z^2}\right] \right\} \tag{6.151}$$

Consequently, the average concentration in the puff is calculated as

$$\bar{c} = \frac{Q_{\mathrm{p}}}{(2\pi)^{3/2}\sigma_x\sigma_y\sigma_z} \exp\left(-\frac{1}{2}\frac{X^2}{\sigma_x^2}\right) \exp\left(-\frac{1}{2}\frac{Y^2}{\sigma_y^2}\right) \left\{ \exp\left[-\frac{1}{2}\frac{(z-h)^2}{\sigma_z^2}\right] + \exp\left[-\frac{1}{2}\frac{(z+h)^2}{\sigma_z^2}\right] \right\} \tag{6.152}$$

A variety of methods to calculate $\sigma_x$, $\sigma_y$, and $\sigma_z$ is provided earlier in this chapter. When no method for $\sigma_x$ is available, it can be assumed equal to $\sigma_y$ as an approximation.

To accurately describe a finite plume as a succession of puffs, a large number of very dilute puffs in short succession should be considered. In that case, the actual concentration is the sum of the concentrations at that location in the individual puffs. This is illustrated in Figure 6.10. The receptor in the figure is located inside puffs 3–5. This is the **direct puff method** of calculating concentrations. For this method to be effective, the spacing between consecutive puff centers should be no more than $\sigma_y$ (Ludwig et al., 1977). This means that each source should emit at least one puff per second to resolve the plumes near the sources in some cases (Scire et al., 2000). It is possible to reduce the total number of puffs required, for example, by merging puffs far from the source that are more closely spaced than necessary (Ludwig et al., 1977) or by defining extra puffs near the receptors (Zannetti, 1981), but the computational requirements of such schemes are still substantial.

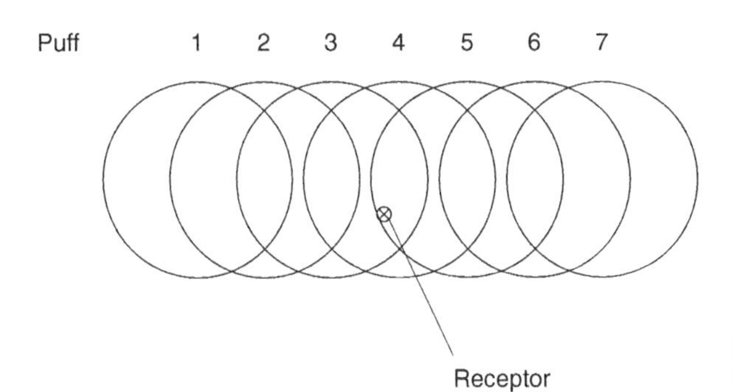


Figure 6.10 Plume as a succession of puffs.

To avoid the weaknesses of direct puff methods, two integrated methods have been proposed and incorporated in MESOPUFF II (Scire et al., 1984a,b) and CALPUFF (Scire et al., 2000). The first method is the integrated puff method. The second is the slug method.

To derive the **integrated puff method**, we make the simplifying assumption that $\sigma_x = \sigma_y$. Hence, the concentration is calculated as

$$\bar{c} = \frac{Q_p}{2\pi\sigma_x\sigma_y} \exp\left[-\frac{1}{2}\frac{(X^2+Y^2)}{\sigma_y^2}\right]\varphi_z \tag{6.153}$$

Second, we make the assumption that the puff size and height do not change during one time increment, that is, $\sigma_y$ and $\varphi_z$ do not change during the time increment. Assume that a puff travels a distance $\Delta s$ during the time increment; then the average concentration $\bar{C}_{ave}$ at the receptor during the time increment is given by

$$\bar{c}_{ave} = \frac{1}{\Delta s}\int_0^{\Delta s} \bar{c}\, ds = \int_0^1 \bar{c}\, dp \tag{6.154}$$

where $dp = ds/\Delta s$, and $p = s/\Delta s$, and $p$ is a dimensionless coordinate of the puff movement.

Assume that a receptor is at the coordinates $(x_r, y_r)$. The puff center moves from $(x_1, y_1)$ to $(x_1 + \Delta x, y_1 + \Delta y)$. At any value of $p$, the coordinates of the puff center are $(x_1 + p\Delta x, y_1 + p\Delta y)$. Hence, eq. (6.153) can be rewritten as

$$\bar{c} = \frac{Q_p}{2\pi\sigma_x\sigma_y} \exp\left[-\frac{1}{2}\frac{(x_1 + p\Delta x - x_r)^2 + (y_1 + p\Delta y - y_r)^2}{\sigma_y^2}\right]\varphi_z \tag{6.155}$$

Equation (6.154) becomes

$$\bar{c}_{ave} = \frac{Q_p\varphi_z}{2\pi\sigma_x\sigma_y}\int_0^1 \exp\left[-\frac{1}{2}\frac{(x_1 + p\Delta x - x_r)^2 + (y_1 + p\Delta y - y_r)^2}{\sigma_y^2}\right] dp \tag{6.156}$$

We can rearrange the terms between brackets in such a way that

$$\overline{c}_{\text{ave}} = \frac{Q_{\text{p}}\varphi_z}{2\pi\sigma_x\sigma_y}\int_0^1 \exp\left[-\frac{1}{2}\left(ap^2 + 2bp + c\right)\right] dp \tag{6.157}$$

where

$$a = \frac{(\Delta x)^2 + (\Delta y)^2}{\sigma_y^2} \tag{6.158}$$

$$b = \frac{\Delta x(x_1 - x_{\text{r}}) + \Delta y(y_1 - y_{\text{r}})}{\sigma_y^2} \tag{6.159}$$

$$c = \frac{(x_1 - x_{\text{r}})^2 + (y_1 - y_{\text{r}})^2}{\sigma_y^2} \tag{6.160}$$

To solve eq. (6.157) in terms of error functions, it is rewritten as

$$\overline{c}_{\text{ave}} = \frac{Q_{\text{p}}\varphi_z}{2\pi\sigma_x\sigma_y}\int_0^1 \exp\left[-\frac{a}{2}\left(p^2 + 2\frac{b}{a}p + \frac{b^2}{a^2} - \frac{b^2}{a^2} + \frac{c}{a}\right)\right] dp \tag{6.161}$$

Hence, the equation becomes

$$\overline{c}_{\text{ave}} = \frac{Q_{\text{p}}\varphi_z}{2\pi\sigma_x\sigma_y}\exp\left(\frac{b^2}{2a} - \frac{c}{2}\right)\int_0^1 \exp\left[-\frac{a}{2}\left(p + \frac{b}{a}\right)^2\right] dp \tag{6.162}$$

Implement the following substitution: $p + b/a = q$; $dp = dq$. It follows that

$$\overline{c}_{\text{ave}} = \frac{Q_{\text{p}}\varphi_z}{2\pi\sigma_x\sigma_y}\exp\left(\frac{b^2}{2a} - \frac{c}{2}\right)\int_{b/a}^{1+b/a} \exp\left(-\frac{a}{2}q^2\right) dq \tag{6.163}$$

From Section B3 (Appendix B) we find

$$\int_0^x \exp(-at^2)dt = \frac{\sqrt{\pi}}{2\sqrt{a}}\text{erf}(\sqrt{a}x) \tag{6.164}$$

Hence, eq. (6.163) becomes

$$\overline{c}_{\text{ave}} = \frac{Q_{\text{p}}\varphi_z}{2\pi\sigma_x\sigma_y}\sqrt{\frac{\pi}{2a}}\exp\left(\frac{b^2}{2a} - \frac{c}{2}\right)\left\{\text{erf}\left[\sqrt{\frac{a}{2}}\left(1 + \frac{b}{a}\right)\right] - \text{erf}\left(\sqrt{\frac{a}{2}}\frac{b}{a}\right)\right\} \tag{6.165}$$

$$\overline{c}_{\text{ave}} = \frac{Q_{\text{p}}\varphi_z}{2\pi\sigma_x\sigma_y}\sqrt{\frac{\pi}{2a}}\exp\left(\frac{b^2}{2a}-\frac{c}{2}\right)\left[\text{erf}\left(\frac{a+b}{\sqrt{2a}}\right)-\text{erf}\left(\frac{b}{\sqrt{2a}}\right)\right] \tag{6.166}$$

In MESOPUFF II, eq. (6.166) is used with dispersion parameters at $p = 0.5$ for all receptor points. In CALPUFF, eq. (6.166) is calculated separately for each receptor point, using the dispersion parameters at the point of closest approach. Based on this approach, the number of puffs needed can be reduced from about one per second to one per hour (Scire et al., 2000).

Close to the source, the assumptions of the integrated puff model do not apply because of the rapid puff growth. Hence, to achieve high accuracy near the source, the **slug method** was developed. A slug is an elongated puff that captures the growth process to some extent. The slug method is also an integrated method, but now the integration is numerical. The slug is the superposition of an infinite number of Gaussian puffs, each containing an infinitesimal amount of pollutant. The slug represents all the pollutant that was emitted during a time step. The leading end of the puff has a larger spread than the trailing end because the leading end is older. The following concentration profile was derived for the slug:

$$\overline{c}(t) = \frac{Q\varphi_z}{2\sqrt{2\pi}u'\sigma_y}\exp\left(-\frac{d_{\text{c}}^2}{2\sigma_y^2}\frac{u^2}{u'^2}\right)\left[\text{erf}\left(\frac{d_{\text{a2}}}{\sqrt{2}\sigma_{y2}}\right)-\text{erf}\left(-\frac{d_{\text{a1}}}{\sqrt{2}\sigma_{y1}}\right)\right] \tag{6.167}$$

where $Q$ (not $Q_{\text{p}}$) is the emission mass flow rate (mg s$^{-1}$), $u'^2 = u^2 + \sigma_{\text{v}}^2$, $d_{\text{c}}$ is the cross-sectional distance from the receptor to the slug axis, $d_{\text{a1}}$ the along-axis distance from the receptor to the leading edge of the slug, $d_{\text{a2}}$ the along-axis distance from the receptor to the trailing edge of the slug, $\sigma_y$, $\sigma_{y1}$, and $\sigma_{y2}$ are the dispersion parameter of the slug at the receptor, the leading edge, and the trailing edge, respectively.

### 6.10.3 Stochastic Puff Models: Parameterization for Instantaneous Puffs

The puff models discussed so far are *average* puffs representing an ensemble average over a certain time. They are assumed to behave in a deterministic fashion. The dispersion parameters used in deterministic puff models are the same as the ones used in plume models. However, parameters have also been proposed for **instantaneous puffs**. Instantaneous puffs can be used in **stochastic puff models**. This type of model is a compromise between a deterministic puff model and a stochastic Lagrangian particle model, aiming to achieve some of the features of a stochastic Lagrangian particle model with less computational effort. In a stochastic puff model, a large number of puffs, each representing an instantaneous pollution cloud, is defined. Each puff moves with the wind, and is allowed to move randomly around the mean wind vector. The average concentration is calculated the same way as a deterministic puff model.

An instantaneous puff model is still only an approximation of the actual behavior of a real pollutant in the atmosphere. Yee et al. (1998) investigated propylene puffs from instantaneous releases and observed a pronounced microstructure over time scales of a few seconds that is not captured in a puff model. The standard deviation of the pollutant concentration due to this microstructure is of the same order of magnitude as the concentration itself.

Another process that is not resolved in instantaneous puff models is **puff slant**. Due to the vertical wind shear (i.e., the increasing wind speed with increasing height), the top of a puff moves faster than the bottom, slanting the puff (DeVito et al., 2009). An example of puff slant will be calculated with a stochastic Lagrangian model in Chapter 9.

Table 6.3 shows a simple instantaneous puff dispersion parameterization based on eqs. (6.168) and (6.169) (Islitzer and Slade, 1968):

$$\sigma_y = ax^b \tag{6.168}$$

$$\sigma_z = cx^d \tag{6.169}$$

Equation (6.168) corresponds well with data of Yee et al. (1998) in neutral conditions but tends to overestimate the plume spread in unstable conditions. DeVito et al. (2009) found that puffs grow linearly with time (< 60 s) with a rate that is relatively insensitive to wind speed. Hence, $a$ and $c$ in eqs. (6.168) and (6.169) may increase with increasing wind speed.

The instantaneous puff dispersion parameters are roughly a third of the corresponding plume parameters. It follows that the maximum concentration in an instantaneous puff is roughly nine times as high as the maximum concentration in a 10-min average plume. Consequently, stochastic puff models would be most valuable in odor nuisance calculations and acute toxicity calculations.

There is more than one way to model stochastic puffs. The simplest method is by randomly defining puffs around the plume axis at each receptor location. The variance of the $y$ and $z$ values of the puff centers are given by

$$\sigma_y^2 = \sigma_{y,c}^2 + \sigma_{y,inst}^2 \tag{6.170}$$

$$\sigma_z^2 = \sigma_{z,c}^2 + \sigma_{z,inst}^2 \tag{6.171}$$

where $\sigma_{y,c}$ and $\sigma_{z,c}$ are the standard deviations of the instantaneous puff center locations in the $x$ and $y$ directions, and $\sigma_{y,inst}$ and $\sigma_{z,inst}$ are the instantaneous puff dispersion parameters. The model of Vanderschaege (1987) is of this type.

**TABLE 6.3 Instantaneous Puff Dispersion Parameters for Use in eqs. (6.12)–(6.13)**

| Conditions | $a$ | $b$ | $c$ | $d$ |
|---|---|---|---|---|
| Unstable | 0.14 | 0.92 | 0.53 | 0.73 |
| Neutral | 0.06 | 0.92 | 0.15 | 0.70 |
| Stable | 0.02 | 0.82 | 0.05 | 0.61 |

A more sophisticated method would be to modify the techniques used in stochastic Lagrangian particle models, so they can be applied to stochastic puffs. A model of this type was developed by de Haan and Rotach (1998). The plume is divided into puffs, and the puff centers are moved on a stochastic trajectory. Stochastic Lagrangian particle models are discussed in Chapter 9.

## 6.11 ADVANCED TOPICS IN METEOROLOGY

A first section on advanced topics in meteorology was included in Chapter 5. It included material that is important for some of the later chapters but not essential for every reader. Some advanced topics could not be included in Chapter 5 because they rely on the concept of the Lagrangian integral time scale. They are included in this section.

### 6.11.1 Spectral Properties of Turbulence

One way to look at a turbulent variable (e.g., $u'$) is by looking at its **spectrum**, or rather its variance spectrum. The variance spectrum is a function of frequency and is closely related to the Fourier transform of the turbulent variable. However, rather than discussing the Fourier transform in detail, we will follow Arya (1999) in adopting a more intuitive approach to spectrum functions.

As defined in Chapter 5, the variance of the wind speed $u$ is given by

$$\sigma_u^2 = \overline{u'^2} \tag{5.137}$$

where $\sigma_u^2$ is twice the $x$ component of the turbulent kinetic energy. The energy represented by $\sigma_u^2$ can be considered as a superposition of oscillations of different frequencies. Each of these oscillations can be isolated by applying a narrow-band filter to the wind speed fluctuation. Define the spectrum function $S'_{uu}(\nu)$ such that $S'_{uu}(\nu)\,d\nu$ is the portion of the wind speed variance associated with all oscillations with frequencies between $\nu$ and $\nu + d\nu$. Hence, by definition,

$$\sigma_u^2 = \int_0^{+\infty} S'_{uu}(\nu)\,d\nu \tag{6.172}$$

The function $S_{uu}(\nu)$ is the function $S'_{uu}(\nu)$ normalized to 1:

$$S_{uu}(\nu) = \frac{S'_{uu}(\nu)}{\sigma_u^2} \tag{6.173}$$

The spectrum functions can also be expressed in terms of angular frequency:

$$S_{uu}(\nu)\,d\nu = S_{uu}(\omega)\,d\omega = S_{uu}(\omega)2\pi\,d\nu \tag{6.174}$$

Hence:

$$S_{uu}(\nu) = 2\pi S_{uu}(\omega) \tag{6.175}$$

Similarly, spectrum functions can be expressed in terms of wave numbers $\kappa = \omega/u$:

$$S_{uu}(\kappa)\, d\kappa = S_{uu}(\omega)\, d\omega = S_{uu}(\omega)u\, d\kappa \tag{6.176}$$

Hence:

$$S_{uu}(\kappa) = uS_{uu}(\omega) \tag{6.177}$$

Spectrum functions are related to autocorrelation functions. The relationships are (Arya, 1999)

$$S_{uu}(\omega) = \frac{2}{\pi}\int_0^{+\infty} R_{uu}(\tau)\cos(\omega\tau)\, d\tau \tag{6.178}$$

$$R_{uu}(\tau) = \int_0^{+\infty} S_{uu}(\omega)\cos(\omega\tau)\, d\omega \tag{6.179}$$

Spectrum functions of atmospheric turbulence span several orders of magnitude, often from $< 0.01$ to $>100\ s^{-1}$. The maximum of $S_{uu}$ is at the lower end of the spectrum (i.e., large turbulent structures).

### 6.11.2 Turbulent Energy Dissipation: Kolmogorov Theory

The kinetic energy contained in turbulent motion is known as **turbulent kinetic energy**. It is usually defined per unit mass of air. It is calculated as (Chapter 5)

$$\overline{e} = \frac{1}{2}\left(\overline{u'^2} + \overline{v'^2} + \overline{w'^2}\right) \tag{5.138}$$

Turbulent kinetic energy is generated by instabilities in large bulk motions of air (e.g., leading to obstacle wakes) and by convective cycling. It follows that turbulence is generated on the low-frequency side of the spectrum (large eddies). At this large scale, viscous effects are negligible because they are driven by velocity gradients, which increase with decreasing eddy size. Viscosity effects occur almost exclusively at the scale of the smallest turbulent eddies (millimeter scale). It follows that turbulence is dissipated on the high-frequency side of the spectrum. Hence, to conserve energy, there is a net transfer of turbulent kinetic energy from the low-frequency side to the high-frequency side of the spectrum (i.e., from large to small eddies). This picture of turbulence is known as the **energy cascade hypothesis**.

From the spectral picture of turbulence and turbulent kinetic energy, we can conclude that there is a frequency range were there is no production or dissipation of kinetic energy, only transfer. This range is known as the **inertial subrange**. Turbulent structures in the 1-cm to 1-m size range are usually in the inertial subrange. These turbulent structures are usually **isotropic**: They have the same properties in all directions. This is because the turbulent eddies were produced from larger eddies, which were themselves produced from even larger eddies. With each transfer, the motion changes direction and eventually "forgets" its initial orientation.

Based on these concepts, Kolmogorov (1941) developed a similarity theory of turbulence. The only variables influencing turbulence in the inertial subrange are the spectrum function $S$ (units m$^3$ s$^{-2}$ when based on $\kappa$), the wave number $\kappa$ (m$^{-1}$), and the **energy dissipation rate** $\varepsilon$. The scale $\eta$ at which the inertial subrange merges into an energy dissipation range is influenced by the kinematic viscosity $\nu$ and the energy dissipation rate $\varepsilon$. Before continuing the derivation of the similarity relationships, we need to properly define the concept of the energy dissipation rate $\varepsilon$.

The energy dissipation rate is the rate of destruction of turbulent kinetic energy. Hence, if the turbulence is perfectly homogeneous and there are no external forces acting upon the turbulent fluid, the energy dissipation rate would be equal to the rate of decrease of turbulent kinetic energy. As turbulent kinetic energy has units m$^2$ s$^{-2}$, the energy dissipation rate has units m$^2$ s$^{-3}$. In an incompressible Newtonian fluid, the energy dissipation rate is given by

$$\varepsilon = \nu\left[\left(\frac{\partial u}{\partial x}\right)^2 + \left(\frac{\partial u}{\partial y}\right)^2 + \left(\frac{\partial u}{\partial z}\right)^2 + \left(\frac{\partial v}{\partial x}\right)^2 + \left(\frac{\partial v}{\partial y}\right)^2 + \left(\frac{\partial v}{\partial z}\right)^2 + \left(\frac{\partial w}{\partial x}\right)^2 + \left(\frac{\partial w}{\partial y}\right)^2 + \left(\frac{\partial w}{\partial z}\right)^2\right] \tag{6.180}$$

It is sometimes more convenient to use numbered dimensions, so an equation like this can be written with summation signs. Hence, eq. (6.180) is usually written as

$$\varepsilon = \nu \sum_{i=1}^{3}\sum_{j=1}^{3}\left(\frac{\partial u_i}{\partial x_j}\right)^2 \tag{6.181}$$

The equation will be derived in Chapter 10. For now we will only discuss $\varepsilon$ based on similarity theory.

As indicated above, the only variables influencing turbulence in the inertial subrange are the spectrum function $S'$ of the turbulent kinetic energy (units m$^3$ s$^{-2}$ when based on $\kappa$), the wave number $\kappa$ (m$^{-1}$), and the energy dissipation rate $\varepsilon$ (m$^2$ s$^{-3}$). These variables incorporate two dimensions, so only one dimensionless number exists connecting them. That number must therefore be constant. Only $S'$ and $\varepsilon$ have seconds in the units, so they must occur as the ratio $S'^3/\varepsilon^2$. To balance meters, we get the following dimensionless number:

$$\pi_1 = \frac{S'^3 \kappa^5}{\varepsilon^2} \tag{6.182}$$

The cubic root of $\pi_1$ is known as the **Kolmogorov constant** $\alpha_k$, or $C_k$. Its value is not known exactly, but it lies in the range 1.53–1.68. A recent determination based on numerical simulations of isotropic turbulence provided an estimate of 1.58 (Donzis and Sreenavasan, 2010).

Based on $\alpha_k$, the spectral function $S'(\kappa)$ can be written as

$$S'(\kappa) = \frac{\alpha_k \varepsilon^{2/3}}{\kappa^{5/3}} \tag{6.183}$$

When the spectral function of the turbulent kinetic energy (or of any velocity variance) is plotted, the inertial subrange is instantly recognizable as the range where $S'$ has a slope of $-\frac{5}{3}$ in log–log scale.

As indicated above, the scale $\eta$ (m) at which the inertial subrange merges into an energy dissipation range, also known as the **Kolmogorov length microscale**, is influenced by the kinematic viscosity $\nu$ (m$^2$ s$^{-1}$) and the energy dissipation rate $\varepsilon$ (m$^2$ s$^{-3}$). Because there is no well-defined scale where the inertial subrange ends, no dimensionless number is defined, and the Kolmogorov length microscale is simply defined as

$$\eta = \nu^{3/4} \varepsilon^{-1/4} \tag{6.184}$$

For instance, with an air viscosity of $1.5 \times 10^{-5}$ m$^2$ s$^{-1}$ and an energy dissipation rate of $10^{-3}$ m$^2$ s$^{-3}$, a Kolmogorov length microscale of $1.36 \times 10^{-3}$ m = 1.36 mm is found.

There are a number of empirical correlations to calculate the energy dissipation rate. In a stable atmosphere, one can use (Stull, 1988)

$$\varepsilon = \frac{3.7 u_*^3}{kL} \tag{6.185}$$

In neutral conditions, the following equation can be used (Stull, 1988):

$$\varepsilon = \frac{u_*^3}{kz} \tag{6.186}$$

In near-neutral conditions in the surface layer, the following equation can be used:

$$\varepsilon = \frac{u_*^3}{kz}\left[1 + 0.5\left(\frac{z}{|L|}\right)^{2/3}\right]^{3/2} \tag{6.187}$$

In unstable conditions, the following equation can be used (Weil, 1990):

$$\varepsilon = \frac{w_*^3}{h_{\text{mix}}}\left\{1.15 \exp\left(-12.5 \frac{z}{h_{\text{mix}}}\right) - 0.2 \exp\left[-50\left(1 - \frac{z}{h_{\text{mix}}}\right)\right] + 0.3\right\} \tag{6.188}$$

Equations (6.185) and (6.188) underestimate the energy dissipation rate near the surface. When these equations predict a lower value of $\varepsilon$ than eq. (6.187), then eq. (6.187) should be used instead.

Another similarity relationship is between the energy dissipation rate, the wind speed variance, and the Lagrangian integral time scale. It is as follows (Tennekes, 1979):

$$C_0\varepsilon = \frac{2\sigma_w^2}{T_{i,L,z}} \tag{6.189}$$

where $C_0$ is known as the **Lagrangian structure function constant**. Estimates of $C_0$ range from 3 to 5 (Sawford, 1985; Wilson and Sawford, 1996; Du, 1997, 1998; Reynolds, 1998; Anfossi et al., 2000). Recent research has not narrowed this range (e.g., Degrazia et al., 2008; Carvalho et al., 2009). Most estimates for $C_0$ are in the $z$ direction. However, Anfossi et al. (2006) estimated $C_0$ in the three directions based on large eddy simulations and found values of 2.5, 4.3, and 4.5 in the $x$, $y$, and $z$ directions, respectively. They also derived the following relationship between $C_0$ and the ratio between the Lagrangian and Eulerian time scales, $\beta$ [see eq. (6.42)]:

$$C_0 = \frac{\pi(1.5\alpha_i\alpha_u)^{3/2}}{\beta i} \tag{6.190}$$

where $\alpha_i = 1$ in the $x$ direction, and $\frac{4}{3}$ in the $y$ and $z$ directions, $\alpha_u = 0.425$, and $i$ is the turbulent intensity. Estimations of the Lagrangian structure function constant will be discussed further in Chapter 9.

Equation (6.189) can be solved for $T_{i,L}$:

$$\boxed{T_{i,L,z} = \frac{2\sigma_w^2}{C_0\varepsilon}} \tag{6.191}$$

Equivalent equations are applicable to the $x$ and $y$ directions. This equation, in combination with eq. (6.68) or (6.95), is probably the most reliable way to calculate Gaussian dispersion.

**Example 6.6.** Recalculate $\sigma_y$ and $\sigma_z$ at 1500 m from the source, under neutral conditions, using the same data as Examples 6.1 and 6.3 (emission height 90 m; $u^* = 0.608\ \mathrm{m\ s^{-1}}$; $u_{90} = 10.34\ \mathrm{m\ s^{-1}}$), but with eq. (6.191) for the Lagrangian integral time scale.

***Solution.*** The turbulent velocities, calculated with eqs. (5.145) and (5.146), are $\sigma_v = 1.133\ \mathrm{m\ s^{-1}}$, and $\sigma_w = 0.775\ \mathrm{m\ s^{-1}}$.

Equation (6.186) can be used for the energy dissipation rate $\varepsilon$ (neutral conditions):

$$\varepsilon = \frac{0.608^3}{0.4 \times 90} = 0.006244\,\mathrm{m^2\,s^{-3}}$$

Equation (6.191) can be used for $T_{i,L}$. Assume a value of 4 for $C_0$ in both $y$ and $z$ directions.

$$T_{i,L,y} = \frac{2\sigma_v^2}{C_0\varepsilon} = \frac{2\times 1.133^2}{4\times 0.006244} = 102.82\ \mathrm{s}$$

$$T_{i,L,z} = \frac{2\sigma_w^2}{C_0\varepsilon} = \frac{2\times 0.775^2}{4\times 0.006244} = 48.13\ \mathrm{s}$$

The Lagrangian integral length scale is

$$L_{L,y} = 10.34\times 102.82 = 1063.1\ \mathrm{m}$$
$$L_{L,z} = 10.34\times 48.13 = 497.7\ \mathrm{m}$$

The recommended equation for $f_y$ and $f_z$ is eq. (6.68). The calculated values are

$$f_y = 1/[1+1500/(2\times 1063.1)]^{0.5} = 0.7657$$
$$f_z = 1/[1+1500/(2\times 497.7)]^{0.5} = 0.6316$$

Based on these values, the following values of $\sigma_y$ and $\sigma_z$ are obtained with eqs. (6.27) and (6.28):

$$\sigma_y = 1.133\times(1500/10.34)\times 0.7657 = 125.9\,\mathrm{m}$$
$$\sigma_z = 0.775\times(1500/10.34)\times 0.6316 = 71.0\,\mathrm{m}$$

With plume-rise-enhanced dispersion, the dispersion parameters become

$$\sigma_y = 125.9^2 + 0.1\times 15^2 = 126.0\,\mathrm{m}$$
$$\sigma_z = 71.0^2 + 0.1\times 15^2 = 71.2\,\mathrm{m}$$

The value for $\sigma_y$ is similar to the uncorrected value calculated in Example 6.1. The value of $\sigma_z$ is greater than the value from Example 6.1. The concentration at the 1500-m distance from the source is

$$c = \frac{100}{2\pi \cdot 10.34 \cdot 126.0 \cdot 71.2} \exp\left(-\frac{1}{2}\frac{0^2}{\sigma_y^2}\right)\left\{\exp\left[-\frac{1}{2}\frac{(0-90)^2}{71.2^2}\right] + \exp\left[-\frac{1}{2}\frac{(0+90)^2}{71.2^2}\right]\right\}$$
$$= 154.4 \times 10^{-6}\,\text{g m}^{-3} = \mathbf{154.4\ \mu g\ m^{-3}}$$

The calculation is included on the enclosed CD, in the Excel file "Example 6.6. Dispersion calculation continuous parameterization.xlsx."

## 6.12 SUMMARY OF THE MAIN EQUATIONS

Formally, the Gaussian plume equation is given by

$$\overline{c} = C_x \varphi_y \varphi_z \tag{6.1}$$

with

$$C_x = \frac{Q}{\overline{u}} \tag{6.2}$$

$$\varphi_y = \frac{1}{\sqrt{2\pi}\sigma_y} \exp\left(-\frac{1}{2}\frac{y^2}{\sigma_y^2}\right) \tag{6.3}$$

In the absence of a temperature inversion, the vertical contribution to eq. (6.1) is given by

$$\varphi_z = \frac{1}{\sqrt{2\pi}\sigma_z}\left\{\exp\left[-\frac{1}{2}\frac{(z-h)^2}{\sigma_z^2}\right] + \exp\left[-\frac{1}{2}\frac{(z+h)^2}{\sigma_z^2}\right]\right\} \tag{6.6}$$

In the presence of a temperature inversion, the equation becomes

$$\varphi_z = \sum_{j=-\infty}^{+\infty}\left(\exp\left\{-\frac{1}{2}\frac{[z-(h+2jh_{mix})]^2}{\sigma_z^2}\right\} + \exp\left\{-\frac{1}{2}\frac{[z+(h+2jh_{mix})]^2}{\sigma_z^2}\right\}\right) \tag{6.8}$$

The most common parameterization for $\sigma_y$ and $\sigma_z$ based on stability classes is

$$\sigma_y = \frac{ax}{(1+bx)^c} \tag{6.10}$$

$$\sigma_z = \frac{dx}{(1+ex)^f} \tag{6.11}$$

Formally, the dispersion parameters can be calculated as

$$\sigma_y = \sigma_v t f_y = \sigma_v \frac{x}{\bar{u}} f_y = i_v x f_y \tag{6.27}$$

$$\sigma_z = \sigma_w t f_z = \sigma_w \frac{x}{\bar{u}} f_z = i_w x f_z \tag{6.28}$$

The Lagrangian autocorrelation function is often described as

$$R_{\mathrm{L}}(\tau) = \exp\left(-\frac{|\tau|}{T_{\mathrm{i,L}}}\right) \tag{6.43}$$

The Lagrangian integral length scale is

$$L_{\mathrm{L},y} = \frac{T_{\mathrm{i,L},y}}{\bar{u}} \tag{6.55}$$

A recommended function for $f_y$ and $f_z$ is

$$f_{x,y,z} = \frac{1}{[1 + x/(2\bar{u}T_{\mathrm{i,L}})]^{0.5}} = \frac{1}{(1 + x/2L_{\mathrm{L}})^{0.5}} \tag{6.68}$$

An equation for $\sigma_y$ based on the autocorrelation function is

$$\sigma_y = \sqrt{2}\frac{\sigma_v}{\bar{u}} L_{\mathrm{L}}^{1/2} \left\{ x - L_{\mathrm{L}} \left[ 1 - \exp\left( -\frac{x}{L_{\mathrm{L}}} \right) \right] \right\}^{1/2} \tag{6.95}$$

A fundamental relation linking $T_{\mathrm{i,L}}$, $K_z$, and $\sigma_w^2$ is

$$T_{\mathrm{i,L}} = \frac{K_z}{\sigma_w^2} \tag{6.105}$$

In neutral conditions, the Lagrangian integral time scale can be calculated as

$$T_{\mathrm{i,L},x,y,z} = \frac{0.5z}{\sigma_w[1 - 15(fz)/u_*]} \tag{6.110}$$

Formally, the calculation in a puff is calculated as

$$\bar{c} = Q_{\mathrm{p}} \varphi_x \varphi_y \varphi_z \tag{6.148}$$

with (no inversion assumed):

$$\varphi_x = \frac{1}{\sqrt{2\pi}\sigma_x}\exp\left(-\frac{1}{2}\frac{X^2}{\sigma_x^2}\right) \tag{6.149}$$

$$\varphi_y = \frac{1}{\sqrt{2\pi}\sigma_y}\exp\left(-\frac{1}{2}\frac{Y^2}{\sigma_y^2}\right) \tag{6.150}$$

$$\varphi_z = \frac{1}{\sqrt{2\pi}\sigma_z}\left\{\exp\left[-\frac{1}{2}\frac{(z-h)^2}{\sigma_z^2}\right]+\exp\left[-\frac{1}{2}\frac{(z+h)^2}{\sigma_z^2}\right]\right\} \tag{6.151}$$

In the integrated puff model the concentration is calculated as

$$\overline{c}_{\text{ave}} = \frac{Q_{\text{p}}\varphi_z}{2\pi\sigma_x\sigma_y}\sqrt{\frac{\pi}{2a}}\exp\left(\frac{b^2}{2a}-\frac{c}{2}\right)\left[\text{erf}\left(\frac{a+b}{\sqrt{2a}}\right)-\text{erf}\left(\frac{b}{\sqrt{2a}}\right)\right] \tag{6.166}$$

In the slug model the concentration is calculated as

$$\overline{c}(t) = \frac{Q\varphi_z}{2\sqrt{2\pi}u'\sigma_y}\exp\left(-\frac{d_{\text{c}}^2}{2\sigma_y^2}\frac{u^2}{u'^2}\right)\left[\text{erf}\left(\frac{d_{\text{a2}}}{\sqrt{2}\sigma_{\text{y2}}}\right)-\text{erf}\left(-\frac{d_{\text{a1}}}{\sqrt{2}\sigma_{\text{y1}}}\right)\right] \tag{6.167}$$

In neutral conditions the energy dissipation rate is calculated as

$$\varepsilon = \frac{u_*^3}{kz} \tag{6.186}$$

An alternative equation for the Lagrangian integral time scale is

$$T_{\text{i,L,z}} = \frac{2\sigma_w^2}{C_0\varepsilon} \tag{6.191}$$

## PROBLEMS

1. In Section 6.5 it was pointed out that eq. (6.20) alone cannot explain the difference between dispersion in rural and urban terrain. Why not?
2. A stack with a height of 50 m emits 20 g s$^{-1}$ of $NO_2$. Plume rise is negligible. The wind speed at a 10-m height is 2.5 m s$^{-1}$. The temperature is 25°C, the solar elevation is 35°, the cloud cover is $\frac{6}{8}$. The terrain is grassland with a surface roughness of 1 cm, an albedo of 0.2, and a Bowen ratio of 1. Based on rawinsonde data taken at the same time a mixing height of 1800 m is estimated.

   Use a continuous parameterization (of your choice but appropriate for the situation) to determine a downwind $NO_2$ concentration plot versus distance from the source on ground level at the plume centerline. Determine the maximum concentration and the downwind concentration where it occurs. Compare with the solution to Problem 2.9. Are the problems comparable?

3. What averaging time is needed to obtain a $\sigma_y$ value twice as large as its 10-min average value?
4. Assuming stability class D, at what distance can we neglect the impact of a 100-m plume rise on $\sigma_y$? What about $\sigma_z$? Assume that we tolerate an error of 10% on the dispersion parameters.
5. Meteorological measurements are conducted simultaneously with plume measurements. The results of the meteorological measurements are $\sigma_v = 0.8\,\text{m s}^{-1}$, $u = 5\,\text{m s}^{-1}$. Meanwhile, at 1000 m from the plume source a $\sigma_y$ value of 107 m is found. Based on this information, estimate $T_{i,L}$ and the energy dissipation rate $\varepsilon$.
6. Draw the autocorrelation coefficient versus correlation time in the range 0–100 s based on the $v$ data below. Estimate the integral time scale.

| $t$ (s) | $v$ (m s$^{-1}$) |
|---|---|
| 0 | 0.0521 |
| 10 | 0.1657 |
| 20 | −0.0752 |
| 30 | −0.1186 |
| 40 | −0.3305 |
| 50 | −0.1730 |
| 60 | −0.8062 |
| 70 | −0.4691 |
| 80 | 0.2718 |
| 90 | 0.3546 |
| 100 | −0.2798 |
| 110 | 0.3458 |
| 120 | −0.1812 |
| 130 | −0.0355 |
| 140 | 0.4130 |
| 150 | 0.1086 |
| 160 | 0.2750 |
| 170 | 0.5721 |
| 180 | 0.8875 |
| 190 | 1.0124 |
| 200 | 0.5581 |

7. Hanna et al. (1990) reported the following dispersion data from the Prairie Grass data set:

| $x$ (m) | $\sigma_y$ (m) |
|---|---|
| 50 | 13.7 |
| 100 | 26 |
| 200 | 49 |
| 400 | 72 |
| 800 | 116 |

The data set was obtained at a wind speed of 3.2 m s$^{-1}$. Use the data to estimate the lateral turbulent velocity and the Lagrangian integral time scale.

**8.** Derive an equation for $\sigma_y$ versus $x$ assuming that the autocorrelation coefficient $R_L(\tau) = 1$ for all $\tau < T_{i,L}$, and $R_L(\tau) = 0$ for all $\tau > T_{i,L}$. Does the equation fulfill the expected limiting behavior of a plume at $x \rightarrow 0$ and at $x \rightarrow +\infty$?

**9.** Calculate $\sigma_z$ of a plume emitted at ground level at night when the wind speed measured at a 10-m height is 7 m s$^{-1}$ and the Obukhov length is 450 m. The surface roughness is 10 cm. Compare your result with the result of Example 6.5 (neutral conditions).

**10.** During an overcast period, isolated puffs are released and measured directly downwind. The release is 10 g of ethylene, emitted quickly at 5 m above the ground without generating plume rise. After a large number of releases, all data are collected for puffs measured 1 km directly downwind at 1 m above the ground, 20 min after the release at a wind speed of 3 m s$^{-1}$ (measured at the release height).

a. What is the highest concentration you expect to be found? Assume that the highest concentration is that at the center of an instantaneous puff. Assume that $\sigma_x = 1.5\sigma_y$.

b. What is the average concentration you expect to be found? Assume that the average is the concentration at the center of an ensemble averaged puff. Again, assume that $\sigma_x = 1.5\sigma_y$. Use an approporiate parameterization of your choice.

**11.** Same as Problem 10, but now measurements are taken for one hour spanning the passage of the puff, and the average is calculated. Use the integrated puff method to calculate the average ethylene concentration expected in this experiment.

## MATERIALS ONLINE

- "Example 6.1. Dispersion calculation stability classes.xlsx": Calculations for Example 6.1.
- "Example 6.2. Dispersion calculation continuous parameterization.xlsx": Calculations for Example 6.2.
- "Example 6.3. Dispersion calculation continuous parameterization.xlsx": Calculations for Example 6.3.
- "Example 6.4. Dispersion calculation continuous parameterization.xlsx": Calculations for Example 6.4.

## REFERENCES

Anfossi D. and Physick W. (2005). Lagrangian particle models. In Zannetti P. (ed.) *Air Quality Modeling. Volume II—Advanced Topics*. Enviro Comp Institute, Fremont, CA, and Air and Waste Management Association, Pittsburgh, PA, pp. 93–161.

Anfossi D., Degrazia G., Ferrero E., Gryning S.E., Morselli M.G., and Trini Castelli S. (2000). Estimation of the Lagrangian structure function constant $C_0$ from surface layer wind data. *Bound. Layer Meteorol.* **95**, 249–270.

Anfossi D., Rizza U., Mangia C., Degrazia G.A., and Pereira Marques Filho E. (2006). Estimation of the ratio between the Lagrangian and Eulerian time scales in an atmospheric boundary layer generated by large eddy simulation. *Atmos. Environ.* **40**, 326–337.

Arya S.P. (1999). *Air Pollution Meteorology and Dispersion*. Oxford University Press, Oxford, UK.

ASME (American Society of Mechanical Engineers) (1973). *Recommended Guide for the Prediction of the Dispersion of Airborne Effluents*, 2nd ed. ASME, New York.

Barad M.L. (1958). *Project Prairie Grass, a Field Program in Diffusion*. Geophysical Research Paper 59, Atmospheric Analysis Laboratory, Air Force Cambridge Research Center, US Air Force, Bedford, MA.

Beychok M.R. (2005). *Fundamentals of Stack Gas Dispersion*, 4th ed. Beychok, Newport Beach, CA.

Briggs G.A. (1973). *Diffusion Estimation of Small Emissions*. Contribution No. 79, Atmospheric Turbulence and Diffusion Laboratory, Oak Ridge, TN.

Briggs G.A. (1982). Similarity forms for ground-source surface-layer diffusion. *Bound. Layer Meteorol.* **23**, 489–502.

Briggs G.A. and McDonald K.R. (1978). *Prairie Grass Revisited: Optimum Indicators of Vertical Spread*. Ninth International NATO/CCMS Technical Meeting, Toronto, Canada.

Briggs G.A., Britter R.E., Hanna S.R., Havens J.A., Robins A.G., and Snyder W.H. (2001). Dense gas vertical dispersion over rough surfaces: Results of wind tunnel studies. *Atmos. Environ.* **35**, 2265–2284.

Britter R.E., Hanna S.R., Briggs G.A., and Robins A. (2003). Short-range vertical dispersion from a ground level source in a turbulent boundary layer. *Atmos. Environ.* **37**, 3885–3894.

Byun D.W. and Ching J.K.S. (1999). *Science Algorithms of the EPA Models-3 Community Multiscale Air Quality (CMAQ) Modeling System*. Report EPA/600/R-99/030, US-EPA, Research Triangle Park, NC.

Carvalho J.C., Rizza U., Lovato R., Degrazia G.A., Filho E.P.M., and Campos C.R.J. (2009). Estimation of the Kolmogorov constant by large-eddy simulation in the stable PBL. *Physica A* **388**, 1500–1508.

Cimorelli A.J., Perry S.G., Venkatram A., Weil J.C., Paine R.J., Wilson R.B., Lee R.F., Peters W.D., Brode R.W., and Paumier J.O. (2004). *AERMOD: Description of Model Formulation*. Report EPA-454/R-03-004, US-EPA, Research Triangle Park, NC.

Deardorff J.W. and Willis G.E. (1975). A parameterization of diffusion into the mixed layer. *J. Appl. Meteorol.* **14**, 1451–1458.

Degrazia G.A. and Anfossi D. (1998). Estimation of the Kolmogorov constant $C_0$ from classical statistical diffusion theory. *Atmos. Environ.* **32**, 3611–3614.

Degrazia G.A., Acevedo O.C., Carvalho J.C., Goulart A.G., Morales O.L.L., Campos Velho H.F., and Moreira D.M. (2005). On the universality of the dissipation rate functional form and of the autocorrelation function exponential form. *Atmos. Environ.* **39**, 1917–1924.

Degrazia G.A., Welter G.S., Wittwer A.R., da Costa Carvalho J., Roberti D.R., Costa Acevedo O., Moraes O.L.L., and de Campos Velho H.F. (2008). Estimation of the Lagrangian Kolmogorov constant from Eulerian measurements for distinct Reynolds number with application to pollution dispersion model. *Atmos. Environ.* **42**, 2415–2423.

de Haan P. and Rotach M.W. (1998). A novel approach to atmospheric dispersion modelling: The puff-particle model. *Quart. J. Roy. Meteorol. Soc.* **124**, 2771–2792.

DeVito T.J., Cao X., Roy G., Costa J.R., and Andrews W.S. (2009). Modelling aerosol concentration distributions from transient (puff) sources. *Can. J. Civil Eng.* **36**, 911–922.

Donzis D.A. and Sreenivasan K.R. (2010). The bottleneck effect and the Kolmogorov constant in isotropic turbulence. *J. Fluid Mech.* **657**, 171–188.

Draxler, R.R. (1976). Determination of atmospheric diffusion parameters. *Atmos. Environ.* **10**, 99–105.

Du S. (1997). Universality of the Lagrangian velocity structure function constant ($C_0$) across different kinds of turbulence. *Bound. Layer Meteorol.* **83**, 207–219.

Du S. (1998). Reply to "Comments on the 'Universality of the Lagrangian velocity structure function constant ($C_0$) across different kinds of turbulence"' by A.M. Reynolds. *Bound. Layer Meteorol.* **89**, 171–172.

EPA (Environmental Protection Agency) (1995). *User's Guide for the Industrial Source Complex (ISC3) Dispersion Models. Volume II—Description of Model Algorithms*. Report EPA-454/B-95-003b, US-EPA, Research Triangle Park, NC.

Gifford F.A. (1961). Use of routine meteorological observations for estimating atmospheric dispersion. *Nuclear Safety* **2**, 47–51.

Hanna S.R. (1981). Lagrangian and Eulerian time-scale relationships in the day-time boundary layer. *J. Appl. Meteorol.* **20**, 242–249.

Hanna S.R. (1982). Applications in air pollution modelling. In Nieuwstadt F.T.M. and van Dop H. (eds.), *Atmospheric Turbulence and Air Pollution Modelling*. Reidel, Dordrecht, The Netherlands.

Hanna S.R. and Britter R.E. (2002). *Wind Flow and Vapor Cloud Dispersion at Industrial and Urban Sites*. AIChE–CPPS, New York.

Hanna S.R., Chang J.S., and Strimaitis D.G. (1990). Uncertainties in source emission rate estimates using dispersion models. *Atmos. Environ.* **24A**, 2971–2980.

Irwin J.S. (1979). *Scheme for Estimating Dispersion Parameters as a Function of Release Height*. EPA-600/4-79-062, US-EPA, Research Triangle Park, NC.

Irwin, J.S. (1983). Estimating plume dispersion—A comparison of several sigma schemes. *J. Climate Appl. Meteorol.* **22**, 92–114.

Islitzer N.F. and Slade D.H. (1968). Diffusion and transport experiments. In Slade D.H. (ed.) *Meteorology and Atomic Energy*. US Department of Energy, Technical Information Center, Oak Ridge, TN, pp. 117–188.

Klug W. (1969). A method for determining diffusion conditions from synoptic observations. *Staub-Reinhalt. Luft* **29**, 14–20.

Kolmogorov A.N. (1941). Local structure of turbulence in an incompressible fluid at very high Reynolds numbers. *Doklady Akad. Nauk. SSSR* **30**, 299–303.

Lamb R.G. (1978). A numerical simulation of dispersion from an elevated point source in the convective boundary layer. *Atmos. Environ.* **12**, 1297–1304.

Ludwig F.L., Gasiorek L.S., and Ruff R.E. (1977). Simplification of a Gaussian puff model for real-time minicomputer use. *Atmos. Environ.* **11**, 431–436.

Martin D.O. (1976). Comment on the change of concentration standard deviations with distance. *J. Air Pollut. Control Assoc.* **26**, 145–146.

McElroy J.L. and Pooler F. (1968). *The St. Louis Dispersion Study*. Report AP-53, US Public Health Service, National Air Pollution Control Administration, Arlington, VA.

Neumann J. (1978). Some observations on the simple exponential function as a Lagrangian velocity correlation function in turbulent diffusion. *Atmos. Environ.* **12**, 1965–1968.

Pasquill F. (1961). Estimation of the dispersion of windborne material. *Meteorol. Mag.* **90**, 33–49.

Pasquill F. (1974). *Atmospheric Diffusion*. Wiley, New York.

Pasquill F. (1979). Atmospheric dispersion modeling. *J. Air Pollut. Control Assoc.* **29**, 117–119.

Reynolds A.M. (1998). Comments on the 'Universality of the Lagrangian velocity structure function constant ($C_0$) across different kinds of turbulence.'" *Bound. Layer Meteorol.* **89**, 161–170.

Sawford B.L. (1985). Lagrangian stochastic simulation of concentration mean and fluctuation fields. *J. Appl. Meteorol.* **24**, 1152–1166.

Scire J.S., Lurmann F.W., Bass A., and Hanna S.R. (1984a). *Development of the MESOPUFF II Dispersion Model*. Report EPA-600/3-84-057, US-EPA, Research Triangle Park, NC.

Scire J.S., Lurmann F.W., Bass A., and Hanna S.R. (1984b). *User's Guide to the MESOPUFF II Model and Related Processor Programs*. Report EPA-600/8-84-013, US-EPA, Research Triangle Park, NC.

Scire J.S., Strimaitis D.G., and Yamartino R.J. (2000). *A User's Guide for the CALPUFF Dispersion Model*. Earth Tech, Concord, MA.

Seinfeld J.H. and Pandis S.N. (2006). *Atmospheric Chemistry and Physics*, 2nd ed. Wiley, Hoboken, NJ.

Smith F.B. (1972). A scheme for estimating the vertical dispersion of a plume from a source near ground level. In *Proceedings of the NATO Expert Panel on Air Pollution Modelling, Oct. 1972*. Report NATO-CCMS-14, North Atlantic Treaty Organization, Brussels.

Stull R.B. (1988). *An Introduction to Boundary Layer Meteorology*. Kluwer Academic, Dordrecht, The Netherlands.

Taylor G.I. (1922). Diffusion by continuous movements. *Proc. London Math. Soc.* **20**, 196–211.

Taylor G.I. (1938). The spectrum of turbulence. *Proc. R. Soc.* **A164**, 476–490.

Tennekes H. (1979). The exponential Lagrangian correlation function and turbulent diffusion in the inertial subrange. *Atmos. Environ.* **13**, 1565–1567.

Turner D.B. (1970). *Workbook of Atmospheric Dispersion Estimates*. US EPA, Washington DC.

Turner D.B. (1994). *Workbook of Atmospheric Diffusion Estimates*, 2nd ed. Lewis Publishers, CRC Press, Boca Raton, FL.

Vanderschaege P. (1987). *Simulatiemodel voor Geurverspreiding in de Atmosfeer*. M.Sc. Thesis, Ghent University, Belgium.

Venkatram A. (1992). Vertical dispersion of ground-level releases in the surface boundary layer. *Atmos. Environ.* **26A**, 947–949.

Venkatram A. and Du S. (1997). An analysis of the asymptotic behavior of cross-wind-integrated ground-level concentrations using Lagrangian stochastic simulation. *Atmos. Environ.* **31**, 1467–1476.

Weber A.H., Irwin J.S., Petersen W.B., Mathis J.J. Jr., and Kahler J.P. (1982). Spectral scales in the atmospheric boundary layer. *J. Appl. Meteorol.* **21**, 1622–1632.

Weil J.C. (1990). A diagnosis of the asymmetry in top-down and bottom-up diffusion using a Lagrangian stochastic model. *J Atmos. Sci.* **47**, 501–515.

Weil J.C., Corio L.A., and Brower R.P. (1997). A PDF dispersion model for buoyant plumes in the convective boundary layer. *J. Appl. Meteorol.* **36**, 982–1003.

Willis G.E. and Deardorff J.W. (1978). A laboratory study of dispersion from an elevated source in a convective mixed layer. *Atmos. Environ.* **12**, 1305–1313.

Willis G.E. and Deardorff J.W. (1981). A laboratory study of dispersion from a source in the middle of the convectively mixed layer. *Atmos. Environ.* **15**, 109–117.

Wilson J.D. (2008). Monin-Obukhov functions for standard deviations of velocity. *Bound. Layer Meteorol.* **129**, 353–369.

Wilson J.D. and Sawford B.L. (1996). Review of Lagrangian stochastic models for trajectories in the turbulent atmosphere. *Bound. Layer Meteorol.* **78**, 191–210.

Yee E., Kosteniuk P.R., and Bowers J.F. (1998). A study of concentration fluctuations in instantaneous clouds dispersing in the atmospheric surface layer for relative turbulent difusion: Basic descriptive statistics. *Bound. Layer Meteorol.* **87**, 409–457.

Zannetti P. (1981). An improved puff algorithm for plume dispersion simulation. *J. Appl. Meteorol.* **20**, 1203–1211.

CHAPTER 7

# PLUME–ATMOSPHERE INTERACTIONS

## 7.1 INTRODUCTION

In Chapter 2, we saw that atmospheric turbulence causes plumes to spread horizontally and vertically and to fluctuate randomly. We have studied plume spreading extensively through the dispersion parameters $\sigma_y$ and $\sigma_z$ and eliminated the fluctuation by taking average concentration values. However, the plume interacts with the atmosphere in different ways as well. Examples are plume rise, plume downwash, and deposition. **Plume rise** is the spontaneous rising of stack gases after they are released to the atmosphere. **Plume downwash** is the effect of building and stack wakes on the movement of plumes. Plumes can get trapped in building wakes, causing high concentrations close to the surface. **Depostion** is the mass transfer of pollutants from the plume to the surface. These processes are not fundamentally linked to any particular model. Usually, plume rise and downwash are used in conjunction with Gaussian plume and puff models. Deposition is commonly used in Gaussian, Eulerian, and stochastic Lagrangian models.

The purpose of this chapter is to understand the plume–atmosphere interactions plume rise, plume downwash, and deposition and to be able to predict them.

## 7.2 PLUME RISE

### 7.2.1 Introduction

Plume rise is caused by two phenomena, **buoyancy** and **momentum**. Most plumes, especially from power generation, are buoyant. They are hotter than the ambient air ($T_s > T_a$) and their average molar mass is fairly close to that of pure air. As a result, they are less dense than the surrounding air and rise. Plumes also leave the stack with a vertical velocity (and hence vertical momentum), and this causes plume rise as well.

In a *stable* atmosphere the plume has an equilibrium height, where the plume stops rising. In an *unstable* atmosphere, the plume can potentially rise until it hits the capping inversion. In practice, this only occurs to parts of the plume caught in

*Air Dispersion Modeling: Foundations and Applications*, First Edition. Alex De Visscher.

updrafts. Once the vertical plume velocity is negligible in comparison with the velocity of updrafts and downdrafts, we can simply assume that plume rise has ended. In a *neutral* atmosphere, a plume can theoretically rise to the inversion layer as well. In practice, the plume rise becomes negligibly slow at large distances from the source. The actual plume rise will depend on how fast momentum and buoyancy are dissipated in the atmosphere.

Plumes can fall as well. **Cold** and **dense plumes** sink because the gravitational force is stronger than the buoyant force. A second cause of plume fall is plume downwash, where obstacles cause plumes to move downward. These are two very different phenomena and will be covered separately.

### 7.2.2 Plume Rise Theory

There are many plume rise models, and their predictions are more than a factor of 10 apart. Selection of an appropriate model is important. There are four model types:

- Purely **empirical** models. These inaccurate, obsolete models should be avoided.
- Models based on **similarity theory** and **dimensional analysis**. These are the most commonly used models.
- Models based on **one-dimensional mass, heat**, and **momentum balances**. This type of model requires numerical integration. It is sometimes used for large, buoyant sources and in combination with downwash models.
- Models based on **computational fluid dynamics** (CFD). These models are computationally very intensive and are mainly used for research purposes.

In the future, CFD may provide us with a full understanding of plume rise. In the mean time, approximations in the fluid dynamics still make this type of model unreliable. Hence, we will consider only models based on dimension analysis (this section) and one-dimensional numerical models (Section 7.2.4).

We consider six variables in the description of plume rise (Briggs, 1965):

Momentum flux parameter: $$F_m = \left(\frac{\overline{\rho_s}}{\overline{\rho}}\right) r_s^2 \overline{w_s}^2 \quad (\mathrm{m^4\,s^{-2}}) \tag{7.1}$$

Buoyancy flux parameter: $$F_b = \left(1 - \frac{\overline{\rho_s}}{\overline{\rho}}\right) g r_s^2 \overline{w_s} \quad (\mathrm{m^4\,s^{-3}}) \tag{7.2}$$

Stability Parameter: $$s = \frac{g}{T}\frac{d\theta}{dz} \quad (\mathrm{s^{-2}}) \tag{7.3}$$

Mean horizontal wind speed $\overline{u}$ $(\mathrm{m\,s^{-1}})$

Time of travel: $t$ (s)

Plume rise: $\Delta z\,(\text{transitional}); \Delta h\,(\text{final})$ (m) where $\overline{\rho_s}$ is stack gas density, $\overline{\rho}$ is atmospheric air density, $r_s$ is stack radius, $\overline{w_s}$ is vertical velocity of the stack gas, $g$ is acceleration due to gravity, $T$ is temperature of the atmosphere, $\theta$ is potential temperature, and $z$ is height. By **transitional plume rise** we mean the height $\Delta z$ a plume has risen by the time it reaches a certain location, whereas **final plume rise** is the height $\Delta h$ a plume has risen by the time it stops rising.

The first variable, the **momentum flux parameter** $F_m$ is a relative measure of the vertical momentum added to the atmosphere per unit time as a result of the plume emission. The second variable, the **buoyancy flux parameter** $F_b$ is a relative measure of the amount of buoyancy added to the atmosphere per unit time as a result of the plume emission. The third variable is the **stability parameter** $s$, discussed in Chapter 5. The remaining three variables are wind speed, time, and plume rise.

The variables have been chosen in such a way that there are only two dimensions, distance and time. Therefore, according to the Buckingham $\pi$ theorem, there exist four (= 6 – 2) dimensionless groups that are linked by a function:

$$\pi_4 = f(\pi_1, \pi_2, \pi_3) \tag{7.4}$$

Measurements of plume rise are not accurate enough to establish a relationship between that many variables. Therefore, some simplifications are needed:

- It is assumed that either $F_m$ or $F_b$ plays a role, but not both (this requirement will be relaxed later on).
- It is assumed that $F_m/\overline{u}$ and $F_b/\overline{u}$ are more relevant than $F_m$ or $F_b$.
- Either stability or time is included, but not both. In other words, we will assume that the impact of stability on transitional plume rise is negligible.

Thanks to these simplifications we can eliminate three variables, so the Buckingham $\pi$ theorem yields the following:

$$\pi_1 = \text{const.} \tag{7.5}$$

Consider the case of transitional plume rise of **buoyancy-dominated plumes**. Eliminating $F_m$, $s$ and combining $F_b$ and $\overline{u}$, the three remaining variables that will make up the dimensionless number are $F_b/\overline{u}$ (m$^3$/s$^2$), $t$ (s), and $\Delta z$ (m). The obvious choice is

$$\pi_1 = \frac{F_b t^2}{\overline{u}(\Delta z)^3} \tag{7.6}$$

Equation (7.6) can be solved for $\Delta z$. The result is

$$\Delta z = \pi_1^{-1/3}\left(\frac{F_b t^2}{\overline{u}}\right)^{1/3} \tag{7.7}$$

From measurement data, $\pi_1^{-1/3}$ has been found to be about 1.6. Because $\Delta z$ is proportional to $t^{2/3}$ in eq. (7.7), this is known as the two-thirds rule of buoyancy-dominated plumes.

Following the same reasoning, the following equation can be derived for transitional plume rise of **momentum-dominated plumes**:

$$\Delta z = \pi_2^{-1/3} \left( \frac{F_{\mathrm{m}} t}{\bar{u}} \right)^{1/3} \tag{7.8}$$

Now $\pi_2^{-1/3}$ has been found to be about 2. Because $\Delta z$ is proportional to $t^{1/3}$ in eq. (7.8), this is known as the one-third rule of momentum-dominated plumes.

Considering that $t = x / \bar{u}$, the following equations can be derived:

Buoyancy dominated:

$$\Delta z = 1.6 \left( \frac{F_{\mathrm{b}} x^2}{\bar{u}^3} \right)^{1/3} \tag{7.9}$$

Momentum dominated:

$$\Delta z = 2 \left( \frac{F_{\mathrm{m}} x}{\bar{u}^2} \right)^{1/3} \tag{7.10}$$

To calculate final plume rise in a ***stable*** atmosphere, the stability parameter $s$ is used in the analysis instead of $t$. Comparing units, it is clear that $t$ is substituted for $s^{-1/2}$ in eqs. (7.7) and (7.8). Hence:

$$\Delta h = \pi_1^{-1/3} \left( \frac{F_{\mathrm{b}}}{\bar{u} s} \right)^{1/3} \tag{7.11}$$

$$\Delta h = \pi_2^{-1/3} \left( \frac{F_{\mathrm{m}}}{\bar{u} s^{1/2}} \right)^{1/3} \tag{7.12}$$

For final plume rise, $\pi_1^{-1/3}$ has been found to be about 2.6, and $\pi_2^{-1/3}$ has been found to be about 1.5.

When the wind speed is very low this approach leads to plume rise tending to infinity, which is unrealistic. In that case, the analysis is made with $F_{\mathrm{b}}$ or $F_{\mathrm{m}}$ instead of $F_{\mathrm{b}}/\bar{u}$ or $F_{\mathrm{m}}/\bar{u}$, and $\bar{u}$ is no longer used as a variable. For final plume rise, the remaining variables are $F_{\mathrm{b}}$ or $F_{\mathrm{m}}$, as well as $s$ and $\Delta h$. The dimensionless numbers are

$$\pi_1 = \frac{F_{\mathrm{b}}}{s^{3/2} (\Delta h)^4} \tag{7.13}$$

for buoyancy-dominated plumes, and

$$\pi_2 = \frac{F_{\mathrm{m}}}{s (\Delta h)^4} \tag{7.14}$$

for momentum-dominated plumes. Hence, the final plume rise heights are given by

$$\Delta h = \pi_1^{-1/4}\left(\frac{F_b^2}{s^3}\right)^{1/8} \tag{7.15}$$

$$\Delta h = \pi_2^{-1/4}\left(\frac{F_m}{s}\right)^{1/4} \tag{7.16}$$

The following values have been put forward for the constants: 5.3 for $\pi_1^{-1/4}$ and 2.4 for $\pi_2^{-1/4}$.

When the atmosphere is near neutral ($s=0$) the above approach would lead to an infinite final plume rise. As discussed before, there is some justification for that, but in practice, plume rise is so slow at large distances from the source that it is more practical to use a constant plume rise representing large distances. Hence, under neutral conditions it is assumed that final plume rise can be based on the variables $F_b$ (or $F_m$), $\bar{u}$, and $\Delta h$. The dimensionless numbers are

$$\pi_1 = \frac{F_b}{\bar{u}^3\,\Delta h} \tag{7.17}$$

for buoyancy-dominated plumes, and

$$\pi_2 = \frac{F_m}{\bar{u}^2(\Delta h)^2} \tag{7.18}$$

for momentum-dominated plumes. Hence, the final plume rise heights are given by

$$\Delta h = \pi_1^{-1}\frac{F_b}{\bar{u}^3} \tag{7.19}$$

$$\Delta h = \pi_2^{-1/2}\frac{F_m}{\bar{u}^2} \tag{7.20}$$

The following values have been put forward for the constants: 400 for $\pi_1^{-1}$ and 3 for $\pi_2^{-1/2}$.

As indicated earlier, there is a practical limit to plume rise even in unstable conditions. Briggs (1975) suggested the following equations for the maximum distance of plume rise:

$$x_f = d_1 F_b^{5/8} \quad \text{for} \quad F_b < 55\,\mathrm{m^4\,s^{-3}} \tag{7.21}$$

$$x_f = d_2 F_b^{2/5} \quad \text{for} \quad F_b > 55\,\mathrm{m^4\,s^{-3}} \tag{7.22}$$

Where $d_1=49\,\mathrm{s^{15/8}\,m^{-3/2}}$ and $d_2=119\,\mathrm{s^{9/5}\,m^{-7/5}}$, $x_f$ is in meters. These equations are not based on dimension analysis and do not convey a fundamental relationship between properties but rather a practical guideline for calculating plume rise. Final plume rise is calculated as plume rise at $x=x_f$. For instance, for buoyancy-dominated plumes [eq. (7.9)], we obtain

$$\Delta h = \frac{c_1 F_b^{3/4}}{\bar{u}} \quad \textit{with} \quad c_1 = 21.4\,\mathrm{s}^{1.25}\,\mathrm{m}^{-1} \qquad F_b < 55\,\mathrm{m}^4\,\mathrm{s}^{-3} \tag{7.23}$$

$$\Delta h = \frac{c_2 F_b^{3/5}}{\bar{u}} \quad \text{with} \quad c_1 = 38.7\,\mathrm{s}^{0.8}\,\mathrm{m}^{-0.4} \qquad F_b > 55\,\mathrm{m}^4\,\mathrm{s}^{-3} \tag{7.24}$$

The final plume rise heights thus obtained are considerably smaller than the plume rise obtained with eq. (7.19) unless $\bar{u} > 6\,\mathrm{m\,s}^{-1}$.

When both **momentum** and **buoyancy** determine plume rise, Briggs (1975) suggested the following equation:

$$\Delta z = \left( \frac{3F_m x}{0.6^2 \bar{u}^2} + \frac{3F_b x^2}{2 \cdot 0.6^2 \bar{u}^3} \right)^{1/3} \tag{7.25}$$

The factor 0.6 in eq. (7.25) is an entrainment rate parameter also encountered in the numerical plume rise model in the next section. The equations derived for transitional plume rise of momentum-dominated and buoyancy-dominated plumes, eqs. (7.9) and (7.10), are limiting cases of eq. (7.25).

Weil (1988) proposed a transitional plume rise equation for stable, windy conditions as an extension of eq. (7.11):

$$\Delta z = 2.66 \left( \frac{F_b}{\bar{u} s} \right)^{1/3} \left[ \frac{0.7 s^{1/2} F_m}{F_b} \sin\left( \frac{0.7 s^{1/2} x}{\bar{u}} \right) + 1 - \cos\left( \frac{0.7 s^{1/2} x}{\bar{u}} \right) \right]^{1/3} \tag{7.26}$$

The distance to final plume rise is given by

$$x_f = \frac{\bar{u}}{0.7 s^{1/2}} \arctan\left( \frac{F_b}{0.7 s^{1/2} F_m} \right) \tag{7.27}$$

Final plume rise is given by eq. (7.11) but with $\pi_1^{-1/3} = 2.66$.

When a plume hits a capping inversion before the final plume rise is reached, it can generally be assumed that the plume will stop rising further. A rule of thumb is that the capping inversion is reached when

$$h_s + \Delta z + 2.15\sigma_z = h_{\mathrm{mix}} \tag{7.28}$$

However, sometimes a plume can penetrate the capping inversion. AERMOD provides a calculation procedure for this situation. It will be discussed in Chapter 14.

Sometimes the wind speed in equations such as eqs. (7.25) and (7.26) is evaluated at the local plume height instead of the source height. In that case, eq. (7.25) is calculated iteratively (Cimorelli et al., 2004; Turner and Schulze, 2007).

**Example 7.1.** A stack with diameter 1.5 m emits 500 $m^3 min^{-1}$ of exhaust gas at a temperature of 100 °C. The ambient temperature is 18 °C and the pressure 98 kPa. The ambient wind speed is 1.2 $m s^{-1}$. The atmosphere is slightly stable, with a potential temperature gradient of 0.001 $K m^{-1}$. Find the most appropriate equation and justify the choice made. Plot the plume rise versus downwind distance up to $x = 200$ m.

***Solution.*** Based on the data, we can calculate all the variables needed to use the plume rise equations.

The stack radius is half of the stack diameter: $r_s = 0.75$ m.

The stack gas flow rate is divided by 60 s $min^{-1}$ and by the cross-sectional stack area ($= \pi r_s^2$) to obtain a vertical stack gas velocity: $w_s = 4.716$ m $s^{-1}$.

Assuming a molar mass of 0.029 kg $mol^{-1}$ for both air and stack gas, the ideal gas law is used to calculate densities of both stack gas and air:

$$\rho_s = \frac{M_{air}\, p}{RT_s} = \frac{0.029 \times 98000}{8.314472 \times 373.15} = 0.916\,\text{kg}\,\text{m}^{-3}$$

$$\rho_a = \frac{M_{air}\, p}{RT_a} = \frac{0.029 \times 98000}{8.314472 \times 291.15} = 1.174\ \text{kg}\ \text{m}^{-3}$$

Based on the potential temperature gradient, the stability parameter is calculated:

$$s = \left(\frac{g}{T}\right)\frac{d\theta}{dz} = \frac{9.80665}{291.15} \times 0.001 = 3.368 \times 10^{-5}\,\text{s}^{-2}$$

The momentum and buoyancy flux parameters are calculated:

$$F_m = (0.916/1.174) \times 0.75^2 \times 4.716^2 = 9.760\,\text{m}^4\text{s}^{-2}$$

$$F_b = (1 - 0.916/1.174) \times 9.80665 \times 0.75^2 \times 4.716 = 5.716\,\text{m}^4\text{s}^{-3}$$

Based on these variables, the plume rise equations are calculated. First, transitional plume rise is calculated with eq. (7.9) (buoyancy dominated) and eq. (7.10) (momentum dominated). It is found that the momentum-dominated plume rise equation dominates in the first 4–5 m from the source, whereas the buoyancy-dominated plume rise dominates further away from the source. It is prudent to use eq. (7.25) for the transitional plume rise calculations.

For the final plume rise equations, a value of 135.5 m is obtained with eq. (7.11) and 16.8 m with eq. (7.12), confirming that the plume rise is buoyancy dominated. We test if these equations overestimate the plume rise due to the low wind speed by using eqs. (7.15) and (7.16), applicable to low wind speeds. Values of 389.7 m and 55.7 m are obtained, respectively, indicating no overestimations with eqs. (7.11)

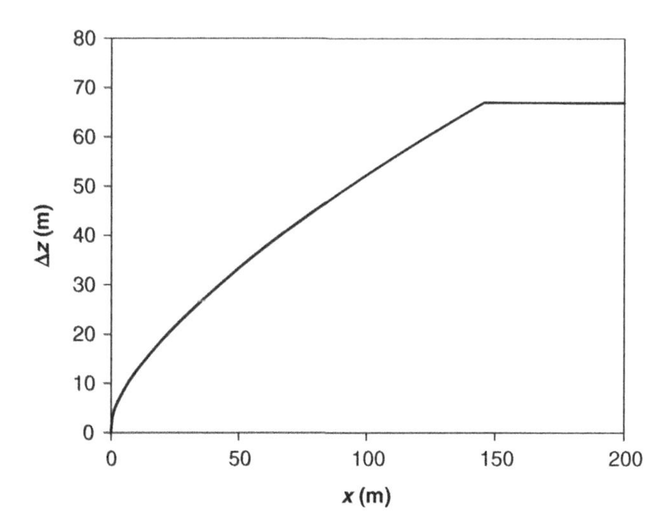


Figure 7.1 Plume rise calculation result.

and (7.12). A similar test is carried out with eqs. (7.19) and (7.20) because eqs. (7.11) and (7.12) might overestimate plume rise due to low stability. Values of 1323.2 and 20.3 m are obtained, respectively, indicating no problem with eqs. (7.11) and (7.12).

On the other hand, a distance to final plume rise of 145.7 m is obtained with eq. (7.21), which indicates a final plume rise of 67.0 m with eq. (7.25). There is no compelling argument for rejecting either value (135.5 or 67.0 m) so the lower value is chosen to ensure that the estimate is conservative. The resulting plume rise profile is shown in Figure 7.1.

The plume rise calculation is on the Excel file "Example 7.1. Plume rise calculations.xlsx."

### 7.2.3 Flare Plume Rise

Flare plumes require a special calculation scheme because the emitted gas is combusted after leaving the "stack" (i.e., the flare tip). As a result, heat is produced, part of which is lost by radiation. Air is entrained by the burning gas, so the gas flow rate at the **flame tip** is larger than the gas flow rate leaving the flare tip. Regulators have proposed several different calculation schemes for incorporating flare combustion in plume rise. They are usually variants of the calculation scheme of Beychok (2005). The basis of Beychok's (2005) calculation is that a buoyant plume starts at the flame tip of the flare. Hence the **theoretical stack height** is introduced as the height of the flame tip above the ground (see Fig. 7.2).

The **flame length** $L_f$ can be estimated based on the heating value $Q_c$ of the flare gas:

$$L_f = 0.00178 Q_c^{0.478} = 2.43 Q_{c,MW}^{0.478} \tag{7.29}$$

with $L_f$ in meters, $Q_c$ in kJ hr$^{-1}$, and $Q_{c,MW}$ in MW.

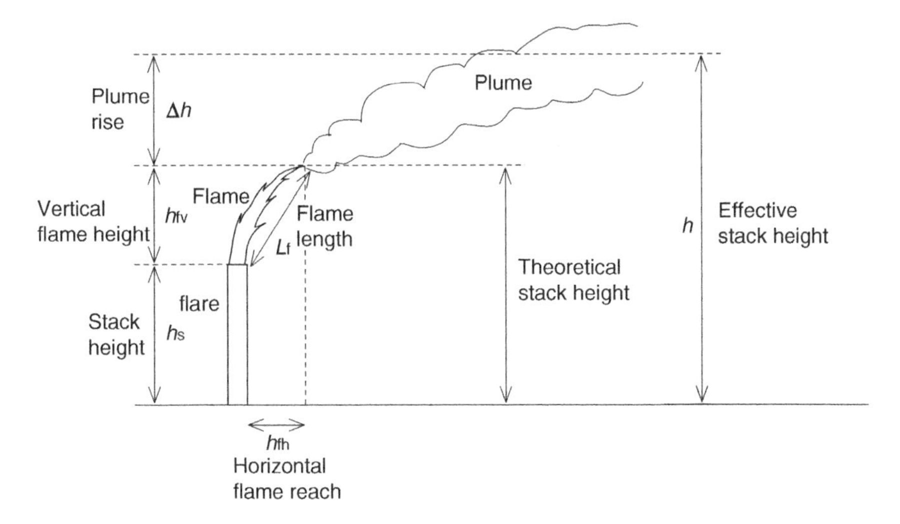


Figure 7.2 Flare stack and flame parameters. Theoretical stack height [after Beychok (2005)].

Because the flare is not straight, a correction is needed to calculate the **vertical flame height** $h_{fv}$. Beychok (2005) suggested the (usually conservative) assumption that the flame is tilted 45°. Hence:

$$h_{fv} = 0.707L_f \tag{7.30}$$

$$h_{fh} = 0.707L_f \tag{7.31}$$

For plume rise equations the **sensible heat emission** $Q_s$ is needed because this determines buoyant behavior of the plume. There is a large uncertainty on the amount of heat lost by radiation in a flare flame, but based on data presented in American Petroleum Institute (API) (2007), 25% seems a reasonable assumption and is recommended by Beychok (2005). Hence:

$$Q_s = 0.75Q_c \tag{7.32}$$

The sensible heat emission can be converted to the buoyancy flux parameter with an approximate equation. Beychok (2005) proposed the following based on work of Briggs:

$$F_b = c_F Q_s \tag{7.33}$$

where $c_F = 8.8\,m^4\,s^{-3}\,MW^{-1}$. The plume rise $\Delta h$ is then calculated assuming that the plume is buoyancy dominated. The effective source height is calculated by

$$h = h_s + h_{fv} + \Delta h \tag{7.34}$$

**Example 7.2.** Calculate the effective source height after final plume rise of a 25-m flare combusting $0.75\,m^3\,s^{-1}$ (at 1 bar and 0 °C) of pure methane (heat of combustion $50{,}000\,kJ\,kg^{-1}$; $M = 0.016\,kg\,mol^{-1}$). Assume a wind speed of $4\,m\,s^{-1}$.

***Solution.*** First, we need the heating value of the methane flow to calculate the flame length. The density of methane at 0 °C and 1 bar is $Mp/RT = 0.6454\,kg\,m^{-3}$.

With this density and a volumetric flow rate of $0.75\,m^3\,s^{-1}$, a mass flow rate of $0.75 \times 0.6454 = 0.4841\,kg\,s^{-1}$ is obtained. With a heat of combustion of $50{,}000\,kJ\,kg^{-1}$, a heating value $Q_c = 24{,}200\,kW = 24.2\,MW$ is obtained.

With a heating value of 24.2 MW, eq. (7.29) is applied to obtain a flame length $L_f = 2.43 \times 24.2^{0.478} = 11.15\,m$. The horizontal flame height $h_{fv} = 11.15/2^{1/2} = 7.88\,m$.

The sensible heat emission is calculated with eq. (7.32): $Q_s = 0.75 \times 24.20 =$ 18.15 MW. This value is used to calculate the buoyancy flux parameter with eq. (7.33):

$$F_b = 8.8 \times 18.15 = 159.7\,m^4 s^{-3}$$

The distance to final plume rise is found with eq. (7.22) ($F_b > 55\,m^4\,s^{-3}$):

$$x_f = 119 \times 159.7^{2/5} = 905.6\,m$$

The plume rise is calculated with eq. (7.9):

$$\Delta h = 1.6 \times (159.7 \times 905.6^2 / 4^3)^{1/3} = 213.15\,m$$

The effective source height at final plume rise is the sum of the flare height, the vertical plume height, and the plume rise. The result is 25 + 7.88 + 203.15 = **236.03 m**.

The calculations are in the Excel file "Example 7.2. Flare plume rise calculation.xls."

An unsolved problem of this approach is that the plume rise equations assume a plume that initially moves vertically. This is not the case with the exhaust gases leaving the flare tip. Hence, the actual plume rise will be less than predicted. Numerical plume rise models may offer a solution to this problem in the future. This type of plume rise model is the subject of the next section.

### 7.2.4 Numerical Plume Rise Calculations

A popular numerical plume rise model is a model incorporated in CALPUFF for large buoyant "area sources" such as forest fires (Scire et al., 2000). The model is also used in the downwash model PRIME (Schulman et al., 2000). It consists of a set of ordinary differential equations that solve the mass, energy, and momentum equations of a plume. This set of differential equations is solved numerically.

Even though the model calculates the plume as a three-dimensional object, it is essentially a one-dimensional model. This is because the independent variable of the

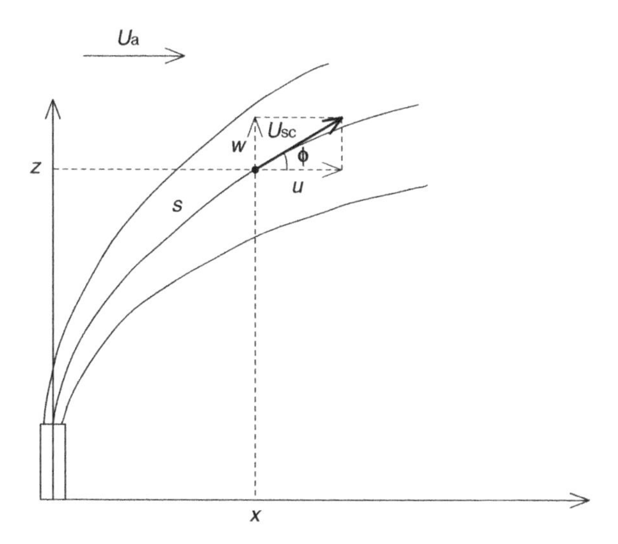


Figure 7.3 Model variables of the numerical plume rise model of Schulman et al. (2000) (Scire et al., 2000).

model is the distance traveled by the plume, $s$. Other variables that define the plume as a three-dimensional object, such as plume location along the $x$ and $z$ direction and plume radius, are calculated as dependent variables. Some of the model variables are defined in Figure 7.3.

The first differential equation of the model is the mass balance, which states that the increase of mass in the plume with $s$ equals the mass mixed into the plume. The model further assumes that there are two driving forces of mass mixing into the plume: shear *along* the plume direction (velocity difference between the plume and ambient air in the direction of the plume axis) and shear *across* the plume direction (the velocity difference projected on a plume cross section). Hence, the mass balance becomes (Schulman et al., 2000):

$$\frac{d}{ds}(\rho U_{sc} r^2) = 2r\alpha\rho_{atm}\left|U_{sc} - U_{atm}\cos\phi\right| + 2r\beta\rho_{atm}\left|U_{atm}\sin\phi\right| \tag{7.35}$$

where $s$ is the distance along the plume centerline path (m), $\rho$ is the density of the plume (kg/m$^3$), $U_{sc}$ is the velocity of the plume centerline (m s$^{-1}$), $r$ is the plume radius (m), $\rho_{atm}$ is the density of the atmosphere (kg m$^{-3}$), $U_{atm}$ is the wind speed of the atmosphere, $\phi$ is the angle between the plume path and the atmospheric streamline, and $\alpha = 0.11$ and $\beta = 0.6$ are empirical parameters defining **entrainment rate** of atmospheric air into the plume resulting from along-plume shear and cross-plume shear.

The term between parentheses in the left-hand side of eq. (7.35) is the mass of the plume per unit length, divided by $\pi$. This is the first dependent variable of the model.

If $u$ and $w$ are the horizontal and vertical component of the plume velocity $U_{sc}$, then the velocities are given by the following equations:

$$\frac{dx}{dt} = u \tag{7.36}$$

$$\frac{dz}{dt} = w \tag{7.37}$$

$$\frac{ds}{dt} = U_{sc} \tag{7.38}$$

By dividing eq. (7.36) by eq. (7.38), we obtain the second differential equation of the model:

$$\frac{dx}{ds} = \frac{u}{U_{sc}} \tag{7.39}$$

The third differential equation is obtained by dividing eq. (7.37) by eq. (7.38):

$$\frac{dz}{ds} = \frac{w}{U_{sc}} \tag{7.40}$$

The velocities will be used to eliminate $\phi$ in eq. (7.35) because the following relations hold:

$$\frac{u}{U_{sc}} = \cos\phi \tag{7.41}$$

$$\frac{w}{U_{sc}} = \sin\phi \tag{7.42}$$

Furthermore, $U_{sc}$ is calculated from $u$ and $w$ as

$$U_{sc} = \sqrt{u^2 + w^2} \tag{7.43}$$

The calculation of $u$ and $w$ themselves will become clear below.

The fourth differential equation is a momentum balance along the wind direction. In this equation, momentum is calculated using the moving ambient air as an observer. Hence, ambient air has zero momentum, and the momentum of the plume is not altered when ambient air is mixed into it. It is also assumed that all the ambient air that causes friction is mixed into the plume, so friction between plume and ambient air layers does not affect the momentum either. The only mechanism that causes a change of the momentum relative to the ambient air is the velocity gradient of ambient air itself: by moving up into faster moving air layers, the plume appears to lose momentum. This leads to the following equation:

$$\frac{d}{ds}\left[\rho U_{sc} r^2 \left(u - U_{atm}\right)\right] = -r^2 \rho w \frac{dU_{atm}}{dz} \tag{7.44}$$

In the vertical direction the same reasoning applies, except that it is assumed that there is no ambient wind speed in the vertical direction. The only mechanism leading to a change in vertical momentum is the difference between gravitational force and buoyant force. This leads to the fifth equation:

$$\frac{d}{ds}\left(\rho U_{sc} r^2 w\right) = g r^2 \left(\rho_{atm} - \rho\right) \tag{7.45}$$

The last equation is the energy balance. The temperature of the ambient air is used as the standard for calculating the enthalpy of the plume. Consequently, air mixed into the plume does not change the enthalpy of the plume. The contributions to an apparent change of the enthalpy are nonadiabatic conditions of the ambient air and radiative heat transfer. Hence, the sixth differential equation is

$$\frac{d}{ds}\left[\rho U_{sc} r^2 (T - T_{atm})\right] = -\left(\frac{dT_{atm}}{dz} + \frac{g}{c_p}\right)\rho w r^2 - \frac{2\varepsilon\sigma}{c_p} r\left(T^4 - T_{atm}^4\right) \tag{7.46}$$

where $T$ is the plume temperature (K), $T_{atm}$ is the temperature of the surrounding air (K), $c_p$ is the heat capacity of air (J kg$^{-1}$ K$^{-1}$), $\varepsilon$ is the emissivity of the plume (–), and $\sigma$ is the Stefan–Boltzman constant ($5.67 \times 10^{-8}$ J s$^{-1}$ m$^{-2}$ K$^{-4}$). Schulman et al. (2000) proposed a value of 0.8 for $\varepsilon$. Such a large value is only realistic for very extended plumes, for example, from forest fires. For ordinary plumes it is probably best to set $\varepsilon = 0$.

Hence, there are six dependent variables: $\rho U_{sc} r^2$, $x$, $z$, $\rho U_{sc} r^2 (u - U_a)$, $\rho U_{sc} r^2 w$, and $\rho U_{sc} r^2 (T - T_a)$. What we need is the values of $\rho$, $U_{sc}$, and so forth. Assuming that the conditions of the ambient air ($U_a$, $T_a$) are known throughout the domain, all variables can be calculated. The plume velocity in the $x$ direction, $u$, can be calculated from the values of the first and fourth variables:

$$u = U_a + \frac{\rho U_{sc} r^2 (u - U_a)}{\rho U_{sc} r^2} \tag{7.47}$$

The plume velocity in the $z$ direction, $w$, can be calculated from the first and fifth variables:

$$w = \frac{\rho U_{sc} r^2 w}{\rho U_{sc} r^2} \tag{7.48}$$

and $T$ can be calculated from the first and the sixth variables:

$$T = T_a + \frac{\rho U_{sc} r^2 (T - T_a)}{\rho U_{sc} r^2} \tag{7.49}$$

Once $T$ is known, $\rho$ can be calculated from the ideal gas law, and $r$ can be calculated from the first variable:

$$r = \sqrt{\frac{\rho U_{sc} r^2}{\rho \cdot U_{sc}}} \tag{7.50}$$

where $U_{sc}$ is calculated from eq. (7.43). Hence, the six differential equations are closed.

De Visscher (2009) extended this plume rise model to include flare combustion. A similar plume rise model with buoyancy generated by chemical reaction was developed by Campbell et al. (2009) and Campbell and Cardoso (2010).

**Example 7.3.** A stack with diameter 1.5 m and height 50 m emits 500 $m^3\,min^{-1}$ of exhaust gas at a temperature of 100 °C. The ambient temperature is 18 °C and the pressure is 98 kPa. The ambient wind speed is 1.2 $m\,s^{-1}$, measured at 50-m height. The atmosphere is slightly stable, with a potential temperature gradient of 0.001 $K\,m^{-1}$. Assume that the wind speed profile is a power law with a power law index of 0.2. Calculate the plume rise with the numerical plume rise model. Plot the plume rise versus downwind distance up to $x = 200$ m. Compare with the result of Example 7.1.

***Solution.*** The problem is solved by numerical integration in Matlab. The solution is given in the Matlab files "main.m," "data.m," and "f.m" in the folder "Numerical plume rise model." For a general introduction to solving differential equations in Matlab, see Appendix B, Section B5.

First, the six dependent variables are defined as follows:

$y(1) = \rho U_{sc} r^2$

$y(2) = x$

$y(3) = z$

$y(4) = \rho U_{sc} r^2 (u - U_{atm})$

$y(5) = \rho U_{sc} r^2 w$

$y(6) = \rho U_{sc} r^2 (T - T_{atm})$

Based on the data, the initial conditions are calculated and stored in the vector *y0* in the file "data.m." To that effect, the following variables are calculated:

$$\rho = \frac{M_{air} p}{RT}$$

$$U_{sc} = \frac{Q}{\pi r^2}$$

where $Q$ is the volumetric flow rate in $m^3\,s^{-1}$. The initial value of $u$ is zero; the initial value of $w$ is $U_{sc}$.

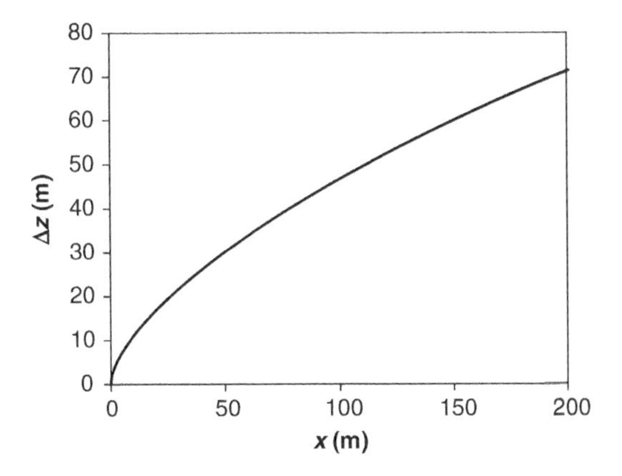


Figure 7.4 Numerical plume rise calculation result.

In the function "f.m" the variables are calculated from $y(1)$ to $y(6)$ as indicated above; $U_{atm}$ is calculated from its value at 50 m with a power law; and $p$ is calculated from its value at 50 m with an equation based on eq. (1.2):

$$p = 98\,\text{kPa} \times \exp[-0.00012(z/\text{m} - 50)]$$

where the factor 1/m is included for unit consistency.

Because this is not a spatially extended plume such as a forest fire, the emissivity $\varepsilon$ is set equal to zero.

The result is shown in Figure 7.4. The plume rise is slightly less than in Example 7.1, possibly because the equations based on dimension analysis do not account for stability in the calculation of transitional plume rise.

The Matlab files that calculate the result are on the website in the "Numerical plume rise model" section.

## 7.3 PLUME DOWNWASH: PRIME (PLUME RISE MODEL ENHANCEMENTS)

### 7.3.1 Introduction

Plume downwash is a process that is caused by the interaction between the atmosphere and the terrain (buildings). When wind moves around a building, additional turbulence is generated in the form of wakes. This is illustrated in Figure 7.5. Air streamlines near the building follow the wakes and are thus deflected. Behind the buildings, the air streamlines move down. Plumes follow the wind streamlines, so they are dragged down behind buildings as well. This process is known as **plume downwash**. If the plumes are sufficiently far from any surrounding buildings, plume downwash will be negligible. As a rule of thumb, if a stack is at least 2.5 times as high as the highest surrounding building, plume downwash will be negligible.

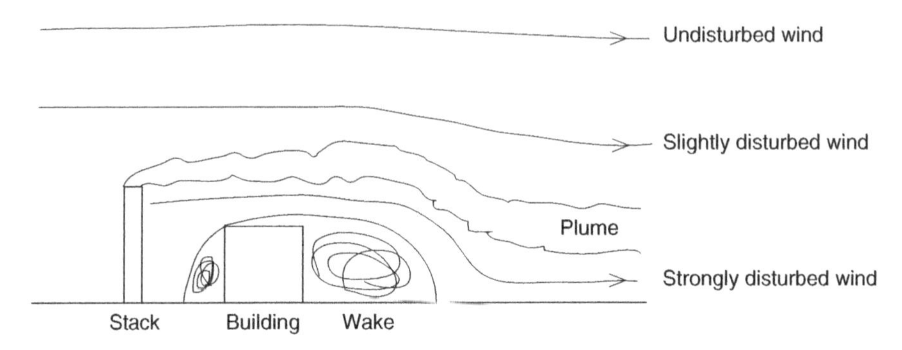


Figure 7.5 Plume downwash.

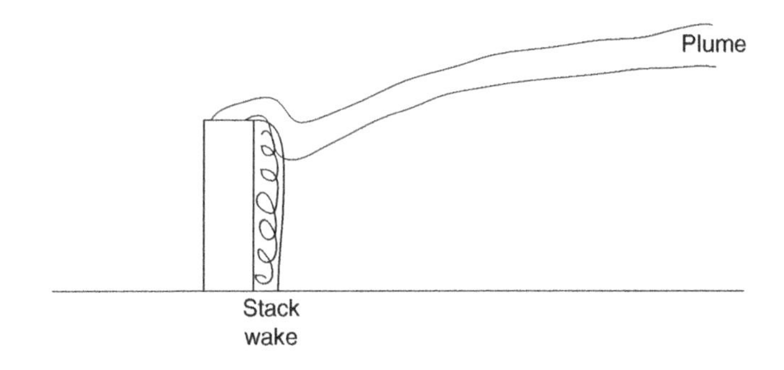


Figure 7.6 Stack downwash.

Stacks have wakes as well. Plumes can get trapped in the wake of the stack itself if the exit speed is insufficient to escape the pressure gradient created by the wake (Figure 7.6). This is known as **stack downwash**. As a rule of thumb, the exit velocity of a plume leaving the stack, $w_s$, should be at least 1.5 times the wind speed at all times. When $w_s < 1.5u$, then stack downwash will generate a negative plume rise, $\Delta h_{sd}$, which can be estimated as follows:

$$\Delta h_{sd} = 2d_s\left(\frac{w_s}{u} - 1.5\right) \tag{7.51}$$

The next sections introduce a downwash model introduced by Schulman et al. (2000) named PRIME (Plume RIse Model Enhancements), based on earlier models and measurements, both in wind tunnels and field tests.

Downwash calculations in PRIME follow three steps. First, the size and location of the building wakes are calculated. Next, the deflection of the streamlines is calculated. Third, the interaction between the plume and the wake is calculated.

### 7.3.2 Wake Size Calculations

An important aspect of PRIME is the calculation of the location of building wakes. In PRIME, building wakes are subdivided in a **recirculating near wake**

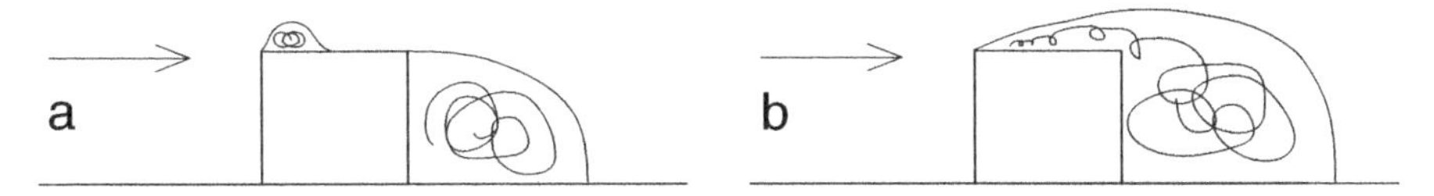


Figure 7.7 Rooftop cavity configurations: (*a*) the rooftop cavity reattaches and (*b*) the rooftop cavity is part of the main cavity.

(or **cavity**) and a **far wake**. Sometimes a separate cavity exists on the rooftop. The following variables are defined:

| | |
|---|---|
| $H_c$ | height of the cavity (depends on location) (m) |
| $H_R$ | height of the cavity at its maximum (m) |
| $W_c$ | half-width of the cavity (m) |
| $L_R$ | length of the cavity (m) |
| $H_W$ | height of the far wake (m) |
| $W_W$ | width of the far wake (m) |
| $H$ | height of the building (m) |
| $W$ | width of the building, projected across the wind flow (m) |
| $L$ | length of the building, projected along the wind direction (m) |
| $x$ | distance from the upwind side of the building (m), not to be confused with the distance from the source |

Furthermore, the **length scale**, $R$ (m), of the building is defined as

$$R = B_S^{2/3} B_L^{1/3} \tag{7.52}$$

with $B_S = \min(H, W)$, and $B_L = \max(H, W)$.

For instance, if a building has a height of 10 m and a width of 20 m, then $R = 10^{2/3} \cdot 20^{1/3} = 12.6$ m.

Depending on the conditions, the cavity on the rooftop can be separate from the main cavity behind the building (when the rooftop cavity reattaches to the roof), or the rooftop cavity can be part of the main cavity. These two situations are illustrated in Figure 7.7. It is assumed that the rooftop cavity reattaches to the roof if

$$L > 0.9R \tag{7.53}$$

When the rooftop cavity reattaches to the roof, then the following equation applies to the maximum height of the cavity:

$$H_R = H \tag{7.54}$$

If the rooftop cavity is part of the main cavity, then the following equation applies:

$$H_R = H + 0.22R \qquad \text{at } x = 0.5R \tag{7.55}$$

The length of the cavity is given by

$$L_R = \frac{1.8W}{(L/H)^{0.3}[1+0.24\ (W/H)]} \quad (0.3 < L/H < 3) \tag{7.56}$$

If $L/H<0.3$, then a value of 0.3 should be used for $L/H$, whereas a value of 3 should be used when $L/H>3$.

When the rooftop cavity reattaches to the roof, $H_c=H$ for $x<L$. For $x>L$, the following equation holds:

$$H_c = H\left[1-\left(\frac{x-L}{L_R}\right)^2\right] \tag{7.57}$$

When the rooftop cavity is part of the main cavity, the following equation holds for $x<0.5R$:

$$H_c = H_R + \frac{4(x-R/2)^2(H-H_R)}{R^2} \tag{7.58}$$

For $x>0.5\ R$, the equation is

$$H_c = H_R\left[1-\frac{(x-R/2)^2}{L+L_R-R/2}\right]^{1/2} \tag{7.59}$$

The width of the cavity is maximal at $x=R$. At $x<R$, the following equation holds:

$$W_c = \frac{W}{2}+\frac{R}{3}-\frac{(x-R)^2}{3R} \tag{7.60}$$

At $x>R$, the cavity width is calculated with

$$W_c = \left(\frac{W}{2}+\frac{R}{3}\right)\sqrt{1-\left(\frac{x-R}{L+L_R-R}\right)^2} \tag{7.61}$$

The height of the far wake is calculated by

$$H_W = 1.2R\left[\frac{x}{R}+\left(\frac{H}{1.2R}\right)^3\right]^{1/3} \tag{7.62}$$

The width of the far wake is calculated by

$$W_W = \frac{W}{2}+\frac{R}{3}\left(\frac{x}{R}\right)^{1/3} \tag{7.63}$$

Of course, the influence of the building is not limited to the near and far wakes. Around the building the wind streamlines are deflected, and plumes will move along these deflected streamlines as well. The next section outlines the PRIME scheme for calculating these deflections.

### 7.3.3 Streamline Deflection Calculation

The wind direction around building wakes is not the same as the undisturbed wind direction. Therefore, adjustments need to be calculated. To that effect, the domain is divided into five sections (Fig. 7.8): (A) upwind, unaffected by the wake ($x<-R$); (B) upwind, affected by the building ($-R<x<0$); (C) downwind, before the wake peak ($0<x<R/2$); (D) downwind, after the wake peak, before the end of the cavity ($R/2<x<L+L_R$); and (E) behind the near wake ($x>L+L_R$).

(A) Upwind, Unaffected by Wake ($x<-R$) The streamline slope is

$$\frac{dz}{dx}=0 \tag{7.64}$$

(B) Upwind, Affected by Building ($-R<x<0$) The streamline slope is

$$\frac{dz}{dx}=\frac{2(H_R-H)(x+R)}{R^2} \qquad z<H \tag{7.65}$$

$$\frac{dz}{dx}=\frac{2(H_R-H)(x+R)}{R^2}\left(\frac{H}{z}\right)^3 \qquad z>H \tag{7.66}$$

However, near the ground the slope is zero:

$$\text{When } R\le H: \frac{dz}{dx}=0 \qquad \text{when} \qquad z<\frac{2H}{3} \tag{7.67}$$

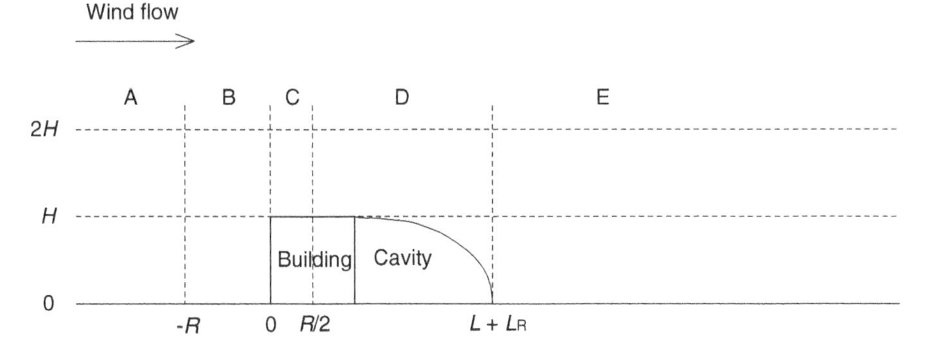


Figure 7.8 Five domains surrounding building wakes for calculating disturbed wind streamlines (adapted from Schulman et al., 2000).

$$\text{When } R > H: \frac{dz}{dx} = 0 \qquad \text{when} \qquad z < \frac{2(2H - R)}{3} \tag{7.68}$$

(C) Downwind, before Wake Peak ($0<x<R/2$) The streamline slope is

$$\frac{dz}{dx} = \frac{-4(H_R - H)(2x/R - 1)}{R} \qquad z < H \tag{7.69}$$

$$\frac{dz}{dx} = \frac{-4(H_R - H)(2x/R - 1)}{R}\left(\frac{H}{z}\right)^3 \qquad z > H \tag{7.70}$$

(D) Downwind, after Wake Peak, before End of Cavity ($R/2<x<L+L_R$) The streamline slope is

$$\frac{dz}{dx} = \frac{(H_R - H)(R - 2x)}{(L + L_R - R/2)^2}\left(\frac{z}{H}\right)^{0.3} \qquad z < H \tag{7.71}$$

$$\frac{dz}{dx} = \frac{(H_R - H)(R - 2x)}{(L + L_R - R/2)^2}\left(\frac{H}{z}\right)^{0.7} \qquad z > H \tag{7.72}$$

(E) Behind the near Wake ($x>L+L_R$)

$$\frac{dz}{dx} = \frac{-(H_R - H)(L + L_R)}{x(L + L_R - R/2)}\left(\frac{z}{H}\right)^{0.3} \qquad z < H \tag{7.73}$$

$$\frac{dz}{dx} = \frac{-(H_R - H)(L + L_R)}{x(L + L_R - R/2)}\left(\frac{H}{z}\right)^{0.7} \qquad z > H \tag{7.74}$$

All these equations apply at $-W/2<y<W/2$. Outside this region, the slope decreases linearly to zero at $y=W/2+R/3$ and $y=-W/2-R/3$.

The main problem with this model is that it is incapable of predicting any wind deflection when the rooftop cavity reattaches to the building. In that case, eq. (7.54) applies, which turns all wind slopes equal to zero.

**Example 7.4.** Calculate wind streamlines around an obstacle with height 10 m, width 20 m, and length 10 m in the downwind direction. Cover the range $x=-20$ m to $x=50$ m, and start from heights 3, 6, 9, 12, and 15 m.

***Solution.*** The solution is calculated in the Matlab files "data.m," "main.m," and "f.m" in the folder "Streamline deflection" on the enclosed CD.

First, all the required variables are calculated. Based on eq. (7.52) we find

$$R = 10^{2/3}\, 20^{1/3} = 12.60\,\text{m}$$

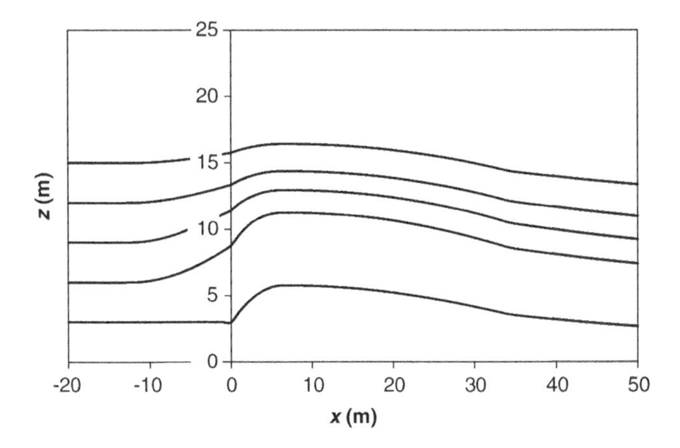


Figure 7.9 Results of streamline deflection calculations.

As the building is shorter than 0.9*R*, the maximum height of the cavity is calculated with eq. (7.55):

$$H_R = 10 + 0.22 \times 12.60 = 12.77\,\text{m}$$

The length of the cavity is calculated with eq. (7.56):

$$L_R = \frac{1.8 \times 20}{1^{0.3}(1 + 0.24 \times 2)} = 24.32\,\text{m}$$

Based on these variables, the differential equations [eqs. (7.64)–(7.74)] are solved numerically. The result is shown in Figure 7.9.

## 7.3.4 Plume–Wake Interaction Calculation

Plumes interact with building wakes in several ways. First, the deflection of the wind streamlines causes a **deflection of the plume**. If a plume has reached final plume height, then the deflections calculated in the previous section can be used directly. However, if the plume is still rising, then the plume trajectory will be a combination of plume rise and plume deflection. Second, plumes trapped in a wake will **slow down**. Third, plumes trapped in a wake will experience an increased turbulence intensity, and hence **increased dispersion;** and fourth, plumes trapped in the cavity will recirculate, causing an **accumulation of pollutants**. Sometimes, only a fraction of a plume gets trapped in a wake. In that case, a distinction must be made between the fraction of the plume that gets trapped and the fraction that does not. Calculations for all of these effects based on the work of Schulman et al. (2000) are described below.

Deflection of a plume can be calculated in conjunction with the numerical plume rise model discussed in Section 7.2.4. To that effect, we calculate the plume height as a function of plume path length *s* by adding the contributions of plume rise and plume deflection:

$$\frac{dz}{ds} = \left(\frac{dz}{ds}\right)_{\text{rise}} + \left(\frac{dz}{dx}\right)_{\text{defl}} \cdot \frac{dx}{ds} \tag{7.75}$$

where $(dz/ds)_{\text{rise}}$ is calculated with the numerical plume rise model:

$$\frac{dz}{ds} = \frac{w}{U_{\text{sc}}} \tag{7.40}$$

and $(dz/dx)_{\text{defl}}$ is calculated as in the previous section; $(dx/ds)$ can also be taken from the numerical plume rise model:

$$\frac{dx}{ds} = \frac{u}{U_{\text{sc}}} \tag{7.39}$$

Building wakes are characterized by a slowing down of the wind (or velocity deficit). On *average*, the slowing down is given by $(x>L)$

$$\frac{\Delta U}{U} = 0.7\left(\frac{L}{x}\right)^{2/3} \tag{7.76}$$

where $U$ is the undisturbed wind speed upwind from the building. As $\Delta U$ and $U$ are only used as a ratio, their individual values are unimportant. The *actual* slowdown in the cavity is more pronounced than the mean velocity deficit:

$$U_{\text{c}} = F_{\text{c}} U_{\text{a}} \tag{7.77}$$

where $U_{\text{a}}$ is the ambient wind speed (i.e., in the absence of the obstacle), and $F_{\text{c}}$ is given by

$$F_{\text{c}} = 1 - \frac{H_{\text{W}}(\Delta U / U)}{0.5(H_{\text{W}} + H_{\text{c}})} \tag{7.78}$$

In the far wake, the slowing effect is dependent on $z$ as well. Hence, the following correction was proposed:

$$U_{\text{W}} = F_{\text{W}} U_{\text{a}} \tag{7.79}$$

where $F_{\text{W}}$ is given by

$$F_{\text{W}} = F_{\text{c}} + \left(\frac{1 - F_{\text{c}}}{H_{\text{W}} - H_{\text{c}}}\right)(z - H_{\text{c}}) \tag{7.80}$$

If there is no cavity at this distance from the building, we use $H_{\text{c}}=0$ in the calculation, and eq. (7.78) simplifies to

$$F_c = 1 - 2\frac{\Delta U}{U} \tag{7.81}$$

Building wakes are also characterized by increased dispersion. The dispersion parameters in the wake are calculated based on turbulence intensity, $i_z$; and $i_z$ is calculated as

$$i_z = i_{z,0}\left\{1 + \frac{1.7 i_{z,N} / i_{z,0} + \Delta U / U}{\left[\frac{(x+R-L)}{R}\right]^{2/3} - \Delta U / U}\right\} \tag{7.82}$$

where $i_{z,N}$=0.06, and $i_{z,0}$ is the turbulence intensity in the absence of the wake (= $\sigma_w/u$). A similar equation is used for $i_y$, but with $i_{y,N}$=0.08. Based on the turbulence intensities, plume growth in wakes can be calculated as follows:

From $x=L$ to $x=L+L_R$, plume growth is given by

$$\frac{d\sigma_y}{dx} = i_y \tag{7.83}$$

$$\frac{d\sigma_z}{dx} = i_z \tag{7.84}$$

For $x>L+L_R$, plume growth is given by

$$\frac{d\sigma_y^2}{dx} = 2\sigma_y(L+L_R)\cdot i_y \cdot \frac{H_W(x)}{H_W(L+L_R)} \tag{7.85}$$

$$\frac{d\sigma_z^2}{dx} = 2\sigma_z(L+L_R)\cdot i_z \cdot \frac{H_W(x)}{H_W(L+L_R)} \tag{7.86}$$

These equations are used up to $x=15R$. Beyond that point, ambient plume growth is used. This can be done by defining a virtual source. As indicated in Section 6.8, a virtual source is an imaginary source located at the point where a plume, calculated with the same equation as is applicable at the receptor, would give the same plume dispersion as the actual plume.

The numerical plume rise model discussed in Section 7.2.4 discribes dispersion (increase of $r$) and plume rise simultaneously. Sometimes, when a numerical plume rise model is used near wakes, the wake creates more entrainment of air into the plume than can be described by the numerical plume rise model. To account for that, the first differential equation of the numerical plume rise model, eq. (7.35), is replaced by

$$\frac{d}{ds}(\rho U_{sc} r^2) = \max\left\{\left[\frac{d}{ds}(\rho U_{sc} r^2)\right]_{entrainment}, 2r\rho_a U_a\left(\frac{dr}{ds}\right)_{wake}\right\} \tag{7.87}$$

where $(dr/ds)_{wake}$ is given by

$$\left(\frac{dr}{ds}\right)_{\text{wake}} = \sqrt{\frac{\pi}{2}}\frac{d\sigma_z}{dx} \tag{7.88}$$

where $\sigma_z$ is calculated in eq. (7.86).

When the plume trajectory and the dispersion parameters are known, **ground-level concentrations** can be calculated. The ground-level concentration of the pollutant in the plume in the near wake, $c_{\text{N}}$, is given by

$$c_{\text{N}} = \frac{BfQ\exp\left[-\frac{1}{2}(y/\sigma_{yc})\right]}{U_H H_{\text{c}} W'_{\text{B}}} \tag{7.89}$$

where $B$=3, a recirculation factor, $f$ is the fraction of the plume that gets trapped in the near wake, $Q$ is the emission rate (mg s$^{-1}$), $U_H$ is the wind speed at height $H$, $\sigma_{yc}$ is a dispersion parameter for the cavity calculated by

$$\sigma_{yc} = \frac{W'_{\text{B}}}{\sqrt{2\pi}} \tag{7.90}$$

and $W'_{\text{B}}$ is a geometric factor given by

$$W'_{\text{B}} = \max\left[\frac{H_{\text{B}}}{3}, \min(W, 3H_{\text{B}})\right] \tag{7.91}$$

Equation (7.89) would lead to unrealistically large values of $c_{\text{N}}$ near the end of the cavity, where $H_{\text{c}}$ approaches 0. Hence, it seems prudent to replace $H_{\text{c}}$ by $H_{\text{R}}$, the maximum cavity height, in eq. (7.89).

The fraction $f$ is based on the overlap between the plume and the cavity, $f_z$:

$$f_z = \text{erf}\left(\frac{H_{\text{c}} - H_{\text{p}}}{\sqrt{2}\sigma_z}\right) \tag{7.92}$$

where $H_{\text{p}}$ is the height of the plume center, and erf the error function:

$$\text{erf}(x) = \frac{2}{\sqrt{\pi}}\int_0^x \exp(-t^2)\, dt \tag{7.93}$$

Values of the error function are tabulated in Section B4 (Appendix B). Factor $f$ is the maximum value of $f_z$ over the interval $x$=$L$ to $x$=$L$+$L_{\text{R}}$. After being trapped in the near wake, this fraction of the plume is transferred to the far wake. The ground-level concentration resulting from this fraction in the far wake, $c_{\text{F}}$, is

$$c_{\text{F}} = \frac{fQ\exp\left[-\frac{1}{2}(y/\sigma_y)^2\right]}{\pi U_{\text{s}}\sigma_y\sigma_z} \tag{7.94}$$

where $\sigma_y$ is the dispersion parameter obtained by taking $\sigma_{yc}$ at $x=L$; $U_s$ is the wind speed at the source height.

The ground-level concentration resulting from the fraction of the plume that is not trapped in the cavity, $c_p$, is given by

$$c_p = \frac{(1-f)Q\exp\left[-\frac{1}{2}(y/\sigma_y)^2\right]\exp\left[\frac{1}{2}(H_p/\sigma_z)^2\right]}{\pi\sigma_y\sigma_z U_s} \tag{7.95}$$

Reflections need to be added to this equation. Here $\sigma_y$ and $\sigma_z$ are calculated assuming a point source.

Wakes have a variable length, ranging from $0.85L_R$ to $1.15L_R$. Hence, it is preferable to account for that effect. The correction is as follows:

$$\begin{aligned}
&c = c_N \quad (L < x < L + 0.85L_R) \\
&c = \lambda c_N + (1-\lambda)c_F \quad (L + 0.85L_R < x < L + L_R) \\
&c = c_p + \lambda c_N + (1-\lambda)c_F \quad (L + L_R < x < L + 1.15L_R) \\
&c = c_p + c_F \quad (x > L + 1.15L_R)
\end{aligned} \tag{7.96}$$

where $\lambda$ is given by

$$\lambda = 1 - \frac{x - L - 0.85L_R}{0.3L_R} \tag{7.97}$$

PRIME is included in several air dispersion models, like AERMOD and CALPUFF.

## 7.4 BEHAVIOR OF DENSER-THAN-AIR PLUMES

The behavior of pollutant plumes heavier than ambient air is fundamentally different from the regular air dispersion when the negative buoyancy is sufficient for the plumes to reach the surface. Such plumes are subjected to what is known as **gravity current**, sometimes referred to as **gravity spread** (Zeman, 1982; Ermak, 1990). Denser-than-air plumes spread on the surface and ramain fairly homogeneous, with a fairly sharp boundary between the plume and the surrounding air, almost as if the plume forms a separate phase. This is especially the case with cold plumes, which are heated intensively from the surface. Hence, the plume behaves like a strongly unstable, well-mixed boundary layer capped by an intense temperature inversion. The plume is gradually diluted by ambient air entrainment until the density difference is sufficiently small to show negligible negative buoyancy. At that point the plume starts behaving like a normal plume.

Zeman (1982) developed a model for the dispersion of denser-than-air releases that formed the basis of the model SLAB (Ermak, 1990). This section will present

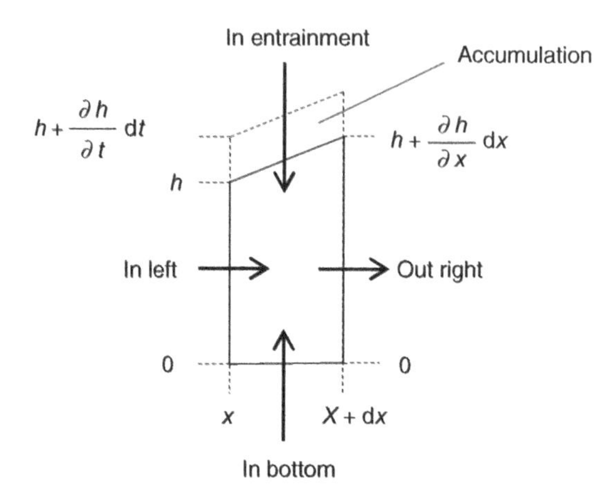


Figure 7.10 Material balance of a denser-than-air plume.

some simplified relationships useful for modeling denser-than-air releases along similar lines as Zeman (1982), without going into a full model description.

We will introduce a material balance, a momentum balance along the wind direction, and a species balance for the description and assume that the plume has a constant (or otherwise known) temperature. A full description would require a heat balance, as denser-than-air plumes are usually much colder than air and exchange much heat with the surroundings.

The **material balance** is based on the differential element shown in Figure 7.10. We can write the equation as

$$\text{Accumulation} = \text{in left} + \text{in bottom} + \text{in entrainment} - \text{out right} \tag{7.98}$$

Assuming constant density, we can write eq. (7.98) as a volume balance: The volume increase of the differential element equals the volumetric flow rate into the element minus the volumetric flow rate out of the volume. Assuming constant width ($y$ direction), this leads to the following equation:

$$\frac{\partial h}{\partial t} \cdot dx = uh + w_b\,dx + w_e\,dx - \left[uh + \frac{\partial (uh)}{\partial x} dx\right] \tag{7.99}$$

where $h$ (m) is the height of the plume, $t$ is time (s), $x$ is the distance in the downwind direction (m), $u$ is the wind speed inside the plume, $w_b$ is the emission expressed as a velocity (m s$^{-1}$), and $w_e$ is the entrainment rate (m s$^{-1}$). Eliminating canceling terms and dividing by $dx$ leads to

$$\boxed{\frac{\partial h}{\partial t} + \frac{\partial (uh)}{\partial x} = w_b + w_e} \tag{7.100}$$

which can also be written as

$$\frac{\partial h}{\partial t} + u\frac{\partial h}{\partial x} + h\frac{\partial u}{\partial x} = w_b + w_e \tag{7.101}$$

In steady-state conditions the equation becomes

$$u\frac{\partial h}{\partial x}+h\frac{\partial u}{\partial x}=w_b+w_e \tag{7.102}$$

The **momentum balance** along the wind direction is as follows:

$$\text{Momentum produced} = \text{force} \tag{7.103}$$

Based on this momentum balance, the following equation can be derived (see Appendix A, Section A 7.1):

$$\boxed{\frac{\partial u}{\partial t}+u\frac{\partial u}{\partial x}=\frac{w_e}{h}\left(\frac{\rho_a u_a}{\rho}-u\right)-\frac{w_b u}{h}-\frac{\beta g}{2\rho}\frac{\partial}{\partial x}\left[h(\rho-\rho_a)\right]-\frac{u_*^2}{h}} \tag{7.104}$$

where subscript $a$ refers to the ambient atmosphere, and $\beta$ is a constant describing the height dependence of $\rho-\rho_a$; $\beta=1$ when $\rho-\rho_a$ is constant, and $\beta=\frac{2}{3}$ when $\rho-\rho_a$ is proportional to $h-x$. Zeman (1982) recommended $\beta=\frac{2}{3}$.

In steady-state conditions the equation becomes

$$u\frac{\partial u}{\partial x}=\frac{w_e}{h}\left(\frac{\rho_a u_a}{\rho}-u\right)-\frac{w_b u}{h}-\frac{\beta g}{2\rho}\frac{\partial}{\partial x}\left[h(\rho-\rho_a)\right]-\frac{u_*^2}{h} \tag{7.105}$$

The species balance can be written as

$$\text{Pollutant accumulation} = \text{pollutant in} - \text{pollutant out} \tag{7.106}$$

This leads to the following equation:

$$\boxed{\frac{\partial q}{\partial t}+u\frac{\partial q}{\partial x}=\frac{w_b}{h}(q_b-q)-\frac{w_e q}{h}} \tag{7.107}$$

where $q$ refers to a mass fraction, and $q_b$ is the mass fraction in the air entering the plume from the surface. Details of the derivation are given in Section A 7.2 (Appendix A).

In steady-state conditions eq. (7.107) becomes

$$u\frac{\partial q}{\partial x}=\frac{w_b}{h}(q_b-q)-\frac{w_e q}{h} \tag{7.108}$$

We can get a qualitative idea of the behavior of dense plumes by considering the steady-state equations [eqs. (7.102), (7.105), and (7.108)] under simplifying conditions. Consider a plume downwind from the pollution source ($w_b=0$), under calm conditions ($u_a=0$). Equation (7.108) then simplifies to

$$u\frac{\partial q}{\partial x}=-\frac{w_e q}{h} \tag{7.109}$$

Not unexpectedly, we find that the concentration must decline as we move downwind with the plume. This is due to the dilution caused by the entrainment ($w_e > 0$). Equation (7.105) simplifies to

$$u\frac{\partial u}{\partial x} = -\frac{uw_e}{h} - \frac{\beta g}{2\rho}\frac{\partial}{\partial x}[h(\rho - \rho_a)] - \frac{u_*^2}{h} \tag{7.110}$$

The first and third terms on the right-hand side are negative. The second term can be either positive or negative due to the opposing effect of an increasing $h$ (see below) and a decreasing $(\rho - \rho_a)$. However, as $(\rho - \rho_a)$ decreases due to dilution, its effect on the overall equation will become negligible, and the velocity is expected to decline.

The material balance under the simplifying conditions is

$$u\frac{\partial h}{\partial x} + h\frac{\partial u}{\partial x} = w_e \tag{7.111}$$

With a positive right-hand side and a velocity that tends to decline, the height of the plume is expected to increase with increasing $x$. It turns out that this is also the case with non-steady-state plumes, as shown in Figure 7.11. As the plume spreads, it tends to have high fronts and a low center.

Calculating density of the plume accurately is tricky because it depends on the pressure, which in turn depends on the density. However, the pressure of the surrounding atmosphere can be used in the calculation of the density with reasonable accuracy. This leads to the following equations:

$$\rho_a = \frac{M_a p}{RT} \tag{7.112}$$

where $M_a$ (0.028964 kg mol$^{-1}$) is the molar mass of the air. The density of the plume is given by

$$\rho = \frac{p\rho_a T_a}{p_a T[1 - q + (M_a / M_c)q]} \approx \frac{\rho_a T_a}{T[1 - q + (M_a / M_c)q]} \tag{7.113}$$

where $M_c$ is the molar mass of the pollutant. In calculational schemes that retain information of the pressure in the previous step, this information can be used to use eq. (7.113) without approximating $p$ by $p_a$.

For the numerical integration of the three differential equations, eq. (7.100), (7.104), and (7.107), for the **non-steady-state** case, either **Eulerian** or the **Lagrangian**

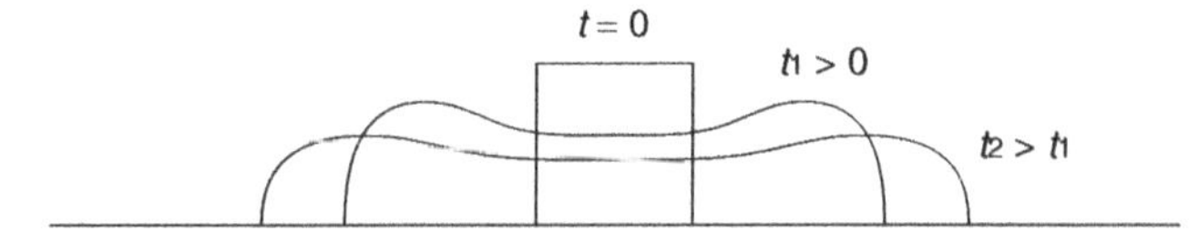


Figure 7.11 Spread of a denser-than-air plume due to gravitational current.

**frame of reference** can be used. Both methods are prone to numerical stability problems in the simulation, but the problem is much more severe in the Eulerian implementation. For that reason, the Lagrangian frame of reference is strongly recommended. In the Lagrangian case, the total differential is defined, for example, for the height $h$:

$$\frac{dh}{dt} = \frac{\partial h}{\partial t} + u\frac{\partial h}{\partial x} \tag{7.114}$$

Hence, the three differential equations become

$$\boxed{\frac{dh}{dt} + h\frac{\partial u}{\partial x} = w_\mathrm{b} + w_\mathrm{e}} \tag{7.115}$$

$$\boxed{\frac{du}{dt} = \frac{w_\mathrm{e}}{h}\left(\frac{\rho_\mathrm{a} u_\mathrm{a}}{\rho} - u\right) - \frac{w_\mathrm{b} u}{h} - \frac{\beta g}{2\rho}\frac{\partial}{\partial x}\left[h(\rho - \rho_\mathrm{a})\right] - \frac{u_*^2}{h}} \tag{7.116}$$

$$\boxed{\frac{dq}{dt} = \frac{w_\mathrm{b}}{h}(q_\mathrm{b} - q) - \frac{w_\mathrm{e} q}{h}} \tag{7.117}$$

In a Lagrangian approach, a number of particles are defined at $t=0$, and followed as a function of time. Equations (7.115)–(7.117) are solved for each particle. The location of each particle can be tracked with the following equation:

$$\frac{dx}{dt} = u \tag{7.118}$$

This approach simplifies the definition of boundary conditions. It suffices to assume a bounding particle that runs out in front of the first plume particle, at the same speed, while $h=0$ and $q=0$ for the bounding particle.

For the solution of the differential equations in the **steady-state** case [eqs. (7.102), (7.105), and (7.108)], the approach is **Eulerian**. The main complication of the steady-state case is that two of the equations are not fully explicit. Equation (7.108) can be solved numerically without complications:

$$\frac{\partial q}{\partial x} = \frac{w_\mathrm{b}}{uh}(q_\mathrm{b} - q) - \frac{w_\mathrm{e} q}{uh} \tag{7.119}$$

We can solve eq. (7.102) for $\partial u/\partial x$:

$$\frac{\partial u}{\partial x} = -\frac{u}{h}\frac{\partial h}{\partial x} + \frac{1}{h}(w_\mathrm{b} + w_\mathrm{e}) \tag{7.120}$$

Equation (7.120) is substituted in eq. (7.105):

$$-\frac{u}{h}\frac{\partial h}{\partial x} + \frac{1}{h}(w_\mathrm{b} + w_\mathrm{e}) = \frac{w_\mathrm{e}}{uh}\left(\frac{\rho_\mathrm{a} u_\mathrm{a}}{\rho} - u\right) - \frac{w_\mathrm{b} u}{uh} - \frac{\beta g}{2u\rho}\frac{\partial}{\partial x}\left[h(\rho - \rho_\mathrm{a})\right] - \frac{u_*^2}{uh} \tag{7.121}$$

Eliminating terms that cancel out and expanding the derivative on the right-hand side leads to

$$-\frac{u}{h}\frac{\partial h}{\partial x}=\frac{w_e}{uh}\left(\frac{\rho_a u_a}{\rho}-2u\right)-\frac{2w_b}{h}-\frac{\beta g}{2u\rho}(\rho-\rho_a)\frac{\partial h}{\partial x}-\frac{\beta g h}{2u\rho}\frac{\partial}{\partial x}(\rho-\rho_a)-\frac{u_*^2}{uh} \quad (7.122)$$

Solving for $\partial h/\partial x$ leads to

$$\frac{\partial h}{\partial x}=\frac{\frac{w_e}{uh}\left(\frac{\rho_a u_a}{\rho}-2u\right)-\frac{2w_b}{h}-\frac{\beta g h}{2u\rho}\frac{\partial}{\partial x}(\rho-\rho_a)-\frac{u_*^2}{uh}}{-\frac{u}{h}+\frac{\beta g}{2u\rho}(\rho-\rho_a)} \quad (7.123)$$

and $\rho$ is calculated with eq. (7.113). However, $\partial(\rho-\rho_a)/\partial x$ is less straightforward. It is calculated as

$$\frac{\partial(\rho-\rho_a)}{\partial x}=\frac{\partial(\rho-\rho_a)}{\partial q}\frac{\partial q}{\partial x} \quad (7.124)$$

where $\partial q/\partial x$ is calculated with eq. (7.119), and $\partial(\rho-\rho_a)/\partial q$ is calculated with

$$\frac{\partial(\rho-\rho_a)}{\partial q}=\frac{T_a(1-M_a/M_c)}{T\left[1-q+(M_a/M_c)q\right]} \quad (7.125)$$

Here it is assumed that $T$ is sufficiently constant not to affect the derivative. If this is not the case, eq. (7.124) needs to be extended to

$$\frac{\partial(\rho-\rho_a)}{\partial x}=\frac{\partial(\rho-\rho_a)}{\partial q}\frac{\partial q}{\partial x}+\frac{\partial(\rho-\rho_a)}{\partial T}\frac{\partial T}{\partial x} \quad (7.126)$$

This case will not be considered here.

Equations (7.119), (7.123), and (7.120) can be integrated numerically if initial conditions, that is, $u$, $h$, and $q$ at $x=0$, are known. We still require an equation for $w_e$. Under certain simplifying conditions (neutral flat plate boundary layer flow) Zeman (1982) suggested:

$$w_e=0.28u_* \quad (7.127)$$

Zeman also suggested the following equation for the friction velocity:

$$u_*=0.08u \quad (7.128)$$

As indicated above, a heat balance was not included here, so the approach is too simple to model cold plumes, such as a boiling propane source. Also, the model discussed here does not contain plume width. In practice, the plume will fan out laterally as well.

**Example 7.5.** An accidental benzene spill generates a benzene-saturated plume at ground level with height 1 m and a horizontal spreading velocity of $0.1\,\mathrm{m\,s^{-1}}$ near the source. The benzene mass fraction of the plume near the source is 0.337. The ambient wind speed is 0. Calculate the spreading of the plume over the next 100 m and/or the first 1000 s away from the source. Assume a temperature of 30 °C for both plume and surroundings and a pressure of 1 bar.

***Solution.*** As we are modeling the plume away from the source, we can assume that $w_b=0$. Benzene has a molar mass of $0.078\,\mathrm{kg\,mol^{-1}}$; air has a molar mass of $0.029\,\mathrm{kg\,mol^{-1}}$.

We will model the plume in steady state and in nonsteady state. In the nonsteady state we will develop both an Eulerian model and a Lagrangian model. We will compare the outputs of the three models.

The steady-state model is programmed in the Matlab files "main.m," "data.m," and "f.m" in the folder "Dense plume steady state." A general overview of how numerical integration of differential equations is handled in Matlab is given in Appendix B, Section B5.

The file "data.m" contains the data for the model, as well as the boundary conditions ($x=0$). We define $y(1)=q$, $y(2)=h$, and $y(3)=u$. A range of 0–100 (m) is chosen for $x$.

The file "main.m" controls the program: It calls the data file and the numerical integration routine. It plots all $y$ values (i.e., $q$, $h$, and $u$) versus $x$ and generates a separate figure for the plume height versus $x$.

The file "f.m" calculates the running variables (e.g., $u_*$, $w_e$, …) and defines the differential equations, based on eq. (7.102) for $h$, eq. (7.105) for $u$, and eq. (7.108) for $q$. The order of the differential equations is important because we need the equation for $q$ to calculate the partial derivative of the density in the equation for $u$, and we need the equation for $u$ to calculate the partial derivative of $u$ in the equation for $h$.

The result of the simulation is shown in Figure 7.12. The plume grows approximately linearly, and dilutes as a result. The velocity increases slightly with distance.

The Eulerian implementation of the non-steady-state model is programmed in the Matlab files "main.m," "data.m," and "f.m" in the folder "Dense plume nonsteady state Eulerian." A brief description of how partial differential equations are converted to ordinary differential equations by finite differences is given in Appendix B, Section B6.

The file "data.m" contains the data for the model, as well as the initial conditions ($t=0$). We define $y(1)=q_1$, $y(2)=h_1$, and $y(3)=u_1$, the value of the variables at node point 1 with $x$ coordinate $x_1$. The $y$ variables continue as $y(4)=q_2$, $y(5)=h_2$, $y(6)=u_2$, and so forth. A range of 0–100 (m) is chosen for $x$, divided into 20 node points with $\Delta x=5\,\mathrm{m}$. The boundary conditions $q=0.337\,\mathrm{kg\,kg^{-1}}$, $h=1\,\mathrm{m}$, and $u=0.1\,\mathrm{m}$ are imposed at $x=0$. Because of the way the spatial derivatives are defined, we only need a boundary condition for $u$ at $x=100\,\mathrm{m}$. The value is assumed to be the same as the value at 95 m.

The file "main.m" controls the program: It calls the data file and the numerical integration routine. It plots all $y$ values versus $x$. However, due to the way the data are organized, very little can be learned from the generated figure. Organizing the data can be done separately in spreadsheet software or added to the code.

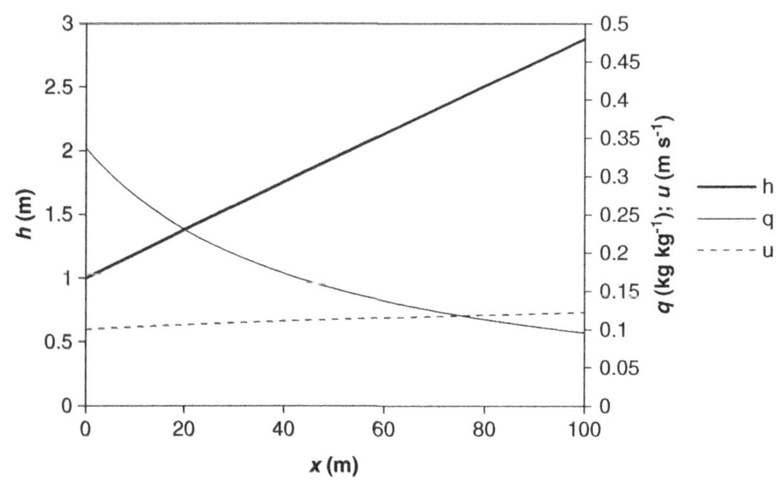


Figure 7.12 Steady-state modeling result, denser-than-air plume.

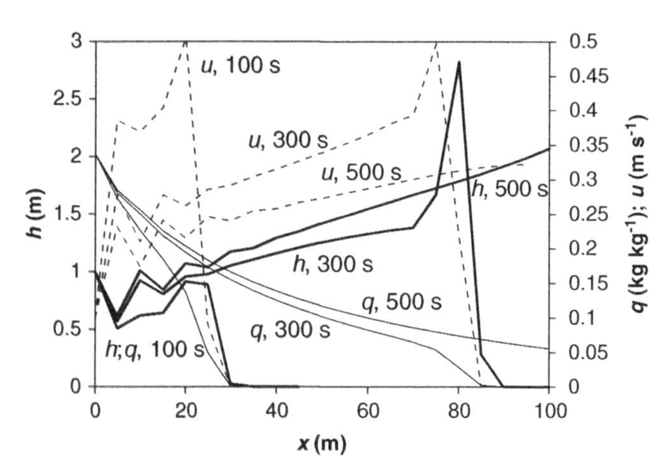


Figure 7.13 Non-steady-state modeling result, denser-than-air plume, Eulerian implementation.

The file "f.m" calculates the running variables (e.g., $u_*$, $w_e$, ...) in each node point and defines the differential equations, based on eq. (7.100) for $h$, eq. (7.104) for $u$, and eq. (7.107) for $q$. For stability reasons the spatial derivatives were defined upstream of the point instead of around the point of interest (see Appendix B), but even that precaution did not result in a stable simulation. Some improvement was achieved by defining a staggered grid with the boundary condition of $u$ at 2.5 m, and grid points for $u$ at 7.5 m, 12.5 m, and the like. This technique is common in Eulerian dispersion models and will be discussed further in Chapter 11. Even so, the algorithm is not stable and diverges completely after about 1200 s.

The result of the simulation is shown in Figure 7.13, with plumes after 100, 300, and 500 s. Comparison with Figure 7.12 makes it clear that the result is very

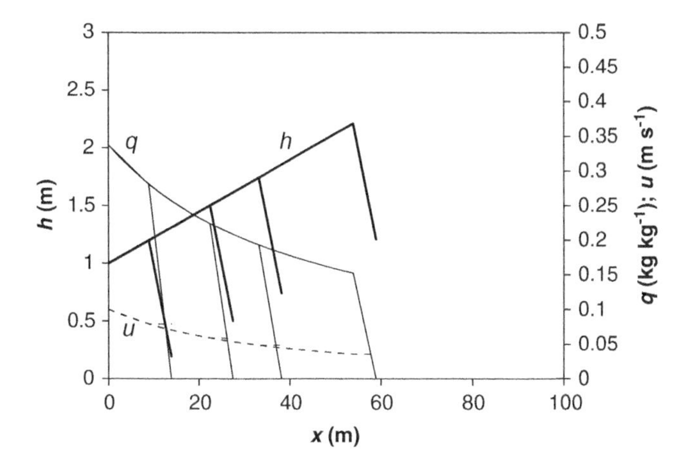


Figure 7.14 Non-steady-state modeling result, denser-than-air plume, Lagrangian implementation. Partly overlapping plumes after 100, 300, 500, and 1000 s.

inaccurate for the plume height and velocity. The algorithm is unstable, leading to oscillating variables, and does not approach the steady-state solution. The benzene mass fraction does not show instabilities, but the model error of the other two variables propagate into the $q$ values as well.

The Lagrangian implementation of the non-steady-state model is programmed in the Matlab files "main.m," "data.m," and "f.m" in the folder "Dense plume non-steady state Lagrangian."

The file "data.m" contains the data for the model as well as the initial conditions ($t=0$). The algorithm must track the $x$ coordinate of each node point as the node points are moving. Hence, we define $y(1)=q_1$, $y(2)=h_1$, $y(3)=u_1$, and $y(4)=x_1$, the value of the variables at node point 1. The $y$ variables continue as $y(5)=q_2$, $y(6)=h_2$, $y(7)=u_2$, $y(8)=x_2$, and so forth.

The file "main.m" controls the program: It calls the data file and the numerical integration routine. It plots all $y$ values versus $x$. However, due to the way the data are organized, very little can be learned from the generated figure. Organizing the data can be done separately in spreadsheet software or added to the code.

The file "f.m" calculates the running variables (e.g., $u_*$, $w_e$, ...) in each node point and defines the differential equations, based on eq. (7.115) for $h$, eq. (7.116) for $u$, eq. (7.117) for $q$, and eq. (7.118) for $x$. For stability reasons the spatial derivatives were defined upstream of the point instead of around the point of interest (see Appendix B), which resulted in a stable simulation. In this implementation, grid nodes were first defined near $x=0$, with narrow spacing at the boundary conditions, but the time derivatives are artificially kept equal to 0 until the next point reaches $x=5$ m. The first point serves as a boundary condition for $u$ and is forced to move at the same velocity as the second point.

The result of the simulation is shown in Figure 7.14, with plumes after 100, 300, 500, and 1000 s. The model is generally consistent with the steady-state solution shown in Figure 7.12 but shows velocities decreasing with $x$, unlike the steady-state result. The reason for the difference is unclear but may be the result of inaccurate finite differencing.

## 7.5 DEPOSITION

So far we assumed that air pollutants are **conservative**, that is, they are not removed from the atmosphere. This is a reasonable assumption for stable gas molecules with low solubility, but particulates and gases with high solubility in water **deposit** on the surface. For a correct description of the air dispersion of such pollutants, the deposition needs to be accounted for.

We distinguish two types of precipitation: **dry deposition**, which is deposition in the absence of rain, and **wet deposition**, which is deposition caused by rain.

The main quantity of interest in deposition is the **deposition flux** $F$ ($\mu g\,m^{-2}\,s^{-1}$), which is the mass of air pollutant deposited to the surface, per unit area of surface, per unit time. In some disciplines, what is defined as a flux here is known as a flux density.

### 7.5.1 Dry Deposition

***7.5.1.1 Theory.*** Dry deposition is the mass transfer of pollutants from the atmosphere to the surface in the absence of precipitation. It refers to both gaseous pollutants and particulate matter (PM).

To quantify dry deposition, the **dry deposition flux** $F$ ($\mu g\,m^{-2}\,s^{-1}$) is usually correlated to the concentration $c$ ($\mu g\,m^{-3}$) at a given reference height (usually $< 10\,m$):

$$F = -v_d\,c \tag{7.129}$$

The minus sign shows that a downward flux is indicated by a negative value. Positive values for downward fluxes can be found in the literature as well.

Note that $c$ has a slight height dependence. Therefore, $v_d$ depends on height as well; $v_d$ ($m\,s^{-1}$) is the **deposition velocity**. As with several other atmospheric variables, this is not an actual velocity but rather a property of the pollutant and the atmosphere that happens to have the units of a velocity. Nevertheless, large particles, whose main deposition mechanism is settling (i.e., falling down), have a settling velocity $v_s$ equal to their deposition velocity.

The **resistance** to deposition $r_t$ is defined as the reciproke ($1/v_d$) of the deposition velocity.

Deposition can also be described as a conventional mass transfer process. In that case, the flux equation is written as

$$F = -v_t\left(c - c_0\right) \tag{7.130}$$

where $v_t$ is the **transfer velocity** and $c_0$ the gas-phase concentration in *equilibrium* with the surface. Equation (7.130) is only needed in cases where there is a clear solubility effect (i.e., re-volatilization) or a clear threshold concentration for deposition.

Dry deposition is usually described in three steps:

- **Aerodynamic transport** from the atmosphere to a thin stagnant air layer at the surface
- **Molecular** or **Brownian transport** through the stagnant (quasi-laminar) air layer
- **Uptake at the surface**

For dry deposition of **gases** these are the only steps that need to be considered. Each step adds a certain **resistance** to the overall resistance of the deposition process. The **total resistance** of the process is calculated as if each individual resistance were an electrical resistance. The resistances are in series:

$$\frac{1}{v_d} = r_t = r_a + r_b + r_c \tag{7.131}$$

where $r_t$ is the total resistance (s m$^{-1}$), $r_a$ (s m$^{-1}$) is the **aerodynamic resistance** associated with mass transfer from the atmosphere to the stagnant air film, $r_b$ (s m$^{-1}$) is the **quasi-laminar layer resistance** associated with mass transfer through the stagnant air film, and $r_c$ (s m$^{-1}$) is the **canopy resistance** or **surface resistance** associated with the actual uptake of the pollutant by the surface.

In the case of **particles** there is an additional mechanism working in parallel with the three resistances: **particle settling**. The fall velocity of a small particle (< 30–80 $\mu$m, depending on the particle density) is given by **Stokes' law**:

$$v_s = \frac{\rho_p d_p^2 g}{18\mu} \tag{7.132}$$

where $\rho_p$ is the particle density (kg m$^{-3}$), $d_p$ is the particle diameter (m), and $\mu$ is the dynamic air viscosity (about $1.8 \times 10^{-5}$ Pa s).

For particles smaller than 5 $\mu$m Stokes' law underestimates the actual settling velocity. This is because air can no longer be considered a continuum at such small scales. The particle falls through the voids between the air molecules. The settling velocity in the Stokes regime is multiplied by a correction factor $C$ known as the **Cunningham correction factor**. It is based on the mean free path of molecules $\lambda$. For air, $\lambda$ is approximated by (Jennings, 1988)

$$\lambda = 1.256469 \frac{\mu}{\sqrt{\rho_f p}} \tag{7.133}$$

where $\mu$ is viscosity in Pa s, $\rho_f$ is air density in kg m$^{-3}$, and $p$ is in Pa. Then $C$ can be calculated as (Allen and Raabe, 1982)

$$C = 1 + 2\frac{\lambda}{d_p}\left[1.257 + 0.4\exp\left(-0.55\frac{d_p}{\lambda}\right)\right] \tag{7.134}$$

Hence, the settling velocity is given by

$$v_s = \frac{\rho_p d_p^2 g}{18\mu}\left\{1 + 2\frac{\lambda}{d_p}\left[1.257 + 0.4\ \exp\left(-0.55\frac{d_p}{\lambda}\right)\right]\right\} \tag{7.135}$$

Equations for the settling velocity of particles outside the Stokes regime are given in Section 7.5.2.1. To incorporate particle settling in the calculation of the dry deposition rate for particles, the following *approximate* equations have been proposed (Slinn et al., 1978; Zhang et al., 2001):

$$v_d = \frac{1}{r_t} = \frac{1}{r_a + r_b} + v_s \tag{7.136}$$

$$v_d = \frac{1}{r_t} = \frac{1}{r_a + r_b + r_a r_b v_s} + v_s \tag{7.137}$$

In the formulations discussed here, canopy resistances are not considered in the dry deposition of particulate matter because the calculation of the quasi-laminar layer resistance, $r_b$, can be assumed to incorporate canopy effects as well. Alternative formulations with a canopy resistance exist as well (Pryor et al., 2008).

Equations (7.136) and (7.137) have the desired properties of a two-resistance model with a parallel third resistance. They predict the correct limiting cases. For instance, if the settling velocity $v_s$ is negligibly small, the equations correctly predict the two-resistance version of eq. (7.131). When settling dominates, that is, when $r_a$ and $r_b$ become infinitely large, the equations predict that deposition rate equals settling velocity.

However, the equations are only approximate because the electrical analogons that were used in the derivation are not entirely applicable. A better approach would be to use a diffusion-and-advection model to derive the equation. The result is the following equation:

$$v_d = \frac{1}{r_t} = \frac{v_s}{1 - \exp\left[-v_s\left(r_a + r_b\right)\right]} \tag{7.138}$$

The derivation is shown in Section A 7.3 (Appendix A). Equation (7.138) also has the desired limiting behavior. When $v_s$ is negibly small, we can use the approximation $\exp(-x) \rightarrow 1 - x$ for $x \rightarrow 0$ to show that the two-resistance model is correctly predicted:

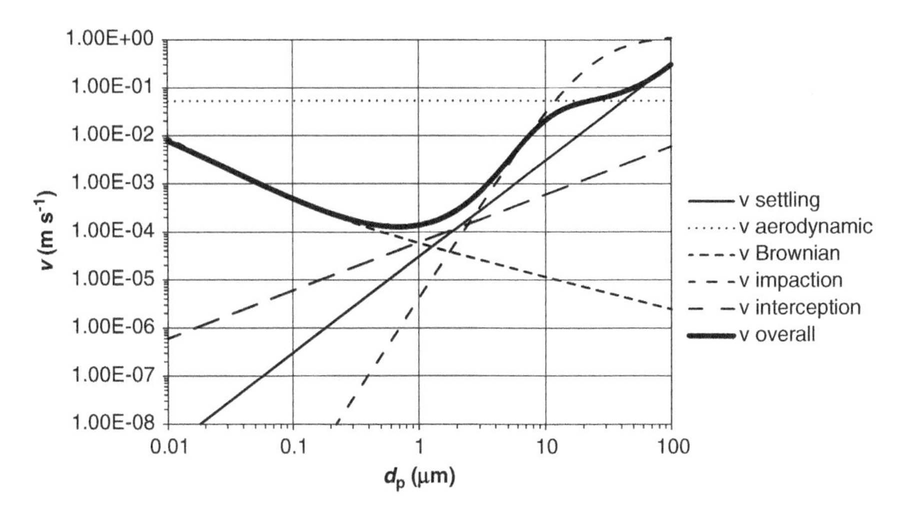


Figure 7.15 Particle dry deposition velocity versus particle diameter.

$$v_d = \frac{v_s}{1-\exp\left[-v_s\left(r_a+r_b\right)\right]} \approx \frac{v_s}{1-\left[1-v_s\left(r_a+r_b\right)\right]} = \frac{v_s}{v_s\left(r_a+r_b\right)} = \frac{1}{r_a+r_b} \tag{7.139}$$

Assuming infinite values of $r_a$ and $r_b$ again leads to the prediction that deposition velocity equals settling velocity. Hence, eq. (7.138) is based on a more realistic model than the standard equations and predicts the same limiting behavior. Fortunately, eqs. (7.136)–(7.138) rarely differ by more than 20%. Equation (7.136) predicts the highest deposition velocities, whereas eq. (7.138) predicts the lowest values.

For particles the deposition velocity depends strongly on particle size. Particles smaller than 0.05 μm deposit rapidly due to Brownian motion (diffusion of particles). Particles between 2 and 20 μm deposit rapidly by interception (the surface is in their path) and impaction (particles are slung to the surface by turbulent air masses). Particles larger than 20 μm deposit rapidly by settling. Between 0.1 and 1 μm diameter all deposition mechanisms are slow. Hence, particles within that range have a low deposition velocity, which can be as low as $10^{-4}$ m s$^{-1}$. This is shown in Figure 7.15. There are some indications that the minimum in the deposition velocity is less pronounced in forest canopies (Pryor et al., 2008).

#### 7.5.1.2 *The Three Resistances.*

In the previous section we established that we need values for three resistances in order to calculate the deposition velocity of gases and two resistances and the settling velocity to calculate the deposition velocity of particulate matter. This section will present practical schemes to calculate these resistances.

The calculation of the aerodynamic resistance, $r_a$, is based on a gradient theory ($K$ theory) of the surface layer. In particular, it is assumed that the turbulent mass diffusivity $K_z$ equals the heat diffusivity (see also Chapter 5). In particular, in a *neutral* atmosphere in the surface layer, we found [eq. (5.192), assuming $z << h_{mix}$ and $\phi_h = 1$]:

$$K_z = ku_* z \tag{7.140}$$

The mass diffusivity is used in the turbulent version of Fick's law to calculate the transfer properties [see eq. (5.176)]:

$$-F = K_z \frac{dc}{dz} \tag{7.141}$$

Substitution leads to

$$-F = ku_* z \frac{dc}{dz} \tag{7.142}$$

Rearranging leads to

$$dc = -\frac{F}{ku_*} \frac{dz}{z} \tag{7.143}$$

We take the **convention** that the quasi-laminar layer starts at a height equal to the roughness length. Hence, integration leads to

$$c_s - c_a = -\frac{F}{ku_*} \ln\left(\frac{z_0}{z}\right) \tag{7.144}$$

In reality, the quasi-laminar layer, which is usually a few millimeters thick, does not correspond with the surface roughness. When there is a pronounced canopy on the surface, the quasi-laminar layer is usually thinner than $z_0$, whereas the quasi-laminar layer is usually thicker than $z_0$ when there is no canopy. The pragmatic approach is to incorporate all resistance related to the atmosphere not covered by $r_a$ as defined here into $r_b$, which is calculated with empirical equations. Appendix C develops a formalism that does not require such conventions.

Compare eq. (7.144) with the resistance theory:

$$-F = \frac{c_a - c_s}{r_a} \tag{7.145}$$

Hence, it is clear that under *neutral* conditions:

$$\boxed{r_a = \frac{1}{ku_*} \ln\left(\frac{z}{z_0}\right)} \tag{7.146}$$

The equation can be rewritten to eliminate the friction velocity. To that effect, consider the following equation [eq. (5.65) with $u_m$ replaced by $u$]:

$$u_* = \frac{ku}{\ln(z/z_0)} \tag{7.147}$$

Substitution into eq. (7.145) leads to

$$r_a = \frac{1}{k^2 u}\left[\ln\left(\frac{z}{z_0}\right)\right]^2 \tag{7.148}$$

where the wind speed is measured at height $z$. Some conclusions follow:

- Turbulent diffusivity increases with height.
- Aerodynamic resistance increases with height.
- Aerodynamic resistance decreases with increasing roughness height.
- Aerodynamic resistance decreases with increasing wind speed.

**Example 7.6.** Calculate $r_a$ at a 10-m height under neutral conditions at a wind speed of 3 m s$^{-1}$ (also measured at a 10-m height), with a roughness length of 0.1 m.

***Solution.***

$$r_a = \frac{1}{(0.4)^2 \cdot 3 \text{ m s}^{-1}}\left[\ln\left(\frac{10}{0.1}\right)\right]^2 = \frac{(4.605)^2}{0.48} \text{ s m}^{-1} = 44.2 \text{ s m}^{-1}$$

Under *stable* conditions, eq. (7.146) can be used very close to the surface, but with rapidly decreasing accuracy away from the surface. Hence, it is better to extend the equation to allow $\phi_h$ to deviate from 1. Equation (5.192) becomes

$$K_z = \frac{ku_* z}{\phi_h(\zeta)} = \frac{ku_* z}{1 + 5z/L} \tag{7.149}$$

Hence, the flux equation (Fick's law) can be written as

$$-F = \frac{ku_* z}{1 + 5z/L}\frac{dc}{dz} \tag{7.150}$$

Rearranging leads to

$$dc = -\frac{F(1 + 5z/L)}{ku_*}\frac{dz}{z} = -\frac{F}{ku_*}\frac{dz}{z} - \frac{5F}{ku_* L}dz \tag{7.151}$$

Integration leads to

$$c_s - c_a = -\frac{F}{ku_*}\ln\left(\frac{z_0}{z}\right) - \frac{5F}{ku_* L}(z - z_0) \tag{7.152}$$

Comparing with eq. (7.145) leads to the following equation for $r_a$ under *stable* conditions:

$$\boxed{r_a = \frac{1}{ku_*}\ln\left(\frac{z}{z_0} + 5\frac{z - z_0}{L}\right)} \tag{7.153}$$

To eliminate friction velocity, we can rearrange eq. (5.81):

$$u_* = \frac{ku}{\ln(z/z_0) + 5\frac{z - z_0}{L}} \tag{7.154}$$

Hence, the aerodynamic resistance can be calculated directly from the wind speed:

$$r_a = \frac{1}{k^2 u}\left[\ln\left(\frac{z}{z_0} + 5\frac{z - z_0}{L}\right)\right]^2 \tag{7.155}$$

**Example 7.7.** Calculate the aerodynamic resistance at a 10-m height under stable conditions with $u = 3\,\mathrm{m\,s^{-1}}$ (measured at 10-m height), a roughness length of 0.1 m, and an Obukhov length of 20 m.

***Solution.***

$$r_a = \frac{1}{(0.4)^2 \cdot 3\,\mathrm{m\,s^{-1}}}\left[\ln\left(\frac{10}{0.1}\right) + 5\frac{10 - 0.1}{20}\right]^2 = \frac{(7.08)^2}{0.48}\,\mathrm{s\,m^{-1}} = 104.4\,\mathrm{s\,m^{-1}}$$

As we see from the example, the aerodynamic resistance increases with stability, that is, mass transfer decreases. This is an expected result.

Under *unstable* conditions, eq. (7.146) can again be used only very close to the surface. Hence we need to extend the equation to a form similar to eq. (7.149), but for unstable conditions:

$$K_z = \frac{ku_* z}{\phi_h(\zeta)} = ku_* z\left(1 - \frac{16z}{L}\right)^{1/2} \tag{7.156}$$

Following the same procedure as outlined for stable conditions, we obtain the following equation for $r_a$:

$$r_{\mathrm{a}} = \frac{1}{ku_*}\left\{\ln\left(\frac{z}{z_0}\right) + \ln\left[\frac{(1+x_0)^2}{(1+x)^2}\right]\right\} \tag{7.157}$$

where

$$x = \left(1 - 16\frac{z}{L}\right)^{1/2} \quad x_0 = \left(1 - 16\frac{z_0}{L}\right)^{1/2} \tag{5.96}$$

**Example 7.8.** Calculate the aerodynmic resistance at a 10-m height under unstable conditions with $u = 3\,\mathrm{m\,s^{-1}}$ (measured at 10-m height), a roughness length of 0.1 m, and an Obukhov length of –10 m. (Additional information: Under these conditions the friction velocity is $0.3402\,\mathrm{m\,s^{-1}}$.)

***Solution.***

$$r_{\mathrm{a}} = \frac{1}{0.4 \cdot 0.3402\,\mathrm{m\,s^{-1}}}\left\{\ln\left(\frac{10}{0.1}\right) + \ln\left[\frac{\left(\sqrt{1+0.1}+1\right)^2}{\left(\sqrt{1+10}+1\right)^2}\right]\right\} = \frac{2.80}{0.1361}\,\mathrm{s\ m^{-1}} = 20.6\ \mathrm{s\ m^{-1}}$$

As we see from the example, the aerodynamic resistance decreases with increasing instability, that is, mass transfer increases. This is also an expected result.

Sometimes $r_a$ is calculated based on the assumption that $K_z = K_m$ [the momentum diffusivity; see eq. (5.174)] (Seinfeld and Pandis, 2006; Holmes et al., 2011). This is an acceptable assumption under stable and neutral conditions, and it is the basis of the calculations in Appendix C. However, this is not an acceptable assumption under unstable conditions.

The aerodynamic resistance is independent of the pollution properties and is even the same for particles and gases. This will not be the case with the other resistances.

To calculate the **quasi-laminar resistance**, we need to distinguish between gases and particles. For **gases**, $r_b$ can be estimated with the following equation (Shepherd, 1974; Slinn et al., 1978; Seinfeld and Pandis, 2006):

$$r_{\mathrm{b}} = \frac{\alpha_{\mathrm{b}}\mathrm{Sc}^{2/3}}{ku_*} \tag{7.158}$$

where $\alpha_b$ is a constant, often taken equal to 2, and Sc is the Schmidt number:

$$\mathrm{Sc} = \frac{\nu}{D} \tag{7.159}$$

where $\nu$ is the kinematic viscosity ($=\mu/\rho$), and $D$ is the molecular diffusivity ($m^2 s^{-1}$). The removal mechanism of gases in the quasi-laminar layer is the same as for particulate matter with deposition rates dominated by Brownian motion (diffusion). Hence, a universal equation for $r_b$ for gases should apply to Brownian motion of particles as well. For that reason, the relevant equations proposed for particles will be included here as well.

Several other values of $\alpha_b$ have been proposed for the calculation of $r_b$. Shepherd (1974) suggested $\alpha_b=10$ for a water surface. Slinn and Slinn (1980) suggested $\alpha_b=4.44$, but they suggested replacing the power of $\frac{2}{3}$ by $\frac{1}{2}$. Zhang et al.'s (2001) value of $\alpha_b$ amounts to $0.4/3=0.133$, and suggested to replace the power of $\frac{2}{3}$ by a number varying from 0.5 to 0.58, depending on the landscape. This leads to extremely low values of the resistance. Based on the wind tunnel measurements at the lower end of the particle diameter range investigated by Möller and Schumann (1970), the original values of $\alpha_b=2$, and a power of $\frac{2}{3}$ seem to best fit the data. On the other hand, field data on deposition of aerosols indicate that wind tunnel experiments underestimate actual deposition rates by an order of magnitude or more (Wesely et al., 1985), which supports the coefficients of Zhang et al. (2001). Weseley et al. (1985) argued that deposition rates are larger when the surface is rough and has a complex fine structure, like lush vegetation.

Owen and Thompson (1963) proposed the following equation:

$$r_b u_* = a\mathrm{Re}_*^m \mathrm{Sc}^n + b \tag{7.160}$$

where $\mathrm{Re}_*$ is the Reynolds number based on surface roughness length and friction velocity:

$$\mathrm{Re}_* = \frac{z_0 u_*}{\nu} \tag{7.161}$$

Fernandez de la Mora and Friedlander (1982) proposed the following coefficients: $a=0.17$, $b=3$, $m=\frac{1}{2}$, and $n=\frac{1}{3}$. With these coefficients, predictions with eq. (7.161) are similar to predictions of eq. (7.158) when the surface is moderately rough (e.g., $z_0=0.1$ m). It predicts that deposition decreases with increasing surface roughness, which may not be consistent with the findings of Wesely et al. (1985), and which is inconsistent with experimental results (Slinn et al., 1978; Sehmel, 1980).

It is clear that a final theory of $r_b$ is not yet available. Pending better data and/or a better theory to put existing data in perspective, the following recommendations can be made. For water and other smooth surfaces, eq. (7.158) can be used without modifications (i.e., $\alpha_b=2$). For rough surfaces, especially vegetation, Zhang et al.'s

(2001) modification is more likely to give accurate results. The following tentative equation provides predictions that are in general agreement with our current understanding of $r_b$:

$$\alpha_b = 0.15(z_0 / \text{m})^{-0.3} \tag{7.162}$$

where the unit m (meter) is included to balance units.

**Example 7.9.** Calculate $r_b$ for a light gas (Sc = 1) and for very fine particulate matter (Sc = 1000). The atmosphere is neutral and the wind speed at a 10-m height is 3 m s$^{-1}$. The roughness length of the surface is 0.1 m. Use eq. (7.158) with the original coefficents and compare with the coefficents of Slinn and Slinn (1980), of Zhang et al. (2001), and with eq. (7.162).

***Solution.*** First, the friction velocity is calculated:

$$u_* = \frac{ku}{\ln(z/z_0)} = \frac{0.4 \cdot 3\text{ms}^{-1}}{\ln(10/0.1)} = 0.2606\,\text{m s}^{-1}$$

For gases:

$$r_b = \frac{2\text{Sc}^{2/3}}{ku_*} = \frac{2 \cdot 1^{2/3}}{0.4 \cdot 0.2606\,\text{m s}^{-1}} = 19.2\,\text{s m}^{-1}$$

For particles:

$$r_b = \frac{2\text{Sc}^{2/3}}{ku_*} = \frac{2 \cdot 1000^{2/3}}{0.4 \cdot 0.2606\,\text{m s}^{-1}} = 1920\,\text{s m}^{-1}$$

The variant of Slinn and Slinn (1980) is

$$r_b = \frac{4.44\text{Sc}^{1/2}}{ku_*}$$

With this equation, values of 42.6 and 1350 s m$^{-1}$ are obtained for gas and particles, respectively.

For the variant of Zhang et al. (2001), we take an average power of 0.54. Hence:

$$r_b = \frac{0.133\text{Sc}^{0.54}}{ku_*}$$

With this equation, values of 1.28 and 53.3 s m$^{-1}$ are obtained for gas and particles, respectively.

Based on eq. (7.162), the calculation of $\alpha_b$ is as follows:

$$\alpha_b = 0.15/0.1^{0.3} = 0.299$$

Based on this value, eq. (7.158) leads to $r_b = 2.87\,\text{s}\,\text{m}^{-1}$ for gas and $287\,\text{s}\,\text{m}^{-1}$ for particles.

We find that for gases the quasi-laminar resistance tends to be smaller than the aerodynamic resistance, whereas for fine particles the quasi-laminar resistance is larger than the aerodynamic resistance. Note that eq. (7.158) is only applicable for extremely fine particulate matter ($D_p < 0.05\,\mu\text{m}$), which has negligible removal efficiency due to impaction and interception.

For **particles**, the quasi-laminar resistance is much more difficult to estimate because $r_b$ covers the effects of Brownian motion, impaction, and interception. Wesely et al. (1985) found that friction velocity and atmospheric stability are the main factors driving $r_b$. Because their work was field based, the particle size was not controlled. However, because of the nature of the aerosol (sulfates), it was assumed that the particle diameter was near the value of minimal deposition rate. They suggested the following equation for $r_b$:

$$r_b = \frac{500 u_*}{u_*^2 + 0.24 w_*^2} \tag{7.163}$$

This equation leads to much larger values than eq. (7.158) or (7.160). Equation (7.163) is meant to incorporate impaction, interception, and diffusion in the diameter range where deposition is minimal.

The calculation can be based on separate calculations for impaction, interception, and diffusion with the following equation (Zhang et al., 2001):

$$r_b = \frac{1}{\varepsilon_0 u_* (E_B + E_{IM} + E_{IN}) R} \tag{7.164}$$

where $\varepsilon_0$ is an empirical constant [a value of 3 was chosen by Zhang et al. (2001)], $E_B$ is the collection efficiency from Brownian motion, $E_{IM}$ is the collection efficiency from impaction, $E_{IN}$ is the collection efficiency from interception, and $R$ is a correction for particles that hit a surface without sticking (= 1 if all particles stick).

In terms of resistances, this equation can be written as

$$\frac{1}{r_b} = \frac{1}{r_{b,B}} + \frac{1}{r_{b,IM}} + \frac{1}{r_{b,IN}} \tag{7.165}$$

For $E_B$ Zhang et al. (2001) suggested the following based on work of Slinn and Slinn (1980) and Slinn (1982):

$$E_B = \text{Sc}^{-\gamma} \tag{7.166}$$

where values of $\gamma$ ranging from $\frac{1}{2}$ for water surfaces to $\frac{2}{3}$ for vegetated surfaces have been proposed. Zhang et al. (2001) proposed values ranging from 0.5 to 0.58. In terms of resistance, the following equation is obtained:

$$r_{b,B} = \frac{Sc^{\gamma}}{3u_* R} \tag{7.167}$$

These equations require a diffusivity for particles, which is taken from Brownian motion theory. The result is (Seinfeld and Pandis, 2006)

$$D = \frac{kTC}{3\pi\mu d_p} \tag{7.168}$$

where $C$ is the Cunningham correction factor [eq. (7.134)], and $k$ is the Boltzmann constant ($1.3806505 \times 10^{-23}$ J K$^{-1}$).

Zhang et al. (2001) reviewed several equations from the literature for $E_{IM}$ (impaction efficiency) and chose the following equation based on Peters and Eiden (1992):

$$E_{IM} = \left(\frac{St}{\alpha + St}\right)^{\beta} \tag{7.169}$$

where *St* is the Stokes number, which has the following form for vegetated surfaces:

$$St = \frac{v_s u_*}{gA} \tag{7.170}$$

where $v_s$ is the settling velocity, and $A$ is the characteristic radius of the particle collector elements (typically 0.002–0.01 m). Smooth surfaces do not have particle collector elements, and the equation proposed by Zhang et al. (2001) to calculate St is

$$St = \frac{v_s u_*^2}{g\nu} \tag{7.171}$$

In eq. (7.169), $\beta = 2$. Depending on the land use category, $\alpha$ ranges from 0.6 to 100. Vegetated surfaces and urban terrain have $\alpha$ values around 1, whereas unvegetated smooth covers including water surfaces have $\alpha$ values of 50–100.

However, the following equation based on Slinn (1982) gives better results:

$$E_{IM} = \frac{St}{\alpha + St^{\beta}} \tag{7.172}$$

The main reason is that Peters and Eiden (1992) used a different definition of St. Instead of eq. (7.170), they used $\mathrm{St} = \rho_\mathrm{p} d_\mathrm{p}^2 u / (9\mu A)$, which equals $\mathrm{St} = 2v_\mathrm{s} u/(gA)$ in the Stokes regime, in the absence of slip. To account for the difference, it is recommended to divide $\alpha$ by 20 if eq. (7.169) is used. Pryor et al. (2008) reviewed other definitions of St used in deposition modeling.

In terms of resistance, eq. (7.169) becomes

$$r_{\mathrm{b,IM}} = \frac{1}{3u_* R}\left(\frac{\alpha + \mathrm{St}}{\mathrm{St}}\right)^{\beta} \tag{7.173}$$

Pleim et al. (1984) used the following equation for $r_{\mathrm{b,IM}}$:

$$r_{\mathrm{b,IM}} = 10^{3/\mathrm{St}} \tag{7.174}$$

The interception collection efficiency, $E_{\mathrm{IN}}$, can be calculated by the following equation modified from Slinn (1982):

$$E_{\mathrm{IN}} = \frac{d_\mathrm{p}}{2A} \tag{7.175}$$

This leads to the following equation for the resistance to deposition by interception:

$$r_{\mathrm{b,IN}} = \frac{1}{3u_* R}\left(\frac{2A}{d_\mathrm{p}}\right) \tag{7.176}$$

Note that eq. (7.175) is different from the equation proposed by Zhang et al. (2001), which is

$$E_{\mathrm{IN}} = \frac{1}{2}\left(\frac{d_\mathrm{p}}{A}\right)^2 \tag{7.177}$$

When eqs. (7.172) and (7.175) are replaced by the original model formulation of Zhang et al. (2001), as is usually done in aerosol models (e.g., Roldin et al., 2011), then impaction and interception are never dominant processes, which is known from experiments to be incorrect (Slinn, 1982).

The fraction of particles that sticks upon hitting a surface, $R$, is given by (Slinn, 1982)

$$R = \exp(-\mathrm{St}^{1/2}) \tag{7.178}$$

An exception is wet surfaces. It is assumed that on wet surfaces $R = 1$ for any particle size. In Figure 7.15, this procedure was used to calculate all velocities.

**Example 7.10.** Calculate the deposition velocity at a reference height of 10 m of particles with diameter 1 μm, density 1000 kg m$^{-3}$, on a surface with roughness 0.1 m and a characteristic radius of particle collector elements of 1 cm. The wind speed at a 10-m height is 3 m s$^{-1}$. Assume neutral conditions, a temperature of 25 °C, a pressure of 1 atm, and a dynamic viscosity of $1.8 \times 10^{-5}$ Pa s. For the purpose of impaction calculations, a rough surface can be assumed. All particles that hit the surface are absorbed.

***Solution.*** First, the settling velocity is calculated. As the particle diameter is less than 5 μm, the Cunningham slip factor needs to be accounted for, which in turn requires the mean free path of air molecules. Equation (7.133) requires the air density, which is calculated with the ideal gas law:

$$\rho_\mathrm{f} = \frac{M_\mathrm{air} p}{RT} = \frac{0.029 \times 101325}{8.314472 \times 298.15} = 1.185\,\mathrm{kg\,m^{-3}}$$

Hence [eq. (7.133)]:

$$\lambda = \frac{0.256469 \times 1.8 \times 10^{-5}}{(1.185 \times 101325)^{0.5}} = 6.526 \times 10^{-8}\,\mathrm{m} = 0.06526\,\mu\mathrm{m}$$

The Cunningham correction factor is calculated as [eq. (7.134)]:

$$C = 1 + 2\left(\frac{0.06526}{1}\right)\left[1.257 + 0.4\exp\left(\frac{-0.55 \times 1}{0.06526}\right)\right] = 1.164$$

Hence, the settling velocity is calculated with eq. (7.132):

$$v_\mathrm{s} = \frac{1000 \times (10^{-6})^2 \times 9.80665}{18 \times 1.8 \times 10^{-5}} = 3.027 \times 10^{-5}\,\mathrm{m\,s^{-1}}$$

For the calculation of the resistances, the friction velocity is calculated from the wind speed with eq. (5.65):

$$u_* = \frac{k u_\mathrm{ref}}{\ln(z_\mathrm{ref}/z_0)} = \frac{0.4 \times 3\,\mathrm{m\,s^{-1}}}{\ln(10/0.1)} = 0.2606\,\mathrm{m\,s^{-1}}$$

The aerodynamic resistance is calculated with eq. (7.146) (neutral atmosphere):

$$r_\mathrm{a} = 1/(0.4 \times 0.2606\,\mathrm{m\,s^{-1}}) \times \ln(10/0.1) = 44.18\,\mathrm{s\,m^{-1}}$$

Because the pollutant is a particle, the quasi-laminar resistance is the result of three resistances in parallel: for Brownian motion, impaction, and interception.

Because the surface is rough, eq. (7.158) will be used, but with a power law index of 0.54 and an $\alpha_b$ value of 0.133. First, the diffusivity of the particles is calculated with eq. (7.168):

$$D = 1.3806505\times10^{-23}\times298.15\times1.164/(3\times3.1415\times1.8\times10^{-5}\times10^{-6}) = 2.825\times10^{-11}\,\mathrm{m^2\,s^{-1}}$$

The Schmidt number is calculated:

$$\mathrm{Sc} = \frac{\nu}{D} = \frac{\mu/\rho_f}{D} = \frac{(1.8\times10^{-5}/1.185)}{2.825\times10^{-11}} = 5.376\times10^5$$

Hence, eq. (7.158) leads to

$$r_{b,B} = 0.133\times(5.376\times10^5)^{0.54}/(0.4\times0.2606) = 1586\,\mathrm{s\,m^{-1}}$$

The resistance to deposition by impaction is calculated with eq. (7.173). However, the Stokes number of Peters and Eiden (1992) will be used in its calculation:

$$\mathrm{St} = \frac{\rho_p d_p^2 u}{9\mu A} = \frac{1000\times(10^{-6})^2\times3}{9\times1.8\times10^{-5}\times0.01} = 1.852\times10^{-3}$$

Hence, with $\alpha = 1$ (rough surface) and $\beta = 2$:

$$r_{b,IM} = 1/(3\times0.2606\times1)\times((1+1.852\times10^{-3})/1.852\times10^{-3}) = 1.744\times10^5\,\mathrm{s\,m^{-1}}$$

Interception resistance is calculated with eq. (7.176):

$$r_{b,IM} = 1/(3\times0.2606\times1)\times(2\times0.01/10^{-6}) = 2.558\times10^4\,\mathrm{s\,m^{-1}}$$

Based on the three quasi-laminar resistances, the overall quasi-laminar resistance can be calculated with the resistances-in-parallel model [eq. (7.165)]:

$$r_b = 1/(1/1586+1/1.744\times10^5+1/2.558\times10^4) = 1488\,\mathrm{s\,m^{-1}}$$

Hence, the deposition velocity is calculated with eq. (7.138):

$$v_d = 3.027\times10^{-5}/(1-\exp(-3.027\times10^{-5}(44.18+1488))) = \mathbf{6.681\times10^{-4}\,m\,s^{-1}}$$

The value is higher than indicated in Figure 7.15 due to the way eq. (7.158) was used to calculate $r_{b,B}$. This indicates that a better understanding of the impact of Brownian motion on dry deposition on various surfaces, or, more general, the impact of the quasi-laminar resistance, is crucial in the development of better dry deposition models.

Hygroscopic particles can accumulate water, so they grow in size. For particles with a diameter of a few micrometers or less, it can be assumed that particles grow to their equilibrium size within the time frame of deposition, whereas larger particles do not reach that equilibrium (Slinn and Slinn, 1980). Sea salt sprays, on the other hand, will shrink until they reach an equilibrium value (de Leeuw et al., 2011).

Because the resistances to deposition are so poorly understood, it is useful to incorporate Brownian motion into the aerodynamic equations, and consider the results. The discussion is limited to neutral conditions. The derivations are given in Appendix C. Incorporating Brownian motion in the form of a diffision constant $D$ into $K_z$ and calculating the resistance all the way to $z=0$ (as opposed to $z=z_0$) lead to

$$r_a + r_b = \frac{1}{ku_*} \ln\left(1 + \frac{ku_* z_{ref}}{D}\right) \tag{7.179}$$

where $z_{ref}$ is the reference height. Note that the resistance all the way to $z=0$ is infinity in the absence of $D$, which is why we needed the assumption that aerodynamic resistance stops at $z=z_0$ in the calculation of $r_a$. Entering molecular diffusion into the model removes that necessity.

Subtracting eq. (7.146) from eq. (7.179) leads to

$$r_b = \frac{1}{ku_*} \ln\left(\frac{z_0}{z_{ref}} + \frac{ku_* z_0}{D}\right) \tag{7.180}$$

This is the resistance based on the assumption that the quasi-laminar layer does not exist as a separate layer, and its value is simply an artefact of the arbitrary choice $z_1=z_0$, along with the effect of $D$ on mass transfer near the surface.

Equation (7.180) yields unrealistically small resistances (a few hundreds of s m$^{-1}$) even for large particles. It follows that there must be a layer close to the surface where eq. (7.140) does not apply. A similar conclusion can also be drawn based on the wind speed profile near a smooth wall.

We define $d$ as the thickness of the layer that violates eq. (7.140). Based on dimension analysis (Appendix C), $d$ is expected to follow a relationship of the form

$$d = \alpha \frac{\nu}{u_*} \tag{C11}$$

where $\alpha$ is an empirical constant. In the case of an aerodynamically smooth water surface, we expect gravity to impact wave formation, which leads to a relationship of the following form:

$$d = \alpha_{\mathrm{w}} \frac{u_*^{(2n-1)/(n+1)} \nu^{1/(n+1)}}{g^{n/(n+1)}} \tag{C54}$$

where $\alpha_{\mathrm{w}}$ and $n$ are empirical constants.

As a first approximation, we assume that the turbulent diffusivity is zero in the layer, and the mass transfer in the layer is governed by molecular diffusion only. As calculated in Appendix C, the resistance of such a system is

$$r_{\mathrm{a}} + r_{\mathrm{b}} = \frac{d}{D} + \frac{1}{ku_*} \ln\left( \frac{ku_* z_{\mathrm{ref}} + D}{ku_* d + D} \right) \tag{7.181}$$

To obtain consistency with the law of the wall for wind speed profile (Chapter 5), a layer thickness ranging from $d = 250\,\mu\mathrm{m}$ to $d = 10\,\mathrm{mm}$ is required, depending on the friction velocity.

In this model, $r_{\mathrm{a}} + r_{\mathrm{b,B}}$ is roughly proportional with $\mathrm{Sc}^{\gamma}$, with $\gamma = 1$, which is outside the range suggested by Hicks (1982) (0.4–0.8), and inconsistent with experimental data of Möller and Schumann (1970), which are consistent with $\gamma = \frac{2}{3}$. Furthermore, when applied to $\mathrm{Sc} = 1$ (gases), the model predicts negative values of $r_{\mathrm{b,B}}$ in some cases, which is inconsistent with the observation that gases also have a quasi-laminar resistance. It follows that the quasi-laminar layer is not entirely free of turbulence.

A model was developed where the quasi-laminar layer had a turbulent diffusivity $K_z$ proportional to $z^2$ (see Appendix C). This model, with $\alpha_{\mathrm{w}} = 718.4$ and $n = -\frac{2}{3}$ in eq. (C54), provided a good fit between the experimental wind speed profile data of Möller and Schumann (1970) and the model. However, for the resistance $r_{\mathrm{a}} + r_{\mathrm{b,B}}$ of fine particles, the model predicts too low a dependence on Sc, and hence too small resistances for particles. In this model, $r_{\mathrm{b,B}}$ is roughly proportional with $\mathrm{Sc}^{\gamma}$, with $\gamma = 0.5$, while the data of Möller and Schumann (1970) indicate $\gamma = \frac{2}{3}$. The equation for $r_{\mathrm{b}}$ and its derivation are given in Appendix C.

The best results were obtained with a model that assumes $K_z$ to be proportional to $z^3$ in the quasi-laminar layer. The equation for $r_{\mathrm{a}} + r_{\mathrm{b}}$ (or $r_{\mathrm{a}} + r_{\mathrm{b,B}}$) is

$$r_{\mathrm{d}} = \frac{1}{3aE_D^{2/3}} \left[ \ln\left( \frac{d + E_D^{1/3}}{E_D^{1/3}} \right) - \frac{1}{2} \ln\left| \frac{d^2 - E_D^{1/3} d + E_D^{2/3}}{E_D^{2/3}} \right| + \frac{3}{\sqrt{3}} \arctan\left( \frac{2d}{\sqrt{3} E_D^{1/3}} \right) \right] + \frac{1}{ku_*} \ln\left( \frac{ku_* z_{\mathrm{ref}} + D}{ku_* d + D} \right) \tag{7.182}$$

where

$$E_D = \frac{3d^2 D}{ku_*} \tag{7.183}$$

This model provided satisfactory agreement with experimental wind speed data of Möller and Schumann (1970) when $\alpha_w = 408.4$ and $n = -\frac{4}{13}$ in eq. (C54). Values of $d$ range from about 1 to 25 mm at smooth water surfaces. In this model, $r_{b,B}$ is roughly proportional with $Sc^{\gamma}$, with $\gamma = 0.67$, consistent with the data of Möller and Schumann (1970), although the resistances are somewhat overestimated. Hence, based on our knowledge of $r_b$ and $r_{b,B}$, we conclude that turbulence in the quasi-laminar layer can best be described by a turbulent diffusion proportional to $z^3$. Note that the considerations that led to the original justification (Slinn et al., 1978) that $\gamma = \frac{2}{3}$ were entirely different from the reasoning applied here. Slinn et al. (1978) considered diffusion in a growing boundary layer, and found a relationslip between resistance and $Sc^{2/3}$ that also incorporates length of the boundary layer.

The model that led to eq. (7.182) is limited to aerodynamically smooth surfaces and is therefore not applicable to the surface of the open sea, which is aerodynamically rough.

For **particulate matter**, due to the way $r_b$ is calculated, the surface resistance $r_c = 0$. The surface resistance for **gases** depends on the nature of the surface. Three types of surface can be defined, each with their own resistance (Seinfeld and Pandis, 2006):

- Water: $r_{cw}$ (water resistance)
- Ground: $r_{cg}$ (ground resistance)
- Vegetation: $r_{cf}$ (foliar resistance)

These resistances work in parallel. Seinfeld and Pandis (2006) suggested that each environment is either a combination of water and vegetation, or ground and vegetation, and that the resistances act in parallel:
For ground + vegetation:

$$\frac{1}{r_c} = \frac{1}{r_{cg}} + \frac{1}{r_{cf}} \tag{7.184}$$

For water + vegetation:

$$\frac{1}{r_c} = \frac{1}{r_{cw}} + \frac{1}{r_{cf}} \tag{7.185}$$

Some of the deposition pathways that make up the surface resistance have transfer resistances that depend on the properties of the depositing molecules. A commonly used technique to model deposition resistances of a wide range of substances is to extablish the resistance for $SO_2$ and for $O_3$, and to develop equations that relate these resistances to the resistance of a molecule with a different set of properties than $SO_2$ and $O_3$ (Wesely, 1989). An interesting approach was adopted by Zhang et al. (2002), who modeled some of the resistances with the following equation:

$$\frac{1}{r_i} = \frac{\alpha_i}{r_{SO_2}} + \frac{\beta_i}{r_{O_3}} \tag{7.186}$$

where subscript $i$ refers to species $i$, and $\alpha_i$ and $\beta_i$ are empirical variables to be determined separately for each species.

The ground resistance and especially the foliar resistance of plant canopies are very complicated and involve many mass flows in series and in parallel. There are different levels of complexities that can be adopted when modeling deposition in such systems. The simplest type of model is the **big leaf models**. As the term indicates, a big leaf model considers only one flux directed to the plants directly, representing the plant by a single leaf. More complicated models are **two-layer models** and **multilayer models**. These models divide the vegetative canopy into two or more layers and consider the fluxes to, across, and between the layers. In fact, these are overlapping concepts because a big leaf model can be a two-layer or multilayer model if only one layer contains leaves.

Because a **leaf area index** (LAI) (area of leaves/area of the ground) is a commonly measured variable in ecology, it is tempting to use it directly in the parameterization of the foliar resistance $r_{cf}$. To understand this, consider the flux to the leaves $F_f$ (in mg per m$^2$ of ground):

$$F_f = \frac{c}{r_{cf}} \tag{7.187}$$

where $c$ is the pollutant concentration at the leaf surface. It would make sense to consider the flux to the leaves in mg per m$^2$ of leaves, $F'_f$:

$$F'_f = \frac{c}{r'_{cf}} \tag{7.188}$$

where $r'_{cf}$ is a leaf area specific resistance. The fluxes are related as follows:

$$F_f = \text{LAI} \cdot F'_f \tag{7.189}$$

Combining eqs. (7.187)–(7.189) leads to

$$r_{cf} = \frac{r'_{cf}}{\text{LAI}} \tag{7.190}$$

Equation (7.190) separates the conditions of the leaves (represented by $r'_{cf}$) from the amount of leaves (represented by LAI). However, $r'_{cf}$ itself depends on LAI, for various reasons. For instance, a smaller fraction of the leaves are sunlit when LAI is large. Hence, eq. (7.190) does not help the parameterization, and is not used in contemporary deposition models.

**TABLE 7.1 Dimensionless Henry Constants for Some Common Air Pollutants**

| Compound | $H$ |
|---|---|
| $SO_2$ | 0.04 |
| $H_2O_2$ | $4\times10^{-7}$ |
| $HNO_3$ | $8\times10^{-8}$ |
| $O_3$ | 2 |
| $NO_2$ | 3.5 |
| PAN | 0.01 |
| HCHO | $4\times10^{-6}$ |

*Source*: From Scire et al. (2000).

We will now discuss the three types of surface resistance. The **water resistance** represents the resistance associated with mass transfer from the water surface to the water bulk. This resistance is dominated by diffusion resistance through a quasi-laminar water layer at the surface not unlike the quasi-laminar layer in the air phase. Hence, what drives the mass transfer is the concentration gradient in the water phase. This gradient can be increased by fast chemical reaction. Hence, we can capture the essence of the water resistance by the following equation (Slinn et al., 1978):

$$r_{cw} = \frac{H}{\alpha k_L} \tag{7.191}$$

where $H$ is the dimensionless Henry constant of the air pollutant, given by

$$H = \left( \frac{c_{air}}{c_{water}} \right)_{equil} \tag{7.192}$$

In words: The Henry constant is the ratio of the concentration of a compound in the air phase and the water phase, at equilibrium; $k_L$ is a mass transfer coefficient; $\alpha$ is the mass transfer enhancement due to aqueous phase reaction; $\alpha = 1$ when the gas is unreactive, and very close to 1 for slowly reactive compounds like $CO_2$. On the other hand $\alpha$ can be as high as 1700 for a fast reacting compound like $SO_2$ at seawater pH (Liss and Slater, 1974).

Some values of the Henry constant at room temperature are given in Table 7.1. More values are given in Section B10 (Appendix B). We find that for highly soluble compounds such as $H_2O_2$ and $HNO_3$ the water resistance is negligible, whereas for poorly soluble compounds such as $O_3$ and $NO_2$ it is the dominant resistance. The water resistance is comparable to the aerodynamic resistance when $H$ is about 0.001.

Because the turbulence in stagnant surface water is generated by the wind, $k_L$ depends mainly on the friction velocity. The following equation has been suggested (Scire et al., 2000):

$$k_{L} = d_{L} u_{*} \tag{7.193}$$

where $d_L$ is an empirical constant. CALPUFF uses a value of $4.8 \times 10^{-4}$ for $d_L$ (Scire et al., 2000). Data presented by Watson et al. (1991) lead to a value of $d_L$ of $1.60 \times 10^{-4}$. This is the recommended value if eq. (7.193) is used. However, there are better parameterizations of $k_L$.

Zhao et al. (2003) compiled a large data set and converted all data to a Schmidt number of 666 (the value of $CO_2$ in water at 20 °C). They obtained the following equation:

$$k_{L} = d_{L} \left( \frac{u_{*}}{1\,\mathrm{m\,s^{-1}}} \right)^{n} \tag{7.194}$$

where $d_L = 1.716 \times 10^{-4}\,\mathrm{m\,s^{-1}}$, and $n = 1.22$. To balance the units $1\,\mathrm{m\,s^{-1}}$ is included in eq. (7.194). However, data sets of this type have a large scatter in both the independent variable ($u_*$) and the dependent variable ($k_L$). Ordinary regression applied to such data underestimates the real slope (or the value of $n$ in a power law relationship). To illustrate that, the data presented by Watson et al. (1991) were analyzed with ordinary regression and with errors-in-variables regression. The former led to values of $d_L = 1.96 \times 10^{-4}\,\mathrm{m\,s^{-1}}$, and $n = 1.31$, whereas the latter led to values of $d_L = 2.58 \times 10^{-4}\,\mathrm{m\,s^{-1}}$, and $n = 1.55$. While these regressions markedly overestimate $k_L$ at high friction velocities as presented by Zhao et al. (2003), they do illustrate the point. In Appendix C some theoretical support for eq. (7.194) with $n = 1.5$ is given.

Zhao et al. (2003) used the data to develop a correlation for $k_L$ based on a breaking-wave parameter. Other formulations for $k_L$ have been suggested as well. It can be assumed that, at least at near-neutral conditions, there is a unique relationship between $u_*$ and the wind speed, provided that the wind speed measurement height is known. Hence, $k_L$ can be correlated to wind speed. For instance, Zhao et al. (2003) proposed the following based on their data set:

$$k_{L} = d_{L,10} \left( \frac{u_{10}}{1\,\mathrm{m\,s^{-1}}} \right)^{n} \tag{7.195}$$

where $d_{L,10} = 2.389 \times 10^{-6}\,\mathrm{m\,s^{-1}}$, $n = 1.35$, and $u_{10}$ is the wind speed at a 10-m height. Again, ordinary regression can cause an underestimated value of $n$. Ordinary regression of the data presented by Watson et al. (1991) leads to $d_{L,10} = 1.32 \times 10^{-6}\,\mathrm{m\,s^{-1}}$, and $n = 1.60$, whereas errors-in-variables regression leads to $d_{L,10} = 6.78 \times 10^{-7}\,\mathrm{m\,s^{-1}}$, and $n = 1.90$. Again, the latter regression overestimates $k_L$ values of Zhao et al. (2003) somewhat at high wind speeds. However, Wanninkhof (1992) found a very similar correlation, with $d_{L,10} = 8.61 \times 10^{-7}\,\mathrm{m\,s^{-1}}$, and $n = 2$. Hence, these are probably among the more reliable relationships and are recommended here, at least for open seawater. They apply to $CO_2$ in water (Sc = 666). Small ponds are less turbulent than the open sea and exhibit lower $k_L$ values.

The mass transfer coefficient of any other compound can be calculated from a reference value $k_{L,ref}$ with the following equation (Schwarzenbach et al., 1993):

$$k_L = k_{L,ref}\left(\frac{D}{D_{ref}}\right)^{\gamma} \tag{7.196}$$

where $\gamma = 0.57 \pm 0.15$ (Holmen and Liss, 1984). Based on the discussion on the quasi-laminar resistance, it seems reasonable to assume that the equation should be rephrased as

$$k_L = k_{L,ref}\left(\frac{Sc_{ref}}{Sc}\right)^{\gamma} \tag{7.197}$$

Hence, the temperature influence on the viscosity of water can be incorporated. This was already pointed out by Holmen and Liss (1984) and applied by Zhao et al. (2003) and others. A typical reference value of Sc is 666, the value for $CO_2$ in water at 20 °C. Note that Schmidt numbers in water are very different from Schmidt numbers in air. The value of $\gamma$ was measured again by Watson et al. (1991), and a value very close to 0.5 was found. The latter value was found on the ocean, whereas the former value (0.57) was obtained in a stirred tank. For this reason, and because 0.5 is within the the confidence interval of the measurements of Holmen and Liss (1984), a value of $\gamma = 0.5$ is recommended. This is also the relationship used by Zhao et al. (2003) to convert all their compiled data to a Schmidt number of 666 ($CO_2$ in water at 20 °C).

As with the quasi-laminar resistance, it is useful to conduct mass transfer calculations in order to gain a further understanding of $k_L$. It will be assumed that a natural water behaves like a mirror image of an atmosphere, with a quasi-laminar resistance at the surface and aerodynamic resistance underneath. Furthermore, it is assumed that the shear stress at the surface is transferred to the liquid, so a friction velocity on the liquid side can be calculated. Because of the square root Sc dependence of the mass transfer, it is assumed that the turbulent diffusivity in the quasi-laminar layer on the liquid side is proportional to the depth squared (see above for a similar discussion on the air side). Thanks to the low value and relatively narrow range of diffusivities of gases in water, a number of simplifications can be made. The full calculation is given in Appendix C. The simplified result is

$$k_L = \left(\frac{\pi}{2}\sqrt{\frac{d_w}{k u_{*w} D_w}}\right)^{-1} \tag{7.198}$$

where $d_w$ is the thickness of the quasi-laminar water layer, given by eq. (C64):

$$d_w = \alpha_w \frac{u_{*w}^{(2n-1)/(n+1)} \nu_w^{1/(n+1)}}{g^{n/(n+1)}} \tag{C64}$$

where $\alpha_w$ and $n$ have the same meaning and values as in the air phase [$\alpha_w = 60$; $n = -\frac{1}{4}$ for experiments of Liss (1973) with an aerodynamically smooth water surface]; and $u_{*w}$ is a water-side friction velocity, given by eq. (C61):

$$u_{*w} = \sqrt{\frac{\rho_{air}}{\rho_{water}}} u_{*,air} = \delta \cdot u_{*,air} \tag{C61}$$

Combining eq. (7.198) with eqs. (C64) and (C61), we can see that $k_L$ is proportional to $u_*^m$, where $m$ equals $(2 - n)/(2+2n)$. Based on experimental data of Liss (1973) in a wind tunnel, a value of $n = -\frac{1}{4}$ was put forward in Appendix C, leading to $m = 1.5$, consistent with calculations made above based on data of Watson et al. (1991).

What is remarkable about the above calculations is that the same set of parameters ($\alpha_w = 60$; $n = -\frac{1}{4}$) apply to both air-side mass transfer and water-side mass transfer. There is a remarkable similarity beween mass transfer on the air side and mass transfer on the water side.

Example values of the diffusivities of various pollutants in air and in water at 25 °C are given in Section B10. The orders of magnitude of diffusivities are $10^{-5}$ $m^2 s^{-1}$ in the air phase and $10^{-9} m^2 s^{-1}$ in the water phase. Estimation methods for diffusivities are discussed by Poling et al. (2000). In the air phase, a diffusivity of $10^{-5} m^2 s^{-1}$ at 25 °C leads to a Sc number of 1.6. In the water phase, a diffusivity of $10^{-9} m^2 s^{-1}$ at 25 °C leads to a Sc number of 900.

In many cases the air pollutant has a significant nonzero background concentration in the water bulk, leading to **bi-directional mass transfer**. The deposition velocity concept [eq. (7.128)] cannot handle this situation, and we will need to use eq. (7.130) instead, which is based on a transfer velocity. This situation is discussed in the next section.

When a pollutant undergoes an extremely fast reaction once dissolved, it is tempting to incorporate the reaction equilibrium into the Henry constant, or into an apparent diffusivity, to incorporate the mass transfer enhancement, rather than using $\alpha$ in eq. (7.191). This is illustrated here for the simplest possible case. Consider the following instantaneous reaction:

$$A \rightarrow B \tag{7.199}$$

with the equilibrium constant:

$$K = \left( \frac{c_{B,water}}{c_{A,water}} \right)_{equil} \tag{7.200}$$

Because both A and B can transport from the interface to the bulk water phase, what we need in eq. (7.191) is an apparent Henry constant $H'$ given by

$$H' = \left( \frac{c_{\text{A,air}}}{c_{\text{A,water}} + c_{\text{B,water}}} \right)_{\text{equil}} = \left( \frac{c_{\text{A,air}}}{c_{\text{A,water}}} \right)_{\text{equil}} \cdot \left( \frac{c_{\text{A,water}}}{c_{\text{A,water}} + c_{\text{B,water}}} \right)_{\text{equil}} = \frac{H}{1+K} \tag{7.201}$$

We find that the more complete the reaction, the lower the apparent Henry constant, and the faster the transfer will be in the water phase. Schwarzenbach et al. (1993) provide a similar argument by calculating an increased apparent diffusivity in the water phase. However, this procedure could lead to incorrect results when very dilute species such as $H^+$ and $OH^-$ are involved, for example, acid–base reactions. For instance, when a volatile acid is dissolved, the acid dissociation will create an accumulation of $H^+$ at the surface, changing the equilibrium. Due to its low concentration, the transfer of $H^+$ away from the surface is limited. Of course, the transfer of the negatively charged base away from the surface will generate a pull on the $H^+$ ions by electrostatic forces, but without more detailed investigations, there is no guarantee that eq. (7.201) will provide an acceptable approximation.

The **ground resistance** is considered *in parallel* with the foliar resistance. Hence, if a pollutant must pass a resistance created by the presence of a vegetative canopy before depositing on the ground, this must be included in the ground resistance. In some models this is expressed as a separate resistance $r_{ac}$ in series with the actual ground resistance $r_{gs}$ (e.g., Wesely, 1989). In some models the ground canopy contribution and the ground surface contribution are combined in a single resistance $r_{cg}$ (e.g., Zhang et al., 2002). However, the latter subdivided the ground resistance in a dry ground and a wet ground contribution, to be considered in parallel:

$$\frac{1}{r_{cg}} = \frac{1 - W_g}{r_{gd}} + \frac{W_g}{r_{gw}} \tag{7.202}$$

where $W_g$ is the wet fraction of the ground (0.9, 0.5, and 0.2 in rainy, dewy, and humid conditions, respectively), $r_{gd}$ is the dry ground resistance, and $r_{gw}$ is the wet ground resistance.

In the model of Wesely (1989), the ground canopy resistance $r_{ac}$ is generally 100–300 s m$^{-1}$ for vegetated land, except in winter, when agricultural land and rangeland can have a ground canopy resistance as low as 10 s m$^{-1}$. Forests have a ground canopy resistance as high as 1000–2000 s m$^{-1}$, whereas water surfaces and deserts have zero ground canopy resistance. The urban value is 100 s m$^{-1}$.

The ground surface contribution to the ground resistance, $r_{gs}$, in the model of Wesely (1989) is pollutant specific. For $SO_2$, $r_{gs}$ is 150–500 s m$^{-1}$ for vegetated and urban surfaces, 0 for water surfaces, and 1000 s m$^{-1}$ for deserts. For $O_3$, it is 150–300 s m$^{-1}$ for vegetated surfaces, 300–600 s m$^{-1}$ for urban surfaces, 400 s m$^{-1}$ for desert surfaces, and 1000–2000 s m$^{-1}$ for water surfaces. An exception is winter (freezing, snow) conditions, when $r_{gs}$ is 3500 s m$^{-1}$ for most surfaces.

For other compounds, deposition on wet surfaces was related to the deposition of $SO_2$ by means of the Henry constant, whereas reactivity of the compound was accounted for by a reactivity factor $f_0$ with a value of 1 ($H_2O_2$), 0.1 (e.g., $NO_2$), or 0 (e.g., $SO_2$) and related to the deposition rate of $O_3$. The equation used was

$$r_{gs} = \frac{1}{\frac{H_{cp}}{10^5 r_{gs,SO_2}} + \frac{f_0}{r_{gs,O_3}}} \tag{7.203}$$

where $H_{cp}$ is the Henry constant expressed as

$$H_{cp} = \left(\frac{c_{water}}{p_i}\right)_{equil} \tag{7.204}$$

where $p_i$ is the partial pressure of the pollutant. The units of $H_{cp}$ in eq. (7.204) is M atm$^{-1}$.

The dry ground resistance $r_{gd}$ in the model of Zhang et al. (2002) is of the same order of magnitude as the $r_{gs}$ values in the model of Wesely (1989), or slightly higher to reflect ground canopy effects. For the wet ground resistance $r_{gw}$ a value of 100 s m$^{-1}$ was used for $SO_2$ in rainy conditions and 200 s m$^{-1}$ for $SO_2$ in dewy conditions. For ozone, the same value as the dry soil resistance was used. Equation (7.186) was used to calculate resistances for other compounds. Typically, $\alpha_i$ reflects the water solubility of the compound (e.g., $\alpha_i = 1$ for $H_2SO_4$, $NH_3$, and $H_2O_2$; 10 for $HNO_3$, 0 for $NO_2$ and $O_3$) whereas $\beta_i$ reflects its reactivity (e.g., $\beta_i = 1$ for $H_2SO_4$ and $H_2O_2$, 10 for $HNO_3$, 0 for $SO_2$ and $NH_3$).

The **foliar resistance** is the most complex of all resistances. Wesely (1989) considered three parallel deposition pathways passing through a total of five resistances. The pathways are the stomata (including a stomatal resistance $r_{st}$ and a mesophyll resistance $r_m$), the cuticle ($r_{cut}$), and the lower canopy (including a canopy resistance $r_{dc}$ and a resistance for leaves, twigs, bark, etc., $r_{cl}$). Hence, the total foliar resistance is given by

$$\frac{1}{r_{cf}} = \frac{1}{r_{st} + r_m} + \frac{1}{r_{cut}} + \frac{1}{r_{dc} + r_{cl}} \tag{7.205}$$

The model of Zhang et al. (2002) also has three parallel pathways, a dry stomatal resistance, a dry cuticle resistance, and a wet cuticle resistance. In their model the overall resistance is

$$\frac{1}{r_{cf}} = \frac{1 - W_{st}}{r_{st} + r_m} + \frac{W_c}{r_{cut,w}} + \frac{1 - W_c}{r_{cut,d}} \tag{7.206}$$

where $W_{st}$ and $W_c$ are wet fraction of the stomata and of the cuticle, respectively. Hence, resistances with the same name have a slightly different meaning in the two models.

In the model of Wesely (1989), the stomatal resistance is calculated with the equation:

$$r_{st} = r_i \left[1 + \left(\frac{200}{R + 0.1}\right)^2\right] \frac{400}{T_s(40 - T_s)} \frac{D_{H_2O}}{D} \tag{7.207}$$

where $R$ is the solar radiation in W m$^{-2}$, and $T_s$ is the surface air tempeature in °C; $r_i$ is a resistance parameter that depends on land use and season. Vegetated areas have a resistance $r_i$ of 140–300 s m$^{-1}$ in spring, 80–150 s m$^{-1}$ in summer, and virtually infinite values the rest of the year. Exceptions are coniferous forests, which have $r_i$ values less than 1000 s m$^{-1}$ throughout the year. Water, deserts, and urban areas have virtually infinite $r_i$ values throughout the year.

The mesophyll resistance $r_m$ in Wesely's (1989) model is calculated with

$$r_m = \frac{1}{H_{cp} / 3000 + 100 f_0} \tag{7.208}$$

According to eq. (7.208), the mesophyll resistance is negligibly small unless a molecule is both poorly soluble and unreactive.

In the model of Wesely (1989), the cuticle resistance $r_{cut}$ is calculated from (see also Walmsley and Wesely, 1996)

$$r_{cut} = \frac{r_{lu}}{10^{-5} H_{cp} + f_0} \tag{7.209}$$

Again, the two terms in the denominator specify the two ways a pollutant can be absorbed in plant matter: by dissolution (hence $H_{cp}$) or by reaction (hence $f_0$). In the presence of vegetation, $r_{lu}$ is 2000–8000 s m$^{-1}$ in spring, 2000–4000 s m$^{-1}$ in summer, and 6000 s m$^{-1}$ to infinity the rest of the year. In the absence of vegetation, $r_{lu}$ is infinity.

The first lower canopy resistance $r_{dc}$ is calculated with the equation:

$$r_{dc} = \frac{100\left[1 + 1000 / (R + 10)\right]}{1 + 1000\theta} \tag{7.210}$$

where $\theta$ is the slope of the terrain in radians.

The second lower canopy resistance is given by

$$r_{cl} = \frac{1}{H_{cp} / 10^5 r_{cl,SO_2} + f_0 / r_{cl,O_3}} \tag{7.211}$$

where the resistances for $SO_2$ and $O_3$ depend on season and vegetation type. The values of $r_{cl,SO_2}$ are generally 2000 s m$^{-1}$ and up in summer and 4000 s m$^{-1}$ and up the rest of the year, except for forests. The value is large enough to generate negligible flux when there is no vegetation. The value for $r_{cl,O_3}$ in vegetated areas is 1000 s m$^{-1}$

in summer and generally lower ($400\,s\,m^{-1}$ and up) the rest of the year. The value is very large when there is no vegetation.

In the model of Zhang et al. (2002), the wet fraction of stomata, $W_{st}$, is 0.5 in rainy and dewy conditions. The wet fraction of the cuticle, $W_c$, is 0.9, 0.7, and 0.2 in rainy, dewy, and humid conditions. Stomatal resistance was based on a model of Brook et al. (1999). Leaf area is divided into sunlit area and shaded area, and a separate resistance is calculated for each of them. Corrections are made for temperature, water vapor deficit, water stress, and diffusivity. The mesophyll resistance, $r_m$, is taken to be $100\,s\,m^{-1}$ for molecules that have both limited solubility and limited oxidizig capacity, and 0 for all other molecules.

The dry cuticle resistance, $r_{cut,d}$, depends on molecule, land use category, and season. A value ranging from 1000 to $9000\,s\,m^{-1}$ was used for both $SO_2$ and $O_3$, except when no leaves are present (e.g., tundra, desert). The wet cuticle resistance, $r_{cut,w}$, is calculated with the following equation:

$$r_{cut,w} = \frac{r_{cut,w0}}{(LAI + 0.1)^{1/3}} \tag{7.212}$$

where $r_{cut,w0}$ is $100\,s\,m^{-1}$ for $SO_2$ under rainy and high humidity conditions, $200\,s\,m^{-1}$ under dewy conditions, $400\,s\,m^{-1}$ for $O_3$ under rainy and high humidity conditions, and $800\,s\,m^{-1}$ for $O_3$ under dewy conditions.

The model of Wesely (1989) was updated and clarified by Walmsley and Wesely (1996). Readers wishing to consult the original model should consult the update as well. The model of Zhang et al. (2002) was updated with additional land use types and other model improvements by Zhang et al. (2003). As with the Wesely model, readers wishing to consult the original model are advised to consult the update as well.

***7.5.1.3 Air–Water Transfer: Equations for Bi-Directional Transfer.*** A framework popular among environmental engineers studying **volatilization of water pollutants** into the air is the so-called **two-film model**, which is derived from chemical engineering. The idea underlying the model is that for a pollutant to be exchanged between an air phase and a water phase, it must travel through two stagnant layers at the air–water interface: one on the water side and one on the air side. Of course, volatilization of a gas from a water body is simply the reverse of a bi-directional deposition process, and calculation schemes of the two processes should be mathematically equivalent. Unfortunately, the relationship between the two-film model and dry deposition theory is a frequently misunderstood subject.

Figure 7.16 illustrates the two-film model with a concentration profile of a pollutant depositing into the water phase. In the steady-state transfer of a conservative (i.e., nonreactive) pollutant from an air phase to a water phase, the flux from the bulk air to the air–water interface must equal the flux form the air–water interface into the bulk water. These fluxes are diffusive, driven by concentration gradients in the air film and the water film. It is assumed that Henry's laws applies at the air–water interface:

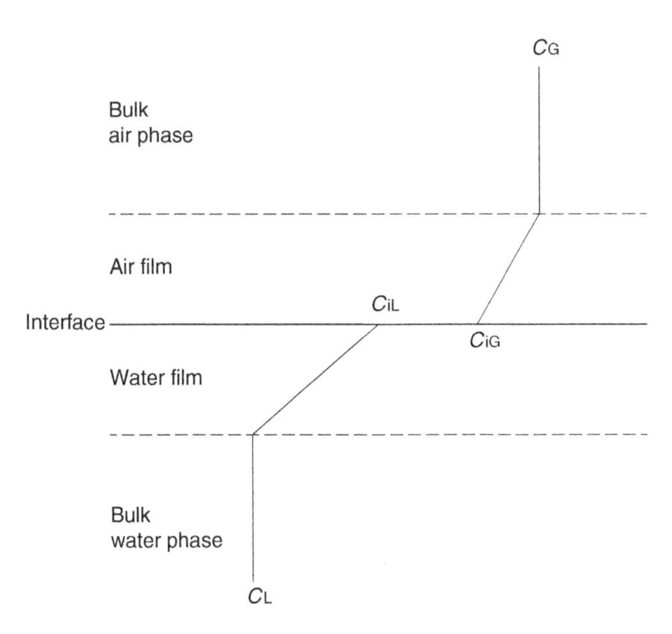


Figure 7.16 Concentration profile of a pollutant near the air–water interface in the two-film model.

$$H = \frac{c_{\text{i,air}}}{c_{\text{i,water}}} \tag{7.213}$$

where $c_{\text{i,air}}$ and $c_{\text{i,water}}$ are the concentrations at the interface, on the air side and the water side, respectively. The stagnant films each have a mass transfer coefficient (or velocity), denoted $k_G$ and $k_L$ for the air (gas) and water (liquid) side, respectively. Hence, the flux across the interface is given by

$$F = k_G(c_{\text{i,air}} - c_{\text{air}}) = k_L(c_{\text{water}} - c_{\text{i,water}}) \tag{7.214}$$

Equation (7.213) can be used to eliminate $c_{\text{i,water}}$:

$$k_G(c_{\text{i,air}} - c_{\text{air}}) = k_L\left(c_{\text{water}} - \frac{c_{\text{i,air}}}{H}\right) \tag{7.215}$$

Equation (7.215) can be solved for $c_{\text{i,air}}$:

$$c_{\text{i,air}} = \frac{k_G c_{\text{air}} + k_L c_{\text{water}}}{k_G + k_L / H} \tag{7.216}$$

Substitution of eq. (7.216) into eq. (7.215), and rearranging, leads to

$$F = \frac{k_G k_L}{H k_G + k_L}(H c_{\text{water}} - c_{\text{air}}) \tag{7.217}$$

Hence, we define the overall mass transfer coefficient $K_G$ as

$$K_G = \frac{k_G k_L}{H k_G + k_L} \tag{7.218}$$

Hence, eq. (7.217) can be rewritten as

$$F = K_G (H c_{\text{water}} - c_{\text{air}}) \tag{7.219}$$

Equation (7.219) describes bi-directional transfer, as it applies to both deposition (when $c_{\text{air}} > c_{\text{water}}$) and volatilization (when $c_{\text{water}} > c_{\text{air}}$).

Note that eq. (7.218) can also be written as

$$\frac{1}{K_G} = \frac{1}{k_L / H} + \frac{1}{k_G} \tag{7.220}$$

This shows that the two-film model can be interpreted as a resistances-in-series model. Provided that the equilibrium gas-phase concentration $Hc_L$ is negligible in comparison with $c_G$, and, as mentioned before, no reaction takes place, we can write

$$r_a + r_b = \frac{1}{k_G} \tag{7.221}$$

$$r_c = \frac{H}{k_L} \tag{7.222}$$

In volatilization studies, $k_G$ and $k_L$ are usually correlated to wind speed instead of friction velocity, and the height of the relevant air-phase concentration for the definition of $k_G$ is usually not specified. However, the experiments are generally conducted in a wind tunnel, where the representative height is typically about 0.1 m (Liss, 1973; Mackay and Yeun, 1983). It is easy to see that a significant portion of the resistance is missed in such experiments. A typical roughness length of a water surface is $10^{-4}$–$10^{-5}$ m. Assuming the higher value, a Schmidt number of 1 and a value $\alpha_b = 2$ in eq. (7.158), the sum of the resistances $r_a + r_b$ is the following for a reference height of 0.1 m:

$$r_a + r_b = \frac{1}{ku_*} \ln\left(\frac{0.1\text{m}}{10^{-4}\text{m}}\right) + \frac{2 \cdot 1}{ku_*} = \frac{22.27}{u_*} \tag{7.223}$$

If the actual reference height is higher, for example, 10 m, then the following resistance is missed in the determination:

$$\Delta r = \frac{1}{ku_*}\ln\left(\frac{10\text{m}}{0.1\text{m}}\right) = \frac{11.51}{u_*} \tag{7.224}$$

About a third of the overall resistance is missed in such experiments. Hence, experimetal determinations of $k_G$ can only be used if a correction is made to account for changes in the reference height. Alternatively, $k_G$ can be calculated from eq. (7.221), where $r_a$ and $r_b$ are calculated as explained in Section 7.5.1.2. Unfortunately, uncorrected values of $k_G$ derived from experiments are commonplace in the volatilization literature.

Another issue with extrapolations from wind tunnel studies is that the water surface in such studies tends to be aerodynamically smooth, whereas many natural waters have an aerodynamically rough water surface.

Another issue is the appropriate correction to account for diffusivity of the pollutant in the gas phase. In the volatilization literature it is common practice to correct for diffusivity in the following fashion (Schwarzenbach et al., 1993):

$$k_G = k_{G,H_2O}\left(\frac{D}{D_{H_2O}}\right)^{2/3} \tag{7.225}$$

However, this correction is only appropriate for the quasi-laminar contribution to the overall air-side resistance to mass transfer (i.e., for $r_b$, not $r_a + r_b$). For gases, the aerodynamic resistance $r_a$, not the quasi-laminar resistance is the dominant resistance to mass transfer. Hence, the dependence of $k_G$ on the diffusivity is much weaker than indicated in eq. (7.225). A power law index of 0.1–0.2 is probably more realistic.

In the liquid phase, the diffusivity of solutes is four orders of magnitude lower than in the gas phase. Hence, resistance to mass transfer on the water side is much more dominated by the quasi-laminar layer than on the air phase and is thus less prone to misinterpretation than the air-side mass transfer; $k_L$ can be calculated with eq. (7.222).

**Example 7.11.** Calculate the deposition flux of ammonia into a lake at 20 °C ($H = 0.00055$; $\nu = 1.5\times10^{-5}\,\text{m}^2\,\text{s}^{-1}$; $D = 2.36\times10^{-5}\,\text{m}^2\,\text{s}^{-1}$; $D_w = 1.74\times10^{-9}\,\text{m}^2\,\text{s}^{-1}$; $\mu_w = 0.001002$ Pa s) containing 25 mg L$^{-1}$ ammonium, of which 0.3% is present in free ammonia form. The air-phase ammonia concentration measured at a 10-m height is 10 ppm. Wind speed at a 10-m height is 5 m s$^{-1}$.

***Solution.*** First, the friction velocity will be calculated from the wind speed at a 10-m height using Charnock's (1955) equation for the surface roughness with the coefficients of Peña and Gryning (2008). This may overestimate the actual surface roughness as a lake will have a smoother surface than the open sea. Because

Charnock's equation contains the friction velocity, the calculation is iterative. Assume as a first approximation a surface roughness of $10^{-5}$ m. Equation (5.65) leads to

$$u_* = 0.4 \times 5 / \ln(10/10^{-5}) = 0.1448\,\mathrm{m\,s^{-1}}$$

Equation (5.66) leads to a new estimate of the surface roughness:

$$z_0 = 0.012 \times 0.1448^2 / 9.80665 = 2.56 \times 10^{-5}\,\mathrm{m}$$

Based on this surface roughness, a new friction velocity is calculated. After a few iterations, a value of $u_* = 0.1574\,\mathrm{m\,s^{-1}}$ and $z_0 = 3.03 \times 10^{-5}$ m are obtained.

Based on this information, $r_a$ and $r_b$ are calculated with eqs. (7.146) and (7.158), respectively. For $r_a$:

$$r_a = 1/(0.4 \times 0.1574)\ln(10/3.03 \times 10^{-5}) = 201.8\,\mathrm{s\,m^{-1}}$$

For $r_b$ we need the Schmidt number of $NH_3$ in the gas phase:

$$\mathrm{Sc} = \frac{\nu}{D} = \frac{1.5 \times 10^{-5}}{2.36 \times 10^{-5}} = 0.6356$$

Hence:

$$r_b = 2 \times 0.6356^{2/3} / (0.4 \times 0.1574) = 23.5\,\mathrm{s\,m^{-1}}$$

$$k_G = (r_a + r_b)^{-1} = 4.439 \times 10^{-3}\,\mathrm{m\,s^{-1}}$$

and $k_L$ is calculated with eq. (7.195) with the coefficients of Wanninkhof (1992):

$$k_L = 8.61 \times 10^{-7} \times 5^2 = 2.153 \times 10^{-5}\,\mathrm{m\,s^{-1}}$$

This value is for $CO_2$ at 20°C, with Sc=666. To correct for the compound, we need the Schmidt number for $NH_3$ in the water phase. We obtain, assuming a water density of 998 kg m$^{-3}$:

$$\mathrm{Sc} = \frac{\nu_w}{D_w} = \frac{\mu_w / \rho_w}{D_w} = \frac{1.002 \times 10^{-3} / 998}{1.74 \times 10^{-9}} = 577$$

Hence, since $K_L$ is inversely proportional to the square root of the Schmidt number:

$$k_L = 2.153\times10^{-5}(666/557)^{0.5} = 2.313\times10^{-5}\,\mathrm{m\,s^{-1}}$$

For the mass transfer calculations we need the concentration of $NH_3$ in the bulk phases, expressed per $m^3$. For the gas phase the ideal gas law is used to convert ppm to mol $m^{-3}$:

$$c_{air} = \left(\frac{10}{10^6}\right)\frac{p}{RT} = \frac{1.01325}{8.314472\times293.15} = 4.157\times10^{-4}\ \mathrm{mol\ m^{-3}}$$

For the water concentration we only consider the free ammonia, which has a concentration of $0.003\times(25\,\mathrm{g\,m^{-3}}/18\,\mathrm{g\,mol^{-1}}) = 4.167\times10^{-3}\,\mathrm{mol\,m^{-3}} = c_{water}$.

For the ammonia flux we need the value of $K_G$, which is calculated with eq. (7.218):

$$\begin{aligned}K_G &= 4.439\times10^{-3}\times2.313\times10^{-5}/(0.00055\times4.439\times10^{-3}+2.313\times10^{-5})\\ &= 4.015\times10^{-3}\,\mathrm{m\,s^{-1}}\end{aligned}$$

Hence, the flux is given by eq. (7.219):

$$F = 4.015\times10^{-3}(0.00055\times4.167\times10^{-3} - 4.157\times10^{-4}) = \mathbf{-1.660\times10^{-6}\,mol\,m^{-2}s^{-1}}$$

The daily mass flux is found by multiplying by the molar mass of ammonia (17 g $\mathrm{mol^{-1}}$) and by 86,400 s $\mathrm{day^{-1}}$:

$$F = -1.660\times10^{-6}\times17\times86,400 = \mathbf{-2.438\,g\,m^{-2}d^{-1}}$$

## 7.5.2 Wet Deposition

Wet deposition is the removal of air pollutants by cloud droplets, rain, and snow. We can distinguish in-cloud scavenging and below-cloud scavenging. Elaborate theories have been developed for wet deposition, but the modeling practice in models such as CALPUFF is fairly simple. This section will first briefly outline some theoretical background on the wet deposition process. This outline will be followed by a description of the actual modeling practice of wet depostion.

### *7.5.2.1 Theoretical Background.*

To understand wet deposition, we need to understand the absorption of pollutants into falling droplets. To start, we will calculate the absorption of a gas into an immobile droplet.

Consider the air around the droplet. We assume spherical symmetry of the pollutant concentration, so the only coordinate needed is the distance from the droplet center, $r$. The domain starts at $r=r_d=d_d/2$, where $d_d$ is the droplet diameter (m). In this coordinate system a positive flux is a flux out of the droplet (to increasing $r$). It is given by Fick's law:

$$F = -D\frac{dc}{dr} \tag{7.226}$$

The rate of mass transfer into the droplet, $\dot{m}$, is given by

$$\dot{m} = -F \cdot 4\pi r^2 \tag{7.227}$$

The minus sign signifies that the mass transfer is defined in the opposite direction as the flux. At steady-state conditions, this mass transfer is constant in space and time. Substituting eq. (7.226) into eq. (7.227) yields

$$\dot{m} = 4\pi r^2 D\frac{dc}{dr} \tag{7.228}$$

Rearranging leads to

$$dc = \frac{\dot{m}dr}{4\pi D r^2} \tag{7.229}$$

Integration of the radius coordinate from $r_d$ to infinity and the concentration from $c_{i,air}$ (interface concentration) to $c_{air}$ (bulk concentration in the air phase) leads to

$$c_{air} - c_{i,air} = \frac{\dot{m}}{4\pi D}\left(-0+\frac{1}{r_d}\right) \tag{7.230}$$

If the solubility of the air pollutant is high, and the droplet is small so mass transfer in the liquid side is fast, then we can assume that the interface concentration is zero. Hence, the mass transfer rate to the droplet is

$$\dot{m} = 4\pi D r_d c_{air} = 2\pi D d_d c_{air} \tag{7.231}$$

The mass flux into the droplet ($-F_d$) is obtained by dividing the mass transfer rate by the droplet surface area:

$$-F_d = \frac{\dot{m}}{\pi d_d^2} = \frac{2D}{d_d}c_{air} \tag{7.232}$$

Define the mass transfer coefficient $K_c$ as the ratio of the mass flux to the droplet to the bulk air concentration (Seinfeld and Pandis, 2006). Hence, eq. (7.232) becomes

$$-F_{\text{d}} = K_{\text{c}}c_{\text{air}} \tag{7.233}$$

where $K_{\text{c}}$ is given by

$$K_{\text{c}} = \frac{2D}{d_{\text{d}}} \tag{7.234}$$

In reality, droplets are not stationary but fall down with a settling velocity $v_{\text{s}}$ relative to the air. Hence, the droplet constantly sweeps through unscrubbed air, enhancing the mass transfer. Furthermore, turbulent air movements can enhance mass transfer as well. This problem is too complicated to solve exactly, but dimension analysis can be used to help us find empirical correlations. The relevant variables are $K_{\text{c}}$ (m s$^{-1}$), $v_{\text{s}}$ (m s$^{-1}$), $D$ (m$^2$ s$^{-1}$), and $d_{\text{d}}$ (m). When there is turbulence, many variables can influence the mass transfer. The most accessible one is the viscosity of the air, which dissipates turbulence. Hence, we add $\nu$ (m$^2$ s$^{-1}$) to the variables. Hence, we have five variables, spanning two dimensions, and we can define three dimensionless numbers. We define the following: $K_{\text{c}}d_{\text{d}}/D$, $\text{Re}=d_{\text{d}}v_{\text{s}}/\nu$, and $\text{Sc} = \nu/D$. Hence, the following relationship applies:

$$\frac{K_{\text{c}}d_{\text{d}}}{D} = f(\text{Re},\text{Sc}) \tag{7.235}$$

Based on eq. (7.239), we can tell immediately that $K_{\text{c}}d_{\text{d}}/D=2$ when $v_{\text{s}}=0$, that is, when $\text{Re}=0$. Bird et al. (1960) proposed the following equation:

$$K_{\text{c}} = \frac{D}{d_{\text{d}}}(2 + 0.6\,\text{Re}^{1/2} \cdot \text{Sc}^{1/3}) \tag{7.236}$$

Hence, when we know the droplet size and the properties of the air and the pollutant, we can calculate the transfer rate constant. The settling velocity of the droplets can be calculated from eq (7.132) for droplets with $d_{\text{d}}$ between 5 and 100 μm. For droplets with diameters between 100 μm and 1.35 mm, the following equation can be used (Theodore and Buonicore, 1988):

$$v_{\text{s}} = \frac{0.153 d_{\text{d}}^{1.14} \rho_{\text{water}}^{0.71} g^{0.71}}{\mu_{\text{air}}^{0.43} \rho_{\text{air}}^{0.29}} \tag{7.237}$$

For larger droplets, the following equation can be used:

$$v_{\text{s}} = \frac{1.74 d_{\text{d}}^{0.5} \rho_{\text{water}}^{0.5} g^{0.5}}{\rho_{\text{air}}^{0.5}} \tag{7.238}$$

The above equations assume that the resistance to mass transfer is on the gas side. To test this assumption, it is useful to consider mass transport in the droplets as well.

This is a non-steady-state diffusion problem. Diffusion inside of a sphere is given by the following partial differential equation:

$$\frac{\partial C_w}{\partial t} = \frac{2D_w}{r}\frac{\partial c_w}{\partial r} + D_w \frac{\partial^2 c_w}{\partial r^2} \tag{7.239}$$

where $c_w$ is the concentration in the droplet phase, $D_w$ is the diffusivity of the pollutant in the droplet, and $r$ is the distance to the droplet center. This equation is derived from Fick's law in Section A7.4 (Appendix A). This equation can be integrated numerically to obtain droplet saturation $S_w$ as a function of time. Droplet saturation is defined as the average concentration of the pollutant concentration in the droplet divided by the concentration in equilibrium with the air-phase concentration at the interface. The other relevant variables of this problem are $D_w$ and $d_d$ (or $r_d$). Hence we have four variables [$S_w$ (dimensionless), $t$ (s), $D_w$ (m$^2$s$^{-1}$), and $r_d$ (m)] spanning two dimensions. This means that two dimensionless variables can be defined: the saturation, $S_w$, and the dimensionless time, $t'=tD_w/r_d^2=4tD_w/d_d^2$. It follows that $S_w$ is a function of $t'$ only:

$$S_w = f(t') \tag{7.240}$$

To find $f$ conveniently, we can nondimensionalize eq. (7.244) [see Section A7.4 (Appendix A)]:

$$\frac{\partial c'_w}{\partial t'} = \frac{2}{r'}\frac{\partial c'_w}{\partial r'} + \frac{\partial^2 c'_w}{\partial r'^2} \tag{7.241}$$

where $c'_w = c_w/c_{i,w}$, $c_{i,w}$ is the concentration of the pollutant at the air–water interface, on the water side, $r'=r/r_d=2r/d_d$, and $t'=t/\tau$, where $\tau=r_d^2/D_w=d_d^2/(4D_w)$. This equation is solved with the following initial and boundary conditions:

$$c'_w = 0 \qquad (t'=0) \tag{7.242}$$

$$\frac{\partial c'_w}{\partial r'} = 0 \qquad (r'=0) \tag{7.243}$$

$$c'_w = 1 \qquad (r'=1) \tag{7.244}$$

Seinfeld and Pandis (2006) provide an analytical solution of this problem as an infinite series. Here the differential equation is integrated numerically. Details are given in Section A7.4 (Appendix A). The saturation is given by the following equation:

$$S_w = \frac{\int_0^1 4\pi r'^2 c' dr}{\int_0^1 4\pi r'^2 dr} = 3\int_0^1 r'^2 c' dr \tag{7.245}$$

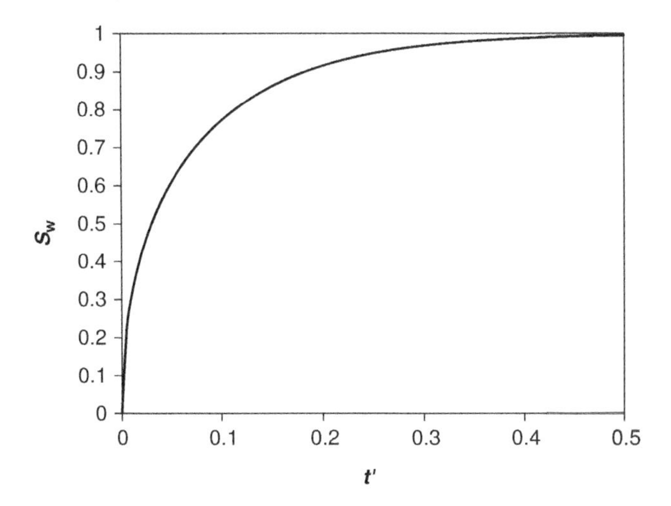


Figure 7.17 Droplet saturation due to wet deposition at constant interface concentration versus dimensionless time.

The result of the simulation is shown in Figure 7.17.

**Example 7.12.** Calculate 90% equilibration times of an air pollutant into a falling droplet: (1) assuming mass transfer limitations in the air phase only and (2) assuming mass transfer limitations in the water phase only. The following data can be assumed: water density 1000 kg m$^{-3}$, air density 1.2 kg m$^{-3}$, droplet diameter 1 mm, air viscosity $1.8\times10^{-5}$ Pa s, pollutant Henry constant 0.001, and pollutant diffusivity $1.5\times10^{-5}$ m$^2$ s$^{-1}$ in air and $1.5\times10^{-9}$ in water.

***Solution.*** For the mass transfer in the air phase the settling velocity of the droplet is needed. This is calculated with eq. (7.237):

$$v_s = 0.153\times0.001^{1.14}\times1000^{0.71}\times9.80665^{0.71}/[(1.8\times10^{-5})^{0.43}\times1.2^{0.29}] = 4.130\,\mathrm{m\,s^{-1}}$$

Hence, the dimensionless numbers can be calculated:

$$\mathrm{Re} = 0.001\times4.13/(1.8\times10^{-5}/1.2) = 275.3$$
$$\mathrm{Sc} = (1.8\times10^{-5}/1.2)/1.5\times10^{-5} = 1$$

Hence, $K_c$ is calculated with eq. (7.236):

$$K_c = (1.5\times10^{-5}/0.001)(2+0.6\times275.3^{1/2}\times1^{1/3}) = 0.1793\,\mathrm{m\,s^{-1}}$$

To allow for equilibration, we need to modify eq. (7.233) to include two-way mass transfer:

$$-F_d = K_c(c_{air} - Hc_w)$$

Combine with eq. (7.232):

$$\frac{\dot{m}}{\pi d_{\mathrm{d}}^{2}} = K_{\mathrm{c}}(c_{\mathrm{air}} - Hc_{\mathrm{w}})$$

The mass accumulation can be written in terms of the concentration in the droplet:

$$\dot{m} = \frac{\pi d_{\mathrm{d}}^{3}}{6}\frac{dc_{\mathrm{w}}}{dt}$$

Hence, the following differential equation describes the accumulation of pollutant assuming only mass transfer limitation in the air phase:

$$\frac{dc_{\mathrm{w}}}{dt} = \frac{6K_{\mathrm{c}}\pi d_{\mathrm{d}}^{2}}{\pi d_{\mathrm{d}}^{3}}(c_{\mathrm{air}} - Hc_{\mathrm{w}}) = \frac{6K_{\mathrm{c}}H}{d_{\mathrm{d}}}\left(\frac{c_{\mathrm{air}}}{H} - c_{\mathrm{w}}\right)$$

This differential equation, with initial condition $c_{\mathrm{w}}=0$ at $t=0$ has the following solution:

$$c_{\mathrm{w}} = \frac{c_{\mathrm{air}}}{H}\left[1 - \exp\left(-\frac{6K_{\mathrm{c}}H}{d_{\mathrm{d}}}t\right)\right]$$

and 90% saturation means that the water-phase concentration equals $0.9c_{\mathrm{air}}/H$. Hence:

$$1 - \exp\left(-\frac{6K_{\mathrm{c}}H}{d_{\mathrm{d}}}t\right) = 0.9$$

Solving for $t$ leads to

$$t = \frac{d_{\mathrm{d}}}{6K_{\mathrm{c}}H}\ln(10)$$

A value of 2.14 s is obtained.

Second, the equilibration time is calculated assuming only mass transfer limitations in the liquid phase. This calculation is based on Figure 7.17. From the figure we can see that a droplet is 90% saturated when $t'$ is about 0.185. Hence, we calculate the actual time based on this dimensionless time:

$$t = \frac{t' r_{\mathrm{d}}^{2}}{D_{\mathrm{w}}}$$

The result is 30.8 s. In practice, there will be some mixing in the droplet due to the air movement around it, but it is clear that the mass transfer is dominated by liquid-phase mass transfer limitations in the current conditions.

Note that the equilibration time in the gas phase is inversely proportional to the Henry constant. Hence, the gas-phase mass transfer becomes relevant for Henry constants on the order of $10^{-4}$ or less (i.e., higher solubility).

The air-side mass transfer has a complex dependence on the droplet size due to the effect of settling but is roughly inversely proportional to the inverse of the droplet size. The water-side mass transfer is inversely proportional to the square of the droplet size. Hence, for droplets of 0.1-mm diameter, the air-side mass transfer becomes limiting in the current conditions.

As with many processes involving aerosol transfer, the wet deposition of partuculate matter involves impaction, interception, and Brownian motion. The collection efficiency of particles by falling droplets is evaluated by comparison with a theoretical maximum collection efficiency. This theoretical maximum is based on the assumption that a falling droplet collects all the particles in its path, including the particles that are less than half of a particle diameter removed from the droplet path. Hence, a droplet sweeps a cylindrical volume with diameter $d_d+d_p$. If the droplets have a settling velocity $v_{s,d}$ and the particles have a settling velocity $v_{s,p}$, then the length of the cylindrical space swept by the droplet is $(v_{s,d} - v_{s,p})t$, where $t$ is time. The volume swept by the droplet per unit time is

$$\frac{\text{Volume swept}}{\text{Time}} = \frac{\pi}{4}(d_d + d_p)(v_{s,d} - v_{s,d}) \qquad (7.246)$$

In reality, not all particles in the volume swept by the droplets are removed from the atmosphere. This is because the air streamlines move around the droplets and tend to deflect the particles around the droplet. Hence, we define the **collision efficiency** $E$ as the fraction of the particles in the swept volume that are collected by the droplet. The collision efficiency depends on the effectiveness of impaction, interception, and Brownian motion as collection mechanisms. Seinfeld and Pandis (2006) recommended the correlation of Slinn (1983) for the collision efficiency:

$$E = \frac{4}{\text{Re}\cdot\text{Sc}}\left(1 + 0.4\,\text{Re}^{1/2}\,\text{Sc}^{1/3} + 0.16\,\text{Re}^{1/2}\,\text{Sc}^{1/2}\right) + 4\phi\left[\frac{1}{\omega} + \left(1 + 2\,\text{Re}^{1/2}\right)\phi\right]$$
$$+\left(\frac{\rho_{\text{water}}}{\rho_p}\right)^{1/2}\left(\frac{\text{St} - S^*}{\text{St} - S^* + \frac{2}{3}}\right)^{3/2} \qquad (7.247)$$

where $\mathrm{Re}=d_\mathrm{d} v_\mathrm{s,d} \rho_\mathrm{air}/(2\mu_\mathrm{air})$ is the Reynolds number based on the *radius* of the droplet; Sc is the Schmidt number of the particle, as defined before $[=\mu_\mathrm{air}/(\rho_\mathrm{air} D)]$; St is the Stokes number, which, in the current context, is defined as $\mathrm{Sc}=2v_\mathrm{s,p}(v_\mathrm{s,d}-v_\mathrm{s,p})/(g\, d_\mathrm{d})$; $\phi$ is the diameter ratio $d_\mathrm{p}/d_\mathrm{d}$; and $\omega$ is the viscosity ratio $\mu_\mathrm{water}/\mu_\mathrm{air}$. $S^*$ is given by

$$S^* = \frac{14.4+\ln(1+\mathrm{Re})}{12+12\ln(1+\mathrm{Re})} \tag{7.248}$$

As with dry deposition, wet deposition of particles is least effective at diameters between 0.1 and 1 μm.

***7.5.2.2 In-Cloud Scavenging.*** In-cloud scavenging, also known as **rainout**, is the collection of pollutants in a cloud. The air inside of a cloud tends to rise because it contains more water vapor than the surrounding air, making it lighter. The rise of air compensates for the settling velocity of the particles. Air pollutants entrained with the rising air can be collected by the cloud droplets. Not all the pollution collected by cloud droplets contributes to wet deposition because up to 90% of the cloud droplets evaporate without raining out, re-releasing the pollutant into the air.

To test if the rate of rainout is limited by precipitation rate alone, or also by the mass transfer rate of pollutants into the cloud droplets, it is useful to calculate the equilibration time of pollutants into the cloud droplets. We follow Seinfeld and Pandis (2006) in adopting the cloud droplet diameter distribution of Battan and Reitan (1957):

$$N(d_\mathrm{d}) = a\exp(-b\cdot d_\mathrm{d}) \tag{7.249}$$

where $a$ ($2.87\times10^8\,\mu\mathrm{m}^{-1}\,\mathrm{m}^{-3}$) and $b$ ($0.265\,\mu\mathrm{m}^{-1}$) are empirical coefficients determined by Levine and Schwartz (1982) in the diameter range from $d_1=5\,\mu\mathrm{m}$ to $d_2=40\,\mu\mathrm{m}$. The total number concentration of droplets in clouds is found by integrating eq. (7.249):

$$N_\mathrm{total} = \int_{d_1}^{d_2} a\exp(-b\cdot d_\mathrm{d})\, dd_\mathrm{d} = \frac{a}{b}[\exp(-b\cdot d_1)-\exp(-b\cdot d_2)] \tag{7.250}$$

A value of $2.88\times10^8\,\mathrm{m}^{-3}$ (droplets per cubic meter) is found. For the total water volume in the cloud, we multiply eq. (7.249) by the volume of one droplet, and integrate:

$$V_\mathrm{total} = \int_{d_1}^{d_2} a\frac{\pi d_\mathrm{d}^3}{6}\exp(-b\cdot d_\mathrm{d})dd_\mathrm{d} \tag{7.251}$$

For the integration we make use of the table of integrals in Section B3 (Appendix B). The result of the integration is

$$V_{\text{total}} = \frac{a\pi}{6}\left[\left(\frac{d_1^3}{b}+\frac{3d_1^2}{b^2}+\frac{6d_1}{b^3}+\frac{6}{b^4}\right)\exp(-b\cdot d_1) - \left(\frac{d_2^3}{b}+\frac{3d_2^2}{b^2}+\frac{6d_2}{b^3}+\frac{6}{b^4}\right)\exp(-b\cdot d_2)\right] \tag{7.252}$$

We obtain a total droplet volume of $1.73\times10^{11}\,\mu\text{m}^3\,\text{m}^{-3}$, or a volume fraction of $1.73\times10^{-7}$ (mass fraction $0.17\,\text{g}\,\text{m}^{-3}$). Water fractions in clouds can be substantially higher, on the order of $1\,\text{g}\,\text{m}^{-3}$ (Giorgi and Chameides, 1985).

To calculate the concentration evolution of a pollutant in the gas phase of a cloud, we need to solve the following material balance:

$$\frac{dc_{\text{air}}}{dt} = F_{\text{d}}A \tag{7.253}$$

where $c_{\text{air}}$ ($\text{mg}\,\text{m}^{-3}$) is the pollutant concentration, $t$ (s) is time, $F_{\text{d}}$ ($\text{mg}\,\text{m}^{-2}\,\text{s}^{-1}$) is the pollutant flux at the droplet surface, and $A$ is the droplet surface area per unit air volume ($\text{m}^2\,\text{m}^{-3}$). Substituting eq. (7.233) ($-F_{\text{d}}=K_{\text{c}}C_{\text{air}}$) into eq. (7.253) and rearranging leads to

$$\frac{dc_{\text{air}}}{c_{\text{air}}} = -K_{\text{c}}A\,dt \tag{7.254}$$

Integration leads to

$$c_{\text{air}} = c_{\text{air},0}(-K_{\text{c}}At) \tag{7.255}$$

The concentration decreases exponentially with time constant $\tau_{\text{c}}=(K_{\text{c}}A)^{-1}$. When there is a distribution of droplet sizes, $K_{\text{c}}A$ is considered in differential form and integrated over a range of droplet diameters. Hence, we obtain

$$\tau_{\text{c}} = \left[\int_{d_1}^{d_2} K_{\text{c}}A_1N(d_{\text{d}})\,dd_{\text{d}}\right]^{-1} \tag{7.256}$$

where $A_1$ is the area of a single droplet with diameter $d_{\text{d}}$, that is, $\pi d_{\text{d}}^2$; $K_{\text{c}}$ is given by eq. (7.231). The integral cannot normally be solved analytically and is solved with Simpson's rule (see Section B4, Appendix B). Assuming a diffusion constant of $10^{-5}$ $\text{m}\,\text{s}^{-1}$, the result is a time constant of 6.2 s. Equation (7.250) becomes

$$C_{\text{air}} = C_{\text{air},0}(-0.161t) \tag{7.257}$$

The concentration decreases by 90% in 14.3 s. Clearly the gas phase can be considered at equilibrium at all times, as large-scale mass transfer into the cloud takes much longer than a few seconds. The calculations are included on the enclosed CD, in the file "Rainout calculation gases.xlsx." Equilibrium partitioning of gaseous air pollutants in clouds is a common assumption in rainout research (Chameides, 1984).

So far we did not yet consider the mass transfer inside the droplets. To make sure that our conclusion is correct, equilibration times will be estimated in the droplet phase as well. Based on Figure 7.17 it is clear that 90% saturation is reached when $t'=0.185$, that is, when $t=0.185d_d^2/(4D_w)$. Assuming a water-phase diffusivity on the order of $10^{-9}\,m^2\,s^{-1}$, the 90% equilibration time ranges from 1.2 ms at $d_d=5\,\mu m$ to 74 ms at $d_d=40\,\mu m$. The droplet side equilibration time is negligible, and the air-phase equilibrium time describes the equilibration dynamics adequately.

A brief calculation based on a simplified version of the model of Giorgi and Chameides (1985) allows us to estimate atmospheric lifetimes of gaseous air pollutants as a result of rainout. The global mean water evaporation rate is about $1.2\times10^{-4}\,m^3\,m^{-2}\,h^{-1}$, that is, every hour a water volume of $1.2\times10^{-4}\,m^3$ is created for every square meter of surface. This water volume can absorb the pollution in $1.2\times10^{-4}/H\,m^3$ of air.

Assuming a mixing height of 1000 m, that is, an air volume of $1000\,m^3$ for every $m^2$ of surface, a fraction $1.2\times10^{-7}/H$ of the air pollutant in the mixing layer is collected for rainout every hour. This leads to an atmospheric residence time of $H/1.2\times10^{-7}$ hours. Insoluble pollutants (e.g., $H=1$) have a negligible rainout rate, whereas a highly soluble pollutant such as $HNO_3$ (Henry constant $8\times10^{-8}$) has a residence time of 0.67 h.

For a low-soluble but reactive pollutant such as $SO_2$, the fate is much more complicated as reaction kinetics in the droplets determine the deposition rate. Examples of droplet-phase chemistry are discussed by Seinfeld and Pandis (2006) and in Chapter 11.

The calculation can be repeated with aerosols. Now $K_c$ can be calculated from the sweep volume of the droplets, the collision efficiency, and the droplet surface area:

$$K_c = \frac{\pi(d_d + d_p)^2(v_{s,d} - v_{s,p})E}{4A_l} \tag{7.258}$$

When the calculation procedure for the equilibration time in the air phase is repeated, a time constant of up to 55 h is obtained for particles with a diameter of 0.6 μm. This calculation is included on the enclosed CD, in the file "Rainout calculation particles.xlsx." It follows that the uptake of particles by cloud water cannot be considered instantaneous. Assuming that the water fraction in clouds is actually $1\,g\,m^{-3}$, and that clouds take up 10% of the mixing layer, we expect a particle time constant of up to 94 h for rainout. As particles do not return to the air phase once collected, no equilibrium considerations are needed.

#### 7.5.2.3 *Below-Cloud Scavenging.*

Below-cloud scavenging is also known as washout. In principle, the mechanisms leading to capture of air pollutants below clouds are the same as in clouds, but some variables, such as contact time and droplet size, are different by orders of magnitude. The main effect of these differences is a much lower effectiveness of washout as a deposition mechanism than rainout. Droplet size is distributed and depends on the rainfall intensity. However, a typical raindrop diameter is 1–2 mm. Equation (7.249) can be used for rain droplets as well, but now $a$ and $b$ are orders of magnitude smaller and depend on the precipitation rate $R$ (mm h$^{-1}$). Sekhon and Srivastava (1971) suggested the following relationships:

$$a = 7\left(\frac{R}{R_1}\right)^{0.37} \tag{7.259}$$

$$b = 0.0038\left(\frac{R}{R_1}\right)^{-0.14} \tag{7.260}$$

where $R_1 = 1$ mm h$^{-1}$; $a$ and $b$ are in μm$^{-1}$ m$^{-3}$ and μm$^{-1}$, respectively. Repeating the calculations from the previous section for $R = 1$ mm h$^{-1}$ leads to time constants of 3.1 and 2000 h for gases and particles, respectively. The calculations are included on the enclosed CD, in the files "Washout calculation gases.xlsx" and "Washout calculation particles.xlsx."

In the case of gases, the above calculation only applies if the pollutant concentration at the air–droplet interface is negligibly small. This means two things:

- The solubility of the gas must be sufficiently large so that all the pollutant molecules reaching the droplet can dissolve.
- The equilibration time due to mass transfer inside the droplet must be negligibly small compared with the fall time of the droplets.

To test the first requirement, consider a rainfall event for 3.1 h at a rainfall intensity of 1 mm h$^{-1}$. This rainwater, the equivalent of a horizontal layer 3.1 mm thick, must be able to absorb all the pollutant present in the boundary layer, which has a thickness on the order of 1 km. Hence, the air volume is 6 orders of magnitude larger than the water volume, and the Henry constant must not be greater than $10^{-6}$ for the dissolution to take place.

To test the second requirement, we return to Figure 7.11. When the droplet diameter is 2 mm (i.e., radius 1 mm), the time needed to reach 90% of saturation is $0.185 \times (0.001\ \text{m})^2/10^{-9}\ \text{m}^2\,\text{s}^{-1} = 185$ s. At this size, the fall velocity is about 4 m s$^{-1}$, so the droplet traverses 1 km in 250 s. We can expect some slowing of the deposition due to nonequilibrium, but the main resistance to mass transfer is on the air side when the Henry constant is as small as assumed here. Because the majority of washout is by droplets smaller than 2 mm, this conclusion probably applies to the overall washout as well.

***7.5.2.4 Practical Calculation Schemes for Wet Deposition.*** While the theories described in Section 7.5.2 provide great insight into how properties of clouds, raindrops, and pollutants affect wet depostion, models such as CALPUFF use much simpler calculation schemes. It is simply assumed that the process is first order in the pollutant concentration and first order in the precipitation rate. This leads to the following equation:

$$c_{\mathrm{air}} = c_{\mathrm{air},0} \exp(-\Lambda t) \tag{7.261}$$

where $\Lambda$, the **scavenging ratio** ($s^{-1}$), is given by

$$\Lambda = \lambda \frac{R}{R_1} \tag{7.262}$$

where $\lambda$ is the **scavenging coefficient** ($s^{-1}$). Default values of the scavenging coefficient as used in CALPUFF are listed in Table 7.2 (Scire et al., 2000); $R$ and $R_1$ are precipitation rate ($\mathrm{mm\,h^{-1}}$) and reference precipitation rate ($1\,\mathrm{mm\,h^{-1}}$), respectively. A drizzle is typically about $0.5\,\mathrm{mm\,h^{-1}}$. Heavy rainfall is typically about $25\,\mathrm{mm\,h^{-1}}$. The values in Table 7.2 are meant to include both in-cloud deposition and below-cloud deposition.

Comparing the values in Table 7.2 with the theory discussed in Section 7.5.2 , it can be observed that the scavenging coefficient of a highly soluble gaseous pollutant such as $HNO_3$ is similar to the value expected from theory ($\lambda = 6\times10^{-5}\,s^{-1}$ in CALPUFF, whereas a time constant of 3.1 h is consistent with a value of $\lambda = 9\times10^{-5}\,s^{-1}$). On the other hand, all other compounds except $NO_x$ have $\lambda$ values several orders of magnitude larger than expected from theory considering only washout. It follows that significant rainout is needed to justify the numbers in Table 7.2. It is surprising that aerosols (represented by $SO_4^{2-}$ and $NO_3^-$ in Table 7.2) have a greater $\lambda$ value than gases, whereas particle removal rates are orders of magnitude smaller than gas removal rates. This may reflect the fact that particles, once collected, do not return to the gas phase (equivalent to zero Henry constant). Based on our rainout time constant estimate for particles in Section 7.5.2.2 we obtain $\Lambda = 2.9\times10^{-6}\,s^{-1}$. Taking the global mean precipitation rate ($0.12\,\mathrm{mm\,h^{-1}}$) as representative, this leads to $\lambda = 2.5\times10^{-5}$, a factor 4 lower than adopted in CALPUFF.

**TABLE 7.2 Default Values of $\lambda$ ($s^{-1}$) as Used in CALPUFF**

| Pollutant | Liquid Precipitation | Frozen Precipitation |
|---|---|---|
| $SO_2$ | $3\times10^{-5}$ | 0 |
| $SO_4^{3-}$ | $1\times10^{-4}$ | $3\times10^{-5}$ |
| $NO_x$ | 0 | 0 |
| $HNO_3$ | $6\times10^{-5}$ | 0 |
| $NO_3^-$ | $1\times10^{-4}$ | $3\times10^{-5}$ |

*Source*: From Scire et al. (2000).

For short precipitation events the following equation is useful:

$$c_{\text{air}}(t+\Delta t) = c_{\text{air}}(t)\exp(-\Lambda \cdot \Delta t) \tag{7.263}$$

Equation (7.261) [or (7.263)] shows concentration decrease due to precipitation only. In practice, the concentration decrease is due to dispersion as well. How that is accounted for is shown in the next section.

### 7.5.3 Gaussian Dispersion Models with Deposition

A Gaussian dispersion model lends itself well to inclusion of a deposition model, provided that the effect of deposition on the ground-level concentration can be introduced conveniently as a correction of the main Gaussian equation. Several procedures have been proposed for that purpose.

In the case of **below-cloud wet deposition**, it can be assumed that all parts of the plume are depleted to an equal extent as a result of precipitation. In that case the concentration can be corrected for precipitation by applying the correction of eq. (7.261) to the source strength:

$$\overline{c} = \frac{Q_0 \exp(-\Lambda t)}{2\pi \overline{u} \sigma_y \sigma_z} \exp\left(-\frac{1}{2}\frac{y^2}{\sigma_y^2}\right)\left\{\exp\left[-\frac{1}{2}\frac{(z-h)^2}{\sigma_z^2}\right] + \exp\left[-\frac{1}{2}\frac{(z+h)^2}{\sigma_z^2}\right]\right\} \tag{7.264}$$

where $t$ is the time the pollutant is subjected to precipitation between emission and reaching the receptor. When there is constant rainfall over the entire trajectory, $t$ is given by

$$t = \frac{x}{\overline{u}} \tag{7.265}$$

In the case of dry deposition the depletion of the pollutant happens at the surface, and there is no longer an exact analytical solution to the problem. Hence, approximations are needed to fit the process of deposition into a Gaussian framework.

The simplest way to approximate the effect of dry deposition on the plume concentration profile is with the **source depletion model** of Chamberlain (1953). This model calculates the deposition from the ground-level concentration but assumes that the effect can be modeled by reducing the source strength $Q$, as if the pollutant is reduced uniformly across the plume. To calculate the reduction of $Q$, a material balance of the plume section from $x$ to $x+dx$ is considered:

$$\text{Pollutant in}_x - \text{pollutant out}_{x+dx} = \text{pollutant deposited} \tag{7.266}$$

The terms on the left-hand side of the equation are straightforward:

$$\text{Pollutant in}_x = Q \tag{7.267}$$

$$\text{Pollutant out}_{x+dx} = Q + \frac{dQ}{dx}dx \tag{7.268}$$

For the deposition term in eq. (7.266) we need the concentration, which can be formally represented by combining eqs. (6.1) and (6.2) (Chapter 6):

$$\overline{c} = \frac{Q}{\overline{u}} \varphi_y \varphi_z \tag{7.269}$$

Hence, the deposition across a differential surface from $x$ to $x+dx$ and from $y$ to $y+dy$ is $F\,dx\,dy$. Based on eq. (7.129) applied to a differential surface $dx\,dy$, we can write

$$-F\,dx\,dy = v_\mathrm{d} \overline{c}\,dx\,dy = v_\mathrm{d} \frac{Q}{\overline{u}} \varphi_y \varphi_z\,dx\,dy \tag{7.270}$$

Integrating from $y=-\infty$ to $y=+\infty$, we obtain the deposition term across a distance increment $dx$ as defined in eq. (7.266). The deposition term becomes

$$-dx \int_{-\infty}^{+\infty} F\,dy = v_\mathrm{d} \frac{Q}{\overline{u}} \varphi_z\,dx \int_{-\infty}^{+\infty} \varphi_y\,dy = v_\mathrm{d} \frac{Q}{\overline{u}} \varphi_z\,dx \tag{7.271}$$

Hence, eq. (7.266) becomes

$$Q - \left( Q + \frac{dQ}{dx} dx \right) = v_\mathrm{d} \frac{Q}{\overline{u}} \varphi_z\,dx \tag{7.272}$$

which reduces to

$$\frac{dQ}{dx} = -v_\mathrm{d} \frac{Q}{\overline{u}} \varphi_z \tag{7.273}$$

Assuming reflection only at the surface, eq. (6.6) can be substituted to obtain

$$\boxed{\frac{dQ}{dx} = -\sqrt{\frac{2}{\pi}} \frac{v_\mathrm{d} Q}{\overline{u} \sigma_z} \exp\left( -\frac{h^2}{2\sigma_z^2} \right)} \tag{7.274}$$

This equation is integrated numerically. This is the procedure adopted in CALPUFF for the calculation of dry deposition. The model can be rewritten as

$$\frac{dQ}{Q} = -\sqrt{\frac{2}{\pi}} \frac{v_\mathrm{d}}{\overline{u} \sigma_z} \exp\left( -\frac{h^2}{2\sigma_z^2} \right) dx \tag{7.275}$$

Integration leads to

$$\ln\left( \frac{Q}{Q_0} \right) = -\sqrt{\frac{2}{\pi}} \frac{v_\mathrm{d}}{\overline{u}} \int_0^x \frac{1}{\sigma_z} \exp\left( -\frac{h^2}{2\sigma_z^2} \right) dx \tag{7.276}$$

Taking the exponential leads to

$$Q = Q_0 \exp\left[-\sqrt{\frac{2}{\pi}}\frac{v_d}{\bar{u}}\int_0^x \frac{1}{\sigma_z}\exp\left(-\frac{h^2}{2\sigma_z^2}\right)dx\right] \tag{7.287}$$

In some special cases, an analytical solution of the integral in eq. (7.277) exists (Yamartino, 2008). The model predicts that the deposition is most pronounced when the source height is low and the atmosphere is stable, leading to prolonged high concentrations at the surface. Unstable atmosphere and high sources lead to less deposition.

Because the source depletion model of Chamberlain (1953) removes pollutant along the entire depth of the plume, it tends to overestimate concentrations at the surface and underestimate concentrations aloft. As a result, it will overestimate deposition and underestimate concentrations far away from the source.

To resolve the issues with the source depletion model, Horst (1977) developed the **surface depletion model**. This model provides a more accurate concentration profile by defining sources of negative emissions at the surface, equal to the local deposition flux.

As a starting point of the derivation of Horst's (1977) surface depletion model, return to eq. (7.269), but with explicit mention of the functional relationships:

$$\bar{c} = \frac{Q}{\bar{u}}\varphi_y(x, y)\varphi_z(x, z, h) \tag{7.278}$$

Consider the deposition flux $F_d$ at location $x=x_d$, $y=y_d$:

$$-F_d(x_d, y_d) = v_d\bar{c}(x_d, y_d) \tag{7.279}$$

The negative emission $Q_d$ associated with the deposition from $x_d$ to $x_d+dx_d$ and from $y_d$ to $y_d+dy_d$ is

$$\frac{\partial^2 Q_d}{\partial x_d \partial y_d}dx_d\, dy_d = v_d\bar{c}(x_d, y_d)\, dx_d\, dy_d \tag{7.280}$$

The negative emission causes a **concentration deficit** $c_d$ downwind at the receptor. The contribution of the negative emission at $x_d$ and $y_d$ and at emission height 0 to the concentration deficit is

$$\frac{\partial^2 \overline{c_d}(x, y)}{\partial x_d\, \partial y_d}dx_d\, dy_d = \frac{(\partial^2 Q_d/\partial x_d\, \partial y_d)dx_d\, dy_d}{\bar{u}}\varphi_y(x-x_d, y-y_d)\varphi_z(x-x_d, z, h=0) \tag{7.281}$$

Substituting $Q_d$ leads to

$$\frac{\partial^2 \overline{c_d}(x, y)}{\partial x_d \partial y_d}dx_d\, dy_d = \frac{v_d\bar{c}(x_d, y_d)dx_d\, dy_d}{\bar{u}}\varphi_y(x-x_d, y-y_d)\varphi_z(x-x_d, z, h=0) \tag{7.282}$$

Integrating eq. (7.282) in the crosswind direction leads to

$$\frac{\partial \overline{c_{\mathrm{d}}}(x, y)}{\partial x_{\mathrm{d}}} dx_{\mathrm{d}} = \frac{v_{\mathrm{d}}}{\bar{u}} \varphi_z (x - x_{\mathrm{d}}, z, h = 0)\, dx_{\mathrm{d}} \cdot \int_{-\infty}^{+\infty} \bar{c}(x_{\mathrm{d}}, y_{\mathrm{d}}) \varphi_y (x - x_{\mathrm{d}}, y - y_{\mathrm{d}})\, dy_{\mathrm{d}} \quad (7.283)$$

Integrating in the downwind direction leads to

$$\overline{c_{\mathrm{d}}}(x, y) = \frac{v_{\mathrm{d}}}{\bar{u}} \cdot \int_0^x \varphi_z (x - x_{\mathrm{d}}, z, h = 0) \int_{-\infty}^{+\infty} \bar{c}(x_{\mathrm{d}}, y_{\mathrm{d}}) \varphi_y (x - x_{\mathrm{d}}, y - y_{\mathrm{d}})\, dy_{\mathrm{d}}\, dx_{\mathrm{d}} \quad (7.284)$$

Hence, the concentration, which is the concentration in the absence of deposition minus the concentration deficit, is given by

$$\bar{c}(x, y) = \frac{Q}{\bar{u}} \varphi_y (x, y) \varphi_z (x, z, h) - \frac{v_{\mathrm{d}}}{\bar{u}} \cdot \int_0^x \varphi_z (x - x_{\mathrm{d}}, z, h = 0) \int_{-\infty}^{+\infty} \bar{c}(x_{\mathrm{d}}, y_{\mathrm{d}}) \varphi_y (x - x_{\mathrm{d}}, y - y_{\mathrm{d}})\, dy_{\mathrm{d}}\, dx_{\mathrm{d}} \quad (7.285)$$

Equation (7.285) is computationally more demanding than the source depletion model, not only because of the double integral but also because the receptor location is in the integrand, so the entire double integral must be recalculated for every receptor location. Fortunately, there is a shortcut (Horst, 1977). The crosswind location dependence of the concentration is $\varphi_y$. Because the deposition rate is proportional to the concentration, it also has a crosswind location dependence $\varphi_y$, which in turn is proportional to the concentration deficit, and consequently to the concentration after dry deposition. Separate the concentration in a factor $\varphi_y$ and a factor independent of $y$:

$$\bar{c}(x, y, z) = \overline{c_{xz}}(x, z) \varphi_y (x, y) \quad (7.286)$$

The meaning of $\overline{c_{xz}}(x, z)$ can be seen from integrating eq. (7.286) in the crosswind direction:

$$\int_{-\infty}^{+\infty} \bar{c}(x, y, z)\, dy = \overline{c_{xz}}(x, z) \int_{-\infty}^{+\infty} \varphi_y (x, y)\, dy = \overline{c_{xz}}(x, z) \quad (7.287)$$

where $\overline{c_{xz}}(x, z)$ is the crosswind integrated concentration. It has units mg m$^{-2}$. We can develop a similar relationship for the deposition flux, as the following relationship holds:

$$-F_{\mathrm{d}}(x, y) = v_{\mathrm{d}} \bar{c}(x, y, z = 0) = v_{\mathrm{d}} \overline{c_{xz}}(x, z = 0) \varphi_y (x, y) = -F_{\mathrm{d},xz}(x) \varphi_y (x, y) \quad (7.288)$$

where $F_{\mathrm{d},xz}$ is given by

$$F_{\mathrm{d},xz}(x) = \int_{-\infty}^{+\infty} F_{\mathrm{d}}(x, y)\, dy \quad (7.289)$$

We now define a crosswind integrated negative emission $Q_{d,xz}$ similar to eq. (7.280):

$$\frac{\partial Q_{d,xz}}{\partial x_d} dx_d = v_d \overline{c_{xz}}(x_d, y_d) dx_d \tag{7.290}$$

We use this emission to calculate its contribution to the crosswind integrated concentration deficit $\overline{c_{d,xz}}$ :

$$\frac{\partial \overline{c_{d,xz}}(x,z)}{\partial x_d} dx_d = \frac{v_d \overline{c_{xz}}(x_d, y_d)\, dx_d}{\overline{u}} \varphi_z(x - x_d, z, h = 0) \tag{7.291}$$

Integrating in the downwind direction leads to

$$\overline{c_{d,xz}}(x,z) = \frac{v_d}{\overline{u}} \int_0^x \overline{c_{xz}}(x_d, y_d) \phi_z(x - x_d, z, h = 0)\, dx_d \tag{7.292}$$

Hence, the crosswind integrated concentration is given by

$$\overline{c_{xz}}(x,z) = \frac{Q}{\overline{u}} \varphi_z(x, z, h) - \frac{v_d}{\overline{u}} \int_0^x \overline{c_{xz}}(x_d, y_d) \varphi_z(x - x_d, z, h = 0) dx_d \tag{7.293}$$

Returning to the local concentration, we obtain

$$\overline{c}(x, y, z) = \frac{Q}{\overline{u}} \varphi_y(x, y) \varphi_z(x, z, h) - \frac{v_d}{\overline{u}} \varphi_y(x, y) \int_0^x \overline{c_{xz}}(x_d, y_d) \varphi_z(x - x_d, z, h = 0)\, dx_d \tag{7.294}$$

The calculation has been reduced to a single integral. The $x$ coordinate of the receptor is still in the integrand, so we still need to recalculate the entire integral for each receptor $x$ coordinate. Yamartino (2008) presented further modifications to reduce computation time based on Laplace transforms.

Due to the computational requirements, the surface depletion model of Horst (1977) has not been popular in air dispersion modeling practice. Hence, a source depletion model with vertical concentration profile correction was developed by Horst (1984). The objective of this model is to achieve accuracies similar to the surface depletion model with the computational needs similar to the source depletion model. In other words, the removal of pollutant from the atmosphere is calculated with eq. (7.274) [or eq. (7.277)], but the concentration profile predicted by the Gaussian equation is corrected.

A correction factor $P(x,z)$ is defined so that the corrected concentration profile is given by

$$\overline{c} = \frac{Q}{\overline{u}} \varphi_y(x, y) \varphi_z(x, z, h) P(x, z) \tag{7.295}$$

The concentration gradient of eq. (7.295) in the absence of correction approaches zero near the ground. This is unrealistic in the presence of dry deposition because the mass flux to the ground is driven by a concentration gradient. That gradient is given by the turbulent version of Fick's law [eq. (5.176)]:

$$F = -K(z)\frac{d\bar{c}}{dz} = -v_{\mathrm{d}}\bar{c}(z_{\mathrm{d}}) \tag{7.296}$$

Rearranging leads to

$$d\bar{c} = v_{\mathrm{d}}\bar{c}(z_{\mathrm{d}})\frac{dz}{K(z)} \tag{7.297}$$

Integration leads to

$$\bar{c}(z) - \bar{c}(z_{\mathrm{d}}) = v_{\mathrm{d}}\bar{c}(z_{\mathrm{d}})\int_{z_{\mathrm{d}}}^{z}\frac{dz}{K(z)} = v_{\mathrm{d}}\bar{c}(z_{\mathrm{d}})R(z, z_{\mathrm{d}}) \tag{7.298}$$

This equation defines the atmospheric resistance, which is essentially part of the aerodynamic resistance. Hence, the relationship between $\bar{c}(z)$ and $\bar{c}(z_{\mathrm{d}})$ is

$$\bar{c}(z) = \bar{c}(z_{\mathrm{d}})\left[1 + v_{\mathrm{d}}R(z, z_{\mathrm{d}})\right] \tag{7.299}$$

Hence, we need to derive a function $P(x,z)$ that will reproduce eq. (7.299) near the surface. To that effect, substitute eq. (7.295) in eq. (7.299):

$$\frac{Q}{\bar{u}}\varphi_{y}(x, y)\varphi_{z}(x, z, h)P(x, z) = \frac{Q}{\bar{u}}\varphi_{y}(x, y)\varphi_{z}(x, z_{\mathrm{d}}, h)P(x, z_{\mathrm{d}})\left[1 + v_{\mathrm{d}}R(z, z_{\mathrm{d}})\right] \tag{7.300}$$

Because the slope near the surface of the uncorrected Gaussian plume model is near zero, we can write

$$\varphi_{z}(x, z, h) \approx \varphi_{z}(x, z_{\mathrm{d}}, h) \tag{7.301}$$

Hence, eq. (7.300) simplifies to

$$P(x, z) = P(x, z_{\mathrm{d}})\left[1 + v_{\mathrm{d}}R(z, z_{\mathrm{d}})\right] \tag{7.302}$$

Hence, we can calculate the correction at any height $z$ if we know the required correction at height $z_{\mathrm{d}}$. The correction $P(x,z)$ must conserve mass when considered over the entire plume cross section. Hence, we require that the pollutant mass flowing through a cross-sectional plane equals $Q$:

$$Q = \int_{z_d}^{+\infty}\int_{-\infty}^{+\infty} uc(x, y, z)\, dy\, dz \tag{7.303}$$

Here we assume that the reference height for defining the deposition velcity is sufficiently small to neglect the pollution flowing underneath. Substituting eq. (7.295) into eq. (7.303), and rearranging the integrals, leads to

$$Q = \frac{Q}{\bar{u}}\bar{u}\int_{-\infty}^{+\infty}\varphi_y(x,y)\,dy\int_{z_d}^{+\infty}\varphi_z(x,z,h)P(x,z)\,dz \tag{7.304}$$

The first integral in eq. (7.304) equals 1 by definition. Substituting eq. (7.302) leads to

$$1 = P(x,z_d)\int_{z_d}^{+\infty}\varphi_z(x,z,h)\left[1+v_d R(z,z_d)\right]dz \tag{7.305}$$

Splitting the integral in two parts allows for a further simplification:

$$1 = P(x,z_d)\left[\int_{z_d}^{+\infty}\varphi_z(x,z,h)dz + v_d\int_{z_d}^{+\infty}\varphi_z(x,z,h)R\left(z,z_d\right)dz\right] \tag{7.306}$$

Again, the first integral is 1 by definition. The equation can now be solved for $P(x,z_d)$:

$$P(x,z_d) = \frac{1}{1+v_d\int_{z_d}^{+\infty}\varphi_z(x,z,h)R(z,z_d)\,dz} \tag{7.307}$$

Substituting eq. (7.302) leads to

$$P(x,z) = \frac{1+v_d R(z,z_d)}{1+v_d\int_{z_d}^{+\infty}\phi_z(x,z,h)R(z,z_d)\,dz} \tag{7.308}$$

The correction depends on the choice made for the resistance $R(z,z_d)$. One possible choice is eq. (7.146), the aerodynamic resistance from deposition theory. However, as Horst (1984) pointed out, this is not necessarily consistent with the growth rate of $\sigma_z$, which could lead to unrealistic results. It is more appropriate to derive a turbulent mass diffusivity $K$ from the growth of $\sigma_z$. Some aspects of this relationship will be discussed in Chapter 11. For now, we will simply take one result from gradient theory:

$$K = u\sigma_z\frac{d\sigma_z}{dx} \tag{7.309}$$

A possible way to derive this result is to consider a non-steady-state diffusion equation in the $z$ direction and substituting the Gaussian plume equation. After substituting $t=x/u$, eq. (7.309) emerges.

For a simple example of an application for eq. (7.309), consider the following equation:

$$\sigma_z = ax \tag{7.310}$$

An example of this functional relationship is Brigg's parameterization for stability classes A and B in rural terrain and stability class C in urban terrain (Briggs, 1973). Application of eq. (7.309) leads to

$$K = ua\sigma_z \tag{7.311}$$

Hence, we obtain the following equation for $R(z,z_d)$ [see eq. (7.298)]:

$$R(z, z_d) = \int_{z_d}^{z} \frac{dz}{ua\sigma_z} = \frac{1}{ua\sigma_z} \int_{z_d}^{z} dz = \frac{z - z_d}{ua\sigma_z} \tag{7.312}$$

However, this leads to a concentration profile that is approximately linear close to the ground, whereas we know from eq. (7.144) that the concentration profile should be approximately logarithmic close to the surface. This is because $K$ is not a constant in vertical transport. Correct limiting behavior of the concentration near $z_d$ is important because the concentration at $z_d$ determines the deposition rate. Hence, Horst (1983) made a number of approximations.

First, $\varphi_z(x,z,h)$ is approximated by $\varphi_z(x,z,h=0)$. This may not seem like an acceptable approximation, but there is only a significant correction when the concentration is near its maximum. At such distances from the source, the approximation is acceptable.

Second, when the source is at ground level, it can be shown that the "average" height of a Gaussian plume particle is

$$\bar{z} = \int_0^{+\infty} \varphi_z \cdot z\, dz = \int_0^{+\infty} \frac{2}{\sqrt{2\pi}\sigma_z} \exp\left(-\frac{1}{2}\frac{z^2}{\sigma_z^2}\right) \cdot z\, dz = \sqrt{\frac{2}{\pi}}\sigma_z \tag{7.313}$$

For this integration the table of integrals in Appendix B is helpful. Hence, eq. (7.312) can be written as

$$R(z, z_d) = \sqrt{\frac{2}{\pi}} \int_{z_d}^{z} \frac{dz'}{ua\bar{z}} \approx \sqrt{\frac{2}{\pi}} \int_{z_d}^{z} \frac{dz'}{uaz'} = \sqrt{\frac{2}{\pi}} \frac{1}{ua} \ln\left(\frac{z}{z_d}\right) \tag{7.314}$$

Since $R(z,z_d)$ will be part of an integral from $z=z_d$ to $z=+\infty$, this change should not have a marked effect on the overall result, however, we do have our desired logarithmic behavior. Substituting eq. (7.314) into eq. (7.308) leads to:

$$P(x, z) = \frac{1 + v_d \sqrt{\frac{2}{\pi}} \frac{1}{ua} \ln\left(\frac{z}{z_d}\right)}{1 + \sqrt{\frac{2}{\pi}} \frac{v_d}{ua} \int_{z_d}^{+\infty} \varphi_z(x, z, h) \ln\left(\frac{z}{z_d}\right) dz} \tag{7.315}$$

When $\varphi_z/z_d$ is sufficiently large, the integral in eq. (7.315) can be approximated with reasonable accuracy by $\ln(2^{1/2}\sigma_z/z_d) - 1$. Hence, the correction $P(x,z)$ becomes

$$P(x,z) = \frac{1 + v_d \sqrt{\frac{2}{\pi}} \frac{1}{ua} \ln\left(\frac{z}{z_d}\right)}{1 + \sqrt{\frac{2}{\pi}} \frac{v_d}{ua} \left[ \ln\left(\frac{\sqrt{2}\sigma_z}{z_d}\right) - 1 \right]} \tag{7.316}$$

The necessary equations for corrections based on some other Briggs equations for $\sigma_z$ will be summarized here. They were taken from EPA (1995).

When $\sigma_z$ is of the form $ax/(1+bx)^{1/2}$, the following equations are obtained:

$$R(z, z_d) = \sqrt{\frac{2}{\pi}} \frac{1}{ua} \left[ \ln\left(\frac{z}{z_d}\right) + \frac{b}{a}\sqrt{\frac{\pi}{2}}(z - z_d) \right] \tag{7.317}$$

$$P(x, z_d) = \left\{ 1 + \sqrt{\frac{2}{\pi}} \frac{v_d}{ua} \left[ \ln\left(\frac{\sqrt{2}\sigma_z}{z_d}\right) - 1 + \frac{b}{a}\frac{\pi}{2}(\sigma_z - z_d) \right] \right\}^{-1} \tag{7.318}$$

From these equations, $P(x,z)$ is obtained with eq. (7.302). This applies to the following case as well.

When $\sigma_z$ is of the form $ax/(1+bx)$, the following equations are obtained:

$$R(z, z_d) = \sqrt{\frac{2}{\pi}} \frac{1}{ua} \left[ \ln\left(\frac{z}{z_d}\right) + \frac{2b}{a}\sqrt{\frac{\pi}{2}}(z - z_d) + \frac{3b^2}{2a^2}\sqrt{\frac{\pi}{2}}(z^2 - z_d^2) \right] \tag{7.319}$$

$$P(x, z_d) = \left\{ 1 + \sqrt{\frac{2}{\pi}} \frac{v_d}{ua} \left[ \ln\left(\frac{\sqrt{2}\sigma_z}{z_d}\right) - 1 + \frac{2b}{a}\frac{\pi}{2}(\sigma_z - z_d) + \frac{3b^2}{2a^2}\frac{\pi}{2}\left(\sigma_z^2 - z_d^2\right) \right] \right\}^{-1} \tag{7.320}$$

When $\sigma_z$ is of the form $ax(1+bx)^{1/2}$, the procedure is more complicated and involves an iterative process. It will not be covered here.

The model as described here does not include the effect of particle settling on plume shape and plume height. Extensions of the above equations to account for settling are given by EPA (1995).

The model of Horst (1983) is used in the air dispersion model ISC3. Doran and Horst (1985) tested several dispersion–deposition models and found that the corrected source depletion model of Horst (1983) best corresponds with the result of field measurements.

**Example 7.13.** Determine a downwind ground-level $SO_2$ concentration profile for the emission conditions given in Example 2.1, but assume a deposition velocity $v_d = 0.002\ \mathrm{m\,s^{-1}}$ (reference height 1 m), and assume the wind speed profile obtained in Example 5.5. The emission is from a coal-fired power plant in rural Pennsylvania and amounts to $100\ \mathrm{g\,s^{-1}}$ $SO_2$ from a stack with height of 75 m. Plume rise is 15 m. Wind speed at a 10-m height is $7\,\mathrm{m\,s^{-1}}$. The weather is overcast. The friction velocity obtained in Example 5.5 is $0.608\,\mathrm{m\,s^{-1}}$. With an infinite value of $L$, this leads to a wind speed of $10.34\,\mathrm{m\,s^{-1}}$ at a 90-m height. Use the Briggs equations for rural terrain for the dispersion parameters. Calculate the apparent emission rate and the ground-level concentration versus downwind distance up to 10 km with the original source depletion model and with Horst's modified source depletion model. Repeat the calculation with an effective source height of 10 m and a deposition velocity of $0.02\,\mathrm{m\,s^{-1}}$.

***Solution.*** First, the source depletion model of Chamberlain (1953) will be used. To that effect, the differential equation for $Q$, eq. (7.279) will be integrated numerically in Matlab, with initial value $Q=100{,}000\,\mathrm{mg\,s^{-1}}$ when $x=0$.
Parameter $\sigma_z$ in the equation is calculated with the Briggs equation for neutral conditions, that is,

$$\sigma_z = \frac{ax}{(1+bx)^{1/2}}$$

with $a=0.06$ and $b=0.0015\,\mathrm{m^{-1}}$.

The above equation causes problems in the first integration step because $\sigma_z=0$ when $x=0$, which leads to infinities in the calculation. This is circumvented by requiring that $dQ/dx=0$ when $\sigma_z=0$.

The numerical integration provides a set of $Q$ values at different distances downwind from the source. These values are used to calculate the ground-level concentration with the basic Gaussian equation with ground-level reflection [eq. (6.7) or eqs.(6.1)–(6.3), (6.6)]. For comparison, the concentration profile in the absence of dry deposition is plotted as well.

The Matlab files with the algortihm are included on the enclosed CD, in the folder "Dispersion and deposition source depletion model." The file "data.m" contains the variables ($u_*$, etc.). The differential equation is defined in the file "f.m." The file "main.m" is the main program, which accesses the other two files, carries out the numerical integration, and uses the result to calculate the concentration profiles. The calculations with the original variables ($h=90\,\mathrm{m}$, $v_d=0.002\,\mathrm{m\,s^{-1}}$) provided results not significantly different from calculations in the absence of deposition and are not shown for that reason. This is not surprising as the time scale for $SO_2$ dry deposition is typically several days. Significant deposition was predicted with a 10-fold deposition velocity and a source height of 10 m. The profile of $Q$ is shown in Figure 7.18, whereas the concentration profile is shown in Figure 7.19.

For the calculations with the source depletion model of Horst (1984), as implemented in ISC3, the surface concentration is corrected by multiplying with $P(x,z_d)$, as obtained with eq. (7.318). Similarly, eq. (7.274) is multiplied by the same factor and integrated numerically. The rest of the procedure is identical to the procedure of the previous calculation.

The Matlab files with the algorithm are included on the enclosed CD, in the folder "Dispersion and deposition modified source depletion model." The structure is the same as the previous calculation. Again the dry deposition was negligible with

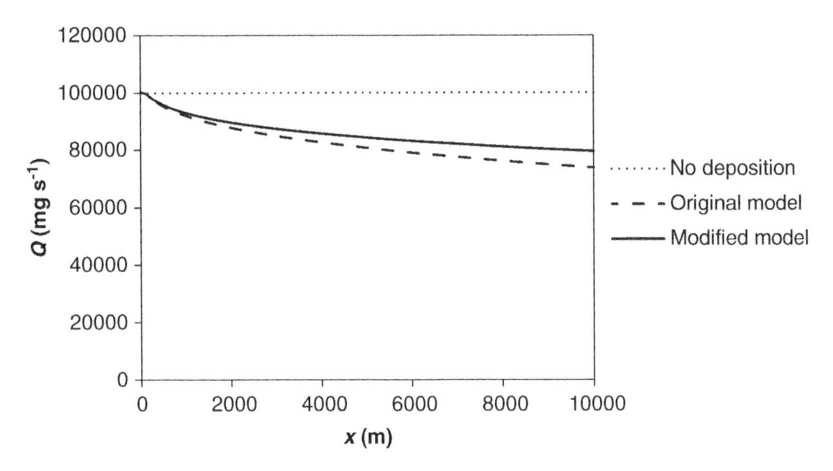


Figure 7.18 Apparent emissions resulting from deposition calculations with two different source depletion models. Source height: 10 m; deposition velocity: 0.02 m $s^{-1}$.

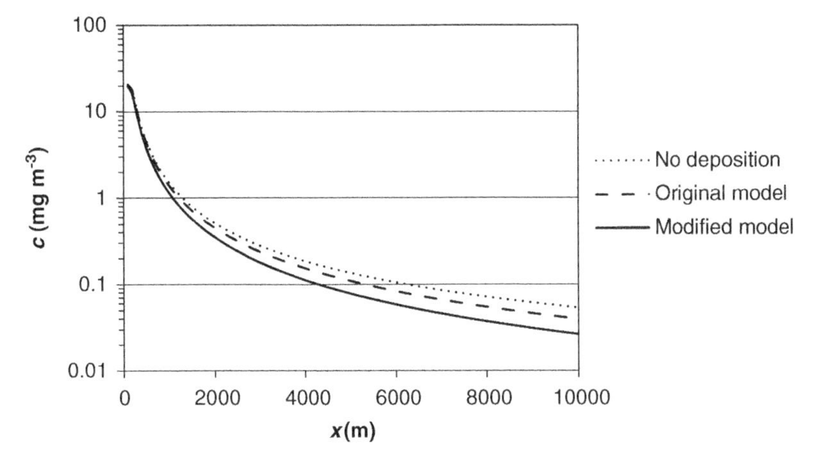


Figure 7.19 Concentration profiles resulting from deposition calculations with two different source depletion models. Source height: 10 m; deposition velocity: 0.02 m $s^{-1}$.

the original variables but significant with the modified variables. The profile of $Q$ is shown in Figure 7.18, along with the results of the previous calculations, whereas the concentration profile is shown in Figure 7.19.

Comparing the results in Figures 7.18 and 7.19, we see that the modified source depletion model predicts lower ground-level concentrations than the original source depletion model. Consequently, the modified source depletion model predicts less deposition than the original source depletion model. Far from the source it is expected that the modified source depletion model will predict higher ground-level concentrations than the original source depletion model. This is not yet the case at 10 km from the source, despite the artificially high deposition velocities in this example.

## 7.6 SUMMARY OF THE MAIN EQUATIONS

Momentum flux parameter: $$F_{\mathrm{m}} = \left(\frac{\overline{\rho_{\mathrm{s}}}}{\overline{\rho}}\right) r_{\mathrm{s}}^{2}\overline{w_{\mathrm{s}}}^{2}\ (\mathrm{m}^{4}\,\mathrm{s}^{-2}) \tag{7.1}$$

Buoyancy flux parameter:

$$F_{\mathrm{b}} = \left(1-\frac{\overline{\rho_{\mathrm{s}}}}{\overline{\rho}}\right) g r_{\mathrm{s}}^{2}\overline{w_{\mathrm{s}}}\ (\mathrm{m}^{4}\,\mathrm{s}^{-3}) \tag{7.2}$$

Stability parameter: $$s = \frac{g}{T}\frac{d\theta}{dz}\ (\mathrm{s}^{-2}) \tag{7.3}$$

Transitional plume rise equations:
Buoyancy dominated:

$$\Delta z = 1.6\left(\frac{F_{\mathrm{b}}x^{2}}{\overline{u}^{3}}\right)^{1/3} \tag{7.9}$$

Momentum dominated:

$$\Delta z = 2\left(\frac{F_{\mathrm{m}}x}{\overline{u}^{2}}\right)^{1/3} \tag{7.10}$$

Final plume rise equations:

$$\Delta h = \pi_{1}^{-1/4}\left(\frac{F_{\mathrm{b}}^{2}}{s^{3}}\right)^{1/8} \tag{7.15}$$

$$\Delta h = \pi_{2}^{-1/4}\left(\frac{F_{\mathrm{m}}}{s}\right)^{1/4} \tag{7.16}$$

where $\pi_{1}^{-1/4} = 5.3$, and $\pi_{2}^{-1/4} = 2.4$. Note that conditions apply. See Section 7.2.2 for alternative equations.
Flame length of a flare:

$$L_{\mathrm{f}} = 0.00178 Q_{\mathrm{c}}^{0.478} = 2.43 Q_{\mathrm{c,MW}}^{0.478} \tag{7.29}$$

Denser-than-air plume with constant width, in Lagrangian coordinates:

$$\frac{dh}{dt} + h\frac{\partial u}{\partial x} = w_{\mathrm{b}} + w_{\mathrm{e}} \tag{7.115}$$

$$\frac{du}{dt} = \frac{w_e}{h}\left(\frac{\rho_a u_a}{\rho} - u\right) - \frac{w_b u}{h} - \frac{\beta g}{2\rho}\frac{\partial}{\partial x}\left[h(\rho - \rho_a)\right] - \frac{u_*^2}{h} \tag{7.116}$$

$$\frac{dq}{dt} = \frac{w_b}{h}(q_b - q) - \frac{w_e q}{h} \tag{7.117}$$

Dry deposition flux equation:

$$F = -v_d c \tag{7.129}$$

Deposition velocity of gases:

$$\frac{1}{v_d} = r_t = r_a + r_b + r_c \tag{7.131}$$

Settling velocity of particles:

$$v_s = \frac{\rho_p d_p^2 g}{18\mu}\left\{1 + 2\frac{\lambda}{d_p}\left[1.257 + 0.4\exp\left(-0.55\frac{d_p}{\lambda}\right)\right]\right\} \tag{7.135}$$

Deposition velocity of particles:

$$v_d = \frac{1}{r_t} = \frac{v_s}{1 - \exp\left[-v_s(r_a + r_b)\right]} \tag{7.138}$$

Aerodynamic resistance, neutral atmosphere:

$$r_a = \frac{1}{ku_*}\ln\left(\frac{z}{z_0}\right) \tag{7.146}$$

Quasi-laminar resistance, gases:

$$r_b = \frac{\alpha_b \mathrm{Sc}^{2/3}}{ku_*} \tag{7.158}$$

Quasi-laminar resistance, particles:

$$\frac{1}{r_b} = \frac{1}{r_{b,B}} + \frac{1}{r_{b,IM}} + \frac{1}{r_{b,IN}} \tag{7.165}$$

Surface (canopy) resistance, ground + vegetation:

$$\frac{1}{r_c} = \frac{1}{r_{cg}} + \frac{1}{r_{cf}} \tag{7.184}$$

For water + vegetation:

$$\frac{1}{r_{\mathrm{c}}}=\frac{1}{r_{\mathrm{cw}}}+\frac{1}{r_{\mathrm{cf}}} \tag{7.185}$$

Liquid-side mass transfer coefficient at a water surface:

$$k_{\mathrm{L}}=d_{\mathrm{L}}\left(\frac{u_*}{1\,\mathrm{m\,s^{-1}}}\right)^n \tag{7.194}$$

$$k_{\mathrm{L}}=d_{\mathrm{L,10}}\left(\frac{u_{10}}{1\,\mathrm{m\,s^{-1}}}\right)^n \tag{7.195}$$

Overall mass transfer coefficient at a water surface:

$$K_{\mathrm{G}}=\frac{k_{\mathrm{G}}k_{\mathrm{L}}}{Hk_{\mathrm{G}}+k_{\mathrm{L}}} \tag{7.218}$$

Material flux at a water surface:

$$F=K_{\mathrm{G}}(Hc_{\mathrm{water}}-c_{\mathrm{air}}) \tag{7.219}$$

Air-side mass transfer coefficient at a water surface:

$$r_{\mathrm{a}}+r_{\mathrm{b}}=\frac{1}{k_{\mathrm{G}}} \tag{7.221}$$

Wet precipitation effect on pollutant concentration:

$$c_{\mathrm{air}}=c_{\mathrm{air,0}}\exp\left(-\Lambda t\right) \tag{7.261}$$

Scavenging ratio:

$$\Lambda=\lambda\frac{R}{R_1} \tag{7.262}$$

Dispersion and wet deposition equation:

$$\overline{c}=\frac{Q_0\exp(-\Lambda t)}{2\pi\overline{u}\sigma_y\sigma_z}\exp\left(-\frac{1}{2}\frac{y^2}{\sigma_y^2}\right)\left\{\exp\left[-\frac{1}{2}\frac{(z-h)^2}{\sigma_z^2}\right]+\exp\left[-\frac{1}{2}\frac{(z+h)^2}{\sigma_z^2}\right]\right\} \tag{7.264}$$

Source depletion model:

$$\frac{dQ}{dx}=-\sqrt{\frac{2}{\pi}}\frac{v_{\mathrm{d}}Q}{\overline{u}\sigma_z}\exp\left(-\frac{h^2}{2\sigma_z^2}\right) \tag{7.274}$$

Concentration profile correction factor for dry deposition:

$$P(x,z)=P(x,z_{\mathrm{d}})[1+v_{\mathrm{d}}R(z,z_{\mathrm{d}})] \tag{7.302}$$

# PROBLEMS

1. Use dimension analysis to derive an equation for the distance to final plume rise ($x_f$) of a buoyancy-dominated plume under windy conditions and neutral atmosphere. Compare with the result that can be obtained by combining the appropriate equations for transitional and final plume rise.
2. Derive an equation for transitional plume rise of a buoyancy-dominated plume under calm conditions and neutral atmosphere. Assume that the wind speed is nonzero but too small to affect plume rise.
3. Derive an equation to calculate $H_{cp}$ [see eq. (7.208)] from the dimensionless Henry constant [eq. (7.194)].
4. Prove eq. (7.306) following the strategy of developing a non-steady-state diffusion equation in the $z$ direction and substituting the Gaussian plume equation.
5. Calculate the plume rise at 1000 m downwind from a source, which is a stack designed to emit $5000\,m^3\,h^{-1}$ with an exit velocity of $5\,m\,s^{-1}$. The actual emission is $7000\,m^3\,h^{-1}$ and has a temperature of 83 °C. The ambient temperature is 12 °C, and the barometric pressure is 93 kPa. The conditions are overcast, and the wind speed measured at a 10-m height is $3.6\,m\,s^{-1}$. The stack height is 55 m.
6. Derive an equation for the limiting wind speed below which it no longer affects final plume rise in buoyancy-dominated plumes.
7. Solve Example 7.2, but with ethylene (molar mass $28\,g\,mol^{-1}$; heat of combustion $47.2\,MJ\,kg^{-1}$) flaring in identical conditions.
8. Solve Problem 5 with a numerical plume rise model. Assume negligible emissivity.
9. Due to an accident in a gas station a gasoline vapor cloud is formed with an initial mass fraction of 0.2, an initial thickness of 2 m, and an initial velocity of $5\,cm\,s^{-1}$. Temperature of plume and surroundings is 25 °C; pressure is 100 kPa. Use octane ($M = 114\,g\,mol^{-1}$). How much time is available to evacuate a community at 250 m from the incident? Do the calculation with (1) a non-steady-state model in Lagrangian coordinates, and (2) with a steady-state model using the downwind plume velocity to estimate travel times. Which of the two solutions would you choose as a basis for evacuation decisions?
10. Estimate the dry deposition velocity of nanoparticles with diameter 10 nm and density $2.5\,g\,cm^{-3}$ into the ocean at a wind speed at a 5-m height of $5\,m\,s^{-1}$. Use 5 m as the reference height. Assume neutral atmosphere.
11. A lake with area $0.8\,km^2$ is contaminated with carbon tetrachloride ($CCl_4$). Assume that the excess $CCl_4$ at the bottom of the lake causes the lake to be constantly saturated with $CCl_4$ (solubility $0.78\,g\,L^{-1}$). The diffusivity of $CCl_4$ is $0.076\,cm^2\,s^{-1}$ in air and $1.04\times10^{-5}$ $cm^2\,s^{-1}$ in water. The dimensionless Henry constant is 1.25. Assume a temperature of 25 °C and a pressure of 1 bar. What is the $CCl_4$ emission from the lake at a wind speed of $4\,m\,s^{-1}$, measured at a 10-m height? Assume approximately zero $CCl_4$ concentration at 10 m height.
12. Adjust eq. (7.279) to incorporate both dry deposition and wet deposition. Recalculate Example 7.13 with the traditional source depletion model assuming $v_d = 0.02\,m\,s^{-1}$ and $\Lambda = 10^{-4}\,s^{-1}$.

## MATERIALS ONLINE

- "Example 7.1. Plume rise calculations.xlsx": Calculations for Example .
- "Example 7.2. Flare plume rise calculations.xlsx": Calculations for Example 7.2.
- Folder "Numerical plume rise model," files "main.m," "data.m," and "f.m": Calculations for Example 7.3.
- Folder "Streamline deflection," files "main.m," "data.m," and "f.m": Calculations for Example 7.4.
- Folder "Dense plume steady state," files "main.m," "data.m," and "f.m": Steady-state calculations for Example 7.5.
- Folder "Dense plume nonsteady state Eulerian," files "main.m," "data.m," and "f.m": Non-steady-state calculations for Example 7.5 (Eulerian implementation).
- Folder "Dense plume nonsteady state Lagrangian," files "main.m," "data.m," and "f.m": Non-steady-state calculations for Example 7.5 (Lagrangian implementation).
- "Rainout calculation gases.xlsx": Calculation of the rainout rate of a gaseous air pollutant.
- "Rainout calculation particles.xlsx": Calculation of the rainout rate of particulates.
- "Washout calculation gases.xlsx": Calculation of the washout rate of a gaseous air pollutant.
- "Washout calculation particles.xlsx": Calculation of the washout rate of particulates.
- Folder "Dispersion and deposition source depletion model," files "main.m," "data.m," and "f.m": Solution of Example 7.13, original source depletion model.
- Folder "Dispersion and deposition modified source depletion model," files "main.m," "data.m," and "f.m": Solution of Example 7.13, modified source depletion model.

## REFERENCES

Allen M.D. and Raabe O.G. (1982). Reevaluation of Millikan's oil drop data for the motion of small particles in air. *J. Aerosol Sci.* **13**, 537–547.

API (2007). *Pressure-Relieving and Depressuring Systems*. ANSI/API Standard 521, 5th ed, Washington, D.C.

Battan L.J. and Reitan C.H. (1957). Droplet size measurements in convective clouds. In *Artificial Stimulation of Rain*. Pergamon Press, New York, pp. 184–191.

Beychok M.R. (2005). *Fundamentals of Stack Gas Dispersion*, 4th ed. Beychok, Newport Beach, CA.

Bird R.B., Steward W.E., and Lightfoot E.N. (1960). *Transport Phenomena*. Wiley, New York.

Briggs G.A. (1965). A plume rise model compared with observations. *J. Air Pollut. Control Assoc.* **15**, 433–438.

Briggs G.A. (1973). *Diffusion estimation of small emissions*. Contribution No. 79, Atmospheric Turbulence and Diffusion Laboratory, Oak Ridge, TN.

Briggs G.A. (1975). Plume rise predictions. In Haugen D.A. (ed.) *Lectures on Air Pollution and Environmental Impact Analyses*. American Meteorological Society, Boston, MA, pp. 59–111.

Brook J., Zhang L., Digiovanni F., and Padro J. (1999). Description and evaluation of a model for routine estimates of air pollutant dry deposition over North America. Part I: Model development. *Atmos. Environ.* **33**, 5037–5051.

Campbell A.N. and Cardoso S.S.S. (2010). Turbulent plumes with internal generation of buoyancy by chemical reaction. *J. Fluid Mech.* **655**, 122–151.

Campbell A.N., Barrie P.J., and Cardoso S.S.S. (2009). Turbulent plumes with internal generation of buoyancy by chemical reaction. *Proceedings 8th World Congress of Chemical Engineering, Montreal, Canada.*

Chamberlain A.C. (1953). *Aspects of Travel and Deposition of Aerosol and Vapor Clouds.* Atomic Energy Research Establishment Report 1261, Her Majesty's Stationary Office, London.

Chameides W.L. (1984). The photochemistry of a remote marine stratiform cloud. *J. Geophys. Res.* **89**, 4739–4755.

Charnock H. (1955). Wind stress on a water surface. *Quart. J. Roy. Meteorol. Soc.* **81**, 639–640.

Cimorelli A.J., Perry S.G., Venkatram A., Weil J.C., Paine R.J., Wilson R.B., Lee R.F., Peters W.D., Brode R.W., and Paumier J.O. (2004). *AERMOD: Description of Model Formulation.* Report EPA-454/R-03-004, US-EPA, Research Triangle Park, NC.

de Leeuw G., Andreas E.L., Anguelova M.D., Fairall C.W., Lewis E.R., O'Dowd C., Schulz M., and Schwartz S.E. (2011). Production flux of sea spray aerosol. *Rev. Geophys.* **49**, RG2001.

De Visscher A. (2009). Extending the PRIME plume rise model to include flare combustion. *Proceedings 8th World Congress of Chemical Engineering, Montreal, Canada.*

Doran J.C. and Horst T.W. (1985). An evaluation of Gaussian plume-depletion models with dual-tracer field measurements. *Atmos. Environ.* **19**, 939–951.

EPA (1995). *User's Guide for the Industrial Source Complex (ISC3) Dispersion Models. Volume II—Description of Model Algorithms.* Report EPA-454/B-95-003b, US-EPA, Research Triangle Park, NC.

Ermak D.L. (1990). *User's Manual for SLAB: An Atmospheric Dispersion Model of Denser-Than-Air Releases.* Livermore, CA, UCRL-MA-105607.

Fernandez de la Mora J. and Friedlander S.K. (1982). Aerosol and gas deposition to fully rough surfaces: Filtration model for blade shaped elements. *Int. J. Heat Mass Transfer*, **25**, 1725–1735.

Giorgi F. and Chameides W.L. (1985). The rainout parameterization in a photochemical model. *J. Geophys. Res.* **90D**, 7872–7880.

Hicks B.B. (1982). Dry deposition. In *Critical Assessment Document on Acid Deposition.* ATDL Contribution File No. 81/24. Atmospheric Turbulence and Diffusion Laboratory, NOAA, Oak Ridge, TN.

Holmen K. and Liss P. (1984). Models for air-water gas transfer: An experimental investigation. *Tellus* **36B**, 92–100.

Holmes H.A., Pardyjak E.R., Perry K.D., and Abbott M.L. (2011). Gaseous dry deposition of atmospheric mercury: A comparison of two surface resistance models for deposition to semiarid vegetation. *J. Geophys. Res.* **116**, D14306.

Horst T.W. (1977). A surface depletion model for deposition from a Gaussian plume. *Atmos. Environ.* **11**, 41–46.

Horst T.W. (1983). A correction to the Gaussian source depletion model. In Pruppacher H.R., Semonin R.G., and Slinn W.G.N. (eds.). *Precipitation Scavenging, Dry Deposition and Resuspension.* Elsevier, Amsterdam, pp. 1205–1218.

Horst T.W. (1984). The modification of plume models to account for dry deposition. *Bound. Layer Meteorol.* **30**, 413–430.

Jennings S.G. (1988). The mean free path in air. *J. Aerosol Sci.* **19**, 159–166.

Levine S.Z. and Schwartz S.E. (1982). In-cloud and below-cloud scavenging of nitric acid vapor. *Atmos. Environ.* **16**, 1725–1734.

Liss P.S. (1973). Processes of gas exchange across an air-water interface. *Deep-Sea Res.* **90**, 221–238.

Liss P.S. and Slater P.G. (1974). Flux of gases across the air-sea interface. *Nature* **247**, 181–184.

Mackay D. and Yeun A.T.K. (1983). Mass transfer coefficient correlations for volatilization of organic solutes from water. *Environ. Sci. Technol.* **17**, 211–217.

Möller U. and Schumann G. (1970). Mechanisms of transport from the atmosphere to the Earth's surface. *J. Geophys. Res.* **75**, 3013–3019.

Owen C.R. and Thompson W.R. (1963). Heat transfer across rough surfaces. *J. Fluid Mech.* **15**, 321–334.

Peña A. and Gryning S.E. (2008). Charnock's roughness length model and non-dimensional wind profiles over the sea. *Bound. Layer Meteorol.* **128**, 191–203.

Peters K. and Eiden R. (1992). Modelling the dry deposition velocity of aerosol particles to a spruce forest. *Atmos. Environ.* **26**, 2555–2564.

Pleim J., Venkatram A., and Yamartino R.J. (1984). *ADOM/TADAP Model Development Program. Volume 4. The Dry Deposition Model.* Ontario Ministry of the Environment, Rexdale, Ontario, Canada.

Poling B.E., Prausnitz J.M., and O'Connell J.P. (2000). *The Properties of Gases and Liquids*, 5th ed. McGraw-Hill, New York.

Pryor S.C., Gallagher M., Sievering H., Larson S.E., Barthelmie R.J., Birsan F., Nemitz E., Rinne J., Kumala M., Grönholm T., Taipale R., and Vesala T. (2008). A review of measurement and modelling results of particle atmosphere-surface exchange. *Tellus* **608**, 42–75.

Roldin P., Swietlicki E., Schurgers G., Arneth A., Lehtinen K.E.J., Boy M., and Kulmala M. (2011). Development and evaluation of the aerosol dynamics and gas phase chemistry model ADCHEM. *Atmos. Chem. Phys.* **11**, 5867–5896.

Schulman L.L., Strimaitis D.G., and Scire J.S. (2000). Development and evaluation of the PRIME plume rise and building downwash model. *J. Air Waste Manage. Assoc.* **50**, 378–390.

Schwarzenbach R.P., Gschwend P.M., and Imboden D.M. (1993). *Environmental Organic Chemistry.* Wiley, New York.

Scire J.S., Strimaitis D.G., and Yamartino R.J. (2000). *A User's Guide for the CALPUFF Dispersion Model.* Earth Tech, Concord, MA.

Sehmel G.A. (1980). Particle and gas dry deposition: A review. *Atmos. Environ.* **14**, 983–1011.

Seinfeld J.H. and Pandis S.N. (2006). *Atmospheric Chemistry and Physics*, 2nd ed. Wiley, Hoboken, NJ.

Sekhon R.S. and Srivastava R.C. (1971). Doppler observations of drop size distributions in a thunderstorm. *J. Atmos. Sci.* **28**, 983–994.

Shepherd J.G. (1974). Measurements of the direct deposition of sulphur dioxide onto grass and water by the profile method. *Atmos. Environ.* **8**, 69–74.

Slinn W.G.N. (1982). Predictions for particle deposition to vegetative canopies. *Atmos. Environ.* **16**, 1785–1794.

Slinn W.G.N. (1983). Precipitation scavenging. In *Atmospheric Science and Power Production.* U.S. Department of Energy, Washington DC.

Slinn S.A. and Slinn W.G.N. (1980), Predictions for particle deposition on natural waters. *Atmos. Environ.* **14**, 1013–1016.

Slinn W.G.N., Hasse L., Hicks B.B., Hogan A.W., Lai D., Liss P.S., Munnich K.O., Sehmel G.A., and Vittori O. (1978). Some aspects of the transfer of atmospheric trace constituents past the air-sea interface. *Atmos. Environ.* **12**, 2055–2087.

Theodore L. and Buonicore A.J. (1988). *Air Pollution Control Equipment for Particulates*, Vol. **1**. CRC Press, Boca Raton, FL.

Turner D.B. and Schulze R.H. (2007). *Practical Guide to Atmospheric Dispersion Modeling.* Trinity Consultants, Dallas, TX, and Air and Waste Management Association, Pittsburgh.

Walmsley J.L. and Wesely M.L. (1996). Modification of coded parameterizations of surface resistances to gaseous dry deposition. *Atmos. Environ.* **30**, 1181–1186.

Wanninkhof R. (1992). Relationship between wind speed and gas exchange over the ocean. *J. Geophys. Res.* **97**, 7373–7382.

Watson A.J., Upstill-Goddard R.C., and Liss P.S. (1991). Air-sea gas exchange in rough and stormy seas measured by a dual-tracer technique. *Nature* **349**, 145–147.

Weil J.C. (1988). Plume rise. In Venkatram A. and Wyngaard J.C. (eds.). *Lectures on Air Pollution Modeling.* Am. Meteorol. Soc., Boston, MA, pp. 119–166.

Wesely M.L. (1989). Parameterization of surface resistances to gaseous dry deposition in regional-scale numerical models. *Atmos. Environ.* **23**, 1293–1304.

Wesely M.L., Cook D.R., Hart R.L., and Speer R.E. (1985). Measurements and parameterization of particulate sulfur dry deposition over grass. *J. Geophys. Res.* **90**, 2131 2143.

Yamartino R.J. (2008). Simulation algorithms in Gaussian plume modeling. In Zannetti P. (ed.). *Air Quality Modeling. Volume III—Special Issues.* Enviro Comp Institute, Fremont, CA, and Air and Waste Management Association, Pittsburgh, pp. 239–280.

Zeman O. (1982). The dynamics and modeling of heavier-than-air, cold gas releases. *Atmos. Environ.* **16**, 741–751.

Zhang L., Gao S, Padro J., and Barrie L. (2001). A size-aggregated dry deposition scheme for an atmospheric aerosol module. *Atmos. Environ.* **35**, 549–560.

Zhang L., Moran M.D., Makar P.A., Brook J.R., and Gong S. (2002). Modelling gaseous dry deposition in AURAMS: A unified regional air-quality modelling system. *Atmos. Environ.* **36**, 537–560.

Zhang L., Brook J.R., and Vet R. (2003). A revised parameterization for gaseous dry deposition in air-quality models. *Atmos. Chem. Phys.* **3**, 2067–2082.

Zhao D., Toba Y., Suzuki Y., and Komori S. (2003). Effect of wind waves on air-sea gas exchange: Proposal of an overall $CO_2$ transfer velocity formula as a function of breaking-wave parameter. *Tellus* **55B**, 478–487.

CHAPTER 8

# GAUSSIAN MODEL APPROACHES IN URBAN OR INDUSTRIAL TERRAIN

## 8.1 INTRODUCTION

Urban and industrial terrains differ from natural terrains in many respects. Because of the large size of the obstacles (buildings), urban terrain is rougher than most natural environments. Furthermore, the heat balance of an urban surface is very different from the heat balance of a rural surface, due to heat production (home heating, machinery) and due to a lower evaporation. These differences have been incorporated into air dispersion models in ways that have been discussed in the preceding chapters. However, there is another difference between urban terrains and most natural terrains that has not been mentioned so far. Due to the *size* of the obstacles in urban terrain, plumes released between obstacles take a substantial time to grow to a size larger than these obstacles and can therefore remain between the obstacles for hundreds of meters downwind. There are some differences between atmospheric turbulence *between* buildings and atmospheric turbulance *above* buildings, and these differences affect plume dispersion. So far we have always made the tacit assumption that we are studying the atmosphere at a height at least twice the average obstacle height. Applying the knowledge obtained so far to plumes released between buildings in an urban or industrial site would lead to substantial errors. Hence, an alternative approach is needed for such cases.

Another issue is the **wind speed profile**. It is generally agreed that the wind speed profile as calculated in similarity theory (see Chapter 5) breaks down at a height about once to twice the average obstacle height, that is, at roughly 10–20 times the roughness length. In urban areas, where surface roughness can be 1 m or more, this poses challenges for the appropriate choice of wind speed in the Gaussian plume equation.

The best way to evaluate dispersion in urban and industrial sites, close to the source, is with wind tunnel studies or with computational fluid dynamics (CFD) (Li et al., 2006). However, this is not always feasible due to time or budget constraints. Hence, we need a technique that is computationally fairly cheap but of reasonable accuracy to deal with such cases.

*Air Dispersion Modeling: Foundations and Applications*, First Edition. Alex De Visscher.

Hanna and Britter (2002) developed a method for calculating air dispersion between buildings with reasonable accuracy, using a Gaussian air dispersion model. This chapter is a modified version of the procedure proposed in Hanna and Britter's book. Readers wishing to apply the ideas outlined in this chapter are encouraged to consult the original approach as well.

The following subjects are covered in this chapter:

- Wind speed profiles in urban terrain. This includes detailed information on estimating roughness length $z_0$ and displacement height $d$, as well as deviations from the logarithmic wind speed profile close to the surface.
- Turbulence between roughness obstacles.
- Dispersion between obstacles and above obstacles. The transition between the two can be described with the virtual source concept or with a numerical modeling approach.

This chapter will mainly focus on neutral conditions.

## 8.2 WIND FLOW AROUND OBSTACLES

The size and some characteristics of building wakes were discussed in detail in Chapter 7 with the objective of understanding the downwash phenomenon. There it was implicitly assumed that each building is isolated, so there is no interaction between the wakes of different buildings. This is not always the case, especially in urban and industrial terrain. The wakes of nearby buildings can interact with each other, and we need to understand some aspects of this interaction if we are to understand turbulence in an urban or industrial environment.

The flow regime of wind between obstacles depends on the *height* of the obstacles defining the roughness $H_r$ and the obstacle *spacing* $S_r$ (the distance between two neighboring obstacle elements). To a lesser extent, it also depends on obstacle width.

We distinguish between three flow regimes. They are illustrated in Figure 8.1. When $S_r/H_r > 3$, then the flow regime is known as **isolated roughness flow**. In this regime, the wakes do not overlap and the impact of each building on the wind flow and turbulence can be considered in isolation. When $S_r/H_r$ is between 1.5 and 3, the obstacle wakes interfere with each other. This is known as **wake interference flow**. The overlapping of the wakes causes extra turbulence. Consequently, the roughness length is usually highest in this regime. When $S_r/H_r$ is less than 1.5, the wakes interact so strongly with each other that stagnant zones are created that do not contribute to turbulence or roughness. In this situation the wind skims over the obstacles, which is known as **skimming flow**. In this regime, the surface roughness is greatly reduced, and instead there is a marked **displacement** of the wind speed profile.

Skimming flow as shown in Figure 8.1 is not common in real terrain (as opposed to wind tunnel studies) because the obstacles are usually not all the same size. Hence, the flow regime is usually wake interference flow even when $S_r/H_r$ is less than 1.5. The result is high roughness and medium displacement of the wind speed profile.

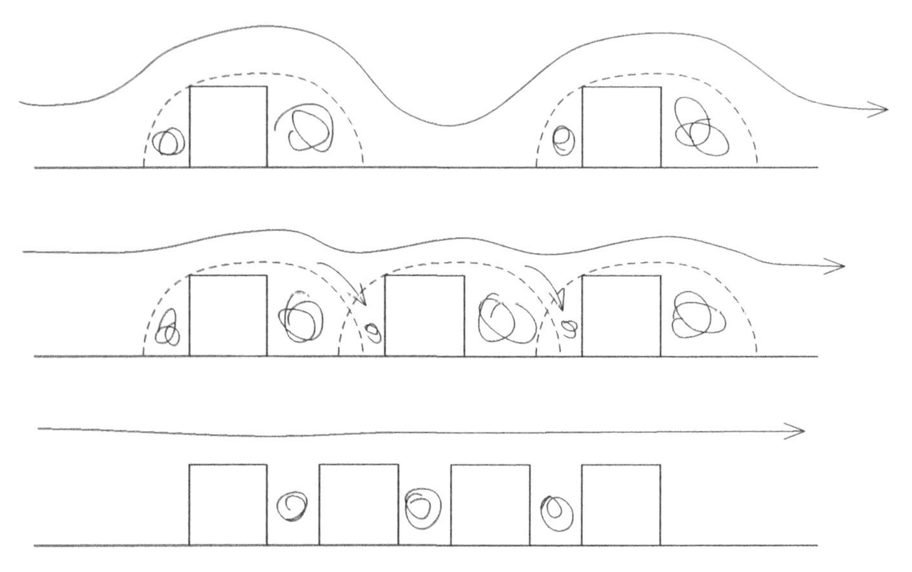

Figure 8.1 Wind flow regimes in urban terrain: isolated roughness flow (*top*), wake interference flow (*middle*), and skimming flow (*bottom*). Based on Oke (1988).

## 8.3 SURFACE ROUGHNESS AND DISPLACEMENT HEIGHT IN URBAN AND INDUSTRIAL TERRAIN

### 8.3.1 Introduction

In Section 8.3, we will focus on the wind speed profile at a sufficient height above the buildings ($z > 2H_r$). Wind speed profiles between buildings will be discussed in Section 8.4. In both sections we will focus on the neutral condition. As we have seen in Chapter 5, the wind speed profile at $z > 2H_r$ under neutral conditions is given by eq. (5.62):

$$u = \frac{u_*}{k} \ln \frac{z}{z_0} \tag{5.64}$$

We have also seen in Chapter 5 that eq. (5.62) loses accuracy when the surface is densely covered with obstacles. In that case, eq. (5.66) can be used:

$$u = \frac{u_*}{k} \ln \frac{z-d}{z_0} \tag{5.68}$$

where $d$ is the **displacement height**. This is relevant in urban environments, which are often densely covered with tall buildings.

**Example 8.1.** Calculate the friction velocity, and the velocity profile up to a 100-m height based on a measured wind speed of 3 m s$^{-1}$ at a 20-m height, (a) assuming a roughness height of 1 m, and (b) assuming a roughness height of 1 m and a displacement height of 5 m. What will be the impact on plume dispersion predictions?

***Solution.*** (a) Solving eq. (5.62) for $u*$ leads to

$$u_* = \frac{uk}{\ln(z / z_0)}$$

For $u = 3$ m s$^{-1}$, $k = 0.4$, $z = 20$ m, and $z_0 = 1$ m, a friction velocity $u* = 0.401$ m s$^{-1}$ is obtained. This value is substituted in eq. (5.62) to obtain wind speeds. The wind speed profile is shown as the dashed line in Figure 8.2.

(b) Solving eq. (5.66) for $u*$ leads to

$$u_* = \frac{uk}{\ln[(z - d) / z_0]}$$

For $u = 3$ m s$^{-1}$, $k = 0.4$, $z = 20$ m, $z_0 = 1$ m, and $d = 5$ m a friction velocity $u* = 0.443$ m s$^{-1}$ is obtained. This value is substituted in eq. (5.66). The wind speed profile is shown as the solid line in Figure 8.2.

Discussion: The agreement between the two calculations is acceptable, with a wind speed difference up to about 10%. However, the slope of the wind speed is severely underestimated without accounting for $d$. The friction velocity is underestimated by about 10% when $d$ is not accounted for. Consequently, the dispersion parameters can be expected to be underestimated by about 10% as well.

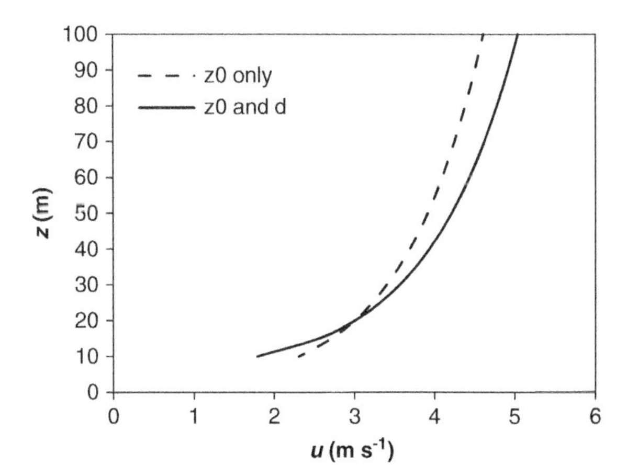


Figure 8.2 Calculated wind speeds from a measured value of 3 m s$^{-1}$ at a 20-m height, with (solid line) and without (dashed line) accounting for displacement height ($z_0 = 1$ m; $d = 5$ m).

We see from the example that ignoring displacement height affects dispersion calculations by about 10%. The deviations will be largest above the wind speed measurement height, where all errors ($u$, $\sigma_y$ and $\sigma_z$) lead to overpredictions of the pollutant concentration. The error can be as large as 30%. In most rural terrains, where roughness elements are much smaller and sparser, the error introduced by ignoring $d$ is usually negligible. In urban terrain the inclusion of $d$ will improve the accuracy of atmospheric property estimates.

## 8.3.2 Determination of $z_0$ and $d$

***8.3.2.1 Measurement Methods*** In principle, the variables $u_*$, $z_0$, and $d$ can be determined by fitting eq. (5.66) to a measured wind speed profile. However, given the relatively minor effect of $d$ on the wind speed profile (see Example 8.1), the similar effect of $z_0$ and $d$ on the wind speed predictions, and the large variability of wind speed, it is nearly impossible to obtain reliable values of $u_*$, $z_0$, and $d$ in this manner. The situation is even worse when deviations from neutral conditions occur, and $L$ becomes a fourth unknown.

The above method can be improved by obtaining an independent measurement of $u_*$. The measurement of $u_*$ is based on eq. (5.167):

$$u_* = \sqrt{\frac{\tau_{xz}}{\rho}} = \sqrt{-\overline{u'w'}} \tag{5.167}$$

where the $x$ axis is chosen along the mean wind direction. The wind speed deviations $u'$ and $w'$ can be measured with a sonic anemometer. This narrows down the number of adjustable parameters to 2 ($z_0$ and $d$). Note, however, that a measurement error on $u_*$ (which itself is variable) will result in a measurement error of $z_0$ and $d$. Again, the possibility of nonneutral conditions can be a confounding factor. In practice, values of $z_0$ and $d$, if critical, should never be based on a single measurement but rather on an extended data set.

***8.3.2.2 Estimation Methods*** *Average* numbers of $z_0$ can be obtained from experience. An example is Table 5.2 (see Section 5.7.2). Additional values for industrial sites are given in Table 8.1 (Hanna and Britter, 2002). Experience is much more limited for values of $d$. As a rule of thumb, $d$ is about 5 times $z_0$.

If the average height of the obstacles defining the surface roughness $H_r$ is known, but nothing else, then the following equations can be used to estimate $z_0$ and $d$ (Hanna and Chang, 1992):

$$z_0 = 0.1H_r \tag{8.1}$$

$$d = 0.5H_r \tag{8.2}$$

When **detailed geometric properties** of the terrain are available, a more refined estimate can be made. However, more refined does not necessarily mean more accurate

**TABLE 8.1 Average Values of the Roughness Length $z_0$ at Industrial Sites**

| Description | $z_0$ (m) |
|---|---|
| Isolated low buildings ($S_r/H_r > 20$) | 0.1 |
| Scattered buildings ($S_r/H_r > 8$–$12$) | 0.25 |
| Moderately covered ($S_r/H_r > 3$–$7$) | 0.5 |
| Densely built area | 1 |
| Chaotic (mix of low rise and high rise) | 2 |

Source: From Hanna and Britter (2002).

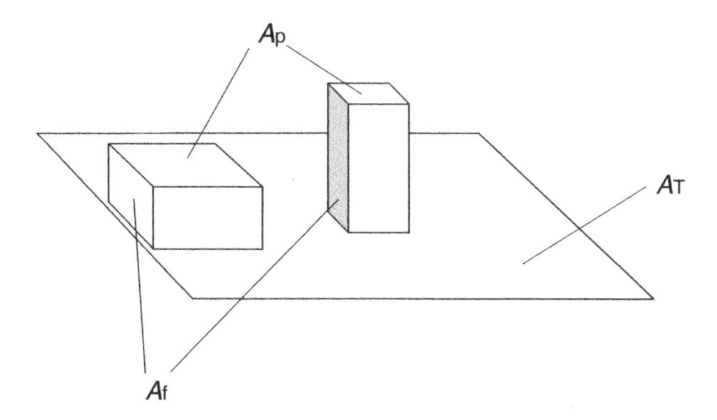


Figure 8.3 Plan area ($A_p$), frontal area ($A_f$) of obstacles on a lot with total area $A_T$.

because the equations outlined below are based on measurement data with considerable scatter. For the estimation, the following geometric properties are defined, based on a representative lot in the terrain (Grimmond and Oke, 1999):

Area $A_p$ is the horizontal area or **plan area** of the obstacles on the lot.
Area $A_f$ is the vertical across-wind area or **frontal area** of the obstacles on the lot.
Area $A_T$ is the **total area** (horizontal) of the lot.

These definitions are illustrated in Figure 8.3. Based on these, the following dimensionless numbers are defined:

$$\lambda_p = \frac{A_p}{A_T} \tag{8.3}$$

$$\lambda_f = \frac{A_f}{A_T} \tag{8.4}$$

Grimmond and Oke (1999) comprehensively reviewed the literature on $z_0$ and $d$ determination and pointed out that a considerable amount of data is based on wind tunnel studies where both the wind and the obstacles are too regular to be representative to real cities. In real cities, skimming is much less common than in wind tunnel experiments. In spite of this, there is an expectation in the literature that the value of $z_0$ should go through a maximum when plotted against $\lambda_p$ or $\lambda_f$. Theoretical justification of this expectation is given in Appendix C. Some commonly used parameterizations were developed by Lettau (1969), Raupach (1994), Bottema (1997), and Macdonald et al. (1998).

From the literature it appears that $\lambda_f$ is a better predictor for $z_0$ and $d$ than $\lambda_p$. Hence, correlations in this chapter will be based on $\lambda_f$. Furthermore, the evidence that $z_0$ passes through a maximum versus $\lambda_f$ in real cities is very scant. In this chapter it will be assumed that there is no such maximum unless there is a specific reason to suspect skimming flow (e.g., a dense collection of identical apartment buildings). The recommendations made here are not the ones of Hanna and Britter (2002). They should be considered *tentative*. For $z_0$:

For $\lambda_f < 0.1$:

$$z_0 = \lambda_f \cdot H_r \tag{8.5}$$

For $\lambda_f > 0.1$:

$$z_0 = 0.1 \cdot H_r \tag{8.6}$$

When there is strong wake interference (chaotic):

$$z_0 = 0.15 \cdot H_r \tag{8.7}$$

When there is strong skimming:

$$z_0 = 0.05 \cdot H_r \tag{8.8}$$

For $d$:

For $\lambda_f < 0.25$ (0.35 in the case of strong skimming):

$$d = 2\lambda_f \cdot H_r \tag{8.9}$$

For $\lambda_f > 0.25$:

$$d = 0.5 \cdot H_r \tag{8.10}$$

For $\lambda_f > 0.35$ in the case of strong skimming:

$$d = 0.7 \cdot H_r \tag{8.11}$$

Readers wishing to apply these rules are recommended to consult the book of Hanna and Britter (2002) and the article of Grimmond and Oke (1999) to get a sense of the uncertainties associated with these estimations. For instance, there is some indication in the data that $z_0$ and $d$ increase more than linearly with $\lambda_f$ when $\lambda_f$ is small (< 0.1). The theoretical description of wind flow in a canopy described in Appendix C confirms this.

***8.3.2.3 Averaging Rules.*** The best way to deal with changing terrain is with the virtual source concept discussed in Section 6.8. However, sometimes only a section of a plume is experiencing changing terrain, so we cannot clearly define terrain properties that apply to the entire plume section. For that situation, Hanna and Britter (2002) proposed an averaging rule. First, a wind direction sector of 30° is defined that approximately coincides with the plume. For each terrain type $i$ the coverage $x_i$ is defined as the fraction of the 30° sector covered by terrain type $i$. Based on the coverages, $z_0$ and $d$ are calculated as the weighted geometric mean. For instance, when there are two terrain types with coverages $x_1$ and $x_2$, and surface roughness lengths $z_{0,1}$ and $z_{0,2}$, the average roughness length is

$$\ln z_0 = \frac{x_1 \ln z_{0,1} + x_2 \ln z_{0,2}}{x_1 + x_2} \tag{8.12}$$

The same rule applies to $d$. For short distances, eq. (8.12) can also be used when the entire plume moves over a terrain with different $z_0$. When plumes move over short terrains, there is insufficient space to fully develop an internal boundary layer and the roughness length will not be well-defined.

**Example 8.2.** Calculate the average surface roughness when a plume moves over a terrain consisting of 40% (or 12°) cut grass ($z_0 = 0.01$ m) and 60% (or 18°) urban terrain ($z_0 = 1$ m).

***Solution.*** Applying eq. (8.12), we find

$$\ln z_0 = \frac{12 \ln 0.01 + 18 \ln 1}{12 + 18} = -1.84$$

Taking the exponential, a value of $z_0 = 0.16$ m is found.

## 8.4 WIND SPEED PROFILES NEAR THE SURFACE: DEVIATIONS FROM SIMILARITY THEORY

### 8.4.1 Theoretical Background

The logarithmic wind speed profile predicted by similarity theory [eq. (5.62) or (5.66)] only applies down to a height of about $2H_r$. Below that height the logarithmic equation underestimates the wind speed. This is illustrated in Figure 8.4. The underestimate is

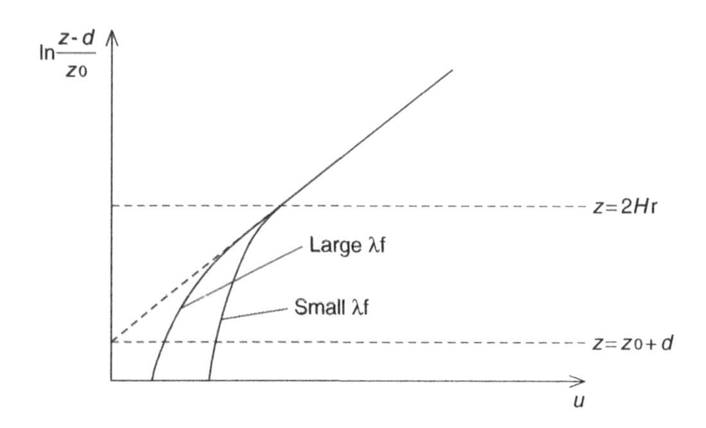


Figure 8.4 Wind speed profiles near the surface: deviation from similarity theory.

most pronounced when $\lambda_f$ is small. Several approaches to estimate the wind speed profile near the surface will be discussed in the next sections. First, some theoretical considerations will be made. A more complete theory is outlined in Appendix C.

Wind speed gradients perpendicular to the wind direction, including vertical profiles, are given by the turbulent version of Newton's viscosity law [eq. (5.174)]:

$$\frac{\tau}{\rho} = \overline{u'w'} = -K_m \frac{\partial \bar{u}}{\partial z} \tag{5.174}$$

Above the canopy, in the surface layer, it can be assumed that the shear stress is approximately constant. This is usually referred to as the "surface" value $\tau_0$, even though it does not strictly apply to the surface when a canopy is present. Assuming that the shear stress $\tau$ equals $\tau_0$, and using the definition of $u_*$ [eq. (5.54) or (5.165)], we obtain

$$\frac{\partial \bar{u}}{\partial z} = \frac{u_*^2}{K_m} \tag{8.13}$$

It follows that a wind speed profile can be calculated from a $K_m$ profile and vice versa. For $z > 2H_r$, under neutral conditions, eq. (5.202) applies

$$K_m = k u_* (z - d) \tag{5.202}$$

Between the roughness defining obstacles, the assumption of constant shear stress can be replaced by the following assumption (see Appendix C):

$$\frac{\partial \tau_{xz}}{\partial z} - \tau_\lambda \frac{\lambda}{H_r} = 0 \tag{C70}$$

where $\tau_\lambda$ is the shear stress on the roughness defining elements, projected on a vertical plane along the wind direction, and $\lambda$ is an area parameter.

Equation (C70) is not universally valid as it ignores the effect of the barometric pressure along the wind direction on the shear stress. However, in most canopies this is not a concern. The result of eq. (C70) is a shear stress decreasing toward the surface. Based on the local shear stress a local friction velocity can be defined as

$$\tau_{xz} = \rho u_{\tau}^{2} \tag{C75}$$

where $u_\lambda < u*$. The lowered local friction velocity leads to a lowered $K_m$ value, which in turn is responsible for the displacement of the wind speed profile expressed in the displacement height $d$. This is discussed in more detail in Appendix C and in Section 8.4.5.

### 8.4.2 Simple Approach

The simplest approach to treat deviations from the logarithmic wind speed profile near the surface is by adopting a constant average value, the **characteristic velocity** $u_c$. It is assumed that $u = u_c$ below a certain height $z_c$, and that $u$ follows the logarithmic equation above $z_c$. The height where the assumed transition from constant to logarithmic occurs is the height where the logarithmic equation predicts $u = u_c$. That height is found by solving eq. (5.66) for height. The result is

$$z_c = d + z_0 \exp\left(\frac{ku_c}{u_*}\right) \tag{8.14}$$

The following equation has been suggested for $u_c$ (Hanna and Britter, 2002):

$$u_c = u_* \sqrt{\frac{2}{\lambda_f}} \tag{8.15}$$

At values of $\lambda_f < 0.15$, the following equation can be used as well (Bentham and Britter, 2003):

$$u_c = u_* \sqrt{\frac{2H_r}{z_0}} \tag{8.16}$$

These equations predict experiments with artificial obstacles very well, but $u_c$ in real situations tends to be lower than in artificial terrains. Further methods to estimate $u_c$ will be discussed in the next sections.

It is clear that a constant wind speed near the surface is physically impossible if we are to maintain the assumption $\tau = \tau_0$. This follows from eq. (8.13): When the velocity gradient is zero, $K_m$ becomes infinity.

### 8.4.3 Exponential Wind Speed Profile

Many authors have proposed models to describe the wind speed profile between canopy elements. The most commonly applied model is the one of Inoue (1963) leading to an exponential velocity profile. Some other models found in the literature are summarized in Section 8.4.5. A new wind speed profile is derived in Appendix C.

Inoue (1963) and Cionco (1965, 1978) derived a wind speed profile near the surface and arrived at the following equation:

$$u = u_{H_r} \exp\left[-a\left(1-\frac{z}{H_r}\right)\right] \tag{8.17}$$

where $a$ in an urban canopy is a wind speed extinction parameter given by (Macdonald, 2000):

$$a = 9.6\lambda_f \tag{8.18}$$

As shown in Appendix C, the dependence of $a$ on $\lambda_f$ may be nonlinear, and it is likely that wind tunnel studies overestimate the wind speed extinction parameter applicable in real urban canopies. Estimates based on the canopy model in Appendix C point at an overestimation of 40%.

In order to use eq. (8.17), a value of $u_{H_r}$, the wind speed at height $H_r$, is still needed. The most straightforward choice is to use the wind speed predicted by the logarithmic equation. This is illustrated in Figure 8.5 for $H_r$ = 10 m and $\lambda_f$ = 0.25. According to the rules of thumb, this leads to $z_0$ = 1 m and $d$ = 5 m. Equation (8.18) leads to $a$ = 2.4. Assuming $u_*$ = 0.5 m s$^{-1}$, the logarithmic wind speed profile starts at $u_{H_r}$ = 2.012 m s$^{-1}$. With these numbers, the wind speed profile is continuous, but the slope of the wind speed is not, leading to a discontinuity of $K_m$. However, the resulting wind speed profile is reasonably accurate.

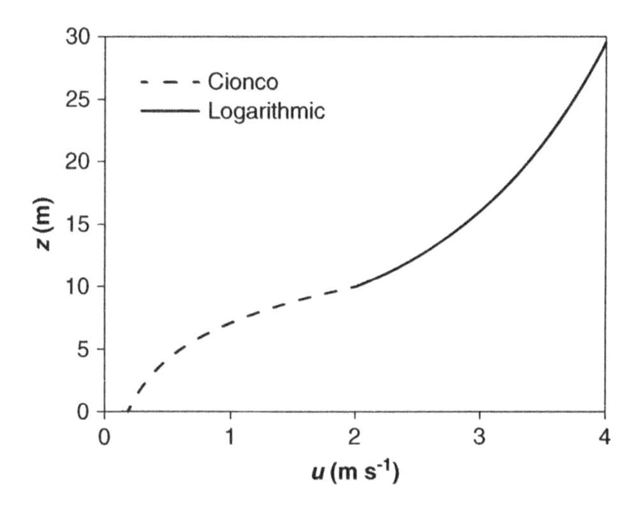


Figure 8.5 Wind speed profile calculated with Cionco's method ($u_*$ = 0.5 m s$^{-1}$, $z_0$ = 1 m, $d$ = 5 m, $\lambda_f$ = 0.25, $a$ = 2.4).

Based on eq. (8.17), an estimate of the characteristic wind speed $u_c$ can be made by simply averaging from $z = 0$ to $z = H_r$. The result is

$$u_c = u_{H_r} \frac{1 - \exp(-a)}{a} \tag{8.19}$$

The equations presented here will not be valid very close to the surface, where a no-slip condition can be expected to hold (i.e., $u = 0$ at $z = 0$), and where viscous forces become dominant. This is the quasi-laminar layer discussed in Section 7.4.1 (dry deposition). A complete theory of the quasi-laminar layer should explain both the quasi-laminar resistance $r_b$ and the wind speed profile near the surface. In Appendix C a wind speed profile is derived that follows the no-slip condition.

### 8.4.4 Junction Methods

Junction methods force both the wind speed and its first derivative with respect to height to be continuous functions in the junction between the logarithmic function and the exponential function. This ensures that the wind speed profile is consistent with a continuous momentum diffusivity profile.

Shinn (1971) proposed a simple junction method for complex forest canopies by using the vertical gradient of $u$ at $z = H_r$ to calculate $a$:

$$\frac{d}{dz}\left\{u_{H_r} \exp\left[-a\left(1 - \frac{z}{H_r}\right)\right]\right\}_{z=H_r} = \frac{d}{dz}\left[\frac{u_*}{k}\ln\left(\frac{z-d}{z_0}\right)\right]_{z=H_r} \tag{8.20}$$

which leads to

$$\frac{u_{H_r} a}{H_r} = \frac{u_*}{k(H_r - d)} \tag{8.21}$$

Solving for $a$ leads to

$$a = \frac{u_*}{k u_{H_r}(1 - d / H_r)} \tag{8.22}$$

This method may overestimate $a$ when $\lambda_f$ is large. An alternative method is proposed here.

So far it was assumed that the junction between the logarithmic wind speed profile and the exponential wind speed profile should be made at the average obstacle height $H_r$. However, allowing the junction to take place at a different height provides an extra degree of freedom, which can be used to ensure that the slope of the two wind speed profiles are the same at the junction. To this effect, eq. (8.17) is rewritten as

$$u = \alpha \exp(\beta z) \tag{8.23}$$

where

$$\beta = \frac{9.6\lambda_f}{H_r} \tag{8.24}$$

and $\alpha$ is determined by assuming that there is a height $z_j$ (junction height) where the prediction of $u$ by eq. (8.23) equals the prediction of $u$ by eq. (5.66), *and* the prediction of the wind speed gradient by eq. (8.23) equals the prediction by eq. (5.66). This leads to the following two equations:

$$u = \frac{u_*}{k}\ln\left(\frac{z_j - d}{z_0}\right) = \alpha\exp(\beta z_j) \tag{8.25}$$

$$\frac{du}{dz} = \frac{u_*}{k(z_j - d)} = \alpha\beta\exp(\beta z_j) \tag{8.26}$$

and $\alpha$ can easily be calculated when $z_j$ is known, for example, by solving eq. (8.25):

$$\alpha = \frac{u_* \ln\left[(z_j - d)/z_0\right]}{k\exp(\beta z_j)} \tag{8.27}$$

Finding $z_j$ is less straightforward. Dividing eq. (8.25) by eq. (8.26) leads to the following implicit equation:

$$(z_j - d)\ln\left(\frac{z_j - d}{z_0}\right) = \frac{1}{\beta} \tag{8.28a}$$

Equation (8.28a) cannot be solved analytically and needs to be solved iteratively. Dividing by $z_0$, it is clear that this equation has only two variables, $\beta z_0$ and $(z_j - d)/z_0$:

$$\frac{z_j - d}{z_0}\ln\left(\frac{z_j - d}{z_0}\right) = \frac{1}{\beta z_0} \tag{8.28b}$$

This relationship is plotted in Figure 8.6. It can be approximated to within 0.0001 for $\beta z_0$ ranging from 0.01 to 1 with the following equation:

$$\frac{z_j - d}{z_0} = -0.311 + 0.8642644(\beta z_0)^{-0.485} + 0.35127972(\beta z_0)^{-0.891} + 0.858735(\beta z_0)^{0.110} \tag{8.28c}$$

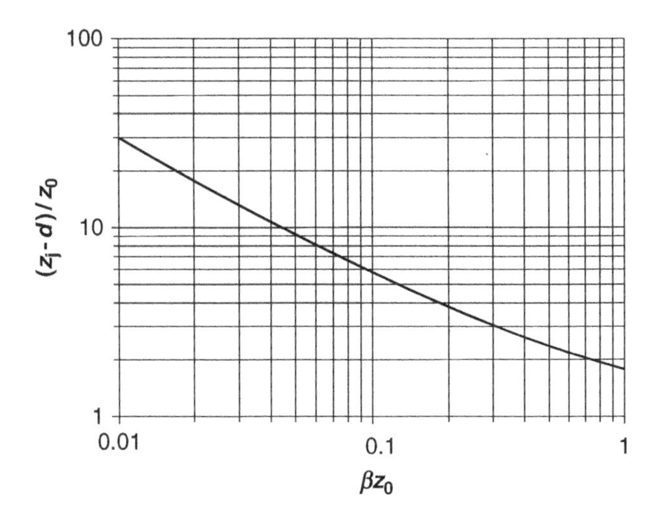


Figure 8.6 Diagram for the calculation of $z_j$.

and $z_j$ is calculated as

$$z_j = z_0 \cdot \frac{z_j - d}{z_0} + d \tag{8.29a}$$

Once $z_j$ is known, $\alpha$ can be calculated with eq. (8.27). The two junction methods are shown in Figure 8.7.

Finally, this junction method also allows to calculate a characteristic wind speed $u_c$ by taking the average of the wind speed from $z = 0$ to $z = H_r$ (assuming $H_r < z_j$):

$$u_c = \frac{\alpha}{H_r \beta}[\exp(H_r \beta) - 1] \tag{8.29b}$$

This equation should be used with care because $z_j$ is not always greater than $H_r$.

### 8.4.5 Other Canopy Wind Speed Descriptions

As indicated before, the most commonly used wind speed profile in canopies is the exponential velocity profile derived by Inoue (1963) and Cionco (1965, 1978). It is based on the assumption that the momentum diffusivity $K_m$ in the canopy is constant. Several other models have been proposed. Most canopy wind speed models were meant to describe plant canopies, but the principles employed are often transferrable to urban terrain. For that reason, the specific application of different models will not be considered here.

Cowan (1968) made the assumption that the momentum diffusivity is proportional to the local wind speed in the canopy and derived the following wind speed profile:

$$u = u_{H_r} \sqrt{\frac{\sinh(\beta z / H_r)}{\sinh(\beta)}} \tag{8.30}$$

where $\beta$ is given by

$$\beta = \frac{2H_r}{(H_r - d)\ln[(H_r - d)/z_0]} \tag{8.31}$$

Cowan (1968) also explored the repercussions of this formulation on mass transfer and its resistance.

A variant of the model of Cionco (1965), also assuming constant $K_m$, leads to (Thom, 1971; Landsberg and James, 1971)

$$u = \frac{u_{H_r}}{[1 + \alpha(1 - z/H_r)]^2} \tag{8.32}$$

where $\alpha$ is a model parameter.

Wang (2012) developed a model for the wind speed profile in canopies that includes the effect of shear on the surface and satisfies the no-slip condition. The resulting wind speed profile contains Bessel functions, making the calculation more onerous. Nonanalytical wind speed models were developed by Coceal and Belcher (2004) and Di Sabatino et al. (2008). These models will not be discussed here.

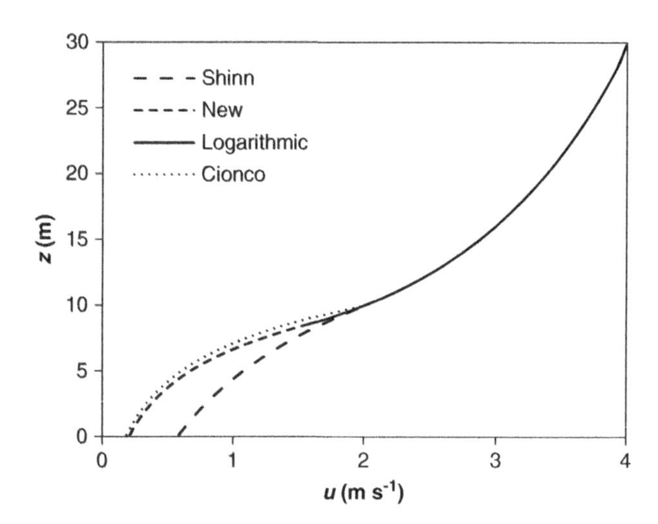


Figure 8.7 Wind speed profile calculated with two junction methods, compared with the Cionco method ($u_* = 0.5$ m s$^{-1}$, $z_0 = 1$ m, $d = 5$ m, $\lambda_f = 0.25$).

In Appendix C a canopy wind speed model is developed that has the main features of similar recent models, including the no-slip condition, but leads to a relatively simple analytical equation. The key assumption is that the height derivative of the regular momentum diffusivity equation is valid within and above the canopy:

$$\frac{dK_{\mathrm{m}}}{dz} = ku_* \tag{C78}$$

Above the canopy, $u*$, the friction velocity, is constant and considered known. Between the canopy elements, a force balance leads to an exponentially decreasing local friction velocity $u_\tau$:

$$u_\tau = u_* \exp\left[-\frac{\gamma\lambda_{\mathrm{f}}}{2}\left(1-\frac{z}{H_{\mathrm{r}}}\right)\right] \tag{C76}$$

Based on these assumptions, and on the following momentum equation:

$$K_{\mathrm{m}}\frac{\partial u}{\partial z} = \frac{\tau_{xz}}{\rho} = u_\tau^2 \tag{C84}$$

the following wind speed profile is derived for $0 < z < H_{\mathrm{r}}$:

$$u = \frac{u_*}{k}\exp\left(-\frac{\gamma\lambda_{\mathrm{f}}}{2}\right)\left[\exp\left(\frac{\gamma\lambda_{\mathrm{f}}}{2}\frac{z}{H_{\mathrm{r}}}\right)-1 + (1-\delta)\ln\left\{\frac{\exp[(\gamma\lambda_{\mathrm{f}}/2)(z/H_{\mathrm{r}})]-(1-\delta)}{\delta}\right\}\right] \tag{C91}$$

where $\gamma$ is a model parameter, estimated at 20, and $\delta$ is given by

$$\delta = \frac{\nu\gamma\lambda_{\mathrm{f}}\exp(\gamma\lambda_{\mathrm{f}}/2)}{2H_{\mathrm{r}}ku_*} \tag{C88}$$

One of the advantages of this model is that it also predicts the displacement height $d$:

$$\frac{d}{H_{\mathrm{r}}} = 1-\frac{2}{\gamma\lambda_{\mathrm{f}}}\left[1-\exp\left(-\frac{\gamma\lambda_{\mathrm{f}}}{2}\right)\right] \tag{C83}$$

The value of $z_0$ is obtained by comparing the model prediction with the logarithmic wind speed profile above the canopy. Equation (C78) ensures that this model has the properties of a junction model as well. It predicts wind speed profiles that are approximately exponential between 0.2 $H_{\mathrm{r}}$ and $H_{\mathrm{r}}$, with wind speed attenuation coefficients that correspond well with the values found by Macdonald (2000). Further details of the model are given in Appendix C.

**Example 8.3.** Chicago has an average obstacle height of about 8 m and an average $\lambda$ of about 0.25 (Grimmond et al., 1998). Estimate the average wind speed 5 m above the ground, $u_5$, using various methods when the wind speed at 20 m above the ground is 4 m s$^{-1}$ at neutral conditions.

***Solution.*** Several methods require an estimate of $z_0$ and $d$. Because we know $\lambda_r$, we use eqs. (8.6) and (8.9), leading to

$$z_0 = 0.1 \times 8\,\mathrm{m} = 0.8\,\mathrm{m}$$
$$d = 2 \times 0.25 \times 8\,\mathrm{m} = 4\,\mathrm{m}$$

This allows us to calculate the friction velocity:

$$u_* = u_{20} \times k / \ln[(20-d)/z_0] = 4 \times 0.4 / \ln[(20-4)/0.8] = 0.5341\,\mathrm{m\,s^{-1}}$$

The simplest estimate of $u_5$ is the characteristic velocity, $u_c$. Hence, eq. (8.15) is used:

$$u_5 \approx u_c = 0.5341 \times (2/0.25)^{0.5} = \mathbf{1.511\ m\,s^{-1}}$$

A more sophisticated estimate of $u_5$ is based on the exponential wind speed profile. The wind speed attenuation factor is calculated with eq. (8.18):

$$a = 9.6 \times 0.25 = 2.4$$

To use the exponential wind speed profile, we need the wind speed at height $H_r$, based on the logarithmic wind speed profile:

$$u_{H_r} = (0.5341/0.4)\ \ln[(8-4)/0.8] = 2.149\ \mathrm{m\ s^{-1}}$$

The wind speed at $z = 5$ m is calculated with eq. (8.17):

$$u_5 = 2.149\ \exp\left[-2.4 \times \left(1 - \frac{5}{8}\right)\right] = \mathbf{0.874\,m\,s^{-1}}$$

It was argued in Section 8.4.3 that eq. (8.18) overestimates $a$ by 40%. In that case, $a = 1.44$, and the following value of $u_5$ is obtained:

$$u_5 = 2.149\ \exp\left[-1.44 \times \left(1 - \frac{5}{8}\right)\right] = \mathbf{1.252\,m\,s^{-1}}$$

We calculate the wind speed with two different junction methods: first the method of Shinn (1971). To that effect, eq. (8.22) is used to calculate $a$:

$$a = 0.5341 \Big/ \left[0.4 \times 2.149 \times \left(1 - \frac{4}{8}\right)\right] = 1.243$$

Equation (8.17) is calculated with the new value of $a$:

$$u_5 = 2.149 \exp\left[-1.243 \times \left(1 - \frac{5}{8}\right)\right] = \mathbf{1.348\,m\,s^{-1}}$$

For the second junction method, we need to determine the height of the junction. To that effect we need $\beta$ [eq. (8.24)]:

$$\beta = 9.6 \times 0.25 / 8 = 0.3\,\mathrm{m}^{-1}$$

Hence, $\beta z_0 = 0.3 \times 0.8 = 0.24$. Equation (8.28c) is used to estimate $(z_j - d)/z_0$:

$$z_0 / z_j - d = -0.311 + 0.8642644 \times 0.24^{-0.485} + 0.35127972 \times 0.24^{-0.891}$$
$$+ 0.858735 \times 0.24^{0.110} = 3.4026$$

Hence:

$$z_j = 0.8 \times 3.4026 + 4 = 6.722\,\mathrm{m}$$

The wind speed at the junction point is calculated with the logarithmic equation:

$$u(z_j) = (0.5341 / 0.4)\ \ln[(6.722 - 4) / 0.8] = 1.635\,\mathrm{m\,s^{-1}}$$

and $\alpha$ can now be calculated with eq. (8.27):

$$\alpha = 0.5341\ \ln(3.4026) / [0.4 \exp(0.24)] = 0.2176\,\mathrm{m\,s^{-1}}$$

Hence, $u_5$ is calculated with eq. (8.23):

$$u_5 = 0.2176\ \exp(0.3 \times 5) = \mathbf{0.9754\,m\,s^{-1}}$$

Repeating the calculation procedure with a 40% lower value of $a$ leads to $u_5$ = **1.258 m s$^{-1}$**.

We will calculate wind speeds with two other canopy models: the Cowan (1968) model and the model of Appendix C. The Cowan model requires parameter $\beta$ as defined by eq. (8.31):

$$\beta = 2 \times 8 / \{(8 - 4)\ \ln[(8 - 4) / 0.8]\} = 2.485$$

Hence, $u_5$ is calculated with eq. (8.30):

$$u_5 = 2.149 \times [\sinh(0.24 \times 5/8) / \sinh(0.24)]^{0.5} = \mathbf{1.323\,m\,s^{-1}}$$

The model in Appendix C requires the variables $\gamma$ (20 is assumed), $\nu$ ($1.5 \times 10^{-5}$ m$^2$ s$^{-1}$), and $\delta$, calculated with eq. (C88):

$$\delta = 1.5\times10^{-5}\times20\times0.25\times\exp(20\times0.25/2)/(2\times8\times0.4\times0.5341) = 2.673\times10^{-4}$$

Hence, $u_5$ is calculated with eq. (C91):

$$u_5 = \left(\frac{0.5341}{0.4}\right)\exp\left(\frac{-20\times0.25}{2}\right)\left\{\exp\left[\left(\frac{20\times0.25}{2}\right)\times\left(\frac{5}{8}\right)\right]-1+\left(1-2.673\times10^{-4}\right)\right.$$
$$\left.\times\ln\left(\frac{\exp\left[\left(\frac{20\times0.25}{2}\right)\times\left(\frac{5}{8}\right)\right]-(1-2.673\times10^{-4})}{2.673\times10^{-4}}\right)\right\} = \mathbf{1.460\,m\,s^{-1}}$$

The difference between the estimates is substantial. The results that are based on actual urban canopies are expected to be more accurate than those based on wind tunnel studies and plant canopies. Unfortunately, this does not leave us with a sufficient basis to make a reliable choice between the above results. However, the calculations converge to a much smaller spread when it is assumed that $a$ is overestimated by 40% in eq. (8.18).

The calculations of this example are included in the file "Example 8.3. Wind speeds in urban canopy.xlsx" on the enclosed CD.

## 8.5 TURBULENCE IN URBAN TERRAIN

Two aspects of turbulence will be discussed in this section: the **turbulence velocity**, and the **Lagrangian integral time**. For both variables, a distinction is made between the turbulence *between* the obstacles and the turbulence *above* the obstacles.

*Above* obstacles in urban terrain, turbulence velocities are essentially the same as above obstacles in other terrains and were discussed in Chapter 5. In neutral and stable conditions, the equations are

$$\frac{\sigma_u}{u_*} = 2.5\left(1-\frac{z}{h_{\text{mix}}}\right)^a \tag{5.144}$$

$$\frac{\sigma_v}{u_*} = 1.9\left(1-\frac{z}{h_{\text{mix}}}\right)^a \tag{5.145}$$

$$\frac{\sigma_w}{u_*} = 1.3\left(1-\frac{z}{h_{\text{mix}}}\right)^a \tag{5.146}$$

where $a$ has a value between 0.5 and 1. Choosing the midpoint value of $a = 0.75$ provides results that are fairly consistent with all the relevant data. Near the surface, $1 - z/h_{\text{mix}} \approx 1$. In unstable conditions, a component due to convective cycling should be added to the wind speed variances (see Chapter 5).

*Between* the obstacles, the turbulent velocities are lower. Measured values vary widely, but Hanna and Britter (2002) recommend the following averages:

$$\frac{\sigma_u}{u_*} = 1.6 \tag{8.33}$$

$$\frac{\sigma_v}{u_*} = 1.4 \tag{8.34}$$

$$\frac{\sigma_w}{u_*} = 1.1 \tag{8.35}$$

The reason for the lower turbulent velocity is the absence of horizontal eddies larger than the spacing between the buildings. Based on experience in the surface layer above the buildings, we can expect these equations to be valid in neutral and stable conditions. In unstable conditions, the turbulent velocity is also determined by the convective velocity scale $w_*$. However, this metric relates to the properties of large convective cycles, which do not fit between the buildings of an urban terrain. Hence, turbulence due to convective cycling can be expected to be greatly attenuated. Hence, it seems reasonable to apply eqs. (8.33)–(8.35) under convective conditions as well.

For the Lagrangian integral time, Hanna and Britter (2002) suggested the following values *above the mean obstacle height*:

$$T_{\mathrm{i,L}} = 1000 \text{ s} \qquad (x \text{ and } y \text{ direction}) \tag{8.36}$$

$$T_{\mathrm{i,L}} = 60 \text{ s} \qquad (z \text{ direction; stable conditions}) \tag{8.37}$$

$$T_{\mathrm{i,L}} = 300 \text{ s} \qquad (z \text{ direction; neutral conditions}) \tag{8.38}$$

$$T_{\mathrm{i,L}} = \infty \qquad (z \text{ direction; unstable conditions}) \tag{8.39}$$

These are typical values based on the Briggs plume dispersion parameterization but do not incorporate the fact that $T_{\mathrm{i,L}}$ tends to decrease with increasing wind speed. Hence, the following alternatives are proposed:

$$T_{\mathrm{i,L}} = \frac{1250 \text{ m}}{u} \qquad (x \text{ and } y \text{ direction}) \tag{8.40}$$

$$T_{\mathrm{i,L}} = \frac{333 \text{ m}}{u} \qquad (z \text{ direction; stable conditions}) \tag{8.41}$$

$$T_{\mathrm{i,L}} = \frac{1667 \text{ m}}{u} \qquad (z \text{ direction; neutral conditions}) \tag{8.42}$$

where the unit meter (m) is included in the balance the units. These equations should lead to fairly accurate results, but they overestimate $T_{\mathrm{i,L}}$ near the ground. Equation (6.191), which calculates $T_{\mathrm{i,L}}$ based on the energy dissipation rate $\varepsilon$,

$$T_{\mathrm{i,L},z} = \frac{2\sigma_w^2}{C_0 \varepsilon} \tag{6.191}$$

is probably the most reliable procedure; and $\varepsilon$ is calculated from eqs. (6.185)–(6.188).

*Between the obstacles*, the Lagrangian integral time depends on how the obstacles affect the wind. In the lateral ($y$) direction, the Lagrangian integral time is limited by the time needed to travel between adjacent buildings at the local wind speed. When no information about building spacing is available, the average obstacle height $H_r$ can be used as a proxy for building spacing. Hence, the following estimate is obtained:

$$T_{i,L} = \frac{H_r}{u} \qquad (y \text{ direction}) \tag{8.43}$$

Wide, flat buildings can increase lateral dispersion because the wind is forced to move around the buildings. In that case, the following equation can be used:

$$T_{i,L} = \frac{W}{u} \qquad (y \text{ direction}) \tag{8.44}$$

where $W$ is the building width. If an area is sparsely built, $H_r$ becomes a poor proxy for building spacing $S$, and it is recommended to use the following equation instead:

$$T_{i,L} = \frac{S}{u} \qquad (y \text{ direction}) \tag{8.45}$$

Hanna and Britter (2002) do not provide an estimate of the Lagrangian integral time in the $z$ direction. Here an estimate will be based on the turbulent mass diffusivity and the turbulence velocity. In Section 6.6.7, the following relationship was developed (Pasquill, 1974):

$$T_{i,L} = \frac{K_z}{\sigma_w^2} \tag{8.46}$$

Equation (8.46) connects the three main characteristics of turbulence. It can also be used above the average obstacle height, in any direction provided that the appropriate $K$ value is known. Above obstacles, eq. (5.198) can be used. Appendix C provides an option for calculating $K$ between obstacles. This would be an alternative way to include an inverse wind speed dependence into eqs. (8.36)–(8.39), as both $K_z$ and $\sigma_w$ tend to be proportional to wind speed.

A value of $T_{i,L}$ for the $x$ direction between buildings will not be provided because it is not needed for plume models, and because the usual formulation of dispersion parameters underestimates $\sigma_x$ between buildings. This is because pollutants trapped in wakes have a disproportionate effect on $\sigma_x$ compared to the other dispersion parameters.

## 8.6 DISPERSION CALCULATIONS IN URBAN TERRAIN NEAR THE SURFACE

### 8.6.1 Introduction

A plume that is emitted between the roughness defining obstacles of an urban or industrial terrain behaves somewhat differently than a plume that is emitted above the roughness defining obstacles.

First of all, the **average plume height** $h_a$, defined as the average height of all pollutant particles in the plume at a given distance from the source, is not the same as the **effective source height** $h$. This is the result of plume reflection, creating asymmetry in the plume. This is obvious when $h = 0$: $h_a$ can never be 0 when $\sigma_z > 0$, so $h_a$ differs from $h$. Hence, as a plume grows, its average height increases, even in the absence of plume rise.

Second, the wind speed profile is very pronounced near the surface. Hence, a plume emitted between the roughness elements will move slowly first and gather speed as it grows and is mixed into faster moving air layers. This mixing into faster moving air layers causes **dilution** of the plume.

Third, while the plume resides between the buildings, it finds itself confined between the buildings, limiting the turbulence by eliminating the largest turbulent eddies. As the plume grows above the buidlings, it is exposed to the full range of turbulent scales. Hence, there will be a drastic change in the growth of the dispersion parameters once the plume starts rising above the buildings. Hanna and Britter (2002) describe this change with a virtual source concept. Here the main focus will be on a formulation in the form of differential equations.

### 8.6.2 Gaussian Model Formulation near the Surface

The ground-level concentration of a pollutant in a plume is calculated with the regular Gaussian equation. To that effect, eq. (6.7) is applied at $z = 0$:

$$c = \frac{Q}{\pi u \sigma_y \sigma_z} \exp\left(-\frac{1}{2}\frac{y^2}{\sigma_y^2}\right) \exp\left(-\frac{1}{2}\frac{h^2}{\sigma_z^2}\right) \tag{8.47}$$

where the symbols have their usual meaning. However, $u$ is not evaluated at the effective source height $h$, but at the average plume height $h_a$.

The average plume height is defined as the average height of pollutant particles in a cross section. The average is weighted using the wind speed as weight. In mathematical terms:

$$h_a = \frac{\int_0^\infty zuc\, dz}{\int_0^\infty uc\, dz} \tag{8.48}$$

High above the ground, where the wind speed $u$ is approximately independent of height, and the plume is approximately symmetrical, this equation leads to $h_a = h$. However, close to the ground $h_a > h$. The extreme case is $h = 0$ and eq. (6.7) simplifies to

$$c = \frac{Q}{\pi u \sigma_y \sigma_z} \exp\left(-\frac{1}{2}\frac{y^2}{\sigma_y^2}\right) \exp\left(-\frac{1}{2}\frac{z^2}{\sigma_z^2}\right) \tag{8.49}$$

This equation is substituted into eq. (8.48), along with a wind speed profile. Depending on the wind speed profile, eq. (8.48) leads to $h_a$ values ranging from $(2/\pi)^{1/2}$ $\sigma_z$ to $(\pi/2)^{1/2}$ $\sigma_z$ (approximately $0.8\sigma_z$ to $1.25\sigma_z$). The former value corresponds with a height-independent wind speed and is the definition of average plume height usually found (see Chapter 7). Derivations of average plume heights are given in Section A8.1 (Appendix A); and $h_a = \sigma_z$ is a reasonable assumption when $h = 0$. Hence, as a rule of thumb, the following equation can be used for the average plume height:

$$h_a = \max(h, \sigma_z) \tag{8.50}$$

Hence, as a plume grows, $h_a$ increases, and the value of $u$ in eq. (8.47) increases, leading to lower values of $c$. This describes the dilution effect of a pollutant mixed into faster moving air layers. What is needed now is a method of calculating dispersion parameters under these changing conditions. This is discussed in the next section.

### 8.6.3 Near Surface Dispersion Parameter Calculation Schemes

Hanna and Britter (2002) recommended the following formalism for the calculation of dispersion parameters [see eqs. (6.27) and (6.28a)]:

$$\sigma_{x,y,z} = \sigma_{u,v,w} t f_{x,y,z} \tag{8.51}$$

where [see eq. (6.68)]:

$$f_{x,y,z} = \frac{1}{(1 + t/2T_{i,L})^{0.5}} \tag{8.52}$$

Combining eqs. (8.51) and (8.52) leads to

$$\sigma_{x,y,z} = \frac{\sigma_{u,v,w} t}{(1 + t/2T_{i,L})^{0.5}} \tag{8.53}$$

Equation (8.53) can be used between obstacles defining the surface roughness, and above, as long as appropriate parameters are used.

Hanna and Britter (2002) proposed an alternative approach for $\sigma_z$ between the roughness elements, based on experimental data as well as a theoretical result that a $K_z$ value increasing proportionally with height leads to linear plume growth. In practice, the $K_z$ value does not increase proportionally with height, as argued in Appendix C.

However, thanks to the small values of $t$ in eq. (8.53), linear plume growth between obstacles is a reasonable assumption.

As indicated above, $\sigma_x$ does not follow eq. (8.53) because pollutants trapped in building wakes cause pronounced plume stretching in the $x$ direction. Hanna and Britter (2002) proposed the following equation for $\sigma_x$ between buildings based on the work of Hanna and Franzese (2000):

$$\sigma_x = A' \frac{u_*}{u_c} x \tag{8.54}$$

where $A'$ is approximately 3.

We still need a procedure to apply eq. (8.53) under changing conditions without creating discontinuities in the plume size. Hanna and Britter (2002) propose to use the virtual source concept (see Section 6.8 for details and an example) to handle the transition from between-building plumes to above-building plumes. However, this approach is only approximate because eq. (8.53) depends on the wind speed through variable $t$ in the numerator. This problem can be solved between the roughness obstacles by using the characteristic wind speed $u_c$, but there is no characteristic wind speed for use above buildings. Hence, an alternative approach based on differential equations is proposed here.

First, we distinguish **real time** $t$ from **apparent time** $t_a$. Real time is the actual time elapsed since the emission of a plume segment. The apparent time is the time needed to grow a plume segment to its current size under current (local) conditions. Real time equals apparent time when the wind speed is constant, and the conditions are constant in time and space. Under changing conditions, eq. (8.53) is rewritten as (considering the $y$ direction as example):

$$\sigma_y = \frac{\sigma_v t_a}{(1 + t_a / 2T_{i,L})^{0.5}} \tag{8.55}$$

The apparent time $t_a$ is calculated by solving eq. (8.55) for $t_a$:

$$t_a = \frac{\sigma_y^2}{4T_{i,L}\sigma_v^2}\left(1 + \sqrt{1 + \frac{16T_{i,L}^2\sigma_v^2}{\sigma_y^2}}\right) \tag{8.56}$$

Equation (8.56) leads to a singularity at $\sigma_y = 0$. However, in this case $t_a = 0$. The differential equation for $\sigma_y$ is obtained by deriving eq. (8.55) with respect to time:

$$\frac{d\sigma_y}{dt} = \frac{d\sigma_y}{dt_a} = \frac{\sigma_v}{(1 + t_a / 2T_{i,L})^{0.5}} - \frac{\sigma_v t_a}{4T_{i,L}(1 + t_a / 2T_{i,L})^{1.5}} \tag{8.57}$$

Substituting eq. (8.55) back into eq. (8.57) leads to

$$\frac{d\sigma_y}{dt} = \frac{\sigma_y}{t_a} - \frac{\sigma_y^3}{4T_{i,L}t_a^2\sigma_v^2} \tag{8.58}$$

Again, a singularity is found when $\sigma_x = 0$. In that case, the following equation holds:

$$\left(\frac{d\sigma_y}{dt}\right)_{t_a=0} = \sigma_v \tag{8.59}$$

Equations equivalent to eqs. (8.56) and (8.58) apply to $\sigma_z$:

$$t_a = \frac{\sigma_z^2}{4T_{i,L}\sigma_w^2}\left(1+\sqrt{1+\frac{16T_{i,L}^2\sigma_w^2}{\sigma_z^2}}\right) \tag{8.60}$$

$$\frac{d\sigma_z}{dt} = \frac{\sigma_z}{t_a} - \frac{\sigma_z^3}{4T_{i,L}t_a^2\sigma_w^2} \tag{8.61}$$

Equations (8.56) and (8.60) provide different values of $t_a$. It is important that the correct values of $t_a$ are used in eqs. (8.58) and (8.61).

We usually want the plume size as a function of location ($x$), not as a function of $t$. Hence, the following equation is added:

$$\frac{dx}{dt} = u \tag{8.62}$$

where $u$ is evaluated at the average plume height $h_a$. Equations (8.58), (8.61), and (8.62) are solved simultaneously with a numerical procedure.

## 8.7 AN EXAMPLE

**Example 8.4.** A truck transporting benzene crashes, creating a crack releasing about 10 liters of liquid benzene (density 879 g $L^{-1}$; molar mass 78 g $mol^{-1}$) onto the street pavement per hour.

The area is a fairly scattered ($\lambda_f = 0.05$) urban–suburban zone with an average building height of 12 m. The weather is overcast. Wind speed at a 30-m height is 6 m $s^{-1}$. Temperature is 12 °C; barometric pressure (sea level) is 101 kPa; altitude is 1000 m.

Assume that the released benzene vaporizes immediately and spreads without buoyancy effects. Calculate the benzene concentration on the plume centerline versus distance from the accident up to 250 m.

***Solution.*** This problem will be solved with a virtual source concept and with the differential equation approach. First, the plume parameters will be calculated.

Estimation of the terrain parameters:

$$z_0 = \lambda_f H_r = 0.05 \times 12\,\text{m} = 0.6\,\text{m} \qquad [\text{eq.}(8.5); \lambda_f < 0.1]$$

$$d = 2\lambda_f H_r = 2 \times 0.05 \times 12\,\text{m} = 1.2\,\text{m} \qquad [\text{eq.}(8.9); \lambda_f < 0.25]$$

Estimation of the wind characteristics: When the weather is overcast, the atmosphere is neutral, and no adjustments for stability are needed. The logarithmic wind speed equation [eq. (5.66)] for neutral conditions is solved for $u_*$ to calculate the friction velocity:

$$u_* = \frac{ku}{\ln[(z-d)/z_0]} \tag{8.63}$$

In this case: $u_* = 0.4 \times 6/\ln(28.8/0.6) = 0.620$ m s$^{-1}$.

We will use the junction method for the wind speed profile. This requires $\alpha$ and $\beta$ for use in eq. (8.23). $\beta$ is calculated with eq. (8.24): $\beta = 9.6 \times 0.05/12$ m $= 0.04$ m$^{-1}$.

For $\alpha$ we need the junction height $z_j$. Based on eq. (8.28c), we find

$$\frac{z_j - d}{z_0} = 15.28$$

From this the value of $z_j$ is calculated with eq. (8.29a): $z_j = 0.6$ m $\times 15.28 + 1.2 = 10.37$ m. Then $\alpha$ is calculated with eq. (8.27): $\alpha = 0.620$ m s$^{-1} \times \ln(15.28)/[0.4 \times \exp(0.04 \times 10.37)] = 2.79$ m s$^{-1}$.

For simplified calculations the characteristic wind speed $u_c$ can be useful. It is calculated with eq. (8.29b): $u_c = 2.79$ m s$^{-1} \times [\exp(12 \times 0.04) - 1]/(12 \times 0.04) = 3.58$ m s$^{-1}$.

Note that, even in approximate calculations, it is not recommended to use $u_c$ in the calculation of the ground-level concentration, only for calculating $\sigma_y$ and $\sigma_z$ values.

The turbulent velocities are calculated from the friction velocity, with eqs. (8.34) and (8.35) between the buildings, and with eqs. (5.145) and (5.146) above the buildings, assuming $1 - z/h_{mix} \approx 1$. The results are $\sigma_v = 0.868$ m s$^{-1}$ and $\sigma_w = 0.682$ m s$^{-1}$ between the buildings, and $\sigma_v = 1.178$ m s$^{-1}$ and $\sigma_w = 0.806$ m s$^{-1}$ above the buildings.

Starting with the virtual source approach, eq. (8.53) is written as a function of the downwind distance by the substitution $t = x/u$. The result is

$$\sigma_{y,z} = \frac{\sigma_{v,w} x / u}{[1 + x/(2uT_{i,L})]^{0.5}} \tag{8.64}$$

We want to use the virtual source concept only at the transition from a between-builing plume and an above-building plume. To avoid having to use a virtual source calculation at every increment of $x$, we assume a constant wind speed in eq. (8.64) at $h_a < H_r$ and at $h_a > H_r$. We use $u_c = 3.58$ m s$^{-1}$ at $h_a < H_r$ and $u_{H_r} = 4.48$ m s$^{-1}$ at $h_a > H_r$.

We also need constant values for $T_{i,L}$ in both ranges. In the $y$ direction between buildings, eq. (8.43) is used, leading to $T_{i,L} = 12$ m/3.58 m s$^{-1}$ = 3.35 s. Above the buildings, eq. (8.40) is used in the $y$ direction, leading to $T_{i,L} = 1250$ m/3.58 m s$^{-1}$ = 279 s. In the $z$ direction between the buildings, we use eq. (8.46), which requires a value of $K_z$. We arbitrarily choose the value at $z = H_r$. Based on eq. (5.198), we find $K_z = 0.4 \times 0.620$ m s$^{-1} \times (12$ m $- 1.2$ m$) = 2.68$ m$^2$ s$^{-1}$. Using eq. (8.46), we find $T_{i,L} = 2.68$ m$^2$ s$^{-1}/(0.682$ m s$^{-1})^2 = 5.76$ s. Above the buildings, we use eq. (8.42) with the wind speed at $z = H_r$, leading to $T_{i,L} = 1667$ m/4.48 m s$^{-1}$ = 372 s.

Based on these data, the equations for $\sigma_y$ and $\sigma_z$, based on eq. (8.64), are

$$\sigma_y = \frac{0.242x}{(1+0.0417x)^{0.5}} \quad \text{(between the buildings)}$$

$$\sigma_y = \frac{0.263x}{(1+0.0004x)^{0.5}} \quad \text{(above the buildings)}$$

$$\sigma_z = \frac{0.190x}{(1+0.0242x)^{0.5}} \quad \text{(between the buildings)}$$

$$\sigma_z = \frac{0.180x}{(1+0.0003x)^{0.5}} \quad \text{(above the buildings)}$$

The equations for between the buildings are applicable until $\sigma_z$ reaches the value of $H_r$, that is,

$$\frac{0.190x}{(1+0.0242x)^{0.5}} = H_r$$

Solving for $x$ leads to

$$x = \frac{0.0242H_r^2 + \sqrt{0.0242^2 H_r^4 + 4\times 0.190^2 H_r^2}}{2\cdot 0.190^2} = 127.5\,\text{m}$$

For $x > 127.5$ m, the above-building equations should be used. However, to avoid discontinuities, $x$ must be adjusted. To that effect, the above procedure is repeated with the above-buildings equation for $\sigma_z$:

$$\frac{0.180x}{(1+0.0003x)^{0.5}} = H_r$$

Solving for $x$ leads to

$$x = \frac{0.0003H_r^2 + \sqrt{0.0003^2 H_r^4 + 4\times 0.180^2 H_r^2}}{2\cdot 0.180^2} = 47.2\,\text{m}$$

It follows that the source appears to be 80.3 m closer than it really is. This translates into the following equation for $\sigma_z$:

$$\sigma_z = \frac{0.180(x-80.3)}{[1+0.0003(x-80.3)]^{0.5}}$$

For $\sigma_y$, we first calculate the value at 127.5 m, which is 12.29 m. Repeating the above procedure with the $\sigma_y$ equations, it is found that the following equation applies for $x > 127.5$ m:

$$\sigma_z = \frac{0.263(x-60.1)}{[1+0.0004(x-60.1)]^{0.5}}$$

The dispersion parameters are shown in Figure 8.8. The downwind concentration profile is shown in Figure 8.9.

The differential equation approach allows us to calculate local conditions in every distance increment, based on the local value of $h_a$. In the $y$ direction $T_{i,L}$ is calculated

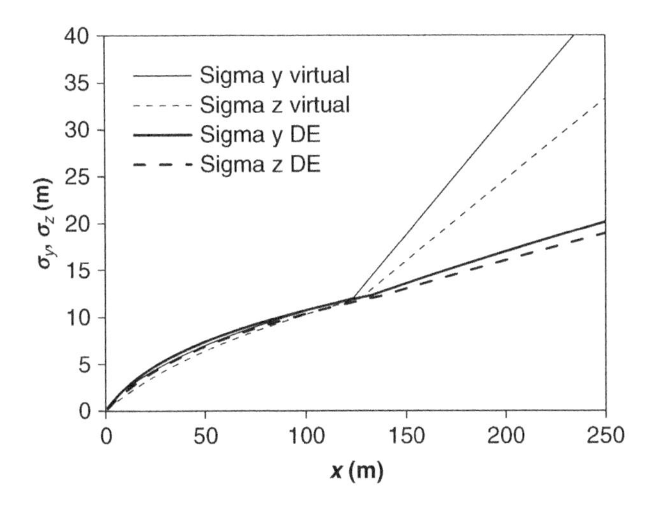


Figure 8.8 $\sigma_y$ and $\sigma_z$ in urban terrain for a plume originating between the roughness elements (virtual = virtual source method; DE = differential equation method).

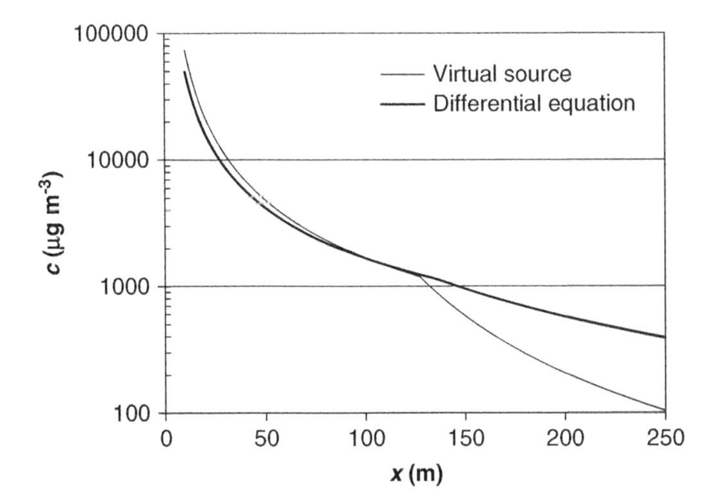


Figure 8.9 Ground-level concentration of the benzene plume versus distance.

from eq. (8.43) between the buildings and from eq. (6.191) above the buildings using a value of the Lagrangian structure function constant $C_0$ in the $y$ direction of 4.3 (Anfossi et al., 2006). For the $z$ direction, $T_{i,L}$ is calculated from eq. (8.46) in all cases. This equation requires knowledge of $K_z$. Above the buildings $K_z$ is calculated from eq. (5.198). Between the buildings, the value obtained with eq. (5.198) with $z = H_r$ is used.

Once the $T_{i,L}$ values are known, the apparent times $t_a$ can be calculated in the $y$ and $z$ directions with eqs. (8.56) and (8.60), respectively. Based on these, the differential calculations are set up with eqs. (8.58) and (8.61). The equations are divided by the wind speed to get differential equations with $x$ as independent variable. For instance, for $\sigma_y$:

$$\frac{d\sigma_y}{dx} = \frac{d\sigma_y}{dt} \cdot \frac{dt}{dx} = \frac{1}{u}\frac{d\sigma_y}{dt} \tag{8.65}$$

where the local wind speed $u$ at height $h_a$ is used.

This numerical integration is implemented in Excel using Euler's rule, and in Matlab with standard Matlab functions (See Appendix B for details of the general procedure).

The result of the calculation is shown in Figure 8.8 for the dispersion parameters. The main difference between the two models is observed above the buildings. The reason is the jump of the $T_{i,L}$ value in the virtual source method, whereas $T_{i,L}$ remains fairly small close to the surface in the differential equation method. Differences close to the source are mainly due to the wind speed profile.

Based on the above information, the ground-level benzene concentration can be calculated with eq. (8.47), assuming $y = 0$ and $z = 0$. The result is shown in Figure 8.9. Both methods provide very similar results. The difference between the two methods is largest at large distances from the source, due to the difference between the dispersion parameters of the two models for plumes above the buildings. As indicated before, the main cause of the difference is the estimate of $T_{i,L}$, which is too large above the buildings when the virtual source method is used.

The solution is given on the enclosed CD, in the folder "Example 8.3. Urban plume dispersion," in the Excel file "Urban Air Dispersion Modeling.xlsx" (virtual source model, and differential equation model; Euler's method), and in the Matlab files "main.m," "f.m," and "data.m" (differential equation model, Runge–Kutta–Fehlberg method).

## 8.8 SUMMARY OF THE MAIN EQUATIONS

Logarithmic wind speed profile, with zero-plane displacement height:

$$u = \frac{u_*}{k}\ln\frac{z-d}{z_0} \tag{5.68}$$

Rule of thumb for $z_0$ and $d$:

$$z_0 = 0.1H_r \tag{8.1}$$

$$d = 0.5H_r \tag{8.2}$$

Frontal area parameter:

$$\lambda_f = \frac{A_f}{A_T} \tag{8.4}$$

Characteristic wind speed between roughness defining obstacles:

$$u_c = u_* \sqrt{\frac{2}{\lambda_f}} \tag{8.15}$$

For $\lambda_f < 0.15$, the following equation can also be used:

$$u_c = u_* \sqrt{\frac{2H_r}{z_0}} \tag{8.16}$$

Exponential wind speed profile:

$$u = u_{H_r} \exp\left[-a\left(1 - \frac{z}{H_r}\right)\right] \tag{8.17}$$

where $a$ is estimated as

$$a = 9.6\lambda_f \tag{8.18}$$

Turbulent velocity between roughness defining obstacles:

$$\frac{\sigma_u}{u_*} = 1.6 \tag{8.33}$$

$$\frac{\sigma_v}{u_*} = 1.4 \tag{8.34}$$

$$\frac{\sigma_w}{u_*} = 1.1 \tag{8.35}$$

Lagrangian integral time scale between obstacles:

$$T_{i,L} = \frac{H_r}{u} \quad (y \text{ direction}) \tag{8.43}$$

$$T_{\mathrm{i,L}} = \frac{K_z}{\sigma_w^2} \quad (8.46)$$

Apparent plume age, $y$ direction:

$$t_{\mathrm{a}} = \frac{\sigma_y^2}{4T_{\mathrm{i,L}}\sigma_v^2}\left(1+\sqrt{1+\frac{16T_{\mathrm{i,L}}^2\sigma_v^2}{\sigma_y^2}}\right) \quad (8.56)$$

Differential equation for $\sigma_y$:

$$\frac{d\sigma_y}{dt} = \frac{\sigma_y}{t_{\mathrm{a}}} - \frac{\sigma_y^3}{4T_{\mathrm{i,L}}t_{\mathrm{a}}^2\sigma_v^2} \quad (8.58)$$

In the $z$ direction:

$$t_{\mathrm{a}} = \frac{\sigma_z^2}{4T_{\mathrm{i,L}}\sigma_w^2}\left(1+\sqrt{1+\frac{16T_{\mathrm{i,L}}^2\sigma_w^2}{\sigma_z^2}}\right) \quad (8.60)$$

$$\frac{d\sigma_z}{dt} = \frac{\sigma_z}{t_{\mathrm{a}}} - \frac{\sigma_z^3}{4T_{\mathrm{i,L}}t_{\mathrm{a}}^2\sigma_w^2} \quad (8.61)$$

# PROBLEMS

1. A plume rises from a source height of 25 m as a buoyancy-dominated plume with buoyancy flux parameter 20 $m^4\ s^{-3}$. The atmosphere is neutral. The wind speed is 3 m $s^{-1}$ at a 10-m height. The average obstacle height is 1 m, with a frontal area parameter of 0.1. Calculate $\sigma_y$ and $\sigma_z$ versus downwind distance using the differential equation formulation, using reasonable assumptions for the input variables. Compare with a continuous dispersion parameterization from Chapter 6.
2. An accidental release of carbon monoxide of 1 kg $s^{-1}$ with an emission height of 5 m and inital spread parameters $\sigma_y = \sigma_z = 1$ m is associated with a lethality at 50 m directly downwind from the source. The average obstacle height is 10 m, and the frontal area parameter is 0.15. Assume a temperature of 20 °C and a pressure of 1 bar. The victim is assumed to breathe air at 1.5 m above the surface. Based on the duration of the exposure, the lethal concentration of CO is assumed to be $300 \pm 100$ ppm. You are appointed as an expert to evaluate if the emission is the likely cause of the death. What is your conclusion?
3. Wind speed measurements are made above a canopy. A friction velocity of 0.52 m $s^{-1}$ is found, a wind speed at a 10-m height of 2.71 m $s^{-1}$, and a wind speed at a 20-m height of 3.91 m $s^{-1}$.

   a. Estimate $z_0$ and $d$.
   b. After reanalysis of the data, a friction velocity of 0.53 m $s^{-1}$ is obtained. Re-estimate $z_0$ and $d$.

4. In an urban canopy with $H_r = 11$ m, $\lambda_f = 0.22$, and a wind speed at $H_r$ of 2.25 m s$^{-1}$, calculate the height where the wind speed is 1.125 m s$^{-1}$, using the main canopy wind speed models. If the wind speed exceeds 1.125 m s$^{-1}$ at ground level, calculate the value of $\lambda_f$ that leads to a wind speed at ground level of 1.125 m s$^{-1}$.
5. Compare eq. (8.15) with eq. (8.16). Under what circumstances or assumptions are the two equations consistent with each other?
6. Calculate for which surface roughness the junction methods predict the same wind speed profile as the regular exponential wind speed profile. Assume $d = 5z_0$ and $\lambda_f = 0.1$.

## MATERIALS ONLINE

- "Example 8.3. Wind speeds in urban canopy.xlsx": Solution to Example 8.3.
- Folder "Example 8.3. Urban plume dispersion." Excel file "Urban Air Dispersion Modeling.xlsx": virtual source solution and differential equation solution (Euler's method) to Example 8.4; Matlab files "main.m," "data.m," and "f.m": differential equation solution to Example 8.4 .

## REFERENCES

Anfossi D., Rizza U., Mangia C., Degrazia G.A., and Pereira Marques Filho E. (2006). Estimation of the ratio between the Lagrangian and Eulerian time scales in an atmospheric boundary layer generated by large eddy simulation. *Atmos. Environ.* **40**, 326–337.

Bentham T. and Britter R. (2003). Spatially averaged flow within obstacle arrays. *Atmos. Environ.* **37**, 2037–2043.

Bottema M. (1997). Urban roughness modelling in relation to pollutant dispersion. *Atmos. Environ.* **31**, 3059–3075.

Cionco R.M. (1965). A mathematical model for air flow in a vegetative canopy. *J. Appl. Meteorol.* **4**, 517–522.

Cionco R.M. (1978). Analysis of canopy index values for various canopy densities. *Bound. Layer Meteorol.* **15**, 81–93.

Coceal O. and Belcher S.E. (2004). A canopy model of mean winds through urban areas. *Q. J. Roy. Meteorol. Soc.* **130**, 1349–1372.

Cowan I.R. (1968). Mass, heat and momentum exchange between stands of plants and their atmospheric environment. *Q. J. Roy. Meteorol. Soc.* **94**, 523–544.

Di Sabatino S., Solazzo E., Paradisi P., and Britter R. (2008). A simple model for spatially-averaged wind profiles within and above an urban canopy. *Bound. Layer Meteorol.* **127**, 131–151.

Grimmond C.S.B. and Oke T.R. (1999). Aerodynamic properties of urban areas derived from analysis of surface form. *J. Appl. Meteorol.* **38**, 1262–1292.

Grimmond C.S.B., King T.S., Roth M., and Oke T.R. (1998). Aerodynamic roughness of urban areas derived from wind observations. *Bound. Layer Meteorol.* **89**, 1–24.

Hanna S.R. and Britter R.E. (2002). *Wind Flow and Vapor Cloud Dispersion at Industrial and Urban Sites*. American Institute of Chemical Engineers, Center for Chemical Process Safety, New York.

Hanna S.R. and Chang J.C. (1992). Boundary layer parameterisations for applied dispersion modelling over urban areas. *Bound. Layer Meteorol.* **58**, 229–259.

Hanna S.R. and Franzese P. (2000). Along wind dispersion—a simple similarity formula compared with observations at 13 field sites and in one wind tunnel. *J. Appl. Meteorol.* **39**, 1700–1714.

Inoue E. (1963). On the turbulent structure of air flow within crop canopies. *J. Meteorol. Soc. Japan* **41**, 317–326.

Landsberg J.J. and James G.B. (1971). Wind profiles in plant canopies: Studies on an analytical model. *J. Appl. Ecol.* **8**, 729–741.

Lettau, H. (1969). Note on aerodynamic roughness parameter estimation on the basis of roughness element description. *J. Appl. Meteorol.* **8**, 828–832.

Li X.X., Liu C.H., Leung D.Y.C., and Lam K.M. (2006). Recent progress in CFD modelling of wind field and pollutant transport in street canyons. *Atmos. Environ.* **40**, 5640–5658.

Macdonald R.W. (2000). Modelling the mean velocity profile in the urban canopy layer. *Bound. Layer Meteorol.* **97**, 25–45.

Macdonald R.W., Griffiths R.F., and Hall D.J. (1998). An improved method for estimation of surface roughness of obstacle arrays. *Atmos. Environ.* **32**, 1857–1864.

Oke T.R. (1988). Street design and urban canopy layer climate. *Energy Build.* **11**, 103–113.

Pasquill F. (1974). *Atmospheric Diffusion*. Wiley, New York.

Raupach M.R. (1994). Simplified expressions for vegetation roughness length and zero-plane displacement as functions of canopy height and area index. *Bound. Layer Meteorol.* **71**, 211–216.

Shinn J.H. (1971). Steady-state two-dimensional flow in forests and the disturbance of surface layer flow by a forest wall. In *R&D Technical Report ECOM-5583*, Atmospheric Sciences Laboratory, White Sands Missile Range, NM.

Thom A.S. (1971). Momentum absorption by vegetation. *Quart. J. Roy. Meteorol. Soc.* **97**, 414–428.

Wang W. (2012). An analytical model for mean wind profiles in sparse canopies. *Bound. Layer Meteorol.* **142**, 383–399.

CHAPTER 9

# STOCHASTIC MODELING APPROACHES

## 9.1 INTRODUCTION

The models discussed in this book so far are **deterministic** models: They do not include any form of randomness other than by defining predictable probability density functions of the wind speed fluctuation, leading to average concentration profiles. In other words, the effect of randomness is described in the model in a nonrandom way. **Stochastic** models are fundamentally different: randomness is built into such models. Of course, computers are deterministic, so the random element in models is only apparent. Apparently random processes in models are also known as **pseudorandom**.

In **stochastic Lagrangian particle** models, the pollutant emission is described by defining a large number of pollutant particles, each representing a small amount of the pollutant. Each particle follows a **random path** around the mean wind speed. To that effect, the trajectory of each particle is divided into a large number of steps, each consisting of a deterministic substep, and a random substep representing a change of the wind speed deviation. Because the wind speed deviation is autocorrelated, the random substeps must be carried out in a way that reproduces this autocorrelation. By following a large number of such particles, it is possible to predict downwind concentrations of the pollutant.

Stochastic Lagrangian particle models are computationally very intensive. However, they have an important advantage over plume and puff models. At large distances from the source, plumes and puffs are too large to be described accurately as a single entity. Different parts of the plume move at different wind speeds and different wind directions and may be subject to different dispersion parameters. It is possible to account for some effects in plume and puff models by plume/puff splitting, stretching, and the like, but this leads to averaging errors, and the result is only approximate. Furthermore, Du (2002) showed using a stochastic Lagrangian particle model that lateral and vertical diffusion are interdependent, something that cannot be modeled within the Gaussian framework. DeVito et al. (2009) showed that real-world puffs are tilted due to the increasing wind speed with height (wind shear), an effect that is not included in Gaussian models.

*Air Dispersion Modeling: Foundations and Applications*, First Edition. Alex De Visscher.

As a result of these approximations, plume models tend to break down at 30–50 km from the source, and puff models at about 200 km from the source. At larger distances, Lagrangian particle models have a clear advantage. Of course, even Lagrangian particle models lose accuracy with increasing distance from the source because the model can only be as accurate as the meteorological model on which it is built. However, it is possible in principle to eliminate averaging errors by defining a sufficient number of pollutant particles.

Examples of stochastic Lagrangian particle models are AUSTAL2000 (Janicke, 2008) and FLEXPART (Stohl et al., 2005).

The objective of this chapter is to outine the basics of stochastic modeling and its application in air dipsersion modeling with stochastic Lagrangian particle models.

## 9.2 FUNDAMENTALS OF STOCHASTIC AIR DISPERSION MODELING

### 9.2.1 Introduction: Properties of the Langevin Equation

Air dispersion is a turbulent version of molecular diffusion and largely follows the same mathematics. Hence, it is no surprise that stochastic air dispersion modeling is largely based on stochastic descriptions of molecular diffusion, which is the theory of Brownian motion.

The first mathematical theory of Brownian motion was developed by Einstein (1905). He combined a theory of osmotic pressure with Stokes' law of drag force on a spherical particle moving in a viscous fluid and derived an equation for the diffusivity of the suspended particle.

A few years later, Langevin (1908) developed a much simpler yet more fundamental theory of Brownian motion that incorporates the inertia of the suspended particle. He showed that Einstein's equation for mean particle displacement versus time is in fact a limiting case for times greater than $10^{-8}$ s, when inertial effects become negligible. Modern stochastic Lagrangian particle theory is a further development of the ideas originally set forth by Langevin. As we will see, the momentum effect that sets apart Langevin's theory from Einstein's theory has its equivalent in the theory of turbulence in the form of autocorrelation, which explains the success of Langevin's train of thought in dispersion modeling. An English translation of Langevin's original paper is published by Lemons and Gythiel (1997).

To fully appreciate the similarities between Brownian motion and the turbulent motion of pollutant particles in a plume, we will first derive some of Langevin's key equations. For simplicity, Langevin (1908) considered only one dimension, with coordinate $x$. Newton's law states:

$$m\frac{d^2x}{dt^2} = F \tag{9.1}$$

where $m$ is the mass of the particle, and $F$ is the force acting on the particle. The force consists of a drag force $F_d$ that attenuates the motion of the particle and a "complementary" (stochastic) force $F_c$ that keeps the Browninan motion going. $F_c$ is a random variable with zero average. In the Stokes regime, $F_d$ is given by

$$F_d = -3\pi\mu d_p \frac{dx}{dt} \tag{9.2}$$

where $\mu$ is the medium viscosity, and $d_p$ is the particle diameter. Hence, eq. (9.1) can be rewritten as

$$m\frac{d^2x}{dt^2} = -3\pi\mu d_p \frac{dx}{dt} + F_c \tag{9.3}$$

In stochastic modeling, a variant of eq. (9.3) is used in direct simulations by defining a series of pseudorandom values for $F_c$. Langevin, however, linked eq. (9.3) to the temperature of the system. First, consider the following mathematical identity:

$$\frac{dx^2}{dt} = 2x\frac{dx}{dt} \tag{9.4}$$

Taking one more derivative leads to

$$\frac{d^2x^2}{dt^2} = 2\frac{dx}{dt}\cdot\frac{dx}{dt} + 2x\frac{d^2x}{dt^2} \tag{9.5}$$

Solving for the second derivative of $x$ leads to

$$\frac{d^2x}{dt^2} = \frac{1}{2x}\frac{d^2x^2}{dt^2} - \frac{1}{x}\left(\frac{dx}{dt}\right)^2 \tag{9.6}$$

Substitution into eq. (9.3) leads to

$$m\left[\frac{1}{2x}\frac{d^2x^2}{dt^2} - \frac{1}{x}\left(\frac{dx}{dt}\right)^2\right] = -3\pi\mu d_p \frac{dx}{dt} + F_c \tag{9.7}$$

Use eq. (9.4) to eliminate $dx/dt$ in the right-hand side of eq. (9.7):

$$\frac{m}{2x}\frac{d^2x^2}{dt^2} - \frac{m}{x}\left(\frac{dx}{dt}\right)^2 = -\frac{3\pi\mu d_p}{2x}\frac{dx^2}{dt} + F_c \tag{9.8}$$

When we consider a large number of particles, and take the average of eq. (9.8) for each particle, then a number of substitutions can be made. First of all, when the average value of $x$ is 0, then the average of $x^2$ is $\sigma_x^2$. Furthermore, the average of $F_c$ is 0 because that was assumed from the onset. Third, $dx/dt$ is the velocity in the $x$ direction, $v_x$. From kinetic gas theory, it is known that the average value of $mv_x^2$ is $kT$, where $k$ is the Boltzmann constant ($1.3806505 \times 10^{-23}\,\mathrm{J\,K^{-1}}$). After multiplying by $x$, these substitutions lead to the following equation:

$$\frac{m}{2}\frac{d^2\sigma_x^2}{dt^2} - kT = -\frac{3\pi\mu d_\mathrm{p}}{2}\frac{d\sigma_x^2}{dt} \tag{9.8}$$

A first integration can be made by substituting the following:

$$z = \frac{d\sigma_x^2}{dt} \tag{9.9}$$

This leads to

$$\frac{m}{2}\frac{dz}{dt} - kT = -\frac{3\pi\mu d_\mathrm{p}}{2} z \tag{9.10}$$

This equation can be rearranged to

$$\frac{dz}{z - 2kT/(3\pi\mu d_\mathrm{p})} = -\frac{3\pi\mu d_\mathrm{p}}{m} dt \tag{9.11}$$

Integration starting from $z=0$ and starting from $t=0$ leads to

$$\ln \frac{z - 2kT/(3\pi\mu d_\mathrm{p})}{2kT/3\pi\mu d_\mathrm{p}} = -\frac{3\pi\mu d_\mathrm{p}}{m} t \tag{9.12}$$

Taking the exponential, and rearranging, leads to

$$z = \frac{2kT}{3\pi\mu d_\mathrm{p}}\left[1 - \exp\left(-\frac{3\pi\mu d_\mathrm{p}}{m} t\right)\right] \tag{9.13}$$

For large values of $t$, the only case considered by Langevin, the exponential can be neglected, and the equation simplifies to

$$z = \frac{2kT}{3\pi\mu d_\mathrm{p}} = \frac{d\sigma_x^2}{dt} \tag{9.14}$$

Integrating from $\sigma_x^2 = 0$ and from $t=0$ leads to

$$\sigma_x^2 = \frac{2kT}{3\pi\mu d_\mathrm{p}} t \tag{9.15}$$

It follows that after a sufficiently long time the $\sigma_x$ of Brownian motion increases proportionally with the square root of the time, just like plumes. At very short times, we can make the following approximation:

$$\exp\left(-\frac{3\pi\mu d_\mathrm{p}}{m} t\right) \approx 1 - \frac{3\pi\mu d_\mathrm{p}}{m} t \tag{9.16}$$

Hence, eq. (9.13) reduces to

$$z = \frac{2kT}{m}t = \frac{d\sigma_x^2}{dt} \tag{9.17}$$

Integrating from $\sigma_x^2 = 0$ and from $t=0$ leads to

$$\sigma_x^2 = \frac{kT}{m}t^2 \tag{9.18}$$

It follows that for sufficiently short times the $\sigma_x$ of Brownian motion increases proportionally with time, just like plumes. The time constant $\tau$ of eq. (9.13) is

$$\tau = \frac{m}{3\pi\mu d_p} \tag{9.19}$$

For typical 1-μm particles, $\tau$ is on the order of $10^{-7}$ s. This can be considered as the Lagrangian integral time of individual particles in Brownian motion in stagnant air.

By integrating eq. (9.13) we can demonstrate the following equation:

$$\sigma_x = \sqrt{at - \frac{a}{b}\left[1 - \exp(-bt)\right]} \tag{9.20}$$

where

$$a = \frac{2kT}{3\pi\mu d_p} \tag{9.21a}$$

$$b = \frac{1}{\tau} \tag{9.21b}$$

Equation (9.20) is mathematically equivalent to eq. (6.95), which describes plume spread in an actual (turbulent) atmosphere based on a statistical argument. Hence, the Lagrangian integral time of large turbulent eddies in the atmosphere is 10 orders of magnitude larger than the diffusion of small particles in stagnant air, but the trends are the same in the two systems. It follows that it will be possible to modify the Langevin equation to suit turbulent dispersion in the atmosphere.

### 9.2.2 Modifying the Langevin Equation for Air Dispersion Modeling: Homogeneous Atmosphere

The main differences between the calculations in the previous section and actual stochastic models are the replacement of molecular properties with turbulent properties, and the use of stochastic numerical methods to solve the equations rather than analytical solutions.

In the following, we will assume that we are considering velocity **fluctuations**, not velocities, and we will consider the crosswind horizontal direction ($y$ direction). Hence, eq. (9.3) can be rewritten as

$$\frac{dv'}{dt} = -\frac{3\pi\mu d_{\mathrm{p}}}{m}v' + \frac{F_{\mathrm{c}}}{m} \tag{9.22}$$

Considering eq. (9.19), this can be written as

$$\frac{dv'}{dt} = -\frac{v'}{\tau} + \frac{F_{\mathrm{c}}}{m} \tag{9.23}$$

and because the time constant in an actual atmosphere is $T_{\mathrm{i,L}}$, this equation can be written as

$$\frac{dv'}{dt} = -\frac{v'}{T_{\mathrm{i,L}}} + \frac{F_{\mathrm{c}}}{m} \tag{9.24}$$

In stochastic modeling, discrete time steps are used, so we need to integrate eq. (9.24). A rigorous argument is given by Legg and Raupach (1982). Here a more intuitive approach will be adopted. The unperturbed differential equation is

$$\frac{dv'}{dt} = -\frac{v'}{T_{\mathrm{i,L}}} \tag{9.25}$$

The solution of this equation is

$$v'(t) = v'(0)\exp\left(-\frac{t}{T_{\mathrm{i,L}}}\right) \tag{9.26}$$

We add an integrated perturbation to this solution:

$$\boxed{v'(t) = v'(0)\exp\left(-\frac{t}{T_{\mathrm{i,L}}}\right) + n(t)} \tag{9.27}$$

Considering the proposed form of the Lagrangian autocorrelation function, eq. (6.43),

$$R_{\mathrm{L}}(\tau) = \exp\left(-\frac{|\tau|}{T_{\mathrm{i,L}}}\right) \tag{6.43}$$

we can write eq. (9.27) as (Smith, 1968)

$$v'(t) = v'(0)R_{\mathrm{L}}(t) + n(t) \tag{9.28}$$

Because the stochastic variable $n(t)$ is a time-integrated form of the perturbation $F_c/m$ in eq. (9.24), the amplitude of $n(t)$ is proportional to $t^{1/2}$. Equation (9.28) is sometimes referred to as an example of a **Markov process**, or as an equation that has the **Markov property**, which means that $v'(t)$ only depends on $v'(0)$, and not on values of $v'$ before that. More information on $n(t)$ can be obtained by taking the square of eq. (9.28) and averaging:

$$\overline{[v'(t)]^2} = \overline{[v'(0)R_L(t)+n(t)]^2} = \overline{[v'(0)R_L(t)]^2} + \overline{[n(t)]^2} + \overline{2v'(0)R_L(t)n(t)} \tag{9.29}$$

Applying averaging rules, and assuming that $n(t)$ is uncorrelated with $v'$:

$$\overline{v'^2(t)} = R_L^2(t)\overline{v'^2(0)} + \overline{n^2(t)} \tag{9.30}$$

When the turbulence is steady state, we can expect the following relationship to apply:

$$\overline{v'^2(t)} = \overline{v'^2(0)} \tag{9.31}$$

Hence, eq. (9.30) can be simplified to

$$\overline{n^2(t)} = \left[1 - R_L^2(t)\right]\overline{v'^2(t)} \tag{9.32}$$

Defining variances, eq. (9.32) can be written as

$$\sigma_n^2 = \left[1 - R_L^2(t)\right]\sigma_v^2 \tag{9.33}$$

or

$$\sigma_n = \sqrt{1 - R_L^2(t)}\,\sigma_v \tag{9.34}$$

To examine the limiting behavior of eq. (9.34) for short time increments, eq. (6.43) is substituted for $R_L(t)$:

$$\boxed{\sigma_n = \sqrt{1 - \exp\left(-\frac{2t}{T_{i,L}}\right)}\,\sigma_v} \tag{9.35}$$

Hence (Hall, 1974):

$$\boxed{v'(t) = v'(0)\exp\left(-\frac{t}{T_{i,L}}\right) + \sqrt{1 - \exp\left(-\frac{2t}{T_{i,L}}\right)}\cdot\sigma_v r(t)} \tag{9.36}$$

where $r(t)$ is a random variable with mean 0 and standard deviation 1.

Considering that $\exp(a) \approx 1 + a$ for small values of $a$, eq. (9.35) can be approximated for short time increments as

$$\sigma_n = \sqrt{\frac{2t}{T_{\mathrm{i,L}}}}\sigma_v \tag{9.37}$$

This confirms the earlier observation that the amplitude (standard deviation) of $n(t)$ should be proportional to the square root of the time increment. Also $n(t)$ is proportional to a stochastic function known as the **Wiener process**, $W(t)$:

$$n(t) = \sqrt{\frac{2}{T_{\mathrm{i,L}}}}\sigma_v W(t) \tag{9.38}$$

where $W(t)$ is a continuous stochastic function with a normal (Gaussian) probability density function with mean 0 and standard deviation $t^{1/2}$. Equation (9.24) can also be written in terms of a Wiener process:

$$dv' = -\frac{v'}{T_{\mathrm{i,L}}}dt + \sqrt{\frac{2\sigma_v^2}{T_{\mathrm{i,L}}}}dW \tag{9.39}$$

We established a practical way to carry out a Lagrangian simulation: eq. (9.27) is calculated recursively with time step $\Delta t$, using a normally distributed random variable with zero average and a standard deviation given by eq. (9.35). The actual location of the particle can be calculated with Euler's method:

$$\boxed{y'(t+\Delta t) = y'(t) + v'(t)\,\Delta t} \tag{9.40}$$

This is illustrated in Figure 9.1 for lateral dispersion with 10 particles ($T_{\mathrm{i,L}} = 100\,\mathrm{s}$, $u = 2\,\mathrm{m\,s^{-1}}$, $\sigma_v = 0.2\,\mathrm{m\,s^{-1}}$, $\Delta t = 10\,\mathrm{s}$). The calculations for this figure are included on the enclosed CD, in the Excel file "Figure 9.1. Stochastic Lagrangian sim 10 particles. xlsx." In the tab Random a set of pseudorandom numbers with a normal distribution is generated. The procedure for the number generation is outlined in Section 9.3. The tab Calculations contains the trajectory calculations. Note that eq. (9.40) above is acceptable in these calculations mainly because the turbulence is assumed to be homogeneous, and only the deviations from average conditions are calculated. In general, Euler's method is not an accurate numerical tool. More sophisticated algorithms for trajectory calculation are summarized by Stohl (1998).

Not all properties of stochastic particle motion can be resolved by the Langevin equation. However, the Langevin equation can be translated into a probabilistic Eulerian form. The result, a partial differential equation involving the probability density function of location and velocity, is known as the **Fokker–Planck equation**. Only the one-dimensional form will be given here. Starting point is the following generalized form of the Langevin equation:

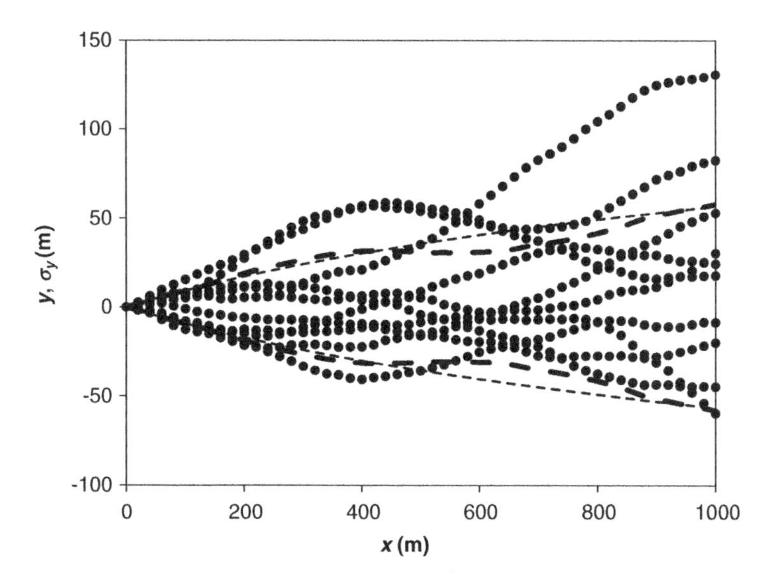


Figure 9.1 Lagrangian particle simulation of lateral dispersion with 10 particles based on eq. (9.28) ($T_{i,L}=100$ s, $u=2$ m s$^{-1}$, $\sigma_v=0.2$ m s$^{-1}$, $\Delta t=10$ s). Dots are particle trajectories, thick dashed lines are $\pm\sigma_y$ of the particles, and thin dashed lines are $\pm\sigma_y$ predicted with the exact solution.

$$dv = a\,dt + b\,dW \tag{9.41}$$

where $a$ and $b$ are coefficients, and $dW$ is the differential Wiener process. The equivalent Fokker–Planck equation can be written as (Weil, 1990)

$$\frac{\partial p}{\partial t} + v\frac{\partial p}{\partial y} = -\frac{\partial}{\partial v}(ap) + \frac{1}{2}\frac{\partial^2}{\partial v^2}(b^2 p) \tag{9.42}$$

where $p$ is the time-dependent probability density function of the location and velocity, that is, $p(y,v,t)\,dy\,dv$ is the probability that an air parcel at time $t$ is located between $y$ and $y+dy$ and has a velocity between $v$ and $v+dv$. Note that $v$ is used in these equations, not $v'$, making the equations more general. Velocity fluctuations can be used as a special case where the average velocity equals zero. However, the term $b\,dW$ describes the diffusion (or dispersion) process, which is fundamentally linked to the velocity fluctuation and the Lagrangian integral time scale, so we can obtain $b$ by comparing eq. (9.41) with eq. (9.39):

$$b = \sqrt{\frac{2\sigma_v^2}{T_{i,L}}} \tag{9.43}$$

Coefficient $a$ in eq. (9.41) depends on the system being modeled and needs to be derived. The procedure to derive $a$ is based on the assumption that an initially well-mixed pollutant should remain well-mixed (the **well-mixed criterion**). It is illustrated in Section A9.1 (Appendix A), where the value of $a$ is derived for a system with homogeneous turbulence and zero average wind speed. The result is

$$a = -\frac{v}{T_{\mathrm{i,L}}} \tag{9.44}$$

This result is expected based on the comparison of eq. (9.39) with eq. (9.41).

### 9.2.3 Langevin Equation in Heterogeneous Atmosphere

The equations discussed in the previous section are only correct when the atmospheric turbulence is homogeneous. This is a reasonable assumption in the $y$ direction but not in the $z$ direction. In the $z$ direction both $T_{\mathrm{i,L}}$ and $\sigma_w$ are height dependent, and a correction to account for this dependence is required.

In the previous section, all spatial variables were expressed in the $y$ direction because it is the direction where homogeneous turbulence is realistic. In this section, spartial variables will be expressed in the $z$ direction because the concepts outlined here are most likely to be applied in that direction.

When $\sigma_w$ is constant and only $T_{\mathrm{i,L}}$ changes with height, as is approximately the case in a neutral atmosphere, the approach of Section 9.2.2 is still valid in principle, as long as sufficiently short time steps are taken to ensure that the $T_{\mathrm{i,L}}$ value of a particle does not change markedly within a single time step (Legg and Raupach, 1982). As a result, very short time steps can be required to resolve dispersion close to the surface, where $T_{\mathrm{i,L}}$ is short (a few seconds or less) and changes strongly with height. Wilson et al. (1981a) developed a rescaling method that transforms the diffusion problem into a mathematically equivalent problem with constant $T_{\mathrm{i,L}}$, thus removing the time step requirement. Good results are obtained when the time step does not exceed $0.1T_{\mathrm{i,L}}$. The rescaling method of Wilson et al. (1981a) is outlined in Section A9.2 (Appendix A).

Note that when $T_{\mathrm{i,L}}$ is variable, it is not strictly an "integral" time scale, as this terminology suggests that a single value represents the entire domain. In the literature it is usually indicated with the Greek letter $\tau$. In this text, the letter $T$ is used to express the conceptual link with the Lagrangian integral time scale.

An issue more important than the variability of $T_{\mathrm{i,L}}$ is the effect of changing $\sigma_w$ on dispersion calculations because it cannot be eliminated by reducing the time step $\Delta t$. To understand the issue, it is useful to return to the molecular diffusion analogy (see also the discussion of Wilson et al., 1981b). In a mixture of two gases with different molar masses, the lighter molecules move more rapidly than the heavier molecules. Hence, when two gases are separated, and then allowed to mix by diffusion, the lighter molecules move into the heavy gas bulk more rapidly than the heavy molecules move into the light gas bulk. Hence, there is an accumulation of molecules in the heavy gas bulk and a depletion of molecules in the light gas bulk, leading to a pressure difference. When the gas can move freely, the pressure must be constant, so an advective gas movement will occur to equalize the pressure. When the gas cannot move freely, for example, because the two gases are separated by a porous plug, the pressure difference can be measured experimentally (Evans et al., 1961).

A similar phenomenon occurs in the atmosphere when there is a (vertical) $\sigma_w$ gradient. Air parcels in zones of high $\sigma_w$ move into zones of low $\sigma_w$ faster than the other way around, leading to an accumulation of air in zones of low $\sigma_w$. This creates a pressure gradient that drives an advective flow from low to high $\sigma_w$. If this effect is not accounted for, a stochastic Lagrangian particle model will predict an accumulation of particles in regions of low $\sigma_w$. In fact, Thomson (1987) showed that a key criterion for a correctly conceived and implemented stochastic Lagrangian particle model is the absence of any accumulation of particles in an initially well-mixed system. This is known as the well-mixed criterion. Attempts to quantify the particle accumulation effect were made by Wilson et al. (1981b) and by Legg and Raupach (1982).

The approach of Wilson et al. (1981b) is a mathematical one: A rescaling method was developed that transforms the diffusion problem into a mathematically equivalent problem with constant $\sigma_w$. However, the transformation entailed a rescaling of the concentration, and when the solution was scaled back to the physical scale, zero flux no longer corresponded with zero concentration gradient, which is physically unrealistic. Hence, Wilson et al. (1981b) introduced a **velocity bias** in the form of an advective flow that ensures that the real concentration gradient is zero under no-flux conditions. The value of the velocity bias according to Wilson et al. (1981b) is

$$\overline{w_{\text{bias}}} = \sigma_w T_{\text{i,L}} \frac{\partial \sigma_w}{\partial z} \tag{9.45}$$

Legg and Raupach (1982) adopted a physical approach to determine the advective flow, which they named **drift velocity**. Based on a Reynolds-averaged momentum equation, they found that a $\sigma_w$ gradient generates a force that acts on each air parcel. By adding this force to the Langevin equation, they derived the following drift velocity:

$$\overline{w_{\text{drift}}} = 2\sigma_w T_{\text{i,L}} \frac{\partial \sigma_w}{\partial z} = T_{\text{i,L}} \frac{\partial \sigma_w^2}{\partial z} \tag{9.46}$$

The drift velocity of Legg and Raupach (1982) is twice as large as the velocity bias of Wilson et al. (1981b). Surprisingly, both groups found that their approach removed the spurious accumulation of particles in zones of low $\sigma_w$ in their calculations, at least in the case of mildly variable $\sigma_w$. Wilson et al. (1983) found that the implementation of Legg and Raupach (1982) introduced a spurious velocity bias approximately equal to half of their drift velocity, which explained the apparent success of their method. In Section 10.2.1 we will see that there is a peculiar inconsistency between the stochastic Lagrangian particle model and the overall mass balance, which introduces the weakness in the Legg and Raupach (1982) model. Wilson et al. (1983) made further improvements to their calculation method based on eq. (9.45), and the validity of their approach was confirmed by Thomson (1987), who provided a more fundamental theoretical framework for the approach based on the Fokker–Planck equation. A simple special case of the Thomson approach is illustrated in Section A9.1 (Appendix A).

The Thomson (1987) approach is the most fundamental procedure to derive stochastic Lagrangian particle models. However, for the sake of intuitive clarity, a more mechanical approach will be adopted here to derive the Wilson et al. (1983) procedure and associated drift velocity. It is a physical approach similar to the one of Legg and Raupach (1982) but based on the much simpler Bernouilli equation. The Bernouilli equation is the foundation of classical hydrodynamics in the absence of friction (viscous) forces and is as follows:

$$\frac{dp}{\rho}+\frac{dw^2}{2}+g\,dz=0 \tag{9.47}$$

In principle, the overall velocity should be used in eq. (9.47) instead of $w$, but when homogeneity is assumed in the horizontal ($x$ and $y$) directions, then the $u$ and $v$ components of the velocity disappear in the differential. We will discard effects of gravity ($g\,dz$) as we are only interested in pressure gradients caused by turbulence for the purpose of this derivation. Hence, taking the ensemble average for a large number of air parcels, eq. (9.47) becomes

$$\frac{d\overline{p}}{\rho}+\frac{d\overline{w^2}}{2}=0 \tag{9.48}$$

Assuming that the drift velocity is small in comparison with the turbulent velocity, the following approximation can be made:

$$\overline{w^2}\approx\overline{w'^2} \tag{9.49}$$

Hence, eq. (9.48) becomes

$$\frac{d\overline{p}}{\rho}+\frac{d\sigma_w^2}{2}=0 \tag{9.50}$$

In terms of a height derivative, this equation becomes

$$\frac{1}{\rho}\frac{\partial\overline{p}}{\partial z}+\frac{1}{2}\frac{\partial\sigma_w^2}{\partial z}=0 \tag{9.51}$$

The left-hand side of eq. (9.51) can be written in terms of a force acting upon an air parcel as follows:

$$\frac{1}{\rho}\frac{\partial\overline{p}}{\partial z}=\frac{A}{A\rho}\frac{\partial\overline{p}}{\partial z}=\frac{\partial F_{\mathrm{d}}}{\partial m}=\frac{F_{\mathrm{d}}}{m} \tag{9.52}$$

where $A$ is the horizontal area of the air parcel; and $F_d$ is a drift force acting upon each air parcel, per unit mass of air. Based on eq. (9.51), the drift force can be written as

$$\frac{F_d}{m} = -\frac{1}{2}\frac{\partial \sigma_w^2}{\partial z} = -\sigma_w \frac{\partial \sigma_w}{\partial z} \tag{9.53}$$

The drift force is added to the Langevin equation [eq. (9.24)] and applied to the vertical direction:

$$\frac{\partial w'}{\partial t} = -\frac{w'}{T_{i,L}} + \frac{F_d}{m} + \frac{F_c}{m} \tag{9.54}$$

Substitution of eq. (9.53) leads to

$$\frac{\partial w'}{\partial t} = -\frac{w'}{T_{i,L}} - \sigma_w \frac{\partial \sigma_w}{\partial z} + \frac{F_c}{m} \tag{9.55}$$

In terms of the Wiener process, eq. (9.55) can be written as

$$dw' = -\left(\frac{w'}{T_{i,L}} + \sigma_w \frac{\partial \sigma_w}{\partial z}\right) dt + \sqrt{\frac{2\sigma_w^2}{T_{i,L}}}\, dW \tag{9.56}$$

Because of the height dependent factor $\sigma_w$ in the drift term in eq. (9.55), the unperturbed equation will contain a height dependence. Not accounting for this height dependence will lead to an incorrect implementation of the Langevin equation (Wilson et al., 1983). Fortunately, the height dependence can be removed with the following substitution:

$$s' = \frac{w'}{\sigma_w} \tag{9.57}$$

Dividing eq. (9.55) by $\sigma_w$ and substituting eq. (9.57) leads to the following equation:

$$\boxed{\frac{ds'}{dt} = -\frac{s'}{T_{i,L}} - \frac{\partial \sigma_w}{\partial z} + \frac{F_c}{\sigma_w m}} \tag{9.58}$$

Remember that the term containing $F_c$ is the stochastic term. In terms of the Wiener process, this equation becomes

$$ds' = -\left(\frac{s'}{T_{i,L}} + \frac{\partial \sigma_w}{\partial z}\right) dt + \sqrt{\frac{2}{T_{i,L}}}\, dW \tag{9.59}$$

The unperturbed version of eq. (9.58) is

$$\frac{ds'}{dt} = -\frac{s'}{T_{i,L}} - \frac{\partial \sigma_w}{\partial z} \tag{9.60}$$

Rearrange to separate variables:

$$\frac{ds'}{s' + T_{i,L}(\partial \sigma_w / \partial z)} = -\frac{dt}{T_{i,L}} \tag{9.61}$$

Integration from time 0 to time $t$ leads to

$$\ln\left[\frac{s'(t) + T_{i,L}(\partial \sigma_w / \partial z)}{s'(0) + T_{i,L}(\partial \sigma_w / \partial z)}\right] = -\frac{t}{T_{i,L}} \tag{9.62}$$

Taking the exponential and rearranging leads to

$$s'(t) = s'(0)\exp\left(-\frac{t}{T_{i,L}}\right) + \left[1 - \exp\left(-\frac{t}{T_{i,L}}\right)\right] \cdot T_{i,L}\frac{\partial \sigma_w}{\partial z} \tag{9.63}$$

Adding the stochastic term leads to (Wilson et al., 1983)

$$\boxed{s'(t) = s'(0)\exp\left(-\frac{t}{T_{i,L}}\right) + \left[1 - \exp\left(-\frac{t}{T_{i,L}}\right)\right] \cdot T_{i,L}\frac{\partial \sigma_w}{\partial z} + n(t)} \tag{9.64}$$

The velocity fluctuation $w'$ can be calculated from $s'$ with the following equation:

$$\boxed{w'(t) = \sigma_w \cdot s'(t)} \tag{9.65}$$

All that remains is a calculation of the standard deviation of $n(t)$. To that effect, eq. (9.64) is squared and averaged. Considering that $n(t)$ and $s'(0)$ have zero mean and are uncorrelated with each other, the following equation is obtained:

$$\overline{s'^2(t)} = \overline{s'^2(0)\exp\left(-\frac{2t}{T_{i,L}}\right)} + \overline{\left[1 - \exp\left(-\frac{t}{T_{i,L}}\right)\right]^2 \cdot T_{i,L}^2\left(\frac{\partial \sigma_w}{\partial z}\right)^2} + \overline{n^2(t)} \tag{9.66}$$

Based on the definition of $s'$, it is clear that its variance must be one. Hence, we can solve eq. (9.66) for $\sigma_n$:

$$\boxed{\sigma_n = \sqrt{1 - \exp\left(-\frac{2t}{T_{i,L}}\right) - \left[1 - \exp\left(-\frac{t}{T_{i,L}}\right)\right]^2 \cdot T_{i,L}^2\left(\frac{\partial \sigma_w}{\partial z}\right)^2}} \tag{9.67}$$

Equation (9.67) is used in conjunction with eqs. (9.64) and (9.65) to calculate the random path of a particle in an atmosphere with variable turbulent velocities.

The third term in the square root is more strongly dependent on $t$ than the other terms and can always be made to be negligible by choosing sufficiently small values of $t$. Hence, the equation can be approximated by

$$\sigma_n = \sqrt{1 - \exp\left(-\frac{2t}{T_{\mathrm{i,L}}}\right)} \tag{9.68}$$

This is the form found by Wilson et al. (1983).

Equations (9.64) and (9.65) were first developed by Wilson et al. (1983) to resolve issues with highly variable turbulence as found in crop canopies. It was set on a more solid theoretical foundation based on the Fokker–Planck equation by Thomson (1987).

The approach discussed in this section applies only to atmospheres with constant density, that is, shallow boundary layers. An approach thar corrects for density changes in deep boundary layers was developed by Stohl and Thomson (1999). An additional correction is needed when the particles are large enough to have a significant settling velocity (Wilson, 2000).

## 9.2.4 Turbulence Data for Stochastic Lagrangian Models

In order to carry out a Lagrangian particle simulation, a number of expressions are needed to calculate at least $\sigma_v$ and $\sigma_w$ at all possible locations, and $T_{\mathrm{i,L}}$ in the $y$ and $z$ directions. In principle, all these variables can be obtained from recommendations made in previous chapters. For the $z$ direction, however, a deviation from these recommendations seems in order.

As a starting point, consider the neutral atmosphere close to the surface. From Chapter 5, eq. (5.149) can be used to obtain

$$\sigma_w = 1.3u_* \tag{9.69}$$

For the calculation of the Lagrangian integral time scale close to the surface, we first consider the turbulent mass diffusivity, based on eq. (5.191):

$$K_z = ku_*z \tag{9.70}$$

Based on relationship (6.105), this can be linked to $T_{\mathrm{i,L}}$ in the $z$ direction:

$$T_{\mathrm{i,L}} = \frac{K_z}{\sigma_w^2} \tag{6.105}$$

Hence, the following equation is obtained:

$$T_{\mathrm{i,L}} = \frac{ku_*z}{1.69u_*^2} = \frac{kz}{1.69u_*} \tag{9.71}$$

The Lagrangian length scale is defined by

$$\Lambda_L = \sigma_w T_{i,L} \tag{9.72}$$

Based on eq. (9.71), it is given by

$$\Lambda_L = \frac{kz\sigma_w}{1.69u_*} = \frac{kz}{1.3} \approx 0.3z \tag{9.73}$$

However, Wilson et al. (1981c) compared stochastic Lagrangian particle modeling results with field data and found that the following equation leads to better predictions for neutral conditions near the surface:

$$\Lambda_L \approx 0.5z \tag{9.74}$$

The reason for the discrepancy is unclear. However, Stull (1988) states that there is evidence to suggest that $K_z \approx 1.35K_m$ under neutral conditions, which would bring $\Lambda_L$ up to slightly above $0.4z$.

Wilson et al. (1981c) proposed correlations for $\Lambda_L$ in nonneutral conditions as well. These correlations are:

For unstable conditions ($L<0$):

$$\Lambda_L = 0.5z\left(1-6\frac{z}{L}\right)^{1/4} \tag{9.75}$$

For stable conditions ($L>0$):

$$\Lambda_L = \frac{0.5z}{1+5(z/L)} \tag{9.76}$$

The choice of the neutral atmosphere value of $\Lambda_L$ is linked to the choice of the Lagrangian structure function constant $C_0$ (see Section 6.11.2). This connection is based on two fundamental relationships for the energy dissipation rate $\varepsilon$. The first one is (Tennekes, 1979)

$$C_0\varepsilon = \frac{2\sigma_w^2}{T_{i,L,z}} \tag{6.189}$$

The second equation is the following finding, valid under neutral conditions (Stull, 1988):

$$\varepsilon = \frac{u_*^3}{kz} \tag{6.186}$$

Combining and solving for $T_{i,L,z}$, leads to

$$T_{i,L,z} = \frac{2\sigma_w^2 kz}{C_0 u_*^3} \tag{9.77}$$

On the other hand, $T_{i,L,z}$ is linked with $K_z$ through eq. (6.105) (see above). This leads to the following relationship:

$$\frac{K_z}{\sigma_w^2} = \frac{2\sigma_w^2 kz}{C_0 u_*^3} \tag{9.78}$$

which leads to the following relationship for $K_z$:

$$K_z = \frac{2\sigma_w^4 kz}{C_0 u_*^3} \tag{9.79}$$

Based on eq. (9.69), the following relationship is obtained:

$$K_z = \frac{2 \cdot 1.3^4}{C_0} k u_* z \tag{9.80}$$

In terms of $\Lambda_L$, eqs. (9.72), (9.77), and (9.69) can be combined to obtain

$$\Lambda_L = \frac{2 \cdot 1.3^3}{C_0} kz \tag{9.81}$$

Note that the factor 1.3 is prone to some uncertainty, and a value of 1.25 is often used. Due to the third power in eq. (9.81), and the fourth power in eq. (9.80), there is some spread in the resulting equations. Nevertheless, when a value of $C_0=5$ is chosen, eq. (9.80) is in general agreement with eq. (9.70), and $K_m=K_z$. Sawford (1985) suggested $C_0=5$, and this value was confirmed by Reynolds (1998). On the other hand, the observations of Wilson et al. (1981c) (eq. 9.74) can be reconciled with eq. (9.81) only if $C_0$ is about 3. This value was confirmed by Du (1997, 1998). Weil (1990), on the other hand, proposed a $C_0$ value as low as 2, based on Anand and Pope (1985). Based on all the available data, Wilson and Sawford (1996) recommended $C_0=3.1$. More recent research has not narrowed the 3–5 range (e.g., Degrazia et al., 2008; Carvalho et al., 2009).

For the *x* and *y* directions there are no clear recommendations for $\Lambda_L$ because $T_{i,L}$ is sufficiently long to be influenced by mesoscale effects that cannot be resolved with a similarity approach. The same mesoscale effects are responsible for the averaging time dependence of lateral dispersion parameters. The reader is referred to Chapter 5 for estimates of $\sigma_u$ and $\sigma_v$ [e.g., eqs. (5.144), (5.145), (5.152), (5.153), and (5.156)]. For $T_{i,L}$, the relations of Weber et al. (1982) (see Section 6.6.7) are useful. Correlations based on $\varepsilon$ (Section 6.11.2) can be used in conjunction with the work of Anfossi et al. (2000), who determined $C_0$ in the $x$, $y$, and $z$ directions (3.2, 4.3, and 4.3, respectively). Note that this suggests the following relationships between the Lagrangian integral time scales, based on the application of eq. (6.191) in different dimensions:

$$\frac{T_{\mathrm{i,L},x}}{T_{\mathrm{i,L},z}} = \frac{4.3}{3.2}\left(\frac{\sigma_u}{\sigma_w}\right)^2 \tag{9.82}$$

$$\frac{T_{\mathrm{i,L},y}}{T_{\mathrm{i,L},z}} = \left(\frac{\sigma_v}{\sigma_w}\right)^2 \tag{9.83}$$

### 9.2.5 Stochastic Lagrangian Particle Modeling of the Convective Boundary Layer

The theory developed so far does not account for the existence of updrafts and downdrafts in a convective boundary layer. A theory of updrafts and downdrafts was developed by Baerentsen and Berkowicz (1984) based on the notion that updrafts and downdrafts are separate entities, each with an average vertical wind speed and a Gaussian wind speed fluctuation. Details of the theory are given in Section 5.13.1.

A simple derivation of a Lagrangian model based on the theory of Baerentsen and Berkowicz (1984) that follows the well-mixed condition is not available. Instead, a new Langevin equation must be derived by imposing the well-mixed condition on the Fokker–Planck equation [eq. (9.42)]. This approach was followed by Luhar and Britter (1989) and by Weil (1990). Weil et al. (1997) applied the Baerentsen and Berkowicz (1984) approach in the development of a Gaussian dispersion model enhancement known for its implementation in AERMOD.

The derivation of the Langevin equation [eq. (9.41)] for convective boundary layers is long and complicated and will not be outlined here. Instead, only the result will be given. As usual, coefficient $b$ is given by

$$b = \sqrt{\frac{2\sigma_w}{T_{\mathrm{i,L}}}} \tag{9.84}$$

The following function for $a$ follows the well-mixed condition:

$$a = -\frac{\sigma_w^2}{T_{\mathrm{i,L}}}\frac{Q}{P_{\mathrm{a}}} + \frac{\phi}{P_{\mathrm{a}}} \tag{9.85}$$

where $Q$, $P_{\mathrm{a}}$, and $\phi$ are given by the following equations:

$$Q = \frac{\lambda_1\left(w - \overline{w_1}\right)}{\sigma_{w1}^2}P_1 + \frac{\lambda_2\left(w - \overline{w_2}\right)}{\sigma_{w2}^2}P_2 \tag{9.86}$$

$$P_a = \lambda_1 P_1 + \lambda_2 P_2 \tag{9.87}$$

$$\begin{aligned}\phi = &-\frac{1}{2}\left(\lambda_1 \frac{\partial \overline{w_1}}{\partial z} + \overline{w_1}\frac{\partial \lambda_1}{\partial z}\right)\operatorname{erf}\left(\frac{w-\overline{w_1}}{\sqrt{2}\overline{w_1}}\right) + \overline{w_1}\left[\lambda_1 \frac{\partial \overline{w_1}}{\partial z}\left(\frac{w^2}{\overline{w_1}^2}+1\right) + \overline{w_1}\frac{\partial \lambda_1}{\partial z}\right]P_1 \\ &+\frac{1}{2}\left(\lambda_2 \frac{\partial \overline{w_2}}{\partial z} + \overline{w_2}\frac{\partial \lambda_2}{\partial z}\right)\operatorname{erf}\left(\frac{w-\overline{w_2}}{\sqrt{2}\overline{w_2}}\right) + \overline{w_2}\left[\lambda_2 \frac{\partial \overline{w_2}}{\partial z}\left(\frac{w^2}{\overline{w_2}^2}+1\right) + \overline{w_2}\frac{\partial \lambda_2}{\partial z}\right]P_2\end{aligned} \tag{9.88}$$

In these equations, $P_1$ and $P_2$ are Gaussian functions given by

$$P_1 = \frac{1}{\sqrt{2\pi}\sigma_{w1}}\exp\left[-\frac{1}{2}\frac{\left(w-\overline{w_1}\right)^2}{\sigma_{w1}^2}\right] \tag{9.89}$$

$$P_2 = \frac{1}{\sqrt{2\pi}\sigma_{w2}}\exp\left[-\frac{1}{2}\frac{\left(w-\overline{w_2}\right)^2}{\sigma_{w2}^2}\right] \tag{9.90}$$

For convenience, Luhar and Britter (1989) provided the following auxiliary equations:

$$\frac{\partial \overline{w_2}}{\partial z} = -\frac{1}{F}\left[\overline{w_2}\frac{\partial \overline{w^3}}{\partial z} + \frac{\partial \sigma_w^2}{\partial z}\left(\frac{\overline{w_2}F}{\sigma_w^2} - 3\sigma_w^2\right)\right] \tag{9.91}$$

$$\frac{\partial \overline{w_1}}{\partial z} = \frac{1}{2\overline{w_2}}\frac{\partial \sigma_w^2}{\partial z} - \frac{\overline{w_1}}{\overline{w_2}}\frac{\partial \overline{w_2}}{\partial z} \tag{9.92}$$

$$\frac{\partial \lambda_1}{\partial z} = \frac{1}{\overline{w_1}+\overline{w_2}}\left(-\lambda_1 \frac{\partial \overline{w_1}}{\partial z} + \lambda_2 \frac{\partial \overline{w_2}}{\partial z}\right) \tag{9.93}$$

$$\frac{\partial \lambda_2}{\partial z} = -\frac{\partial \lambda_1}{\partial z} \tag{9.94}$$

where

$$F = 4\sigma_w^2 \overline{w_2} + \overline{w^3} \tag{9.95}$$

The remaining variables, such as $\lambda_l$ (fractional area in updrafts) and $\lambda_2$ (fractional area in downdrafts) are defined in Section 5.13.1, along with their calculation.

## 9.3 NUMERICAL ASPECTS OF STOCHASTIC MODELING

This section will concentrate on the stochastic aspects of trajectory modeling. The numerical aspects of deterministic trajectory modeling are reviewed by Stohl (1998).

The main numerical challenge in stochastic modeling is the generation of pseudorandom numbers with a Gaussian distribution. This can be carried out in two steps:

- A set of pseudorandom numbers with a homogeneous distribution (top hat distribution) is generated, usually in the range 0–1.
- The set of numbers is converted to a normally distributed set by means of the inverse cumulative normal distribution, for example, with mean 0 and standard deviation 1. In other words: If $r$ is the random number from the previous set, then the number in the normal distribution that is greater than a fraction $r$ of the normally distributed numbers is obtained in the second set.

These steps are straightforward in Microsoft Excel and some other modeling software that have built-in functions for both. For instance, the Excel function RAND() generates a pseudorandom number between 0 and 1 with a homogeneous distribution. The Excel function NORMINV($r$,0,1) turns $r$ into a random number with normal distribution, with mean 0 and standard deviation 1. Hence, NORMINV(RAND(), $a$, $b$) produces a number from a normal distribution with mean $a$ and standard deviation $b$. Note that Park and Miller (1988) and Press et al. (1992) strongly discourage the use of *any* prepackaged random generator without knowledge of the algorithm implemented, as many are flawed, and some badly so. Fortunately, the situation has improved much since these authors wrote those warnings, but flawed methods are still in use.

Generating pseudorandom numbers is less straightforward when programming languages without built-in pseudorandom number generators are used. Press et al. (1992, 2007) provide a number of algorithms. The one discussed here is known as the Minimal Standard algorithm (Schrage, 1979; Park and Miller, 1988). It is considered obsolete in the literature on random number generation and inadequate for intensive Monte Carlo simulations (Press et al., 2007). However, thanks to its simplicity, and the modest numerical requirements of the examples presented in this chapter, it is an acceptable choice for demonstrating the inner workings of stochastic Lagrangian particle models here, especially as this type of model is a fairly forgiving application of random number generation. Nevertheless, readers wishing to develop their own models for real-world applications are recommended to investigate more sophisticated random number generators.

First, a random set of whole numbers is generated from a seed $P_0$ with the algorithm:

$$R_{n+1} = a \cdot \mathrm{mod}(P_n, q) - r \cdot \mathrm{int}\left(\frac{P_n}{q}\right) \tag{9.96}$$

where

$$a = 16\,807$$
$$m = 2\,147\,483\,647$$
$$q = 127\,773$$
$$r = 2\,836$$

This algorithm sometimes generates a negative number. However, the actual set of numbers, $P$, is obtained with the following rules:

$$P_{n+1} = R_{n+1} \quad \text{if } R_{n+1} > 0 \tag{9.97}$$

$$P_{n+1} = m + R_{n+1} \quad \text{if } R_{n+1} < 0 \tag{9.98}$$

This algorithm generates a highly irregular series of numbers ranging from 1 to $m-1$. Hence, the set of numbers defined by

$$p_n = \frac{P_n}{m} \tag{9.99}$$

is a pseudorandom set of numbers between 0 and 1 that satisfies the requirements with sufficient accuracy. The period of the algorithm is $m-1$, that is, it rotates through all the numbers from 1 to $m-1$ in an apparently random order. The algorithm is not recommended for applications requiring more than $10^8$ random numbers (Press et al., 1992), which limits the application to short-term single-source calculations. Current software applications tend to have longer periods. For instance, at the time of this writing the algorithm underlying the RAND() function in Excel is the Wichman–Hill algorithm (Whichmann and Hill, 1982), which has a period of about $6.95 \times 10^{12}$. The programming language Python has a random generator based on the Mersenne Twister algorithm of Matsumoto and Nishimura (1998), with a period of about $4.31 \times 10^{6001}$, which can be considered infinitely long for practical purposes. A practical way to increase the period of random number generation is by using three different generators with different periods. The three generators are combined based on the property that, assuming that $a$, $b$, and $c$ are homogeneously distributed random numbers with range 0–1, the function $a+b+c \bmod 1$ is also homogeneously distributed with range 0–1.

Step 2 of the algorithm is slightly different from the suggested procedure above because there is a numerically cheaper way to generate uncorrelated normally distributed numbers known as the Box–Muller method (Press et al., 1992). To that effect, two consecutive numbers $p_{2n-1}$ and $p_{2n}$ are used to calculate two uncorrelated numbers $N_{2n-1}$ and $N_{2n}$:

$$N_{2n-1} = \sqrt{-2\ln(p_{2n-1})}\cos(2\pi \cdot p_{2n}) \tag{9.100}$$

$$N_{2n} = \sqrt{-2\ln(p_{2n-1})}\sin(2\pi \cdot p_{2n}) \tag{9.101}$$

The set $N$ is normally distributed with mean 0 and standard deviation 1. If a different standard deviation is needed, then all numbers in the set are simply multiplied by the desired standard deviation.

When execution speed is important, eqs. (9.100) and (9.101) are not recommended because of the slow execution of the sin and cos functions. An alternative is provided by Leva (1992).

A second issue is the proper choice of the time increment. When the turbulence is homogeneous, or the Wilson et al. (1981a) methodology is used to make the system homogeneous in the case of varying $T_{i,L}$ and constant $\sigma_w$, a time step equal to 0.1 times $T_{i,L}$ is sufficiently short to achieve good accuracy (Wilson et al., 1981a). When $T_{i,L}$ is variable, and the Wilson et al. (1981a) methodology is not used, sufficiently short time steps must be used to ensure that the $T_{i,L}$ experienced by a particle does not change markedly within a single time step (Legg and Raupach, 1982). When both $\sigma_w$ and $T_{i,L}$ are variable, the following time step is recommended (Thomson, 1987):

$$\Delta t = \min\left(0.025 T_{i,L}, \frac{0.1\sigma_w}{|a|}, \frac{0.01\sigma_w}{w(\partial\sigma_w / \partial z)}\right) \tag{9.102}$$

where $a$ refers to the coefficient in eq. (9.42). This equation may not adequately account for changes in $T_{i,L}$.

In the vertical direction, particles are usually *contained*, either by the surface, the mixing height, or both, and boundary conditions are needed to ensure containment. In the case of homogeneous Gaussian turbulence this is straightforward. According to Wilson and Flesch (1993), perfect reflection, that is, changing the sign of the velocity component perpendicular to the reflecting surface, is exactly valid in such cases.

However, in the vertical direction turbulence is never homogeneous, and the velocity distribution can be skewed or not centered around zero velocity. In that case, perfect reflection would change the velocity distribution, which leads to a violation of the well-mixed condition. Several researchers have proposed a boundary condition to minimize the effect on the velocity distribution. Defining $P(w)$ as the probability density function of the velocity, Wilson and Sawford (1996) suggested that the boundary condition should solve the following equation:

$$\frac{P(w^+)dw^+}{\int_0^\infty P(w)dw} = \frac{P(w^-)dw^-}{\int_{-\infty}^0 P(w)dw} \tag{9.103}$$

where $w^-$ and $w^+$ are the vertical wind speed component before and after reflection, respectively.

Weil (1990) suggested the following:

$$\frac{\int_0^{w^+} P(w)dw}{\int_0^\infty P(w)dw} = \frac{\int_{w^-}^0 P(w)dw}{\int_{-\infty}^0 P(w)dw} \tag{9.104}$$

So far it was assumed that all particles reaching the surface reflect, that is, there is no deposition. In the presence of deposition, a fraction of the particles hitting the surface should be taken out of the atmosphere at random. The question arises what fraction corresponds with a given deposition velocity $v_d$. Wilson et al. (1989) suggested the following relationship:

$$\frac{1-R}{1+R} = \sqrt{\frac{\pi}{2}} \frac{v_d}{\sigma_w} \tag{9.105}$$

where $R$ is the reflection probability.

Wilson and Flesch (1993) discovered that some turbulence parameterizations in stochastic Lagrangian particle models create conditions where the particles never reach the surface. This is obvious from eq. (9.105): If the modeled value of $\sigma_w$ tends to 0 upon reaching the surface, the deposition velocity is zero as well, regardless of the chosen reflection probability. This is only possible when the particles do not reach the wall. A similar phenomenon can be expected when there is a large positive vertical gradient of $\sigma_w$ near the surface. From eq. (9.45) it is clear that large upward drift velocities are generated in such conditions, driving the particles back into the bulk of the atmosphere.

Wilson (2000), Boehm and Aylor (2005), and Bocksel and Loth (2006) proposed adjustments to incorporate particle settling into stochastic Lagrangian particle models. The well-mixed condition cannot be used as a criterion to test such models.

Once the calculation of the particle trajectory is completed, a concentration profile needs to be inferred from the particle locations. The simplest way to achieve this is the **particle-in-cell method**. The domain is divided into cubic cells with size $h$ (m) and the number of particles in each cell, $n$, is counted. The concentration in the cell is estimated as

$$c = \frac{mn}{h^3} \tag{9.106}$$

where $m$ is the pollutant mass represented by one particle. When only one or two dimenstions are considered, then the power of $h$ must be changed to 1 or 2, and $m$ must be defined appropriately to ensure unit consistency. In other words, $m$ must have units of μg $m^{-2}$ for one dimension and μg $m^{-1}$ for two dimensions.

The particle-in-cell method is not the most accurate estimation method because it forces each cell to have a homogeneous concentration, and it involves no smoothing of statistical fluctuations. The **kernel method** (Lorimer, 1986; Lorimer and Ross, 1986) is more accurate. In this method each particle contributes toward the concentration estimate, but with a weight that vanishes rapidly with distance from the point of interest. The concentration is calculated as

$$c = \frac{m}{h^3} \sum_{i=1}^{n} k\left(\frac{d_i}{h}\right) \tag{9.107}$$

where $n$ is the number of particles in the domain, $h$ is the resolution (m), $d_i$ is the distance between particle $i$ and the point of interest, and $k$ is a function with integral

1 over the computational domain. The Gaussian distribution is usually used as the kernel function:

$$k(x) = \pi^{-n_d/2} \exp(-x^2) \tag{9.108}$$

where $n_d$ is the number of dimensions considered in the estimate of the concentration. Again, $m$ must be defined appropriately to ensure unit consistency.

Lorimer (1986) recommended the following kernel function for the three-dimensional case:

$$k(x) = \pi^{-3/2}(2.5 - x^2)\exp(-x^2) \tag{9.109}$$

When concentrations near the domain boundary are calculated, the kernel function needs to be renormalized to ensure that the integral of $k(x)$ over the computational domain remains equal to 1. When the terrain is flat and the boundary is the surface, eq. (9.108) can be normalized as follows:

$$k(x) = \frac{2\pi^{-n_d/2}}{1 + \mathrm{erf}(z_i/h)} \exp(-x^2) \tag{9.110}$$

where $z_i$ is the height of the point of interest above the surface. If the surface is not flat, it is recommended to do the calculation in terrain following coordinates.

**Example 9.1.** Calculate the concentration profile associated with Figure 9.1, at a 1000-m distance from the source. It can be assumed that the release rate is 1 g s$^{-1}$, the wind speed is 2 m s$^{-1}$, the particles are well-mixed in the $z$ direction, with a mixing height of 20 m. The particle locations in the $y$ directions (in m) are −19.8552, −59.6014, 17.7876, 30.2642, 52.9285, 24.3665, −44.7198, 82.5435, 130.6672, and −8.4718.

***Solution.*** The mass of a particle must reflect the fact that the data will be analyzed in one dimension ($y$ direction). A three-dimensional data analysis with a release of 10 particles per second would lead to a mass of $10^6$ μg s$^{-1}$/10 particles s$^{-1}$ = $10^5$ μg particle$^{-1}$. Because the particles are well-mixed over a 20-m mixing height, it is as if a single, 20-m high cell is defined, and the particle mass is per 20 m height, or $10^5$ μg particle$^{-1}$/20 m = 5000 μg particle$^{-1}$ m$^{-1}$. Because the 10 particles shown in Figure 9.1 are assumed to represent 1 second of emission, it is as if we defined a single cell with a length 2 m ($u$ = 2 m s$^{-1}$). Hence: $m$ = 5000 μg particle$^{-1}$ m$^{-1}$/2 m = 2500 μg particle$^{-1}$ m$^{-2}$.

For the particle-in-cell method, the number of particles in 50-m cells is determined. Smaller cells lead to increased fluctuation, whereas larger cells lead to decreased resolution. The result is shown in Figure 9.2. The theoretical solution expected with the Gaussian plume model is shown in the figure as well. For the kernel method, the Gaussian equation with $h$ = 25 m is used.

Both methods lead to a reasonable accuracy, as judged by comparison with the Gaussian result. Higher accuracy is possible by increasing the number of particles.

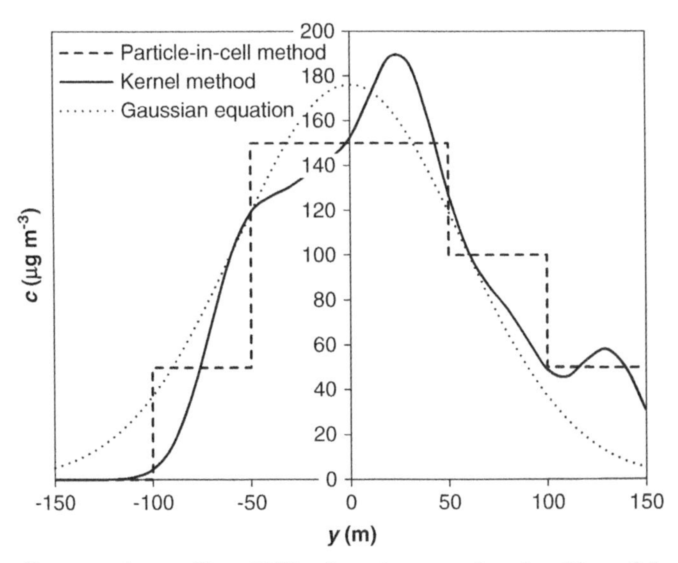


Figure 9.2 Concentration profile at 1000 m from the source based on Figure 9.1.

As can be seen from Example 9.1, a reasonable accurary is possible with 10 particles. To achieve high accuracy, on the order of 30 particles is needed. For a two-dimensional calculation we would need this number in the $y$ and $z$ directions, that is, about 100–1000 particles. This is also the number of particles needed if we wish to reuse the same particle paths at different times to construct a three-dimensional plume. For a three-dimensional instantaneous puff, we need a distribution in the $x$ direction as well, so 1000–30,000 particles are needed. For the full three-dimensional Lagrangian calculation, without reuse of the same particle paths at different times, but with a continuous release of particles, we need a set of particles to fill two dimensions every 1–10 s. This leads to 36,000–3.6 million particles for a one-hour calculation. In three dimensions, the lower estimate is generally sufficient for most applications.

## 9.4 STOCHASTIC LAGRANGIAN CALCULATION EXAMPLES

**Example 9.2.** To illustrate the principle of stochastic Lagrangian particle modeling, we will complete the calculation for a single particle in the simplest case of homogeneous Gaussian turbulence. As indicated before, homogeneous Gaussian turbulence is an acceptable approximation in the $y$ direction (crosswind). Assume the following conditions:

Wind speed: $u = 2\,\mathrm{m\,s^{-1}}$

Turbulent velocity: $\sigma_v = 0.2\,\mathrm{m\,s^{-1}}$

Lagrangian time scale: $T_{i,L}$: 100 s

***Solution.*** We choose a time step $\Delta t$ of 10 s. In homogeneous turbulence, this is sufficiently short. Hence, we find the following intermediary results:

$$\exp\left(-\frac{\Delta t}{T_{\mathrm{i,L}}}\right) = 0.904837$$

$$\sigma_n = \sqrt{1-\exp\left(-\frac{2t}{T_{\mathrm{i,L}}}\right)}\sigma_v = 0.085151\,\mathrm{m\,s^{-1}}$$

For the random number generation the method of eqs. (9.96)–(9.101) is used with a seed of 80,000,000.

We need initial conditions. As initial crosswind location, $y=0$ at $t=0$ is assumed. As initial crosswind velocity fluctuation, $v'$, at $t=0$, the first random number $N_1$ (= 1.9941239) is multiplied by $\sigma_v$, leading to $v'=0.398825\,\mathrm{m\,s^{-1}}$.

Consecutive velocity fluctuations are calculated with eq. (9.36), where $r(t)$ of step $n$ equals $N_{n+1}$. For the first step this leads to $v'=0.398825\times0.904837+0.085151\times 0.6681220=0.417763\,\mathrm{m\,s^{-1}}$.

Crosswind locations are calculated with eq. (9.40), based on the average velocity fluctuation. Hence, for the first step: $y=0+[(0.398825+0.417763)/2]\times 10=4.08294\,\mathrm{m}$.

Further steps are shown in Table 9.1. Note that one of the $R$ values is negative, which invokes eq. (9.98). The velocity fluctuations show the expected autocorrelation: Consecutive values are similar, but the similarity disappears as the time lag increases. This results into a relatively smooth but irregular path, similar to the trajectories in Figure 9.1. At a downwind distance of 400 m a value of $y'=18.5\,\mathrm{m}$ is obtained. More particles can easily be modeled by continuing the random number series or by changing the seed. Under identical conditions, $\sigma_y$ is about 30 m at a downwind distance of 400 m.

The calculations for this example are on the enclosed CD, in the file "Example 9.2. Stochastic Lagrangian particle homogeneous.xlsx."

**Example 9.3.** Calculate the dispersion in two dimensions from a continuous line source at 1 m above the ground, with an emission of $1\,\mathrm{mg\,s^{-1}\,m^{-1}}$ (i.e., per meter of source). Assume neutral conditions, $u_*=0.25\,\mathrm{m\,s^{-1}}$, $z_0=0.1\,\mathrm{m}$, $d=0.4\,\mathrm{m}$, $H_r=1\,\mathrm{m}$ (average obstacle height). For $z<H_r$, an exponential wind speed profile can be assumed:

$$u = u_{H_r}\exp\left[-a\left(1-\frac{z}{H_r}\right)\right] \tag{8.17}$$

with $a=1$. Furthermore, assume $K_z=ku_*z$ when $z>H_r$, and $K_z=ku_*H_r$ when $z<H_r$. And $\sigma_w=1.3\,u_*$ can be assumed throughout.

Take a snapshot of a continuous plume 60 s after the start of the emission and compare with a snapshot of a puff taken 60 s after the emission. Assume that transfer

**TABLE 9.1 Stochastic Lagrangian Model Calculations in Lateral Direction: $u = 2\,m\,s^{-1}$, $\sigma_v = 0.2\,m\,s^{-1}$, $T_{i,L} = 100\,s$ (seed = 80,000,000).**

| $R_n$ | $P_n$ | $p_n$ | $N_n$ | $t$ (s) | $x$ (m) | $v'$ (m s$^{-1}$) | $y'$ (m) |
|---|---|---|---|---|---|---|---|
| 235236978 | 235236978 | 0.1095408 | 1.9941239 | 0 | 0 | 0.398825 | 0.00000 |
| 110495119 | 110495119 | 0.0514533 | 0.6681220 | 10 | 20 | 0.417763 | 4.08294 |
| 1665594025 | 1665594025 | 0.7756027 | −0.6724450 | 20 | 40 | 0.320748 | 7.77550 |
| 1189439530 | 1189439530 | 0.5538759 | −0.2367438 | 30 | 60 | 0.270066 | 10.72956 |
| −15089213 | 2132394434 | 0.9929735 | 0.0987392 | 40 | 80 | 0.252773 | 13.34376 |
| 1946151102 | 1946151102 | 0.9062472 | −0.0659779 | 50 | 100 | 0.223101 | 15.72313 |
| 638143857 | 638143857 | 0.2971589 | −0.9114586 | 60 | 120 | 0.124258 | 17.45992 |
| 750471481 | 750471481 | 0.3494655 | 1.2634159 | 70 | 140 | 0.220015 | 19.18129 |
| 1002722336 | 1002722336 | 0.4669290 | −0.5584578 | 80 | 160 | 0.151524 | 21.03898 |
| 1450123143 | 1450123143 | 0.6752662 | −1.1005822 | 90 | 180 | 0.043389 | 22.01355 |
| 427754598 | 427754598 | 0.1991888 | 0.1739577 | 100 | 200 | 0.054072 | 22.50085 |
| 1643762077 | 1643762077 | 0.7654364 | −1.7879439 | 110 | 220 | −0.103319 | 22.25462 |
| 1479593131 | 1479593131 | 0.6889892 | 0.4725243 | 120 | 240 | −0.053251 | 21.47176 |
| 1808604104 | 1808604104 | 0.8421969 | −0.7223434 | 130 | 260 | −0.109692 | 20.65705 |
| 1725636290 | 1725636290 | 0.8035620 | −0.6470374 | 140 | 280 | −0.154350 | 19.33684 |
| 1002473295 | 1002473295 | 0.4668130 | 0.1369104 | 150 | 300 | −0.128003 | 17.92507 |
| 1559458350 | 1559458350 | 0.7261794 | 0.6378658 | 160 | 320 | −0.061507 | 16.97752 |
| 1926060426 | 1926060426 | 0.8968918 | −0.4827459 | 170 | 340 | −0.096760 | 16.18619 |
| 129689956 | 129689956 | 0.0603916 | 2.3692451 | 180 | 360 | 0.114192 | 16.27335 |
| 3188787 | 3188787 | 0.0014849 | 0.0221054 | 190 | 380 | 0.105208 | 17.37035 |
| 2054335581 | 2054335581 | 0.9566246 | 0.2970744 | 200 | 400 | 0.120492 | 18.49885 |

in the $z$ direction is governed by random motions only, whereas transfer in the $x$ direction is governed by the mean wind speed only.

***Solution.*** To determine the wind speed profile, $u(H_r)$ is calculated first:

$u(H_r) = (0.25/0.4)\ \ln[(1-0.4)/0.1] = 1.120\,m\,s^{-1}$

When $z > 1$ m, the wind speed is calculated with

$u = (0.25/0.4)\ \ln[(z-0.4)/0.1]$

When $z < 1$ m, the wind speed is calculated with

$u = 1.120\ \exp[-(1-z)]$

In the last two equations $z$ is in meters.

The calculation of $K_z$ is given in the problem statement: $\sigma_w = 1.3 \times 0.25 = 0.325\,m\,s^{-1}$. Based on $K_z$ and $\sigma_w$. $T_{i,L}$ is calculated as

$$T_{i,L} = \frac{K_z}{\sigma_w^2} \tag{6.105}$$

The lowest expected value is $0.4 \times 0.25 \times 1/0.325^2 = 0.947$ s. This means that time steps no greater than approximately 0.025 s are needed to ensure numerical accuracy.

To ensure low statistical fluctuation, we will consider at least 30 particles per second. Given that we need about 40 time steps per second, we will release one new particle with every time step in the plume calculation. This leads to a total emission of $40 \times 60 = 2400$ particles. In the puff calculation we will release 2400 particles simultaneously.

The regular Langevin equation [eq. (9.37)] will be used for the calculation of the update of the wind speed fluctuation. Particle trajectories are tracked, and when $z<0$, then the sign of both $z$ and $w$ of that particle is changed. The Minimal Standard technique is used for random number generation, with a seed of 1,234,567,890. Because the procedure calculates two random numbers in each cycle, only half of the numbers is used, leaving the other numbers available for improvements to the model without changing the existing random steps if so desired.

The solution is implemented in the Matlab file "example_9_3.m" on the enclosed CD. The result of the continuous plume calculation is shown in Figure 9.3. We see the normal plume spread with reflection on the surface, but at large distances the plume seems to lift off the ground. This is not a real process. Instead, higher particles move faster due to vertical wind shear and reach further than lower particles. The effect would disappear if longer simulation times were chosen.

The puff calculation is implemented in the Matlab file "example_9_3_2.m" on the enclosed CD. The result of the continuous plume calculation is shown in Figure 9.4. In Figure 9.4 it is clear that what looked like plume rise or lofting in Figure 9.3 is actually the result of puff tilting. Note that the puff shows a pronounced spread in the $x$ direction, in spite that $u'=0$ was assumed in the calculation. The plume spread along the wind direction is also due to wind shear: Particles are exchanged between slower moving low layers and faster moving high layers.

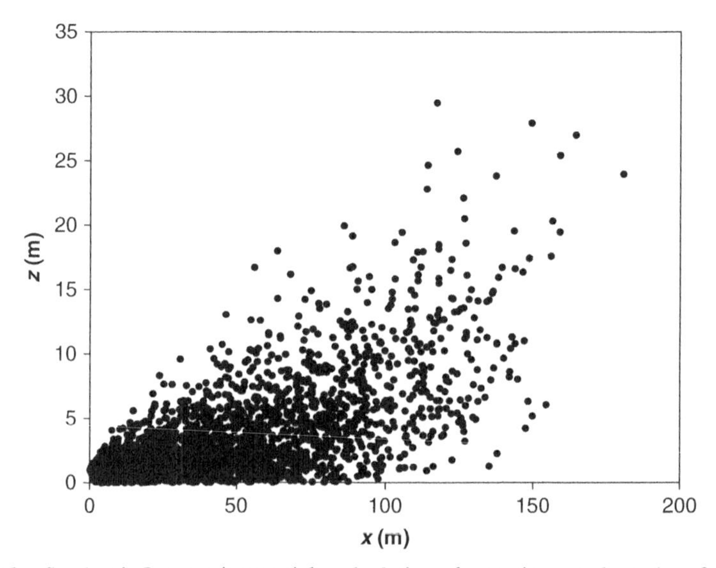


Figure 9.3 Stochastic Lagrangian particle calculation of a continuous plume ($u_* = 0.25\,\mathrm{m\,s^{-1}}$, emission height 1 m, $z_0 = 0.1$ m, $d = 0.4$ m, 40 particles $\mathrm{s^{-1}}$, 60 s emission time).

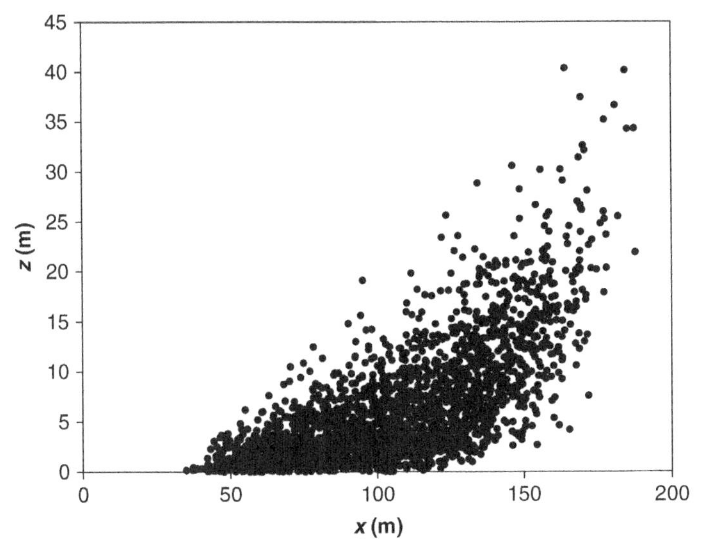


Figure 9.4 Stochastic Lagrangian particle calculation of a puff ($u_* = 0.25\,\mathrm{m\,s^{-1}}$, emission height 1 m, $z_0 = 0.1$ m, $d = 0.4$ m, 2400 particles $\mathrm{s^{-1}}$, $t = 60$ s after emission).

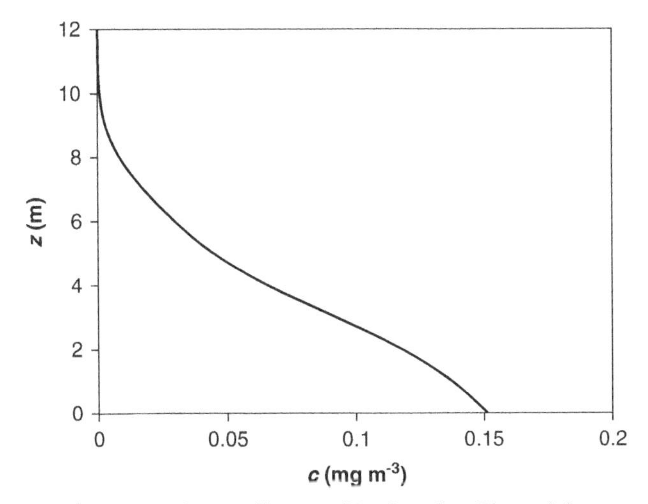


Figure 9.5 Vertical concentration profile at $x = 35$ m based on Figure 9.3.

Because not all the particles emitted at $t = 0$ have reached $x = 40$ m, the plume is not fully developed except in the first 35 m. For that reason a vertical concentration profile for the plume is only calculated at 35 m from the source. The calculation was made with eq. (9.110), using two dimensions, and $h = 2$ m. The result is shown in Figure 9.5. The profile is smooth indicating that a sufficient number of particles were released to resolve the concentration with a resolution of 2 m. However, with $h = 1$ m the profile would not be smooth. The concentration gradient at the surface is not zero. This may be the result of artifacts of the kernel technique or because the time step was still too long. Just above the obstacles, the $T_{i,L}$ value of a particle can change by about 1% in a single step, which could potentially lead to a slight accumulation of particles in the canopy layer ($z < 1$ m).

## 9.5 SUMMARY OF THE MAIN EQUATIONS

The original Langevin equation (i.e., for microscopic particles in stagnant fluid) is

$$m\frac{d^2x}{dt^2} = -3\pi\mu d_p \frac{dx}{dt} + F_c \tag{9.3}$$

Particle dispersion of the original Langevin equation is

$$\sigma_x = \sqrt{at - \frac{a}{b}\left[1 - \exp(-bt)\right]} \tag{9.20}$$

where

$$a = \frac{2kT}{3\pi\mu d_p} \tag{9.21a}$$

$$b = \frac{1}{\tau} \tag{9.21b}$$

Integrated Langevin equation in homogeneous turbulence is

$$v'(t) = v'(0)\exp\left(-\frac{t}{T_{i,L}}\right) + \sqrt{1 - \exp\left(-\frac{2t}{T_{i,L}}\right)} \cdot \sigma_v r(t) \tag{9.36}$$

Langevin equation in homogeneous turbulence based on the Wiener process is

$$dv' = -\frac{v'}{T_{i,L}}dt + \sqrt{\frac{2\sigma_v^2}{T_{i,L}}}dW \tag{9.39}$$

Particle trajectory calculation is

$$y'(t + \Delta t) = y'(t) + v'(t)\,\Delta t \tag{9.40}$$

Fokker–Planck equation is

$$\frac{\partial p}{\partial t} + v\frac{\partial p}{\partial y} = -\frac{\partial}{\partial v}(ap) + \frac{1}{2}\frac{\partial^2}{\partial v^2}(b^2 p) \tag{9.42}$$

where $b$ is given by

$$b = \sqrt{\frac{2\sigma_v^2}{T_{i,L}}} \tag{9.43}$$

Langevin equation for nonhomogeneous turbulence is

$$ds' = -\left(\frac{s'}{T_{\mathrm{i,L}}} + \frac{\partial \sigma_w}{\partial z}\right) dt + \sqrt{\frac{2}{T_{\mathrm{i,L}}}} dW \tag{9.59}$$

where

$$s' = \frac{w'}{\sigma_w} \tag{9.57}$$

The integrated Langevin equation in nonhomogeneous turbulence is

$$s'(t) = s'(0) \exp\left(-\frac{t}{T_{\mathrm{i,L}}}\right) + \left[1 - \exp\left(-\frac{t}{T_{\mathrm{i,L}}}\right)\right] \cdot T_{\mathrm{i,L}} \frac{\partial \sigma_w}{\partial z} + n(t) \tag{9.64}$$

$$w'(t) = \sigma_w \cdot s'(t) \tag{9.65}$$

with

$$\sigma_n = \sqrt{1 - \exp\left(-\frac{2t}{T_{\mathrm{i,L}}}\right) - \left[1 - \exp\left(-\frac{t}{T_{\mathrm{i,L}}}\right)\right]^2 \cdot T_{\mathrm{i,L}}^2 \left(\frac{\partial \sigma_w}{\partial z}\right)^2} \tag{9.67}$$

The maximum recommended step size for a stochastic trajectory calculation is

$$\Delta t = \min\left(0.025 T_{\mathrm{i,L}}, \frac{0.1\sigma_w}{|a|}, \frac{0.01\sigma_w}{w(\partial \sigma_w / \partial z)}\right) \tag{9.102}$$

The kernel method for concentration calculations from stochastic particle assemblies is

$$c = \frac{m}{h^3} \sum_{i=1}^{n} k\left(\frac{d_i}{h}\right) \tag{9.107}$$

## PROBLEMS

1. Derive a one-dimensional Langevin equation for homogeneous turbulence in the $y$ direction with nonzero average wind speed $\overline{v} = v_0$, based on the Fokker–Planck equation (see Section A 9.1, Appendix A).
2. Solve Example 9.3, but add a random motion in the $x$ direction. Make plausible assumptions about dispersion in the $x$ direction.
3. Solve Example 9.3, but use the equation $\sigma_w = 1.3\, u_* \exp(-0.01\, z)$, where $z$ is in meters.
4. Solve Example 9.3, but use the Wilson et al. (1981a) technique (see Section A 9.2, Appendix A) to account for changes in $T_{\mathrm{i,L}}$.
5. Solve Example 9.3 for a point source with emission 25 mg s$^{-1}$. Plot a top view and a side view of the resulting plume.

## MATERIALS ONLINE

- "Figure 9.1. Stochastic Lagrangian sim 10 particles.xlsx": Calculation of a stochastic Lagrangian particle model in homogeneous atmosphere with 10 particles.
- "Example 9.1. Particle concentration calculations.xlsx": Calculation of a concentration profile from the particle locations at $x = 1000\,\text{m}$ in Figure 9.1.
- "Example 9.2. Stochastic Lagrangian particle homogeneous.xlsx": Solution of Example 2.
- "Example_9_3.m": Matlab file for the solution of Example 9.3 (continuous emission).
- "Example_9_3_2.m": Matlab file for the solution of Example 9.3 (puff emission).

## REFERENCES

Anand M.S. and Pope S.B. (1985). Diffusion behind a line source in grid turbulence. In Bradbury J.T.S., Durst F., Launder B.E., Schmidt F.W. and Whitelaw J.H. (eds.). *Turbulent Shear Flows*. Springer, New York, pp. 46–61.

Anfossi D., Degrazia G., Ferrero E., Gryning S.E., Morselli M.G., and Trini Castelli S. (2000). Estimation of the Lagrangian structure function constant $C_0$ from surface layer wind data. *Bound. Layer Meteorol.* **95**, 249–270.

Baerentsen J.H. and Berkowicz R. (1984). Monte Carlo simulation of plume dispersion in the convective boundary layer. *Atmos. Environ.* **18**, 701–712.

Bocksel T.L. and Loth E. (2006). Stochastic modeling of particle diffusion in a turbulent boundary layer. *Int. J. Multiph. Flow* **32**, 1234–1253.

Boehm M.T. and Aylor D.E. (2005). Lagrangian stochastic modeling of heavy particle transport in the convective boundary layer. *Atmos. Environ.* **39**, 4841–4850.

Carvalho J.C., Rizza U., Lovato R., Degrazia G.A., Filho E.P.M., and Campos C.R.J. (2009). Estimation of the Kolmogorov constant by large-eddy simulation in the stable PBL. *Physica A* **388**, 1500–1508.

Degrazia G.A., Welter G.S., Wittwer A.R., da Costa Carvalho J., Roberti D.R., Costa Acevedo O., Moraes O.L.L., and de Campos Velho H.F. (2008). Estimation of the Lagrangian Kolmogorov constant from Eulerian measurements for distinct Reynolds number with application to pollution dispersion model. *Atmos. Environ.* **42**, 2415–2423.

DeVito T.J., Cao X, Roy G., Costa J.R., and Andrews W.S. (2009). Modelling aerosol concentration distributions from transient (puff) sources. *Can. J. Civil Eng.* **36**, 911–922.

Du S. (1997). Universality of the Lagrangian velocity structure function constant ($C_0$) across different kinds of turbulence. *Bound. Layer Meteorol.* **83**, 207–219.

Du S. (1998). Reply to "Comments on the 'Universality of the Lagrangian velocity structure function constant ($C_0$) across different kinds of turbulence'" by A.M. Reynolds. *Bound. Layer Meteorol.* **89**, 171–172.

Du S. (2002). On the inter-dependency between lateral diffusion and vertical diffusion in the atmospheric surface layer. *Atmos. Environ.* **36**, 3049–3054.

Einstein A. (1905). Über die von der molekularkinetischen Theorie der Wärme geforderte Bewegung von in ruhenden Flüssigkeiten suspendierten Teilchen. *Ann. Phys.* **17**, 549–560.

Evans R.B. III, Watson G.M., and Mason E.A. (1961). Gaseous diffusion in porous media at uniform pressure. *J. Chem. Phys.* **35**, 2076–2083.

Hall C.D. (1974). The simulation of particle motion in the atmosphere by a numerical random walk model. *Quart. J. Roy. Meteorol. Soc.* **101**, 235–244.

Janicke (2008). *AUSTAL2000. Program Documentation of Version 2.4.4.4*. Janicke Consulting, Dunum, Germany.

Langevin P. (1908). Sur la théorie du mouvement brownien. *C. R. Acad. Sci. (Paris)* **146**, 530–533.

Legg B.J. and Raupach M.R. (1982). Markov-chain simulation of particle dispersion in inhomogeneous flows: The mean drift velocity induced by a gradient in Eulerian velocity variance. *Bound. Layer Meteorol.* **24**, 3–13.

Lemons D.S. and Gythiel A. (1997). Paul Langevin's 1908 paper "On the theory of Brownian motion" ["Sur la théorie du mouvement brownien," *C. R. Acad. Sci. Paris 146*, 530–533 (1908)]. *Am. J. Phys.* **65**, 1079–1081.

Leva J.L. (1992). A fast normal random number generator. *ACM Trans. Math. Softw.* **18**, 449–453.

Lorimer G.S. (1986). The kernel method for air quality modelling—I. Mathematical foundation. *Atmos. Environ.* **20**, 1447–1452.

Lorimer G.S. and Ross D.G. (1986). The kernel method for air quality modelling—II. Comparison with analytical solutions. *Atmos. Environ.* **20**, 1773–1780.

Luhar A. and Britter R.E. (1989). A random walk model for dispersion in inhomogeneous turbulence in a convective boundary layer. *Atmos. Environ.* **23**, 1911–1924.

Matsumoto M. and Nishimura T. (1998). Mersonne Twister: A 623-dimensionally equidistributed uniform pseudorandom number generator. *ACM Trans. Model. Comp. Sim.* **8**, 3–30.

Park S.K. and Miller K.W. (1988). Random number generators—Good ones are hard to find. *Commun. ACM* **31**, 1192–1201.

Press W.H., Flannery B.P., Teukolsky S.A., and Vetterling W.T. (1992). *Numerical Recipes in C. The Art of Scientific Computing*, 2nd edn. Cambridge University Press, Cambridge, UK.

Press W.H., Teukolsky S.A., Vetterling W.T., and Flannery B.P. (2007). *Numerical Recipes*, 3rd edn. Cambridge University Press, Cambridge, UK.

Reynolds A.M. (1998). Comments on the "Universality of the Lagrangian velocity structure function constant ($C_0$) across different kinds of turbulence." *Bound. Layer Meteorol.* **89**, 161–170.

Sawford B.L. (1985). Lagrangian stochastic simulation of concentration mean and fluctuation fields. *J. Appl. Meteorol.* **24**, 1152–1166.

Schrage L. (1979). A more portable Fortran random generator. *ACM Trans. Math. Softw.* **5**, 132–138.

Smith F.B. (1968). Conditioned particle motion in a homogeneous turbulent field. *Atmos. Environ.* **2**, 491–508.

Stohl A. (1998). Computation, accuracy and applications of trajectories—A review and bibliography. *Atmos. Environ.* **32**, 947–966.

Stohl A. and Thomson D.J. (1999). A density correction for Lagrangian particle dispersion models. *Bound. Layer Meteorol.* **90**, 155–167.

Stohl A., Forster C., Frank A., Seibert P., and Wotawa G. (2005). Technical note: The Lagrangian particle dispersion model FLEXPART version 6.2. *Atmos. Chem. Phys.* **5**, 2461–2474.

Stull R.B. (1988). *An Introduction to Boundary Layer Meteorology*. Kluwer Academic, Dordrecht, The Netherlands.

Tennekes H. (1979). The exponential Lagrangian correlation function and turbulent diffusion in the inertial subrange. *Atmos. Environ.* **13**, 1565–1567.

Thomson D.J. (1987). Criteria for the selection of stochastic models of particle trajectories in turbulent flows. *J. Fluid Mech.* **180**, 529–556.

Weber A.H., Irwin J.S., Petersen W.B., Mathis J.J., Jr., and Kahler J.P. (1982). Spectral scales in the atmospheric boundary layer. *J. Appl. Meteorol.* **21**, 1622–1632.

Weil J.C. (1990). A diagnosis of the asymmetry in top-down and bottom-up diffusion using a Lagrangian stochastic model. *J Atmos. Sci.* **47**, 501–515.

Weil J.C., Corio L.A., and Brower R.B. (1997). A PDF dispersion model for buoyant plumes in the convective boundary layer. *J. Appl. Meteorol.* **36**, 982–1003.

Wichmann B.A. and Hill I.D. (1982). Algorithm 183: An efficient and portable pseudo-random generator. *Appl. Stat.* **31**, 188–190.

Wilson J.D. (2000). Trajectory models for heavy particles in atmospheric turbulence: Comparison with observations. *J. Appl. Meteorol.* **39**, 1894–1912.

Wilson J.D. and Flesch T.K. (1993). Flow boundaries in random-flight dispersion models: Enforcing the well-mixed condition. *J. Appl. Meteorol.* **32**, 1695–1707.

Wilson J.D. and Sawford B.L. (1996). Review of Lagrangian stochastic models for trajectories in the turbulent atmosphere. *Bound. Layer Meteorol.* **78**, 191–210.

Wilson J.D., Thurtell G.W., and Kidd G.E. (1981a). Numerical simulation of particle trajectories in inhomogeneous turbulence. I: Systems with constant turbulent velocity scale. *Bound. Layer Meteorol.* **21**, 295–313.

Wilson J.D., Thurtell G.W., and Kidd G.E. (1981b). Numerical simulation of particle trajectories in inhomogeneous turbulence. II: Systems with variable turbulent velocity scale. *Bound. Layer Meteorol.* **21**, 423–441.

Wilson J.D., Thurtell G.W. and Kidd G.E. (1981c). Numerical simulation of particle trajectories in inhomogeneous turbulence, III: Comparison of predictions with experimental data for the atmospheric surface layer. *Bound. Layer Meteorol.* **21**, 443–463.

Wilson J.D., Legg B.J. and Thomson D.J. (1983). Calculation of particle trajectories in the presence of a gradient in turbulent-velocity variance. *Bound. Layer Meteorol.* **27**, 163–169.

Wilson J.D., Ferrandino F.J. and Thurtell G.W. (1989). A relationship between deposition velocity and trajectory reflection probability for use in stochastic Lagrangian dispersion models. *Agric. Forest Meteorol.* **47**, 139–154.

CHAPTER 10

# COMPUTATIONAL FLUID DYNAMICS AND METEOROLOGICAL MODELING

## 10.1 INTRODUCTION

So far in this book we have always assumed that wind speed and turbulence profiles, or, in three dimensions, wind speed and turbulence fields, are available or obtainable from similarity calculations. In practice, this is not always the case. Local wind speed profiles may be disturbed by obstacles, and the model approaches we discussed in Chapters 5 and 8 may not always be satisfactory. We may need weather data in remote or rugged terrain, where representative data from a meteorological station is not available, depriving us of the imput data needed for similarity calculations. Or the modeling domain may be too large to be covered by measured data alone, necessitating a prognostic meteorological calculation.

In all these cases the answer to the problem is moving closer to the actual physics of atmospheric turbulence. *In principle*, direct simulation of atmospheric dynamics is possible with the **Navier–Stokes** equation. However, direct simulation is impossible *in practice* with current computer technology in all but the simplest cases because the scales at which turbulent motions exist cover many orders of magnitude. The smallest turbulence scale is the scale at which viscosity dissipates turbulent energy: on the order of a few millimeters and smallest near the surface. The Kolmogorov length scale [eq. (6.184)] is an estimation of this scale. The largest turbulence scale is the size of the planetary boundary layer: up to a few kilometers, 5–6 orders of magnitude larger than the smaller scale. In the horizontal directions, the largest turbulence scale overlaps with the smallest scale at which weather patterns induce variability in the wind speeds, so there is no set boundary beyond which the effect on the simulation becomes negligible. But even considering 5 orders of magnitude in each dimension necessitates a grid with $10^{15}$ nodes for a direct simulation to be successful. Such a fine grid would necessitate a time step smaller than 0.001 s, so at least 100 million time steps are required to calculate a single day.

*Air Dispersion Modeling: Foundations and Applications*, First Edition. Alex De Visscher.

These issues force turbulence modelers and meteorologists to take a step away from reality and use empirical models to emulate the smallest scales as they cannot be resolved by direct simulation. The two main approaches to achieve this are Reynolds-averaged Navier–Stokes (RANS) and large eddy simulation (LES). Numerical modeling based on the Navier–Stokes equation is termed **computational fluid dynamics (CFD)**. The applications of CFD range from resolving small-scale turbulence around obstacles to global climate modeling.

The Navier–Stokes equation is usually insufficient to model a turbulent system. Material and energy balances are usually required as well. CFD packages usually allow the user to solve these equations simultaneously with the Navier–Stokes equation.

The objective of this chapter is to familiarize the reader with computational fluid dynamics for air quality purposes. The different approaches will be discussed with their advantages and disadvantages.

## 10.2 CFD MODEL FORMULATION: FUNDAMENTALS

### 10.2.1 The Navier–Stokes Equation

The basis of CFD is the Navier–Stokes equation, a momentum balance for fluids (liquids and gases). It is usually written in vector form, but in this book it will be written in Cartesian form, with one equation for each dimension. For the derivation, we start from Newton's law:

$$\text{Mass} \times \text{acceleration} = \text{force} \tag{10.1}$$

We *follow* a fluid element with mass $m$ as it is subjected to a gravity force $F_g$, a Coriolis force $F_c$, a pressure force $F_p$, and a viscous shear force $F_s$:

$$m \times \text{acceleration} = F_g + F_c + F_p + F_s \tag{10.2}$$

We will discuss the terms in eq. (10.2) one by one, and find equations to describe them for a fluid element with density $\rho$, located at a point with Cartesian coordinates $(x,y,z)$, moving at a velocity with components $(u,v,w)$, and with dimensions $dx$ in the $x$ direction, $dy$ in the $y$ direction, and $dz$ in the $z$ direction.

The mass of the fluid element is

$$m = \rho\, dx\, dy\, dz \tag{10.3}$$

The acceleration of the fluid element is the total derivative of the fluid velocity with respect to time. As the fluid velocity depends on location as well as time, we can write the acceleration as

$$\text{Acceleration} = \frac{du(x,y,z,t)}{dt} = \frac{\partial u}{\partial t} + \frac{\partial u}{\partial x}\frac{\partial x}{\partial t} + \frac{\partial u}{\partial y}\frac{\partial y}{\partial t} + \frac{\partial u}{\partial z}\frac{\partial z}{\partial t}$$
$$= \frac{\partial u}{\partial t} + u\frac{\partial u}{\partial x} + v\frac{\partial u}{\partial y} + w\frac{\partial u}{\partial z} \tag{10.4}$$

in the $x$ direction. The acceleration in the $y$ and $z$ directions is immediately obvious.

The gravity force acts down (i.e., is negative) in the $z$ direction. It equals mass times acceleration of gravity:

$$F_{g,z} = -\rho\, dx\, dy\, dz \cdot g \tag{10.5}$$

The Coriolis force is an apparent force that is felt by all objects moving in a noninertial frame of reference, in this case a rotating Earth. This is easiest to visualize as an object thrown in the north direction ($v>0$) on the Northern Hemisphere of a globe (Fig. 10.1). At the start of its trajectory, the object moves east with Earth. As it is thrown north, it moves toward a more slowly moving Earth, which is an effect of the Earth's curvature. Hence, from the perspective of Earth, the object accelerates in the east direction ($a_x>0$), which is felt as an apparent force, the Coriolis force. Based on this picture, the following equation can be derived for the Coriolis force in the $x$ direction:

$$F_{c,x} = \rho\, dx\, dy\, dz \cdot 2\Omega \sin\varphi \cdot v = \rho\, dx\, dy\, dz \cdot fv \tag{10.6}$$

where $\Omega$ is the angular velocity of Earth ($7.292\times10^{-5}$ rad s$^{-1}$ or simply $7.292\times10^{-5}$ s$^{-1}$), $\varphi$ is the latitude (radians: 1 rad = 57.29578° = 57° 17′ 44.8″), and $f$ is the Coriolis parameter (s$^{-1}$). From eq. (10.6), it is clear that

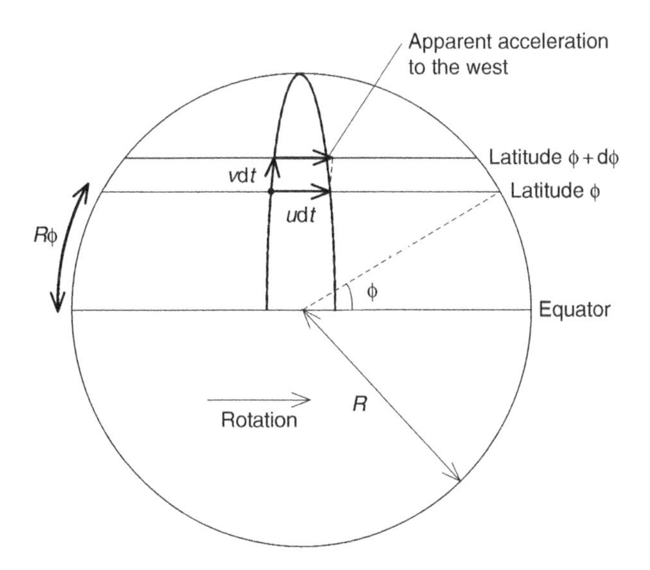


Figure 10.1 Object moving north on a rotating Earth: Coriolis effect.

$$f = 2\Omega \sin \varphi \tag{10.7}$$

An intuitive derivation of eq. (10.6) is given in Section A10.1 (Appendix A).

The existence of a Coriolis force for an object moving east or west is less obvious. However, from Figure 10.2 it can be seen that the plane that connects the movement of the object to the center of Earth is not stationary from the perspective of Earth. For an object moving east ($u>0$), there is an acceleration to the south ($a_y<0$). Another way of stating this is that an object moving east (with the rotation of Earth) experiences an increased centrifugal force, which has a southern component. The Coriolis force is as follows:

$$F_{c,y} = -\rho\, dx\, dy\, dz \cdot fu \tag{10.8}$$

Equation (10.8) is essentially the same as eq. (10.6), but due to the choice of the $x$ and $y$ axes, an acceleration to the right is positive when the object moves parallel with the $y$ axis, in the direction of increasing $y$, and negative when the object moves parallel with the $x$ axis, in the direction of increasing $x$.

The pressure term in eq. (10.2) is the result of a pressure differential. For instance, consider the pressures in the $x$ direction (Fig. 10.3). Considering that force equals pressure multiplied by area, one obtains

$$F_{p,x} = p\, dy\, dz - \left( p + \frac{\partial p}{\partial x} dx \right) dy\, dz = -\frac{\partial p}{\partial x} dx\, dy\, dz \tag{10.9}$$

The application to the $y$ and $z$ directions is obvious.

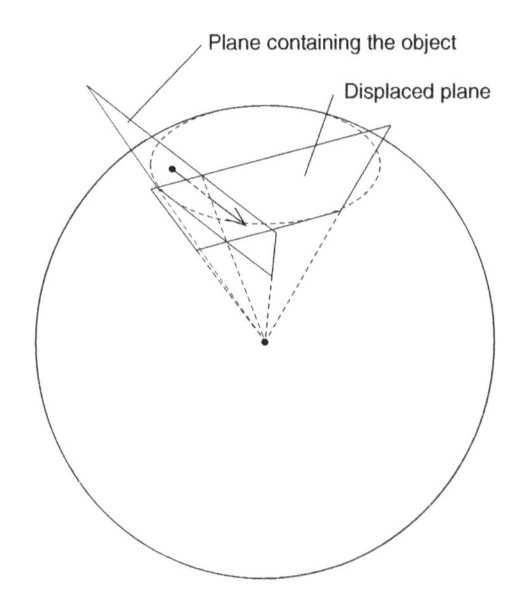


Figure 10.2 Object moving east on a rotating Earth: Coriolis effect.

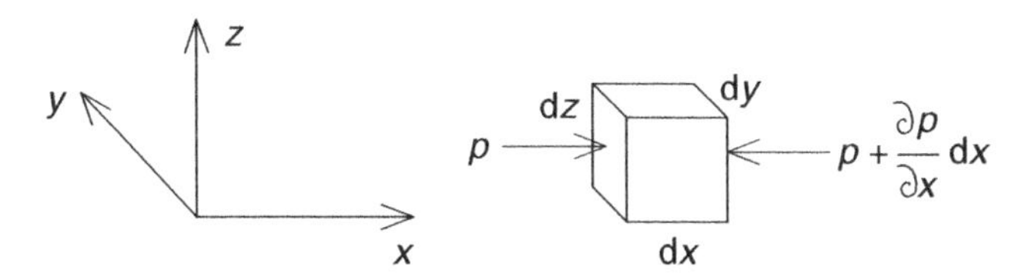


Figure 10.3 Pressure forces in a fluid element *dx/dy/dz* in the *x* direction.

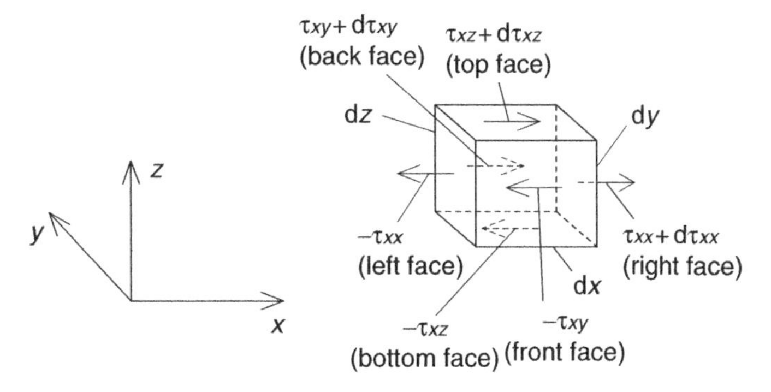


Figure 10.4 Viscous forces in a fluid element *dx/dy/dz* in the *x* direction.

The viscous stress term is less straightforward because it involves forces acting on every face of the fluid element. This is shown in Figure 10.4. In the figure, the stresses $\tau_{xy}$ and $\tau_{xz}$ are **shear stresses**: stresses resulting in forces acting upon a plane parallel to the force. This is in contrast with pressure, which results in a force normal to the plane. As with pressure, shear stress has units Pa [force (N) per unit area (m$^2$)]; and $\tau_{xx}$ is a normal stress.

The viscous force term can be written as

$$F_{s,x} = -\tau_{xx}\, dy\, dz + \left(\tau_{xx} + \frac{\partial \tau_{xx}}{\partial x} dx\right) dy\, dz - \tau_{xy}\, dx\, dz + \left(\tau_{xy} + \frac{\partial \tau_{xy}}{\partial y} dy\right) dx\, dz$$
$$- \tau_{xz}\, dx\, dy + \left(\tau_{xz} + \frac{\partial \tau_{xz}}{\partial z} dz\right) dx\, dy \tag{10.10}$$

This equation reduces to

$$F_{s,x} = \frac{\partial \tau_{xx}}{\partial x} dx\, dy\, dz + \frac{\partial \tau_{xy}}{\partial y} dx\, dy\, dz \frac{\partial \tau_{xz}}{\partial z} dx\, dy\, dz \tag{10.11}$$

Viscous stresses are driven by velocity gradients. The relationship between velocity gradient and viscous stress in a fluid is straightforward when two conditions are met:

- The fluid is incompressible. This is an acceptable assumption in the atmosphere because pressure changes are small over the scale where viscous forces are relevant (millimeters to centimeters).

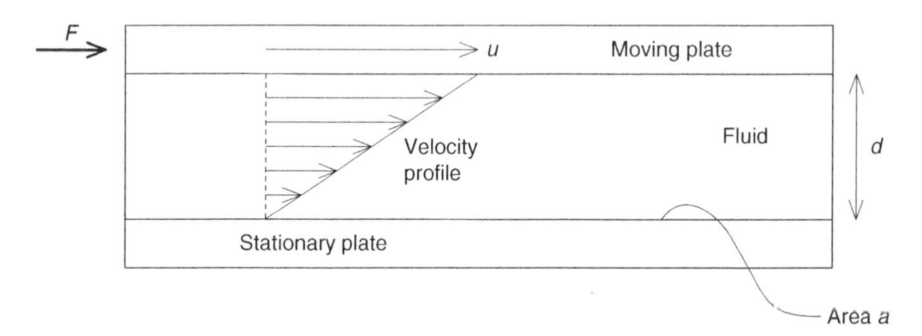


Figure 10.5 Viscous forces in a fluid enclosed between a stationary plate and a moving plate.

- The fluid is **Newtonian**. For organized incompressible flow this means that the shear stress is proportional to the velocity gradient. An example of organized straight flow with a velocity gradient is shown in Figure 10.5.

Newton's viscosity law for an **incompressible gas** is as follows:

$$\tau_{xz} = \mu\left(\frac{\partial u}{\partial z} + \frac{\partial w}{\partial x}\right) \tag{10.12}$$

where $\mu$ is the viscosity (Pa s). A similar equation applies to $\tau_{xy}$ and $\tau_{xx}$. For the latter, the two terms between brackets are identical. To visualize eq. (10.12), we can identify the following terms in Figure 10.5:

$$\tau_{xz} = \frac{F}{A} \tag{10.13}$$

$$\frac{\partial u}{\partial z} = \frac{u}{d} \tag{10.14}$$

Figure 10.5 explains only the first term between brackets in eq. (10.12). This is because the second term is zero in this case. However, it can be seen intuitively that a second term must exist. Figure 10.6 considers the torques around the center of a fluid element that are generated by the shear stresses $\tau_{xz}$ and $\tau_{zx}$. It is clear from Figure 10.6 that the stresses generate a rotation of the fluid element unless $\tau_{xz} = \tau_{zx}$. As no such rotation is observed in Figure 10.5, it follows that the two shear stresses must indeed be the same. Equation (10.12) guarantees this and is consistent with eqs. (10.13) and (10.14) in the absence of a vertical velocity.

Viscous forces are friction forces. It is intuitively clear that intermolecular forces will cause a force that resists lateral and vertical velocity gradients. Less intuitively obvious is the fact that velocity gradients along the flow direction leads to friction forces as well. This is because there is no simple vizualization similar to Figure 10.5 for this stress component. However, the need to include a $\tau_{xx}$ term in the viscous force will become clear from the turbulent analogy when we discuss the Reynolds-averaged Navier–Stokes equation.

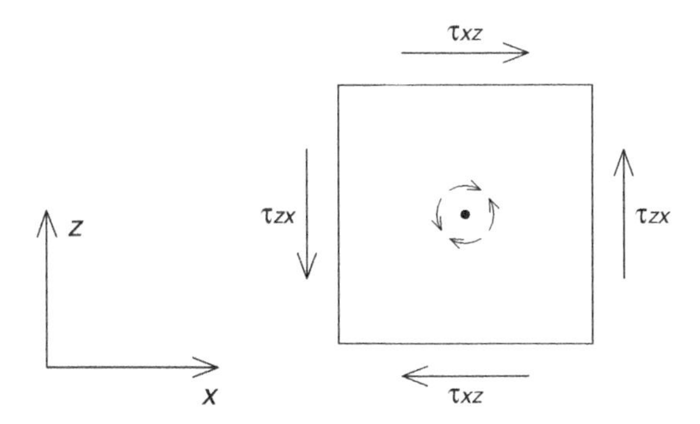


Figure 10.6 Stresses in a fluid element to balance torque.

Substituting eqs. (10.3)–(10.6),(10.8) and (10.9) and (10.11), as well as equivalent equations for the $y$ and $z$ directions, in eq. (10.2) for the $x$, $y$, and $z$ directions, and dividing by $\rho\, dx\, dy\, dz$, leads to the following three equations:

$$\left(\frac{\partial u}{\partial t}+u\frac{\partial u}{\partial x}+v\frac{\partial u}{\partial y}+w\frac{\partial u}{\partial z}\right)=fv-\frac{1}{\rho}\frac{\partial p}{\partial x}+\frac{1}{\rho}\left(\frac{\partial \tau_{xx}}{\partial x}+\frac{\partial \tau_{xy}}{\partial y}+\frac{\partial \tau_{xz}}{\partial z}\right) \quad (10.15)$$

$$\left(\frac{\partial v}{\partial t}+u\frac{\partial v}{\partial x}+v\frac{\partial v}{\partial y}+w\frac{\partial v}{\partial z}\right)=-fu-\frac{1}{\rho}\frac{\partial p}{\partial y}+\frac{1}{\rho}\left(\frac{\partial \tau_{yx}}{\partial x}+\frac{\partial \tau_{yy}}{\partial y}+\frac{\partial \tau_{yz}}{\partial z}\right) \quad (10.16)$$

$$\left(\frac{\partial w}{\partial t}+u\frac{\partial w}{\partial x}+v\frac{\partial w}{\partial y}+w\frac{\partial w}{\partial z}\right)=-g-\frac{1}{\rho}\frac{\partial p}{\partial z}+\frac{1}{\rho}\left(\frac{\partial \tau_{zx}}{\partial x}+\frac{\partial \tau_{zy}}{\partial y}+\frac{\partial \tau_{zz}}{\partial z}\right) \quad (10.17)$$

Substituting eq. (10.12) leads to

$$\frac{\partial u}{\partial t}+u\frac{\partial u}{\partial x}+v\frac{\partial u}{\partial y}+w\frac{\partial u}{\partial z}=fv-\frac{1}{\rho}\frac{\partial p}{\partial x}+\frac{\mu}{\rho}\left(2\frac{\partial^2 u}{\partial x^2}+\frac{\partial^2 u}{\partial y^2}+\frac{\partial^2 v}{\partial x\partial y}+\frac{\partial^2 u}{\partial z^2}+\frac{\partial^2 w}{\partial x\partial z}\right) \quad (10.18)$$

$$\frac{\partial v}{\partial t}+u\frac{\partial v}{\partial x}+v\frac{\partial v}{\partial y}+w\frac{\partial v}{\partial z}=-fu-\frac{1}{\rho}\frac{\partial p}{\partial y}+\frac{\mu}{\rho}\left(\frac{\partial^2 v}{\partial x^2}+\frac{\partial^2 u}{\partial x\partial y}+2\frac{\partial^2 v}{\partial y^2}+\frac{\partial^2 v}{\partial z^2}+\frac{\partial^2 w}{\partial y\partial z}\right) \quad (10.19)$$

$$\frac{\partial w}{\partial t}+u\frac{\partial w}{\partial x}+v\frac{\partial w}{\partial y}+w\frac{\partial w}{\partial z}=-g-\frac{1}{\rho}\frac{\partial p}{\partial z}+\frac{\mu}{\rho}\left(\frac{\partial^2 w}{\partial x^2}+\frac{\partial^2 u}{\partial x\partial z}+\frac{\partial^2 w}{\partial y^2}+\frac{\partial^2 v}{\partial y\partial z}+2\frac{\partial^2 w}{\partial z^2}\right) \quad (10.20)$$

The factor $\mu/\rho$ can also be written as $\nu$, the kinematic viscosity ($m^2s^{-1}$). As we will see in the next section, the following equation applies in an **incompressible fluid:**

$$\frac{\partial u}{\partial x}+\frac{\partial v}{\partial y}+\frac{\partial w}{\partial z}=0 \quad (10.21)$$

Taking the partial derivative with respect to $x$ leads to

$$\frac{\partial}{\partial x}\left(\frac{\partial u}{\partial x}+\frac{\partial v}{\partial y}+\frac{\partial w}{\partial z}\right)=0 \quad (10.22)$$

Hence:

$$\frac{\partial^2 u}{\partial x^2}+\frac{\partial^2 v}{\partial x \partial y}+\frac{\partial^2 w}{\partial x \partial z}=0 \tag{10.23}$$

Similar equations apply in the $y$ and $z$ directions. Substituting into eqs. (10.18)–(10.20), and introducing the kinematic viscosity, leads to

$$\boxed{\frac{\partial u}{\partial t}+u\frac{\partial u}{\partial x}+v\frac{\partial u}{\partial y}+w\frac{\partial u}{\partial z}=fv-\frac{1}{\rho}\frac{\partial p}{\partial x}+\nu\left(\frac{\partial^2 u}{\partial x^2}+\frac{\partial^2 u}{\partial y^2}+\frac{\partial^2 u}{\partial z^2}\right)} \tag{10.24}$$

$$\boxed{\frac{\partial v}{\partial t}+u\frac{\partial v}{\partial x}+v\frac{\partial v}{\partial y}+w\frac{\partial v}{\partial z}=-fu-\frac{1}{\rho}\frac{\partial p}{\partial y}+\nu\left(\frac{\partial^2 v}{\partial x^2}+\frac{\partial^2 v}{\partial y^2}+\frac{\partial^2 v}{\partial z^2}\right)} \tag{10.25}$$

$$\boxed{\frac{\partial w}{\partial t}+u\frac{\partial w}{\partial x}+v\frac{\partial w}{\partial y}+w\frac{\partial w}{\partial z}=-g-\frac{1}{\rho}\frac{\partial p}{\partial z}+\nu\left(\frac{\partial^2 w}{\partial x^2}+\frac{\partial^2 w}{\partial y^2}+\frac{\partial^2 w}{\partial z^2}\right)} \tag{10.26}$$

These are the Navier–Stokes equations.

*The same equations* would have been obtained if the stresses had been calculated with equations of the following form:

$$\tau_{xz}=\mu\frac{\partial u}{\partial z} \tag{10.27}$$

Hence, **kinematically**, an incompressible fluid behaves as if shear stresses follow eq. (10.27). **Dynamically**, this is not correct, as this approach will predict incorrect stresses.

Equations (10.24)–(10.26) are a form of the Navier–Stokes equations. These equations can be used for **direct simulations of turbulence**. While this is potentially the most fundamental way to model turbulence, it is by no way the most efficient way to do so, and except in very simple cases the computational resources are not available to model turbulence accurately with eqs. (10.24)–(10.26). The reason is obvious: Turbulence is not included at all in eqs. (10.24)–(10.26), whereas viscosity, which is usually not a relevant variable in atmospheric turbulence, features prominently in eqs. (10.24)–(10.26). The only way the Navier–Stokes equations can model turbulence accurately is by directly simulating the turbulent kinetic energy cascade all the way to the scale where the energy is dissipated by viscous forces. This means that the grid cell size must be on the order of 1 mm to simulate atmospheric turbulence directly with any degree of accuracy.

There are two ways to make the Navier–Stokes equations more efficient. The most efficient, but potentially least accurate, method is the **Reynolds-averaged Navier–Stokes (RANS)** method (Section 10.3). In this method Reynolds-averaging equations are used to remove all turbulence from the wind speeds. The wind speed

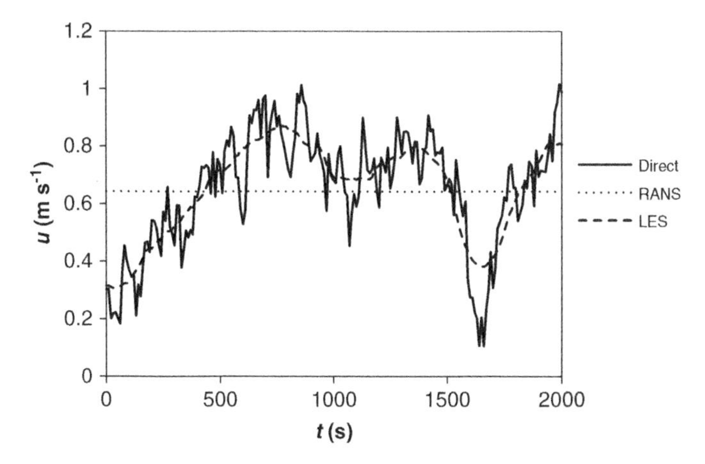


Figure 10.7 Wind speed features resolved by a Reynolds-averaged Navier–Stokes (RANS) method, a large eddy simulation (LES) method, and a direct simulation.

fluctuations ($u'$, etc.), which contain all the turbulence, are parameterized in terms of average variables. RANS can be used as a steady-state method or a non-steady-state method. However, any transient behavior resulting from turbulence cannot be resolved in a RANS model. Hence, many turbulence phenomena cannot be modeled with a RANS method simply because they are inherently nonsteady state. An example is the cyclic structures in some building wakes.

The second, potentially more accurate, but less efficient, method to model turbulence is the **large eddy simulation (LES)** method (Section 10.4). This method involves simulating the large eddies of a turbulent system directly, while parameterizing only smaller eddies as turbulence is dissipated down the energy cascade. This is a non-steady-state method.

The difference between direct simulation, RANS, and LES is illustrated in Figure 10.7. The figure shows the expected outcome of each simulation in terms of wind speed variability. RANS can only reproduce the Reynolds average wind speed. LES can resolve the large-scale features of the wind speed but not the small-scale features. Only a direct simulation can resolve the features of the wind speed at all scales. Note that the actual features of turbulence cannot be predicted because the system is chaotic. Only the statistical features of the turbulence can be resolved by a simulation.

### 10.2.2 The Material and Energy Balance

The Navier–Stokes equations by themselves cannot predict turbulence or dispersion without additional equations. The choice of additional equation depends on the problem to be solved and the assumptions that are made in the solution.

In general, a **mass balance** can be used to provide additional information. Consider the masses flowing in and out of an incremental volume $dx\ dy\ dz$ in an Eulerian frame of reference. The following equation is obtained:

$$\frac{\partial \rho}{\partial t} dx\, dy\, dz = \rho u\, dy\, dz - \left[ \rho u + \frac{\partial(\rho u)}{\partial x} dx \right] dy\, dz + \rho v\, dx\, dz - \left[ \rho v + \frac{\partial(\rho v)}{\partial y} dy \right] dx\, dz$$
$$+ \rho w\, dx\, dy - \left[ \rho w + \frac{\partial(\rho w)}{\partial z} dz \right] dx\, dy \qquad (10.28)$$

which simplifies to

$$\frac{\partial \rho}{\partial t} = \frac{\partial(\rho u)}{\partial x} + \frac{\partial(\rho v)}{\partial y} + \frac{\partial(\rho w)}{\partial z} \qquad (10.29)$$

Equation (10.29) is also known as the **continuity equation**. When the fluid is incompressible, this equation can be simplified to

$$\boxed{\frac{\partial u}{\partial x} + \frac{\partial v}{\partial y} + \frac{\partial w}{\partial z} = 0} \qquad (10.30)$$

In the atmosphere, incompressible flow applies with reasonable accuracy. The requirements for this assumption are that the wind speed is much smaller than the speed of sound, and that the vertical air movements are less than one kilometer (Byun et al., 2003).

In non-steady-state conditions the following equation also applies to an incompressible fluid:

$$\frac{\partial}{\partial t}\left( \frac{\partial u}{\partial x} + \frac{\partial v}{\partial y} + \frac{\partial w}{\partial z} \right) = 0 \qquad (10.31)$$

By rearrangement of the differentials, the following equation is obtained:

$$\frac{\partial}{\partial x}\frac{\partial u}{\partial t} + \frac{\partial}{\partial y}\frac{\partial v}{\partial t} + \frac{\partial}{\partial z}\frac{\partial w}{\partial t} = 0 \qquad (10.32)$$

When eq. (10.29) (compressible flow) is used, we need an equation to relate density to the other variables. First, density is related to total concentration $c_{tot}$ (total number of moles per unit volume) with the molar mass $M_{air}$:

$$\rho = M_{air} c_{tot} \qquad (10.33)$$

In eq. (10.33) care must be taken to maintain unit consistency. Usually this means using air molar mass in kg mol$^{-1}$ (see Table 1.1, Chapter 1). Total concentration is related to pressure with the **ideal gas law**:

$$c_{tot} = \frac{n}{V} = \frac{p}{RT} \qquad (10.34)$$

To model the dispersion of pollutants, we need a **material balance for the pollutant** as a separate entity. A brief derivation of the material balance is given here. A more thorough discussion of the derivation is given in Chapter 11. In direct simulations, the mass transfer mechanisms to be accounted for are advection and molecular diffusion. The following equation is a generic material balance:

$$\begin{aligned}\text{Accumulation} &= \text{advection in} - \text{advection out} + \text{diffusion in} \\ &\quad - \text{diffusion out} + \text{source}\end{aligned} \tag{10.35}$$

Each term will be calculated in mg s$^{-1}$, for a differential volume $dx\,dy\,dz$. The accumulation term is

$$\text{Accumulation} = \frac{\partial c}{\partial t}dx\,dy\,dz \tag{10.36}$$

where $c$ is the pollutant concentration (mg m$^{-3}$). The mass transfer terms will first be shown in one dimension. The total mass transfer is the sum of the contributions of the three dimensions. For advective transfer in the $x$ direction the equation is

$$\text{Advection in} - \text{advection out} = cu\,dy\,dz - \left(cu + \frac{\partial cu}{\partial x}dx\right)dy\,dz \tag{10.37}$$

which simplifies to

$$\text{Advection in} - \text{advection out} = -\frac{\partial cu}{\partial x}dy\,dz \tag{10.38}$$

For diffusion, Fick's law can be used [eq. (5.171)]. Hence, the following equation applies for diffusive transfer into the volume:

$$\text{Diffusion in} - \text{diffusion out} = -D\frac{\partial c}{\partial x}dy\,dz - \left[-D\frac{\partial c}{\partial x} - \frac{\partial}{\partial x}\left(D\frac{\partial c}{\partial x}\right)dx\right]dy\,dz \tag{10.39}$$

The molecular diffusivity, $D$ (m$^2$ s$^{-1}$), can generally be approximated as a constant. Hence, eq. (10.30) reduces to

$$\text{Diffusion in} - \text{diffusion out} = D\frac{\partial^2 c}{\partial x^2}dx\,dy\,dz \tag{10.40}$$

The source term (or sink term when negative) includes chemical reaction, deposition, and the like. It is given by

$$\text{Source} = S\,dx\,dy\,dz \tag{10.41}$$

where $S$ is the rate of production of the pollutant (mg m$^{-3}$ s$^{-1}$).

Applying all the terms in eq. (10.35) in three dimensions, and dividing by $dx\,dy\,dz$, leads to the following equation:

$$\boxed{\frac{\partial c}{\partial t} = -\frac{\partial cu}{\partial x} - \frac{\partial cv}{\partial y} - \frac{\partial cw}{\partial z} + D\left(\frac{\partial^2 c}{\partial x^2} + \frac{\partial^2 c}{\partial y^2} + \frac{\partial^2 c}{\partial z^2}\right) + S} \tag{10.42}$$

A similar equation can be developed for the **humidity** $h$.

If deviations from neutral conditions need to be accounted for, as is usually the case, a **heat balance**, or more accurately, a **thermodynamic energy balance**, is required. In generic form, this equation reads like

$$\begin{aligned}\text{Accumulation} &= \text{convection in} - \text{convection out} + \text{conduction in} \\ &\quad - \text{conduction out} + \text{radiation in} - \text{radiation out}\end{aligned} \tag{10.43}$$

When an air parcel moves in the vertical direction, the parcel is compressed or decompressed, leading to an enthalpy change. Normally, this effect needs to be accounted for in eq. (10.43), but this can be avoided by using the **potential temperature** in eq. (10.43) instead of the actual temperature.

Bearing the above in mind, the enthalpy accumulation in a differential volume is

$$\text{Accumulation} = \frac{\partial \rho c_p \theta}{\partial t} dx\, dy\, dz \tag{10.44}$$

where $c_p$ is the specific heat of air ($1006\,\text{J}\,\text{kg}^{-1}\,\text{K}^{-1}$), and $\theta$ is the potential temperature (K). The other terms will be outlined in the $x$ direction. The equivalent terms for other directions are obvious.

The convective terms are

$$\text{Convection in} - \text{convection out} = \rho c_p \theta u\, dy\, dz - \left( \rho c_p \theta u + \frac{\partial \rho c_p \theta u}{\partial x} dx \right) dy\, dz \tag{10.45}$$

Assuming constant density (incompressible flow) and constant specific heat, this equation reduces to

$$\text{Convection in} - \text{convection out} = -\rho c_p \frac{\partial \theta u}{\partial x} dx\, dy\, dz \tag{10.46}$$

Heat conduction can be described by Fourier's law [eq. (5.170)]. Hence, the conduction terms can be described as

$$\text{Conduction in} - \text{conduction out} = -\lambda \frac{\partial \theta}{\partial x} dy\, dz - \left[ -\lambda \frac{\partial \theta}{\partial x} - \frac{\partial}{\partial x} \left( \lambda \frac{\partial \theta}{\partial x} dx \right) \right] dy\, dz \tag{10.47}$$

Assuming constant thermal conductivity, $\lambda$ ($\text{J}\,\text{m}^{-1}\,\text{K}^{-1}$), this equation reduces to

$$\text{Conduction in} - \text{conduction out} = \lambda \frac{\partial^2 \theta}{\partial x^2} dx\, dy\, dz \tag{10.48}$$

The radiation terms are based on radiative heat fluxes, $R_x$, $R_y$, and $R_z$ ($\text{J}\,\text{m}^{-2}\,\text{s}^{-1}$). Hence, in the $x$ direction:

$$\text{Radiation in} - \text{radiation out} = R_x\, dy\, dz - \left( R_x + \frac{\partial R_x}{\partial x} dx \right) dy\, dz \tag{10.49}$$

which simplifies to

$$\text{Radiation in} - \text{radiation out} = -\frac{\partial R_x}{\partial x} dx\, dy\, dz \tag{10.50}$$

Applying the above equations in the three-dimensional heat balance, and dividing by $\rho c_p\, dx\,dy\,dz$ leads to the following:

$$\frac{\partial\theta}{\partial t} = -\left(\frac{\partial\theta u}{\partial x}+\frac{\partial\theta v}{\partial y}+\frac{\partial\theta w}{\partial z}\right)+\frac{\lambda}{\rho c_p}\left(\frac{\partial^2\theta}{\partial x^2}+\frac{\partial^2\theta}{\partial y^2}+\frac{\partial^2\theta}{\partial z^2}\right)-\frac{1}{\rho c_p}\left(\frac{\partial R_x}{\partial x}+\frac{\partial R_y}{\partial y}+\frac{\partial R_z}{\partial z}\right) \tag{10.51}$$

If there is any evaporation, condensation, or heat of reaction in the atmosphere, then corresponing terms need to be added to eq. (10.51).

For the equations in this section the same comments apply as for the Navier–Stokes equations: They will only give correct results in the atmosphere if all levels of turbulence are simulated directly. An example is eq. (10.42), which is the actual pollutant dispersion equation in direct CFD simulations. This equation only allows for mass transfer by advection and molecular diffusion, whereas transfer in the cross-wind and vertical directions is almost entirely due to turbulence. It follows that the same kind of parameterizations that are necessary to solve the Navier–Stokes equations effectively for the atmosphere will be required for the material and heat balance as well. This will be covered in Section 10.3.3.

## 10.3 REYNOLDS-AVERAGED NAVIER–STOKES (RANS) TECHNIQUES

### 10.3.1 Averaging the Navier–Stokes Equations

In this section we will use **Reynolds decomposition** (see Section 5.11.1) to separate the turbulent part of the fluid motion in the Navier–Stokes equation from the time-averaged part. In the next section we will find out how these turbulent parts can be parameterized based on our knowledge of the properties of turbulence.

Consider the Navier–Stokes equations [eqs. (10.24)–(10.26)], and apply Reynolds averaging on all variables:

$$\begin{aligned}&\frac{\partial(\bar u+u')}{\partial t}+(\bar u+u')\frac{\partial(\bar u+u')}{\partial x}+(\bar v+v')\frac{\partial(\bar u+u')}{\partial y}+(\bar w+w')\frac{\partial(\bar u+u')}{\partial z}\\&\quad=f(\bar v+v')-\frac{1}{(\bar\rho+\rho')}\frac{\partial(\bar p+p')}{\partial x}\\&\quad+\frac{\mu}{(\bar\rho+\rho')}\left[\frac{\partial^2(\bar u+u')}{\partial x^2}+\frac{\partial^2(\bar u+u')}{\partial y^2}+\frac{\partial^2(\bar u+u')}{\partial z^2}\right]\end{aligned} \tag{10.52}$$

$$\begin{aligned}&\frac{\partial(\bar v+v')}{\partial t}+(\bar u+u')\frac{\partial(\bar v+v')}{\partial x}+(\bar v+v')\frac{\partial(\bar v+v')}{\partial y}+(\bar w+w')\frac{\partial(\bar v+v')}{\partial z}\\&\quad=-f(\bar u+u')-\frac{1}{(\bar\rho+\rho')}\frac{\partial(\bar p+p')}{\partial y}\\&\quad+\frac{\mu}{(\bar\rho+\rho')}\left[\frac{\partial^2(\bar v+v')}{\partial x^2}+\frac{\partial^2(\bar v+v')}{\partial y^2}+\frac{\partial^2(\bar v+v')}{\partial z^2}\right]\end{aligned} \tag{10.53}$$

$$\frac{\partial(\overline{w}+w')}{\partial t}+(\overline{u}+u')\frac{\partial(\overline{w}+w')}{\partial x}+(\overline{v}+v')\frac{\partial(\overline{w}+w')}{\partial y}+(\overline{w}+w')\frac{\partial(\overline{w}+w')}{\partial z}$$
$$=-g-\frac{1}{(\overline{\rho}+\rho')}\frac{\partial(\overline{p}+p')}{\partial z}$$
$$+\frac{\mu}{(\overline{\rho}+\rho')}\left[\frac{\partial^2(\overline{w}+w')}{\partial x^2}+\frac{\partial^2(\overline{w}+w')}{\partial y^2}+\frac{\partial^2(\overline{w}+w')}{\partial z^2}\right] \tag{10.54}$$

We use **Reynolds averaging** (see Section 5.11.1) to take the time average of eqs. (10.52)–(10.54). The following rules can be applied:

- $\overline{f'}=0$ for any variable $f$.
- However, $\overline{f'g'}\neq 0$.
- $\overline{fg}=\overline{(\overline{f}+f')(\overline{g}+g')}=\overline{f}\cdot\overline{g}+\overline{f'}\cdot\overline{g}+\overline{f}\cdot\overline{g'}+\overline{f'g'}=\overline{f}\cdot\overline{g}+\overline{f'g'}$.
- Pressure and density fluctuate much less than velocity. Their fluctuations can be neglected.

Using these rules, the Reynolds-averaged Navier-Stokes equations can be written as

$$\frac{\partial\overline{u}}{\partial t}+\overline{u}\frac{\partial\overline{u}}{\partial x}+\overline{u'\frac{\partial u'}{\partial x}}+\overline{v}\frac{\partial\overline{u}}{\partial y}+\overline{v'\frac{\partial u'}{\partial y}}+\overline{w}\frac{\partial\overline{u}}{\partial z}+\overline{w'\frac{\partial u'}{\partial z}}$$
$$=f\overline{v}-\frac{1}{\overline{\rho}}\frac{\partial\overline{p}}{\partial x}+\frac{\mu}{\overline{\rho}}\left(\frac{\partial^2\overline{u}}{\partial x^2}+\frac{\partial^2\overline{u}}{\partial y^2}+\frac{\partial^2\overline{u}}{\partial z^2}\right) \tag{10.55}$$

$$\frac{\partial\overline{v}}{\partial t}+\overline{u}\frac{\partial\overline{v}}{\partial x}+\overline{u'\frac{\partial v'}{\partial x}}+\overline{v}\frac{\partial\overline{v}}{\partial y}+\overline{v'\frac{\partial v'}{\partial y}}+\overline{w}\frac{\partial\overline{v}}{\partial z}+\overline{w'\frac{\partial v'}{\partial z}}$$
$$=-f\overline{u}-\frac{1}{\overline{\rho}}\frac{\partial\overline{p}}{\partial y}+\frac{\mu}{\overline{\rho}}\left(\frac{\partial^2\overline{v}}{\partial x^2}+\frac{\partial^2\overline{v}}{\partial y^2}+\frac{\partial^2\overline{v}}{\partial z^2}\right) \tag{10.56}$$

$$\frac{\partial\overline{w}}{\partial t}+\overline{u}\frac{\partial\overline{w}}{\partial x}+\overline{u'\frac{\partial w'}{\partial x}}+\overline{v}\frac{\partial\overline{w}}{\partial y}+\overline{v'\frac{\partial w'}{\partial y}}+\overline{w}\frac{\partial\overline{w}}{\partial z}+\overline{w'\frac{\partial w'}{\partial z}}$$
$$=-g-\frac{1}{\overline{\rho}}\frac{\partial\overline{p}}{\partial z}+\frac{\mu}{\overline{\rho}}\left(\frac{\partial^2\overline{w}}{\partial x^2}+\frac{\partial^2\overline{w}}{\partial y^2}+\frac{\partial^2\overline{w}}{\partial z^2}\right) \tag{10.57}$$

The Reynolds-averaged Navier–Stokes equations are very similar to the original Navier–Stokes equations, but with three extra terms representing covariances. We have seen in Section 5.11.4 that covariances represent turbulent fluxes. However, before attepting to interpret these fluxes, we will combine the Reynolds-averaged Navier–Stokes equations with the mass balance. For incompressible flow we have found eq. (10.30). Reynolds decomposition leads to

$$\frac{\partial(\overline{u}+u')}{\partial x}+\frac{\partial(\overline{v}+v')}{\partial y}+\frac{\partial(\overline{w}+w')}{\partial z}=0 \tag{10.58}$$

Reynolds averaging leads to

$$\frac{\partial \overline{u}}{\partial x}+\frac{\partial \overline{v}}{\partial y}+\frac{\partial \overline{w}}{\partial z}=0 \tag{10.59}$$

Subtract eq. (10.59) from eq. (10.58) to obtain a mass balance for the turbulent part of the wind movement:

$$\frac{\partial u'}{\partial x}+\frac{\partial v'}{\partial y}+\frac{\partial w'}{\partial z}=0 \tag{10.60}$$

Multiply eq. (10.60) by $u'$ and take the average:

$$\overline{u'\frac{\partial u'}{\partial x}}+\overline{u'\frac{\partial v'}{\partial y}}+\overline{u'\frac{\partial w'}{\partial z}}=0 \tag{10.61}$$

Equation (10.61) is added to eq. (10.55):

$$\begin{aligned}&\frac{\partial \overline{u}}{\partial t}+\overline{u}\frac{\partial \overline{u}}{\partial x}+\overline{u'\frac{\partial u'}{\partial x}}+\overline{u'\frac{\partial u'}{\partial x}}+\overline{v}\frac{\partial \overline{u}}{\partial y}+\overline{v'\frac{\partial u'}{\partial y}}+\overline{u'\frac{\partial v'}{\partial y}}+\overline{w}\frac{\partial \overline{u}}{\partial z}+\overline{w'\frac{\partial u'}{\partial z}}+\overline{u'\frac{\partial w'}{\partial z}}\\&=f\overline{v}-\frac{1}{\overline{\rho}}\frac{\partial \overline{p}}{\partial x}+\frac{\mu}{\overline{\rho}}\left(\frac{\partial^2 \overline{u}}{\partial x^2}+\frac{\partial^2 \overline{u}}{\partial y^2}+\frac{\partial^2 \overline{u}}{\partial z^2}\right)\end{aligned} \tag{10.62}$$

The velocity covariance terms can be rewritten as follows:

$$\begin{aligned}&\frac{\partial \overline{u}}{\partial t}+\overline{u}\frac{\partial \overline{u}}{\partial x}+\frac{\partial \overline{u'^2}}{\partial x}+\overline{v}\frac{\partial \overline{u}}{\partial y}+\frac{\partial \overline{u'v'}}{\partial y}+\overline{w}\frac{\partial \overline{u}}{\partial z}+\frac{\partial \overline{u'w'}}{\partial z}\\&=f\overline{v}-\frac{1}{\overline{\rho}}\frac{\partial \overline{p}}{\partial x}+\frac{\mu}{\overline{\rho}}\left(\frac{\partial^2 \overline{u}}{\partial x^2}+\frac{\partial^2 \overline{u}}{\partial y^2}+\frac{\partial^2 \overline{u}}{\partial z^2}\right)\end{aligned} \tag{10.63}$$

Comparing with eq. (5.166), it is clear that the new terms are turbulent stress terms. For that reason, they are moved to the right-hand side and combined with the viscous stress terms:

$$\begin{aligned}&\frac{\partial \overline{u}}{\partial t}+\overline{u}\frac{\partial \overline{u}}{\partial x}+\overline{v}\frac{\partial \overline{u}}{\partial y}+\overline{w}\frac{\partial \overline{u}}{\partial z}\\&=f\overline{v}-\frac{1}{\overline{\rho}}\frac{\partial \overline{p}}{\partial x}\\&+\frac{1}{\overline{\rho}}\left(\mu\frac{\partial^2 \overline{u}}{\partial x^2}-\overline{\rho}\frac{\partial \overline{u'^2}}{\partial x}+\mu\frac{\partial^2 \overline{u}}{\partial y^2}-\overline{\rho}\frac{\partial \overline{u'v'}}{\partial y}+\mu\frac{\partial^2 \overline{u}}{\partial z^2}-\overline{\rho}\frac{\partial \overline{u'w'}}{\partial z}\right)\end{aligned} \tag{10.64}$$

Similar equations can be developed for the $y$ and $z$ directions:

$$\frac{\partial \bar{v}}{\partial t}+\bar{u}\frac{\partial \bar{v}}{\partial x}+\bar{v}\frac{\partial \bar{v}}{\partial y}+\bar{w}\frac{\partial \bar{v}}{\partial z} = -f\bar{u}-\frac{1}{\bar{\rho}}\frac{\partial \bar{p}}{\partial y} + \frac{1}{\bar{\rho}}\left(\mu\frac{\partial^2 \bar{v}}{\partial x^2}-\bar{\rho}\frac{\partial \overline{u'v'}}{\partial x}+\mu\frac{\partial^2 \bar{v}}{\partial y^2}-\bar{\rho}\frac{\partial \overline{v'^2}}{\partial y}+\mu\frac{\partial^2 \bar{v}}{\partial z^2}-\bar{\rho}\frac{\partial \overline{v'w'}}{\partial z}\right) \tag{10.65}$$

$$\frac{\partial \bar{w}}{\partial t}+\bar{u}\frac{\partial \bar{w}}{\partial x}+\bar{v}\frac{\partial \bar{w}}{\partial y}+\bar{w}\frac{\partial \bar{w}}{\partial z} = -g-\frac{1}{\bar{\rho}}\frac{\partial \bar{p}}{\partial z}+\frac{1}{\bar{\rho}}\left(\mu\frac{\partial^2 \bar{w}}{\partial x^2}-\bar{\rho}\frac{\partial \overline{u'w'}}{\partial x}+\mu\frac{\partial^2 \bar{w}}{\partial y^2}-\bar{\rho}\frac{\partial \overline{v'w'}}{\partial y}+\mu\frac{\partial^2 \bar{w}}{\partial z^2}-\bar{\rho}\frac{\partial \overline{w'^2}}{\partial z}\right) \tag{10.66}$$

Equations (10.64)–(10.66) are the same as eqs. (10.24)–(10.26), except that the stress terms include both viscous and turbulent stress. This explains how Reynolds averaging can make turbulence modeling more efficient: It provides the tools to incorporate turbulent stress explicitly in the Navier–Stokes equations. However, we pay a price for that efficiency: We have introduced six new variables: the covariances $\overline{u'v'}$, $\overline{u'w'}$, and $\overline{v'w'}$ and the variances $\overline{u'^2}$, $\overline{v'^2}$, and $\overline{w'^2}$. We cannot derive equations to describe these variables from first principles. This is known as the **turbulence closure problem**.

Equations (10.64)–(10.66) also allow us to understand why there should be a viscous stress $\tau_{xx}$ that is not a shear stress. An air parcel exchanges material with the surroundings whenever $u' \neq 0$. This represents momentum that is exchanged with the surroundings, and this momentum change is felt as a force. On a molecular scale the same phenomenon occurs with individual molecules moving into and out of the air parcel, where it is felt as a viscous force. For an observer moving with the air parcel, the force associated with momentum transfer in the direction of the bulk movement is felt as a normal force.

An additional simplification is possible in eq. (10.66) if it is assumed that the average pressure follows the hydrostatic equation [eq. (5.2)]:

$$\frac{1}{\bar{\rho}}\frac{\partial \bar{p}}{\partial z}=-g \tag{10.67}$$

This reduces eq. (10.66) to

$$\frac{\partial \bar{w}}{\partial t}+\bar{u}\frac{\partial \bar{w}}{\partial x}+\bar{v}\frac{\partial \bar{w}}{\partial y}+\bar{w}\frac{\partial \bar{w}}{\partial z} = \frac{1}{\bar{\rho}}\left(\mu\frac{\partial^2 \bar{w}}{\partial x^2}-\bar{\rho}\frac{\partial \overline{u'w'}}{\partial x}+\mu\frac{\partial^2 \bar{w}}{\partial y^2}-\bar{\rho}\frac{\partial \overline{v'w'}}{\partial y}+\mu\frac{\partial^2 \bar{w}}{\partial z^2}-\bar{\rho}\frac{\partial \overline{w'^2}}{\partial z}\right) \tag{10.68}$$

Often it is assumed that turbulence is homogeneous in the $x$ and $y$ directions. When that is the case, the partial derivatives with respect to $x$ and $y$ in the viscosity and turbulent dissipation terms can be omitted. This is an important simplification because it reduces the number of new variables to only three: the covariances $\overline{u'w'}$ and $\overline{v'w'}$ and the variance $\overline{w'^2}$. Furthermore, if we simply assume $\overline{w} = 0$, we eliminate one equation and the variance $\overline{w'^2}$.

Note that eq. (10.60), which is a mass balance for the turbulent part of the wind flow, is not usually accounted for in stochastic Lagrangian particle models. This weakness in stochastic Lagrangian particle models is not normally problematic, except when a Reynolds-averaged Navier–Stokes equation is used to calculate drift velocities. It is this flaw that caused Legg and Raupach (1982) to predict drift velocities twice as large as the "velocity bias" of Wilson et al. (1981).

## 10.3.2 Closing the Navier–Stokes Equations

***10.3.2.1 Introduction: Notation.*** In the previous section we laid the foundation to greatly improve the efficiency of computational fluid dynamics by Reynolds averaging the Navier–Stokes equations. The price to pay for increased efficiency is the introduction of six new variables: three variances and three covariances. In this section we will consider ways to estimate these variables in terms of average variables.

So far, all equations in this book were written in full, without summations or Einstein's index notation. This made some equations rather long, but allowed the reader to see each variable in every dimension as a separate term or factor. In this section the complexity of the equations is such that the added size of the equations is no longer justified by the convenicence of having every term of the equation in full, and summations will be used. However, the index notation will not be used, and every summation will be indicated with a summation sign in the equations. The notation changes as follows:

$$x_1 = x \tag{10.69}$$

$$x_2 = y \tag{10.70}$$

$$x_3 = z \tag{10.71}$$

$$u_1 = u \tag{10.72}$$

$$u_2 = v \tag{10.73}$$

$$u_3 = w \tag{10.74}$$

For instance, the Reynolds-averaged Navier–Stokes equations, eqs. (10.64)–(10.66), are now written as

$$\frac{\partial \overline{u}_1}{\partial t} + \sum_{i=1}^{3} \overline{u}_i \frac{\partial \overline{u}_1}{\partial x_i} = f\overline{u}_2 - \frac{1}{\overline{\rho}}\frac{\partial \overline{p}}{\partial x_1} + \frac{1}{\overline{\rho}}\left(\mu \sum_{i=1}^{3} \frac{\partial^2 \overline{u}_1}{\partial x_i^2} - \overline{\rho} \sum_{i=1}^{3} \frac{\partial \overline{u_1' u_i'}}{\partial x_i}\right) \tag{10.75}$$

$$\frac{\partial \overline{u}_2}{\partial t} + \sum_{i=1}^{3} \overline{u}_i \frac{\partial \overline{u}_2}{\partial x_i} = -f\overline{u}_1 - \frac{1}{\overline{\rho}}\frac{\partial \overline{p}}{\partial x_2} + \frac{1}{\overline{\rho}}\left(\mu \sum_{i=1}^{3} \frac{\partial^2 \overline{u}_2}{\partial x_i^2} - \overline{\rho} \sum_{i=1}^{3} \frac{\partial \overline{u_2' u_i'}}{\partial x_i}\right) \tag{10.76}$$

$$\frac{\partial \overline{u}_3}{\partial t}+\sum_{i=1}^{3}\overline{u}_i\frac{\partial \overline{u}_3}{\partial x_i}=-g-\frac{1}{\overline{\rho}}\frac{\partial \overline{p}}{\partial x_3}+\frac{1}{\overline{\rho}}\left(\mu\sum_{i=1}^{3}\frac{\partial^2 \overline{u}_3}{\partial x_i^2}-\overline{\rho}\sum_{i=1}^{3}\frac{\partial \overline{u_3' u_i'}}{\partial x_i}\right) \tag{10.77}$$

In this section we will start from the Reynolds decomposed, nonaveraged Navier–Stokes equations [eqs. (10.52)–(10.54)], so it is useful to have these equations available in the new notation. They are as follows:

$$\frac{\partial(\overline{u}_1+u_1')}{\partial t}+\sum_{i=1}^{3}(\overline{u}_i+u_i')\frac{\partial(\overline{u}_1+u_1')}{\partial x_i}$$
$$=f(\overline{u}_2+u_2')-\frac{1}{\overline{\rho}+\rho'}\frac{\partial(\overline{p}+p')}{\partial x_1}+\frac{\mu}{\overline{\rho}+\rho'}\left[\sum_{i=1}^{3}\frac{\partial^2(\overline{u}_1+u_1')}{\partial x_i^2}\right] \tag{10.78}$$

$$\frac{\partial(\overline{u}_2+u_2')}{\partial t}+\sum_{i=1}^{3}(\overline{u}_i+u_i')\frac{\partial(\overline{u}_2+u_2')}{\partial x_i}$$
$$=-f(\overline{u}_1+u_1')-\frac{1}{\overline{\rho}+\rho'}\frac{\partial(\overline{p}+p')}{\partial x_2}+\frac{\mu}{\overline{\rho}+\rho'}\left[\sum_{i=1}^{3}\frac{\partial^2(\overline{u}_2+u_2')}{\partial x_i^2}\right] \tag{10.79}$$

$$\frac{\partial(\overline{u}_3+u_3')}{\partial t}+\sum_{i=1}^{3}(\overline{u}_i+u_i')\frac{\partial(\overline{u}_3+u_3')}{\partial x_i}$$
$$=-g-\frac{1}{\overline{\rho}+\rho'}\frac{\partial(\overline{p}+p')}{\partial x_3}+\frac{\mu}{\overline{\rho}+\rho'}\left[\sum_{i=1}^{3}\frac{\partial^2(\overline{u}_3+u_3')}{\partial x_i^2}\right] \tag{10.80}$$

***10.3.2.2 First-Order Closure: Gradient Transport Theory.*** Closure techniques for the Navier–Stokes equations are classified by the **order**, which defines the highest order moments that are resolved directly, without parameterization. First-order closure means that only average properties are resolved directly, and variances and covariances are parameterized.

Gradient transport theory is one of the older closure techniques for computational fluid dynamics and is rendered obsolete by more powerful higher-order techniques such as the $k$-$\varepsilon$ technique. However, it is useful to start here because gradient transport theory lays the foundation for higher-order techniques. It dates back to Boussinesq's idea that turbulence increases the apparent viscosity of a fluid (see Chapter 5). The turbulent form of Newton's viscosity law was given in eq. (5.174):

$$\frac{\tau}{\rho}=\overline{u'w'}=-K_{\mathrm{m}}\frac{\partial \overline{u}}{\partial z} \tag{5.174}$$

However, we saw in Section 10.2.1 that Newton's viscosity law requires a second term [eq. (10.12)]. The same reasoning applies to turbulent viscosity. Hence, eq. (5.174) must be extended to

$$\frac{\tau}{\rho} = \overline{u'w'} = -K_{\mathrm{m}}\left(\frac{\partial \bar{u}}{\partial z} + \frac{\partial \bar{w}}{\partial x}\right) \tag{10.81}$$

A similar equation applies in other dimensions. The general form of eq. (10.81) is

$$\overline{u_i' u_j'} = \overline{u_j' u_i'} = -K_{\mathrm{m}}\left(\frac{\partial \bar{u}_i}{\partial x_j} + \frac{\partial \bar{u}_j}{\partial x_i}\right) \tag{10.82}$$

These covariances can be substituted in the Reynolds-averaged Navier–Stokes equations [eqs. (10.75)–(10.77)]. Assuming constant $K_{\mathrm{m}}$ values, one obtains

$$\frac{\partial \bar{u}_1}{\partial t} + \sum_{i=1}^{3} \bar{u}_i \frac{\partial \bar{u}_1}{\partial x_i} = f\bar{u}_2 - \frac{1}{\bar{\rho}}\frac{\partial \bar{p}}{\partial x_1} + \frac{1}{\bar{\rho}}\left[\mu \sum_{i=1}^{3} \frac{\partial^2 \bar{u}_1}{\partial x_i^2} + \bar{\rho}\sum_{i=1}^{3} K_{\mathrm{m}} \frac{\partial}{\partial x_i}\left(\frac{\partial \bar{u}_1}{\partial x_i} + \frac{\partial \bar{u}_i}{\partial x_1}\right)\right] \tag{10.83}$$

$$\frac{\partial \bar{u}_2}{\partial t} + \sum_{i=1}^{3} \bar{u}_i \frac{\partial \bar{u}_2}{\partial x_i} = -f\bar{u}_1 - \frac{1}{\bar{\rho}}\frac{\partial \bar{p}}{\partial x_2} + \frac{1}{\bar{\rho}}\left[\mu \sum_{i=1}^{3} \frac{\partial^2 \bar{u}_2}{\partial x_i^2} + \bar{\rho}\sum_{i=1}^{3} K_{\mathrm{m}} \frac{\partial}{\partial x_i}\left(\frac{\partial \bar{u}_2}{\partial x_i} + \frac{\partial \bar{u}_i}{\partial x_2}\right)\right] \tag{10.84}$$

$$\frac{\partial \bar{u}_3}{\partial t} + \sum_{i=1}^{3} \bar{u}_i \frac{\partial \bar{u}_3}{\partial x_i} = -g - \frac{1}{\bar{\rho}}\frac{\partial \bar{p}}{\partial x_3} + \frac{1}{\bar{\rho}}\left[\mu \sum_{i=1}^{3} \frac{\partial^2 \bar{u}_3}{\partial x_i^2} + \bar{\rho}\sum_{i=1}^{3} K_{\mathrm{m}} \frac{\partial}{\partial x_i}\left(\frac{\partial \bar{u}_3}{\partial x_i} + \frac{\partial \bar{u}_i}{\partial x_3}\right)\right] \tag{10.85}$$

Applying the same reasoning as in the direct Navier–Stokes simulation, it is possible to rearrange the partial differentials in the last term of eq. (10.85), in a way that isolates the three terms of the incompressible continuity equation [eq. (10.60)]. Hence, for incompressible simulations these terms can be eliminated, and the above equations are simplified as follows:

$$\frac{\partial \bar{u}_1}{\partial t} + \sum_{i=1}^{3} \bar{u}_i \frac{\partial \bar{u}_1}{\partial x_i} = f\bar{u}_2 - \frac{1}{\bar{\rho}}\frac{\partial \bar{p}}{\partial x_1} + \left(\frac{\mu}{\bar{\rho}} + K_{\mathrm{m}}\right)\sum_{i=1}^{3} \frac{\partial^2 \bar{u}_1}{\partial x_i^2} \tag{10.86}$$

$$\frac{\partial \bar{u}_2}{\partial t} + \sum_{i=1}^{3} \bar{u}_i \frac{\partial \bar{u}_2}{\partial x_i} = -f\bar{u}_1 - \frac{1}{\bar{\rho}}\frac{\partial \bar{p}}{\partial x_2} + \left(\frac{\mu}{\bar{\rho}} + K_{\mathrm{m}}\right)\sum_{i=1}^{3} \frac{\partial^2 \bar{u}_2}{\partial x_i^2} \tag{10.87}$$

$$\frac{\partial \bar{u}_3}{\partial t} + \sum_{i=1}^{3} \bar{u}_i \frac{\partial \bar{u}_3}{\partial x_i} = -g - \frac{1}{\bar{\rho}}\frac{\partial \bar{p}}{\partial x_3} + \left(\frac{\mu}{\bar{\rho}} + K_{\mathrm{m}}\right)\sum_{i=1}^{3} \frac{\partial^2 \bar{u}_3}{\partial x_i^2} \tag{10.88}$$

Closure of the Navier–Stokes equations requires knowledge of $K_{\mathrm{m}}$.

In fact, any property $\xi$ of the atmosphere can be dispersed by turbulence, so the gradient theory is not limited to the covariance between two velocity fluctuations. The transport of a scalar property $\xi$ is given by

$$\overline{u_j' \xi'} = -K \frac{\partial \bar{\xi}}{\partial x_j} \tag{10.89}$$

The subscript $m$ is no longer used in eq. (10.89) because the transport no longer refers to momentum. Under neutral and stable conditions, it is reasonable to assume that the $K$ values are the same, irrespective of what is being transported. This is not the case in unstable conditions, so we can expect first-order closure models to perform poorly in unstable conditions.

Assuming that the atmospheric turbulence is homogeneous in the $x$ and $y$ directions, eq. (10.82) closes the Navier–Stokes equations provided that an estimate of $K_m$ is available. This brings us back to Prantl's mixing length theory (see Chapter 5), which will be outlined in more detail here.

Assume that a parcel of air moves up the boundary layer over a distance $z'$, and the boundary layer has a positive wind speed gradient. The air parcel moves into faster moving air, so it will have a negative wind speed fluctuation. In equation form:

$$u' = -\frac{\partial \overline{u}}{\partial z} z' \tag{10.90}$$

For the air parcel to move up, it must have a positive vertical velocity fluctuation. Hence, we can assume there is a relationship bewteen $u'$ and $w'$:

$$w' = -cu' \tag{10.91}$$

However, this is only valid if the wind speed gradient is positive. If it is negative, the $u'$ and $w'$ are positively correlated:

$$w' = cu' \tag{10.92}$$

The above three equations can be combined as

$$w' = c\left|\frac{\partial \overline{u}}{\partial z}\right| z' \tag{10.93}$$

Hence, the velocity covariance can be obtained by combining eq. (10.90) with eq. (10.93), and taking the average:

$$\overline{u'w'} = -c\overline{z}'^2 \left|\frac{\partial \overline{u}}{\partial z}\right| \cdot \frac{\partial \overline{u}}{\partial z} \tag{10.94}$$

Comparing with eq. (5.174), we find the following equation for $K_m$:

$$K_m = c\overline{z}'^2 \left|\frac{\partial \overline{u}}{\partial z}\right| \tag{10.95}$$

The fact that eq. (5.174), not eq. (10.81), is used here is a weakness of the method. Combining $c\overline{z}'^2$ into a single length variable $l^2$, we obtain

$$K_m = l^2 \left|\frac{\partial \overline{u}}{\partial z}\right| \tag{10.96}$$

It follows that the model is closed if we can find a calculation method for $l$ that does not involve second-order moments. To that effect, consider the wind speed profile from similarity theory under neutral conditions [eq. (5.64)]:

$$u = \frac{u_*}{k} \ln \frac{z}{z_0} \tag{5.64}$$

Hence:

$$\left| \frac{\partial \overline{u}}{\partial z} \right| = \frac{u_*}{kz} \tag{10.97}$$

Substituting into eq. (10.96) leads to

$$K_{\mathrm{m}} = \frac{l^2 u_*}{kz} \tag{10.98}$$

On the other hand, an equation for $K_{\mathrm{m}}$ at neutral conditions near the surface is known from similarity theory [eq. (5.187)]:

$$K_{\mathrm{m}} = k u_* z \tag{5.187}$$

Combining eq. (10.98) with eq. (5.187), and solving for $l$ leads to

$$l = kz \tag{10.99}$$

and $k$ is the von Karman constant (0.4). Hence, the model is closed. Equations (10.96) and (10.99) define $K_{\mathrm{m}}$, which can be substituted in the Reynolds-averaged Navier–Stokes equations [eqs. (10.86)–(10.88)].

Many formulations for first-order closure have been proposed for nonneutral conditions (Stull, 1988). Most of these rely on the Monin–Obukhov length, which is based on $u_*$, which, in turn, is based on the velocity covariances we are trying to calculate. Hence, the calculation becomes iterative.

***10.3.2.3 Turbulent Kinetic Energy Equation.*** One of the measures of turbulence discussed is the turbulent kinetic energy per unit mass of air [eq. (5.138)]:

$$\overline{e} = \frac{1}{2}\left(\overline{u'^2} + \overline{v'^2} + \overline{w'^2}\right) = \frac{1}{2}\sum_{i=1}^{3} \overline{u_i'^2} \tag{10.100}$$

The turbulent kinetic energy turns out to be a good basis for turbulent closure modeling. In particular, the relationship between the turbulent kinetic energy and the turbulent energy dissipation rate $\varepsilon$ (see Section 6.11.2) gives rise to a class of closure models known as **$k$-$\varepsilon$ models** (where $k$ is a different notation for turbulent kinetic energy).

The $k$-$\varepsilon$ models will be discussed in the next section. In this section we will derive a prognostic equation for the turbulent kinetic energy based on the Navier–Stokes equations. We will mainly focus on the $x$ direction for the derivation and only present the main points in the derivation for the $y$ and $z$ components in the turbulent kinetic energy.

Starting from the Reynolds-decomposed, nonaveraged Navier–Stokes equation for the $x$ direction, eq. (10.78), expanding the terms between brackets on the left-hand side and ignoring density fluctuations leads to

$$\frac{\partial \overline{u}_1}{\partial t}+\frac{\partial u_1'}{\partial t}+\sum_{i=1}^{3}\overline{u}_i\frac{\partial \overline{u}_1}{\partial x_i}+\sum_{i=1}^{3}\overline{u}_i\frac{\partial u_1'}{\partial x_i}+\sum_{i=1}^{3}u_i'\frac{\partial \overline{u}_1}{\partial x_i}+\sum_{i=1}^{3}u_i'\frac{\partial u_1'}{\partial x_i}$$
$$=f\left(\overline{u}_2+u_2'\right)-\frac{1}{\overline{\rho}}\frac{\partial\left(\overline{p}+p'\right)}{\partial x_1}+\frac{\mu}{\overline{\rho}}\left[\sum_{i=1}^{3}\frac{\partial^2\left(\overline{u}_1+u_1'\right)}{\partial x_i^2}\right] \tag{10.101}$$

Subtracting the Reynolds-averaged Navier–Stokes equation for the $x$ direction, eq. (10.75), leads to

$$\frac{\partial u_1'}{\partial t}+\sum_{i=1}^{3}\overline{u}_i\frac{\partial u_1'}{\partial x_i}+\sum_{i=1}^{3}u_i'\frac{\partial \overline{u}_1}{\partial x_i}+\sum_{i=1}^{3}u_i'\frac{\partial u_1'}{\partial x_i}=fu_2'-\frac{1}{\overline{\rho}}\frac{\partial p'}{\partial x_1}+\frac{\mu}{\overline{\rho}}\sum_{i=1}^{3}\frac{\partial^2 u_1'}{\partial x_i^2}+\sum_{i=1}^{3}\frac{\partial \overline{u_1'u_i'}}{\partial x_i} \tag{10.102}$$

Equation (10.102) is a prognostic equation for the velocity fluctuation in the $x$ direction. The derivation of the equivalent equation for the $y$ direction is straightforward:

$$\frac{\partial u_2'}{\partial t}+\sum_{i=1}^{3}\overline{u}_i\frac{\partial u_2'}{\partial x_i}+\sum_{i=1}^{3}u_i'\frac{\partial \overline{u}_1}{\partial x_i}+\sum_{i=1}^{3}u_i'\frac{\partial u_2'}{\partial x_i}=-fu_1'-\frac{1}{\overline{\rho}}\frac{\partial p'}{\partial x_2}+\frac{\mu}{\overline{\rho}}\sum_{i=1}^{3}\frac{\partial^2 u_2'}{\partial x_i^2}+\sum_{i=1}^{3}\frac{\partial \overline{u_2'u_i'}}{\partial x_i} \tag{10.103}$$

The equation for the $z$ direction must be derived more carefully because density fluctuations, which have a negligible direct effect on the momentum of an air parcel, cause a nonnegligible buoyancy effect, especially in unstable conditions. Accounting for buoyancy effects of density fluctuations while ignoring all other effects of density fluctuations is known as the **Boussinesq approximation**. A convenient way to introduce the Boussinesq approximation is by multiplying eq. (10.80) by $(\overline{\rho}+\rho')/\overline{\rho}$ to bring forward the density effect in the gravity term:

$$\left(1+\frac{\rho'}{\overline{\rho}}\right)\frac{\partial(\overline{u}_3+u_3')}{\partial t}+\left(1+\frac{\rho'}{\overline{\rho}}\right)\sum_{i=1}^{3}(\overline{u}_i+u_i')\frac{\partial(\overline{u}_3+u_3')}{\partial x_i}$$
$$=-\left(\frac{\overline{\rho}+\rho'}{\overline{\rho}}\right)g-\frac{1}{\overline{\rho}}\frac{\partial(\overline{p}+p')}{\partial x_3}+\frac{\mu}{\overline{\rho}}\left[\sum_{i=1}^{3}\frac{\partial^2(\overline{u}_3+u_3')}{\partial x_i^2}\right] \tag{10.104}$$

The density factors in the left-hand side of eq. (10.104) can be approximated by 1. In the gravity term, the density fluctuation depends on the temperature and pressure fluctuation. Because the temperature fluctuation is more than 10 times larger than the pressure fluctuation (Garratt, 1994), we can ignore the latter. Hence, we can set density inversely proportional to the potential temperature, which leads to

$$\frac{\bar{\rho}+\rho'}{\bar{\rho}}=\frac{\bar{\theta}}{\bar{\theta}+\theta'}=\frac{\bar{\theta}}{\theta}=\frac{\theta-\theta'}{\theta} \tag{10.105}$$

Hence, eq. (10.104) can be written as

$$\frac{\partial(\bar{u}_3+u_3')}{\partial t}+\sum_{i=1}^{3}\left(\bar{u}_i+u_i'\right)\frac{\partial(\bar{u}_3+u_3')}{\partial x_i}$$
$$=-\left(\frac{\theta-\theta'}{\theta}\right)g-\frac{1}{\bar{\rho}}\frac{\partial(\bar{p}+p')}{\partial x_3}+\frac{\mu}{\bar{\rho}}\left[\sum_{i=1}^{3}\frac{\partial^2(\bar{u}_3+u_3')}{\partial x_i^2}\right] \tag{10.106}$$

Subtracting the Reynolds-averaged equation, eq. (10.77), leads to

$$\frac{\partial u_3'}{\partial t}+\sum_{i=1}^{3}\bar{u}_i\frac{\partial u_3'}{\partial x_i}+\sum_{i=1}^{3}u_i'\frac{\partial \bar{u}_3}{\partial x_i}+\sum_{i=1}^{3}u_i'\frac{\partial u_3'}{\partial x_i}$$
$$=\frac{\theta'}{\theta}g-\frac{1}{\bar{\rho}}\frac{\partial p'}{\partial x_3}+\frac{\mu}{\bar{\rho}}\sum_{i=1}^{3}\frac{\partial^2 u_3'}{\partial x_i^2}+\sum_{i=1}^{3}\frac{\partial\overline{u_3'\,u_i'}}{\partial x_i} \tag{10.107}$$

where the actual potential temperature in the gravity term can be approximated by the average potential temperature:

$$\frac{\partial u_3'}{\partial t}+\sum_{i=1}^{3}\bar{u}_i\frac{\partial u_3'}{\partial x_i}+\sum_{i=1}^{3}u_i'\frac{\partial \bar{u}_3}{\partial x_i}+\sum_{i=1}^{3}u_i'\frac{\partial u_3'}{\partial x_i}=\frac{\theta'}{\bar{\theta}}g-\frac{1}{\bar{\rho}}\frac{\partial p'}{\partial x_3}+\frac{\mu}{\bar{\rho}}\sum_{i=1}^{3}\frac{\partial^2 u_3'}{\partial x_i^2}+\sum_{i=1}^{3}\frac{\partial\overline{u_3'\,u_i'}}{\partial x_i} \tag{10.108}$$

Back to the $x$ direction, eq. (10.102) is multiplied by $2u_1'$:

$$2u_1'\frac{\partial u_1'}{\partial t}+\sum_{i=1}^{3}2u_1'\,\bar{u}_i\frac{\partial u_1'}{\partial x_i}+\sum_{i=1}^{3}2u_1'\,u_i'\frac{\partial \bar{u}_1}{\partial x_i}+\sum_{i=1}^{3}2u_1'\,u_i'\frac{\partial u_1'}{\partial x_i}$$
$$=f\cdot 2u_1'\,u_2'-\frac{1}{\bar{\rho}}2u_1'\frac{\partial p'}{\partial x_1}+\frac{\mu}{\bar{\rho}}\sum_{i=1}^{3}\frac{2u_1'\,\partial^2 u_1'}{\partial x_i^2}+\sum_{i=1}^{3}\frac{2u_1'\,\partial\overline{u_1'\,u_i'}}{\partial x_i} \tag{10.109}$$

The factor $2u_1'$ is incorporated in the partial derivative whenever possible:

$$\frac{\partial {u_1'}^2}{\partial t}+\sum_{i=1}^{3}\bar{u}_i\frac{\partial u_1'^2}{\partial x_i}+\sum_{i=1}^{3}2u_1'\,u_i'\frac{\partial \bar{u}_1}{\partial x_i}+\sum_{i=1}^{3}u_i'\frac{\partial {u_1'}^2}{\partial x_i}$$
$$=f\cdot 2u_1'\,u_2'-\frac{1}{\bar{\rho}}2u_1'\frac{\partial p'}{\partial x_1}+\frac{\mu}{\bar{\rho}}\sum_{i=1}^{3}\frac{2u_1'\,\partial^2 u_1'}{\partial x_i^2}+\sum_{i=1}^{3}\frac{2u_1'\,\partial\overline{u_1'\,u_i'}}{\partial x_i} \tag{10.110}$$

The average of eq. (10.110) is taken, and the Reynolds averaging rules are applied:

$$\frac{\partial \overline{u_1'^2}}{\partial t}+\sum_{i=1}^{3}\overline{u}_i\frac{\partial \overline{u_1'^2}}{\partial x_i}+\sum_{i=1}^{3}2\overline{u_1'\,u_i'}\frac{\partial \overline{u}_1}{\partial x_i}+\sum_{i=1}^{3}\overline{u_i'\frac{\partial u_1'^2}{\partial x_i}}$$
$$=f\cdot 2\overline{u_1'\,u_2'}-\frac{1}{\overline{\rho}}\overline{2u_1'\frac{\partial p'}{\partial x_1}}+\frac{\mu}{\overline{\rho}}\sum_{i=1}^{3}\overline{2u_1'\frac{\partial^2 u_1'}{\partial x_i^2}} \tag{10.111}$$

The same procedure is applied to the $y$ and $z$ directions:

$$\frac{\partial \overline{u_2'^2}}{\partial t}+\sum_{i=1}^{3}\overline{u}_i\frac{\partial \overline{u_2'^2}}{\partial x_i}+\sum_{i=1}^{3}2\overline{u_2'\,u_i'}\frac{\partial \overline{u}_2}{\partial x_i}+\sum_{i=1}^{3}\overline{u_i'\frac{\partial u_2'^2}{\partial x_i}}$$
$$=-f\cdot 2\overline{u_2'u_1'}-\frac{1}{\overline{\rho}}\overline{2u_2'\frac{\partial p'}{\partial x_2}}+\frac{\mu}{\overline{\rho}}\sum_{i=1}^{3}\overline{2u_2'\frac{\partial^2 u_2'}{\partial x_i^2}} \tag{10.112}$$

$$\frac{\partial \overline{u_3'^2}}{\partial t}+\sum_{i=1}^{3}\overline{u}_i\frac{\partial \overline{u_3'^2}}{\partial x_i}+\sum_{i=1}^{3}2\overline{u_3'\,u_i'}\frac{\partial \overline{u}_3}{\partial x_i}+\sum_{i=1}^{3}\overline{u_i'\frac{\partial u_3'^2}{\partial x_i}}$$
$$=\frac{2g}{\overline{\theta}}\overline{u_3'\,\theta'}-\frac{1}{\overline{\rho}}\overline{2u_3'\frac{\partial p'}{\partial x_3}}+\frac{\mu}{\overline{\rho}}\sum_{i=1}^{3}\overline{2u_3'\frac{\partial^2 u_3'}{\partial x_i^2}} \tag{10.113}$$

To obtain an equation for the turbulent kinetic energy, eqs. (10.111)–(10.113) are summed and divided by 2:

$$\frac{1}{2}\sum_{j=1}^{3}\frac{\partial \overline{u_j'^2}}{\partial t}+\frac{1}{2}\sum_{j=1}^{3}\sum_{i=1}^{3}\overline{u}_i\frac{\partial \overline{u_j'^2}}{\partial x_i}+\sum_{j=1}^{3}\sum_{i=1}^{3}\overline{u_j'\,u_i'}\frac{\partial \overline{u}_j}{\partial x_i}+\frac{1}{2}\sum_{j=1}^{3}\sum_{i=1}^{3}\overline{u_i'\frac{\partial u_j'^2}{\partial x_i}}$$
$$=\frac{g}{\overline{\theta}}\overline{u_3'\,\theta'}-\frac{1}{\overline{\rho}}\sum_{j=1}^{3}\overline{u_j'\frac{\partial p'}{\partial x_j}}+\frac{\mu}{\overline{\rho}}\sum_{j=1}^{3}\sum_{i=1}^{3}\overline{u_j'\frac{\partial^2 u_j'}{\partial x_i^2}} \tag{10.114}$$

The turbulent kinetic energy [eq. (10.103)] is substituted in eq. (10.117) whenever applicable:

$$\frac{\partial \overline{e}}{\partial t}+\sum_{i=1}^{3}\overline{u}_i\frac{\partial \overline{e}}{\partial x_i}+\sum_{j=1}^{3}\sum_{i=1}^{3}\overline{u_j'\,u_i'}\frac{\partial \overline{u}_j}{\partial x_i}+\sum_{i=1}^{3}\overline{u_i'\frac{\partial e}{\partial x_i}}$$
$$=\frac{g}{\overline{\theta}}\overline{u_3'\,\theta'}-\frac{1}{\overline{\rho}}\sum_{j=1}^{3}\overline{u_j'\frac{\partial p'}{\partial x_j}}+\frac{\mu}{\overline{\rho}}\sum_{j=1}^{3}\sum_{i=1}^{3}\overline{u_j'\frac{\partial^2 u_j'}{\partial x_i^2}} \tag{10.115}$$

Equation (10.118) has two terms containing third-order moments. The one on the left-hand side can be reworked by considering the continuity equation assuming constant density [eq. (10.60)]:

$$\sum_{i=1}^{3}\frac{\partial u_i'}{\partial x_i}=0 \tag{10.116}$$

Multiplying eq. (10.116) by $u_j'^2$ and summing over $j$ leads to

$$\sum_{j=1}^{3}\sum_{i=1}^{3} u_j'^2 \frac{\partial u_i'}{\partial x_i} = 0 \tag{10.117}$$

Now consider the following partial derivative:

$$\sum_{j=1}^{3}\sum_{i=1}^{3} \frac{\partial u_j'^2 u_i'}{\partial x_i} = \sum_{j=1}^{3}\sum_{i=1}^{3} u_j'^2 \frac{\partial u_i'}{\partial x_i} + \sum_{j=1}^{3}\sum_{i=1}^{3} u_i' \frac{\partial u_j'^2}{\partial x_i} \tag{10.118}$$

The first term in the right-hand side of eq. (10.118) is the same as the left-hand side of eq. (10.117). Therefore, it equals zero. In the other terms the summations can be rearranged:

$$\sum_{i=1}^{3} \frac{\partial u_i' \sum_{j=1}^{3} u_j'^2}{\partial x_i} = \sum_{i=1}^{3} u_i' \frac{\partial \sum_{j=1}^{3} u_j'^2}{\partial x_i} \tag{10.119}$$

In both terms the turbulent kinetic energy can be recognized. Hence, dividing by 2:

$$\sum_{i=1}^{3} \frac{\partial u_i' e}{\partial x_i} = \sum_{i=1}^{3} u_i' \frac{\partial e}{\partial x_i} \tag{10.120}$$

Reynolds averaging leads to the last term on the left-hand side of eq. (10.115). Substitution leads to

$$\begin{aligned} &\frac{\partial \bar{e}}{\partial t} + \sum_{i=1}^{3} \bar{u}_i \frac{\partial \bar{e}}{\partial x_i} + \sum_{j=1}^{3}\sum_{i=1}^{3} \overline{u_j' u_i'} \frac{\partial \bar{u}_j}{\partial x_i} + \sum_{i=1}^{3} \frac{\partial \overline{u_i' e}}{\partial x_i} \\ &= \frac{g}{\bar{\theta}} \overline{u_3' \theta'} - \frac{1}{\bar{\rho}} \sum_{j=1}^{3} \overline{u_j' \frac{\partial p'}{\partial x_j}} + \frac{\mu}{\bar{\rho}} \sum_{j=1}^{3}\sum_{i=1}^{3} \overline{u_j' \frac{\partial^2 u_j'}{\partial x_i^2}} \end{aligned} \tag{10.121}$$

This is a prognostic equation for the turbulent kinetic energy. The last term is a viscous energy dissipation term. A closer look at this term leads to some useful insights. Consider the following second partial derivative:

$$\frac{\partial^2 u_j'^2}{\partial x_i^2} = \frac{\partial}{\partial x_i}\left(\frac{\partial u_j'^2}{\partial x_i}\right) = \frac{\partial}{\partial x_i}\left(2u_j' \frac{\partial u_j'}{\partial x_i}\right) = 2\left(\frac{\partial u_j'}{\partial x_i}\right)^2 + 2u_j' \frac{\partial^2 u_j'}{\partial x_i^2} \tag{10.122}$$

Take the average:

$$\frac{\partial^2 \overline{u_j'^2}}{\partial x_i^2} = 2\overline{\left(\frac{\partial u_j'}{\partial x_i}\right)^2} + 2\overline{u_j' \frac{\partial^2 u_j'}{\partial x_i^2}} \tag{10.123}$$

There is a fundamental difference between the left-hand side and the terms in the right-hand side that makes the former several orders of magnitude smaller than the latter. In the left-hand side the average is taken first and the derivative is taken next. The average variable changes much more slowly than the variable itself. In this case, the average is a velocity variance, typically on the order of $1\,m^2 s^{-2}$, which changes over hundreds of meters. Its gradient will vary over similar distances. Hence, the left-hand side is on the order of $10^{-4} s^{-2}$. In the right-hand side we are dealing with actual variables that fluctuate over a few centimeters or less, so the terms on the right-hand side are on the order of $1000\,s^{-2}$ or larger. It follows that the left-hand side is negligibly small, and the right-hand side consists of two approximately equal terms of opposite sign:

$$\overline{\left(\frac{\partial u'_j}{\partial x_i}\right)^2} = -\overline{u'_j \frac{\partial^2 u'_j}{\partial x_i^2}} \tag{10.124}$$

Substitution into eq. (10.121) leads to

$$\frac{\partial \bar{e}}{\partial t} + \sum_{i=1}^{3} \bar{u}_i \frac{\partial \bar{e}}{\partial x_i} + \sum_{j=1}^{3}\sum_{i=1}^{3} \overline{u'_j u'_i} \frac{\partial \bar{u}_j}{\partial x_i} + \sum_{i=1}^{3} \frac{\partial \overline{u'_i e}}{\partial x_i} = \frac{g}{\bar{\theta}} \overline{u'_3 \theta'} - \frac{1}{\bar{\rho}} \sum_{j=1}^{3} \overline{u'_j \frac{\partial p'}{\partial x_j}} - \frac{\mu}{\bar{\rho}} \sum_{j=1}^{3}\sum_{i=1}^{3} \overline{\left(\frac{\partial u'_j}{\partial x_i}\right)^2} \tag{10.125}$$

Before discussing the meaning of the last term in eq. (10.125), we will make one more change, based on the following partial derivative:

$$\sum_{j=1}^{3} \overline{\frac{\partial u'_j p'}{\partial x_j}} = \sum_{j=1}^{3} \overline{u'_j \frac{\partial p'}{\partial x_j}} + \sum_{j=1}^{3} \overline{p' \frac{\partial u'_j}{\partial x_j}} = \sum_{j=1}^{3} \overline{u'_j \frac{\partial p'}{\partial x_j}} + \overline{p' \sum_{j=1}^{3} \frac{\partial u'_j}{\partial x_j}} \tag{10.126}$$

The last term in eq. (10.126) contains the continuity equation, which is zero. Hence, that term is zero, and the equation reduces to

$$\sum_{j=1}^{3} \overline{\frac{\partial u'_j p'}{\partial x_j}} = \sum_{j=1}^{3} \frac{\partial \overline{u'_j p'}}{\partial x_j} = \sum_{j=1}^{3} \overline{u'_j \frac{\partial p'}{\partial x_j}} \tag{10.127}$$

Hence, eq. (10.125) can be written as

$$\frac{\partial \bar{e}}{\partial t} + \sum_{i=1}^{3} \bar{u}_i \frac{\partial \bar{e}}{\partial x_i} + \sum_{j=1}^{3}\sum_{i=1}^{3} \overline{u'_j u'_i} \frac{\partial \bar{u}_j}{\partial x_i} + \sum_{i=1}^{3} \frac{\partial \overline{u'_i e}}{\partial x_i} = \frac{g}{\bar{\theta}} \overline{u'_3 \theta'} - \frac{1}{\bar{\rho}} \sum_{j=1}^{3} \frac{\partial \overline{u'_j p'}}{\partial x_j} - \frac{\mu}{\bar{\rho}} \sum_{j=1}^{3}\sum_{i=1}^{3} \overline{\left(\frac{\partial u'_j}{\partial x_i}\right)^2} \tag{10.128}$$

To understand the last term in eq. (10.128), consider the energy dissipated in Couette flow, a system consisting of a liquid moving between a stationary plate and a moving plate (Fig. 10.5). To keep such a system moving at a velocity $u$, a force $F$ must be maintained on the upper plate. The energy $E$ used to maintain the flow equals force multiplied by displacement:

$$E = Fut \tag{10.129}$$

where $t$ is the time. This energy used equals the energy dissipated by viscous forces. The energy dissipated per unit time and per unit mass, $\varepsilon$, is given by

$$\varepsilon = \frac{Fu}{m} \tag{10.130}$$

If the system is a fluid element $dx\,dy\,dz$, then its mass is

$$m = \rho\, dx\, dy\, dz \tag{10.131}$$

Taking the underlying fluid as a reference, $u$ is the *relative* velocity between the top and bottom plates. Hence, assuming that the velocity profile in Figure 10.5 is in the $z$ direction:

$$u = \frac{\partial u_1}{\partial x_3} dx_3 \tag{10.132}$$

The force $F$ is the shear stress, calculated with Newton's viscosity law, multiplied by the area:

$$F = \mu \frac{\partial u_1}{\partial x_3} dx_1\, dx_2 \tag{10.133}$$

The velocity gradient also generates a force perpendicular to the flow. However, a force perpendicular to the flow does not dissipate energy and will not be considered here. Hence, the energy dissipated per unit time and per unit mass is

$$\varepsilon = \frac{\mu(\partial u_1/\partial x_3)^2\, dx_1\, dx_2\, dx_3}{\rho\, dx_1\, dx_2\, dx_3} = \frac{\mu}{\rho}\left(\frac{\partial u_1}{\partial x_3}\right)^2 \tag{10.134}$$

In a real atmosphere, the energy dissipation rate is a time average of the contributions in all directions. Hence:

$$\varepsilon = \frac{\mu}{\overline{\rho}} \sum_{j=1}^{3} \sum_{i=1}^{3} \overline{\left(\frac{\partial u_j'}{\partial x_i}\right)^2} \tag{10.135}$$

We can rewrite eq. (10.135) as

$$\boxed{\frac{\partial \overline{e}}{\partial t} + \sum_{i=1}^{3} \overline{u}_i \frac{\partial \overline{e}}{\partial x_i} + \sum_{j=1}^{3}\sum_{i=1}^{3} \overline{u_j'\, u_i'} \frac{\partial \overline{u}_j}{\partial x_i} + \sum_{i=1}^{3} \frac{\partial \overline{u_i'\, e}}{\partial x_i} = \frac{g}{\overline{\theta}} \overline{u_3'\, \theta'} - \frac{1}{\overline{\rho}} \sum_{j=1}^{3} \frac{\partial \overline{u_j' p'}}{\partial x_j} - \varepsilon} \tag{10.136}$$

Equation (10.136) links the turbulent kinetic energy to the energy dissipation rate. It is a key equation in turbulence closure.

***10.3.2.4 One-and-Half-Order Closure: k-ε Models.*** The prognostic relationship between turbulent kinetic energy and energy dissipation rate derived in the previous section provides a powerful tool to close the Reynolds-averaged Navier–Stokes equation. The class of models based on this tool is known as $k$-$\varepsilon$ models, based on the alternative symbol $k$ for turbulent kinetic energy. Based on the notation used in this book, they could be called $\overline{e}$-$\varepsilon$ models. The $k$-$\varepsilon$ models are sometimes described as one-and-half-order closure models because some of the second-order moments occurring in the model are calculated prognostically in terms of third-order moments, whereas other second-order moments are parameterized.

The first step in deriving a $k$-$\varepsilon$ model is parameterizing all the higher-order moments in eq. (10.136) based on eq. (10.89):

$$\overline{u_i' u_j'} = -K\left(\frac{\partial \overline{u}_j}{\partial x_i} + \frac{\partial \overline{u}_i}{\partial x_j}\right) \tag{10.137}$$

$$\overline{u_i' e} = -K\frac{\partial \overline{e}}{\partial x_i} \tag{10.138}$$

$$\overline{u_3' \theta} = -K\frac{\partial \overline{\theta}}{\partial x_3} \tag{10.139}$$

Note that eq. (10.138) is a parameterization of a third-order moment. The other moments are second order. With this parameterization the pressure–wind speed covariance could be resolved with a similar term. However, Launder and Spalding (1974) proposed

$$\overline{u_i'\left(e + \frac{p'}{\overline{\rho}}\right)} = -K\frac{\partial \overline{e}}{\partial x_i} \tag{10.140}$$

This transforms eq. (10.136) into the following form:

$$\frac{\partial \overline{e}}{\partial t} + \sum_{i=1}^{3} \overline{u}_i \frac{\partial \overline{e}}{\partial x_i} + \sum_{j=1}^{3}\sum_{i=1}^{3} K\left(\frac{\partial \overline{u}_j}{\partial x_i} + \frac{\partial \overline{u}_i}{\partial x_j}\right)\left(\frac{\partial \overline{u}_j}{\partial x_i}\right) + \sum_{i=1}^{3} \frac{\partial}{\partial x_i}\left(K\frac{\partial \overline{e}}{\partial x_i}\right) = \frac{g}{\overline{\theta}} K \frac{\partial \overline{\theta}}{\partial x_3} - \varepsilon \tag{10.141}$$

In neutral conditions, the potential temperature gradient is zero, and eq. (10.141) reduces to

$$\frac{\partial \overline{e}}{\partial t} + \sum_{i=1}^{3} \overline{u}_i \frac{\partial \overline{e}}{\partial x_i} + \sum_{j=1}^{3}\sum_{i=1}^{3} K\left(\frac{\partial \overline{u}_j}{\partial x_i} + \frac{\partial \overline{u}_i}{\partial x_j}\right)\left(\frac{\partial \overline{u}_j}{\partial x_i}\right) + \sum_{i=1}^{3} \frac{\partial}{\partial x_i}\left(K\frac{\partial \overline{e}}{\partial x_i}\right) = -\varepsilon \tag{10.142}$$

Because momentum and kinetic energy do not necessarily have the same turbulent diffusivity, a fitting parameter $\sigma_k$ was added to eq. (10.142) (Jones and Launder, 1972). Hence:

$$\boxed{\frac{\partial \bar{e}}{\partial t}+\sum_{i=1}^{3}\bar{u}_i\frac{\partial \bar{e}}{\partial x_i}+\sum_{j=1}^{3}\sum_{i=1}^{3}K\left(\frac{\partial \bar{u}_j}{\partial x_i}+\frac{\partial \bar{u}_i}{\partial x_j}\right)\left(\frac{\partial \bar{u}_j}{\partial x_i}\right)+\sum_{i=1}^{3}\frac{\partial}{\partial x_i}\left(\frac{K}{\sigma_k}\frac{\partial \bar{e}}{\partial x_i}\right)=-\varepsilon} \tag{10.143}$$

Launder and Spalding (1974) proposed to model the energy dissipation rate with an almost identical equation:

$$\boxed{\frac{\partial \varepsilon}{\partial t}+\sum_{i=1}^{3}\bar{u}_i\frac{\partial \varepsilon}{\partial x_i}+\frac{c_1\varepsilon}{\bar{e}}\sum_{j=1}^{3}\sum_{i=1}^{3}K\left(\frac{\partial \bar{u}_j}{\partial x_i}+\frac{\partial \bar{u}_i}{\partial x_j}\right)\left(\frac{\partial \bar{u}_j}{\partial x_i}\right)+\sum_{i=1}^{3}\frac{\partial}{\partial x_i}\left(\frac{K}{\sigma_\varepsilon}\frac{\partial \varepsilon}{\partial x_i}\right)=-\frac{c_2\varepsilon^2}{\bar{e}}} \tag{10.144}$$

Jones and Launder (1972) approximated eq. (10.137) by the following:

$$\overline{u_i'\,u_j'}=-K\frac{\partial \bar{u}_j}{\partial x_i} \tag{10.145}$$

and found slightly different formulations as a result.

What is still needed to close the model is a parameterization of the turbulent diffusivity $K$. It is assumed that $K$ only depends on $\bar{e}$ and $\varepsilon$. Hence, dimension analysis dictates that the relationship is as follows:

$$K=c_\mu\frac{\bar{e}^2}{\varepsilon} \tag{10.146}$$

The following parameters were found to lead to accurate results, provided that the Reynolds number is not too small (fully turbulent flow):

$\sigma_k=1$

$\sigma_\varepsilon=1.3$

$c_1=1.55$

$c_2=2$

$c_\mu=0.09$

The Reynolds-averaged Navier–Stokes equations [eqs. (10.86)–(10.88)] can be used with $K_m=K$.

The method described here does not provide initial conditions of $\bar{e}$ and $\varepsilon$. It is recommended to use arbitrary but plausible values based on fully turbulent conditions, and to model sufficiently long for the effect of the initial conditions to fade out.

The above is only one example of a $k$-$\varepsilon$ model. Jones and Launder (1972) found that corrections to the model were needed near the wall (low Reynolds number). Furthermore, many variants have been proposed for the formulation and the parameter values (e.g., Mellor, 1973; Duynkerke, 1988; Abe et al., 1994; Shih et al., 1995).

A different class of RANS closure is the class of Reynolds stress models. In this type of models, the Reynolds stresses are calculated with differential equations similar to the ones for $\overline{e}$ and $\varepsilon$ (e.g., Launder et al., 1975). Due to the additional equations, the computational needs for this type of model is substantially larger than the computational needs of $k$-$\varepsilon$ models. Reynolds stress models are rarely used in air dispersion modeling.

To maintain consistency with the rest of this book, it is useful to translate the main equations back into the notation without summation signs. For the turbulence kinetic energy and its dissipation rate, the following equations have been found [eqs. (10.143) and (10.144)]:

$$\frac{\partial \overline{e}}{\partial t}+\overline{u}\frac{\partial \overline{e}}{\partial x}+\overline{v}\frac{\partial \overline{e}}{\partial y}+\overline{w}\frac{\partial \overline{e}}{\partial z}$$
$$+K\left[2\left(\frac{\partial \overline{u}}{\partial x}\right)^2+2\left(\frac{\partial \overline{v}}{\partial y}\right)^2+2\left(\frac{\partial \overline{w}}{\partial z}\right)^2+\left(\frac{\partial \overline{u}}{\partial y}+\frac{\partial \overline{v}}{\partial x}\right)^2+\left(\frac{\partial \overline{u}}{\partial z}+\frac{\partial \overline{w}}{\partial x}\right)^2+\left(\frac{\partial \overline{v}}{\partial z}+\frac{\partial \overline{w}}{\partial y}\right)^2\right]$$
$$+\frac{\partial}{\partial x}\left(\frac{K}{\sigma_k}\frac{\partial \overline{e}}{\partial x}\right)+\frac{\partial}{\partial y}\left(\frac{K}{\sigma_k}\frac{\partial \overline{e}}{\partial y}\right)+\frac{\partial}{\partial z}\left(\frac{K}{\sigma_k}\frac{\partial \overline{e}}{\partial z}\right)=-\varepsilon \tag{10.147}$$

$$\frac{\partial \varepsilon}{\partial t}+\overline{u}\frac{\partial \varepsilon}{\partial x}+\overline{v}\frac{\partial \varepsilon}{\partial y}+\overline{w}\frac{\partial \varepsilon}{\partial z}$$
$$+\frac{c_1\varepsilon K}{\overline{e}}\left[2\left(\frac{\partial \overline{u}}{\partial x}\right)^2+2\left(\frac{\partial \overline{v}}{\partial y}\right)^2+2\left(\frac{\partial \overline{w}}{\partial z}\right)^2+\left(\frac{\partial \overline{u}}{\partial y}+\frac{\partial \overline{v}}{\partial x}\right)^2+\left(\frac{\partial \overline{u}}{\partial z}+\frac{\partial \overline{w}}{\partial x}\right)^2+\left(\frac{\partial \overline{v}}{\partial z}+\frac{\partial \overline{w}}{\partial y}\right)^2\right]$$
$$+\frac{\partial}{\partial x}\left(\frac{K}{\sigma_\varepsilon}\frac{\partial \varepsilon}{\partial x}\right)+\frac{\partial}{\partial y}\left(\frac{K}{\sigma_\varepsilon}\frac{\partial \varepsilon}{\partial y}\right)+\frac{\partial}{\partial z}\left(\frac{K}{\sigma_\varepsilon}\frac{\partial \varepsilon}{\partial z}\right)=-\frac{c_2\varepsilon^2}{\overline{e}} \tag{10.148}$$

All the closure schemes discussed in this chapter determine the values of the higher-order moments based on local conditions. They are known as **local closure** methods. There is also a class of closure methods that consider the entire atmosphere in the deterimation of a local flux. These methods are known as **non-local closure** methods (e.g., Blackadar, 1978; Stull, 1984). They will be discussed in Chapter 11.

### 10.3.3 Reynolds-Averaged Material and Energy Balances

As indicated earlier, the Navier–Stokes equations rarely define the entire problem to be solved. Material and energy balances are usually needed as well. Hence, when a Reynolds-averaged Navier–Stokes equation is used, Reynolds-averaged material and energy balances are needed as well.

The Reynolds-averaged overall mass balance (continuity equation) of an incompressible fluid was discussed earlier [eq. (10.59)].

For the ideal gas law, eqs. (10.33) and (10.34) can be combined as follows:

$$p\frac{M_{\text{air}}}{R} = \rho T \tag{10.149}$$

Reynolds decompostion, followed by Reynolds averaging, leads to

$$\bar{p}\frac{M_{\text{air}}}{R} = \bar{\rho}\cdot\bar{T} + \overline{\rho' T'} \tag{10.150}$$

Reynolds decomposition and averaging of the material balance [eq. (10.42)] leads to the following equation:

$$\begin{aligned}\frac{\partial \bar{c}}{\partial t} = &-\frac{\partial \bar{c}\cdot\bar{u}}{\partial x} - \frac{\partial \overline{c'u'}}{\partial x} - \frac{\partial \bar{c}\cdot\bar{v}}{\partial y} - \frac{\partial \overline{c'v'}}{\partial y} - \frac{\partial \bar{c}\cdot\bar{w}}{\partial z} - \frac{\partial \overline{c'w'}}{\partial z} - \frac{\partial cw}{\partial z} \\ &+ D\left(\frac{\partial^2 \bar{c}}{\partial x^2} + \frac{\partial^2 \bar{c}}{\partial y^2} + \frac{\partial^2 \bar{c}}{\partial z^2}\right) + \bar{S}\end{aligned} \tag{10.151}$$

First-order closure of eq. (10.151) is straightforward: The second-order moments in the equation are material fluxes, which can be calculated with a turbulent version of Fick's law. For instance:

$$\overline{c'u'} = -K_x \frac{\partial \bar{c}}{\partial x} \tag{10.152}$$

where $K_x$ can be calculated as in Section 10.2.2.2.

For higher-order closure, an argument similar to the one that led to the turbulence kinetic energy equation can be developed. A more detailed discussion of the Reynolds-averaged material balance is included in Chapter 11.

The Reynolds-averaged heat balance follows the same principles as the Reynolds-averaged material balance. Based on eq. (10.51), the following equation is obtained:

$$\begin{aligned}\frac{\partial \bar{\theta}}{\partial t} = &-\left(\frac{\partial \bar{\theta}\cdot\bar{u}}{\partial x} + \frac{\partial \overline{\theta' u'}}{\partial x} + \frac{\partial \bar{\theta}\cdot\bar{v}}{\partial y} + \frac{\partial \overline{\theta' v'}}{\partial y} + \frac{\partial \bar{\theta}\cdot\bar{w}}{\partial z} + \frac{\partial \overline{\theta' w'}}{\partial z}\right) \\ &+ \frac{\lambda}{\rho c_p}\left(\frac{\partial^2 \bar{\theta}}{\partial x^2} + \frac{\partial^2 \bar{\theta}}{\partial y^2} + \frac{\partial^2 \bar{\theta}}{\partial z^2}\right) - \frac{1}{\rho c_p}\left(\frac{\partial \overline{R_x}}{\partial x} + \frac{\partial \overline{R_y}}{\partial y} + \frac{\partial \overline{R_z}}{\partial z}\right)\end{aligned} \tag{10.153}$$

Again, first-order closure can be achieved with a $K$ theory to describe the fluxes. For instance,

$$\overline{c'\theta'} = -K_{\text{h}} \frac{\partial \bar{\theta}}{\partial x} \tag{10.154}$$

Higher-order closure is achieved in a manner similar to the higher-order closure of the Navier–Stokes equations. Details will not be discussed here.

## 10.4 LARGE EDDY SIMULATION (LES)

### 10.4.1 Introduction

In the previous section it was shown how Reynolds averaging can greatly reduce the computational needs for solving the Navier–Stokes equations. However, in the process all turbulent features have been removed from the model: It is as if we resolved an equivalent laminar flow problem with a highly location-dependent viscosity. However, not all turbulent processes behave like a laminar process with added randomness. Furthermore, some turbulent structures are so large that we want to resolve them explictly. An example is convective cycling, where the updrafts and downdrafts are part of the atmosphere's turbulence. A RANS model that removes such features by Reynolds averaging them is too incomplete to be of any use for air dispersion modeling. Furthermore, there is a more fundamental reason why RANS is inappropriate in the case of convective cycling: Reynolds averaging assumes that the fluctuations are small compared to the average behavior of the system. In the case of convective cycling, where the velocity of the updrafts can be of the same order of magnitude as the horizontal wind speed, this condition is not met. Hence, we need a more refined framework to satisfactorily model the convective boundary layer. Such a framework is large eddy simulation (LES)

Like RANS, LES is based on averaging, but instead of averaging all turbulence, only the turbulent motions smaller than a given size (e.g., the cell size of the computational grid) are averaged, whereas any turbulence that can be resolved on the grid is simulated directly. It follows that the parameterization of the higher order moments will depend on an arbitrarily chosen resolution such as the grid cell size. How that is achieved is subject of the next section.

### 10.4.2 Turbulence Modeling in LES

In LES, the derivation of the Navier–Stokes equation for the mean wind speed is essentially the same as for the Reynolds-averaged Navier–Stokes equation shown in Section 10.3.1. Hence, eqs. (10.64)–(10.66) can be used for LES, but the meaning of the variables is different. Instead of averaging out all the turbulence, we define average properties as if we are passing the turbulence through a filter. For instance, to define average wind speed in a point $(x,y,z)$, we could define a volume $V$ around the point, and calculate the average as

$$\overline{u} = \frac{1}{V}\int_0^V u\,dV \tag{10.155}$$

where $V$ can be the grid cell size $(\Delta x \Delta y \Delta z)$, or larger, but never smaller. When the grid cell size is chosen as the averaging volume, the turbulence closure can be described as **subgrid scale modeling**.

The size of $V$ should not be smaller than the grid cell size because all the turbulence in the atmosphere must be either simulated directly or modeled in the closure model. If $V$ is smaller than the grid cell size, then the part of the turbulence at a scale larger than $V$, but smaller than the grid cell size, is not resolved. In its simplest form, LES uses concepts from Kolmogorov's energy cascade theory in the closure technique, so $V$ should be smaller than the largest eddies in the inertial subrange, where the main energy transfer is from larger to smaller eddies, in these types of LES closure.

When LES closure is based on Kolmogorov's energy cascade theory, the starting point is the Kolmogorov microscale $\eta$, which is a measure of the smallest eddies in turbulent flow (see Section 6.11.2). Based on dimension analysis, assuming that the turbulent kinetic energy dissipation rate $\varepsilon$ depends only on kinematic viscosity $\nu$ and on $\eta$, $\varepsilon$ can be calculated based on eq. (6.184):

$$\varepsilon = \frac{\nu^3}{\eta^4} \tag{10.156}$$

Similarly, the subgrid kinetic energy dissipation rate $\varepsilon_s$ must be based on the turbulent viscosity $K_m$ and the averaging length scale $l$ as

$$\varepsilon_s = \frac{K_m^3}{c^4 l^4} \tag{10.157}$$

where $c$ is a dimensionless fitting constant, and $l$ is given by

$$l = V^{1/3} \tag{10.158}$$

In the case $V = \Delta x \, \Delta y \, \Delta z$, $l$ is given by

$$l = (\Delta x \, \Delta y \, \Delta z)^{1/3} \tag{10.159}$$

On the other hand, $\varepsilon_s$ can be defined in a manner similar to $\varepsilon$ [eq. (10.135)]. With summations (Sorbjan, 2005):

$$\varepsilon_s = \frac{K_m}{2} \sum_{j=1}^{3} \sum_{i=1}^{3} \left( \frac{\partial \overline{u}_i}{\partial x_j} + \frac{\partial \overline{u}_j}{\partial x_i} \right)^2 \tag{10.160}$$

Hence, combining eqs. (10.157) and (10.160), $K_m$ is calculated as

$$K_m = c^2 l^2 \sqrt{\frac{1}{2} \sum_{j=1}^{3} \sum_{i=1}^{3} \left( \frac{\partial \overline{u}_i}{\partial x_j} + \frac{\partial \overline{u}_j}{\partial x_i} \right)^2} \tag{10.161}$$

Under neutral conditions, $c = 0.1$ provides good results. Under unstable conditions, a value of $c = 0.21$ should be used (Deardorff, 1971). The subgrid mass and heat diffusivity $K_h$ is not always the same as the subgrid momentum diffusivity $K_m$. Deardorff (1972) suggested

$$K_h = c_K K_m \tag{10.162}$$

where $c_K$=3, at least in unstable conditions, except close to the surface, where $c_K$ is close to 1.

Because of the variability of $c_K$ near the surface, the above approach to LES closure leads to unrealistic results near boundaries. To resolve that problem, Deardorff (1973, 1974) developed a LES closure scheme based on a prognostic equation for all second-order moments. The third-order moments were parameterized. Later, Deardorff (1980) developed a closure scheme based on an equation for subgrid kinetic energy. The equation was as follows:

$$\frac{\partial \bar{e}}{\partial t} = -\sum_{i=1}^{3} \frac{\partial}{\partial x_i}(\bar{u}_i \cdot \bar{e}) - \sum_{i=1}^{3}\sum_{j=1}^{3} \overline{u_i' u_j'} \frac{\partial \bar{u}_i}{\partial x_j} + \frac{g}{\bar{\theta}} \overline{x_3' \theta'} - \sum_{i=1}^{3} \frac{\partial}{\partial x_i} \overline{u_i' \left( e + \frac{p'}{\bar{\rho}} \right)} - \varepsilon \tag{10.163}$$

The velocity covariances were parameterized as follows:

$$\overline{u_i' u_j'} = -K_\mathrm{m} \left( \frac{\partial \bar{u}_i}{\partial x_j} + \frac{\partial \bar{u}_j}{\partial x_i} \right) \tag{10.164}$$

whereas the velocity variances were parameterized as

$$\overline{u_i' u_i'} = -2K_\mathrm{m} \frac{\partial \bar{u}_i}{\partial x_i} + \frac{2}{3}\bar{e} \tag{10.165}$$

The third-order moment is parameterized as

$$\overline{u_i' \left( e + \frac{p'}{\bar{\rho}} \right)} = -2K_\mathrm{m} \frac{\partial \bar{e}}{\partial x_i} \tag{10.166}$$

The turbulent momentum diffusivity $K_\mathrm{m}$ was calculated by

$$K_\mathrm{m} = 0.1 l_\mathrm{s} \sqrt{\bar{e}} \tag{10.167}$$

where $l_\mathrm{s}$ is given by

$$l_\mathrm{s} = \min \left( l, 0.76 \sqrt{\frac{\bar{e}}{\frac{g}{\bar{\theta}} \cdot \frac{\partial \bar{\theta}}{\partial z}}} \right) \tag{10.168}$$

where $l$ is given by eq. (10.159). Turbulent mass and heat diffusivity were calculated as

$$K_\mathrm{h} = \left( 1 + \frac{2 l_\mathrm{s}}{l} \right) K_\mathrm{m} \tag{10.169}$$

The kinetic energy dissipation rate $\varepsilon$ was calculated as

$$\varepsilon = \frac{c\bar{e}^{3/2}}{l_s} \tag{10.170}$$

where

$$c = 0.19 + 0.51\frac{l_s}{l} \tag{10.171}$$

As with RANS, many varieties of LES exist. Some examples of such varieties were presented by Moeng (1984) and by Mason (1989).

## 10.5 NUMERICAL METHODS IN CFD

The Navier–Stokes equations and the material and energy balances are usually solved by finite differences on a grid. There are many ways a set of differential equations can be implemented on a grid, and some implementations are more likely to lead to numerical instability than others. An important principle to bear in mind in modeling transport is to *keep the variables separate from their fluxes*. To understand how that works, consider a variable $a$ that is transported by an advective process and a diffusive process. The advective process is proportional to the gradient of $a$, whereas the diffusive process is proportional to the gradient of the flux of $a$, $F_a$. A simple differential equation of this form would be

$$\frac{\partial a}{\partial t} = -u\frac{\partial a}{\partial x} - \frac{\partial F_a}{\partial x} \tag{10.172}$$

where the flux is given by a gradient theory:

$$F_a = -D\frac{\partial a}{\partial x} \tag{10.173}$$

Consider a grid in the $x$ direction with spacing $\Delta x$, where both $a$ and $F_a$ are defined at each grid point. The gradients of $a$ and $F_a$ at grid point $n$ are approximated as finite differences:

$$\frac{\partial a_n}{\partial x} = \frac{a_{n+1} - a_{n-1}}{2\Delta x} \tag{10.174}$$

$$\frac{\partial F_{a_n}}{\partial x} = \frac{F_{a_{n+1}} - F_{a_{n-1}}}{2\Delta x} \tag{10.175}$$

Each flux is also calculated as a finite difference:

$$F_{a_{n+1}} = -D\frac{a_{n+2} - a_n}{2\Delta x} \tag{10.176}$$

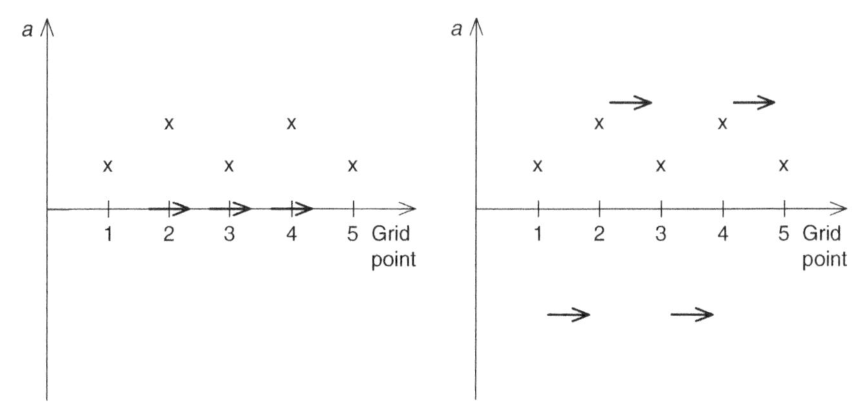


Figure 10.8 Finite differences ($a$) on a regular grid and ($b$) on a staggered grid. ×: $a$; arrows: $F_a$.

$$F_{a_{n-1}} = -D\frac{a_n - a_{n-2}}{2\Delta x} \tag{10.177}$$

Hence, the change of $a$ is calculated as

$$\frac{\partial a_n}{\partial t} = -u\frac{a_{n+1} - a_{n-1}}{2\Delta x} - D\frac{a_{n+2} - 2a_n + a_{n-2}}{(2\Delta x)^2} \tag{10.178}$$

Equation (10.178) has an important destabilizing feature: It never calculates the difference between two adjacent values of $a$. The repercussion of this is clear from Figure 10.8. In the figure, the variable $a$ alternates between a high and a low value, a feature that is sometimes induced by the finite differencing of the boundary condition. In the regular grid, where the variable and its flux are not separated, both $a$ and $F_a$ have zero gradient in the finite difference scheme, so eq. (10.178) predicts zero time derivative. The finite difference scheme does not "feel" the alternating pattern and cannot dampen it by diffusive forces. The alternating grid pattern, on the other hand, corresponds with the following equation:

$$\frac{\partial a_n}{\partial t} = -u\frac{a_{n+1} - a_{n-1}}{2\Delta x} - D\frac{a_{n+1} - 2a_n + a_{n-1}}{(\Delta x)^2} \tag{10.179}$$

Now the finite difference scheme leads to fluxes away from the high values of $a$, and toward the low values of $a$, leading to a dampening of the alternating pattern. The alternating grid pattern is much more likely to lead to a stable calculation scheme than the regular grid pattern. However, even eq. (10.179) does not always lead to a stable calculation either. More robust algorithms are needed to guarantee stability.

Based on the above comparison, the grid is usually built as follows. Consider a grid consisting of cubic cells. The center of the grid cell contains the pressure, concentration, and temperature. The wind speeds are defined in the centers of the faces of the cube: $u$ at the front and back faces, $v$ at the left and right faces, and $w$ at the top and bottom faces. Stresses, if defined as separate entities, are defined at the centers of the edges of the cube (Sorbjan, 2005).

For the numerical integration of the time, a Runge–Kutta routine was proposed by Sorbjan (2005) and an Adams–Bashford method was proposed by Deardorff (1980).

Traditionally, regular grids were used for finite differencing. However, irregular grids have gained popularity in recent years. Irregular grids, for instance, using triangular prisms instead of cubes, are more flexible because the grid can be made to consist of small triangles where high resolution is required and large triangles where low resolution is adequate. Irregular grids tend to be more accurate and efficient than regular grids (Sarma, 2008). However, the mathematics underlying such finite differences is more complicated than in the case of regular grids.

## 10.6 METEOROLOGICAL MODELING

The domain of a CFD calculation can be very small, for example, in case details of the circumstances of emission need to be modeled or to study dispersion of pollutants in street canyons. Of an entirely different scale is meteorological modeling to generate the wind speeds and stability conditions necessary to run Gaussian, Lagrangian, or Eulerian air dispersion models. The domain of meteorological model can be as large as an entire continent, or even the entire globe. Due to the scale, new levels of complexity are introduced in this type of model, such as cloud physics, water evaporation, short wave (incoming) and long wave (outgoing) radiation, and the like.

Furthermore, approximations such as incompressible flow or the Boussinesq approximation no longer hold in all circumstances of meteorological modeling. The most contemporary meteorological models consider compressible flow. This can cause problems when such a model is linked to an Eulerian model that assumes incompressible flow, as the result will be a violation of mass conservation (Byun et al., 2003).

A related problem occurs when concentration profiles over large height differences are modeled. Using $c$, in mg $m^{-3}$, to express concentration, the material balance usually predicts that the concentration at steady state is independent of height. Summing up a meterial balance for every compound in air, it follows that the total concentration is predicted to be independent of height. However, from the barometric equation we can calculate that the density, and hence the total concentration, decreases with height. It follows that the material balance does not predict concentrations accurately when the domain is large enough to contain pronounced density variations. To solve this problem, the material balance is often expressed in terms of the pollutant mass fraction $q$ (e.g., in μg $kg^{-1}$). This is explained in more detail in Chapter 11.

Another issue of meteorological modeling is to ensure that the model is as consistent as reasonably possible with the available meteorological measurement data (data assimilation). The meteorological model MM5, for instance, uses a relaxation technique that "nudges" the model results toward the measurement data with a user-defined time constant and radius of influence.

Large-scale meteorological models have a domain reaching up to high altitudes. Hence, pressure and density vary substantially within the domain. Meteorological models often use pressure or a variable linked to pressure as vertical coordinate instead of height. An example of a variable linked to pressure is the dimensionless pressure $\sigma$:

$$\sigma = \frac{p - p_{\text{top}}}{p_{\text{surf}} - p_{\text{top}}} \tag{10.180}$$

where $p_{\text{top}}$ is the pressure at the top of the computational domain, and $p_{\text{surf}}$ is the pressure at the surface. Hence, by definition $\sigma = 0$ at the top, and $\sigma = 1$ at the surface. The advantage of this approach is that the variable $\sigma$ is proportional to the mass of air considered (Pleim and Chang, 1992).

## 10.7 SUMMARY OF THE MAIN EQUATIONS

Newton's viscosity law (incompressible fluid):

$$\tau_{xz} = \mu\left(\frac{\partial u}{\partial z} + \frac{\partial w}{\partial x}\right) \tag{10.12}$$

Navier–Stokes equations for incompressible flow of a Newtonian fluid:

$$\frac{\partial u}{\partial t} + u\frac{\partial u}{\partial x} + v\frac{\partial u}{\partial y} + w\frac{\partial u}{\partial z} = fv - \frac{1}{\rho}\frac{\partial p}{\partial x} + \nu\left(\frac{\partial^2 u}{\partial x^2} + \frac{\partial^2 u}{\partial y^2} + \frac{\partial^2 u}{\partial z^2}\right) \tag{10.24}$$

$$\frac{\partial v}{\partial t} + u\frac{\partial v}{\partial x} + v\frac{\partial v}{\partial y} + w\frac{\partial v}{\partial z} = -fu - \frac{1}{\rho}\frac{\partial p}{\partial y} + \nu\left(\frac{\partial^2 v}{\partial x^2} + \frac{\partial^2 v}{\partial y^2} + \frac{\partial^2 v}{\partial z^2}\right) \tag{10.25}$$

$$\frac{\partial w}{\partial t} + u\frac{\partial w}{\partial x} + v\frac{\partial w}{\partial y} + w\frac{\partial w}{\partial z} = -g - \frac{1}{\rho}\frac{\partial p}{\partial z} + \nu\left(\frac{\partial^2 w}{\partial x^2} + \frac{\partial^2 w}{\partial y^2} + \frac{\partial^2 w}{\partial z^2}\right) \tag{10.26}$$

Continuity equation, incompressible flow:

$$\frac{\partial u}{\partial x} + \frac{\partial v}{\partial y} + \frac{\partial w}{\partial z} = 0 \tag{10.30}$$

Material balance:

$$\frac{\partial c}{\partial t} = -\frac{\partial cu}{\partial x} - \frac{\partial cv}{\partial y} - \frac{\partial cw}{\partial z} + D\left(\frac{\partial^2 c}{\partial x^2} + \frac{\partial^2 c}{\partial y^2} + \frac{\partial^2 c}{\partial z^2}\right) + S \tag{10.42}$$

Energy balance:

$$\frac{\partial \theta}{\partial t} = -\left(\frac{\partial \theta u}{\partial x} + \frac{\partial \theta v}{\partial y} + \frac{\partial \theta w}{\partial z}\right) + \frac{\lambda}{\rho c_p}\left(\frac{\partial^2 \theta}{\partial x^2} + \frac{\partial^2 \theta}{\partial y^2} + \frac{\partial^2 \theta}{\partial z^2}\right) - \frac{1}{\rho c_p}\left(\frac{\partial R_x}{\partial x} + \frac{\partial R_y}{\partial y} + \frac{\partial R_z}{\partial z}\right) \tag{10.51}$$

Reynolds-averaged Navier-Stokes equations (RANS):

$$\begin{aligned}\frac{\partial \bar{u}}{\partial t} + \bar{u}\frac{\partial \bar{u}}{\partial x} + \bar{v}\frac{\partial \bar{u}}{\partial y} + \bar{w}\frac{\partial \bar{u}}{\partial z} &= f\bar{v} - \frac{1}{\bar{\rho}}\frac{\partial \bar{p}}{\partial x} \\ &+ \frac{1}{\bar{\rho}}\left(\mu\frac{\partial^2 \bar{u}}{\partial x^2} - \bar{\rho}\frac{\partial \overline{u'^2}}{\partial x} + \mu\frac{\partial^2 \bar{u}}{\partial y^2} - \bar{\rho}\frac{\partial \overline{u'v'}}{\partial y} + \mu\frac{\partial^2 \bar{u}}{\partial z^2} - \bar{\rho}\frac{\partial \overline{u'w'}}{\partial z}\right)\end{aligned} \tag{10.64}$$

$$\begin{aligned}\frac{\partial \bar{v}}{\partial t} + \bar{u}\frac{\partial \bar{v}}{\partial x} + \bar{v}\frac{\partial \bar{v}}{\partial y} + \bar{w}\frac{\partial \bar{v}}{\partial z} &= -f\bar{u} - \frac{1}{\bar{\rho}}\frac{\partial \bar{p}}{\partial y} \\ &+ \frac{1}{\bar{\rho}}\left(\mu\frac{\partial^2 \bar{v}}{\partial x^2} - \bar{\rho}\frac{\partial \overline{u'v'}}{\partial x} + \mu\frac{\partial^2 \bar{v}}{\partial y^2} - \bar{\rho}\frac{\partial \overline{v'^2}}{\partial y} + \mu\frac{\partial^2 \bar{v}}{\partial z^2} - \bar{\rho}\frac{\partial \overline{v'w'}}{\partial z}\right)\end{aligned} \tag{10.65}$$

$$\begin{aligned}\frac{\partial \bar{w}}{\partial t} + \bar{u}\frac{\partial \bar{w}}{\partial x} + \bar{v}\frac{\partial \bar{w}}{\partial y} + \bar{w}\frac{\partial \bar{w}}{\partial z} &= -g - \frac{1}{\bar{\rho}}\frac{\partial \bar{p}}{\partial z} \\ &+ \frac{1}{\bar{\rho}}\left(\mu\frac{\partial^2 \bar{w}}{\partial x^2} - \bar{\rho}\frac{\partial \overline{u'w'}}{\partial x} + \mu\frac{\partial^2 \bar{w}}{\partial y^2} - \bar{\rho}\frac{\partial \overline{v'w'}}{\partial y} + \mu\frac{\partial^2 \bar{w}}{\partial z^2} - \bar{\rho}\frac{\partial \overline{w'^2}}{\partial z}\right)\end{aligned} \tag{10.66}$$

Reynolds-averaged Navier–Stokes equations with turbulence closure:

$$\frac{\partial \bar{u}_1}{\partial t} + \sum_{i=1}^{3}\bar{u}_i\frac{\partial \bar{u}_1}{\partial x_i} = f\bar{u}_2 - \frac{1}{\bar{\rho}}\frac{\partial \bar{p}}{\partial x_1} + \left(\frac{\mu}{\bar{\rho}} + K_{\mathrm{m}}\right)\sum_{i=1}^{3}\frac{\partial^2 \bar{u}_1}{\partial x_i^2} \tag{10.86}$$

$$\frac{\partial \bar{u}_2}{\partial t} + \sum_{i=1}^{3}\bar{u}_i\frac{\partial \bar{u}_2}{\partial x_i} = -f\bar{u}_1 - \frac{1}{\bar{\rho}}\frac{\partial \bar{p}}{\partial x_2} + \left(\frac{\mu}{\rho} + K_{\mathrm{m}}\right)\sum_{i=1}^{3}\frac{\partial^2 \bar{u}_2}{\partial x_i^2} \tag{10.87}$$

$$\frac{\partial \bar{u}_3}{\partial t} + \sum_{i=1}^{3}\bar{u}_i\frac{\partial \bar{u}_3}{\partial x_i} = -g - \frac{1}{\bar{\rho}}\frac{\partial \bar{p}}{\partial x_3} + \left(\frac{\mu}{\bar{\rho}} + K_{\mathrm{m}}\right)\sum_{i=1}^{3}\frac{\partial^2 \bar{u}_3}{\partial x_i^2} \tag{10.88}$$

Energy dissipation rate (definition):

$$\varepsilon = \frac{\mu}{\bar{\rho}}\overline{\sum_{j=1}^{3}\sum_{i=1}^{3}\left(\frac{\partial u_j'}{\partial x_i}\right)^2} \tag{10.135}$$

Prognostic equation for the energy dissipation rate:

$$\frac{\partial \bar{e}}{\partial t}+\sum_{i=1}^{3}\bar{u}_i\frac{\partial \bar{e}}{\partial x_i}+\sum_{j=1}^{3}\sum_{i=1}^{3}\overline{u_j'u_i'}\frac{\partial \bar{u}_j}{\partial x_i}+\sum_{i=1}^{3}\frac{\partial \overline{u_i'e}}{\partial x_i}=\frac{g}{\bar{\theta}}\overline{u_3'\theta'}-\frac{1}{\bar{\rho}}\sum_{j=1}^{3}\frac{\partial \overline{u_j'p'}}{\partial x_j}-\varepsilon \qquad (10.136)$$

The $k$-$\varepsilon$ model:

$$\frac{\partial \bar{e}}{\partial t}+\sum_{i=1}^{3}\bar{u}_i\frac{\partial \bar{e}}{\partial x_i}+\sum_{j=1}^{3}\sum_{i=1}^{3}K\left(\frac{\partial \bar{u}_j}{\partial x_i}+\frac{\partial \bar{u}_i}{\partial x_j}\right)\left(\frac{\partial \bar{u}_j}{\partial x_i}\right)+\sum_{i=1}^{3}\frac{\partial}{\partial x_i}\left(\frac{K}{\sigma_k}\frac{\partial \bar{e}}{\partial x_i}\right)=-\varepsilon \qquad (10.143)$$

$$\frac{\partial \varepsilon}{\partial t}+\sum_{i=1}^{3}\bar{u}_i\frac{\partial \varepsilon}{\partial x_i}+\frac{c_1\varepsilon}{\bar{e}}\sum_{j=1}^{3}\sum_{i=1}^{3}K\left(\frac{\partial \bar{u}_j}{\partial x_i}+\frac{\partial \bar{u}_i}{\partial x_j}\right)\left(\frac{\partial \bar{u}_j}{\partial x_i}\right)+\sum_{i=1}^{3}\frac{\partial}{\partial x_i}\left(\frac{K}{\sigma_\varepsilon}\frac{\partial \varepsilon}{\partial x_i}\right)=-\frac{c_2\varepsilon^2}{\bar{e}} \qquad (10.144)$$

## REFERENCES


Abe K., Kondoh T., and Nagano Y. (1994). A new turbulence model for predicting fluid flow and heat transfer in separating and reattaching flows. I. Flow field calculations. *Int. J. Heat Mass Transfer* **37**, 139–151.

Blackadar A.K. (1978). Modeling pollutant transfer during daytime convection. In *Preprints, Fourth Symposium in Atmospheric Turbulence, Diffusion, and Air Quality, Reno*. American Meteorological Society, Boston, pp. 443–447.

Byun D.W., Lacser A., Yamartino R., and Zannetti P. (2003). Eulerian dispersion models. In Zannetti P. (ed.). *Air Quality Modeling. Volume I. Fundamentals*. Enviro Comp Institute, Fremont, CA, and Air and Waste Management Association, Pittsburgh, pp. 213–291.

Deardorff J.W. (1971). On the magnitude of the subgrid scale eddy coefficient. *J. Comput. Phys.* **7**, 120–133.

Deardorff J.W. (1972). Numerical investigation of neutral and unstable planetary boundary layers. *J. Atmos. Sci.* **29**, 91–115.

Deardorff J.W. (1973). The use of subgrid transport equations in a three-dimensional model of atmospheric turbulence. *J. Fluids Eng.* **95**, 429–438.

Deardorff J.W. (1974). Three-dimensional numerical study of the height and mean structure of a heated planetary boundary layer. *Bound. Layer Meteorol.* **7**, 81–106.

Deardorff J.W. (1980). Stratocumulus-capped mixed layers derived from a three-dimensional model. *Bound. Layer Meteorol.* **18**, 495–527.

Duynkerke P.G. (1988). Application of the $E$-$\varepsilon$ turbulence closure model to the neutral and stable atmospheric boundary layer. *J. Atmos. Sci.* **45**, 865–880.

Garratt J.R. (1994). *The Atmospheric Boundary Layer*. Cambridge University Press, Cambridge, UK.

Jones W.P. and Launder B.E. (1972). The prediction of laminarization with a two-equation model of turbulence. *Int. J. Heat Mass Transfer* **15**, 301–314.

Launder B.E. and Spalding D.B. (1974). The numerical computation of turbulent flows. *Comp. Meth. Appl. Mech. Eng.* **3**, 269–289.

Launder B.E., Reece G.J., and Rodi W. (1975). Progress in the development of a Reynolds-stress turbulence closure. *J. Fluid Mech.* **68**, 537–566.

Legg B.J. and Raupach M.R. (1982). Markov-chain simulation of particle dispersion in inhomogeneous flows: The mean drift velocity induced by a gradient in Eulerian velocity variance. *Bound. Layer Meteorol.* **24**, 3–13.

Mason P.J. (1989). Large-eddy simulation of the convective atmospheric boundary layer. *J. Atmos. Sci.* **46**, 1492–1516.

Mellor G.L. (1973). Analytic prediction of the properties of stratified planetary surface layers. *J. Atm. Sci.* **30**, 1061–1069.
Moeng C.H. (1984). A large-eddy simulation model for the study of planetary boundary-layer turbulence. *J. Atmos. Sci.* **41**, 2052–2062.
Pleim J.E. and Chang J.S. (1992). A non-local closure model for vertical mixing in the convective boundary layer. *Atmos. Environ.* **26A**, 965–981.
Sarma A. (2008). Meteorological modeling for air quality applications. In Zannetti P. (ed.). *Air Quality Modeling. Volume III. Special Issues*. Enviro Comp Institute, Fremont, CA, and Air and Waste Management Association, Pittsburgh, pp. 131–168.
Shih T.H., Liou W.W., Shabbir A., Yang Z., and Zhu J. (1995). A new $k$-$\varepsilon$ eddy viscosity model for high Reynolds number turbulent flows. *Computers Fluids* **24**, 227–238.
Sorbjan, Z. (2005). Large-eddy simulations of the atmospheric boundary layer. In Zannetti P. (ed.). *Air Quality Modeling. Volume II. Advanced Topics*. Enviro Comp Institute, Fremont, CA, and Air and Waste Management Association, Pittsburgh, pp. 11–82.
Stull R.B. (1984). Transilient turbulence theory. Part 1: The concept of eddy mixing across finite distances. *J. Atmos. Sci.* **41**, 3351–3367.
Stull R.B. (1988). *An Introduction to Boundary Layer Meteorology*. Kluwer Academic, Dordrecht, The Netherlands.
Wilson J.D., Thurtell G.W., and Kidd G.E. (1981). Numerical simulation of particle trajectories in inhomogeneous turbulence, II: Systems with variable turbulent velocity scale. *Bound. Layer Meteorol.* **21**, 423–441.

CHAPTER 11

# EULERIAN MODEL APPROACHES

## 11.1 INTRODUCTION

The previous chapter was an outline of computational fluid dynamics (CFD), a powerful simulation technique based on the Navier–Stokes equations that is sometimes used in air dispersion modeling. However, in and of itself, it is not a dispersion model. A material balance must be solved simultaneously with the Navier–Stokes equations to obtain an air dispersion model. The result is a type of model that differs from all dispersion models discussed in this book so far. How that works out in practice is the subject of the present chapter, in particular in the case where the CFD calculation is a large-scale meteorological model calculation. This type of dispersion model is known as an **Eulerian dispersion model**. As we have seen before, the Eulerian frame of reference is fixed in space. In an Eulerian dispersion model we calculate the transfer of pollutants on a grid that is fixed to Earth. The main advantage of Eulerian dispersion models is that rigorous chemical models can be incorporated seamlessly in the algorithm, in contrast with other dispersion models, which can only include simple chemistry or where the incorporation is difficult. An advantage the Eulerian models have in common with Lagrangian particle models is that the plume is not modeled as a single entity, so that transport over long distances can be modeled with relatively minor loss of accuracy. However, the main disadvantages of Eulerian models are their lack of resolution and their high computational needs. The issues are essentially the same issues that plague large-scale CFD models: Too many grid points are needed to run an Eulerian dispersion model with high accuracy, and the computational resources needed increase rapidly with increasing resolution. Another issue is that the physics underlying Eulerian disperison models, gradient transport theory ($K$ theory), breaks down at short distances from the source, where the size of the largest turbulent eddies is not negligible in comparison with the distance between the source and the receptor. Whereas Lagrangian particle models are well equipped to deal with dispersion on a local scale, Eulerian models only work well on a regional scale. Eulerian dispersion models usually require more computational resources than Lagrangian particle models.

Given the advantages and disadvantages, Eulerian dispersion models have a very specific niche: that of modeling secondary pollutants such as ozone. Hence, to

*Air Dispersion Modeling: Foundations and Applications*, First Edition. Alex De Visscher.

fully understand this class of models, it is important to consider atmospheric chemistry in much more detail than the outline in Chapter 3.

The objective of this chapter is an in-depth study of Eulerian dispersion modeling including both the physics and the chemistry. At the end of this chapter, the reader will be familiar with the way ozone pollution can be predicted with Eulerian dispersion models.

## 11.2 GOVERNING EQUATIONS OF EULERIAN DISPERSION MODELS

The equations governing air dispersion in an Eulerian frame of reference are essentially the equations governing computational fluid dynamics (the Navier–Stokes equation, the continuity equation, the material balance) and have been discussed in Chapter 10. However, whereas Chapter 10 discussed the Navier–Stokes equations in great detail and payed less attention to the material balance, the main emphasis in this chapter will be on the material balance.

The Navier–Stokes equations form the momentum balance for the calculation of the flow field that forms the basis of the Eulerian model. They are as follows:

$$\frac{\partial u}{\partial t}+u\frac{\partial u}{\partial x}+v\frac{\partial u}{\partial y}+w\frac{\partial u}{\partial z}=fv-\frac{1}{\rho}\frac{\partial p}{\partial x}+\nu\left(\frac{\partial^2 u}{\partial x^2}+\frac{\partial^2 u}{\partial y^2}+\frac{\partial^2 u}{\partial z^2}\right) \tag{11.1}$$

$$\frac{\partial v}{\partial t}+u\frac{\partial v}{\partial x}+v\frac{\partial v}{\partial y}+w\frac{\partial v}{\partial z}=-fu-\frac{1}{\rho}\frac{\partial p}{\partial y}+\nu\left(\frac{\partial^2 v}{\partial x^2}+\frac{\partial^2 v}{\partial y^2}+\frac{\partial^2 v}{\partial z^2}\right) \tag{11.2}$$

$$\frac{\partial w}{\partial t}+u\frac{\partial w}{\partial x}+v\frac{\partial w}{\partial y}+w\frac{\partial w}{\partial z}=-g-\frac{1}{\rho}\frac{\partial p}{\partial z}+\nu\left(\frac{\partial^2 w}{\partial x^2}+\frac{\partial^2 w}{\partial y^2}+\frac{\partial^2 w}{\partial z^2}\right) \tag{11.3}$$

Once the wind speed components $u$, $v$, and $w$ in directions $x$, $y$, and $z$ are known, they can be used to calculate the transport of pollutants with a **material balance**. As explained in Section 10.2.2, a generic material balance of a differential volume $dx\,dy\,dz$ is as follows:

$$\text{Accumulation} = \text{advection in} - \text{advection out} + \text{diffusion in} - \text{diffusion out} + \text{source} \tag{11.4}$$

The mass (in mg) of pollutant in the differential volume is $c_i\,dx\,dy\,dz$, where $c_i$ is the mass concentration of pollutant $i$ (mg m$^{-3}$). Hence, the accumulation of pollutant in the differential volume is

$$\text{Accumulation} = \frac{\partial c_i}{\partial t}dx\,dy\,dz \tag{11.5}$$

The transfer terms will be discussed in the $x$ direction only. The equivalent terms for the $y$ and $z$ directions are straightforward.

Advective transport into the differential volume is the mass of pollutant that is carried into the volume by airflow per unit time. Hence, it equals the concentration of pollutant multiplied by the volumetric flow rate of air into the differential volume. The volumetric flow rate is the wind speed in the direction considered, multiplied by the cross-sectional area of the volume element. Hence:

$$\text{Advection in} = c_i u \, dy \, dz \tag{11.6}$$

The advective flow of pollutant $i$ out of the differential volume is the same but based on the conditions at the far side of the differential volume. The result is

$$\text{Advection out} = \left( c_i u + \frac{\partial c_i u}{\partial x} dx \right) dy \, dz \tag{11.7}$$

Both concentration and wind speed can change with location. The importance of this for the material balance will be discussed later. Based on eqs. (11.6) and (11.7), the following net advection is obtained:

$$\text{Advection in} - \text{advection out} = -\frac{\partial c_i u}{\partial x} dx \, dy \, dz \tag{11.8}$$

The diffusion terms can be calculated with Fick's law. In the most general case, multi-component diffusion should be calculated with the Stefan–Maxwell equations. These equations describe the concentration gradient corresponding with a set of fluxes as follows (Jaynes and Rogowski, 1983):

$$-\frac{p}{RT} \frac{\partial y_i}{\partial x} = \sum_{\substack{j=1 \\ j \neq i}}^{n} \frac{N_i y_j - N_j y_i}{D_{ij}} \tag{11.9}$$

where $y_i$ is the mole fraction of compound $i$, $N_i$ is the flux of $i$ in a fixed frame of reference, and $D_{ij}$ is the binary diffusivity of a mixture of compounds $i$ and $j$. However, it can be shown that eq. (11.9) is a diffusion and advection equation if all but one compound in the mixture (in this case air) only has a trace concentration. To that effect, eq. (11.9) is rewritten to separate air (subscript $a$) from the other compounds. The ideal gas law is also introduced to eliminate $p/RT$:

$$-c_{\text{tot}} \frac{\partial y_i}{\partial x} = N_i \sum_{\substack{j=1 \\ j \neq i \\ j \neq \text{a}}}^{n} \frac{y_j}{D_{ij}} + \frac{N_i y_\text{a}}{D_{i\text{a}}} - y_i \sum_{\substack{j=1 \\ j \neq i \\ j \neq \text{a}}}^{n} \frac{N_j}{D_{ij}} - \frac{y_i N_\text{a}}{D_{i\text{a}}} \tag{11.10}$$

where $c_{\text{tot}}$, as well as the other concentrations in the derivation of eq. (11.12), are amount concentrations (mol m$^{-3}$). If temperature and pressure vary only slowly with distance, then the left-hand side of eq. (11.10) is the concentration gradient of $i$. The first and third terms are assumed to be negligibly small in comparison with the second and fourth terms because of the dominance of air in the mixture. It is further assumed that $y_\text{a} \approx 1$, and $N_\text{a} \approx N_{\text{tot}}$, where $N_{\text{tot}}$ is the total flux. Hence:

$$-c_{\text{tot}}\frac{\partial y_i}{\partial x}=-\frac{\partial c_i}{\partial x}=\frac{N_i}{D_{ia}}-\frac{y_i N_{\text{tot}}}{D_{ia}} \tag{11.11}$$

The total flux equals total concentration times wind speed. Hence, rearranging leads to

$$N_i=-D_{ia}\frac{\partial c_i}{\partial x}+uy_i c_{\text{tot}}=-D_{ia}\frac{\partial c_i}{\partial x}+uc_i \tag{11.12}$$

Equation (11.12) states that the total flux of $i$ equals the sum of a Fickian diffusion term and an advective term; and $D_{ia}$ is the binary diffusivity constant of pollutant $i$ in air. In this section, $D_{ia}$ will be abbreviated as $D_i$. Hence, it is established that Fick's law can be used to describe the diffusive terms:

$$\text{Diffusion in}=-D_i\frac{\partial c_i}{\partial x}dydz \tag{11.13}$$

Diffusion of $i$ at the far end of the differential volume is

$$\text{Diffusion out}=\left[-D_i\frac{\partial c_i}{\partial x}-\frac{\partial}{\partial x}\left(D_i\frac{\partial c_i}{\partial x}\right)dx\right]dydz \tag{11.14}$$

The net diffusion into the differential volume is

$$\text{Diffusion in}-\text{diffusion out}=-D_i\frac{\partial c_i}{\partial x}dy\,dz-\left[-D_i\frac{\partial c_i}{\partial x}-\frac{\partial}{\partial x}\left(D_i\frac{\partial c_i}{\partial x}\right)dx\right]dy\,dz \tag{11.15}$$

which simplifies as

$$\text{Diffusion in}-\text{diffusion out}=\frac{\partial}{\partial x}\left(D_i\frac{\partial c_i}{\partial x}\right)dxdydz \tag{11.16}$$

The diffusivity can be considered constant except over very large distances. Hence, eq. (11.16) can be approximated as

$$\text{Diffusion in}-\text{diffusion out}=D_i\frac{\partial^2 c_i}{\partial x^2}dxdydz \tag{11.17}$$

Finally, the source/sink term will simply be described as

$$\text{Source}=S_i\,dxdydz \tag{11.18}$$

where $S_i$ is the mass of $i$ that is produced per unit volume. When $S_i$ is negative, then $i$ is removed from the atmosphere. Source/sink terms can include emission, deposition, reaction, and the like. In fact, not all possible sources and sinks are included in this term. Dry deposition and emission can be introduced in the model as boundary conditions as well.

If the source/sink term refers to a reaction, then the following equation links source (in mg $m^{-3}$ $s^{-1}$) to reaction rate $r_i$ (in mol $m^{-3}$ $s^{-1}$):

$$S_i = 10^3 M_i r_i \tag{11.19}$$

where $M_i$ is the molar mass of $i$ (in g $mol^{-1}$) and the factor $10^{-3}$ has units of mg $g^{-1}$. Reaction rate is denoted with lowercase here to distinguish from the reaction rate $R_i$ in units more common in atmospheric chemistry, molecules $cm^{-3}$ $s^{-1}$. The relationship is

$$r_i = \frac{10^6 R_i}{N_A} \tag{11.20}$$

where $r_i$ is reaction rate in molecules $cm^{-3}$ $s^{-1}$, the factor $10^6$ has the units $cm^3$ $m^{-3}$, and $N_A$ is Avogadro's number ($6.0221415 \times 10^{23}$ molecules $mol^{-1}$).

Combining all terms, also in the $y$ and $z$ directions, into the material balance [eq. (11.4)], and dividing by $dx\,dy\,dz$ leads to

$$\boxed{\frac{\partial c_i}{\partial t} = -\frac{\partial c_i u}{\partial x} - \frac{\partial c_i v}{\partial y} - \frac{\partial c_i w}{\partial z} + D_i\left(\frac{\partial^2 c_i}{\partial x^2} + \frac{\partial^2 c_i}{\partial y^2} + \frac{\partial^2 c_i}{\partial z^2}\right) + S_i} \tag{11.21}$$

Equation (11.21) is sometimes used with the wind speeds outside the derivatives in the advective terms. To see the repercussions of this, expand the differentials:

$$\frac{\partial c_i}{\partial t} = -u\frac{\partial c_i}{\partial x} - c_i\frac{\partial u}{\partial x} - v\frac{\partial c_i}{\partial y} - c_i\frac{\partial v}{\partial y} - w\frac{\partial c_i}{\partial z} - c_i\frac{\partial w}{\partial z} + D_i\left(\frac{\partial^2 c_i}{\partial x^2} + \frac{\partial^2 c_i}{\partial y^2} + \frac{\partial^2 c_i}{\partial z^2}\right) + S_i \tag{11.22}$$

Bringing the wind speeds out of the derivatives in the advective terms generates three additional terms, which can be grouped as the pollutant concentration multiplied by the wind speed divergence:

$$\frac{\partial c_i}{\partial t} = -u\frac{\partial c_i}{\partial x} - v\frac{\partial c_i}{\partial y} - w\frac{\partial c_i}{\partial z} - c_i\left(\frac{\partial u}{\partial x} + \frac{\partial v}{\partial y} + \frac{\partial w}{\partial z}\right) + D_i\left(\frac{\partial^2 c_i}{\partial x^2} + \frac{\partial^2 c_i}{\partial y^2} + \frac{\partial^2 c_i}{\partial z^2}\right) + S_i \tag{11.23}$$

The wind speed divergence is zero when the flow is incompressible. In that case, the following equation is an acceptable material balance:

$$\frac{\partial c_i}{\partial t} = -u\frac{\partial c_i}{\partial x} - v\frac{\partial c_i}{\partial y} - w\frac{\partial c_i}{\partial z} + D_i\left(\frac{\partial^2 c_i}{\partial x^2} + \frac{\partial^2 c_i}{\partial y^2} + \frac{\partial^2 c_i}{\partial z^2}\right) + S_i \tag{11.24}$$

However, eq. (11.24) must not be used whenever compressible wind flow was assumed in the calculation of the wind field. Using eq. (11.24) with compressible wind flow leads to a violation of the conservation of mass. However, there is a variant of eq. (11.24) based on mass fractions that can be used in combination with compressible wind flow without violating mass conservation and that has additional advantages. To that effect, consider the pollution mass fraction defined as

$$q_i = \frac{c_i}{\rho} \tag{11.25}$$

The units of $q_i$ can be described as milligrams of pollutant per kilogram of polluted air. Mass fraction is often used in the modeling of water transport in the atmosphere. Substitution into eq. (11.21) leads to

$$\frac{\partial \rho q_i}{\partial t} = -\frac{\partial \rho q_i u}{\partial x} - \frac{\partial \rho q_i v}{\partial y} - \frac{\partial \rho q_i w}{\partial z} + D_i \left( \frac{\partial^2 \rho q_i}{\partial x^2} + \frac{\partial^2 \rho q_i}{\partial y^2} + \frac{\partial^2 \rho q_i}{\partial z^2} \right) + S_i \tag{11.26}$$

Expand the derivatives in the advective terms by considering the variables as $\rho u \times q_i$, and so forth. The result is

$$q_i \frac{\partial \rho}{\partial t} + \rho \frac{\partial q_i}{\partial t} = -\rho u \frac{\partial q_i}{\partial x} - q_i \frac{\partial \rho u}{\partial x} - \rho v \frac{\partial q_i}{\partial y} - q_i \frac{\partial \rho v}{\partial y} - \rho w \frac{\partial q_i}{\partial z} - q_i \frac{\partial \rho w}{\partial z}$$
$$+ D_i \left( \frac{\partial^2 \rho q_i}{\partial x^2} + \frac{\partial^2 \rho q_i}{\partial y^2} + \frac{\partial^2 \rho q_i}{\partial z^2} \right) + S_i \tag{11.27}$$

To simplify eq. (11.27), consider the non-steady-state overall mass balance of a differential volume:

$$\frac{\partial \rho}{\partial t} + \frac{\partial \rho u}{\partial x} + \frac{\partial \rho v}{\partial y} + \frac{\partial \rho w}{\partial z} = 0 \tag{11.28}$$

Based on this mass balance, it is clear that the first term in the left-hand side of eq. (11.27) cancels three terms in the right-hand side, simplifying eq. (11.27) to

$$\rho \frac{\partial q_i}{\partial t} = -\rho u \frac{\partial q_i}{\partial x} - \rho v \frac{\partial q_i}{\partial y} - \rho w \frac{\partial q_i}{\partial z} + D_i \left( \frac{\partial^2 \rho q_i}{\partial x^2} + \frac{\partial^2 \rho q_i}{\partial y^2} + \frac{\partial^2 \rho q_i}{\partial z^2} \right) + S_i \tag{11.29}$$

Considering that the diffusive terms in this equation were approximations based on the fact that the diffusivity changes negligibly, and that diffusivity is inversely proportional with pressure and more than proportional with absolute temperature, it is clear that a further approximation is possible based on the assumption that density changes do not affect the diffusion process. Hence, the density can be taken out of the derivatives in the diffusion terms and canceled with the other densities. Hence:

$$\frac{\partial q_i}{\partial t} = -u \frac{\partial q_i}{\partial x} - v \frac{\partial q_i}{\partial y} - w \frac{\partial q_i}{\partial z} + D_i \left( \frac{\partial^2 q_i}{\partial x^2} + \frac{\partial^2 q_i}{\partial y^2} + \frac{\partial^2 q_i}{\partial z^2} \right) + \frac{S_i}{\rho} \tag{11.30}$$

In fact, the diffusion terms in eq. (11.30) are more accurate than the diffusion terms in the preceding equations. To see this, consider a zero diffusive flux condition over large vertical distances, where density and total concentration are not constant.

It is expected that zero diffusive flux is associated with zero gradient of $q_i$, but not with zero gradient of $c_i$. Hence, eq. (11.30) is more consistent with the physics of large-scale vertical diffusion than the equations preceding it.

The above equations can only be used in laminar flow, or when all scales of turbulence are resolved in a direct simulation based on the Navier–Stokes equation. In practice, this is not a realistic situation. We can only simulate the largest scales of turbulence at best, whereas the smaller scales need to be modeled indirectly. The process of separating some or all of the turbulence from the rest of the calculation (e.g., by **Reynolds averaging**) generates more unknowns than equations. This is known as the **closure problem**. The unknowns are variances and covariances, represent turbulent fluxes, and are usually resolved by assuming a functional dependence from the **gradient of the average properties**.

The **Reynolds-averaged Navier–Stokes equations** were derived in Chapter 10. The result is as follows:

$$\frac{\partial \bar{u}}{\partial t}+\bar{u}\frac{\partial \bar{u}}{\partial x}+\bar{v}\frac{\partial \bar{u}}{\partial y}+\bar{w}\frac{\partial \bar{u}}{\partial z}$$
$$= f\bar{v}-\frac{1}{\bar{\rho}}\frac{\partial \bar{p}}{\partial x}+\frac{1}{\bar{\rho}}\left(\mu\frac{\partial^2 \bar{u}}{\partial x^2}-\bar{\rho}\frac{\partial \overline{u'^2}}{\partial x}+\mu\frac{\partial^2 \bar{u}}{\partial y^2}-\bar{\rho}\frac{\partial \overline{u'v'}}{\partial y}+\mu\frac{\partial^2 \bar{u}}{\partial z^2}-\bar{\rho}\frac{\partial \overline{u'w'}}{\partial z}\right) \tag{11.31}$$

$$\frac{\partial \bar{v}}{\partial t}+\bar{u}\frac{\partial \bar{v}}{\partial x}+\bar{v}\frac{\partial \bar{v}}{\partial x}+\bar{w}\frac{\partial \bar{v}}{\partial x}$$
$$=-f\bar{u}-\frac{1}{\bar{\rho}}\frac{\partial \bar{p}}{\partial y}+\frac{1}{\bar{\rho}}\left(\mu\frac{\partial^2 \bar{v}}{\partial x^2}-\bar{\rho}\frac{\partial \overline{u'v'}}{\partial x}+\mu\frac{\partial^2 \bar{v}}{\partial y^2}-\bar{\rho}\frac{\partial \overline{v'^2}}{\partial y}+\mu\frac{\partial^2 \bar{v}}{\partial z^2}-\bar{\rho}\frac{\partial \overline{v'w'}}{\partial z}\right) \tag{11.32}$$

$$\frac{\partial \bar{w}}{\partial t}+\bar{u}\frac{\partial \bar{w}}{\partial x}+\bar{v}\frac{\partial \bar{w}}{\partial x}+\bar{w}\frac{\partial \bar{w}}{\partial x}$$
$$=-g-\frac{1}{\bar{\rho}}\frac{\partial \bar{p}}{\partial z}+\frac{1}{\bar{\rho}}\left(\mu\frac{\partial^2 \bar{w}}{\partial x^2}-\bar{\rho}\frac{\partial \overline{u'w'}}{\partial x}+\mu\frac{\partial^2 \bar{v}}{\partial y^2}-\bar{\rho}\frac{\partial \overline{v'w'}}{\partial y}+\mu\frac{\partial^2 \bar{w}}{\partial z^2}-\bar{\rho}\frac{\partial \overline{w'^2}}{\partial z}\right) \tag{11.33}$$

Equations (11.31)–(11.33) have more than three unknowns, so a number of additional equations are needed to close the model. In general, the additional equations are of the form:

$$\overline{u'_i u'_j} = -K_{\mathrm{m}}\left(\frac{\partial \overline{u_i}}{\partial x_j}+\frac{\partial \overline{u_j}}{\partial x_i}\right) \tag{11.34}$$

where $u_1 = u$, $u_2 = v$, $u_3 = w$, $x_1 = x$, $x_2 = y$, and $x_3 = z$.

Reynolds averaging for transport calculations involving turbulent flow follows the following principles, also discussed in Chapter 10:

- $\overline{f'} = 0$ for any variable $f$.
- However, $\overline{f'g'} \neq 0$.

- $\overline{fg} = \overline{(\overline{f}+f')(\overline{g}+g')} = \overline{f}\cdot\overline{g} + \overline{f'}\cdot\overline{g} + \overline{f}\cdot\overline{g'} + \overline{f'g'} = \overline{f}\cdot\overline{g} + \overline{f'g'}\cdot$
- Pressure and density fluctuate much less than velocity. Their fluctuations can be neglected. An exception is the effect of density fluctuation on buoyancy.

First, Reynolds decomposition is applied to the material balance [eq. (11.21)]:

$$\frac{\partial(\overline{c}_i + c'_i)}{\partial t} = -\frac{\partial(\overline{c}_i + c'_i)(\overline{u}+u')}{\partial x} - \frac{\partial(\overline{c}_i + c'_i)(\overline{v}+v')}{\partial y} - \frac{\partial(\overline{c}_i + c'_i)(\overline{w}+w')}{\partial z} + D_i\left[\frac{\partial^2(\overline{c}_i + c'_i)}{\partial x^2} + \frac{\partial^2(\overline{c}_i + c'_i)}{\partial y^2} + \frac{\partial^2(\overline{c}_i + c'_i)}{\partial z^2}\right] + (\overline{S}_i + S'_i) \tag{11.35}$$

Expanding the terms between brackets in the advection terms and Reynolds averaging, applying the rules above, leads to

$$\frac{\partial \overline{c_i}}{\partial t} = -\frac{\partial \overline{c_i}\cdot\overline{u}}{\partial x} - \frac{\partial \overline{c'_i u'}}{\partial x} - \frac{\partial \overline{c_i}\cdot\overline{v}}{\partial y} - \frac{\partial \overline{c'_i v'}}{\partial y} - \frac{\partial \overline{c_i}\cdot\overline{w}}{\partial z} - \frac{\partial \overline{c'_i w'}}{\partial z} + D_i\left(\frac{\partial^2 \overline{c_i}}{\partial x^2} + \frac{\partial^2 \overline{c_i}}{\partial y^2} + \frac{\partial^2 \overline{c_i}}{\partial z^2}\right) + \overline{S_i} \tag{11.36}$$

The turbulent portions of the advection terms are turbulent diffusion terms. They can be closed using a gradient transport theory ($K$ theory):

$$\overline{c'_i u'} = -K_x \frac{\partial \overline{c_i}}{\partial x} \tag{11.37}$$

$$\overline{c'_i v'} = -K_y \frac{\partial \overline{c_i}}{\partial y} \tag{11.38}$$

$$\overline{c'_i w'} = -K_z \frac{\partial \overline{c_i}}{\partial z} \tag{11.39}$$

The calculation of the $K$ values is discussed in the next section.

As indicated before, eq. (11.39) predicts that zero vertical flux corresponds with zero concentration gradient. Applying this to all atmospheric compounds and summing the results, it follows that eq. (11.39) predicts constant total concentration in the atmosphere, that is, constant density. This is justified in shallow air layers, but in deep atmospheric boundary layers there is a pronounced decrease of the air density with height that is not accounted for in eq. (11.39). This problem does not occur when the concentration is expressed as a mass fraction. Hence, a material balance based on concentrations contains a bias that does not exist in a material balance based on mass balances. This is why contemporary Eulerian dispersion models usually use mass fractions instead of concentrations as the variable.

## 11.3 CLOSING THE MATERIAL BALANCE FOR TURBULENT MOTION

### 11.3.1 Local Closure

In the previous section the Reynolds-averaged material balance was developed [eq. (11.36)]. Introducing a turbulent version of Fick's law [eqs. (11.37)–(11.39)] into the material balance leads to the following equation:

$$\frac{\partial \overline{c_i}}{\partial t} = -\frac{\partial \overline{c_i} \cdot \overline{u}}{\partial x} - \frac{\partial \overline{c_i} \cdot \overline{v}}{\partial y} - \frac{\partial \overline{c_i} \cdot \overline{w}}{\partial z} + \frac{\partial}{\partial x}\left[\left(D_i + K_x\right)\frac{\partial \overline{c_i}}{\partial x}\right]$$
$$+ \frac{\partial}{\partial y}\left[\left(D_i + K_y\right)\frac{\partial \overline{c_i}}{\partial y}\right] + \frac{\partial}{\partial z}\left[\left(D_i + K_z\right)\frac{\partial \overline{c_i}}{\partial z}\right] + \overline{S_i} \tag{11.40}$$

In the special case where the diffusivities (molecular and turbulent) are constant, they can be taken out of the derivative in the diffusion terms. However, this is not always justified, especially in the $z$ direction, as $K_z$ increases markedly with height.

When the $K$ values are parameterized based on average properties without considering higher-order moments, the closure technique is known as **first-order closure**. What follows is a brief overview of first-order closure methods for the material balance.

In the $x$ and the $y$ direction, the $K$ values are poorly understood. In Section 6.6.7 a useful equation was derived that can be used in this context [eq. (6.105)]. It leads to the following estimates for $K_x$ and $K_y$:

$$K_x = \sigma_u^2 T_{\mathrm{i,L},x} \tag{11.41}$$

$$K_y = \sigma_v^2 T_{\mathrm{i,L},y} \tag{11.42}$$

Equation (6.191) can be used to estimate $T_{\mathrm{i,L},x}$ and $T_{\mathrm{i,L},y}$ from the energy dissipation rate $\varepsilon$ when no other formulations for $T_{\mathrm{i,L},x}$ and $T_{\mathrm{i,L},y}$ are available (Tennekes, 1979). Applied to the $x$ and $y$ directions, this leads to

$$T_{\mathrm{i,L},x} = \frac{2\sigma_u^2}{C_0 \varepsilon} \tag{11.43}$$

$$T_{\mathrm{i,L},y} = \frac{2\sigma_v^2}{C_0 \varepsilon} \tag{11.44}$$

where $C_0$ is the Lagrangian structure function constant. Anfossi et al. (2006) estimated $C_0 = 2.5$ and 4.3 in the $x$ and $y$ directions, respectively. Formulations for $\varepsilon$ are given in Section 6.6.7 [eqs. (6.185)–(6.188)].

These equations lead to overestimations of dispersion near the source because of the non-Fickian nature of turbulent diffusion. However, this is an inherent weakness of Eulerian models and is not linked with any parameterization of $K$.

The numerical solution of the material balance usually leads to substantial **numerical dispersion**, a mathematical artifact that takes on the form of an artificial diffusion. In some cases where large-scale simulations with grid cells in the kilometer range are conducted, one can assume that numerical dispersion exceeds actual dispersion, and it is best to assign $K_x$ and $K_y$ a value of zero. When this is not the case, it is best to estimate numerical dispersion and subtract it from the actual dispersion in the form of a reduced $K$ value. In the $z$ direction, grid cell spacing is usually much smaller than in the $x$ direction, and numerical dispersion is usually less dominant. Numerical dispersion and ways to deal with it are discussed in Section 11.5.

In the vertical direction, the above approach can be used as well. However, a more straightforward closure technique is based on eq. (5.192) (Chapter 5). This is based on the observation that the turbulent heat diffusivity equals the turbulent mass diffusivity. The equation is as follows in unstable and neutral conditions (Nieuwstadt, 1984):

$$K_z = \frac{k u_* z \left(1 - z/h_{\text{mix}}\right)^{1.5}}{\varphi_{\text{h}}(\zeta)} \tag{11.45}$$

The variables are defined in Chapter 5. In principle, this equation is only valid under stable and neutral conditions. For unstable conditions, the following equation [eq. (5.193)] has been proposed (Byun and Ching, 1999):

$$K_z = k w_* z \left(1 - \frac{z}{h_{\text{mix}}}\right) \tag{11.46}$$

However, this would predict $K_z$ values smaller than the neutral values under unstable but very nearly neutral conditions, which is unrealistic. Hence, it is recommended to add eq. (11.45) applied to neutral conditions to eq. (11.46) to avoid that situation. The following equation approaches eq. (11.46) in unstable conditions and approaches eq. (11.45) in unstable, near-neutral conditions:

$$K_z = k u_* z \left(1 - \frac{z}{h_{\text{mix}}}\right) \left[\left(1 - \frac{z}{h_{\text{mix}}}\right)^{1.5} - \frac{h_{\text{mix}}}{kL}\right]^{1/3} \tag{11.47}$$

See Section 5.11.5 for further discussion on this issue.

Above the mixing layer, in the free atmosphere, the following equation has been proposed (Byun et al., 2003):

$$K_z = K_0 + S \frac{Ri_{\text{c}} - Ri_{\text{B}}}{Ri_{\text{c}}} l^2 \tag{11.48}$$

where $K_0$ is a background diffusivity of 1 m$^2$ s$^1$, $S$ is the vertical gradient of the horizontal wind speed [in both $x$ and $y$ directions, i.e., $(u^2 + v^2)^{1/2}$], $Ri_{\text{c}}$ is a critical bulk

Richardson number set at 0.25, $l$ is a length scale set at 40 m, and $Ri_B$ is the bulk Richardson number, defined as

$$Ri_B = \frac{g}{\theta_0 S^2} \frac{\Delta\theta}{\Delta z} \tag{11.49}$$

It is recommended to ensure a smooth transition between the boundary layer $K_z$ value and the free atmosphere value.

**Higher-order closure models**, which are based on prognostic equations to resolve some of the higher-order moments in the Navier–Stokes equations and the material balance, can be applied as well. This has been discussed in detail in Chapter 10 for the Navier–Stokes equations, and the principle is the same for the material balance. For these closure methods it is important to maintain consistency between the closure of the Navier–Stokes equations and the material balance.

The most straightforward higher-order closure method is based on the turbulent kinetic energy and the kinetic energy dissipation rate ($k$-$\varepsilon$ method). This is sometimes referred to as a one-and-half-order method as some but not all of the second-order moments are resolved. This type of turbulence closure was discussed in Section 10.3.2.4. The result was as follows (eq. 10.146):

$$K = c_\mu \frac{\overline{e}^2}{\varepsilon} \tag{10.146}$$

Accurate results were obtained with $c_\mu = 0.09$ (Jones and Launder, 1972). However, $K$ refers to the turbulent momentum diffusivity $K_m$ (or turbulent kinematic viscosity). In stable and neutral conditions, $K_m$ can be assumed to be equal to $K_z$, but this is not the case in unstable conditions, where $K_z$ exceeds $K_m$. Deardorff (1972) suggested that the turbulent heat diffusivity $K_h$ is three times $K_m$ in unstable conditions, and it is generally assumed that $K_z = K_h$. This leaves open the question of how to assign $K_z$ in slightly unstable conditions. Comparing eqs. (5.185) and (5.189) in Section 5.11.5,

$$K_m = \frac{ku_* z(1 - z/h_{mix})^{1.5}}{\phi_m(\zeta)} \tag{5.185}$$

$$K_h = \frac{ku_* z(1 - z/h_{mix})^{1.5}}{\phi_h(\zeta)} \tag{5.189}$$

with $\zeta = z/L$, it is clear that the following equation is valid in neutral and stable conditions:

$$\frac{K_h}{K_m} = \frac{\phi_m(\zeta)}{\phi_h(\zeta)} \tag{11.50}$$

If it is *assumed* that this equation is also valid in unstable conditions, and we assume the following equations for $\phi_m(\zeta)$ and $\phi_h(\zeta)$ in unstable conditions,

$$\phi_m(\zeta) = (1 - 16\zeta)^{-1/4} \tag{5.77}$$

$$\phi_{h}(\zeta)=\phi_{m}^{2}(\zeta)=(1-16\zeta)^{-1/2} \tag{5.92}$$

then the following relationship is obtained between $K_z$ and $K_m$:

$$K_{z}=\left(1-16\frac{z}{L}\right)^{1/4}K_{m} \tag{11.51}$$

This relationship is consistent with Deardorff's (1972) assumption that $K_h$ ($= K_z$) $= 3K_m$ when $z = 5L$ and sets $K_z$ between $2K_m$ and $4K_m$ in a wide height range from $-0.9375L$ to $-15.9375L$. A comprehensive list of $K_z$ formulations is given by Byun et al. (2003).

Fully second-order and higher-order closure techniques exist as well. However, such models are very complex, and the limited added accuracy does not warrant the added complexity and computational needs. They are rarely used in air quality modeling (Byun et al., 2003).

### 11.3.2 Nonlocal Closure

The problem with all closure techniques discussed in the previous section is that the covariances are evaluated using only the local conditions. However, convective cycles directly transport air masses across the entire boundary layer, so within the averaging times relevant in Eulerian dispersion models, it is as if the fluxes are directly impacted by the conditions of the entire boundary layer. This has led to the formulation of **nonlocal closure techniques**.

With the exception of a study of Stull and Driedonks (1987), which uses turbulent kinetic energy closure, nonlocal closure models are generally first-order. The first attempt to develop a nonlocal vertical dispersion scheme is by Blackadar (1978). This model is one of the options in the meteorological model MM5. It is only applicable to a convective atmosphere, so a different model is always needed to model neutral or stable conditions.

To apply the model of Blackadar (1978), consider the mixing layer as having $n$ layers, with layer 1 at the bottom and layer $n$ at the top. Each layer has a height $h_i$ so that

$$h_{mix}=\sum_{i=1}^{n}h_{i} \tag{11.52}$$

where $h_{mix}$ is the height of the mixing layer. Each layer has a potential temperature $\theta_i$. The bottom layer receives a sensible heat flux $H$ (J s$^{-1}$ m$^{-2}$) from the surface, and in steady-state conditions this heat flux is lost to the layers aloft. It is assumed that the layers aloft do not exchange heat or mass with each other, only with layer 1. Assume that the mass flow rate of air exchanged between layer 1 and layer $i$ per unit area of surface (i.e., the mass flux) is $Q_{mi}$ (kg m$^{-2}$ s$^{-1}$). Further, assume that these flow rates are proportional to the mass of layer $i$:

$$Q_{mi}=M_{u}\rho_{i}h_{i} \tag{11.53}$$

where $M_u$ is an exchange rate constant (kg kg$^{-1}$ s$^{-1}$ or simply s$^{-1}$). In principle, the exchange rate constants to different layers could be different, but this is usually not assumed in this model. Rate constant $M_u$ is a variable that needs to be determined by solving the sensible heat balance in the bottom layer. Under steady-state conditions the sensible heat received from the surface equals the difference between the sensible heat sent to the layers aloft and the sensible heat received from the layers aloft. In equation form:

$$H = c_p \theta_1 \sum_{i=2}^{n} Q_i - c_p \sum_{i=2}^{n} \theta_i Q_i \tag{11.54}$$

Substituting eq. (11.53), this equation can be written as

$$H = c_p M_u \sum_{i=2}^{n} \rho_i h_i (\theta_1 - \theta_i) \tag{11.55}$$

Hence:

$$M_u = \frac{H}{c_p \sum_{i=2}^{n} \rho_i h_i (\theta_1 - \theta_i)} \tag{11.56}$$

Assuming only vertical material transfer, the pollutant concentration change in each layer can be calculated based on these exchange rates. In layer 1, the material balance involves a dry deposition rate and material exchange with all other layers:

$$\rho_1 h_1 \frac{\partial q_1}{\partial t} = -v_d q_1 \rho_1 - q_1 M_u \sum_{i=2}^{n} \rho_i h_i + M_u \sum_{i=2}^{n} q_i \rho_i h_i \tag{11.57}$$

Hence, the concentration change with time in layer 1 is

$$\frac{\partial q_1}{\partial t} = -\frac{v_d}{h_1} q_1 + \frac{M_u}{\rho_1 h_1} \sum_{i=2}^{n} \rho_i (q_i - q_1) h_i \tag{11.58}$$

In the layers aloft, the only material transfers are by exchange with layer 1. Hence:

$$\rho_i h_i \frac{\partial q_i}{\partial t} = M_u \rho_i h_i (q_1 - q_i) \tag{11.59}$$

or simply

$$\frac{\partial q_i}{\partial t} = M_u (q_1 - q_i) \tag{11.60}$$

The densities in eq. (11.58) cancel each other out in the case of a shallow convective boundary layer, and although they are not negligible in deep convective boundary layers, they are usually ignored (e.g., Byun and Ching, 1999; Byun et al., 2003).

Instead, a correction $h_1/h_i$ is applied to eq. (11.60). This correction conserves mass when density differences are neglected in the special case where each layer has the same mass. This is how the Blackadar (1978) scheme is usually implemented. If eqs. (11.58) and (11.60) are used, mass conservation is inherent to the model formulation.

In Eulerian models it is customary to calculate the concentration changes due to horizontal dispersion, vertical dispersion, advection, and chemical reaction separately. Hence, the above equations can be incorporated conveniently into an Eulerian model.

The model of Blackadar (1978) is only a crude approximation of atmosphere dynamics. In reality, all layers exchange mass with each other. Pleim and Chang (1992) developed a more realistic model, the **Asymmetrical Convective Model (ACM)**, by assuming that only the upward material flow follows the Blackadar scheme, whereas the downward flow is from each layer to the next, that is, layer $n$ only sends air to layer $n$–1; layer $n$–1 sends air to layer $n$–2, and so forth.

In this model a distinction is made between the upward exchange rate constant $M_u$, which is constant for every layer, and the downward exchange rate constant for layer $i$, $M_{di}$. The upward mass flow rate is the same as in the Blackadar (1978) model [eq. (11.53)]:

$$Q_{mui} = M_u \rho_i h_i \tag{11.61}$$

The downward mass flow rate is based on a vertical mass balance. The mass flow rate $Q_{mdi}$ moving from layer $i$ to layer $i$–1 equals the sum of the mass flow rates moving up to the layers $i$ through $n$. Hence:

$$Q_{mdi} = M_u \sum_{j=i}^{n} \rho_j h_j \tag{11.62}$$

Pleim and Chang (1992) use a nondimensionalized pressure as vertical coordinate:

$$\sigma = \frac{p - p_{top}}{p_{surf} - p_{top}} = \frac{p - p_{top}}{p^*} \tag{11.63}$$

where $p_{top}$ is the pressure at the top of the domain (or the top of the convective boundary layer), $p_{surf}$ is the pressure at the surface, and $p^*$ is defined as $p_{surf} - p_{top}$. The differential of the new coordinate is

$$d\sigma = \frac{dp}{p^*} = -\frac{\rho g dz}{p^*} \tag{11.64}$$

Hence, the height of each layer, $h_i$, can be linked to the associated $\sigma$ decrement $\Delta\sigma_i$:

$$\rho_i h_i = \frac{p^*}{g} \Delta\sigma_i \tag{11.65}$$

The upward and downward mass flow rate are thus calculated as

$$Q_{\mathrm{mu}i} = \frac{M_{\mathrm{u}} p^*}{g} \Delta\sigma_i \tag{11.66}$$

$$Q_{\mathrm{md}i} = \frac{M_{\mathrm{u}} p^*}{g} \sum_{j=i}^{n} \Delta\sigma_j \tag{11.67}$$

As with the Blackadar (1978) model, $M_{\mathrm{u}}$ is determined with a heat balance of the bottom layer. The sensible heat flux $H$ given off by the soil equals the difference between the sensible heat of the upward airflows leaving layer 1 and the sensible heat of the downward airflow received by layer 1:

$$H = \theta_1 c_p \frac{M_{\mathrm{u}} p}{g} \sum_{i=2}^{n} \Delta\sigma_i - \theta_2 c_p \frac{M_{\mathrm{u}} p}{g} \sum_{j=2}^{n} \Delta\sigma_j \tag{11.68}$$

Solving for $M_{\mathrm{u}}$ leads to

$$M_{\mathrm{u}} = \frac{H}{(\theta_1 - \theta_2) c_p (p/g) \sum_{i=2}^{n} \Delta\sigma_i} \tag{11.69}$$

The material balances are calculated in a manner similar to the Blackadar (1978) model. For the bottom layer, the mass accumulation of pollutant depends on the deposition rate, the pollutant mass sent to the layers aloft, and on the pollutant mass received from layer 2. Hence:

$$\rho_1 h_1 \frac{\partial q_1}{\partial t} = -v_{\mathrm{d}} q_1 \rho_1 - q_1 \sum_{i=2}^{n} Q_{\mathrm{mu}i} + q_2 Q_{\mathrm{md}2} \tag{11.70}$$

After substitution of eqs. (11.75)–(11.67), the following equation is obtained:

$$\frac{p^*}{g} \Delta\sigma_1 \frac{\partial q_1}{\partial t} = -v_{\mathrm{d}} q_1 \rho_1 - q_1 \sum_{i=2}^{n} \frac{M_{\mathrm{u}} p^*}{g} \Delta\sigma_i + q_2 \frac{M_{\mathrm{u}} p^*}{g} \sum_{i=2}^{n} \Delta\sigma_i \tag{11.71}$$

Equation (11.71) is divided by $p^* \Delta\sigma_1 / g$ to obtain

$$\frac{\partial q_1}{\partial t} = -\frac{v_{\mathrm{d}} q_1 \rho_1 g}{p^* \Delta\sigma_1} + (q_2 - q_1) \frac{M_{\mathrm{u}}}{\Delta\sigma_1} \sum_{i=2}^{n} \Delta\sigma_i \tag{11.72}$$

For layer 2 and up, the material balance is established by the flows received from the bottom layer and the layer just above, and by the flow leaving to the layer just below. Hence:

$$\rho_i h_i \frac{\partial q_i}{\partial t} = q_1 Q_{\mathrm{mu}i} + q_{i+1} Q_{\mathrm{md}i+1} - q_i Q_{\mathrm{md}i} \tag{11.73}$$

Substitution of eqs. (11.65)–(11.67) into eq. (11.73) leads to

$$\frac{p^*}{g}\Delta\sigma_i\frac{\partial q_i}{\partial t}=q_1\frac{M_{\text{u}}p^*}{g}\Delta\sigma_i+q_{i+1}\frac{M_{\text{u}}p^*}{g}\sum_{j=i+1}^{n}\Delta\sigma_j-q_i\frac{M_{\text{u}}p^*}{g}\sum_{j=i}^{n}\Delta\sigma_j \tag{11.74}$$

Equation (11.74) is divided by $p^*\Delta\sigma_i/g$ to obtain

$$\frac{\partial q_i}{\partial t}=q_1M_{\text{u}}+q_{i+1}\frac{M_{\text{u}}}{\Delta\sigma_i}\sum_{j=i+1}^{n}\Delta\sigma_j-q_i\frac{M_{\text{u}}}{\Delta\sigma_i}\sum_{j=i}^{n}\Delta\sigma_j \tag{11.75}$$

Ad hoc theories such as the ones above have been generalized into a more general theory known as **transilient turbulence theory** (Stull, 1984, 1988). The models of Blackadar (1978) and of Pleim and Chang (1992) assume that air is exchanged only between certain pairs of layers and not between others. In reality, each two pairs of layers will exchange some air. This is the assumption underlying transilient turbulence theory.

Because of the large number of gas exchanges, transilient turbulence theory is best described in matrix form. For the sake of simplicity, consider local gradient transport theory first, based on a turbulent version of Fick's law (eq. 5.176):

$$J=\overline{c'w'}=-K_z\frac{\partial\bar{c}}{\partial z} \tag{5.176}$$

Assume that $F_{ii+1}$ is the flux between layer $i$ and layer $i$+1. Hence, remembering that $c_i=q_i\rho_i$:

$$F_{ii+1}=-K_{zii+1}\rho_{ii+1}\frac{q_{i+1}-q_i}{\Delta z_{ii+1}} \tag{11.76}$$

The factor $\rho_{ii+1}$ (the average density of the air between the centers of layers $i$ and $i$+1) was put outside the differential because in a compressible atmophere the diffusive flux is zero when $q$ is constant, not when $c$ is constant; $K_{zii+1}$ is the local turbulent diffusivity between layer $i$ and layer $i$+1, and $z_{ii+1}$ is the distance between the centers of layers $i$ and $i$+1. Considering only vertical mass transport, a material balance of horizontal layer $i$ is

$$h_i\rho_i\frac{\partial q_i}{\partial t}=F_{i-1i}-F_{ii+1} \tag{11.77}$$

Applying eq. (11.76) to resolve the fluxes in eq. (11.77) leads to

$$h_i\rho_i\frac{\partial q_i}{\partial t}=-K_{zi-1i}\rho_{i-1i}\frac{q_i-q_{i-1}}{\Delta z_{i-1i}}+K_{zii+1}\rho_{ii+1}\frac{q_{i+1}-q_i}{\Delta z_{ii+1}} \tag{11.78}$$

Dividing eq. (11.78) by $h_i\rho_i$, and rearranging, leads to

$$\frac{\partial q_i}{\partial t} = \frac{K_{zi-1i}\rho_{i-1i}}{h_i\rho_i\,\Delta z_{i-1i}} q_{i-1} + \left( -\frac{K_{zi-1i}\rho_{i-1i}}{h_i\rho_i\,\Delta z_{i-1i}} - \frac{K_{zii+1}\rho_{ii+1}}{h_i\rho_i\,\Delta z_{ii+1}} \right) q_i + K_{zii+1}\rho_{ii+1} \frac{K_{zii+1}\rho_{ii+1}}{h_i\rho_i\,\Delta z_{ii+1}} q_{i+1} \quad (11.79)$$

Some simplifications are possible when the density is constant and the layers have the same thickness $\Delta z = h_i$. Then the material balance becomes

$$\frac{\partial q_i}{\partial t} = \frac{K_{zi-1i}}{(\Delta z)^2} q_{i-1} - \frac{K_{zi-1i} + K_{zii+1}}{(\Delta z)^2} q_i + \frac{K_{zii+1}}{(\Delta z)^2} q_{i+1} \quad (11.80)$$

For convenience, we will apply the matrix notation with eq. (11.80), although in most cases the full eq. (11.79) needs to be applied. For the first and last layers we will assume no material exchange across the domain boundaries (i.e., no depostion and full reflection at the top of the atmosphere). Hence:

$$\frac{\partial q_1}{\partial t} = -\frac{K_{z12}}{(\Delta z)^2} q_1 + \frac{K_{z12}}{(\Delta z)^2} q_2 \quad (11.81)$$

$$\frac{\partial q_n}{\partial t} = \frac{K_{zn-1n}}{(\Delta z)^2} q_{n-1} - \frac{K_{zn-1n}}{(\Delta z)^2} q_n \quad (11.82)$$

All the equations [eqs. (11.86)–(11.88)] can be combined in a single matrix equation:

$$\begin{pmatrix} \frac{\partial q_1}{\partial t} \\ \frac{\partial q_2}{\partial t} \\ \frac{\partial q_3}{\partial t} \\ \frac{\partial q_4}{\partial t} \end{pmatrix} = \begin{pmatrix} -\frac{K_{z12}}{(\Delta z)^2} & \frac{K_{z12}}{(\Delta z)^2} & 0 & 0 \\ \frac{K_{z12}}{(\Delta z)^2} & -\frac{K_{z12}+K_{z23}}{(\Delta z)^2} & \frac{K_{z23}}{(\Delta z)^2} & 0 \\ 0 & \frac{K_{z23}}{(\Delta z)^2} & -\frac{K_{z23}+K_{z34}}{(\Delta z)^2} & \frac{K_{z34}}{(\Delta z)^2} \\ 0 & 0 & \frac{K_{z34}}{(\Delta z)^2} & -\frac{K_{z34}}{(\Delta z)^2} \end{pmatrix} \begin{pmatrix} q_1 \\ q_2 \\ q_3 \\ q_4 \end{pmatrix} \quad (11.83)$$

The $4 \times 4$ matrix is a material balance matrix $\mathbf{M}$ where the nondiagonal elements $M_{ij}$ (units in $s^{-1}$) refer to the rate at which mass is transferred from layer $j$ to layer $i$. The diagonal elements $M_{ii}$ refer to the rate at which material is lost to other layers. The $M_{ii}$ elements maintain mass conservation. In case each layer has the same total mass, the sum of the elements in each column must be zero. Likewise, the sum of the elements in each row must be zero.

In local diffusion only adjacent layers exchange material, and as a result only the main diagonal and the two diagonals around it have nonzero elements. This is known as a **tridiagonal matrix**. Diffusion is nonlocal if material is exchanged between nonneighboring layers, that is, if the material balance matrix is not tridiagonal.

For instance, consider the Blackadar (1978) model. Equations (11.58) and (11.60) can be written in matrix form as

$$\begin{pmatrix} \dfrac{\partial q_1}{\partial t} \\ \dfrac{\partial q_2}{\partial t} \\ \dfrac{\partial q_3}{\partial t} \\ \dfrac{\partial q_4}{\partial t} \end{pmatrix} = \begin{pmatrix} -\dfrac{M_u \rho_2 h_2}{\rho_1 h_1} - \dfrac{M_u \rho_3 h_3}{\rho_1 h_1} - \dfrac{M_u \rho_4 h_4}{\rho_1 h_1} & \dfrac{M_u \rho_2 h_2}{\rho_1 h_1} & \dfrac{M_u \rho_3 h_3}{\rho_1 h_1} & \dfrac{M_u \rho_4 h_4}{\rho_1 h_1} \\ M_u & -M_u & 0 & 0 \\ M_u & 0 & -M_u & 0 \\ M_u & 0 & 0 & -M_u \end{pmatrix} \begin{pmatrix} q_1 \\ q_2 \\ q_3 \\ q_4 \end{pmatrix} \tag{11.84}$$

Note that the deposition term in eq. (11.58) is not included in eq. (11.84). If each layer has the same mass, eq. (11.84) can be simplified as

$$\begin{pmatrix} \dfrac{\partial q_1}{\partial t} \\ \dfrac{\partial q_2}{\partial t} \\ \dfrac{\partial q_3}{\partial t} \\ \dfrac{\partial q_4}{\partial t} \end{pmatrix} = \begin{pmatrix} -3M_u & M_u & M_u & M_u \\ M_u & -M_u & 0 & 0 \\ M_u & 0 & -M_u & 0 \\ M_u & 0 & 0 & -M_u \end{pmatrix} \begin{pmatrix} q_1 \\ q_2 \\ q_3 \\ q_4 \end{pmatrix} \tag{11.85}$$

The simplest numerical solver of differential equations is Euler's method. This involves discretizing the derivatives:

$$\begin{pmatrix} \dfrac{q_1(t+\Delta t) - q_1(t)}{\Delta t} \\ \dfrac{q_2(t+\Delta t) - q_2(t)}{\Delta t} \\ \dfrac{q_3(t+\Delta t) - q_3(t)}{\Delta t} \\ \dfrac{q_4(t+\Delta t) - q_4(t)}{\Delta t} \end{pmatrix} = \begin{pmatrix} -3M_u & M_u & M_u & M_u \\ M_u & -M_u & 0 & 0 \\ M_u & 0 & -M_u & 0 \\ M_u & 0 & 0 & -M_u \end{pmatrix} \begin{pmatrix} q_1(t) \\ q_2(t) \\ q_3(t) \\ q_4(t) \end{pmatrix} \tag{11.86}$$

Rearranging leads to

$$\begin{pmatrix} q_1(t+\Delta t) \\ q_2(t+\Delta t) \\ q_3(t+\Delta t) \\ q_4(t+\Delta t) \end{pmatrix} = \begin{pmatrix} q_1(t) \\ q_2(t) \\ q_3(t) \\ q_4(t) \end{pmatrix} + \Delta t \begin{pmatrix} -3M_u & M_u & M_u & M_u \\ M_u & -M_u & 0 & 0 \\ M_u & 0 & -M_u & 0 \\ M_u & 0 & 0 & -M_u \end{pmatrix} \begin{pmatrix} q_1(t) \\ q_2(t) \\ q_3(t) \\ q_4(t) \end{pmatrix} \tag{11.87}$$

which can be written as

$$\begin{pmatrix} q_1(t+\Delta t) \\ q_2(t+\Delta t) \\ q_3(t+\Delta t) \\ q_4(t+\Delta t) \end{pmatrix} = \begin{pmatrix} 1-3M_u\Delta t & M_u\Delta t & M_u\Delta t & M_u\Delta t \\ M_u\Delta t & 1-M_u\Delta t & 0 & 0 \\ M_u\Delta t & 0 & 1-M_u\Delta t & 0 \\ M_u\Delta t & 0 & 0 & 1-M_u\Delta t \end{pmatrix} \begin{pmatrix} q_1(t) \\ q_2(t) \\ q_3(t) \\ q_4(t) \end{pmatrix} \tag{11.88}$$

The $4 \times 4$ matrix in eq. (11.88) is known as the **transilient matrix**. Fiedler and Moeng (1985) developed a transilient matrix for a convective mixing layer by requiring that upward and downward turbulent diffusion calculations with the transilient theory correspond with predictions of a large eddy simulation. Stull and Driedonks (1987) developed a way to construct the transilient matrix based on turbulent kinetic energy calculations.

## 11.4 ATMOSPHERIC CHEMISTRY

### 11.4.1 Introduction

The main advantage of Eulerian models is that this class of models can accommodate any set of chemical reactions, even involving complex kinetics. Conceptually, including chemical reactions in an Eulerian model means entering the kinetics in the source/sink term in eq. (11.21) by means of eqs. (11.19) and (11.20). While this is simple conceptually, the implementation of complicated kinetic models is computationally expensive: in a model such as CMAQ, 50–90 % of the CPU time is needed to calculate the numerical integrations involved in resolving the chemical kinetics (Byun and Ching, 1999).

The main application of Eulerian air quality models with chemistry is the calculation of ozone pollution. This is a photochemical process, so any succesful Eulerian model should include some photochemical reactions.

Some of the main reactions in photochemical smog have already been shown in Chapter 3 and will be reviewed briefly here. However, a detailed discussion of atmospheric ozone chemistry models requires knowledge of chemical and photochemical kinetics and will be deferred to a later section. Stratospheric reactions are outside the realm of air dispersion modeling and will not be discussed in this book.

In the absence of hydrocarbons, ozone is produced and destroyed in a cycle involving NO and $NO_2$:

$$NO_2 + UV \rightarrow NO + O(^3P) \tag{3.10}$$

$$O(^3P) + O_2 \rightarrow O_3 \tag{3.11}$$

$$NO + O_3 \rightarrow NO_2 + O_2 \tag{3.12}$$

where $O(^3P)$ denotes the oxygen atom in the ground (triplet) state. A number of additional reactions become relevant when nonmethane hydrocarbons are present:

$$O_3 + UV \rightarrow O_2 + O(^1D) \tag{3.13}$$

$$H_2O + O(^1D) \rightarrow 2HO^{\cdot} \tag{3.14}$$

$$HO^{\cdot} + RH \rightarrow H_2O + R^{\cdot} \tag{3.15}$$

$$R^{\cdot} + O_2 \rightarrow R\text{-}O\text{-}O^{\cdot} \tag{3.16}$$

$$R\text{-}O\text{-}O^{\cdot} + NO \rightarrow R\text{-}O^{\cdot} + NO_2 \tag{3.17}$$

In the above reactions, $O(^1D)$ denotes the oxygen atom in the excited (singlet) state, and RH stands for a hydrocarbon. Species such as $R^{\cdot}$ and $R\text{-}O^{\cdot}$ have an **unpaired electron**. They are much more reactive than molecules that have only paired electrons. They are referred to as **radicals**. Due to their high reactivity, radicals usually have extremely low concentrations.

Some of these reactions, such as eq. (3.16), are so fast that they can be considered *instantaneous*. Simulating them directly would not alter the end result of the calculation, but it would increase the computation time of the model by an order of magnitude or more. Hence, one of the main challenges in atmospheric chemistry modeling is finding the appopriate assumptions and approximations that retain the salient features of the real chemistry, without creating an insurmountable computational demand.

## 11.4.2 Introduction to Chemical Kinetics

In atmospheric chemistry it is customary to express concentrations in molecules $cm^{-3}$, not in mol $m^{-3}$. This convention will be followed in the current chapter as well. The relation between the number concentration $C$ in molecules $cm^{-3}$ and the amount concentration $c$ in mol $m^3$ is

$$c_i = \frac{10^6 C_i}{N_A} \tag{11.89}$$

where $N_A$ is Avogadro's constant ($6.0221415 \times 10^{23}$ molecules $mol^{-1}$). This is one of the few occasions in this book where SI units will not be used.

In complex reaction kinetics we distinguish between the reaction rate of a **compound** and the reaction rate of a **reaction**. We will start with the latter. Consider the following chemcial reaction:

$$aA + bB \rightarrow cC + dD \tag{11.90}$$

Assuming this is reaction number $i$, the reaction rate $R_i$ of this reaction is given by

$$R_i = -\frac{1}{Va}\frac{\partial N_{Ai}}{\partial t} = -\frac{1}{Vb}\frac{\partial N_{Bi}}{\partial t} = \frac{1}{Vc}\frac{\partial N_{Ci}}{\partial t} = \frac{1}{Vd}\frac{\partial N_{Di}}{\partial t} \tag{11.91}$$

where $V$ is the volume of the gas phase under consideration (cm$^3$), and $N_{Ai}$ is the number of molecules of A reacting (i.e., forming) in reaction $i$. The rate of a reaction is positive if it runs from left to right and negative if it runs from right to left. When a reaction has a positive reaction rate, the number of molecules of **reactants** (A and B) reacting is negative (A and B are consumed), whereas the number of molecules of **reaction products** (C and D) is positive.

If $i$ is the only reaction that involves A, then the reaction rate of A is defined as

$$R_A = -aR_i = \frac{1}{V}\frac{\partial N_{Ai}}{\partial t} \tag{11.92}$$

To define the reaction rate of A when A is involved in multiple reactions, it is useful to define new stoichiometric coefficients defined by considering all compounds involved in the reaction as reaction products:

$$0 = \nu_A A + \nu_B B + \nu_C C + \nu_D D \tag{11.93}$$

Hence, $\nu_A = -a$, $\nu_B = -b$, $\nu_C = c$, and $\nu_D = d$. Define $\nu_{Ai}$ as more generally the stoichiometric coefficient of A in reaction $i$, then the reaction rate of A in a system with multiple reactions is

$$R_A = \sum_{i=1}^{n} \nu_{Ai} R_i \tag{11.94}$$

In atmospheric chemistry, the number of reactions $n$ can run in the hundreds.

We need a relationship that allows the calculation of the reaction rate based on the conditions of the reaction. This relationship is the reaction kinetics. The relevant conditions for the reaction rate in gas phase are temperature, pressure, and concentration of the reactants. The effect of reactant **concentration** on the reaction rate can be described with the concept of the **rate constant** and the **reaction order**. For instance, if the reaction $i$ described by eq. (11.90) can be assumed to be irriversible (i.e., goes from left to right only), then the reaction kinetics can usually be described as

$$R_i = k_i C_A^{n_{iA}} C_B^{n_{iB}} \tag{11.95}$$

where $k_i$ is the rate constant of the reaction, $n_{iA}$ is the **order** of the reaction in A, and $n_{iB}$ is the order of the reaction in B. The overall order of the reaction is the sum of the orders of the reactor to the individual reactants. If the reaction is an **elementary reaction** (i.e., the reaction occurs in one step), then the order of the reaction equals the **molecularity** of the reaction, that is, $n_{iA} = a$ and $n_{iB} = b$. However, this is *not* the case if the mechanism of the reaction can be subdivided into more than one smaller reactions. For instance, termolecular reactions (i.e., reactions that involve three reactant molecules) are usually not elementary because the likelihood of three molecules colliding simultaneously is small.

To illustrate the principles involved, consider the following reaction (Atkinson et al., 2004):

$$O^{\cdot} + O_3 \rightarrow 2O_2 \tag{11.96}$$

This reaction is first order in the oxygen atom concentration, first order in the ozone concentration, and has a rate constant of $8.0\times10^{-15}$ cm$^3$ molecule$^{-1}$ s$^{-1}$ at 25 °C. Hence, the reaction rate is given by

$$R_i = k_i C_{\mathrm{O}} C_{\mathrm{O_3}} \tag{11.97}$$

where $k_i = 8.0\times10^{-15}$ cm$^3$ molecule$^{-1}$ s$^{-1}$. Assuming a concentration of $10^{12}$ molecules cm$^{-3}$ for ozone, and a concentration of $10^4$ molecules cm$^{-3}$ for oxygen atoms, a reaction rate of $8.0\times10^{-15}$ cm$^3$ molecule$^{-1}$ s$^{-1}\times10^{12}$ molecules cm$^{-3}\times10^4$ molecules cm$^{-3}$ = 80 molecules cm$^{-3}$ s$^{-1}$ is obtained. If this were the only reaction to take place in the atmosphere, the reaction rates of oxygen atoms and ozone would be −80 molecules cm$^{-3}$ s$^{-1}$, whereas the reaction rate of oxygen gas would be 160 molecules cm$^{-3}$ s$^{-1}$.

A special case of reaction kinetics that deserves further attention is the case of first-order kinetics. There are few atmospheric reactions that have true and irreversible first-order kinetics, but second-order reactions often behave like a first-order reaction because one of the reactants has a constant concentration. An example is the reaction of eq. (3.16). Even complex reactive systems can have an overall behavior that is nearly first-order in typical atmospheric conditions. For now we will simply consider a theoretical case A → B. The reaction kinetics is

$$R_1 = k_1 C_{\mathrm{A}} \tag{11.98}$$

In the absence of other reactions, of mass transfer mechanisms, and of pressure or temperature changes (i.e., at constant volume), the concentration change of A can be calculated with the following equation, which is based on eq. (11.92):

$$\frac{dC_{\mathrm{A}}}{dt} = -k_1 C_{\mathrm{A}} \tag{11.99}$$

Integration of eq. (11.99) leads to the following equation:

$$C_{\mathrm{A}} = C_{\mathrm{A,0}} \exp(-k_1 t) \tag{11.100}$$

where $C_{\mathrm{A,0}}$ is the initial concentration of A. Two characteristic times are often defined in association with a first-order reaction. These are the **time constant** $\tau$ and the **half-life** $t_{1/2}$:

$$\tau = \frac{1}{k_1} \tag{11.101}$$

$$t_{1/2} = \frac{\ln(2)}{k_1} \tag{11.102}$$

It can easily be shown that the concentration $C_{\mathrm{A}}$ reduces from $C_{\mathrm{A,0}}$ to $e^{-1}C_{\mathrm{A,0}} \approx 0.35C_{\mathrm{A,0}}$ in a time $\tau$ and from $C_{\mathrm{A,0}}$ to $0.5C_{\mathrm{A,0}}$ in a time $t_{1/2}$.

The influence of **temperature** on the reaction rate is expressed as a temperature relationship of the rate constant. The most frequently used temperature relationship is the **Arrhenius** equation:

$$k_i = A\exp\left(-\frac{E}{RT}\right) \quad (11.103)$$

where $A$ is a constant known as the **frequency factor**, and $E$ is a constant known as the **activation energy**. And $A$ has the same units as $k_i$, whereas $E$ has units J mol$^{-1}$. For instance, the reaction in eq. (11.96) has a frequency factor of $8.0\times10^{-12}$ cm$^3$ molecule$^{-1}$ s$^{-1}$ and an activation energy of 17,130 J mol$^{-1}$. Often $E/R$ is presented in the literature instead of $E$.

Equation (11.103) usually has a limited temperature range. For instance, in the case of reaction (11.96) the temperature range for the above coefficients is 200–400 K.

For liquid-phase reactions and high-temperature gas-phase reactions $E$ is positive and expresses the energy needed for a molecule to overcome the barrier to chemical reaction. A positive activation energy means that the rate constant increases with increasing temperature, and the increase is more pronounced with higher activation energies. In gas phase at low temperature, on the other hand, there are many reactions that have a rate constant decreasing with increasing temperature. In this case the reaction is nonelementary, with an intermediate that destabilizes rapidly with increasing temperature, so the overall reaction rate appears to decrease with increasing temperature. The practical way to deal with such cases is to adopt a negative activation energy that fits the rate constant data well.

Sometimes the temperature range of eq. (11.103) is extended by modifying the relationship as follows:

$$k_i = A\left(\frac{T}{T_0}\right)^n \exp\left(-\frac{E}{RT}\right) \quad (11.104)$$

where $n$ is an empirical constant; and $T_0$ is a reference temperature used to balance units and to give $A$ the meaning of an apparent frequency factor. For instance, the reaction

$$2\mathrm{HO}^{\cdot} \rightarrow \mathrm{H_2O} + \mathrm{O}^{\cdot} \quad (11.105)$$

follows eq. (11.104) with $A = 6.2\times10^{-14}$ cm$^3$ molecule$^{-1}$ s$^{-1}$, $n = 2.6$, $T_0 = 298$ K, and $E/R = -945$ K. The temperature range is 200–350 (Atkinson et al., 2004). An Arrhenius equation can describe the relatively strong negative temperature dependence of the rate constant at 200–250 K and the weak temperature dependence of the rate constant at 250–350 K, but not both. Hence, an Arrhenius equation does not adequately describe the rate constant of this reaction in all conditions relevant for tropospheric and stratospheric chemistry.

In most cases gas-phase reaction kinetics has no **pressure dependence** other than through the values of the concentrations of the reactants. However, some combination reactions have an overall order of 3 at low pressure and an overall order of 2 at high pressure. Conseqently, there is an intermediate pressure range where eq. (11.95) is an inadequate description of the reaction kinetics. This phenomenon is known as **falloff**. An example of such a reaction is

$$2\mathrm{HO}^{\cdot} \rightarrow \mathrm{H_2O_2} \tag{11.106}$$

The mechanism of this type of reactions is as follows (Atkinson et al., 2004):

$$\mathrm{A} + \mathrm{B} \rightarrow \mathrm{AB}^* \tag{11.107}$$

$$\mathrm{AB}^* \rightarrow \mathrm{A} + \mathrm{B} \tag{11.108}$$

$$\mathrm{AB}^* + \mathrm{M} \rightarrow \mathrm{AB} + \mathrm{M} \tag{11.109}$$

where M is usually either $N_2$ or $O_2$. At low pressure most intermediates AB* decompose, and the overall reaction rate is proportional to the concentration of M. At high pressure every intermediate that is produced reacts further to AB, and the actual concentration of M no longer matters.

A simple method to predict rate constants in the falloff pressure range was developed by Troe (1979). The equation is

$$k_i = \frac{k_0 k_\infty}{k_0 + k_\infty} F \tag{11.110}$$

where $k_0$ is the apparent second-order rate constant in the low-pressure limit (expressed as a third-order rate constant multiplied by the total number concentration), $k_\infty$ is the rate constant in the high-pressure limit, and $F$ is a broadening factor that is a measure of the width of the transition region between the low-pressure and high-pressure limits. $F$ is calculated as

$$\log F = \frac{\log F_c}{1 + \left[\dfrac{\log\left(k_0 / k_\infty\right)}{0.75 - 1.27 \log F_c}\right]^2} \tag{11.111}$$

In eqs. (11.110) and (11.111), $k_0$, $k_\infty$, and $F_c$ are empirical parameters, and log is the logarithm with base 10. Sander et al. (2006) uses an alternative formulation.

For example, in the kinetics of the reaction in eq. (11.106), it was observed that $k_0 = 6.9 \times 10^{-31}$[M] cm$^3$ molecule$^{-1}$ s$^{-1}$, where [M] is the total concentration in molecules cm$^{-3}$; $k_\infty = 2.6 \times 10^{-11}$ cm$^3$ molecule$^{-1}$ s$^{-1}$, and $F_c = 0.5$. At 298.15 K and 100,000 Pa, the total concentration is $2.43 \times 10^{19}$ molecules cm$^{-3}$, leading to a $k_0$ value of $1.68 \times 10^{-11}$ cm$^3$ molecule$^{-1}$ s$^{-1}$. Based on these values it is found that $F$ = 0.510, which leads to $k_i = 5.19 \times 10^{-12}$ cm$^3$ molecule$^{-1}$ s$^{-1}$. The transition pressure is very wide: to ensure accurate rate constants to within 10%, the limiting rate constants should only be used below 100 Pa at the lower pressure limit and above 100 MPa at the higher pressure limit.

Authoritative reviews of rate constants for atmospheric reactions are maintained by the IUPAC Subcommittee on Data Evaluation and published at regular intervals (Baulch et al., 1980, 1982, 1984; Atkinson et al., 1989, 1992, 1997a, 1997b, 1999, 2000, 2004, 2006, 2007). Useful data are also maintained by NASA's Jet Propulsion Laboratory (Sander et al., 2006; 2011).

### 11.4.3 Introduction to Photochemical Kinetics

One way to look at a photochemical reaction is as a reaction that has a photon as a reactant. The photon is consumed in the reaction like any other reactant. However, there are two fundamental differences between photons and actual reactants:

- Sometimes photons are consumed without leading to any chemical reaction. This will lead to the definition of **quantum yield**.
- The photons move at light speed. This means that we will need to define a variable expressing some kind of photon flux to replace the number concentration that is used in chemical kinetics. That variable is the **photon fluence rate** and its wavelength derivative, the **spectral photon fluence rate**, known mostly as the **spectral actinic flux**.

If a surface above Earth's atmosphere is oriented toward the sun, it receives about 1370 W $m^{-2}$ of light. An energy flux like this is formally defined in photochemistry as the **irradiance** ($E$). Due to absorption, the light hitting Earth's surface is less intense: Here the irradiance on a surface pointed directly at the sun is about 1000 W $m^{-2}$ (0.1 W $cm^{-2}$). In photochemistry, the preferred length unit is in centimeters, so the irradiance is expressed as 0.1 W $cm^{-2}$.

Light is emitted at a range of wavelengths, some of which are photochemically active and some of which are not. Hence the **spectral irradiance** $E_\lambda(\lambda)$ (W $m^{-2}$ $nm^{-1}$) is defined such that the irradiance between wavelength $\lambda$ and $\lambda + d\lambda$ is $E_\lambda(\lambda)\, d\lambda$. Hence:

$$E = \int_0^\infty E_\lambda(\lambda)\, d\lambda \tag{11.112}$$

The number of photochemical events in the atmosphere does not depend on the energy received but on the number of photons. Hence, the **spectral photon irradiance** $E_{\mathrm{p},\lambda}$ (photons $cm^{-2}$ $s^{-1}$ $nm^{-1}$ or simply $cm^{-2}$ $s^{-1}$ $nm^{-1}$) is defined as the number of photons incident on a surface, per unit area, per unit time, and per unit wavelength. To convert, the energy of one photon $e_\mathrm{p}$ (J $photon^{-1}$ or simply J) is needed. It is given by

$$e_\mathrm{p} = h\nu \tag{11.113}$$

where $h$ is the Planck constant ($6.6260693\times10^{-34}$ J s), and $\nu$ ($s^{-1}$) is the frequency of the light; also $\nu$ is linked to the wavelength through the light speed $c$ ($2.99792458\times10^{8}$ m $s^{-1}$):

$$\nu = \frac{c}{\lambda} \tag{11.114}$$

The value of $\lambda$ in eq. (11.114) is in meters. If a value in nanometers is used, it must be divided by $10^9$ to maintain unit consistency. Hence, $E_{\mathrm{p},\lambda}$ can be calculated from $E_\lambda$ as

$$E_{\mathrm{p},\lambda} = \frac{\lambda(\mathrm{nm})E_\lambda}{10^9 hc} \tag{11.115}$$

The **photon irradiance** $E_p$ is the integral of the spectral photon irradiance over the entire wavelength spectrum:

$$E_p = \int_0^\infty E_{p,\lambda}(\lambda)d\lambda \qquad (11.116)$$

If all the light from the sun reached Earth directly without reflections or scattering, then the spectral photon irradiance, along with the (unique) direction of the light, would be sufficient to calculate photochemical reaction rates. However, this is not the case. Why that is important can be seen from considering the Beer–Lambert law of light absorption, which states that the absorbance of a system, $A(\lambda)$, as defined in the equation, is proportional to the path length $l$:

$$A(\lambda) = \log\left(\frac{E_{p,\lambda}^0}{E_{p,\lambda}}\right) = \varepsilon(\lambda)cl \qquad (11.117)$$

where $\varepsilon(\lambda)$ is the **molar decadic absorption coefficient**, and $c$ is the molar concentration. When light absorption (and hence photochemical activity) is considered in a horizontal air layer, then light rays coming from different directions travel different distances through the air layer and experience different absorptions. Hence, two systems experiencing the same spectral photon irradiance may experience different photochemical activities. The spectral photon irradiance does not contain the information needed to calculate photochemical reaction rates. We need a different measure of light intensity that is directly related to photochemical activity. That measure is the spectral actinic flux $I(\lambda)$.

The term "spectral actinic flux" is not formally defined in the IUPAC glossary of terms used in photochemistry (Braslavsky, 2007). No term in the glossary defines the same concept, but from related definitions it is clear that the IUPAC term would be "spectral photon fluence rate" $E_{p,\lambda,o}$, with units of photons $cm^{-2}$ $s^{-1}$ $nm^{-1}$ or simply $cm^{-2}$ $s^{-1}$ $nm^{-1}$. Given this omission in the IUPAC glossary, and the fact that the term actinic flux is commonly used in atmospheric chemistry literature (e.g., Byun and Ching, 1999; Pun et al., 2005; Seinfeld and Pandis, 2006), the term actinic flux will be adopted in this text.

The spectral actinic flux can be defined as the number of photons entering a small sphere around the point of interest per unit time, divided by the cross-sectional area of the sphere. Because it is a spectral quantity, its units are photons $cm^{-2}$ $s^{-1}$ $nm^{-1}$. The spectral actinic flux is always greater than the spectral photon irradiance. The two properties are the same when all the light rays are perpendicular to the surface of interest and go in the same direction (i.e., not back and forth).

The spectral actinic flux multiplied by the number concentration of the absorbing molecule is proportional to the number of photons absorbed per unit time, volume, and wavelength. For a justification of this, see Seinfeld and Pandis (2006). We will only consider the relationship based on an analysis of the units. Consider $a_p(\lambda)$ the rate of absorption of photons (photons $cm^{-3}$ $s^{-1}$ $nm^{-1}$). It is related to the spectral actinic flux as

$$a_p(\lambda) = \sigma(\lambda)CI(\lambda) \qquad (11.118)$$

With $C$ (number concentration) in molecules $cm^{-3}$, the variable $\sigma(\lambda)$ must have units of $cm^2$ $molecule^{-1}$ or simply $cm^2$ to balance the units. The variable $\sigma(\lambda)$ is termed the **absorption cross section**. It is a measure of the ability of a molecule to absorb photons. It usually depends on the wavelength very strongly. Values are typically on the order of $10^{-20}$ $cm^2$, with a range of several orders of magnitude.

Not all absorbed photons cause chemical reactions. The fraction of photons that cause a chemical reaction is termed the **quantum yield** $\phi(\lambda)$. By definition, it is a number between 0 (no reaction) and 1 (all photons cause reaction). It also depends on the wavelength and is usually small at long wavelengths and approaches 1 when the photon energy exceeds the energy needed to induce the reaction. The reaction rate per nanometer of absorbed light spectrum is obtained by multiplying $a_p(\lambda)$ by $\phi(\lambda)$.

Because the properties we considered so far are spectral properties, we need to integrate across wavelength to obtain the photochemical reaction rate:

$$R_i = C\int_0^{\infty} \phi(\lambda)\sigma(\lambda)I(\lambda)\, d\lambda \tag{11.119}$$

Hence, the photochemical reaction $i$ is first order in the reacting molecule, with rate constant $j_i$:

$$j_i = \int_0^{\infty} \phi(\lambda)\sigma(\lambda)I(\lambda)\, d\lambda \tag{11.120}$$

and the reaction kinetics can be written as

$$R_i = j_i C \tag{11.121}$$

For tropospheric chemistry, the relevant wavelength window is mostly between 300 and 400 nm. Light with a wavelength shorter than 300 nm barely reaches the troposphere and only needs to be considered in the stratosphere. Most photochemical reactions in the atmosphere have low quantum yields at wavelengths longer than 400 nm.

**Example 11.1** Calculate the photchemical rate constant of ozone photolysis to the singlet oxygen atom, $O(^1D)$, based on the data in Table 11.1. Also calculate the photchemical rate constant of ozone photolysis to the triplet oxygen atom, $O(^3P)$, assuming that the combined quantum yield of both reactions is 1.

***Solution.*** The values in Table 11.1 are averages over 10-nm intervals, not point values. Hence, for calculating the integral we simply calculate the product $\phi(\lambda)\sigma(\lambda)$ $I(\lambda)$ in every interval and multiply by 10 nm. The sums of these contributions are $j_1$ and $j_2$. This is shown in Table 11.1.

**TABLE 11.1 Properties of Photolysis of Ozone to O($^1$D) ($j_1$) and to O($^3$P) ($j_2$)[a]**

| $\lambda$ (nm) | $\sigma$ ($10^{-20}$ cm$^2$) | $\phi_1$ | $I$ ($10^{12}$ photons cm$^{-2}$ s$^{-1}$ nm$^{-1}$) | contribution to $j_1$ (s$^{-1}$) | contribution to $j_2$ (s$^{-1}$) |
|---|---|---|---|---|---|
| 300 | 40 | 0.9 | 3.6 | $1.296\times10^{-5}$ | $1.44\times10^{-7}$ |
| 310 | 11 | 0.6 | 41 | $2.706\times10^{-5}$ | $1.80\times10^{-6}$ |
| 320 | 3 | 0.17 | 91 | $4.64\times10^{-6}$ | $2.27\times10^{-6}$ |
| 330 | 0.7 | 0.08 | 151 | $8.46\times10^{-6}$ | $9.72\times10^{-7}$ |
| 340 | 0.15 | 0.04 | 160 | $9.60\times10^{-7}$ | $2.30\times10^{-7}$ |
| 350 | 0.03 | 0 | 178 | 0 | $5.34\times10^{-8}$ |
| 360 | 0.007 | 0 | 184 | 0 | $1.29\times10^{-8}$ |
| 370 | 0.0015 | 0 | 226 | 0 | $3.39\times10^{-9}$ |
| 380 | 0.0006 | 0 | 219 | 0 | $1.31\times10^{-9}$ |
| 390 | 0.0007 | 0 | 222 | 0 | $1.55\times10^{-9}$ |
| 400 | 0.0012 | 0 | 300 | 0 | $3.60\times10^{-9}$ |
| | | | | sum: $4.56\times10^{-5}$ | sum: $5.5\times10^{-6}$ |

[a] Values are averages of 10-nm intervals.

For the photolysis to O($^1$D), a rate constant of $4.56\times10^{-5}$ s$^{-1}$, or 0.164 h$^{-1}$, is obtained. The rate constant for photolysis to O($^3$P) is estimated at $5.5\times10^{-6}$ s$^{-1}$, or 0.020 h$^{-1}$. The overall photolysis rate of ozone is 0.184 h$^{-1}$.

### 11.4.4 Gas-Phase Reactions in Tropospheric Chemistry

As indicated in the example in the preceding section, tropospheric photochemistry mostly takes place in the near-ultraviolet region. They are the primary drivers of photochemical smog and tropospherical ozone formation. The following photochemical reactions initiate the formation of photochemical smog:

$$O_3 + h\nu \rightarrow O_2 + O(^1D) \qquad \text{(Reaction 1)} \qquad (11.122)$$

$$O_3 + h\nu \rightarrow O_2 + O(^3P) \qquad \text{(Reaction 2)} \qquad (11.123)$$

$$NO_2 + h\nu \rightarrow NO + O(^3P) \qquad \text{(Reaction 3)} \qquad (11.124)$$

The first two reactions were the subject of the example in the preceding section. In typical sunny conditions the reactions take a few hours to complete. The reaction products are oxygen gas and either a singlet oxygen atom or a triplet oxygen atom. The singlet form has a higher energy state and is more reactive than the triplet form. This has important repercussions for our discussion further down. The rate constants obtained depend strongly on the conditions (cloudiness, solar elevation) but the values obtained in the previous section are representative: $j_1 = 4.5\times10^{-5}$ s$^{-1}$; $j_2 = 0.5\times10^{-5}$ s$^{-1}$.

Reaction 3 has a much higher rate constant due to a larger absorption cross section and a wider wavelength window reaching up to about 420 nm. A typical photochemical rate constant for $NO_2$ is $j_3 = 8\times10^{-3}$ s$^{-1}$. Hence, once $NO_2$ is formed in sunny conditions, it reacts back to NO in a few minutes, producing $O(^3P)$.

Reactions 1 and 2 are much slower than reaction 3, and will be ignored for now. Assuming that $O_3$, $NO_2$, NO, and $O(^3P)$ are the only substances in the atmosphere besides $O_2$ and $N_2$, two more reactions are needed to start making calculations. These reactions are

$$O(^3P) + O_2 \rightarrow O_3 \qquad \text{(Reaction 4)} \qquad (11.125)$$

$$NO + O_3 \rightarrow NO_2 + O_2 \qquad \text{(Reaction 5)} \qquad (11.126)$$

Reactions 3, 4, and 5 form a cycle that constantly interconverts $O_3$ into $O_2$ and back and interconverts NO into $NO_2$ and back. This system is too simple to provide an accurate prediction of what happens in a real atmosphere, but it allows us to introduce some concepts and approaches in atmospheric chemistry.

The rate constants of reactions 4 and 5 at 25 °C are (Atkinson et al., 2004)

$$k_4 = 6.0\times10^{-34}[O_2] + 5.6\times10^{-34}[N_2]\,\text{cm}^6\,\text{molecule}^{-2}\,\text{s}^{-1} \qquad (11.127)$$

$$k_5 = 1.8\times10^{-14}\,\text{cm}^3\,\text{molecule}^{-1}\,\text{s}^{-1} \qquad (11.128)$$

where a compound between square brackets is its concentration in molecules cm$^{-3}$. Assuming 25 °C and 100 kPa, the total concentration can be calculated with the ideal gas law. The result is 40.34 mol m$^{-3}$ = $2.43\times10^{19}$ molecules cm$^{-3}$. Assuming 78% $N_2$ and 21% $O_2$, and substituting in eq. (11.127) leads to

$$k_4 = 1.37\times10^{-14}\,\text{cm}^3\,\text{molecule}^{-1}\,\text{s}^{-1} \qquad (11.129)$$

One of the reactants in reaction 4 is oxygen, which can be assumed to have a constant concentration. Hence, we can describe this reaction with an **apparent first-order rate constant** in $O(^3P)$ by multiplying eq. (11.129) by the oxygen concentration. This situation is described as **pseudo-first-order kinetics**. Apparent first-order constants will be indicated with a prime:

$$k'_4 = 7.0\times10^4\,\text{s}^{-1} \qquad (11.130)$$

If reaction 4 were the only reaction, it would run to completion in a few tens of microseconds ($\tau$ = $1.4\times10^{-5}$ s). In a system where $O(^3P)$ is formed in only one reaction, and destroyed in only one reaction, we can combine the two, as they are only a few tenths of microseconds apart. Hence, an overall reaction representing both reaction 3 and reaction 4 is

$$NO_2 + O_2 + h\nu \rightarrow NO + O_3 \qquad \text{(Reactions 3+4)} \qquad (11.131)$$

Because this is not an elementary reaction, the order does not need to be the same as the molecularity. Because reaction 4 always follows reaction 3 closely, the kinetics of reaction 3 determines the overall kinetics of reaction 3+4. We say that reaction 3 is the **rate-limiting step** in the mechanism, whereas reaction 4 is an *instantaneous* reaction. The chemistry in Eulerian models such as CMAQ contains many nonelementary reactions. Reaction 3+4 is *not* one of them because $O(^3P)$ takes part in too many reactions. Here, we will assume that reaction 3+4 is first order in $NO_2$, with rate constant $8\times10^{-3}$ $s^{-1}$.

The overall reaction rate of $NO_2$ is determined by the rate of reaction 3+4, which destroys $NO_2$, and by the rate of reaction 5, which produces it:

$$r_{NO_2} = -r_{3+4} + r_5 = -j_3[NO_2] + k_5[NO][O_3] \tag{11.132}$$

The reaction rates of $O_3$ and NO are the same but opposite in sign. Because the reactions are relatively fast, the chemical species evolve to a constant concentration within a few minutes. This is the **steady-state** concentration. A steady state is not the same as a thermodynamic equilibrium. The ultraviolet light constantly pushes the reaction away from the equilibrium. The steady state can be calculated by requiring that the reaction rate equals zero. Hence:

$$-j_3[NO_2] + k_5[NO][O_3] = 0 \tag{11.133}$$

It follows that the steady-state $[NO_2]/[NO]$ ratio is determined by the photolysis rate constant $j_3$ (and hence the amount of sunlight) and the ozone concentration:

$$\frac{[NO_2]}{[NO]} = \frac{k_5[O_3]}{j_3} \tag{11.134}$$

Assuming 15 ppb of ozone, a typical background value, and using the total concentration of $2.43\times10^{19}$ molecules $cm^{-3}$ calculated above, this is consistent with an ozone number concentration of $3.64\times10^{11}$ molecules $cm^{-3}$. In our example, this leads to a steady-state $[NO_2]/[NO]$ ratio of 0.82 (i.e., [NO] = 5.5 ppm, $[NO_2]$ = 4.5 ppm). In heavily polluted air with 60 ppb ozone, the ratio is 3.3 (i.e., [NO] = 2.3 ppm, $[NO_2]$ = 7.7 ppm). Considering that the $[NO_2]/[NO]$ ratio emitted by combustion processes is usually less than 0.1 (Cooper and Alley, 2011), it follows that a large portion of NO is converted to $NO_2$, removing $O_3$ in the process (known as **$O_3$ titration**). If these three reactions were the only ones occurring, then NO emissions would lead to a decline of the ozone concentration. This is sometimes observed in industrial areas at high latitudes where $j_3$ is always low (hence a high ratio $[NO_2]/[NO]$), and there is insufficient sunlight for other photochemical processes to occur.

Returning to reaction 4, we can estimate the steady-state concentration of $O(^3P)$ by requiring that $r_3 = r_4$. This leads to the following equation:

$$j_3[NO_2] = k_4'[O(^3P)] \tag{11.135}$$

which leads to the following concentration ratio:

$$\frac{[O(^3P)]}{[NO_2]} = \frac{j_3}{k_4'} \tag{11.136}$$

Under the conditions assumed here, the concentration ratio $[O(^3P)]/[NO_2]$ is $1.1\times10^{-7}$. Assuming an $NO_2$ concentration of 10 ppb, we find an $O(^3P)$ concentration as low as $1.1\times10^{-6}$ ppb, or $2.8\times10^4$ molecules $cm^{-3}$. Highly reactive species like this have an extremely low concentration. This makes the numerical simulation of atmospheric chemical reactions extremely difficult.

The first missing ingredient to make this model more realistic is water vapor. In the presence of water, three chemical reactions and one photochemical reaction need to be included:

$$H_2O + O(^1D) \rightarrow 2HO^{\cdot} \qquad \text{(Reaction 6)} \tag{11.137}$$

$$2HO^{\cdot} \rightarrow H_2O + O^{\cdot} \qquad \text{(Reaction 7)} \tag{11.138}$$

$$2HO^{\cdot} + M \rightarrow H_2O_2 + M \qquad \text{(Reaction 8)} \tag{11.139}$$

$$H_2O_2 + h\nu \rightarrow 2HO^{\cdot} \qquad \text{(Reaction 9)} \tag{11.140}$$

Reactions 6 and 7 have the following rate constants at 25 °C:

$$k_6 = 2.2\times10^{-10}\ \text{cm}^3\,\text{molecule}^{-1}\,\text{s}^{-1} \tag{11.141}$$

$$k_7 = 1.48\times10^{-12}\ \text{cm}^3\,\text{molecule}^{-1}\,\text{s}^{-1} \tag{11.142}$$

Reaction 8 is in the falloff range. Its rate constant at 1 bar pressure was calculated in Section 11.4.2:

$$k_8 = 1.68\times10^{-11}\ \text{cm}^3\,\text{molecule}^{-1}\,\text{s}^{-1} \tag{11.143}$$

Reaction 8 is the dominant $^{\cdot}OH$ recombination reaction. Based on data reviewed by Sander et al. (2006), the photolysis rate constant of $H_2O_2$ (reaction 9) is of the order of magnitude of $j_9 = 8\times10^{-6}$ $s^{-1}$. If photolysis were the only removal mechanism, $H_2O_2$ would remain in the air for several days. In this time other removal mechanisms, such as dry deposition, will occur.

At this level of complexity, the oxygen species $O(^1D)$ can be removed from the reaction scheme in a manner analogous to $O(^3P)$, by combining reactions 1 and 6 into a single reaction with rate constant $j_1 = 4.5\times10^{-5}$ $s^{-1}$, which is first order in ozone. Similarly, reactions 7 and 8 can be combined to a reaction with rate constant $k_{7+8} = 1.83\times10^{-11}$ $cm^3$ $molecule^{-1}$ $s^{-1}$. Ignoring the photolysis of $H_2O_2$ as a source of $^{\cdot}OH$ radicals, the steady-state condition for $^{\cdot}OH$ radicals is obtained by requiring that its reaction rate equals zero. Noting the stoichiometric factor 2 in both reactions, one obtains

$$2j_1[O_3]-2k_{7+8}[HO^\cdot]^2=0 \tag{11.144}$$

Solving for the hydroxyl radical concentration leads to

$$[HO^\cdot]=\sqrt{\frac{j_1[O_3]}{k_{7+8}}} \tag{11.145}$$

Again, using an ozone number concentration of $3.64\times10^{11}$ molecules cm$^{-3}$ (15 ppb) and the $j_1$ value used previously, a hydroxyl radical concentration of $9.47\times10^{8}$ molecules cm$^{-3}$ (0.039 ppb) is obtained. In polluted air (60 ppb ozone), the hydroxyl radical concentration doubles. Of course, reactions 7 and 8 are not the only removal mechanism of $HO^\cdot$ in polluted air. However, these values establish a baseline that will be compared with later calculations.

For the sake of completeness, the $O(^1D)$ concentration can be calculated using the same procedure as used for $O(^3P)$. Assuming a water mole fraction of 0.02 (63% relative humidity at 25 °C and 100 kPa), a number concentration of 0.153 molecules cm$^{-3}$ is obtained.

$HO^\cdot$ radicals are scavenged by organic pollutants. Propane will be used as an example here. Propane reacts with $HO^\cdot$ by abstraction of a hydrogen atom:

$$HO^\cdot+C_3H_8\rightarrow H_2O+{}^\cdot C_3H_7 \qquad \text{(Reaction 10)} \tag{11.146}$$

This is a second-order reaction with rate constant $k_{10}=1.1\times10^{-12}$ cm$^3$ molecule$^{-1}$ s$^{-1}$. If this is the dominant scavenging mechanism for $HO^\cdot$ radicals, its reaction rate must be equal to the $HO^\cdot$ production rate, $2r_{1+6}$, where the factor 2 is due to the fact that two $HO^\cdot$ radicals are produced in reaction 6. Hence:

$$2j_1[O_3]-k_{10}[HO^\cdot][C_3H_8]=0 \tag{11.147}$$

Solving for the hydroxyl radical concentration leads to

$$[HO^\cdot]=\frac{2j_1[O_3]}{k_{10}[C_3H_8]} \tag{11.148}$$

Assuming a propane concentration of 50 ppb ($1.21\times10^{12}$ molecules cm$^{-3}$) and 60 ppb ozone, a $HO^\cdot$ concentration of $9.82\times10^{7}$ molecules cm$^{-3}$ is obtained, a factor 20 lower than expected in the absence of organic molecules. Comparing $r_{10}$ with $r_{7+8}$ under these conditions, it is clear that less than 0.3% of the hydroxyl radicals react by recombining. Hence, the assumption leading to eq. (11.148) is justified.

In an actual atmospheric chemistry model, approximations such as eq. (11.147) would not be made because there is always a large number of reactions consuming $HO^\cdot$ radicals, none of which is dominant.

Rate constant $k_{10}[HO^{\cdot}]$ is a pseudo-first-order degradation rate constant of propane. Under the current conditions it has a value of $1.1 \times 10^{-4}$ s$^{-1}$, or 0.39 h$^{-1}$. Propane degrades photochemically in a few hours.

For simplicity, it will be assumed here that reaction 10 leads to *i*-propyl radicals only, without *n*-propyl radical formation. Alkyl radicals react with oxygen immediately:

$$i\text{-}^{\cdot}C_3H_7 + O_2 \rightarrow i\text{-}C_3H_7OO^{\cdot} \qquad \text{(Reaction 11)} \qquad (11.149)$$

At 25 °C, this reaction has a rate constant of $k_{11} = 1.1 \times 10^{-11}$ cm$^3$ molecule$^{-1}$ s$^{-1}$. Assuming the oxygen concentration constant at $5.10 \times 10^{18}$ molecules cm$^{-3}$, this leads to a pseudo-first-order rate constant of $k'_{11} = 5.61 \times 10^7$ s$^{-1}$. The reaction goes to completion in a fraction of a microsecond ($\tau = 1.8 \times 10^{-8}$ s) and can be considered instantaneous. Reactions 10 and 11 can be combined as

$$HO^{\cdot} + C_3H_8 + O_2 \rightarrow H_2O + i\text{-}C_3H_7OO^{\cdot} \qquad \text{(Reactions 10+11)} \qquad (11.150)$$

The peroxy radical reacts with nitrogen oxide. The main pathway (96%) is the formation of an oxy radical:

$$i\text{-}C_3H_7OO^{\cdot} + NO \rightarrow i\text{-}C_3H_7O^{\cdot} + NO_2 \qquad \text{(Reaction 12)} \qquad (11.151)$$

The other 4% is the formation of $i\text{-}C_3H_7ONO_2$. The rate constant of reaction 12 is $k_{12} = 9.4 \times 10^{-12}$ cm$^3$ molecule$^{-1}$ s$^{-1}$ (Atkinson et al., 2006). $i\text{-}C_3H_7OO^{\cdot}$ can also react with $NO_2$, forming $i\text{-}C_3H_7O_2NO_2$.

The further degradation of $i\text{-}C_3H_7O^{\cdot}$ will be discussed further down. First, the effect of reaction 12 on the $NO_x$ dynamics will be discussed. The reaction rate of $NO_2$ is

$$r_{NO_2} = r_5 + r_{12} - r_3 \qquad (11.152)$$

Furthermore, if $i\text{-}C_3H_7OO^{\cdot}$ is only produced in reaction 10+11, and it is only consumed in reaction 12, then these two reactions have the same rate. Hence:

$$r_{NO_2} = r_5 + r_{10} - r_3 \qquad (11.153)$$

Substituting the reaction kinetics into eq. (11.153) leads to

$$r_{NO_2} = k_5[NO][O_3] + k_{10}[HO^{\cdot}][C_3H_8] - j_3[NO_2] \qquad (11.154)$$

The reaction rate of $NO_2$ is zero in steady-state conditions. Substituting $[NO_x] = [NO] + [NO_2]$ to eliminate $[NO_2]$ and solving for [NO] leads to

$$[NO] = -\frac{j_3[NO_x] - k_{10}[HO^{\cdot}][C_3H_8]}{k_5[O_3] + j_3} \qquad (11.155)$$

Entering all the values obtained above for an $O_3$ concentration of 60 ppb, and an $NO_x$ concentration of 10 ppb, an NO concentration of $5.29 \times 10^{10}$ molecules cm$^{-3}$ is obtained, or 2.18 ppb. Hence, the $NO_2$ concentration is 7.82 ppb, leading to a $[NO_2]/[NO]$ ratio of 3.59. This ratio is not very different from the ratio in the absence of propane, but the effect on the ozone reaction rate is profound.

In the absence of propane, the ozone concentration maintains steady state because reaction 3+4 and reaction 5 are balanced through the NO–$NO_2$ cycle. When reaction 12 converts an additional NO to $NO_2$, the ozone reaction rates are no longer balanced. Ignoring $r_2$, which has not been accounted for in any of the above, the reaction rate of ozone is

$$r_{O_3} = -r_1 + r_{3+4} - r_5 \tag{11.156}$$

Substituting the kinetics of each reaction in eq. (11.156) leads to the following:

$$r_{O_3} = -k_1[O_3] + j_3[NO_2] - k_5[NO][O_3] \tag{11.157}$$

Substituting all the rate constants and concentrations leads to a net ozone reaction rate of $6.56 \times 10^7$ molecules cm$^{-3}$ s$^{-1}$. This amounts to $2.70 \times 10^{-3}$ ppb s$^{-1}$, or 9.72 ppb h$^{-1}$. The net ozone production rate is almost 10 ppb per hour. Hence, a substantial amount of ozone can accumulate in a day if there are no ozone removal mechanisms other than reactions 1, 2, and 5. This imbalance is what causes elevated ozone concentrations in photochemical smog.

Sometimes, the photochemical degradation of $NO_2$ cannot keep up with the production of $NO_2$ through reaction 12. This manifests itself in eq. (11.155), which becomes negative in this case. What happens then is an accumulation of peroxyl radicals until alternative reaction pathways become dominant. One such pathway is the Russell mechanism:

$$2i\text{-}C_3H_7OO^{\cdot} \rightarrow 2i\text{-}C_3H_7O^{\cdot} + O_2 \qquad \text{(Reaction 13)} \tag{11.158}$$

A similar reaction takes place in parallel, leading to complete molecules instead of radicals:

$$2i\text{-}C_3H_7OO^{\cdot} \rightarrow i\text{-}C_3H_7OH + CH_3C(O)CH_3 + O_2 \qquad \text{(Reaction 14)} \tag{11.159}$$

where the O between brackets forms a carbonyl group (C=O) with the carbon preceding it. With $k_{13} = 5.6 \times 10^{-16}$ cm$^3$ molecule$^{-1}$ s$^{-1}$ and $k_{14} = 4.4 \times 10^{-16}$ cm$^3$ molecule$^{-1}$ s$^{-1}$, it is found that $k_{13+14} = 1.0 \times 10^{-15}$ cm$^3$ molecule$^{-1}$ s$^{-1}$ (Atkinson et al., 2006).

For short-chain hydrocarbons, with up to three carbon atoms, the main reaction of the oxyalkyl radical produced in reactions 12 and 13 is abstraction of the hydrogen atom on the carbon that has the oxy radical (Atkinson, 1997a):

$$i\text{-}C_3H_7O^{\cdot} + O_2 \rightarrow CH_3C(O)CH_3 + HO_2^{\cdot} \qquad \text{(Reaction 15)} \tag{11.160}$$

Hence the main product is acetone. The rate constant of this reaction is $k_{15} = 7\times10^{-15}$ $cm^3$ $molecule^{-1}$ $s^{-1}$ (Atkinson et al., 2006). Molecules with four or more carbon atoms in succession undergo isomerizations, broadening the product spectrum (Atkinson, 1997b).

Side reactions of reaction 15 are the following:

$$i\text{-}C_3H_7O^{\cdot} + NO \rightarrow i\text{-}C_3H_7ONO \quad \text{(Reaction 16)} \tag{11.161}$$

$$i\text{-}C_3H_7O^{\cdot} + NO \rightarrow CH_3C(O)CH_3 + HNO \quad \text{(Reaction 17)} \tag{11.162}$$

where $k_{16+17} = 4\times10^{-11}$ $cm^3$ $molecule^{-1}$ $s^{-1}$ (Atkinson et al., 2006). Reaction 16 is the dominant path of the two (80%). Because NO concentrations in the atmosphere are more than six orders of magnitude lower than the $O_2$ concentration, reaction 15 is the dominant reaction.

All reaction products undergo further reactions. Here are some rules that apply to the degradation of organic compounds:

- For alkanes the main first step in the degradation is hydrogen abstraction by the hydroxyl radical. For alkenes the main first step is addition of a hydroxyl radical to the double bond, creating an OH group and an unpaired electron on adjacent carbon atoms ($\beta$-hydroxyalkyl radicals).
- $\beta$-Hydroxyalkoxy radicals, which are expected to form from $\beta$-hydroxyalkyl radicals, eliminate one of the functional groups on the carbon that has the OH group, whereas the O atom with the unpaired electron forms an aldehyde or ketone group (carbonyl group, C=O).
- When the alkyl radical formed by additon or abstraction is an $\alpha$-hydroxyalkyl radical (i.e., where one of the carbons has an OH group and an unpaired electron), the dominant reaction is abstraction of the hydrogen atom on the OH group, with the formation of a carbonyl group.

The reactive species produced in atmospheric chemistry undergo a multitude of reactions, which modify the trends discussed above. Here are the main ones. They are by no means the only ones that take place.

$$O^{\cdot}(^1D) + O_2 \rightarrow O^{\cdot}(^3P) + O_2 \quad \text{(Reaction 18)} \tag{11.163}$$

$$O^{\cdot} + O_3 \rightarrow 2O_2 \quad \text{(Reaction 19)} \tag{11.164}$$

$$HO^{\cdot} + H_2O_2 \rightarrow HO^{\cdot}_2 + H_2O \quad \text{(Reaction 20)} \tag{11.165}$$

$$HO^{\cdot} + O_3 \rightarrow HO^{\cdot}_2 + O_2 \quad \text{(Reaction 21)} \tag{11.166}$$

$$HO^{\cdot}_2 + O_3 \rightarrow HO^{\cdot} + 2O_2 \quad \text{(Reaction 22)} \tag{11.167}$$

$$HO^{\cdot}_2 + HO^{\cdot} \rightarrow H_2O + O_2 \quad \text{(Reaction 23)} \tag{11.168}$$

$$2HO^{\cdot}_2 \rightarrow H_2O_2 + O_2 \qquad \text{(Reaction 24)} \qquad (11.169)$$

$$2HO^{\cdot}_2 + M \rightarrow H_2O_2 + O_2 + M \qquad \text{(Reaction 25)} \qquad (11.170)$$

The rate constants for these reactions at 25 °C are $8.0\times10^{-15}$ cm$^3$ molecule$^{-1}$ s$^{-1}$ for reaction 19, $1.7\times10^{-12}$ cm$^3$ molecule$^{-1}$ s$^{-1}$ for reaction 20, $7.3\times10^{-14}$ cm$^3$ molecule$^{-1}$ s$^{-1}$ for reaction 21, $2.0\times10^{-15}$ cm$^3$ molecule$^{-1}$ s$^{-1}$ for reaction 22, $1.1\times10^{-10}$ cm$^3$ molecule$^{-1}$ s$^{-1}$ for reaction 23, and $1.6\times10^{-12}$ cm$^3$ molecule$^{-1}$ s$^{-1}$ for reaction 24. Reaction 25 is a third-order version of reaction 24. These two reactions are not modeled as falloff but rather as two parallel reactions. The third-order rate constant of reaction 25 is $5.2\times10^{-32}$ cm$^6$ molecule$^{-2}$ s$^{-1}$ when $N_2$ is the third molecule and $4.5\times10^{-32}$ cm$^6$ molecule$^{-2}$ s$^{-1}$ when $O_2$ is the third molecule.

An even larger number of reactions are involved in $NO_x$ and $NO_y$ reactions. The main ones are

$$O^{\cdot} + NO + M \rightarrow NO_2 + M \qquad \text{(Reaction 26)} \qquad (11.171)$$

$$O^{\cdot} + NO_2 \rightarrow O_2 + NO \qquad \text{(Reaction 27)} \qquad (11.172)$$

$$O^{\cdot} + NO_2 + M \rightarrow NO_3 + M \qquad \text{(Reaction 28)} \qquad (11.173)$$

$$O^{\cdot} + NO_3 \rightarrow NO_2 + O_2 \qquad \text{(Reaction 29)} \qquad (11.174)$$

$$O(^1D) + N_2 \rightarrow O(^3P) + N_2 \qquad \text{(Reaction 30)} \qquad (11.175)$$

$$HO^{\cdot} + NO + M \rightarrow HNO_2 + M \qquad \text{(Reaction 31)} \qquad (11.176)$$

$$HO^{\cdot} + NO_2 + M \rightarrow HNO_3 + M \qquad \text{(Reaction 32)} \qquad (11.177)$$

$$HO^{\cdot} + HNO_2 + \rightarrow H_2O + NO_2 \qquad \text{(Reaction 33)} \qquad (11.178)$$

$$HO^{\cdot} + HNO_3 + \rightarrow H_2O + NO_3 \qquad \text{(Reaction 34)} \qquad (11.179)$$

$$HO^{\cdot} + NO_3 + \rightarrow HO^{\cdot}_2 + NO_2 \qquad \text{(Reaction 35)} \qquad (11.180)$$

$$HO_2^{\cdot} + NO \rightarrow HO^{\cdot} + NO_2 \qquad \text{(Reaction 36)} \qquad (11.181)$$

$$HO_2^{\cdot} + NO_2 + M \rightarrow HNO_4 + M \qquad \text{(Reaction 37)} \qquad (11.182)$$

$$HNO_4 + M \rightarrow HO_2^{\cdot} + NO_2 + M \qquad \text{(Reaction 38)} \qquad (11.183)$$

$$2NO + O_2 \rightarrow 2NO_2 \qquad \text{(Reaction 39)} \qquad (11.184)$$

$$NO + NO_3 \rightarrow 2NO_2 \qquad \text{(Reaction 40)} \qquad (11.185)$$

$$NO_2 + O_3 \rightarrow NO_3 + O_2 \quad \text{(Reaction 41)} \quad (11.186)$$

$$NO_2 + NO_3 + M \rightarrow N_2O_5 + M \quad \text{(Reaction 42)} \quad (11.187)$$

$$N_2O_5 + M \rightarrow NO_2 + NO_3 + M \quad \text{(Reaction 43)} \quad (11.188)$$

$$N_2O_5 + H_2O \rightarrow 2HNO_3 \quad \text{(Reaction 44)} \quad (11.189)$$

$$N_2O_5 + 2H_2O \rightarrow 2HNO_3 + H_2O \quad \text{(Reaction 45)} \quad (11.190)$$

The rate constants of these reactions are shown in Table 11.2.

### 11.4.5 Chemistry of Aerosol Formation in the Troposphere

Aerosols can be both a primary pollutant and a secondary pollutant. Hence, to correctly predict aerosol concentrations, it is necessary to account for chemical reactions that lead to aerosol formation. The main source of secondary aerosol formation is the oxidation of $SO_2$ to sulfates, and this will be the main topic of this section. There are three main mechanisms of $SO_2$ oxidation to $SO_3$: gas-phase oxidation by hydroxyl radicals, oxidation by hydrogen peroxide in the droplet phase, and oxidation by ozone in the droplet phase. In the droplet phase, the acids $H_2SO_3$ and $H_2SO_4$ are formed. The relative importance of these three pathways depends on the presence, the amount, and the size of cloud water droplets and on the concentration of hydrogen peroxide in the air.

Other sources of aerosols are organic in origin, such as long-chain hydrocarbons or alkene oligomerization products forming a solid phase or dissolving in droplets.

In this section a brief outline of the most important aerosol-forming reactions will be given. A more comprehensive overview of the chemistry in atmospheric water droplets is given by Seinfeld and Pandis (2006). A comprehensive simulation model of chemical reactions in atmospheric water droplets is given by Pandis and Seinfeld (1989).

***11.4.5.1 Gas-Phase Reactions: $SO_2$ Oxidation.*** The gas-phase reaction mechanism for the formation of sulfates is simple when $SO_2$ is the only source of atmospheric sulfur:

$$O^{\cdot} + SO_2 + M \rightarrow SO_3 + M \quad \text{(Reaction 46)} \quad (11.191)$$

$$HO^{\cdot} + SO_2 + M \rightarrow HSO^{\cdot}_3 + M \quad \text{(Reaction 47)} \quad (11.192)$$

$$HSO^{\cdot}_3 + O_2 \rightarrow HO^{\cdot}_2 + SO_3 \quad \text{(Reaction 48)} \quad (11.193)$$

**TABLE 11.2 Rate Constants of Atmospheric Reactions Involving $NO_x$**

| Reaction Number | Rate Constant ($cm^3$ $molecule^{-1}$ $s^{-1}$ except if indicated otherwise) |
|---|---|
| 26 | $k_0 = 1.0\times10^{-31}[N_2]$ |
| | $k = 3.0\times10^{-11}$ |
| | $F_c = 0.85$ |
| 27 | $1.0\times10^{-11}$ |
| 28 | $k_0 = 1.3\times10^{-31}[N_2]$ |
| | $k = 2.3\times10^{-11}$ |
| | $F_c = 0.6$ |
| 29 | $1.7\times10^{-11}$ |
| 30 | $2.6\times10^{-11}$ |
| 31 | $k_0 = 7.4\times10^{-31}[N_2]$ |
| | $k = 3.3\times10^{-11}$ |
| | $F_c = 0.81$ |
| 32 | $k_0 = 3.3\times10^{-30}[N_2]$ |
| | $k = 4.1\times10^{-11}$ |
| | $F_c = 0.4$ |
| 33 | $6.0\times10^{-12}$ |
| 34 | $1.5\times10^{-13}$ |
| 35 | $2.0\times10^{-11}$ |
| 36 | $8.8\times10^{-12}$ |
| 37 | $k_0 = 1.8\times10^{-31}[N_2]$ |
| | $k = 4.7\times10^{-12}$ |
| | $F_c = 0.6$ |
| 38 | $k_0 = 1.3\times10^{-20}[N_2]$ ($s^{-1}$) |
| | $k = 0.25$ ($s^{-1}$) |
| | $F_c = 0.6$ |
| 39 | $2.0\times10^{-38}$ ($cm^6$ $molecule^{-2}$ $s^{-1}$) |
| 40 | $2.6\times10^{-11}$ |
| 41 | $3.5\times10^{-17}$ |
| 42 | $k_0 = 3.6\times10^{-30}[N_2]$ |
| | $k = 1.9\times10^{-12}$ |
| | $F_c = 0.35$ |
| 43 | $k_0 = 1.2\times10^{-19}[N_2]$ ($s^{-1}$) |
| | $k = 0.069$ ($s^{-1}$) |
| | $F_c = 0.35$ |
| 44 | $2.5\times10^{-22}$ |
| 45 | $1.8\times10^{-39}$ ($cm^6$ $molecule^{-2}$ $s^{-1}$) |

Source: From Atkinson et al. (2004).

Reaction 46 is third order, with rate constant $k_{46} = 1.4\times10^{-33}$ $cm^6$ $molecule^{-2}$ $s^{-1}$ if the third molecule is $N_2$; reaction 47 is a falloff reaction with $k_0 = 4.5\times10^{-31}[N_2]$ $cm^3$ $molecule^{-1}$ $s^{-1}$, $k_\infty = 1.3\times10^{-12}$ $cm^3$ $molecule^{-1}$ $s^{-1}$, and $F_c = 0.525$; reaction 48 has a rate constant $k_{48} = 4.3\times10^{-13}$ $cm^3$ $molecule^{-1}$ $s^{-1}$.

Considering the conditions found in the previous section ($O^{\bullet}$ concentration $2.8\times10^4$ molecules cm$^{-3}$; $HO^{\bullet}$ concentration $9.82\times10^7$ molecules cm$^{-3}$), we find that reaction 46 has a negligible impact (apparent first-order rate constant on the order of $10^{-9}$ s$^{-1}$), whereas reaction 47 ($k_{47} = 7.95\times10^{-13}$ cm$^3$ molecule$^{-1}$ s$^{-1}$) leads to an apparent first-order rate constant of $7.8\times10^{-5}$ s$^{-1}$, or **0.28 h$^{-1}$** for $SO_2$. Reaction 48 is practically instantaneous.

***11.4.5.2 Introduction to Thermodynamic Equilibria.*** To fully understand aerosol formation, we need to consider phase transfer to the atmospheric water phase and aqueous-phase chemical reactions. Both involve thermodynamic equilibria that need to be accounted for. Hence, a brief introduction on chemical thermodynamics will be given here.

Chemical thermodynamics can describe **chemical equilibria**, which include reaction equilibria, phase equilibria, and combinations of both. An example of a combination of both is the following:

$$SO_3(g) + H_2O(l) \rightarrow 2H^+(aq) + SO_4^{2-}(aq) \tag{11.194}$$

As eq. (11.194) shows, the phase is indicated with every species in case a phase change is involved. In this case a distinction is made between a **liquid** phase (indicating that the species, water, is the solvent) and an **aqueous** phase (indicating that the species is a solute). The **equilibrium condition** of a chemical reaction is given by the **mass action law**:

$$K = \frac{a^2(H^+)a\left(SO_4^{2-}\right)}{a(SO_3)a\left(H_2O\right)} \tag{11.195}$$

The structure of eq. (11.195) is linked with the structure of eq. (11.194): The reactants are found in the denominator, the reaction products in the numerator, and the power of each factor represents the stoichiometric factor.

In eq. (11.195), $a(i)$ refers to the **activity** of species $i$, which has a different meaning in each phase. If we assume that all solutes are at *low concentration*, all solvents are *nearly pure*, and all gases are at *low pressure*, then we assume that the phases are *ideal*, that is, they follow idealized simplified thermodynamic relationships. The definitions of activity in ideal phases are the following:

- For **ideal gases** the activity is the partial pressure divided by 1 bar to cancel out the units.
- For **ideal liquids** the activity is the mole fraction.
- For **ideal solutes** (aqueous solution) the activity is the molality (mol solute per kg solvent), which can be approximated by the concentration (mol per liter solution).

Hence, the equilibrium condition can be written as

$$K = \frac{\left[\mathrm{H}^+\right]^2\left[\mathrm{SO_4}^{2-}\right]}{p(\mathrm{SO_3})x(\mathrm{H_2O})} \quad (11.196)$$

where the square brackets denote concentrations, and $K$ is the **equilibrium constant**. It can be calculated from the **standard Gibbs free energy** of the reaction, $\Delta_r G^\circ$ a thermodynamic variable:

$$K = \exp\left(\frac{-\Delta_r G^\circ}{RT}\right) \quad (11.197)$$

The Gibbs free energy of reaction, in turn, depends on the standard **enthalpy** of reaction, $\Delta_r H^\circ$, and the standard **entropy** of reaction, $\Delta_r S^\circ$:

$$\Delta_r G^\circ = \Delta_r H^\circ - T\Delta_r S^\circ \quad (11.198)$$

The enthalpy is a measure of energy that expresses itself as heat in the case of constant pressure. The entropy is a measure of the degradation of the quality of the energy, which manifests itself as heat divided by temperature of the transfer of the heat. The word "standard" refers to a state that, again, depends on the phase:

- For gases, the standard state is a partial pressure of 1 bar [the old standard, 1 atm, or 1.01325 bar, was replaced in 1982 (Mills et al., 1993) but is still frequently used], assuming ideal gas.
- For liquids, the standard state is the pure liquid.
- For solutes, the standard state is a molality of 1 mol kg$^{-1}$ assuming ideal conditions (approximately 1 mol L$^{-1}$).

The standard enthalpy of reaction is calculated form the standard enthalpy of **formation**, $\Delta_f H^\circ$, a property tabulated in many handbooks and resources for a temperature of 25 °C:

$$\Delta_r H^\circ = 2\Delta_f H^\circ(\mathrm{H^+(aq)}) + \Delta_f H^\circ\left(\mathrm{SO_4^{2-}(aq)}\right) - \Delta_f H^\circ(\mathrm{SO_3(g)}) - \Delta_f H^\circ(\mathrm{H_2O}(\ell)) \quad (11.199)$$

As with the equilibrium constant, the structure of eq. (11.199) is linked with the structure of eq. (11.194): Reactants are subtracted, reaction products are added, and the coefficients are the stoichiometric factors.

For the standard entropy of reaction, the absolute entropies can be used for the calculation, again a property tabulated at 25 °C:

$$\Delta_r S^\circ = 2S^\circ(\mathrm{H^+(aq)}) + S^\circ\left(\mathrm{SO_4^{2-}(aq)}\right) - S^\circ(\mathrm{SO_3(g)}) - S^\circ(\mathrm{H_2O}(\ell)) \quad (11.200)$$

Good references for thermodynamic data are the CODATA key values for thermodynamics published by Cox et al. (1989) and available online at many websites; the *CRC Handbook of Chemistry and Physics* (Haynes, 2011), the NIST Chemistry Webbook (http://webbook.nist.gov/chemistry), and most textbooks on chemical thermodynamics.

**Example 11.2.** Calculate the equilibrium constant of the reaction

$$H_2O(\ell) \rightarrow H^+(aq) + OH^-(aq)$$

at 25 °C. Use the following thermodynamic properties from Cox et al. (1989):

$H_2O(\ell)$: $\Delta_f H^\circ = -285.83$ kJ mol$^{-1}$; $S^\circ = 69.95$ J mol$^{-1}$ K$^{-1}$
$H^+(aq)$: $\Delta_f H^\circ = 0$; $S^\circ = 0$ (by convention)
$OH^-(aq)$: $\Delta_f H^\circ = -230.015$ kJ mol$^{-1}$; $S^\circ = -10.90$ J mol$^{-1}$ K$^{-1}$

***Solution.*** The standard enthalpy of reaction is

$$\Delta_r H^\circ = 0 + (-230.015 \text{ kJ mol}^{-1}) - (-285.83 \text{ kJ mol}^{-1}) = 55.815 \text{ kJ mol}^{-1} = 55{,}815 \text{ J mol}^{-1}$$

$$\Delta_r S^\circ = 0 + (-10.90 \text{ J mol}^{-1} \text{ K}^{-1}) - 69.95 \text{ J mol}^{-1} \text{ K}^{-1} = -80.85 \text{ J mol}^{-1} \text{ K}^{-1}$$

Hence,

$$\Delta_r G^\circ = 55{,}815 \text{ J mol}^{-1} - 298.15 \text{ K} \times (-80.85 \text{ J mol}^{-1} \text{ K}^{-1}) = 79{,}920 \text{ J mol}^{-1}$$

$$K = \exp(-79{,}920 \text{ J mol}^{-1}/(8.314472 \text{ J mol}^{-1} \text{ K}^{-1} \times 298.15 \text{ K})) = \mathbf{9.97 \times 10^{-15}}$$

In the atmosphere it is usually possible to assume ideal gas conditions. In cloud water and dilute aerosols, it can be assumed that the water activity approaches 1, and nonionic solutes behave like ideal solutes. The same assumption is usually made for ionic solutes, but this is not always justified. For **nonideal solutes** the activity is given by

$$a(i) = \gamma_i m_i \approx \gamma_i [i] \tag{11.201}$$

where square brackets denote concentration. Coefficient $\gamma_i$ is the **activity coefficient**, a measure of the nonideality of the solute, and is 1 (ideal solution) by definition at the low concentration limit. Hence, we should modify eq. (11.196) as

$$K = \frac{\gamma_{H^+}^2 [H^+]^2 \gamma_{SO_4^{2-}} \left[SO_4{}^{2-}\right]}{p(SO_3)} \tag{11.202}$$

A convenient model for the estimation of activity coefficients in dilute to moderately concentrated solutions (< 1 mol kg$^{-1}$) is the specific ion theory (SIT) (e.g., DeVisscher et al., 2012). The following simplified version provides an estimate of the activity coefficient in dilute solutions ($I < 0.05$ mol kg$^{-1}$):

$$\log(\gamma_i) = -\frac{Az_i^2\sqrt{I}}{1+1.5\sqrt{I}} \tag{11.203}$$

where $A$ is the Debye–Hückel constant (0.5115 at 25 °C), $z_i$ is the charge number of the ion (1 for $H^+$, 2 for $SO_4^{2-}$), and $I$ is the **ionic strength**, defined as

$$I = 0.5\sum_{i=1}^{n} z_i^2 m_i \tag{11.204}$$

where the summation is over all ionic species.

If we accept an error tolerance of 10% resulting from assuming ideal solutions, we obtain a maximum ionic strength of $1.07\times10^{-2}$ mol kg$^{-1}$ for ions with charge number 1, and a maximum ionic strength of $5.35\times10^{-4}$ mol kg$^{-1}$ for ions with charge number 2. However, if we aim for an accuracy within 10% by using eq. (11.196) instead of eq. (11.202), the ionic strength should be no more than $2.33\times10^{-4}$ mol kg$^{-1}$. This condition is not always fulfilled. For simplicity, we will assume ideal solution here.

Equation (11.203) is valid to within a few percents for ionic strengths up to 0.05 mol kg$^{-1}$. At higher concentrations an ion-specific approach is needed. The most comprehensive and accurate approach for ionic strengths up to 6 mol kg$^{-1}$ is the Pitzer formalism (Pitzer, 1973, 1986, 1991; Pitzer and Mayorga, 1973, 1974; Pitzer and Kim, 1974). However, in highly concentrated solutions, like the liquid film surrounding sulfate particles as water condenses on the particle surface (Clegg and Brimblecombe, 1988a, 1988b), the Pitzer formalism can be inadequate. In such cases, a mole-fraction-based version of Pitzer's approach (Clegg and Pitzer, 1992, 1994; Clegg et al., 1992, 1994) provides better results.

One type of equilibrium constant is a solubility constant, which describes the transfer to a dissolved state. For instance, for $CO_2$ from the gas phase the reaction is

$$CO_2(g) \rightarrow CO_2(aq) \tag{11.205}$$

with equilibrium constant:

$$K_s = \frac{\gamma_{CO_2}[CO_2(aq)]}{p(CO_2)} \tag{11.206}$$

A related property is the **Henry constant**, a measure of the solubility of a gas at the low-pressure limit:

$$k_H = \lim_{p\to0}\frac{p(CO_2)}{x(CO_2(aq))} \tag{11.207}$$

where $x(CO_2(aq))$ is the mole fraction. Unlike equilibrium constants, which are made dimensionless by dividing each factor by its standard state, the Henry constant retains its units, usually bar. The relationship between the Henry constant and the solubility constant is

$$k_H = \frac{55.508\,\text{bar}}{K_s} \tag{11.208}$$

Equation (11.208) refers to $K_s$ based on a 1 mol kg$^{-1}$ standard state but can be used with reasonable accuracy in the concentration (mol L$^{-1}$) scale. In envionmental engineering, it is often customary to express the Henry constant as follows:

$$H_{cc} = \lim_{[CO_2(g)]\to 0} \frac{[CO_2(g)]}{[CO_2(aq)]} \tag{11.209}$$

where square brackets denote concentration in mol L$^{-1}$. The relationship with the solubility constant is

$$H_{cc} = \frac{100}{RTK_s} \tag{11.210}$$

where the factor 100 combines the conversion of pressure from bar to Pa and the conversion of the gas-phase concentration from mol m$^{-3}$ to mol L$^{-1}$.

**Example 11.3.** Convert the Henry constant of $CO_2$ ($k_H$ = 1642 bar) to a solubility constant $K_s$ and to a dimensionless Henry constant $H_{cc}$.

***Solution.*** Solving eq. (11.208) for $K_s$, we obtain $K_s$ = 55.508 bar/1642 bar = 0.0338. Applying eq. (11.210), we obtain $H_{cc}$ = 100/8.314472/298.15/0.0338 = 1.19.

There are two ways to calculate equilibria in complex systems. The first is to define all the relevant equilibrium constants, as well as material balances, and solve the set of equations with the concentrations, pressures, and the like as unknowns.

The second way to calculate equilibria in complex systems is based on the thermodynamic condition that the equilibrium is the state with the lowest Gibbs free energy. Hence, the total Gibbs free energy is calculated and minimized, with the material balances as constraints. The Gibbs free energy of an individual compound in a single phase, $G_i$, is

$$G_i = G_i^\circ + RT\ln(a(i)) \tag{11.211}$$

The total Gibbs free energy is the sum of the Gibbs free energy of all compounds in all phases.

***11.4.5.3 Pollutant Equilibria in Small Droplets*** In the previous section the basic thermodynamic relationships needed to calculate equilibria were outlined. In this section these equations will be applied to calculate pollutant partitioning between the atmospheric gas and liquid phases. We will concentrate on $CO_2$, which is a background compound that is always present, on $SO_2$ and on $SO_3$.

The Henry constant of $CO_2$ is highly temperature dependent, so it is useful to take the temperature effect into account. The following equation applies (De Visscher et al., 2012):

$$\ln\left(\frac{k_H}{p_{v,w}}\right) = \frac{A}{T_r} + \frac{B\theta^{0.355}}{T_r} + CT_r^{-0.41}\exp(\theta) + D \tag{11.212}$$

where $p_{v,w}$ is the vapor pressure of water (bar), $T_r$ is the reduced temperature ($T/T_c$), with $T_c$ the critical temperature, 647.096 K, and $\theta = 1 \times T_r$. The coefficients are:

$A = -9.14122$

$B = 2.81920$

$C = 11.28516$

$D = -0.80660$

The vapor pressure of water can be calculated with an equation proposed by Wagner and Pru$\beta$ (2002):

$$p_{v,w} = p_c \exp\left[\frac{T_c}{T}\left(a_1\theta + a_2\theta^{1.5} + a_3\theta^3 + a_4\theta^{3.5} + a_5\theta^4 + a_6\theta^{7.5}\right)\right] \tag{11.213}$$

where $p_c$ is the critical pressure of water (220.64 bar), and the coefficients are

$a_1 = -7.85951783$

$a_2 = 1.84408259$

$a_3 = -11.7866497$

$a_4 = 22.6807411$

$a_5 = -15.9618719$

$a_6 = 1.80122502$

Equation (11.212) is an extended form of an equation proposed by Fernández-Prini et al. (2003). The values of the Henry constant in its different forms at 25 °C were given in the example in the previous section. For $K_c$ a value of 0.0338 was found.

Considering a $CO_2$ concentration of 380 ppm, and a barometric pressure of 1 bar, the $CO_2$ partial pressure is $380 \times 10^{-6}$ bar. A slightly lower value will be found when one accounts for the diluting effect of water vapor. This effect will be ignored here. Substituting $p(CO_2)$ into the definition of $K_c$ and solving for $[CO_2(aq)]$ leads to $[CO_2(aq)] = 380 \times 10^{-6} \times 0.0338 = 1.28 \times 10^{-5}$, with concentration units mol L$^{-1}$.

The next step is to take the dissociation of $CO_2$ into account. The relevant reactions are

$$CO_2(aq) + H_2O(l) \rightarrow H^+(aq) + HCO_3^-(aq) \tag{11.214}$$

$$HCO_3^-(aq) \rightarrow H^+(aq) + CO_3^{2-}(aq) \tag{11.215}$$

$$H_2O(l) \rightarrow H^+(aq) + OH^-(aq) \tag{11.216}$$

The equilibrium constants of these reactions are $K_1 = 10^{-6.347}$ for eq. (11.214), $K_2 = 10^{-10.329}$ for eq. (11.215), and $K_w = 10^{-14}$ for eq. (11.216). These values are also temperature dependent. Recommendations for temperature relationships are given by De Visscher et al. (2012) for $K_1$ and $K_2$ and by Marshall and Franck (1981) and Bandura and Lvov (2006) for $K_w$.

To calculate the $CO_2$–carbonate speciation in a droplet, four unknowns need to be calculated: $[H^+(aq)]$, $[OH^-(aq)]$, $[HCO_3^-(aq)]$, and $[CO_3^{2-}(aq)]$. The equilibrium requirements [eqs. (11.214)–(11.216)] provide only three relationships, so a fourth relationship is needed. That fourth relationship is the requirement that the droplet must be electrically neutral, so all positively charged ions need to be balanced by negatively charged ions. This leads to the following four relationships:

$$[H^+(aq)] = [OH^-(aq)] + [HCO_3^-(aq)] + 2\left[CO_3^{2-}(aq)\right] \tag{11.217}$$

$$[HCO_3^-(aq)] = \frac{K_1[CO_2(aq)]}{[H^+(aq)]} \tag{11.218}$$

$$\left[CO_3^{2-}(aq)\right] = \frac{K_2[HCO_3^-(aq)]}{[H^+(aq)]} = \frac{K_1K_2[CO_2(aq)]}{[H^+(aq)]^2} \tag{11.219}$$

$$[OH^-(aq)] = \frac{K_w}{[H^+(aq)]} \tag{11.220}$$

Substituting eqs. (11.218)–(11.220) into eq. (11.217) leads to

$$[H^+(aq)] = \frac{K_w}{[H^+(aq)]} + \frac{K_1[CO_2(aq)]}{[H^+(aq)]} + 2\frac{K_1K_2[CO_2(aq)]}{[H^+(aq)]^2} \tag{11.221}$$

This equation can be written as a polynome equation:

$$[H^+(aq)]^3 - (K_w + K_1[CO_2(aq)])[H^+(aq)] - 2K_1K_2[CO_2(aq)] = 0 \tag{11.222}$$

This equation can be solved algebraically or iteratively. The solution is $[H^+(aq)] = 2.40 \times 10^{-6}$ mol L$^{-1}$. Since pH is defined as

$$\mathrm{pH} = -\log(a(\mathrm{H}^+(\mathrm{aq}))) \approx -\log([\mathrm{H}^+(\mathrm{aq})]) \tag{11.223}$$

it follows that the pH of a water solution in equilibrium with atmospheric $CO_2$ is 5.62. Applying eqs. (11.218)–(11.220) leads to

$$\left[\mathrm{HCO_3}^-(\mathrm{aq})\right] = 2.40\times10^{-6}\,\mathrm{mol\,L^{-1}}$$
$$\left[\mathrm{CO_3}^{2-}(\mathrm{aq})\right] = 4.68\times10^{-11}\,\mathrm{mol\,L^{-1}}$$
$$\left[\mathrm{OH}^-(\mathrm{aq})\right] = 4.16\times10^{-9}\,\mathrm{mol\,L^{-1}}$$

The ionic strength of this solution is $2.40\times10^{-6}$ mol L$^{-1}$. Hence, the assumption of ideal solution is justified.

Next, the impact of $SO_2$ dissolution on the composition of the atmospheric water phase will be calculated. The reactions are essentially the same as in the $CO_2$–carbonate system, with $CO_2$, $HCO_3^-$, and $CO_3^{2-}$ replaced by $SO_2$, $HSO_3^-$, and $SO_3^{2-}$. The equilibrium constants will be denoted $K_{s,S}$, $K_{1,S}$, and $K_{2,S}$, for the solubility constant, the first and second acid dissociation constants of $SO_2$–sulfurous acid, respectively. Values at 25 °C are $K_{s,S}$ = 1.386 (Wilhelm et al., 1977), $K_{1,S} = 10^{-1.886}$ and, $K_{2,S} = 10^{-7.180}$ (Seinfeld and Pandis, 2006). The dimensionless Henry constant is 0.0291.

Assume an $SO_2$ concentration of 10 ppb (partial pressure $10^{-8}$ bar). Based on the Henry constant, an equilibrium $SO_2$ concentration of $1.386\times10^{-8}$ mol L$^{-1}$ is obtained. Based on this information, a new equilibrium can be calculated. The additional equilibrium relationships imposed by the presence of the $SO_2$–sulfhite system are

$$[\mathrm{HSO_3}^-(\mathrm{aq})] = \frac{K_{1,S}[\mathrm{SO_2}(\mathrm{aq})]}{[\mathrm{H}^+(\mathrm{aq})]} \tag{11.224}$$

$$\left[\mathrm{SO_3}^{2-}(\mathrm{aq})\right] = \frac{K_{2,S}[\mathrm{HSO_3}^-(\mathrm{aq})]}{\left[\mathrm{H}^+(\mathrm{aq})\right]} = \frac{K_{1,S}K_{2,S}[\mathrm{SO_2}(\mathrm{aq})]}{[\mathrm{H}^+(\mathrm{aq})]^2} \tag{11.225}$$

The charge balance is extended with two extra terms:

$$\begin{aligned}[\mathrm{H}^+(\mathrm{aq})] = [\mathrm{OH}^-(\mathrm{aq})] + [\mathrm{HCO_3}^-(\mathrm{aq})] + 2\left[\mathrm{CO_3}^{2-}(\mathrm{aq})\right] + [\mathrm{HSO_3}^-(\mathrm{aq})] \\ + 2\left[\mathrm{SO_3}^{2-}(\mathrm{aq})\right]\end{aligned} \tag{11.226}$$

Substituting the equilibrium relations and writing as a polynome equation lead to

$$\begin{aligned}&[\mathrm{H}^+(\mathrm{aq})]^3 - (K_w + K_1[\mathrm{CO_2}(\mathrm{aq})] + K_{1,S}[\mathrm{SO_2}(\mathrm{aq})])[\mathrm{H}^+(\mathrm{aq})] \\ &-(2K_1K_2[\mathrm{CO_2}(\mathrm{aq})] + 2K_{1,S}K_{2,S}[\mathrm{SO_2}(\mathrm{aq})]) = 0\end{aligned} \tag{11.227}$$

By trial and error, it can be established that the equilibrium $H^+$ concentration is slightly above $10^{-5}$ mol L$^{-1}$. By an iterative solution, the solution is found to be $1.37\times10^{-5}$ mol L$^{-1}$ (pH 4.86). The ion concentrations are

$$\left[\mathrm{HCO_3^-(aq)}\right] = 4.20\times10^{-7}\,\mathrm{mol\,L^{-1}}$$

$$\left[\mathrm{CO_3^{2-}(aq)}\right] = 1.44\times10^{-12}\,\mathrm{mol\,L^{-1}}$$

$$\left[\mathrm{HSO_3^-(aq)}\right] = 1.32\times10^{-5}\,\mathrm{mol\,L^{-1}}$$

$$\left[\mathrm{SO_3^{2-}(aq)}\right] = 6.34\times10^{-8}\,\mathrm{mol\,L^{-1}}$$

$$\left[\mathrm{OH^-(aq)}\right] = 7.30\times10^{-10}\,\mathrm{mol\,L^{-1}}$$

The total sulfur content is $1.32\times10^{-5}$ mol L$^{-1}$. Substituing this value back into the definition of the solubility constant or the Henry constant provides the **apparent solubility constant** or the **apparent Henry constant**, an adjusted solubility or Henry constant that accounts for speciation of the compound in the water phase. The value obtained for $SO_2$ is $K_s = 1320$, or $H_{cc} = 3.05\times10^{-5}$. Due to its acid nature, $SO_2$ will reduce the concentration of bicarbonate and carbonate in the droplet. On the other hand, the $SO_2$ sulfate system will be relatively unaffected by a $CO_2$–carbonate system.

A similar calculation can be made for $SO_3$. Due to the strongly acid nature of $H_2SO_4$, it is clear that the apparent solubility constant of $SO_3$ will be extremely large. This will have a tendency to decrease the pH, and hence the apparent solubility of $SO_2$ in atmospheric water droplets. A similar effect applies to $HNO_3$, which is formed from $NO_x$ chemistry in the atmosphere. Ammonia ($NH_3$), on the other hand, will tend to increase the pH of atmospheric water due to its alkaline nature. Both $NH_3$ and $HNO_3$ tend to occur predominantly to exclusively in the atmospheric water phase (Seinfeld and Pandis, 2006).

***11.4.5.4 $SO_2$ Oxidation in the Atmospheric Water Phase*** In this section we will consider the situation of an atmosphere as discussed earlier, with 10 ppb $SO_2$, 60 ppb ozone, and $9.82\times10^7$ molecules cm$^{-3}$ hydroxyl radicals. In Section 11.4.5.1 it was established that the apparent $SO_2$ oxidation rate constant in the gas phase is 0.28 h$^{-1}$. In Section 11.4.5.2 the speciation of $SO_2$ in water droplets under these conditions was established.

It will be assumed that the $H_2O_2$ concentration is 1 ppb. We will calculate $SO_2$ oxidation in cloud droplets by ozone and by hydrogen peroxide and compare the resulting rates with the gas-phase rates. As an extreme case, we will consider a water content typical of clouds, 1 g m$^{-3}$, or a volume fraction $10^{-6}$.

Ozone has a solubility constant $K_s$ of 0.011. Hence, at a partial pressure of $6\times10^{-8}$ bar (60 ppb), the equilibrium ozone concentration in a droplet is $6.6\times10^{-10}$ mol L$^{-1}$. The oxidation kinetics of $S^{IV}$ (as in $SO_2$) to $S^{VI}$ (as in $SO_3$) is as follows (Hoffmann and Calvert, 1985):

$$r_{S^{IV}} = \left(k_0\left[\mathrm{SO_2(aq)}\right] + k_1\left[\mathrm{HSO_3^-(aq)}\right] + k_2\left[\mathrm{SO_3^{2-}(aq)}\right]\right)[\mathrm{O_3(aq)}] \tag{11.228}$$

where $k_0 = 2.4\times10^4$ L mol$^{-1}$ s$^{-1}$, $k_1 = 3.7\times10^5$ L mol$^{-1}$ s$^{-1}$, and $k_2 = 1.5\times10^9$ L mol$^{-1}$ s$^{-1}$. Entering the concentrations in eq. (11.228) leads to a reaction rate of $6.6\times10^{-8}$ mol L$^{-1}$ s$^{-1}$. With a water volume fraction of $10^{-6}$, the apparent $SO_2$ oxidation rate in the atmosphere as a whole resulting from the ozone-mediated oxidation in the water phase is $10^{-6}\times6.6\times10^{-8}$ mol L$^{-1}$ s$^{-1}$ = $6.6\times10^{-14}$ mol L$^{-1}$ s$^{-1}$. To translate this into an apparent first-order rate constant, the $SO_2$ concentration in the gas phase in mol L$^{-1}$ is calculated. Based on the ideal gas law, a value of $4.03\times10^{-7}$ mol m$^{-3}$ = $4.03\times10^{-10}$ mol L$^{-1}$ is obtained. Substitution into the first-order rate law, and solving for the rate constant, leads to

$$k' = \frac{r_{SO_2}}{c_{SO_2}} = \frac{6.6\times10^{-14}\,\mathrm{mol\,L^{-1}\,s^{-1}}}{4.03\times10^{-10}\,\mathrm{mol\,L^{-1}}} = 1.64\times10^{-4}\,\mathrm{s^{-1}} = 0.589\,\mathrm{h^{-1}} \quad (11.229)$$

A rate constant of 0.589 h$^{-1}$ is obtained. It follows that, at least in clouds with a water content of 1 g m$^{-3}$, oxidation by ozone in the water phase is a more important $SO_2$ oxidation mechanism than gas-phase oxidation by hydroxyl radicals. However, as $H_2SO_4$ accumulates in the droplets, the speciation of the $SO_2$–sulfite system moves towards $SO_2$, which reduces the apparent solubility and oxidation rate of $SO_2$. Furthermore, it remains to be determined if the mass transfer rates from the atmosphere to the droplet phase are sufficiently fast to maintain these reaction rates.

The oxidation by $H_2O_2$ has a very different dynamic than the oxidation by $O_3$, as $H_2O_2$ is less reactive but more soluble than $O_3$. With a solubility constant $K_s = 10^5$ mol L$^{-1}$ s$^{-1}$ and a partial pressure of $10^{-9}$ bar (1 ppb), the liquid-phase $H_2O_2$ concentration is $10^{-4}$ mol L$^{-1}$.

The reaction kinetics of the oxidation of $S^{IV}$ to $S^{VI}$ by $H_2O_2$ is as follows (Hoffmann and Calvert, 1985):

$$r_{S^{IV}} = \frac{k\left[\mathrm{H^+(aq)}\right]\left[\mathrm{H_2O_2(aq)}\right]\left[\mathrm{HSO_3^-(aq)}\right]}{1+\mathrm{K}\left[\mathrm{H^+(aq)}\right]} \quad (11.230)$$

where $k = 7.5\times10^7$ L$^2$ mol$^{-2}$ s$^{-1}$, and $K = 13$ L mol$^{-1}$. At the pH considered here (4.86), the denominator of eq. (11.230) is approximately 1. The reaction rate is $1.36\times10^{-6}$ mol L$^1$ s$^1$. At a water volume fraction of $10^6$ this leads to an apparent reaction rate in the atmosphere of $1.36\times10^{-12}$ mol L$^{-1}$ s$^{-1}$. Substituting into the first-order rate law leads to an apparent rate constant of $3.36\times10^{-3}$ s$^{-1}$ = **12.1 h$^{-1}$**. It follows that, unless the water fraction in the atmosphere is much smaller than 1 g m$^{-1}$, the oxidation by $H_2O_2$ in the droplet phase dominates $SO_2$ oxidation to $SO_3$. Again, this assumes the absence of significant mass transfer limitations in and around droplets.

There are several other reactions that can oxidize $SO_2$ into $SO_3$ in the droplet phase. The main one is catalyzed oxidation with oxygen. These reactions will not be discussed here. The reader is referred to Seinfeld and Pandis (2006). In the remainder of this section, we will investigate if mass transfer limitations will slow down the reactions described above.

In Chapter 7 is was determined that the 90% equilibration time of mass transfer *inside* droplets is given by the equation

$$t_{90\%} = \frac{0.185 d_d^2}{4 D_w} \quad (11.231)$$

where $d_d$ is the droplet diameter and $D_w$ is the diffusivity of the solute in water. Assuming a diffusivity of $10^{-9}$ m$^{-2}$ s$^{-1}$, the 90% equilibration time ranges from 1.2 ms in 5-μm droplets to 74 ms in 40-μm droplets. This is a typical size range in clouds (Seinfeld and Pandis, 2006). These values will be compared with reaction time constants.

What is needed for this comparison is the time constant of the **limiting substance**, that is, the substance that will show mass transfer limitations first. When all diffusivities are the same, and the stoichiometric coefficients are all equal to one, the limiting substance is simply the substance present at the lowest concentration.

For the oxidation by ozone, the limiting substance is ozone. Dividing the reaction rate by the ozone concentration leads to a rate constant of $6.6\times10^{-8}$ mol L$^{-1}$ s$^{-1}$/ $6.6\times 10^{-10}$ mol L$^{-1}$ = 100 s$^{-1}$, or a time constant of 1/100 s$^{-1}$ = 10 ms. The reaction is kinetically controlled in small cloud droplets and mass transfer limited in large droplets.

For the oxidation by $H_2O_2$, the limiting substance is sulfur. As acid–base reactions are usually very fast, we can assume that the three sulfur species will contribute to mass transfer, whether they are effectively oxidized or not. Hence, the relevant rate constant is $1.36\times10^{-6}$ mol L$^{-1}$ s$^{-1}$/$1.32\times10^{-5}$ mol L$^{-1}$ = 0.103 s$^{-1}$, and the time constant is 1/0.103 s$^{-1}$ = 9.71 s, much longer than the equilibration time for mass transfer. Hence, there is no mass transfer limitation in the droplets for $SO_2$ oxidation by $H_2O_2$.

To investigate the impact of mass transfer from the bulk air to the droplets (mass transfer limitation *outside* the droplets), we can use some other equations used to describe wet deposition (Chapter 7). Equation (7.233) describes the mass (or molar) flux $F_d$ toward a droplet in terms of a mass transfer coefficient $K_c$ in the limit of zero concentration inside the droplet:

$$-F_d = K_c c_{air} \tag{7.233}$$

where $c_{air}$ is the pollutant concentration in the bulk air phase. When the pollutant concentration in the droplet, and hence on the air–droplet interface is not zero, eq. (7.233) must be extended as

$$-F_d = K_c (c_{air} - c_{i,air}) \tag{11.232}$$

where $c_{i,air}$ is the concentration in the air phase at the air–droplet interface. See also Example 7.12 for related calculations, and $K_c$ can be calculated with eq. (11.236):

$$K_c = \frac{D}{d_d}(2 + 0.6\,\mathrm{Re}^{1/2}\cdot \mathrm{Sc}^{1/3}) \tag{7.236}$$

where $D$ is the diffusivity of the pollutant in the air phase, Re is the Reynolds number (Re = $d_d\, v_s/v$) and Sc is the Schmidt number (Sc = $v/D$); $v$ is the kinematic viscosity of the air, and $v_s$ is the settling velocity of the droplet. For the latter, we can apply eq. (7.132) to droplets smaller than about 100 μm:

$$v_s = \frac{\rho_w d_d g}{18\mu} \tag{11.233}$$

where $\rho_w$ is the density of water, and $\mu$ is the dynamic viscosity of air. For the calculations, a kinematic viscosity of $1.5\times10^{-5}$ m$^2$ s$^{-1}$ is assumed, a dynamic viscosity of $1.8\times10^{-5}$ Pa s, a Sc number of 1 (hence $D = 1.5\times10^{-5}$ m$^2$ s$^{-1}$), and $\rho_w = 1000$ kg m$^{-3}$. For droplets with diameters 5 and 40 μm, a settling velocity of $7.57\times10^{-4}$ m s$^{-1}$ and $4.84\times10^{-2}$ m s$^{-1}$ is obtained, respectively. This leads to $K_c$ values of 6.03 and 0.807 m s$^{-1}$ for 5-μm and 40-μm droplets, respectively.

To determine if the external mass transfer limits the oxidation process, we will derive an equilibration time constant associated with mass transfer from the bulk air to the droplet, and compare it with the time constant of reaction, for the limiting compound. The compound selected is $H_2O_2$ because it has the lowest concentration in the atmosphere, and it is associated with the fastest oxidation of $S^{IV}$ to $S^{VI}$. The molar flow of compound into the droplet is the flux multiplied by the droplet area:

$$-\pi d_d^2 F_d = K_c \pi d_d^2 (c_{air} - c_{i,air}) \tag{11.234}$$

The amount of compound accumulating in the droplet is the droplet volume multiplied by the concentration increase rate:

$$-\pi d_d^2 F_d = \frac{1}{6}\pi d_d^3 \frac{dc_w}{dt} \tag{11.235}$$

Equation (11.234) is set equal to eq. (11.235):

$$\frac{1}{6}\pi d_d^3 \frac{dc_w}{dt} = K_c \pi d_d^2 (c_{air} - c_{i,air}) \tag{11.236}$$

For the sake of the current calculation it is assumed that the water-phase concentration is linked to the interface concentration on the air side by means of Henry's law (i.e., it is assumed that mass transfer in the water phase is not limiting). Equation (11.209) is most suited for that purpose. For the current problem, this translates into

$$H_{cc} = \frac{c_{i,air}}{c_w} \tag{11.237}$$

Using eq. (11.237) to eliminate $c_w$ from eq. (11.236), and rearranging, leads to

$$\frac{dc_{i,air}}{dt} = \frac{6H_{cc}K_c}{d_d}(c_{air} - c_{i,air}) \tag{11.238}$$

Solving this differential equation leads to

$$c_{i,air} = c_{air}\left[1 - \exp\left(-\frac{6H_{cc}K_c}{d_d}t\right)\right] \tag{11.239}$$

The mass transfer is a first-order process with a rate constant $k' = 6H_{cc}K_c/d_d$, and a time constant of $d_d/(6H_{cc}K_c)$. The values of the time constant obtained for $H_2O_2$ transfer into 5-μm droplets and 40-μm droplets is 0.343 and 19.9 s, respectively.

The time constant of disappearance of $H_2O_2$ can be calculated based on the apparent first-order rate constant of S(IV) oxidation in the $H_2O_2$ concentration, that is, the reaction rate divided by the $H_2O_2$ concentration. The result is $1.36\times10^{-6}$ mol $L^{-1}$ $s^{-1}/10^{-4}$ mol $L^{-4} = 0.0136$ $s^{-1}$. Hence, the time constant of $H_2O_2$ reaction is 1/0.0136 $s^{-1}$ = 73.7 s, greater than the time constant of mass transfer in all droplets. The mass transfer from the bulk air to the droplet surface is not limiting.

As a general conclusion we can state that the gas-phase oxidation of $SO_2$ to $SO_3$ by hydroxyl radicals is the main mechanism in the absence of fog or clouds. However, in the presence of clouds, the main mechanism is oxidation in the water droplets by $H_2O_2$. However, as the $H_2O_2$ concentration in the atmosphere is generally lower than the $SO_2$ concentration, this only applies until the atmospheric $H_2O_2$ is depleted. After that, oxidation by ozone in the droplets is dominant when the water content is high (> 0.5 g $m^{-3}$). In thin clouds or fog, oxidation by ozone occurs but the dominant mechanism is oxidation by hydroxyl radicals. These conclusions are in general agreement with the more detailed model of Seigneur and Saxena (1988). Detailed criteria on when droplet size effects are important in aqueous-phase atmospheric chemistry are given by Fahey and Pandis (2001).

In this section it was assumed that the equilibration between the gas side and the liquid side at the interface is instantaneous. This is not always the case because not every molecule hitting an interface moves to the other phase. This aspect of mass transfer will not be discussed here.

***11.4.5.5 Aerosol-Forming Organic Reactions.*** A large fraction of the particulate matter in the atmosphere is organic in orgin. The main class in organic aerosols is soot. As soot is mainly formed *before* emission, it will not be discussed here.

Atmospheric reactions can convert gas-phase pollutants into aerosols by *increasing their solubility*, and by *decreasing their vapor pressure*. The first mechanism will cause pollutants to transfer into water droplets, whereas the second mechanism will cause the pollutants to form a separate solid or liquid phase, or adsorb on a solid surface.

As an example of the first mechanism, consider the atmospheric chemistry of propane discussed in Section 11.4.4. Some of the reaction products of atmospheric propane oxidation are isopropanol and acetone, both of which are more soluble than propane.

Consider 1 $m^3$ of atmosphere, a volume fraction $f_w$ of which is liquid water, and the remaining $(1-f_w)$ gas. The concentration of the pollutant in air and water phase is $c_{air}$ and $c_w$, respectively. The total number of moles of pollutant in 1 $m^3$ of atmosphere, $n$, is

$$n = n_{air} + n_w = c_{air}(1-f_w) + c_w f_w \tag{11.240}$$

It is assumed that Henry's law applies to the pollutant partitioning between the gas phase and the droplet phase. The dimensionless Henry constant $H_{cc}$ is the most convenient form here. Using $H_{cc}$ to eliminate $c_{air}$ leads to

$$n = n_{air} + n_w = H_{cc} c_w (1-f_w) + c_w f_w \tag{11.241}$$

Hence, the fraction of polltutant that is found in the droplet phase, $f$, is

$$f = \frac{n_{\mathrm{w}}}{n_{\mathrm{air}} + n_{\mathrm{w}}} = \frac{f_{\mathrm{w}}}{H_{cc}(1 - f_{\mathrm{w}}) + f_{\mathrm{w}}} \tag{11.242}$$

Using the Henry constants proposed by Sander (1999) at 25 °C, and converting to $H_{cc}$ values, leads to $H_{cc} = 27.2$ for propane, $1.3 \times 10^{-3}$ for acetone, and $3.1 \times 10^{-4}$ for isopropanol. Substitution into eq. (11.242) leads to $f = 3.7 \times 10^{-8}$ for propane, $7.4 \times 10^{-4}$ for acetone, and $3.2 \times 10^{-3}$ for isopropanol. Further oxidation steps increase the solubility further. Eventually, the majority of the reaction products end up in the droplet phase.

A decrease of the vapor pressure is typical when alkenes are present in the atmosphere. Organic radicals can add to the double bond of the alkane, producing a larger radical. For instance, in the presence of ethylene, the following reaction is possible:

$$i - \mathrm{C_3H_7O^{\cdot}} + \mathrm{C_2H_4} \rightarrow \mathrm{C_3H_7OCH_2C^{\cdot}H_2} \tag{11.243}$$

With each addition, the vapor pressure of the resulting molecules becomes smaller, until condensation occurs. It is unlikely that a single compound accumulates to sufficient concentrations to form a pure organic liquid. It is more likely that a group of compounds form an organic liquid mixture. The criterion for condensation is the following:

$$\sum_{i=1}^{n} \frac{p_i}{p_{\mathrm{v},i}} > 1 \tag{11.244}$$

where $p_i$ is the partial pressure of compound $i$, and $p_{\mathrm{v},i}$ is the vapor pressure of compound $i$. This equation is based on Raoult's law and assumes that the organic molecules form an ideal liquid mixture.

Organic pollutants, individually or collectively, do not need to fulfill eq. (11.244) to enter the aerosol phase. This is because organic pollutants can *adsorb* on particle surfaces. Adsorption is an interaction between a gas or liquid compound and a solid surface, causing molecules of the compound to accumulate at the surface. Adsorption depends mainly on the area of the solid surface, and on the ratio $p_i/p_{\mathrm{v},i}$. Hence, atmospheric reactions that decrease the vapor pressure will increase the amount of adsorption taking place.

## 11.5 NUMERICAL ASPECTS OF EULERIAN DISPERSION MODELING

An Eulerian dispersion model is a very complex set of ordinary and partial differential equations, which has to be solved numerically. Each partial differential equation can be converted into a set of ordinary differential equations, so that ultimately what needs to be solved is a very large set of ordinary differential equations. Many solution techniques have been proposed to solve sets of ordinary differential equations. Some of the more common ones are described in detail by Press et al. (2007). None of the conventional solvers are adequate for the numerical problem of an Eulerian dispersion model. This is because each process—advection, dispersion, and chemical reaction—have different numerical properties and require different types of solvers to provide a solution of acceptable accuracy with an acceptable amount of computational power. Hence, each process is solved separately, with its own numerical solver. This is known as **operator splitting**. In the next sections, some general principles of the solution techniques for each process are outlined.

### 11.5.1 Advection

Advection, the movement of a pollutant with the mean wind speed, is difficult to model accurately because of two reasons: **numerical instability** and **numerical dispersion**. Numerical instability is a tendency of numerical errors to accumulate and grow and can often be diagnosed as negative pollutant concentrations. Numerical dispersion is a tendency of pollutant plumes to become wider, even when no dispersion term is included in the material balance. Both can be illustrated with the following numerical scheme. Consider a one-dimensional material balance with only an advection term:

$$\frac{\partial q}{\partial t} = -u\frac{\partial q}{\partial x} \tag{11.245}$$

where $q$ is the pollutant mass fraction. Replace the time derivative by an explicit finite difference (Euler's method) and the space derivative by a finite difference:

$$\frac{\partial q_n}{\partial t} = \frac{q_n^{t+\Delta t} - q_n^t}{\Delta t} \tag{11.246}$$

$$\frac{\partial q_n}{\partial x} = \frac{q_{n+1}^t - q_{n-1}^t}{2\Delta x} \tag{11.247}$$

where the subscript is the node point number, the superscript is the time, $\Delta x$ is the node spacing, and $\Delta t$ is the time increment. The material balance becomes

$$\frac{q_n^{t+\Delta t} - q_n^t}{\Delta t} = -u\frac{q_{n+1}^t - q_{n-1}^t}{2\Delta x} \tag{11.248}$$

which leads to

$$q_n^{t+\Delta t} = q_n^t - u\frac{q_{n+1}^t - q_{n-1}^t}{2\Delta x}\Delta t \tag{11.249}$$

In eq. (11.249), $u\Delta t/\Delta x$ is a dimensionless number, the **Courant number**. Table 11.3 shows the first few time steps for a calculation of eq. (11.249) on a grid with eight internal node points, a Courant number of 0.5 ($u = 1$; $\Delta x = 1$; $\Delta t = 0.5$), and $q = 0$ at the external node points (boundary condition). The lack of stability of this scheme is clear from the negative numbers in the result, which is also illustrated in Figure 11.1. Another sign of instability is the occurrence of rippling: the production of secondary

**TABLE 11.3 Advection of Plume Modeled with Simple Explicit Finite Difference Scheme [eq. (11.249)][a].**

| $t$ | $q_1$ | $q_2$ | $q_3$ | $q_4$ | $q_5$ | $q_6$ | $q_7$ | $q_8$ |
|---|---|---|---|---|---|---|---|---|
| 0 | 0.000 | 0.000 | 1.000 | 1.000 | 0.000 | 0.000 | 0.000 | 0.000 |
| 0.5 | 0.000 | −0.250 | 0.750 | 1.250 | 0.250 | 0.000 | 0.000 | 0.000 |
| 1 | 0.063 | −0.438 | 0.375 | 1.375 | 0.563 | 0.063 | 0.000 | 0.000 |
| 1.5 | 0.172 | −0.516 | −0.078 | 1.328 | 0.891 | 0.203 | 0.016 | 0.000 |
| 2 | 0.301 | −0.453 | −0.539 | 1.086 | 1.172 | 0.422 | 0.066 | 0.004 |

[a] $u = 1$; $\Delta x = 1$; $\Delta t = 0.5$. Courant number = 0.5. $q_0 = q_9 = 0$.

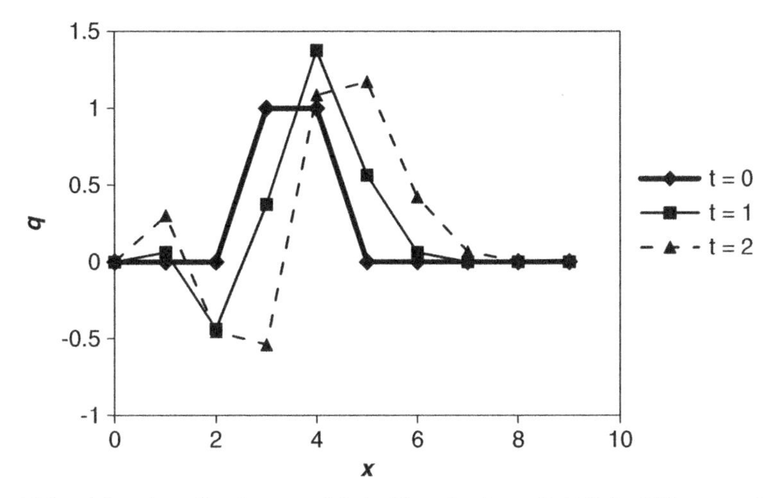


Figure 11.1 Advection of a plume modeled with a simple explicit finite difference scheme. $u = 1$; $\Delta x = 1$; $\Delta t = 0.5$. Courant number = 0.5. $q_0 = q_9 = 0$.

plumes. Further steps reveal that the secondary peak in node point 1 grows larger, spreads to node point 2 at $t = 3$, and detaches from the boundary at $t = 5$. Numerical schemes designed to increase the stability will tend to take away pollutant where it is too high, and add pollutant where it is too low. However, without a memory of previous concentrations, an algorithm has no way of knowing what is too high or too low, and will reduce gradients indiscriminately. In other words: To increase stability, numerical dispersion is introduced in the calculation.

A typical way to achieve the increased numerical stability at the expense of numerical dispersion is by using the knowledge that pollutant entering grid cell $n$ in the case of a positive wind speed $u$ is the pollution located on the right side of cell $n-1$, and the pollution located on the right side of cell $n$ leaves that cell. The fraction of the cell that moves to the next cell is the Courant number $\beta$. For instance, if it is assumed that all cells have a homogeneous concentration, then the above reasoning leads to the following material balance for cell $n$:

$$q_n^{t+\Delta t} = q_n^t - u\frac{q_n^t - q_{n-1}^t}{\Delta x}\Delta t \tag{11.250}$$

**TABLE 11.4 Advection of Plume Modeled with eq. (11.250)[a]**

| $t$ | $q_1$ | $q_2$ | $q_3$ | $q_4$ | $q_5$ | $q_6$ | $q_7$ | $q_8$ |
|---|---|---|---|---|---|---|---|---|
| 0 | 0.000 | 0.000 | 1.000 | 1.000 | 0.000 | 0.000 | 0.000 | 0.000 |
| 0.5 | 0.000 | 0.000 | 0.500 | 1.000 | 0.500 | 0.000 | 0.000 | 0.000 |
| 1 | 0.000 | 0.000 | 0.250 | 0.750 | 0.750 | 0.250 | 0.000 | 0.000 |
| 1.5 | 0.000 | 0.000 | 0.125 | 0.500 | 0.750 | 0.500 | 0.125 | 0.000 |
| 2 | 0.000 | 0.000 | 0.063 | 0.313 | 0.625 | 0.625 | 0.313 | 0.063 |

[a] $u = 1$; $\Delta x = 1$; $\Delta t = 0.5$. Courant number = 0.5. $q_0 = q_9 = 0$.

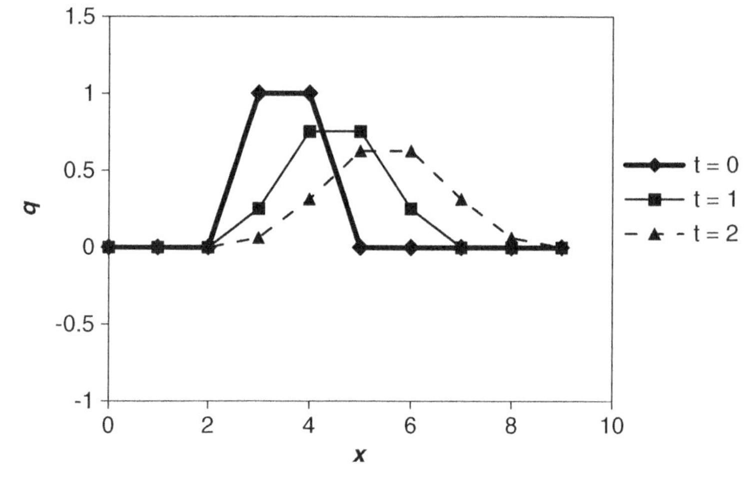


Figure 11.2 Advection of a plume modeled with eq. (11.241). $u = 1$; $\Delta x = 1$; $\Delta t = 0.5$. Courant number = 0.5. $q_0 = q_9 = 0$.

The first steps of this procedure are shown in Table11.4 and Figure11.2. It is clear that the ghost peaks and oscillations have disappeared, but the plume spreads even though no such plume spread was built into the material balance. While eq. (11.249) is preferable to eq. (11.250), it is by no means a satisfactory approach.

Several numerical methods based on similar principles have been proposed to achieve an accurate and stable calculation of advection. Byun and Ching (1999) describe the following methods, which have been incorporated in CMAQ: the piecewise parabolic method of Colella and Woodward (1984), the method of Bott (1989), and the cubic scheme of Yamartino (1993).

In the piecewise parabolic method of Colella and Woodward (1984), it is assumed that the concentration in each grid cell is parabolic. Furthermore, it is assumed that the concentration $q_n$ is the average concentration in the cell, and that the concentrations at the edges of the cells, $q_{L,n}$ and $q_{R,n}$ are given by

$$q_{L,n} = \frac{7}{12}(q_{n-1} + q_n) - \frac{1}{12}(q_{n+1} - q_{n-2}) \tag{11.251}$$

$$q_{R,n} = \frac{7}{12}(q_n + q_{n+1}) - \frac{1}{12}(q_{n+2} - q_{n-1}) \quad (11.252)$$

When $q_n$ is a minimum or a maximum, the concentration profile is assumed to be homogeneous. When the edge values are such that the parabolic concentration profile is not monotonic, one of the edge values is chosen so that the concentration gradient at the other edge is zero.

In the Bott (1989) method, the concentration profile within each cell is a second- or fourth-order polynome constructed such that it describes the correct average concentration of the cell itself and the surrounding two (second-order) or four (fourth-order) cells. Further precautions are taken to ensure that the amount of pollutant leaving a cell on the downwind side is positive and less than the pollutant contained in the cell.

The Yamartino (1993) method is based on a cubic spline, which is a smooth piecewise cubic function with a smooth derivative. The determination of the derivatives at the node points is based on the idea that the concentration difference $q_{n+1} - q_{n-1}$ not only depends on the derivative at node point $n$ but also on the derivatives at node points $n-1$ and $n+1$. Hence, a set of linear equations, to be solved simulataneously, is constructed to determine the derivatives. The cubic spline of each cell is then constructed based on the concentration and derivative of the cell and of the adjacent cells. Yamartino (1993) provided various corrections to ensure that the scheme remains stable, including a small degree of dispersion. This scheme is popular thanks to its reliability, but its main disadvantage, as with all schemes outlined above, is its reliance on ad hoc corrections to improve the performance of the core computation, which is of relatively low accuracy.

A powerful technique that has been proposed to calculate advection in Eulerian models is the Accurate Space Derivative (ASD) scheme of Gazdag (1973). The ASD approach follows a strategy unlike the methods described previously: to achieve the highest possible accuracy with the core calculation, without ad hoc corrections. To this effect, the space derivatives are determined by means of a Fourier transform of the entire domain. As the entire domain contributes to the calculation of the derivative, the accuracy of the derivatives is far superior to any other technique. Furthermore, the accuracy of the time derivative is improved by writing its truncation error in terms of spatial derivatives using the material balance, a technique originally proposed by Lantz (1971) that will be discussed in more detail below. The ASD method is the most accurate calculation method available to date, and was used by Byun and Ching (1999) as a benchmark against which to test other methods. The disadvantages of Gazdag's (1973) method, however, are the need for a Fourier transform, which means high computational needs and the need for complex numbers, and the tendency of instabilities to spread rapidly across the entire domain, rendering the entire calculation useless. Avoiding instabilities is thus much more critical in the ASD method than in any other scheme.

Because the Gazdag (1973) method is not a popular one, the technique of Lantz (1971) to write truncation errors in the time derivative as space derivatives has not received the attention it deserves. It was originally proposed for reservoir simulations in petroleum engineering and has been applied in the simulation of solute transport in groundwater (Notordamojo et al., 1991; Moldrup et al., 1994) and in the simulation of gas transport in soils (Moldrup et al., 1996). The principle behind it is mostly known under the term "numerical dispersion".

A concentration change over time can be written exactly as a Taylor series:

$$q^{t+\Delta t} = q^t + \frac{\partial q}{\partial t}\Delta t + \frac{\partial^2 q}{\partial t^2}\frac{(\Delta t)^2}{2!} + \frac{\partial^3 q}{\partial t^3}\frac{(\Delta t)^3}{3!} + \cdots \tag{11.253}$$

When a finite number of terms in the right-hand side of eq. (11.253) is kept, discarding the higher-order terms, the equation is *truncated*. The combined value of the discarded terms is the **truncation error**. The first discarded term is a good measure of the truncation error. When only the first two terms are kept (as is usually the case in advection modeling), then the truncation error is proportional to the square of the time increment. The truncated equation has first-order accuracy because the highest-order term that is kept is the first-order term in the Taylor series.

To increase the accuracy of the numerical integration, we can keep three terms instead of two. Hence, eq. (11.253) becomes

$$q^{t+\Delta t} \approx q^t + \frac{\partial q}{\partial t}\Delta t + \frac{\partial^2 q}{\partial t^2}\frac{(\Delta t)^2}{2!} \tag{11.254}$$

To resolve the second-order term, the material balance [eq. (11.245)] can be used:

$$\frac{\partial^2 q}{\partial t^2} = \frac{\partial}{\partial t}\frac{\partial q}{\partial t} = \frac{\partial}{\partial t}\left(-u\frac{\partial q}{\partial x}\right) \tag{11.255}$$

Expanding the time derivative:

$$\frac{\partial^2 q}{\partial t^2} = -\frac{\partial u}{\partial t}\frac{\partial q}{\partial x} - u\frac{\partial}{\partial t}\left(\frac{\partial q}{\partial x}\right) \tag{11.256}$$

Switching the derivatives in the second term in the right-hand side leads to

$$\frac{\partial^2 q}{\partial t^2} = -\frac{\partial u}{\partial t}\frac{\partial q}{\partial x} - u\frac{\partial}{\partial x}\left(\frac{\partial q}{\partial t}\right) \tag{11.257}$$

The material balance [eq. (11.245)] can be used once again to resolve the time derivative on the righthand side:

$$\frac{\partial^2 q}{\partial t^2} = -\frac{\partial u}{\partial t}\frac{\partial q}{\partial x} + u\frac{\partial}{\partial x}\left(u\frac{\partial q}{\partial x}\right) = -\frac{\partial u}{\partial t}\frac{\partial q}{\partial x} + \frac{\partial u}{\partial x}\frac{\partial q}{\partial x} + u^2\frac{\partial^2 q}{\partial x^2} \tag{11.258}$$

The first two terms are associated with changes of the wind speed. The third term, however, is mathematically the same as a dispersion term. Not accounting for this term [i.e., using only two terms in eq. (11.253)] leads to a spurious negative diffusivity that destabilizes the calculation by sharpening slopes and creating ripples.

Hence, a more accurate calculation scheme is obtained by substituting eq. (11.258) into eq. (11.254):

$$q^{t+\Delta t} \approx q^t + \frac{\partial q}{\partial t}\Delta t + \left( -\frac{\partial u}{\partial t}\frac{\partial q}{\partial x} + u\frac{\partial u}{\partial x}\frac{\partial q}{\partial x} + u^2\frac{\partial^2 q}{\partial x^2} \right)\frac{(\Delta t)^2}{2!} \tag{11.259}$$

Substituting the material balance in the second term of eq. (11.259) leads to an improved version of eq. (11.249), an equation with **numerical dispersion correction**:

$$q^{t+\Delta t} \approx q^t - u\frac{\partial q}{\partial x}\Delta t + \left( -\frac{\partial u}{\partial t}\frac{\partial q}{\partial x} + u\frac{\partial u}{\partial x}\frac{\partial q}{\partial x} + u^2\frac{\partial^2 q}{\partial x^2} \right)\frac{(\Delta t)^2}{2!} \tag{11.260}$$

In eq. (11.260) the simplest way to calculate the partial derivatives is as follows:

$$\frac{\partial q_n}{\partial x} = \frac{q_{n+1} - q_{n-1}}{\Delta x} \tag{11.261}$$

$$\frac{\partial^2 q_n}{\partial x^2} = \frac{q_{n+1} - 2q_n + q_{n-1}}{(\Delta x)^2} \tag{11.262}$$

The results of this calculation scheme are shown in Table 11.5 and in Figure 11.3 for the same conditions as the previous calculation schemes. Clearly a major part of the instability has been removed without introducing numerical dispersion. The main reason why there is still instability in the calculation is because of the low accuracy of eqs. (11.261) and (11.262). The instability can be removed by replacing these equations by more accurate ones. A procedure for generating finite difference schemes with higher-order accuracy is given by Fornberg (1988). Additional improvements of the accuracy can be obtained by adding higher–order terms to eq. (11.254). Following this strategy offers the best prospects for faster and more accurate numerical schemes in future Eulerian dispersion models.

### 11.5.2 Diffusion/Dispersion

Molecular diffusion and turbulent diffusion or dispersion are mathematically equivalent when a gradient theory is used and will be discussed collectively as "dispersion" here. Modeling dispersion in the horizontal directions ($x$ and $y$) is very different from modeling dispersion in the $z$ direction. Due to the closer spacing of the grid in the $z$ direction, numerical dispersion is less of a concern in this direction. On the other hand, nonlocal dispersion effects are more pronounced in the $z$ direction than in the $x$ and $y$ directions. For these reasons, horizontal and vertical dispersion are usuallly treated separately.

The appropriate quantification of dispersion in the **horizontal directions** is not well understood and often smaller than the numerical dispersion of the algorithms

**TABLE 11.5 Advection of Plume Modeled with Finite Difference Method Corrected for Numerical Dispersion [eqs. (11.260)–(11.262)][a].**

| $t$ | $q_1$ | $q_2$ | $q_3$ | $q_4$ | $q_5$ | $q_6$ | $q_7$ | $q_8$ |
|---|---|---|---|---|---|---|---|---|
| 0 | 0.000 | 0.000 | 1.000 | 1.000 | 0.000 | 0.000 | 0.000 | 0.000 |
| 0.5 | 0.000 | −0.125 | 0.625 | 1.125 | 0.375 | 0.000 | 0.000 | 0.000 |
| 1 | 0.016 | −0.172 | 0.281 | 1.031 | 0.703 | 0.141 | 0.000 | 0.000 |
| 1.5 | 0.033 | −0.158 | 0.018 | 0.791 | 0.896 | 0.369 | 0.053 | 0.000 |
| 2 | 0.045 | −0.108 | −0.145 | 0.488 | 0.923 | 0.606 | 0.178 | 0.020 |

[a] $u = 1$; $\Delta x = 1$; $\Delta t = 0.5$. Courant number = 0.5. $q_0 = q_9 = 0$.

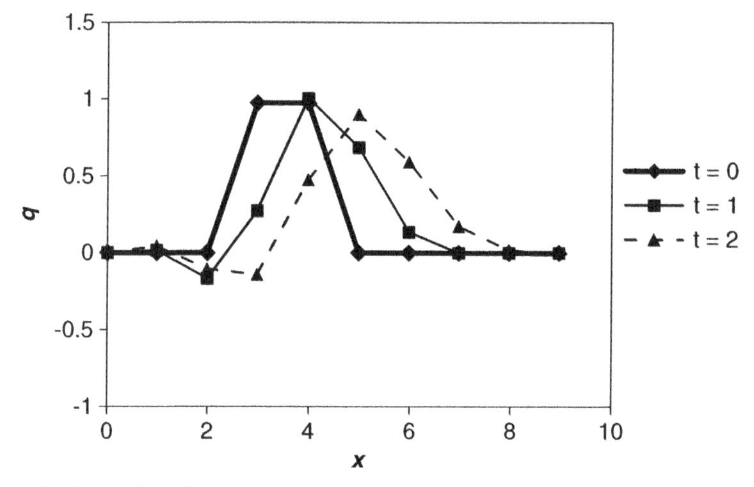


Figure 11.3 Advection of a plume modeled with a finite difference method corrected for numerical dispersion [eqs. (11.251)–(11.253)]. $u = 1$; $\Delta x = 1$; $\Delta t = 0.5$. Courant number = 0.5. $q_0 = q_9 = 0$.

used to model advection. Hence, horizontal dispersion is often ignored in Eulerian dispersion models. When the turbulent diffusivity does not change much in the horizontal direction, which is usually a reasonable assumption, then eq. (11.40) can be adapted as

$$\frac{\partial q}{\partial t} = (D + K_x)\frac{\partial^2 q}{\partial x^2} + (D + K_y)\frac{\partial^2 q}{\partial y^2} \tag{11.263}$$

For the spatial second derivatives, the same equations can be used as discussed in the previous section [e.g., eq. (11.262) and higher-order variants of Fornberg (1988)].

Modeling dispersion in the **vertical direction** is very closely linked with the model used, as the Blackadar (1978) model and its successors already include a material balance. Hence, no new methods than the ones already described are needed.

### 11.5.3 Chemical Reaction Kinetics

The simulation of the chemical reactions is the most time-consuming part of an Eulerian dispersion calculation. This is partly due to the complexity of the chemistry, with typically about 100 chemical and photochemical reactions to be accounted for, and partly due to the nature of radical reactions.

What is typical about radical reactions is that the reactions *producing* radicals are very slow, whereas reactions *consuming* radicals are very fast. How that affects a numerical simulation will be illustrated below. For simplified chemical mechanisms it is often possible to remove the fast reaction dynamics with the pseudo-steady-state approximation, as illustrated in Section 11.4, but this is usually not possible with complex reaction mechanisms, where the reactive species undergo many reactions in parallel. Hence, efficient algorithms are needed that allow for the direct simulation of such chemical systems.

To illustrate the problem, consider the following two reactions:

$$\mathrm{A} \rightarrow \mathrm{B} \tag{11.264}$$

$$\mathrm{B} + \mathrm{C} \rightarrow \mathrm{D} \tag{11.265}$$

Assume that the first reaction is slow, with a rate constant $k_1 = 10^{-5}$ (units are omitted as we are considering only the numerical aspect of this system as a mathematical problem). The second reaction is fast, with rate constant $k_2 = 10^5$. As an approximate solution, we can assume that every molecule of B reacts with C immediately, so the chemical mechanism can be simplified as

$$\mathrm{A} + \mathrm{C} \rightarrow \mathrm{D} \tag{11.266}$$

which is first order in A and zero order in C, with rate constant $k_1$. The concentration of B can be calculated by assuming that the rate of the two reactions of eqs. (11.264) and (11.265) are the same:

$$k_1[\mathrm{A}] = k_2[\mathrm{B}][\mathrm{C}] \tag{11.267}$$

Solving for [B] leads to

$$[\mathrm{B}] = \frac{k_1[\mathrm{A}]}{k_2[\mathrm{C}]} \tag{11.268}$$

Assuming [A] = [C] = 1, the concentration [B] is $10^{-10}$. Solving this problem is now reduced to solving the material balance for one of the species other than B:

$$\frac{\partial[\mathrm{A}]}{\partial t} = -k_1[\mathrm{A}] \tag{11.269}$$

**TABLE 11.6 Numerical Solution of Dynamics of Chemical Reactions [eqs. (11.264) and(11.265)] with eq. (11.271) Assuming a Pseudo-Steady-State Concentration of B[a].**

| $t$ | [A] numerical | [A] exact |
|---|---|---|
| 0 | 1 | 1 |
| 1000 | 0.9900 | 0.9900 |
| 2000 | 0.9801 | 0.9802 |
| 3000 | 0.9703 | 0.9704 |
| 4000 | 0.9606 | 0.9608 |
| 5000 | 0.9510 | 0.9512 |
| 6000 | 0.9415 | 0.9418 |
| 7000 | 0.9321 | 0.9324 |
| 8000 | 0.9227 | 0.9231 |
| 9000 | 0.9135 | 0.9139 |
| 10000 | 0.9044 | 0.9048 |

[a] $k_1 = 10^{-5}$; $k_2 = 10^5$; $\Delta t = 1000$. Exact solution is eq. (11.270).

which can be solved analytically:

$$[\mathrm{A}] = [\mathrm{A}]_0 \exp(-k_1 t) \tag{11.270}$$

where $[\mathrm{A}]_0$ is the concentration of A at time zero. Because [B] is negligibly small, and we chose $[\mathrm{A}]_0 = [\mathrm{C}]_0$, we can state that [C] = [A]. Assuming $[\mathrm{D}]_0 = 0$, it follows that $[\mathrm{D}] = [\mathrm{A}]_0 - [\mathrm{A}]$.

For the sake of comparison, we will solve this problem numerically assuming pseudosteady state and without this assumption. With the pseudo-steady state approximation, we only need a numerical solution for eq. (11.269). The simplest method is Euler's method:

$$[\mathrm{A}]^{t+\Delta t} = [\mathrm{A}]^t - k_1 [\mathrm{A}]^t \Delta t \tag{11.271}$$

The result of the calculation is shown in Table 11.6 with a time step $\Delta t = 1000$. The numerical solution is almost identical to the analytical solution, indicating that this mechanism poses no numerical problems.

However, if we wish to solve the problem without pseudo-steady-state approximation, we need to solve eq. (11.269) along with two other differential equations:

$$\frac{\partial[\mathrm{B}]}{\partial t} = k_1[\mathrm{A}] - k_2[\mathrm{B}][\mathrm{C}] \tag{11.272}$$

$$\frac{\partial[\mathrm{C}]}{\partial t} = -k_2[\mathrm{B}][\mathrm{C}] \tag{11.273}$$

**TABLE 11.7 Numerical Solution of Dynamics of Chemical Reactions [eqs. (11.264) and (11.265)] with eq. (11.271) Assuming No Pseudo-Steady-State Concentration of B[a].**

| $t$ | [A] numerical | [A] exact | [B] numerical | [B] steady state | [C] numerical | [D] numerical | [D] exact |
|---|---|---|---|---|---|---|---|
| 0 | 1 | 1 | 0 | $1.00\times10^{-10}$ | 1.0000 | 0 | 0 |
| $1\times10^{-4}$ | 1 | 1 | $1.00\times10^{-9}$ | $1.00\times10^{-10}$ | 1.0000 | 0 | $1.00\times10^{-9}$ |
| $2\times10^{-4}$ | 1 | 1 | $-8.00\times10^{-9}$ | $1.00\times10^{-10}$ | 1.0000 | $1.00\times10^{-8}$ | $2.00\times10^{-9}$ |
| $3\times10^{-4}$ | 1 | 1 | $7.30\times10^{-8}$ | $1.00\times10^{-10}$ | 1.0000 | $-7.00\times10^{-8}$ | $3.00\times10^{-9}$ |
| $4\times10^{-4}$ | 1 | 1 | $-6.56\times10^{-7}$ | $1.00\times10^{-10}$ | 1.0000 | $6.60\times10^{-7}$ | $4.00\times10^{-9}$ |
| $5\times10^{-4}$ | 1 | 1 | $5.90\times10^{-6}$ | $1.00\times10^{-10}$ | 1.0000 | $-5.90\times10^{-6}$ | $5.00\times10^{-9}$ |
| $6\times10^{-4}$ | 1 | 1 | $-5.31\times10^{-5}$ | $1.00\times10^{-10}$ | 0.9999 | $5.32\times10^{-5}$ | $6.00\times10^{-9}$ |
| $7\times10^{-4}$ | 1 | 1 | $4.78\times10^{-4}$ | $1.00\times10^{-10}$ | 1.0005 | $-4.78\times10^{-4}$ | $7.00\times10^{-9}$ |
| $8\times10^{-4}$ | 1 | 1 | $-4.31\times10^{-3}$ | $1.00\times10^{-10}$ | 0.9957 | $4.31\times10^{-3}$ | $8.00\times10^{-9}$ |
| $9\times10^{-4}$ | 1 | 1 | $3.86\times10^{-2}$ | $1.00\times10^{-10}$ | 1.0386 | $-3.86\times10^{-2}$ | $9.00\times10^{-9}$ |
| $10\times10^{-4}$ | 1 | 1 | $-3.62\times10^{-1}$ | $1.00\times10^{-10}$ | 0.6379 | $3.62\times10^{-1}$ | $1.00\times10^{-8}$ |

[a] $k_1 = 10^{-5}$; $k_2 = 10^5$; $\Delta t = 1000$. Exact solution is based on eq. (11.270); steady-state solution of [B] is based on eq. (11.268).

The implementation of eqs. (11.272) and (11.273) with Euler's method leads to the following equations:

$$[\mathrm{B}]^{t+\Delta t} = [\mathrm{B}]^t + \left(k_1[\mathrm{A}]^t - k_2[\mathrm{B}]^t[\mathrm{C}]^t\right)\Delta t \tag{11.274}$$

$$[\mathrm{C}]^{t+\Delta t} = [\mathrm{C}]^t - k_2[\mathrm{B}]^t[\mathrm{C}]^t\,\Delta t \tag{11.275}$$

We will assume that the initial concentration of B is 0. [D] is calculated with a species balance: Because D is produced out of C, and C does not take part in any other reactions, we find

$$[\mathrm{D}] = [\mathrm{C}]_0 - [\mathrm{C}] \tag{11.276}$$

This time a time step of $10^{-4}$ is chosen in the numerical integration. The result is shown in Table 11.7. In spite of the very short time increments, the concentration of the intermediate B is not calculated accurately, which leads to inaccurate calculations of C and D as well. Even though the time constant of the overall process is $1/k_1 = 10^5$, the time constant of the dynamics of B is $1/(k_2[\mathrm{C}]) = 10^{-5}$ initially, and an explicit numerical integration scheme like Euler's method requires smaller time increments than the time constant to produce acceptable results. It follows that about a billion time increments are needed to resolve this chemical system. When a system of differential equations has time constants that are several orders of magnitude apart, we say that it is a **stiff set of differential equations**.

In this particular system there is a very simple solution to the numerical problem posed by Euler's method. When [B] is evaluated at time $t + \Delta t$ every time it occurs, the problem disappears. Equations (11.274) and (11.275) are reformulated as follows:

$$[B]^{t+\Delta t} = [B]^t + \left(k_1[A]^t - k_2[B]^{t+\Delta t}[C]^t\right)\Delta t \tag{11.277}$$

$$[C]^{t+\Delta t} = [C]^t - k_2[B]^{t+\Delta t}[C]^t\,\Delta t \tag{11.278}$$

The implicit equation (11.277) must be solved for $[B]^{t+\Delta t}$:

$$[B]^{t+\Delta t} = \frac{[B]^t + k_1[A]^t\,\Delta t}{1 + k_2[C]^t\,\Delta t} \tag{11.279}$$

Hence, the set of differential equations is solved with eqs. (11.271), (11.279), and (11.278) (in that order). The result is exactly the same as the pseudo-steady-state calculation (Table 11.6) and is not shown for that reason. The values of [B] found by the algorithm are also the values of the steady-state solution. Hence, an implicit method can solve stiff sets of differential equations with an accuracy and efficiency (in number of steps needed) comparable to a pseudo-steady-state approximation. The required number of time increments is reduced from 1 billion to 10.

In practice, it is not usually feasible to find an exact implicit algorithm for all species in a reaction mechanism. In fact, even the simple problem outlined above would have been much more complicated if we had decided to derive implicit equations for all variables. Hence, a more systematic approach is required.

A well-known algorithm for the solution of stiff sets of ordinary differential equations is the algorithm of Gear (1971), an example of a backward differentiation algorithm. In this algorithm a first estimate of the concentrations at time $t + \Delta t$ is made, and this estimate is improved iteratively. In itself, the Gear algorithm is often insufficiently efficient to resolve the chemistry in an Eulerian dispersion model. For that reason, additional improvements were made to the algorithm by Jacobson and Turco (1994) and by Jacobson (1995) by improving the matrix operations to avoid excess calculations. This type of improved Gear algorithm is used in CMAQ. An alternative option in CMAQ is the use of an iterative quasi-steady-state approximation (Byun and Ching, 1999).

### 11.5.4 Boundary Conditions

Boundary conditions are needed for the advection and diffusion processes. For horizontal advection and diffusion, an upwind boundary condition can be imposed in the form of a background concentration. At the downwind boundary, a boundary condition can be imposed by requiring that the flux entering the last cell equals the flux leaving that cell.

For vertical diffusion, a boundary condition is needed at the surface and at the top of the domain. At the surface, the boundary condition will be imposed by the dry deposition assumption. This leads to a relationship between the concentration and the mass flux, which can be substituted in the material balance and solved for the concentration. At the top of the domain, total reflection can be imposed, that is, zero vertical concentration gradient. In nonlocal closure models, the boundary conditions are usually included in the model formulation. For a realistic boundary condition it is necessary that the domain of the computational grid includes the entire mixing layer.

### 11.5.5 Plume-in-Grid Modeling

The main disadvantage of Eulerian dispersion models is their low resolution. This is a particular weakness in the presence of large point sources. The normal approach in an Eulerian model is to dilute the emission immediately over an entire grid cell. But with horizontal grid spacings of 1–20 km, it can take hours for a plume to grow to that size. If a plume contains large amounts of $NO_x$, and the ambient air contains hydrocarbons, it can take hours before sufficient mixing has occurred to initiate the production of significant amounts of ozone. Due to the simulated mixing, Eulerian models will predict an immediate ozone production. To avoid such inaccuracies, the plume-in-grid approach has been introduced.

The main issue to be solved in a plume-in-grid submodel is the fact that a plume is a Lagrangian entity whereas an Eulerian grid does not recognize plumes. Hence, plume development is calculated separately from the Eulerian grid calculations. Another issue is the incorporation of complex, nonlinear chemistry into the plume, with different reactions dominating in different parts of the plume. For instance, in the example of a large $NO_x$ containing plume in a VOC containing background, ozone concentrations in the center of the plume will decline due to ozone titration by NO. At the plume edges, where $NO_x$ has mixed with VOCs, the ozone concentration will rise.

To solve these issues, the Eulerian model CMAQ implements a procedure proposed by Gillani (1986). In this approach the plume cross section is represented by a rectangular slab divided into 10 vertical sections of equal width. The width of the sections increase proportionally with the plume width. Hence, pollutant concentrations and the chemistry can be evaluated separately in each section. The center of the cross section is the plume center, which moves with the wind speed averaged over the cross section. When the plume has reached sufficient size and maturity (dilution), its contents are incorporated into the Eulerian grid.

## 11.6 SUMMARY OF THE MAIN EQUATIONS

Material balance (direct simulation):

$$\frac{\partial c_i}{\partial t} = -\frac{\partial c_i u}{\partial x} - \frac{\partial c_i v}{\partial y} - \frac{\partial c_i w}{\partial z} + D_i\left(\frac{\partial^2 c_i}{\partial x^2} + \frac{\partial^2 c_i}{\partial y^2} + \frac{\partial^2 c_i}{\partial z^2}\right) + S_i \qquad (11.21)$$

Material balance (direct simulation) in terms of mass fraction:

$$\frac{\partial q_i}{\partial t} = -u\frac{\partial q_i}{\partial x} - v\frac{\partial q_i}{\partial y} - w\frac{\partial q_i}{\partial z} + D_i\left(\frac{\partial^2 q_i}{\partial x^2} + \frac{\partial^2 q_i}{\partial y^2} + \frac{\partial^2 q_i}{\partial z^2}\right) + \frac{S_i}{\rho} \qquad (11.30)$$

Material balance (Reynolds averaged, closed):

$$\frac{\partial \overline{c_i}}{\partial t} = -\frac{\partial \overline{c_i} \cdot \overline{u}}{\partial x} - \frac{\partial \overline{c_i} \cdot \overline{v}}{\partial y} - \frac{\partial \overline{c_i} \cdot \overline{w}}{\partial z} + \frac{\partial}{\partial x}\left[(D_i + K_x)\frac{\partial \overline{c_i}}{\partial x}\right] + \frac{\partial}{\partial y}\left[(D_i + K_y)\frac{\partial \overline{c_i}}{\partial y}\right] + \frac{\partial}{\partial z}\left[(D_i + K_z)\frac{\partial \overline{c_i}}{\partial z}\right] + \overline{S_i} \tag{11.40}$$

Blackadar nonlocal scheme:

$$\frac{\partial q_1}{\partial t} = -\frac{v_\mathrm{d}}{h_1} q_1 + \frac{M_u}{\rho_1 h_1} \sum_{i=2}^{n} \rho_i (q_i - q_1) h_i \tag{11.58}$$

$$\frac{\partial q_i}{\partial t} = M_u (q_1 - q_i) \tag{11.60}$$

where

$$M_\mathrm{u} = \frac{H}{c_p \sum_{i=2}^{n} \rho_i h_i (\theta_1 - \theta_i)} \tag{11.56}$$

In transilient matrix form (assuming layers of constant mass):

$$\begin{pmatrix} q_1(t+\Delta t) \\ q_2(t+\Delta t) \\ q_3(t+\Delta t) \\ q_4(t+\Delta t) \end{pmatrix} = \begin{pmatrix} 1-3M_\mathrm{u}\,\Delta t & M_\mathrm{u}\,\Delta t & M_\mathrm{u}\,\Delta t & M_\mathrm{u}\,\Delta t \\ M_\mathrm{u}\,\Delta t & 1-M_\mathrm{u}\,\Delta t & 0 & 0 \\ M_\mathrm{u}\,\Delta t & 0 & 1-M_\mathrm{u}\,\Delta t & 0 \\ M_\mathrm{u}\,\Delta t & 0 & 0 & 1-M_\mathrm{u}\,\Delta t \end{pmatrix} \begin{pmatrix} q_1(t) \\ q_2(t) \\ q_3(t) \\ q_4(t) \end{pmatrix} \tag{11.88}$$

Time constant:

$$\tau = \frac{1}{k_1} \tag{11.101}$$

Half-life:

$$\tau = \frac{\ln(2)}{k_1} \tag{11.102}$$

Arrhenius equation:

$$k_i = A \exp\left(-\frac{E}{RT}\right) \tag{11.103}$$

Extended Arrhenius equation:

$$k_i = A\left(\frac{T}{T_0}\right)^n \exp\left(-\frac{E}{RT}\right) \tag{11.104}$$

Falloff equation:

$$k_i = \frac{k_0 k_\infty}{k_0 + k_\infty} F \tag{11.110}$$

with

$$\log F = \frac{\log F_c}{1 + \left[\dfrac{\log(k_0 / k_\infty)}{0.75 - 1.27 \log F_c}\right]^2} \tag{11.111}$$

Photolytical reaction rate:

$$R_i = C \int_0^\infty \phi(\lambda)\sigma(\lambda)I(\lambda)\, d\lambda \tag{11.119}$$

Equilibrium constant calculation:

$$K = \exp\left(\frac{-\Delta_r G^\circ}{RT}\right) \tag{11.197}$$

Gibbs free energy calculation:

$$\Delta_r G^\circ = \Delta_r H^\circ - T\, \Delta_r S^\circ \tag{11.198}$$

## PROBLEMS

**1.** Using the actinic fluxes in Table 11.1, calculate the photochemical rate constant of $H_2O_2$ photolysis [reaction 9; eq. (11.140)]. Assume a quantum yield of 1 and the following absorption cross section (averages of 10 nm intervals):

| $\lambda$ (nm) | $\sigma$ ($10^{-20}$ cm$^2$) |
|---|---|
| 300 | 0.68 |
| 310 | 0.39 |
| 320 | 0.22 |
| 330 | 0.13 |
| 340 | 0.07 |
| 350 | 0.04 |
| 360+ | 0 |

## REFERENCES

Anfossi D., Rizza U., Mangia C., Degrazia G.A., and Pereira Marques Filho E. (2006). Estimation of the ratio between the Lagrangian and Eulerian time scales in an atmospheric boundary layer generated by large eddy simulation. *Atmos. Environ.* **40**, 326–337.

Atkinson R. (1997a). Atmospheric reactions of alkoxy and $\beta$-hydroxyalkoxy radicals. *Int. J. Chem. Kinet.* **29**, 99–111.

Atkinson R. (1997b). Gas-phase tropospheric chemistry of volatile organic compounds: 1. Alkanes and alkenes. *J. Phys. Chem. Ref. Data* **26**, 215–290.

Atkinson R., Baulch D.L., Cox R.A., Hampson R.F. Jr., Kerr R.F., and Troe J. (1989). Evaluated kinetic and photochemical data for atmospheric chemistry: Supplement III, IUPAC Subcommittee on Gas Kinetic Data Evaluation for Atmospheric Chemistry. *J. Phys. Chem. Ref. Data* **18**, 881–1097.

Atkinson R., Baulch D.L., Cox R.A., Hampson R.F. Jr., Kerr R.F., and Troe J. (1992). Evaluated kinetic and photochemical data for atmospheric chemistry: Supplement IV, IUPAC Subcommittee on Gas Kinetic Data Evaluation for Atmospheric Chemistry. *J. Phys. Chem. Ref. Data* **21**, 1125–1568.

Atkinson R., Baulch D.L., Cox R.A., Hampson R.F. Jr., Kerr R.F., Rossi M.J., and Troe J. (1997a). Evaluated kinetic and photochemical data for atmospheric chemistry: Supplement V, IUPAC Subcommittee on Gas Kinetic Data Evaluation for Atmospheric Chemistry. *J. Phys. Chem. Ref. Data* **26**, 521–1011.

Atkinson R., Baulch D.L., Cox R.A., Hampson R.F. Jr., Kerr R.F., Rossi M.J., and Troe J. (1997b). Evaluated kinetic and photochemical data for atmospheric chemistry: Supplement VI, IUPAC Subcommittee on Gas Kinetic Data Evaluation for Atmospheric Chemistry. *J. Phys. Chem. Ref. Data* **26**, 1329–1499.

Atkinson R., Baulch D.L., Cox R.A., Hampson R.F. Jr., Kerr R.F., Rossi M.J., and Troe J. (1999). Evaluated kinetic and photochemical data for atmospheric chemistry: Supplement VII, IUPAC Subcommittee on Gas Kinetic Data Evaluation for Atmospheric Chemistry. *J. Phys. Chem. Ref. Data* **28**, 191–393.

Atkinson R., Baulch D.L., Cox R.A., Hampson R.F. Jr., Kerr R.F., Rossi M.J., and Troe J. (2000). Evaluated kinetic and photochemical data for atmospheric chemistry: Supplement VIII, IUPAC Subcommittee on Gas Kinetic Data Evaluation for Atmospheric Chemistry. *J. Phys. Chem. Ref. Data* **29**, 167–266.

Atkinson R., Baulch D.L., Cox R.A., Crowley J.N., Hampson R.F., Hynes R.G., Jenkin M.E., Rossi M.J., and Troe J. (2004). Evaluated kinetic and photochemical data for atmospheric chemistry: Volume I – Gas phase reactions of $O_x$, $HO_x$, $NO_x$ and $SO_x$ species. *Atmos. Chem. Phys.* **4**, 1461–1738.

Atkinson R., Baulch D.L., Cox R.A., Crowley J.N., Hampson R.F., Hynes R.G., Jenkin M.E., Rossi M.J., and Troe J. (2006). Evaluated kinetic and photochemical data for atmospheric chemistry: Volume II – Gas phase reactions of organic species. *Atmos. Chem. Phys.* **6**, 3625–4055.

Atkinson R., Baulch D.L., Cox R.A., Crowley J.N., Hampson R.F., Hynes R.G., Jenkin M.E., Rossi M.J., and Troe J. (2007). Evaluated kinetic and photochemical data for atmospheric chemistry: Volume III – Gas phase reactions of inorganic halogens. *Atmos. Chem. Phys.* **7**, 981–1191.

Bandura A.V. and Lvov S.N. (2006). The ionization constant of water over wide ranges of temperature and density. *J. Phys. Chem. Ref. Data* **35**, 15–30.

Baulch D.L., Cox R.A., Hampson R.F. Jr., Kerr J.A., Troe J., and Watson R.T. (1980). Evaluated kinetic and photochemical data for atmospheric chemistry, CODATA Task Group on Chemical Kinetics. *J. Phys. Chem. Ref. Data* **9**, 295–471.

Baulch D.L., Cox R.A., Crutzen P.J., Hampson R.F. Jr., Kerr J.A., Troe J., and Watson R.T. (1982). Evaluated kinetic and photochemical data for atmospheric chemistry: Supplement I, CODATA Task Group on Chemical Kinetics. *J. Phys. Chem. Ref. Data* **11**, 327–496.

Baulch D.L., Cox R.A., Hampson R.F. Jr., Kerr J.A., Troe J., and Watson R.T. (1984). Evaluated kinetic and photochemical data for atmospheric chemistry: Supplement II, CODATA Task Group on Chemical Kinetics. *J. Phys. Chem. Ref. Data* **13**, 1259–1380.

Blackadar A.K. (1978). Modeling pollutant transfer during daytime convection. In *Preprints, Fourth Symposium in Atmospheric Turbulence, Diffusion, and Air Quality, Reno*. Am. Meteorol. Soc, pp. 443–447.

Bott A. (1989). A positive definite advection scheme obtained by nonlinear renormalization of the advective fluxes. *Mon. Weather Rev.* **117**, 1006–1015.

Braslavsky S.E. (2007). Glossary of terms used in photochemistry, 3rd ed. *Pure Appl. Chem.* **79**, 293–465.

Byun D.W. and Ching J.K.S. (1999). *Science Algorithms of the EPA Models3 Community Multiscale Air Quality (CMAQ) Modeling System*. Report EPA/600/R-99/030, US-EPA.

Byun D.W., Lacser A., Yamartino R., and Zannetti P. (2003). Eulerian dispersion models. In Zannetti P. (ed.). *Air Quality Modeling. Volume I—Fundamentals*. Enviro Comp Institute, Fremont, CA, and Air and Waste Management Association, Pittsburgh, pp. 213–291.

Clegg S.L. and Brimblecombe P. (1988a). Equilibrium partial pressures of strong acids over concentrated saline solutions. I. $HNO_3$. *Atmos. Environ.* **22**, 91–100.

Clegg S.L. and Brimblecombe P. (1988b). Equilibrium partial pressures of strong acids over concentrated saline solutions. II. HCl. *Atmos. Environ.* **22**, 117–129.

Clegg S.L. and Pitzer K.S. (1992). Thermodynamics of multicomponent, miscible, ionic solutions: Generalized equations for symmetrical electrolytes. *J. Phys. Chem.* **96**, 3513–3520.

Clegg S.L. and Pitzer K.S. (1994). Thermodynamics of multicomponent, miscible, ionic solutions: Generalized equations for symmetrical electrolytes. *Erratum J. Phys. Chem.* **98**, 1368.

Clegg S.L., Pitzer K.S., and Brimblecombe P. (1992). Thermodynamics of multicomponent, miscible, ionic solutions. 2. Mixtures including unsymmetrical electrolytes. *J. Phys. Chem.* **96**, 9470–9479.

Clegg S.L., Pitzer K.S., and Brimblecombe P. (1994). Thermodynamics of multicomponent, miscible, ionic solutions. 2. Mixtures including unsymmetrical electrolytes. Erratum *J. Phys. Chem.* **98**, 1368.

Colella P. and Woodward P.R. (1984). The piecewise parabolic method (PPM) for gas-dynamical simulations. *J. Comp. Phys.* **54**, 174–201.

Cooper C.D. and Alley F.C. (2011). *Air Pollution Control*. 4th ed., Waveland Press, Long Grove, IL.

Cox, J.D., Wagman, D.D., and Medvedev, V.A. (1989). *CODATA Key Values for Thermodynamics*. Hemisphere, New York.

Deardorff J.W. (1972). Numerical investigation of neutral and unstable planetary boundary layers. *J. Atmos. Sci.* **29**, 91–115.

De Visscher A., Vanderdeelen J., Königsberger E., Churgalov B.R., Ichikuni M., and Tsurumi M. (2012). IUPAC-NIST Solubility Data Series. 93. Alkaline earth carbonates in aqueous systems. Part 1. Introduction, Be and Mg. *J. Phys. Chem. Ref. Data* **41**, paper 013105.

Fahey K.M. and Pandis S.N. (2001). Optimizing model performance: Variable size resolution in cloud chemistry modeling. *Atmos. Environ.* **35**, 4471–4478.

Fernández-Prini R., Alvarez J.L., and Harvey A.H. (2003). Henry's constants and vapour-liquid distribution constants for gaseous solutes in $H_2O$ and $D_2O$ at high temperatures. *J. Phys. Chem. Ref. Data* **32**, 903–916.

Fiedler B.H. and Moeng C.H. (1985). A practical integral closure model for mean vertical transport of a scalar in a convective boundary layer. *J. Atmos. Sci.* **42**, 359–363.

Fornberg B. (1988). Generation of finite difference formulas on arbitrarily spaced grids. *Math. Comp.* **51**, 699–706.

Gazdag J. (1973). Numerical convective schemes based on accurate computation of space derivatives. *J. Comp. Phys.* **13**, 100–113.

Gear C.W. (1971). The automatic integration of ordinary differential equations. *Comm. ACM* **14**, 176–179.

Gillani N.V. (1986). *Ozone Formation in Pollutant Plumes: A Reactive Plume Model with Arbitrary Crosswind Resolution*. EPA-600/3-86-051, Research Triangle Park, NC.

Haynes W.M. (2011). *The CRC Handbook of Chemistry and Physics*, 92nd ed. CRC Press, Boca Raton, FL.

Hoffmann M.R. and Calvert J.G. (1985). *Chemical Transformation Modules for Eulerian Acid Deposition Models, Vol. 2. The Aqueous-Phase Chemistry*. EPA/600/3-85/017, U.S. Environmental Protection Agency, Research Triangle Park, NC.

Jacobson M. (1995). Computation of global photochemistry with SMVGEAR II. *Atmos. Environ.* **29**, 2541–2546.

Jacobson M. and Turco R.P. (1994). SMVGEAR: A sparse-matrix, vectorized Gear code for atmospheric models. *Atmos. Environ.* **28**, 273–284.

Jaynes D.B. and Rogowski, A S. (1983). Applicability of Fick's Law to gas diffusion. *Soil Sci. Soc. Am. J.* **47**, 425–430.

Jones W.P. and Launder B.E. (1972). The prediction of laminarization with a two-equation model of turbulence. *Int. J. Heat Mass Transfer* **15**, 301–314.

Lantz R.B. (1971). Quantitative evaluation of numerical diffusion (truncation error), *Soc. Petrol. Eng. J.* **11**, 315–320.

Marshall W.L. and Franck E.U. (1981). Ion product of water substance, 0–1000°C, 1–10,000 bars. New international formulation and its background. *J. Phys. Chem. Ref. Data* **10**, 295–304.

Mills I., Cvitaš T., Homann K., Kallay N., and Kuchitsu K. (1993). *Quantities, Units and Symbols in Physical Chemistry ("Green Book")*, 2nd ed. IUPAC and Blackwell Science, Oxford, UK.

Moldrup P., Yamaguchi T., Rolston D.E., Vestergaard K., and J. A. Hansen. (1994). Removing numerically induced dispersion from finite difference models for solute and water transport in unsaturated soils. *Soil Sci.* **157**, 153–161.

Moldrup P., Kruse C.W., Yamaguchi T., and Rolston D.E. (1996). Modeling diffusion and reaction in soils: I. A diffusion and reaction corrected finite difference calculation scheme. *Soil Sci.* **161**, 347–354.

Nieuwstadt F.T.M. (1984). The turbulent structure of the stable, nocturnal boundary layer. *J. Atmos. Chem.* **41**, 2202–2216.

Notodarmojo S., Ho G.E., Scott W.D., and Davis G.B. (1991). Modelling phosphorus transport in soils and groundwater with two-consecutive reactions. *Water Res.* **25**, 1205–1216.

Pandis S.N. and Seinfeld J.H. (1989). Sensitivity analysis of a chemical mechanism for aqueous-phase atmospheric chemistry. *J. Geophys. Res.* **94**, 1105–1126.

Pitzer K.S. (1973). Thermodynamics of electrolytes. I. Theoretical basis and general equations. *J. Phys. Chem.* **77**, 268–277.

Pitzer, K.S. (1986). Theoretical considerations of solubility with emphasis on mixed aqueous electrolytes. *Pure Appl. Chem.* **58**, 1599–1610.

Pitzer, K.S. (1991). Ion interaction approach: Theory and data correlation. In Pitzer, K.S. (ed.). *Activity Coefficients in Electrolyte Solutions*, 2nd ed. CRC Press, Boca Raton, FL.

Pitzer K.S. and Kim J.J. (1974). Thermodynamics of electrolytes. IV. Activity and osmotic coefficients for mixed electrolytes. *J. Am. Chem. Soc.* **96**, 5701–5707.

Pitzer K.S. and Mayorga G. (1973). Thermodynamics of electrolytes. II. Activity and osmotic coefficients for strong electrolytes with one or both ions univalent. *J. Phys. Chem.* **77**, 2300–2308.

Pitzer K.S. and Mayorga G. (1974). Thermodynamics of electrolytes. III. Activity and osmotic coefficients for 2-2 electrolytes. *J. Solut. Chem.* **3**, 539–546.

Pleim J.E. and Chang J.S. (1992). A non-local closure model for vertical mixing in the convective boundary layer. *Atmos. Environ.* **26A**, 965–981.

Press W.H., Teukolsky S.A., Vetterling W.T., and Flannery B.P. (2007). *Numerical Recipes*, 3rd ed. Cambridge University Press, Cambridge, UK.

Pun B.K., Seigneur C., and Cichelsen H. (2005). Atmospheric transformations. In Zannetti P. (ed.). *Air Quality Modeling. Volume II—Advanced Topics*. Enviro Comp Institute, Fremont, CA, and Air and Waste Management Association, Pittsburgh, pp. 163–232.

Sander R. (1999). *Compilation of Henry's Law Constants for Inorganic and Organic Species of Potential Importance in Environmental Chemistry*. Version 3. http://www.mpch-mainz.mpg.de/~sander/res/henry.html.

Sander S.P., Friedl R.R., Ravishankara A.R., Golden D.M., Kolb C.E., Kurylo M.J., Molina M.J., Moortgat G.K., Keller-Rudek H., Finlayson-Pitts B.J., Wine P.H., Huie R.E., and Orkin V.L. (2006). *Chemical kinetics and photochemical data for use in atmospheric studies*. Evaluation number 15. JPL Publication 06-2, NASA, Jet Propulsion Laboratory, Pasadena, CA.

Sander S.P., Friedl R.R., Abbatt J.P.D., Barker J.R., Burkholder J.B., Golden D.M., Kolb C.E., Kurylo M.J., Moortgat G.K., Wine P.H., Huie R.E., and Orkin V.L. (2011). *Chemical kinetics and photochemical data for use in atmospheric studies*. Evaluation number 17. JPL Publication 10-6, NASA, Jet Propulsion Laboratory, Pasadena, CA.

Seigneur C. and Saxena P. (1988). A theoretical investigation of sulfate formation in clouds. *Atmos. Environ.* **22**, 101–115.

Seinfeld J.H. and Pandis S.N. (2006). *Atmospheric Chemistry and Physics*, 2nd ed. Wiley, Hoboken, NJ.

Stull R.B. (1984). Transilient turbulence theory. Part 1: The concept of eddy mixing across finite distances. *J. Atmos. Sci.* **41**, 3351–3367.

Stull R.B. (1988). *An Introduction to Boundary Layer Meteorology*. Kluwer Academic, Dordrecht, The Netherlands.

Stull R.B. and Driedonks A.G.M. (1987). Applications of the transilient turbulence parameterization to atmospheric boundary layer simulations. *Bound. Layer Meteorol.* **40**, 209–239.

Tennekes H. (1979). The exponential Lagrangian correlation function and turbulent diffusion in the inertial subrange. *Atmos. Environ.* **13**, 1565–1567.

Troe J. (1979). Predictive possibilities of unimolecular rate theory. *J. Phys. Chem.* **83**, 114–126.

Wagner W. and Pru$\beta$ A. (2002). The IAWPS formulation 1995 for the thermodynamic properties of ordinary water substance for general and scientific use. *J. Phys. Chem. Ref. Data* 31, 387–535.

Wilhelm E., Battino R., and Wilcock R.J. (1977). Low-pressure solubility of gases in liquid water. *Chem. Rev.* **77**, 219–262.

Yamartino R.J. (1993). Nonnegative, conserved scalar transport using grid-cell-centered, spectrally constrained Blackman cubics for applications on a variable-thickness mesh. *Mon. Weather Rev.* **121**, 753–763.

CHAPTER 12

# PRACTICAL ASPECTS OF AIR DISPERSION MODELING

## 12.1 INTRODUCTION

The first 11 chapters discussed the theoretical knowledge needed to develop and run air dispersion models. However, the air dispersion modeler often runs into problems of a more mundane nature when solving a problem. For example: Where exactly on the map is the source? How do I estimate the emission? Where do I find representative meteorological data for my problem? These are challenges that are not tied specifically to an air dispersion model or to a specific branch of air pollution meteorology.

Nevertheless, the air dispersion modeler needs to solve these problems in order to be successful. The objective of this chapter is to discuss practical issues that are not specifically tied to one air dispersion model or another.

## 12.2 SOURCE CHARACTERIZATION AND SOURCE MODELING

The prediction of an air dispersion model can only be as accurate as the emission data fed into the model. This goes without saying, but the quality of the source data is often not questioned in the course of a modeling project. The requirement for data depends on the problem to be solved. When the modeler is only interested in the impact of a well-defined set of emission sources, then it suffices to characterize only this set of sources. However, very often the question at hand is whether or not the addition of a new source will lead to exceedences of the air quality standards, given the emission sources that are already in the region. In that case, all sources that may affect the local air quality need to be included, and the quality of their characterization will affect the quality of the result. Hence, it is often necessary to estimate third-party emissions.

The best quality data is obtained when they are based on **direct measurements**. Point sources are usually measured with a CEMS (Continuous Emission Monitoring System). A CEMS is an instrument that measures the flow velocity of the gas in a stack (e.g., with pitot tubes) and simultaneously measures the concentration

*Air Dispersion Modeling: Foundations and Applications*, First Edition. Alex De Visscher.

of the compound of interest in the stack gas (e.g., with an infrared analyzer). The emission is calculated as

$$Q = Awc \tag{12.1}$$

where $Q$ is the emission rate (g s$^{-1}$), $A$ is the cross-sectional area of the stack (m$^2$), $w$ is the velocity of the stack gas (m s$^{-1}$), and $c$ is the pollutant concentration (g m$^{-3}$). Usually, the emission is variable and the average is calculated:

$$Q = \frac{1}{\Delta t} \int_{t}^{t+\Delta t} Awc\, dt \tag{12.2}$$

With modern data loggers averaging is a simple operation.

When measured emission data cannot be obtained, a second line of approach is to consult the **emission database** of the authorities in charge. For the United States, emission databases are maintained by the EPA and available at http://www.epa.gov/ttn/chief/eiinformation.html. In Canada, emission databases are maintained by Environment Canada and available at http://www.ec.gc.ca/inrp-npri/.

One of the issues of using a government database is that they are inherently incomplete. Only emissions greater than a certain threshold value are required by law to be reported, and for many pollutants thousands of small sources that individually remain below the radar contribute substantially to the overall emission and hence on the air quality. When using emission databases, a decision must be made on a case-by-case basis whether the coverage of the database is sufficient to ensure good quality of the modeling result.

When measurements and inventories are insufficient to cover all the sources in an air dispersion modeling domain, the use of **emission factors** can improve the coverage. An emission factor is a number that relates an activity to the emission it causes. The estimation of an emission with an emission factor is as follows:

$$Q = \mathrm{Act} \cdot \mathrm{EF}\left(1 - \frac{\mathrm{ER}}{100}\right) \tag{12.3}$$

where Act is a quantification of the activity leading to the emission, EF is the emission factor, and ER is the emission reduction efficiency (in %) of any air pollution control technology in place. A comprehensive overview of emission factors is given on the EPA website at http://www.epa.gov/ttn/chief/ap42/index.html.

**Example 12.1.** According to the EPA (1995) the emission factor of a pulverized coal-fired power plant with a dry bottom, wall-fired furnace running on bituminous coal has a PM10 emission factor of 1.15$A$ kg tonne$^{-1}$, where $A$ stands for percent ash content (Note that ton in the original reference refers to short tons, not metric tonnes). Estimate the PM10 emission of a power plant consuming 180 tonne bituminous coal per hour with ash content 7%, with an electrostatic precipitator eliminating 99.5% of the PM10 produced.

***Solution.*** The emission factor is $1.15A = 1.15 \times 7 = 8.05\,\text{kg tonne}^{-1}$. The activity rate here is 180 tonne $\text{h}^{-1}$. Application of eq. (12.3) yields

$$Q = 180 \times 8.05 \times (1 - 99.5/100) = \mathbf{7.245\,kg\,h^{-1}}$$

No matter what efforts are made to ensure that the emissions included in an air dispersion model are complete, there are cases where the gaps in the data are too big to ensure accurate results. In such cases, the modeler may have to resort to calculating relative measures of air pollution impact. One such measure is the **intake fraction**, which is defined as the fraction of an emitted air polluted that is breathed in by receptors (Bennett et al., 2002). The intake fraction is much less sensitive to the quality of the emission data set than the predicted pollutant concentration and can be used to compare sources and estimate impacts on a scenario basis (Zhou et al., 2003; Carella and Mudu, 2009). It is also a useful policy-making tool, for example, by applying the concept to subpopulations instead of an entire populations (Fraser, 2011). The intake fraction $iF$ is calculated as follows:

$$iF = \frac{P \cdot c \cdot \text{BR}}{Q} \tag{12.4}$$

where $P$ is the population (number of people), $c$ is the pollutant ground-level concentration, BR is the breathing rate of an average person, and $Q$ is the emission rate. When the population is spread out over a large geographical area, the following equation is useful:

$$iF = \frac{\int_{\text{Area}} P_{\text{d}} \cdot c \cdot \text{BR}\, d\text{Area}}{Q} \tag{12.5}$$

where $P_{\text{d}}$ is the population density.

The intake fraction is not a useful indicator of impact when the impact of interest has a strongly nonlinear relationship with the concentration. An example is the acute toxic effect of $H_2S$.

## 12.3 COORDINATE SYSTEMS

One of the issues of expressing geographical data such as meteorological data or air dispersion modeling output is the representation of a sphere on a flat surface. This is realized with a **projection**. Any map on a flat surface will distort the geography to some extent, and the larger the area mapped, the larger the distortion. In itself, distortion is not a major issue. However, it becomes problematic when data from different sources are combined in an air dispersion calculation, and different projections were used to determine coordinates. Projection inconsistency can cause receptors or terrain to move by several hundred meters relative to the source. Hence, it is important to pay attention to the type and specification of the projection used.

The two main projections used in air dispersion modeling are the **Lambert conformal conic (LCC) projection** and the **universal transverse Mercator (UTM)** projection. The shape of Earth that was assumed in the projection is referred to as the **datum**.

In the Lambert conformal conic projection the projection is on to a cone that has the same axis as Earth and that intersects Earth at two parallels, the standard reference latitudes. The shape of the projection depends mainly on the two parallels chosen.

To calculate the coordinates in an LCC projection from the coordinates on a spherical datum with radius $a=6370\,\text{km}$, latitude $\varphi$, longitude $\lambda$ (positive in the Western Hemisphere, negative in the Eastern Hemisphere), standard latitudes $\varphi_1$ and $\varphi_2$, and latitude and longitude of origin $\varphi_{\text{ORI}}$ and $\lambda_{\text{ORI}}$ (the latitude and longitude where $y=0$ and $x=0$), the following procedure can be used (Scire et al., 2000). First, the cone constant, $\sin\varphi_0$, is calculated:

$$\sin\varphi_0 = \frac{\ln\left(\dfrac{\cos\varphi_1}{\cos\varphi_2}\right)}{\ln\left\{\dfrac{\tan[(90-\varphi_1)/2]}{\tan[(90-\varphi_2)/2]}\right\}} \tag{12.6}$$

Next, the auxiliary function $\psi$ is calculated:

$$\psi = \frac{\left[a\dfrac{(\cos\varphi_1)}{(\sin\varphi_0)}\right]}{\{\tan[(90-\varphi_1)/2]\}^{\sin\varphi_0}} \tag{12.7}$$

Based on these variables, two polar radii are calculated:

$$\rho = \psi\left[\tan\left(\frac{90-\varphi}{2}\right)\right]^{\sin\varphi_0} \tag{12.8}$$

$$\rho_{\text{ORI}} = \psi\left[\tan\left(\frac{90-\varphi_{\text{ORI}}}{2}\right)\right]^{\sin\varphi_0} \tag{12.9}$$

Next, the polar angle $\theta$:

$$\theta = (\lambda_{\text{ORI}} - \lambda)\sin\varphi_0 \tag{12.10}$$

Finally, the $x$ and $y$ coordinates are calculated as

$$x = \rho\sin\theta \tag{12.11}$$

$$y = \rho_{\text{ORI}} - \rho\cos\theta \tag{12.12}$$

The main advantage of the LCC projection is that it conserves angles, and it can project a major portion of the globe, at least in principle. However, distortions increase with increasing size of the map, and size is not conserved.

The UTM projection is based on very different principles and is in many ways complementary to the LCC. It also conserves angles and distorts size and shape, but the distortion is minimal in the entire domain. UTM is not capable of projecting a major portion of the globe on a single plane because it divides the globe into 60 zones, strips of 6° longitude, from pole to pole, and projects each strip on a separate plane. As a result, the UTM can be cumbersome when very large areas, spanning several zones, are considered. The zones are numbered, starting from 1 at 180–174°W, and moving east. The contiguous United States are contained within Zones 10 (parts of California, Oregon, and Washington) to 19 (New England). The center of the zone (i.e., at the equator, at 3° latitude from both sides) the coordinates are $x=500\,\text{km}$, $y=0$.

The projection of the zone is on a cylinder wrapped around Earth so that the intersecting circle is at the center of the zone. The resulting map is shrunk by 0.04% to improve accuracy of the map, leading to features being slightly too small near the center latitude, and too large near the side latitudes.

## 12.4 DATA HANDLING

The main issue with data handling in air dispersion modeling is the sheer volume of data that is used. Consider a 100×100 grid with 10 vertical layers. At each modeled time (usually every hour) 100,000 wind speeds are considered in $x$, $y$, and $z$ directions and 10,000 concentrations for each compound (assuming only ground-level concentrations). After 1 year of modeling, the number of data runs in the billions of wind speeds. Data mining can be a challenge with such a deluge of information, so the modeler is advised to use visualization software. Some commercial versions of air dispersion models, such as AERMOD View and CALPUFF View (Lakes Environmental), come with visualization software included. The versions referred to on the EPA website, however, do not. CALPUFF, for instance, requires the third-party software SURFER (Golden Software, Inc.) for the visualization to be fully functional.

Finding the necessary data to run an air dispersion model can also be a challenge. Meteorological and terrain data for the United States are collected on the TRC website at http://www.src.com/datasets/datasets_main.html. Outside the United States the availability of data is less systematic. Some regulators provide recommended 5-year meteorological data, which is an excellent way to make regulatory air dispersion modeling more objective. An example is the province of Alberta, Canada (http://www.albertamm5data.com).

## 12.5 MODEL VALIDATION

Model validation is not a common practice in air dispersion modeling because the number of data sets of adequate quality is very limited. A repository of validation data sets can be found at http://www.jsirwin.com. Some of these are very old and have been used in the development of current air dispersion parameterization schemes, so they are not genuine validation data sets.

Sometimes attempts are made to validate a real-time air dispersion model by measuring pollutant concentrations locally, where the maximum concentration is predicted. This can be challenging when the plume is narrow and meanders strongly. When the plume is missed in the measurement, it seems that the model overestimates concentrations, whereas in fact it missed the location of the concentration peaks. As indicated in Chapter 1, air dispersion models usually predict concentration **histograms** more accurately than the concentration at any given place or time. Therefore, a comparison of histograms is usually a better test of a model's success than a direct comparison of concentrations.

## REFERENCES

Bennett D.H., McKone T.E., Evans J.S., Nazaroff W.W., Margni D., Jolliet O., and Smith K.R. (2002). Defining intake fraction. *Environ. Sci. Technol.* **36**, 206A–211A.

Carella B. and Mudu P. (2009). Exposure to air pollution: An intake fraction application in Turin Province. *Arch. Environ. Occup. Health* **64**, 156–163.

EPA (Environmental Protection Agency) (1995). *Compilation of Air Pollutant Emission Factors. Volume I: Stationary Point and Area Sources.*, 5th ed. US-EPA Report AP-42.

Fraser S. (2011). *Estimating Intake Fraction by Loose Coupling an Air Dispersion Model and a Geospatial Information System (GIS).* M.Sc. Thesis, University of Calgary, Canada.

Scire J.S., Robe F.R., Fernau M.E., and Yamartino R.J. (2000). *A User's Guide for the CALMET Meteorological Model.* Earth Tech, Concord, MA.

Zhou Y., Levy J.I., Hammitt J.K., and Evans J.S. (2003). Estimating population exposure to power plant emissions using CALPUFF: A case study in Beijing, China. *Atmos. Environ.* **37**, 815–826.

CHAPTER 13

# ISC3 AND SCREEN3: A DETAILED DESCRIPTION

## 13.1 INTRODUCTION

ISC3 and SCREEN3 are products of a previous generation of air dispersion modeling. They are relatively simple Gaussian plume models based on stability classes and show the deficiencies of such a modeling approach. They are not adequate for detailed modeling, but they are useful quick estimation methods for worst-case scenarios; in other words, they are good screening models.

ISC3 is available on the EPA website at http://www.epa.gov/ttn/scram/dispersion_alt.htm. Source code, executable file, and documentation can be downloaded there. SCREEN3 can be found at http://www.epa.gov/ttn/scram/dispersion_screening.htm.

ISC3 uses user-defined meteorological data and can handle multiple sources. It has a short-term variant and a long-term variant. The short-term version uses meteorological data versus time (i.e., as a time series), whereas the long-term model uses meteorological data in histogram form (EPA, 1995b).

SCREEN3 is a screening version of ISC3. It looks at only one source and runs a simulation with all realistic combinations of wind speed and stability class. At a series of distances from the source the highest concentration is selected, and from that information the maximum overall concentration is selected (EPA, 1995a).

The objective of this chapter is to provide a detailed overview of ISC3 and SCREEN3.

## 13.2 ISC3 MODEL DESCRIPTION

ISC stands for Industrial Source Complex. It is a Gaussian plume model based on stability class parameterizations, and for that reason it is no longer a recommended model. Some of the features of ISC3 is its capability to model multiple sources: area, volume, pit, and flare sources, dry and wet deposition, and inversion layers.

The basis of the model is the Gaussian dispersion model, eq. (6.1), with a decay term $D$ for reactivity effects.

*Air Dispersion Modeling: Foundations and Applications*, First Edition. Alex De Visscher.

$$\bar{C} = C_x \varphi_y \varphi_z D \tag{13.1}$$

where $C_x$ is the downwind dilution term. Instead of eq. (6.2), the following equation is used in ISC3:

$$C_x = \frac{Q}{u_s} \tag{13.2}$$

where $Q$ is the emission, and $u_s$ is the wind speed at the *real* source height, not at the effective source height. This choice was made in order to obtain conservative estimates of the ground-level concentration. This was mandated by the EPA at the time ISC3 was the EPA-preferred model. More recent models do not use this assumption, and it is no longer mandated by EPA.

The crosswind dilution factor is calculated by eq. (6.3):

$$\varphi_y = \frac{1}{\sqrt{2\pi}\sigma_y} \exp\left(-\frac{1}{2}\frac{y^2}{\sigma_y^2}\right) \tag{6.3}$$

where $\sigma_y$ is the lateral dispersion parameter, and $y$ is the crosswind distance from the source. For rural terrain, the calculation of $\sigma_y$ is based on eqs. (6.17) and (6.18) (EPA, 1995b) with the coefficients given in Table A6.5 (Appendix A, Section A6.1):

$$\sigma_y = 465.11628 x \tan(\theta) \tag{6.17}$$

where

$$\theta = 0.017453293\left[c - d \ln(x)\right] \tag{6.18}$$

Note that $x$ is in kilometers with these coefficients, whereas $\sigma_z$ is in meters.

For urban terrain, the Briggs (1973) equation [eq. (6.10)] is used with coefficients based on the McElroy and Pooler (1968) data (see Table 6.2, Chapter 6):

$$\sigma_y = \frac{ax}{(1+bx)^c} \tag{6.10}$$

In the absence of dry deposition, the vertical dilution term is given by eq. (6.8), that is, it accounts for reflection on both the surface and on an inversion layer:

$$\varphi_z = \sum_{j=-\infty}^{+\infty}\left(\exp\left\{-\frac{1}{2}\frac{\left[z-(h+2jh_{\text{mix}})\right]^2}{\sigma_z^2}\right\} + \exp\left\{-\frac{1}{2}\frac{\left[z+(h+2jh_{\text{mix}})\right]^2}{\sigma_z^2}\right\}\right) \tag{6.8}$$

An exception is made for stable conditions, where no upper boundary is assumed. Hence, in stability classes E and F the vertical dilution term is eq. (6.6):

$$\varphi_z = \frac{1}{\sqrt{2\pi}\sigma_z}\left\{\exp\left[-\frac{1}{2}\frac{(z-h)^2}{\sigma_z^2}\right]+\exp\left[-\frac{1}{2}\frac{(z+h)^2}{\sigma_z^2}\right]\right\} \tag{6.6}$$

In unstable and neutral conditions, eq. (6.8) is only used for $\sigma_z$ values up to $1.6h_{mix}$. When $\sigma_z$ exceeds $h_{mix}$, the well-mixed assumption is made in the vertical direction [eq. (6.9)]:

$$\varphi_z = \frac{1}{h_{mix}} \tag{6.9}$$

When the effective source height exceeds the height of the mixing layer, it is assumed that the entire layer penetrates the stable layer aloft. Hence, the ground-level concentration is assumed zero in those conditions.

In the presence of dry deposition, the calculation of the vertical dilution factor is more complicated. This will be discussed further down.

The vertical dispersion parameters for rural terrain are calculated with eq. (6.13), but with several sets of parameters depending on the downwind distance:

$$\sigma_z = cx^d \tag{6.13}$$

The coefficients are given in Table A6.6 (Section A6.1, Appendix A). Note that $x$ is in kilometers with these coefficients, whereas $\sigma_z$ is in meters.

For urban terrain the Briggs (1973) parameterization is used [eq. (6.11)]:

$$\sigma_z = \frac{dx}{(1+ex)^f} \tag{6.11}$$

with coefficients based on the data of McElroy and Pooler (1968). The coefficients are given in Table 6.2 (Chapter 6).

The decay term $D$ in eq. (13.1) is the result of chemical reaction. It is assumed to be the result of first-order reaction (exponential decay). The rate constant is user defined, but for $SO_2$ in urban terrain, a default value of 0.0000481 $s^{-1}$ is used.

The **wind speed profile** is calculated with the power law [eq. (2.3)]:

$$u_2 = u_1\left(\frac{z_2}{z_1}\right)^p \tag{2.3}$$

where $u_1$ and $u_2$ are wind speeds 1 and 2, and $z_1$ and $z_2$ are heights 1 and 2. The values of $p$ depend on the stability class and on the terrain. ISC3 uses somewhat different values than the ones recommended in Chapter 2. The $p$ values used in ISC3 are shown in Table 13.1.

**TABLE 13.1 Values of *p* for Use in eq. (2.3) to Predict Wind Speed Profiles**

| Stability Class | *p* for Rural Terrain | *p* for Urban Terrain |
|---|---|---|
| A | 0.07 | 0.15 |
| B | 0.07 | 0.15 |
| C | 0.10 | 0.20 |
| D | 0.15 | 0.25 |
| E | 0.35 | 0.30 |
| F | 0.55 | 0.30 |

*Source*: From EPA (1995b).

The stack height is corrected for **stack tip downwash**. No correction is needed when the stack gas exit velocity $w_s$ exceeds $1.5u_s$, where $u_s$ is the wind speed at the stack height. When $w_s < 1.5u_s$, eq. (7.51) is used:

$$\Delta h_{sd} = 2d_s\left(\frac{w_s}{u} - 1.5\right) \tag{7.51}$$

where $d_s$ is the stack tip diameter, and $u$ is evaluated at the real stack height. $\Delta h_{sd}$ (<0) is added to the stack height.

ISC3 calculates **plume rise** based on either the momentum flux parameter $F_m$ or the buoyancy flux parameter $F_b$:

$$\text{Momentum flux parameter}: F_m = \left(\frac{\overline{\rho_s}}{\overline{\rho}}\right) r_s^2 \overline{w_s}^2 \qquad (\text{m}^4\text{s}^{-2}) \tag{7.1}$$

$$\text{Buoyancy flux parameter}: F_b = \left(1 - \frac{\overline{\rho_s}}{\overline{\rho}}\right) g r_s^2 \overline{w_s} \qquad (\text{m}^4\text{s}^{-3}) \tag{7.2}$$

The actual algorithm uses temperatures instead of densities as input. However, if the gas is ideal, and the stack gas has the same molar mass as ambient air, the equations are equivalent.

Whether momentum or buoyancy dominates, plume rise depends on the temperature difference between the stack gas and the ambient air, $\Delta T = T_s - T_a$. Based on a comparison of momentum-dominated and buoyancy-dominated plume rise, a crossover temperature difference $\Delta T_c$ is defined. In unstable or neutral conditions, the following equations are used for $\Delta T_c$:

For $F_b < 55$ m$^4$ s$^{-3}$:

$$\Delta T_c = 0.0297 T_s \frac{w_s^{1/3}}{d_s^{2/3}} \tag{13.3}$$

For $F_b > 55$ m$^4$ s$^{-3}$:

$$\Delta T_c = 0.00575 T_s \frac{w_s^{2/3}}{d_s^{1/3}} \tag{13.4}$$

In stable conditions, the crossover temperature difference is based on the stability parameter [eq. (5.48)]:

$$s = \frac{g}{T_a} \frac{d\theta}{dz} \tag{5.48}$$

where default values are used for the potential temperature gradient:
For stability class E:

$$\frac{d\theta}{dz} = 0.02\,\mathrm{K\,m^{-1}} \tag{13.5}$$

For stability class F:

$$\frac{d\theta}{dz} = 0.035\,\mathrm{K\,m^{-1}} \tag{13.6}$$

The crossover temperature difference is

$$\Delta T_c = 0.019582\, T_s\, w_s \sqrt{s} \tag{13.7}$$

When $\Delta T > \Delta T_c$, the plume rise is buoyancy dominated; when $\Delta T < \Delta T_c$, the plume rise is momentum dominated. Transitional plume rise is calculated with eq. (7.9) for buoyancy-dominated plumes:

$$\Delta z = 1.6 \left( \frac{F_b x^2}{\overline{u}^3} \right)^{1/3} \tag{7.9}$$

For momentum-dominated plumes, the calculation of transitional plume rise depends on the atmospheric conditions. In unstable or neutral conditions, an equation similar to eq. (7.10) is used:

$$\Delta z = \left( \frac{3 F_m x}{\beta_j^2 u_s^2} \right)^{1/3} \tag{13.8}$$

where $\beta_j$ is the jet entrainment parameter, defined as

$$\beta_j = \frac{1}{3} + \frac{u_s}{v_s} \tag{13.9}$$

In stable conditions, the transitional plume rise of buoyancy-dominated plumes is calculated with

$$\Delta z = \left[ 3F_m \frac{\sin\left(x\sqrt{s}/u_s\right)}{\beta_j^2 u_s \sqrt{s}} \right] \tag{13.10}$$

The distance to final plume rise for buoyancy-dominated plumes is calculated with eq. (7.21) or eq. (7.22) when the atmosphere is unstable or neutral:

$$x_f = d_1 F_b^{5/8} \quad \text{for} \quad F_b < 55\,\text{m}^4\text{s}^{-3} \tag{7.21}$$

$$x_f = d_2 F_b^{2/5} \quad \text{for} \quad F_b > 55\,\text{m}^4\text{s}^{-3} \tag{7.22}$$

where $d_1 = 49\ \text{s}^{15/8}\ \text{m}^{-3/2}$, and $d_2 = 119\ \text{s}^{9/5}\ \text{m}^{-7/5}$; and $x_f$ is in meters. When final plume rise is reached, $x$ is replaced by $x_f$ in eq. (7.9).

In stable conditions, the distance to final plume rise of buoyancy-dominated plumes is

$$x_f = 2.0715 \frac{u_s}{\sqrt{s}} \tag{13.11}$$

This procedure reproduces eq. (7.11) for final plume rise.

For momentum-dominated plume rise in unstable or neutral conditions, eqs. (7.21) and (7.22) ase used for the distance to final plume rise, as with buoyany-dominated plumes. In stable conditions, the distance to final plume rise is calcuated with the following equation:

$$x_f = 0.5\pi \frac{u_s}{\sqrt{s}} \tag{13.12}$$

Final plume rise of momentum-dominated plumes is calculated directly, with the following equations. In unstable and neutral conditions:

$$\Delta h = \frac{3 d_s w_s}{u_s} \tag{13.13}$$

In stable conditions:

$$\Delta h = 1.5 \left( \frac{F_m}{u_s \sqrt{s}} \right)^{1/3} \tag{7.12}$$

ISC3 contains algorithms for **building downwash**, which can potentially change plumes in two ways:

- It reduces the effective source height.
- It increases plume dispersion by trapping in the building wake.

The downwash calculation scheme in ISC3 is very similar to the algorithm used in CALPUFF and will not be discussed here. See Chapter 15 for details.

Plume rise *enhances* plume dispersion. The enhanced plume dispersion is calculated with equations that are almost identical to eqs. (6.21) and (6.22):

$$\sigma_y = \sqrt{\sigma_{y,0}^2 + \left(\frac{\Delta h}{3.5}\right)^2} \tag{13.14}$$

$$\sigma_z = \sqrt{\sigma_{z,0}^2 + \left(\frac{\Delta h}{3.5}\right)^2} \tag{13.15}$$

ISC3 calculates **dry deposition** with eq. (7.128):

$$F = -v_d\, c \tag{7.129}$$

where $F$ is the flux (positive flux is up, hence the minus sign), $c$ is the pollutant concentration, and $v_d$ is the dry depostion velocity; $v_d$ is calculated with eq. (7.137):

$$v_d = \frac{1}{r_t} = \frac{1}{r_a + r_b + r_a r_b v_s} + v_s \tag{7.137}$$

where $r_a$ is the aerodynamic resistance, $r_b$ is the quasi-laminar resistance, and $v_s$ is the particle settling velocity.

Resistance $r_a$ is calculated with a variant of eq. (7.153) in a stable atmosphere:

$$r_a = \frac{1}{ku_*} \ln\left(\frac{z}{z_0} + 4.7\frac{z - z_0}{L}\right) \tag{13.16}$$

Unfortunately, the aerodynamic resistance in unstable conditions is not based on eq. (7.157) but on the assumption $K_z = K_m$. Hence, the following equation is used:

$$r_a = \frac{1}{ku_*} \ln\left[\frac{(x-1)(x_0+1)}{(x_0-1)(x+1)}\right] \tag{13.17a}$$

where

$$x = \left(1 - 16\frac{z}{L}\right)^{1/2} \qquad x_0 = \left(1 - 16\frac{z_0}{L}\right)^{1/2} \tag{5.96}$$

which is mathematically equivalent with eq. (7.157).

Resistance $r_b$ is calculated with the following equation, which covers Brownian motion and impaction:

$$r_b = \frac{1}{u_*(\text{Sc}^{-2/3} + 10^{-3/\text{St}})} \tag{13.17b}$$

where Sc is the Schmidt number:

$$\text{Sc} = \frac{\nu}{D} \tag{7.159}$$

and St is the Stokes number:

$$\text{St} = \frac{v_s u_*^2}{g\nu} \tag{7.171}$$

Velocity $v_s$ is calculated in the Stokes regime with a slip correction factor:

$$v_s = \frac{(\rho_p - \rho_a) d_p^2 g}{18\mu} C \tag{13.18}$$

where $\rho_p$ is the particle density, $\rho_a$ is the air density, $d_p$ is the particle diameter, $g$ is the acceleration due to gravity, and $\mu$ is the dynamic viscosity of air ($1.81 \times 10^{-5}$ Pa s); and $C$ is the slip correction factor:

$$C = 1 + 2\frac{\lambda}{d_p}\left[1.257 + 0.4\exp\left(-0.55\frac{d_p}{\lambda}\right)\right] \tag{7.134}$$

where $\lambda$ is the mean free path of air molecules ($6.5 \times 10^{-8}$ m was used). For the diffusivity of the particles, the following equation was used:

$$D = \left(8.09 \times 10^{-20}\,\text{m}^3\,\text{s}^{-1}\,\text{K}^{-1}\right)\frac{T_a C}{d_p} \tag{13.19}$$

Units are included in eq. (13.19) to ensure that the units of $D$ are m$^2$ s$^{-1}$.

A value of $10^{-4}$ m$^2$ s$^{-1}$ is added to $v_d$ to include phoretic effects. This is only important for particles between 0.1 and 1 μm in size.

**Wet deposition** is calculated in ISC3 based on a scavenging ratio, $\Lambda$, which is calculated with the following equation:

$$\Lambda = \lambda \frac{R}{R_1} \quad (7.262)$$

where $\lambda$ is a user-defined scavenging coefficient, $R$ is the precipitation rate, and $R_1 = 1$ mm h$^{-1}$.

For the impact of dry deposition on the downwind concentration, the corrected source depletion model of Horst (1983) was used. Details of this procedure were given in Section 7.5.3 and will not be repeated here. In summary, the vertical term is adjusted in three ways:

- A constant term (<1) is multiplied to represent overall depletion of the pollutant (source depletion).
- A height-dependent term is multiplied to account for depletion at the surface. The term is less than 1 near the surface and greater than 1 far from the surface. The vertically integrated concentration is not affected by this term.
- The plume height is adjusted to account for the settling velocity of the particles.

Wet deposition is accounted for in the concentration calculations by multiplying the emission by an exponential factor:

$$Q = Q_0 \exp(-\Lambda t) \quad (13.20)$$

where $t$ is the travel time of the pollutant since its emission.

Three types of terrain are distinguished in ISC3: simple terrain, complex terrain and intermediate terrain.

**Simple terrain** is terrain lower than the release height. In elevated simple terrain, no adjustments are made.

**Complex terrain** is assumed when the terrain height is above the plume centerline height. In complex terrain, the following assumptions are made:

- The plume center does not follow the terrain in stable conditions and is located midway between a terrain following $z$ coordinate and a nonterrain following $z$ coordinate in neutral and unstable conditions. However, the plume centerline height is assumed never to be lower than 10 m above the terrain.
- The mixing height follows the terrain.
- The wind speed profile follows the terrain.
- The plume concentration is averaged over 22.5° sectors.

**Intermediate terrain** is terrain above the release height but below the plume centerline height. In intermediate terrain, calculations are made for simple terrain and for complex terrain, and the highest value is taken.

ISC3 can handle **volume sources**, which are sources that start with an initial value of $\sigma_y$ and $\sigma_z$. The virtual source concept is used to calculate the dispersion, that is, the distance needed to reach the initial $\sigma_y$ and $\sigma_z$ values are added to the actual downwind distance from the source in the calculation of $\sigma_y$ and $\sigma_z$. This distance can be different for the $y$ and $z$ directions.

ISC3 can also calculate **area sources**. These are sources with a rectangular shape. They are calculated by numerical integration across the rectangular area.

An additional type of source that is defined in ISC3 is an **open-pit source**. An example of an open pit is a surface mine. Two adjustments are accounted for when modeling this type of source:

- Only a fraction of the particulate matter escapes the pit.
- The emission occurs mainly on the upwind side of the pit. This is the result of recirculation of air in the pit. This emission is modeled as an area source.

## 13.3 SCREEN3 MODEL DESCRIPTION

SCREEN3 is a screening version of ISC3. It uses the same dispersion parameters as ISC3 and several other features, such as area and volume sources. However, it does not allow for multiple sources, and it does not include atmospheric chemistry or deposition.

SCREEN3 is not based on actual measured or modeled meteorological data. Instead, a fixed set of meteorological conditions is used, and from this set the worst-case ground-level concentration is selected. The set of meteorological conditions is shown in Table 13.2.

In stability classes A–D, the **mixing height** is calculated with the following equation:

**TABLE 13.2 Set of Meteorological Conditions Used in SCREEN3: Stability Class and Wind Speed at 10 m Height**

| Stability Class $u_{10}$ (m s$^{-1}$) | 1 | 1.5 | 2 | 2.5 | 3 | 3.5 | 4 | 4.5 | 5 | 8 | 10 | 15 | 20 |
|---|---|---|---|---|---|---|---|---|---|---|---|---|---|
| A | • | • | • | • | • | | | | | | | | |
| B | • | • | • | • | • | • | • | • | • | | | | |
| C | • | • | • | • | • | • | • | • | • | • | • | | |
| D | • | • | • | • | • | • | • | • | • | • | • | • | • |
| E | • | • | • | • | • | • | • | • | | | | | |
| F | • | • | • | • | • | • | • | | | | | | |

$$h_{mix} = 320u_{10} \tag{13.21}$$

**Plume rise** and **stack tip downwash** are calculated as in ISC3. **Plume rise induced dispersion** is incorporated as well. SCREEN3 also includes a building downwash procedure.

SCREEN3 has a fumigation option. Two types of fumigation are included: inversion breakup fumigation and shoreline fumigation. For the latter, the following conservative estimate is made for the thermal internal boundary layer height:

$$h_T = 6x^{0.5} \tag{13.22}$$

where both $h_T$ and $x$ are in meters.

## REFERENCES

Briggs G.A. (1973). *Diffusion Estimation of Small Emissions*. Contribution No. 79, Atmospheric Turbulence and Diffusion Laboratory, Oak Ridge, TN.

EPA (1995a). *SCREEN3 Model User's Guide*. Report EPA-454/B-95-004, US-EPA, Research Triangle Park, NC.

EPA (1995b). *User's Guide for the Industrial Source Complex (ISC3) Dispersion Models. Volume II – Description of Model Algorithms*. Report EPA-454/B-95-003b, US-EPA, Research Triangle Park, NC.

Horst T.W. (1983). A correction to the Gaussian source-depletion model. In: Pruppacher H.R., Semonin R.G. and Slinn W.G.N. (eds.) *Precipitation Scavenging, Dry Deposition and Resuspension*. Elsevier, NY.

McElroy J.L. and Pooler F. (1968). *The St. Louis Dispersion Study*. Report AP-53, US Public Health Service, National Air Pollution Control Administration.

CHAPTER 14

# AERMOD AND AERMET: A DETAILED DESCRIPTION

## 14.1 INTRODUCTION

AERMOD is a sophisticated Gaussian plume model based on continuous parameterization of atmospheric dispersion. AERMET is the meteorological preprocessor of AERMOD (Cimorelli et al., 2004). These models have greatly increased in importance in recent years, as EPA decided in 2005 to replace ISC3 by AERMOD as the preferred air dispersion model (EPA, 2005). AERMOD and AERMET have become the workhorses of the air dispersion modeler.

Continuous parameterization is the main reason why AERMOD has superseded ISC3, but it is not the only reason. AERMOD implements a bi-Gaussian vertical dispersion function in the convective boundary layer, thus accounting explicitly for updrafts and downdrafts in such atmospheres. This feature makes AERMOD vastly superior to ISC3 when modeling dispersion in the daytime atmosphere.

However, the increased level of sophistication of air dispersion in AERMOD comes at a price. Thanks to the relatively simple equations used by ISC3 to calculate dispersion parameters, it was possible to use the relatively accurate corrected source depletion model described by Horst (1984) to calculate dry deposition. AERMOD's choice to implement more sophisticated schemes for the calculation of dispersion parameters necessitated the choice of simpler and less accurate calculation schemes for dry deposition. Fortunately, within the distances appropriate for AERMOD (< 50 km) dry deposition is usually limited, and the loss of accuracy of the deposition scheme will be more than compensated by the increased accuracy of the dispersion scheme.

AERMOD, including the preprocessor AERMET, is available on the EPA website at http://www.epa.gov/ttn/scram/dispersion_prefrec.htm. Source code, executable file, documentation, and extensive supporting material can be downloaded there.

This chapter contains a thorough overview of the model formulation of AERMOD and AERMET.

*Air Dispersion Modeling: Foundations and Applications*, First Edition. Alex De Visscher.

## 14.2 DESCRIPTION OF AERMET

AERMET is the meteorological preprocessor of AERMOD. The function of AERMET is to calculate the heat balance of the surface, determine whether the atmosphere is stable or convective (i.e., whether the heat flows from the atmosphere to the surface or vice versa), and calculate the friction velocity and the convective velocity scale, as well as the Obukhov length. The algorithms employed for these calculations will be outlined in this section.

The first decision that needs to be made is whether the surface transfers heat to the atmosphere or the other way around. To that effect, AERMET calculates the minimum solar angle needed to generate sufficient heat for the surface to transfer heat to the atmosphere. The basis of this calculation is the following equation (Kasten and Czeplak, 1980; Holtslag and van Ulden, 1983):

$$R = (990 \sin \phi - 30)(1 - 0.75n^{3.4}) \tag{5.54}$$

where $R$ is the solar radiation flux received by the surface (W m$^{-2}$), $\varphi$ is the solar elevation (angle of the sun above the horizon), and $n$ the fractional cloud cover (a number between 0 and 1, usually given in steps of $\frac{1}{8}$). Equations for the solar elevation are given in Section A5.2 (Appendix A).

However, the distinction between stable and unstable conditions depends on the net radiation, not on the solar radiation. The following equation was put forward by Holtslag and van Ulden (1983):

$$R_\mathrm{N} = \frac{R(1-r) + 60n - 5.67 \times 10^{-8} T^4 + 5.31 \times 10^{-13} T^6}{1.12} \tag{5.55}$$

where $r$ is the albedo of the surface, $n$ is the fractional cloud cover (0–1, usually in steps of $\frac{1}{8}$), and $T$ is the temperature in K. Substitution of eq. (5.54) into eq. (5.55) leads to

$$R_\mathrm{N} = \frac{(990 \sin \varphi - 30)(1 - 0.75n^{3.4})(1-r) + 60n - 5.67 \times 10^{-8} T^4 + 5.31 \times 10^{-13} T^6}{1.12} \tag{14.1}$$

The atmosphere is neutral when $R_\mathrm{N}=0$. Hence, solving eq. (14.1) for the solar elevation assuming $R_\mathrm{N}=0$ provides the required criterion:

$$\sin \varphi_\mathrm{crit} = \frac{1}{990}\left[\frac{-60n + 5.67 \times 10^{-8} T^4 - 5.31 \times 10^{-13} T^6}{(1 - 0.75n^{3.4})(1-r)} + 30\right] \tag{14.2}$$

where $\varphi_\mathrm{crit}$ is the critical solar elevation for transition from stable to unstable conditions.

The albedo itself is a function of $\varphi$:

$$r = r' + (1 - r')\exp\left[-0.1\varphi - 0.5(1 - r')^2\right] \tag{14.3}$$

with $r'$ the albedo at 90°; and $\varphi$ is in degrees. A figure of this dependency is given in Section A5.3 (Appendix A). Combining eq. (14.2) with eq. (14.3) leads to an implicit equation, which can be solved iteratively to obtain $\varphi_{\text{crit}}$. The value is usually around 13° for clear and partly cloudy conditions and about 23° for overcast conditions.

The sensible heat flux $q$ from the surface to the atmosphere is calculated from the net radiation with eq. (5.53):

$$q = \frac{(1 - C_{\text{G}})R_{\text{N}}}{1 + 1/B} \tag{5.53}$$

where $C_{\text{G}}$ is the fraction of the net radiation that goes to the ground (=0.1 in AERMOD), and $B$ is the Bowen ratio, that is, the ratio between sensible heat flux and latent heat flux.

The further calculation of the meteorological parameters depends on whether the boundary layer is convective or stable, that is, whether $\varphi$ is greater than or less than $\varphi_{\text{crit}}$.

When the atmosphere is convective, the variables $u_*$ (friction velocity) and $L$ (Obukhov length) are calculated with an iterative procedure. First, an initial estimate of $u_*$ is made based on a measured wind speed based on eq, (5.64):

$$u = \frac{u_*}{k}\ln\frac{z}{z_0} \tag{5.64}$$

Solving for $u_*$ leads to eq. (5.65):

$$u_* = \frac{ku_{\text{m}}}{\ln(z_{\text{m}}/z_0)} \tag{5.65}$$

where $u_{\text{m}}$ is the measured wind speed and $z_{\text{m}}$ is the wind speed measurement height.

Next, an estimate of $L$ is made with eq. (5.71):

$$L = -\frac{\rho_0 c_p T_0 u_*^3}{kgq} \tag{5.71}$$

where $q$ is the value obtained from eq. (5.53). Further iterations of $u_*$ are based on eqs. (5.82) and (5.83):

$$u = \frac{u_*}{k}\left\{\ln\frac{z}{z_0} + \ln\left[\frac{(n_0^2+1)(n_0+1)^2}{(n^2+1)(n+1)^2}\right] + 2[\arctan(n) - \arctan(n_0)]\right\} \tag{5.82}$$

with

$$n_0 = \left(1 - 16\frac{z_0}{L}\right)^{1/4} \quad n = \left(1 - 16\frac{z}{L}\right)^{1/4} \tag{5.83}$$

Note that the arctan function in eq. (5.82) has radians as units, not degrees. Solving eq. (5.82) for $u_*$ leads to the following equation:

$$u_* = \frac{ku}{\ln\frac{z}{z_0} + \ln\left(\frac{(n_0^2+1)(n_0+1)^2}{(n^2+1)(n+1)^2}\right) + 2(\arctan(n) - \arctan(n_0))} \tag{14.4}$$

where $u_*$ and $L$ are calculated by using eqs. (14.4) and (5.71) successively until convergence is reached.

In the convective boundary layer, the convective velocity scale $w_*$ is calculated based on its definition:

$$w_* = \left(\frac{gqh_{\text{mix}}}{T_0 \rho c_p}\right)^{1/3} \tag{5.150}$$

The calculation of the mixing layer height is discussed further down.

At night time, the above procedure for calculating $u_*$ and $L$ does not lead to accurate results, and a different approach is used, based on the theory in Section 5.9. Based on observations of the friction temperature in the nocturnal boundary layer, the following equation was put forward for the friction velocity:

$$u_* = \frac{C_D u}{2}\left[1 + \sqrt{1 - \left(\frac{2u_0}{\sqrt{C_D}u}\right)^2}\right] \tag{5.107}$$

where

$$C_D = \frac{k}{\ln(z/z_0)} \tag{5.104}$$

$$u_0^2 = \frac{5g\theta_*(z - z_0)}{T_0} \tag{5.105}$$

and the friction temperature $\theta_*$ is calculated with (van Ulden and Holtslag, 1985)

$$\theta_* = 0.09\,(1 - 0.5n^2) \tag{5.100}$$

Equation (5.107) only has a real solution when the wind speed exceeds a minimum value given by

$$u_{\text{min}} = \frac{2u_0}{\sqrt{C_{\text{D}}}} \tag{5.108}$$

This is because eq. (5.100) is not valid at low wind speeds. At the minimum wind speed the friction velocity is given by eq. (5.107):

$$u_{*\text{min}} = \frac{C_{\text{D}} u_{\text{min}}}{2} = \sqrt{C_{\text{D}}}\, u_0 \tag{5.109}$$

When the wind speed is less than $u_{\text{min}}$, the friction velocity and the friction temperature are calculated as follows:

$$u_* = u_{*\text{min}} \frac{u}{u_{\text{min}}} \tag{5.110}$$

$$\theta_* = 0.09(1 - 0.5n^2)\frac{u}{u_{\text{crit}}} \tag{5.111}$$

Once $u_*$ and $\theta_*$ are known, $L$ is calculated with

$$L = \frac{u_*^2 T_0}{kg\theta_*} \tag{5.89}$$

The sensible heat flux is calculated by solving eq. (5.88) for $q$:

$$q = -\rho_0 c_p u_* \theta_* \tag{14.5}$$

An alternative formulation for $u^*$ was in review as of June 2013. When the fractional cloud cover $n$ is not known, AERMET allows for an alternative calculation of $u_*$, $\theta_*$, and $L$ based on potential temperature values at two different heights. The procedures are shown in Section A 5.9 (Appendix A) and will not be repeated here.

The calculation of the mixing layer height depends on the type of atmosphere (convective or stable). In the convective boundary layer, a heat balance approach is used to calculate the mixing layer height. The relevant equation is

$$1.4 \int_{t_{\text{sunrise}}}^{t} q(t)\,dt = \rho c_p \int_0^{h_{\text{mix}}} [\theta(h_{\text{mix}}) - \theta(z)]\,dz \tag{5.122}$$

This equation is solved for $h_{\text{mix}}$.

Equation (5.122) will underestimate the mixing layer height near the onset of convective conditions because the atmosphere is still dominated by mechanical mixing. For that reason, the calculation procedure for stable atmosphere (see below) is also followed in unstable conditions, and the highest value of $h_{\text{mix}}$ is adopted.

In a stable boundary layer, it is assumed that eq. (5.115) predicts an equilibrium value of the mixing layer height. Hence:

$$h_{\mathrm{mix,e}} = Bu_*^{3/2} \tag{14.6}$$

where $B = 230\,\mathrm{s}^{1.5}\,\mathrm{m}^{0.5}$.

Further, it is assumed that the actual mixing layer height approaches the equilibrium value with time constant $\tau$ given by

$$\tau = \frac{h_{\mathrm{mix}}}{\beta_\tau u_*} \tag{14.7}$$

where $\beta_\tau = 2$. The actual mixing layer height at time $t + \Delta t$ is calculated from the value at time $t$, the equilibrium value and the time constant with the following equation:

$$h_{\mathrm{mix}}^{t+\Delta t} = h_{\mathrm{mix}}^{t} \exp\left(-\frac{\Delta t}{\tau}\right) + h_{\mathrm{mix,e}}^{t+\Delta t}\left[1 - \exp\left(-\frac{\Delta t}{\tau}\right)\right] \tag{14.8}$$

## 14.3 DESCRIPTION OF AERMOD

AERMOD consists of a meteorological interface and an actual dispersion model. The meteorological interface consists of the calculation of the wind speed profile, the potential temperature profile, and the profile of the vertical and lateral dispersion parameters. These calculations are outlined in Sections 14.3.1–14.3.3. The actual dispersion model covers the actual calculation of the concentration and includes issues such as plume rise, building downwash, plume dispersion modifiers, and plume meandering effects. These calculations are covered in the remainder of Section 14.3.

### 14.3.1 Wind Speed Profile

In unstable conditions, the wind speed profile is calculated with eq. (5.82) (see above). However, in stable conditions, the wind speed profile is not calculated with eq. (5.81) because of its lack of accuracy in extremely stable conditions. Instead, the following equation is used:

$$u = \frac{u_*}{k}\left[\ln\frac{z}{z_0} - 17\exp\left(-0.29\frac{z}{L}\right) + 17\exp\left(-0.29\frac{z_0}{L}\right)\right] \tag{14.9}$$

At low values of $z/L$ (slightly stable conditions) this equation reduces to eq. (5.81).

The above equations are only used from $7z_0$ to $h_{\mathrm{mix}}$. Below $7z_0$ the following equation is used:

$$u = \frac{z}{7z_0} u_{7z_0} \tag{14.10}$$

where $u_{7z_0}$ is the wind speed at height $7z_0$. At heights exceeding the mixing height $h_{mix}$, the wind speed at $z = h_{mix}$ is used.

AERMOD has no independent calculation scheme of the height dependence of the wind direction. However, it does accept measured wind directions at different heights, and interpolates linearly between them. Below the lowest measurement height and above the heighest measurement height the wind direction is assumed to be constant.

### 14.3.2 Potential Temperature Profile

In AERMOD the potential temperature profile is calculated in two steps. First a profile of the potential temperature gradient is generated. Then, this gradient is integrated to obtain potential temperatures.

In the **convective boundary layer** it is assumed that the potential temperature gradient is zero above the surface layer:

$$\frac{\partial \theta}{\partial z} = 0 \qquad (z < h_{mix}) \tag{14.11}$$

This is a reasonable assumption far above the surface but not near the surface, so it is important to use measured temperatures at sufficient height for the values to be representative. An example of a potential temperature profile in a convective boundary layer is given in Figure 5.5. Clearly the assumption of zero potential temperature gradient is valid in the mixing layer, except close to the surface.

For $z > h_{mix}$ an early morning value based on a **morning temperature sounding** is used, up to a height $h_{mix} + 500\,\text{m}$. Above that, the following equation is used:

$$\frac{\partial \theta}{\partial z} = 0.005\ \text{Km}^{-1} \qquad (z > h_{mix} + 500\ \text{m}) \tag{14.12}$$

In the **stable boundary layer** the following two equations are used at heights below 100 m (Dyer, 1974; Panofsky and Dutton, 1984):

$$\frac{\partial \theta}{\partial z} = \frac{\theta_*}{k \cdot 2\text{m}} \left(1 + \frac{10\text{m}}{L}\right) \qquad (z < 2\text{m}) \tag{14.13}$$

$$\frac{\partial \theta}{\partial z} = \frac{\theta_*}{kz} \left(1 + \frac{5z}{L}\right) \qquad (2\text{m} < z < 100\text{m}) \tag{14.14}$$

where the unit m in eq. (14.13) is to balance units. Above a 100-m height the following equation is used (Stull, 1983; van Ulden and Holtslag, 1985):

$$\frac{\partial\theta}{\partial z}=\left(\frac{\partial\theta}{\partial z}\right)_{z=100\,\mathrm{m}}\exp\left(-\frac{z-100\,\mathrm{m}}{0.44z_{i\theta}}\right)\quad(z>100\,\mathrm{m}) \tag{14.15}$$

where $z_{i\theta}=\max(h_{\mathrm{mix}},100\,\mathrm{m})$. The potential temperature gradient should be at least $0.002\,\mathrm{K\,m^{-1}}$. If lower values are predicted, then $0.002\,\mathrm{K\,m^{-1}}$ is used instead.

Separate rules apply when a measured value of the temperature gradient is available (Cimorelli et al., 2004).

To calculate the potential temperature profile from the potential temperature gradient profile, a reference potential temperature value $\theta_{\mathrm{ref}}$ at a reference height $z_{\mathrm{ref}}$ is calculated from the temperature $T_{\mathrm{ref}}$ measured at $z_{\mathrm{ref}}$:

$$\theta_{\mathrm{ref}}=T_{\mathrm{ref}}+\frac{g}{c_p}z_{\mathrm{msl}} \tag{14.16}$$

where $z_{\mathrm{msl}}$ is a height given by

$$z_{\mathrm{msl}}=z_{\mathrm{ref}}+z_{\mathrm{base}} \tag{14.17}$$

where $z_{\mathrm{base}}$ is a user-defined base height. From the reference potential temperature, the potential temperature profile is built up layer by layer based on the following incremental equation:

$$\theta_{z+\Delta z}=\theta_z+\overline{\frac{\partial\theta}{\partial z}}\Delta z \tag{14.18}$$

where $\Delta z$ is the height increment (negative for heights below the reference height).

### 14.3.3 Profile of Vertical Turbulence

In AERMOD the vertical velocity variance $\left(\sigma_w^2\right)$ of the convective boundary layer is the sum of a convective term and a mechanical term:

$$\sigma_w^2=\sigma_{wc}^2+\sigma_{wm}^2 \tag{14.19}$$

This approach was also followed in Section 5.11.3. Note that the velocity variance, not the turbulent velocity, is the additive property. Hence, with respect to turbulent velocity we get the equivalent of eq. (5.156):

$$\sigma_w=\sqrt{\sigma_{wc}^2+\sigma_{wm}^2} \tag{14.20}$$

For the calculation of the convective portion of the vertical wind speed variance, equations similar to eqs. (5.152)–(5.154) are used, but with different coefficients:

$$\sigma_{wc}^2=1.6\left(\frac{z}{h_{\mathrm{mix,c}}}\right)^{2/3}w_*^2\qquad(z<0.1h_{\mathrm{mix}}) \tag{14.21}$$

$$\sigma_{wc}^2 = 0.35w_*^2 \qquad (0.1h_{mix} < z < h_{mix}) \tag{14.22}$$

$$\sigma_{wc}^2 = 0.35w_*^2 \exp\left(-6\frac{z-h_{mix,c}}{h_{mix,c}}\right) \qquad (z > h_{mix}) \tag{14.23}$$

where $h_{mix,c}$ is the mixing layer height in convective conditions, and $w_*$ was defined earlier [eq. (5.150)].

The calculation of the mechanical portion of the vertical wind speed variance is calculated in two parts: one term, $\sigma_{wml}^2$, arising from turbulence in the mixing layer and one term, $\sigma_{wmr}^2$, arising from residual turbulence above the mixing layer:

$$\sigma_{wm}^2 = \sigma_{wml}^2 + \sigma_{wmr}^2 \tag{14.24}$$

The term $\sigma_{wml}^2$ is calculated with eq. (5.146) using $a=0.5$ within the mixing layer. Above the mixing layer, $\sigma_{wml}^2$ equals 0:

$$\sigma_{wml}^2 = 1.3u_*\left(1-\frac{z}{h_{mix}}\right)^{1/2} \qquad (z < h_{mix}) \tag{14.25}$$

$$\sigma_{wml}^2 = 0 \qquad (z > h_{mix}) \tag{14.26}$$

The residual term is calculated as follows:

$$\sigma_{wmr}^2 = 0.02\frac{z}{h_{mix}} \qquad (z < h_{mix}) \tag{14.27}$$

$$\sigma_{wmr}^2 = 0.02 \qquad (z > h_{mix}) \tag{14.28}$$

The calculation of the vertical wind speed variance in the stable boundary layer is essentially the same as in the convective boundary layer, except that $\sigma_{wc}=0$ under stable conditions.

### 14.3.4 Profile of Horizontal Turbulence

The calculation of the horizontal wind speed variance follows a pattern similar to the calculation of the vertical wind speed variance. In **unstable conditions** the calculation is based on the following equation:

$$\sigma_v^2 = \sigma_{vc}^2 + \sigma_{vm}^2 \tag{14.29}$$

where $\sigma_{vc}^2$ is calculated as follows within the mixed layer:

$$\sigma_{vc}^2 = 0.35w_*^2 \qquad (z < h_{mix}) \tag{14.30}$$

Above the mixed layer the calculation scheme depends on the value of $\sigma_{vc}^2$ within the mixed layer. If $0.35w_*^2 < 0.25\,\mathrm{m^2\,s^{-2}}$, $\sigma_{vc}^2$ is calculated as follows:

$$\sigma_{vc}^2 = 0.35w_*^2 \qquad (z > h_{mix}) \tag{14.31}$$

If $0.35w_*^2 > 0.25\,\mathrm{m^2\,s^{-2}}$, $\sigma_{vc}^2$ is calculated as follows:

$$\sigma_{vc}^2 = 0.35w_*^2 - \left(\frac{0.35w_*^2 - 0.25\mathrm{m^2\,s^{-2}}}{0.2h_{mix,c}}\right)(z - h_{mix,c}) \quad (h_{mix} < z < 1.2h_{mix}) \tag{14.32}$$

$$\sigma_{vc}^2 = 0.25\,\mathrm{m^2\,s^{-2}} \qquad (z > 1.2h_{mix}) \tag{14.33}$$

The calculation of $\sigma_{vm}^2$ depends on the **mechanically** generated mixing height:

$$\sigma_{vm}^2 = \sigma_{v0}^2 = 3.6u_*^2 \qquad (z = 0) \tag{14.34}$$

$$\sigma_{vm}^2 = \sigma_{vmh}^2 = \min\left(0.25\mathrm{m^2\,s^{-2}}, \sigma_{v0}^2\right) \quad (z > h_{mix,m}) \tag{14.35}$$

$$\sigma_{vm}^2 = \sigma_{v0}^2 - \left(\frac{\sigma_{v0}^2 - \sigma_{vmh}^2}{h_{mix,m}}\right)z \qquad (0 < z < h_{mix,m}) \tag{14.36}$$

In **stable conditions** the calculation scheme for the horizontal wind speed variance is the same as for convective conditions, except that $\sigma_{vc}^2 = 0$.

### 14.3.5 Calculation of Effective Variables

Plumes are spatially extended, but the calculation of plume variables can usually take only a single value of the atmospheric conditions, although these conditions vary within the plume. This can be a source of error in dispersion calculations. The error can be reduced by using a suitable averaging procedure to ensure that the plume variables as calculated in the model are based on atmospheric conditions that are representative for the calculation. AERMOD stands out among air dispersion models for the efforts made to calculate such average variables or **effective** variables.

In AERMOD the procedure for calculating effective variables can be summarized as follows:

- The plume centroid height $H_p$ as defined below is calculated.
- Values of $\sigma_w$ and $u$ at $H_p$ are used to calculate $\sigma_z$.
- The height interval from $H_p$ to the receptor, or from $H_p$, $2.15\sigma_z$ in the direction of the receptor, whichever comes first, is considered.
- The variable of interest is integrated across the height interval defined in the previous step, and divided by the height of the interval, to obtain the effective variable.

For instance, if the receptor (height $z_r$) is at less than $2.15\sigma_z$ away from $H_p$, then the effective wind speed $\tilde{u}$ is calculated as

$$\tilde{u} = \frac{1}{H_\mathrm{p} - z_\mathrm{r}} \int_{z_\mathrm{r}}^{H_\mathrm{p}} u(z)\,dz \tag{14.37}$$

The plume centroid height mentioned in the first step is calculated as follows. If the distance to final plume rise $x_\mathrm{f}$ has not yet been reached, the plume centroid height is taken equal to the effective plume height as obtained by adding the transitional plume rise to the stack height:

$$H_\mathrm{p} = h_\mathrm{s} + \Delta z \tag{14.38}$$

When the distance to the source exceeds $x_\mathrm{f}$, the calculation depends on the atmospheric conditions. In a stable atmosphere, $H_\mathrm{p}$ stays constant and is equal to the effective plume height based on final plume rise:

$$H_\mathrm{p} = h_\mathrm{s} + \Delta h \tag{14.39}$$

In an unstable atmosphere, plume spreading is so prounounced that the plume centroid height moves even after $x_\mathrm{f}$ is reached. Hence, a second downwind distance $x_\mathrm{m}$ is calculated, an estimate of the distance to complete vertical mixing:

$$x_\mathrm{m} = \frac{\overline{u} h_\mathrm{mix}}{\overline{\sigma_w}} \tag{14.40}$$

where $\overline{u}$ and $\overline{\sigma_w}$ are *average* values, not to be confused with effective values:

$$\overline{u} = \frac{1}{h_\mathrm{mix}} \int_0^{h_\mathrm{mix}} u\,dz \tag{14.41}$$

$$\overline{\sigma_w} = \frac{1}{h_\mathrm{mix}} \int_0^{h_\mathrm{mix}} \sigma_w\,dz \tag{14.42}$$

When the distance to the source exceeds $x_\mathrm{m}$, the plume is assumed to be well mixed across the mixing layer, and the plume centroid height is, therefore,

$$H_\mathrm{p} = \frac{h_\mathrm{mix}}{2} \tag{14.43}$$

Between $x_\mathrm{f}$ and $x_\mathrm{m}$ the plume centroid height is calculated as a linear interpolation between the limiting values:

$$H_\mathrm{p} = H_\mathrm{p}(x_\mathrm{f}) + \left[\frac{h_\mathrm{mix}}{2} - H_\mathrm{p}(x_\mathrm{f})\right] \frac{x - x_\mathrm{f}}{x_\mathrm{m} - x_\mathrm{f}} \tag{14.44}$$

where $H_\mathrm{p}(x_\mathrm{f})$ is the value of $H_\mathrm{p}$ at $x = x_\mathrm{f}$.

### 14.3.6 Complex Terrain in AERMOD

When a landscape feature such as a hill is on the path of a plume, the following events are possible in AERMOD:

- The plume travels over the hill (i.e., it follows the terrain).
- The plume travels "through" the hill (i.e., it moves horizontally).
- Part of the plume travels over the hill while the rest travels through it.

In reality, the plume can follow only a horizontal path *around* a landscape feature, not through it. However, such a process is difficult to include in a plume concept and is not included in AERMOD.

The basis of the calculation of plume behavior in complex terrain is the concept of the **dividing streamline height** $H_c$. This concept is discussed in Section 5.13.3. In stable conditions, an air parcel can only overcome the stratification of the atmosphere to move over a hill if it has sufficient kinetic energy to overcome the buoyant forces. With increasing height, an air parcel has more kinetic energy available (higher wind speed) and requires less energy as it requires less vertical movement to reach the hill top. Hence, there is a height where the air parcel has just enough kinetic energy to reach the hill top. This height is the dividing streamline height. In Section 5.13.3 an approximate derivation links the dividing streamline height to the Brunt–Väisälä frequency $N_{BV}$ [eq. (5.257)]:

$$N_{BV} = \sqrt{\frac{g}{\theta}\frac{d\theta}{dz}} \tag{5.257}$$

Assuming constant $N_{BV}$, it was shown that the dividing streamline height is given by eq. (5.265). AERMOD uses a more sophisticated equation that incorporates the height dependence of the Brunt–Väisälä frequency:

$$\frac{1}{2}u^2(H_c) = \int_{H_c}^{h_c} N_{BV}^2(h_c - z)\,dz \tag{14.45}$$

where $u(H_c)$ is the wind speed at height $H_c$. In AERMOD, $h_c$ is the height of the terrain that dominates the flow in the vicinity of the receptor, usually the height of the nearest hill.

In unstable and neutral conditions the Brunt–Väisälä frequency is not defined because the buoyancy does not counteract the vertical movement of the air parcel. In these conditions the dividing streamline height is ground level, and the entire plume has sufficient kinetic energy to reach the top of the hill.

When the dividing streamline height is known, $\varphi$ is defined as the fraction of the plume that is below $H_c$, that is, the fraction of the plume that has insufficient kinetic energy to reach the top of the hill. Based on $\varphi$, the plume is divided into two parts:

- A part with fraction $f$ that moves horizontally on an absolute altitude scale (i.e., which does not follow the terrain)

- A part with fraction $(1 - f)$ that moves horizontally in terrain following coordinates (i.e., which moves up and down with the landscape)

Fraction $f$ is given by the following equation:

$$f = 0.5\,(1+\varphi) \tag{14.46}$$

A consequence of this calculation scheme is that half of the plume will not follow the landscape ($f$=0.5) even if it is entirely capable of reaching the top of the hill ($\varphi = 0$). If the plume is entirely incapable of reaching the top of the hill ($\varphi = 1$), then the plume will not follow the landscape at all ($f$=1).

For the implementation of the terrain effects in the concentration calculation, AERMOD calculates each pollutant concentration twice: once in terrain-following coordinates and once in a coordinate system that does not follow the terrain. The concentration generated by AERMOD is a weighted average of these two concentrations.

Assume that a receptor is located at a downwind distance $x_r$ from the source and at a lateral distance $y_r$ from the source. It is located at a height $z_r$ above the base of the stack emitting the plume and at a height $z_p$ above the local ground. The concentration obtained by applying the Gaussian plume equation to the receptor coordinates $(x_r, y_r, z_r)$ is $c(x_r, y_r, z_r)$. The concentration obtained by applying the Gaussian plume equation to the receptor coordinates $(x_r, y_r, z_p)$ is $c(x_r, y_r, z_p)$. The overall concentration calculated by AERMOD is

$$c = fc\,(x_r, y_r, z_r) + (1-f)\,c(x_r, y_r, z_p) \tag{14.47}$$

### 14.3.7 Concentration Predictions in AERMOD

In the previous section the impact of complex terrain on the concentration predictions in AERMOD was outlined. We have seen that AERMOD calculates concentrations in two different coordinate systems (one that follows the landscape and one that does not follow the landscape) and takes a weighted average. In this section the concentration calculations given a certain coordinate system will be outlined.

The concentration calculation in AERMOD follows the general pattern of plume modeling:

$$\bar{C} = C_x \varphi_y \varphi_z \tag{6.1}$$

where $C_x$ is the dilution factor due to wind:

$$C_x = \frac{Q}{\bar{u}} \tag{6.2}$$

and $\varphi_y$ and $\varphi_z$ are horizontal and vertical dispersion factors.

However, in AERMOD the value of $\bar{u}$ (here the time-averaged wind speed) in eq. (6.2) is replaced by $\tilde{u}$, so the following equation is obtained:

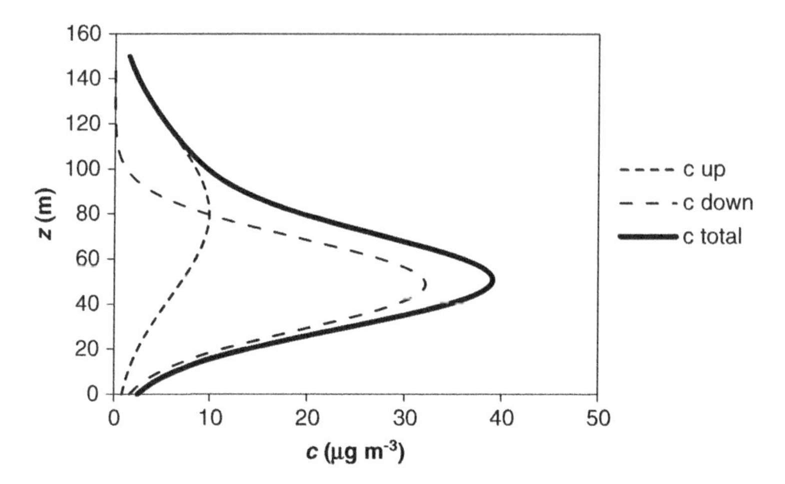


Figure 14.1 Vertical concentration profile resulting from a Gaussian concentration profile representing updrafts and a Gaussian concentration profile representing downdrafts.

$$C_x = \frac{Q}{\tilde{u}} \tag{14.48}$$

In the convective boundary layer the existence of convective cycles is accounted for in a number of ways. As discussed in Sections 5.12.1 and 5.13.1, convective cycles consist of a relatively small but strong updraft and a more extended but weaker downdraft. A small fraction of the plume will get caught in updrafts and rise considerably in comparison with the average plume particle. A larger fraction of the plume will get caught in downdrafts and sink slightly in comparison with the average plume particle.

The main repercussion of this plume behavior is that the vertical concentration profile of the pollutant is not Gaussian but can be described by the superposition of two Gaussian curves. This is illustrated in Figure 14.1. The result is a skewed concentration profile.

The behavior of convective cycles near the capping inversion also affects plume behavior. A plume particle that hits the inversion layer after being caught in an updraft will be transported along the inversion layer before descending in a downdraft. This leads to a delay of the downward movement in comparison with the reflection against the inversion layer.

In addition to the effects of convective cycling, plume behavior can also deviate from simple Gaussian spreading when part of the plume penetrates the capping inversion. This section of the plume behaves very differently from the fraction of the plume that stays in the mixing layer. It will be subject to less turbulence and hence less dispersion.

These three effects are dealt with in AERMOD by considering three sources: the direct source, an indirect source representing reflection on the capping inversion, and a penetrated source representing plume particles trapped in the capping inversion. This is illustrated in Figure 14.2.

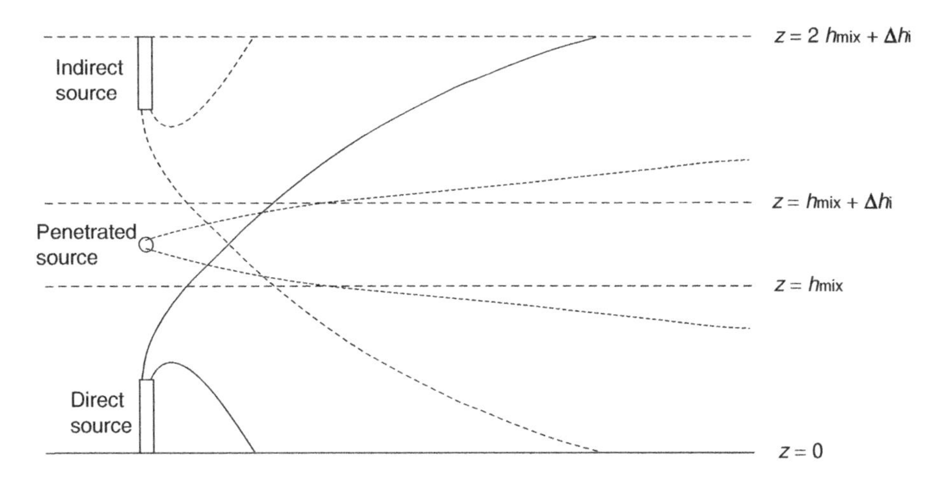


Figure 14.2 Real and virtual plumes in the convective boundary layer in AERMOD.

The **direct source** contribution to the vertical dispersion factor $\varphi_z$ in eq. (6.1), $\varphi_{zd}$, is given by the following equation:

$$\phi_{zd} = \frac{\lambda_1 f_p}{\sqrt{2\pi}\sigma_{z1}} \exp\left[-\frac{(z-\psi_{d1})^2}{2\sigma_{z1}^2}\right] + \frac{\lambda_2 f_p}{\sqrt{2\pi}\sigma_{z2}} \exp\left[-\frac{(z-\psi_{d2})^2}{2\sigma_{z2}^2}\right] \tag{14.49}$$

where $f_p$ is the fraction of the plume that is not trapped in the capping inversion (see below). The variables $\psi_{d1}$ and $\psi_{d2}$ are effective source heights corresponding with updrafts and downdrafts, respectively. They are given by the following equations:

$$\psi_{d1} = h_s + \Delta h_d + \frac{\overline{w_1} x}{u} \tag{14.50}$$

$$\psi_{d2} = h_s + \Delta h_d + \frac{\overline{w_2} x}{u} \tag{14.51}$$

where $h_s$ is the source height, $\Delta h_d$ is the direct plume rise, and $u$ is the wind speed at the stack height. Other variables are defined as in Section 5.13.1. Whenever there is a choice between alternatives in that section, AERMOD chooses the option of Weil et al. (1997).

For clarity, reflection on the surface is not included in eq. (14.49). However, it is included in AERMOD, so in practice eq. (14.49) is replaced by what is in principle an infinite sum of reflection terms. An exception is the first reflection on the inversion layer, which is handled by the indirect source term.

In eq. (14.49), the factor $f_p$ is calculated based on a plume rise consideration. The following is an estimate of the plume rise into a stable environment:

$$\Delta h_{eq} = \left[2.6^3 P_S + \left(\frac{2}{3}\right)^3\right]^{1/3} \Delta h_h \tag{14.52}$$

where $P_s$ is a penetration parameter, defined by:

$$P_s = \frac{F_b}{uN_h^2 \, \Delta h_h^3} \tag{14.53}$$

where $F_b$ is the buoyancy flux parameter. Assuming an emission with the same molar mass as air, and ideal gas behavior, eq. (7.2) can be converted to

$$F_b = g w_s r_s^2 \frac{\Delta T}{T_s} \tag{14.54}$$

where $w_s$ is the exit velocity of the stack gas, $r_s$ is the stack radius, $\Delta T$ is the difference between the stack gas temperature and the ambient temperature, and $T_s$ is the temperature of the stack gas. Variable $N_h$ is defined by

$$N_h = \sqrt{\frac{g}{\theta_{h_{mix}}} \left( \frac{\partial \theta}{\partial z} \right)_{z > h_{mix}}} \tag{14.55}$$

where $\theta_{h_{mix}}$ is the potential temperature evaluated at $z = h_{mix}$; and $\Delta h_h$ is defined as the distance between the stack tip and the inversion layer:

$$\Delta h_h = h_{mix} - h_s \tag{14.56}$$

The calculation of $f_p$ depends on the relative values of $\Delta h_h$ and $\Delta h_{eq}$:

$$\text{If } \Delta h_h < 0.5 \Delta h_{eq} : f_p = 0 \tag{14.57}$$

$$\text{If } \Delta h_h > 1.5 \Delta h_{eq} : f_p = 1 \tag{14.58}$$

$$\text{If } 0.5 \Delta h_{eq} < \Delta h_h < 1.5 \Delta h_{eq} : f_p = \frac{\Delta h_h}{\Delta h_{eq}} - 0.5 \tag{14.59}$$

The dispersion parameters in eq. (14.49) are discussed in the next section.

The **indirect source** contribution to the vertical dispersion factor $\varphi_z$ in eq. (6.1), $\varphi_{zi}$, is given by the following equation:

$$\varphi_{zi} = \frac{\lambda_1 f_p}{\sqrt{2\pi}\sigma_{z1}} \exp\left[ -\frac{(z - \psi_{r1} - 2h_{mix})^2}{2\sigma_{z1}^2} \right] + \frac{\lambda_2 f_p}{\sqrt{2\pi}\sigma_{z2}} \exp\left[ -\frac{(z - \psi_{r2} - 2h_{mix})^2}{2\sigma_{z2}^2} \right] \tag{14.60}$$

where

$$\psi_{r1} = \psi_{d1} - \Delta h_i \tag{14.61}$$

$$\psi_{r2} = \psi_{d2} - \Delta h_i \tag{14.62}$$

and $\Delta h_i$ is given by

$$\Delta h_i = \frac{x}{u_p}\sqrt{\frac{2F_b h_{mix}}{\alpha_r u_p r_y r_z}} \tag{14.63}$$

where $u_p$ is the wind speed used for calculating plume rise (see further), $F_b$ is the buoyancy flux parameter (see above), $\alpha_r = 1.4$, and $r_y r_z$ is given by

$$r_y r_z = r_h^2 + \frac{a_e \lambda_y^{3/2}}{4}\frac{w_*^2 x^2}{u_p^2} \tag{14.64}$$

where

$$r_h = \beta_2 (h_{mix} - h_s) \tag{14.65}$$

where $\beta_2 = 0.4$, $\lambda_y = 2.3$, and $\alpha_e = 0.1$. Again, AERMOD includes additional reflections on the surface that are not included in eq. (14.60). The dispersion parameters in eq. (14.60) are discussed in the next section.

The **penetrated source** contribution to the vertical dispersion factor $\varphi_z$ in eq. (6.1), $\varphi_{zp}$, is given by the following equation:

$$\varphi_{zp} = \frac{1-f_p}{\sqrt{2\pi}\sigma_{zp}}\exp\left[-\frac{(z-h_{ep})^2}{2\sigma_{zp}^2}\right] \tag{14.66}$$

where the calculation of $h_{ep}$ depends on the degree of plume penetration into the capping inversion. When the penetration is full (i.e., $f_p = 0$), then $h_{ep}$ is calculated as follows:

$$h_{ep} = h_s + \Delta h_{eq} \tag{14.67}$$

When the penetration is not full, $h_{ep}$ is calculated as

$$h_{ep} = \frac{h_s + h_{mix}}{2} + 0.75\Delta h_{eq} \tag{14.68}$$

AERMOD also includes reflections of the penetrated plume on the surface and a soft boundary located 2.15 standard deviations above the penetrated plume center. The dispersion parameters in eq. (14.66) are discussed in the next section.

The vertical dispersion factor $\varphi_z$ is calculated as

$$\varphi_z = \varphi_{zd} + \varphi_{zi} + \varphi_{zp} \tag{14.69}$$

In the stable boundary layer, the calculation of $\varphi_z$ is much more straightforward because the dispersion can be considered Gaussian. Hence, eq. (6.6) applies:

$$\varphi_z = \frac{1}{\sqrt{2\pi}\sigma_z}\left\{\exp\left[-\frac{1}{2}\frac{(z-h)^2}{\sigma_z^2}\right]+\exp\left[-\frac{1}{2}\frac{(z+h)^2}{\sigma_z^2}\right]\right\} \tag{6.6}$$

In addition, AERMOD considers reflections on a soft upper boundary similar to the penetrated source in the convective boundary layer. See Cimorelli et al. (2004) for details.

The *lateral* plume dispersion factor $\varphi_y$ is more straightforward than the *vertical* plume dispersion factor because the dispersion can be considered Gaussian in this direction, irrespective of the conditions of the boundary layer. Hence, eq. (6.3) applies:

$$\varphi_y = \frac{1}{\sqrt{2\pi}\sigma_y}\exp\left(-\frac{1}{2}\frac{y^2}{\sigma_y^2}\right) \tag{6.3}$$

At large distances, plume meandering due to mesoscale wind direction changes can increase plume dispersion. AERMOD includes a procedure to account for such effects. However, except for low wind speeds, this procedure is only invoked at distances were the use of AERMOD is not recommended. See Cimorelli et al. (2004).

## 14.3.8 Dispersion Parameters in AERMOD

Plume dispersion parameters in AERMOD include plume-rise-induced dispersion. Hence, Eqs. (6.21) and (6.22) apply:

$$\sigma_y = \sqrt{\sigma_{y,0}^2 + 0.08\,(\Delta h)^2} \tag{6.21}$$

$$\sigma_z = \sqrt{\sigma_{z,0}^2 + 0.08(\Delta h)^2} \tag{6.22}$$

where $\sigma_{y,0}$ and $\sigma_{z,0}$ are the plume dispersion parameters in the absence of plume rise, and $\Delta h$ is the plume rise.

In the case of **lateral dispersion** the same equation is used for both the convective and the stable boundary layer:

$$\sigma_{y,0} = \frac{\tilde{\sigma}_v x}{\tilde{u}\left[1+78 z_{PG}\tilde{\sigma}_v x/\left(z_{max}\tilde{u}h_{mix}\right)\right]^{0.3}} \tag{14.70}$$

where $z_{PG}$=0.46 m, $z_{max}$ is given by

$$z_{max} = \max(z_s, z_{PG}) \tag{14.71}$$

where the definition of $z_s$ depends on the conditions. In the stable boundary layer, $z_s$ is the effective source height $h$, given by the sum of the actual source hight and the plume rise. In the case of the convective boundary layer, $z_s$ is the actual source height, for the penetrated source, $z_s$ equals $h_{ep}$ [eqs. (14.67)–(14.68)].

The power 0.3 in the denominator of eq. (14.70) does not permit us to define a Lagrangian integral time scale. Nevertheless, a comparison with eq. (6.68) provides insight into the structure of eq. (14.70):

$$f_{x,y,z} = \frac{1}{[1+x/(2\bar{u}T_{\mathrm{i,L}})]^{0.5}} = \frac{1}{[1+x/(2L_{\mathrm{L}})]^{0.5}} \tag{14.68}$$

If a Lagrangian integral time were to be defined based on eq. (14.70), it would increase with increasing emission height. In Section 6.11.2 we have seen that the Lagrangian integral time scale should indeed increase with increasing height, at least in the surface layer. Most Gaussian dispersion models do not account for this effect. Even the popular Briggs (1973) equations [eqs. (6.10) and (6.11)] do not account for this effect. This is one of the main reasons for AERMOD's success.

The **vertical dispersion** parameter depends on the state of the boundary layer (stable or convective). In the stable boundary layer, the dispersion parameter depends on two components: a near-surface component $\sigma_{z,g}$ and an elevated component $\sigma_{z,e}$. The calculations is as follows for an effective source height $h < h_{\mathrm{mix}}$:

$$\sigma_{z,0} = \left(1 - \frac{h}{h_{\mathrm{mix}}}\right)\sigma_{z,g} + \frac{h}{h_{\mathrm{mix}}}\sigma_{z,e} \tag{14.72}$$

When the effective source height exceeds the mixing layer height ($h > h_{\mathrm{mix}}$), the calculation is as follows:

$$\sigma_{z,0} = \sigma_{z,e} \tag{14.73}$$

where

$$\sigma_{z,e} = \frac{\tilde{\sigma}_w x}{\tilde{u}\left(1 + \frac{\tilde{\sigma}_w x}{2\tilde{u}}\left(\frac{1}{0.36h} + \frac{N_{\mathrm{BV}}}{0.27\tilde{\sigma}_w}\right)\right)^{1/2}} \tag{14.74}$$

$$\sigma_{z,g} = \sqrt{\frac{2}{\pi}}\frac{u_* x}{\tilde{u}\,[1+0.7(x/L)]^{1/3}} \tag{14.75}$$

In the convective boundary layer we need two dispersion parameters: one for the updrafts ($\sigma_{z,0,1}$) and one for the downdrafts ($\sigma_{z,0,2}$). Part of the calculations were already discussed in Section 5.13.1. Variables that are not defined in the current section can be found there. Again, the parameters depend on an elevated component and a near-surface component, but this time it is the variances that are added:

$$\sigma_{z,0,1}^2 = \sigma_{z,e,1}^2 + \sigma_{z,g}^2 \tag{14.76}$$

$$\sigma_{z,0,2}^2 = \sigma_{z,e,2}^2 + \sigma_{z,g}^2 \tag{14.77}$$

with

$$\sigma_{z,e,1} = \alpha_b \frac{\sigma_{w1} x}{\tilde{u}} \tag{14.78}$$

$$\sigma_{z,e,2} = \alpha_b \frac{\sigma_{w2} x}{\tilde{u}} \tag{14.79}$$

The variable $\alpha_b$ is defined as follows:

$$\alpha_b = \min\left(0.6 + \frac{4H_p}{h_{mix}}, 1\right) \tag{14.80}$$

where $H_p$ is the plume centroid height, defined in eqs. (14.38), (14.39), (14.43), and (14.44). The following equations have been found in Section 5.13.1:

$$\sigma_{w1} = R\overline{w_1} \tag{5.218}$$

$$\sigma_{w2} = -R\overline{w_2} \tag{5.219}$$

$$\overline{w_1} = \frac{S\sigma_w \gamma_1}{2}\left(1 + \sqrt{1 + \frac{4}{\gamma_1^2 \gamma_2 S^2}}\right) \tag{5.222}$$

$$\overline{w_2} = \frac{S\sigma_w \gamma_1}{2}\left(1 + \sqrt{1 + \frac{4}{\gamma_1^2 \gamma_2 S^2}}\right) \tag{5.223}$$

$$S = 0.105 \frac{w_*^3}{\sigma_w^3} \tag{5.214}$$

$$\gamma_1 = \frac{1 + R^2}{1 + 3R^2} \tag{5.226}$$

$$\gamma_2 = 1 + R^2 \tag{5.227}$$

where $R$ was chosen equal to 2 in AERMOD.

The near-surface component of the dispersion parameter is calculated as follows when $H_p < 0.1 h_{mix}$:

$$\sigma_{z,g} = b_c \left(1 - 10\frac{H_p}{h_{mix}}\right)\left(\frac{u_*}{\tilde{u}}\right)^2 \frac{x^2}{|L|} \tag{14.81}$$

where $b_c = 0.5$. When $H_p > 0.1 h_{mix}$, the equation is

$$\sigma_{z,g} = 0 \tag{14.82}$$

### 14.3.9 Plume Rise Calculations in AERMOD

In the convective boundary layer the transitional plume rise is calculated with the Briggs (1975) equation for plumes governed by both buoyancy and momentum:

$$\Delta z = \left( \frac{3F_{\mathrm{m}}x}{0.6^2 \overline{u}^2} + \frac{3F_{\mathrm{b}}x^2}{2 \cdot 0.6^2 \overline{u}^3} \right)^{1/3} \tag{7.25}$$

The wind speed used in eq. (7.25) is the wind speed at the stack tip. Other variables in eq. (7.25) are the buoyancy flux parameter $F_{\mathrm{b}}$ [see eq. (14.54)] and the momentum flux parameter $F_{\mathrm{m}}$:

$$F_{\mathrm{m}} = w_{\mathrm{s}}^2 r_{\mathrm{s}}^2 \frac{T}{T_{\mathrm{s}}} \tag{14.83}$$

Based on Briggs (1975), the following equations are used for the maximum distance of plume rise:

$$x_{\mathrm{f}} = d_1 F_{\mathrm{b}}^{5/8} \quad \text{for} \quad F_{\mathrm{b}} < 55\,\mathrm{m}^4\mathrm{s}^{-3} \tag{7.21}$$

$$x_{\mathrm{f}} = d_2 F_{\mathrm{b}}^{2/5} \quad \text{for} \quad F_{\mathrm{b}} < 55\,\mathrm{m}^4\mathrm{s}^{-3} \tag{7.22}$$

where $d_1 = 49\,\mathrm{s}^{15/8}\,\mathrm{m}^{-3/2}$, and $d_2 = 119\,\mathrm{s}^{9/5}\,\mathrm{m}^{-7/5}$; and $x_{\mathrm{f}}$ is in meters.

In the stable boundary layer the plume rise is calculated with eq. (7.26):

$$\Delta z = 2.66 \left( \frac{F_{\mathrm{b}}}{\overline{u}s} \right)^{1/3} \left[ \frac{0.7s^{1/2} F_{\mathrm{m}}}{F_{\mathrm{b}}} \sin\left( \frac{0.7s^{1/2}x}{\overline{u}} \right) + 1 - \cos\left( \frac{0.7s^{1/2}x}{\overline{u}} \right) \right]^{1/3} \tag{7.26}$$

where $s$ is the stability parameter, or the square of the Brunt–Väisälä frequency. Define $N'$ as

$$N' = 0.7 N_{\mathrm{BV}} \tag{14.84}$$

Hence, eq. (7.26) can also be written as

$$\Delta z = 2.66 \left( \frac{F_{\mathrm{b}}}{\overline{u} N_{\mathrm{BV}}^2} \right)^{1/3} \left[ \frac{N' F_{\mathrm{m}}}{F_{\mathrm{b}}} \sin\left( \frac{N'x}{\overline{u}} \right) + 1 - \cos\left( \frac{N'x}{\overline{u}} \right) \right]^{1/3} \tag{14.85}$$

A problem that arises in a stable atmosphere is that the wind speed is strongly height dependent, and the wind speed at the real source height may not be representative for the average wind speed during plume rise. For that reason, an iterative procedure is adopted:

- Calculate $\Delta z$ with the values of $N_{\mathrm{BV}}$ and the wind speed at stack height $h_{\mathrm{s}}$.
- Evaluate $N_{\mathrm{BV}}$ and the wind speed at height $h_{\mathrm{s}}$ and at height $h_{\mathrm{s}} + \Delta z/2$ and take the average.

- Calculate $\Delta z$ with the new values of $N_{BV}$ and the wind speed.
- If the new value of $\Delta z$ differs from the old value by more than an acceptable margin, repeat step 2.

The distance to final plume rise is given by

$$x_f = \frac{\overline{u}}{N'} \arctan\left( \frac{F_b}{N'F_m} \right) \tag{14.86}$$

Final plume rise is given by eq. (7.11) but with $\pi_1^{-1/3} = 2.66$:

$$\Delta h = 2.66 \left( \frac{F_b}{\overline{u} N_{BV}^2} \right)^{1/3} \tag{14.87}$$

When the atmosphere is close to neutral, the Brunt–Väisälä frequency approaches zero, and the final plume rise approaches infinity. Therefore, a different equation is needed. AERMOD uses the following equation in those conditions:

$$\Delta z = 1.2 L_n^{3/5} (h_s + 1.2 L_n)^{2/5} \tag{14.88}$$

where $L_n$ is a neutral length scale, calculated as

$$L_n = \frac{F_b}{u u_*^2} \tag{14.89}$$

Equation (14.88) is used whenever it predicts lower values than eq. (14.85) or eq. (14.87) (whichever is applicable).

An additional problem arises when the wind speed approaches zero. In that case, plume rise becomes infinitely large. In that case, eq. (7.15) can be used:

$$\Delta h = \pi_1^{-1/4} \left( \frac{F_b^2}{s^3} \right)^{1/8} \tag{7.15}$$

However, instead of using 5.3 for $\pi_1^{-1/4}$, as in Chapter 7, a value of 4 is used by AERMOD.

In exceptional circumstances it is possible that both $N_{BV}$ and the wind speed approach zero. In that case all equations designed for stable conditions overestimate the plume rise. Hence, eq. (7.25), which is valid for the convective boundary layer, is used.

## REFERENCES

Briggs G.A. (1973). *Diffusion Estimation of Small Emissions.* Contribution No. 79, Atmospheric Turbulence and Diffusion Laboratory, Oak Ridge, TN.

Briggs G.A. (1975). Plume rise predictions. In Haugen D.A. (ed.). *Lectures on Air Pollution and Environmental Impact Analyses.* American Meteorological Society, Boston, pp. 59–111.

Cimorelli A.J., Perry S.G., Venkatram A., Weil J.C., Paine R.J., Wilson R.B., Lee R.F., Peters W.D., Brode R.W., and Paumier J.O. (2004). *AERMOD: Description of Model Formulation*. Report EPA-454/R-03-004, US-EPA, Research Triangle Park, NC.

Dyer A.J. (1974). A review of flux-profile relationships. *Bound. Layer Meteorol.* **7**, 363–372.

EPA (Environmental Protection Agency). (2005). 40 CFR Part 51. Revision to the guideline on air quality models: Adoption of a preferred general purpose (flat and complex terrain) dispersion model and other revisions; Final rule. *Federal Register* **70(216)**, 68218–68261 (http://www.epa.gov/scram001/guidance/guide/appw_05.pdf).

Holtslag A.A.M. and van Ulden A.P. (1983). A simple scheme for daytime estimates of the surface fluxes from routine weather data. *J. Clim. Appl. Meteorol.* **22**, 517–529.

Horst T.W. (1984). The modification of plume models to account for dry deposition. *Bound. Layer Meteorol.* **30**, 413–430.

Kasten F. and Czeplak G. (1980). Solar and terrestrial radiation dependent on the amount and type of cloud. *Solar Energy* **24**, 177–189.

Panofsky H.A. and Dutton J.A. (1984). *Atmospheric Turbulence: Models and Methods for Engineering Applications*. Wiley, New York.

Stull, R.B. (1983). A heat flux history length scale for the nocturnal boundary layer. *Tellus* **35A**, 219–230.

van Ulden A.P. and Holtslag A.A.M. (1985). Estimation of atmospheric boundary layer parameters for diffusion applications. *J. Climate Appl. Meteorol.* **24**, 1196–1207.

Weil J.C., Corio L.A., and Brower R.P. (1997). A PDF dispersion model for buoyant plumes in the convective boundary layer. *J. Appl. Meteorol.* **36**, 982–1003.

CHAPTER 15

# CALPUFF AND CALMET: A DETAILED DESCRIPTION

## 15.1 INTRODUCTION

CALPUFF is a Gaussian puff model that was developed in the late 1980s (Scire et al., 2000a). The meteorological preprocessor of CALPUFF is CALMET (Scire et al., 2000b). The CALPUFF/CALMET modeling system is one of the most commonly used models, as it is one of the EPA-recommended models. In particular, EPA recommends CALPUFF for applications where the distance between the source and the receptor exceeds 50 km.

The quality of a meteorological preprocessor is one of the main determinants of the overall quality of the air dispersion model, and this is particularly true for the CALPUFF/CALMET modeling system in a wide range of conditions. The main purpose of CALMET is to obtain the best possible meteorological data based on the available information. In particular, CALMET can receive measured data, modeled data (i.e., generated by a meteorological model like MM5 or WRF), or both. When a high-resolution terrain data set is available, CALMET is capable of using this information to estimate local deviations from meteorological data measured or modeled at a coarser resolution.

CALPUFF is a relatively old model that has undergone many modifications and improvements over the years, and many of the older model versions are still included in the model as model options. Some of these options are no longer recommended, so deviations from the default options should be chosen with care.

The official versions of CALPUFF and CALMET are available on the TRC website at http://www.src.com/calpuff/calpuff1.htm. Additional information is available on the EPA website at http://www.epa.gov/scram001/dispersion_prefrec.htm.

This chapter contains a thorough overview of the model formulation of CALPUFF and CALMET. After reading this chapter, the reader should have a good understanding of how these models work and make rational choices among the various options.

*Air Dispersion Modeling: Foundations and Applications*, First Edition. Alex De Visscher.

## 15.2 DESCRIPTION OF CALMET

### 15.2.1 Coordinate System

CALMET uses a coordinate system that follows the landscape. This can be expressed as follows:

$$Z = z - h_t \tag{15.1}$$

where $Z$ is the terrain-following vertical coordinate, $z$ is the vertical coordinate with respect to an absolute (Cartesian) altitude scale, and $h_t$ is the altitude of the terrain. The vertical wind speed in terrain-following coordinates is as follows:

$$W = w - u\frac{\partial h_t}{\partial x} - v\frac{\partial h_t}{\partial y} \tag{15.2}$$

where $W$ is the vertical wind speed in terrain-following coordinates, $w$ the vertical wind speed in Cartesian coordinates, and $u$ and $v$ are the horizontal wind components in the $x$ and $y$ directions. The $x$ coordinate runs from west to east; the $y$ coordinate runs from south to north.

The domain is divided into grid cells, and the grid points for the horizontal wind speed components $u$ and $v$ are defined in the centers of the cells. The vertical wind speed component, however, is defined at the center of the top and bottom faces of the cells. The origin of the grid is the southernmost, westernmost point in the grid, at ground level, so that the first grid point for $u$ and $v$ is displaced one-half of the cell size east, north, and upward. The first grid point for $w$ is defined at ground level and has zero value. In the $z$ direction each layer has a different thickness to ensure sufficient resolution near the ground while covering the entire mixing layer. The heights of the layer boundaries are defined by the user.

### 15.2.2 Introduction to Wind Field Calculations in CALMET

The main function of CALMET is to generate wind fields of sufficient accuracy to resolve local features that affect the movement of air pollutants. This section briefly outlines the procedure followed to generate these wind fields. Sections 15.2.3–15.2.5 describe the algorithms in more detail.

The wind field is generated in three steps:

- **Wind field initial guess determination.** This can be based on three types of data: upper air measurements (e.g., rawinsonde data), ground-level measurements, or a prognostic model (e.g., MM5 output).
- **Step 1 formulation.** In this step the initial guess is refined with mostly physics-based adjustments, such as kinematic effects, slope effects, and mass conservation.
- **Step 2 formulation.** In this step further refinements are made based on observational data (if available) and with mathematical adjustments, such as interpolation and smoothing. This step is completed with additional mass conservation adjustments.

### 15.2.3 Initial Guess Wind Field Generation in CALMET

The first step in the development of gridded wind speed data in CALMET is the establishment of an initial guess. There are several ways to achieve this based on the data provided by the user.

When only a single wind speed is known, CALMET can use a spatially constant wind speed provided by the user. If a measured time series of upper-air data (rawinsonde) is available, CALMET can vertically average each measurement and interpolate in time between measurements. These options do not exploit the benefits of CALPUFF over other models. Unless the terrain is complex, or the wind speed is variable, the modeler is better off using AERMOD if these options are to be used.

A better alternative is when multiple upper-air data sets are available. In that case, wind speeds in each grid cell can be generated with the following equation:

$$u = \frac{\left(u_1 / r_1^2\right) + \left(u_2 / r_2^2\right) + \cdots}{\left(1 / r_1^2\right) + \left(1 / r_2^2\right) + \cdots} \tag{15.3}$$

where $u_1$, $u_2$, ... are the wind speeds at station 1, 2, ..., and $r_1$, $r_2$, ... are the distances from the grid point to station 1, 2, ... . This way spatial variability is introduced in the initial guess wind field, but the temporal resolution is poor because upper-air data are usually obtained only once or twice a day.

To overcome the weak time resolution of upper-air data, and to allow a larger number of measured wind speeds to be used in an initial guess, CALPUFF is capable of using a measured wind speed near the surface at hourly intervals and extrapolate it vertically using similarity theory. The input data needed for the similarity theory are discussed in a later section.

When a combination of upper-air data sets and ground-level data sets are used, the user may prefer to use predominantly ground-level data in the lower layers of the domain and mainly upper-air data in the higher layers. Especially in valleys were the local conditions may deviate from the upper part of the planetary boundary layer, this approach is desired. To accommodate this, CALPUFF contains a user-defined variable, BIAS, in each vertical layer of the grid, which controls the weighing of the data. The weights of the ground-level data are multiplied by a factor of 1 when BIAS is between –1 and 0 and by a factor of (1 – BIAS) when BIAS is between 0 and 1. The weights of the upper-air data is multiplied by a factor of (1 + BIAS) when BIAS is between –1 and 0 and by a factor of 1 when BIAS is between 0 and 1. Consequently, upper-air data are ignored completely when BIAS = –1, whereas ground-level data are ignored completely when BIAS = 1. When BIAS = 0 both ground-level data and upper-air data are used in an unbiased manner.

An initial wind field can also be obtained from an external simulation with a prognostic meteorological model such as MM5 (Penn State/NCAR Mesoscale Model) or CSUMM (Colorado State University Mesoscale Model). In this case the user has the option of skipping the Step 1 formulation and proceeding directly to the step 2 formulation for processing the meteorogical model data. The grid used in the meteorological model does not need to be the same as the grid in CALMET. Usually the meteorological model will use a much coarser grid than CALMET.

### 15.2.4 Step 1: Wind Field Formulation in CALMET

As indicated before, the step 1 formulation consists of the treatment of kinematic effects, slope flow, blocking effects, and divergence minimization.

***15.2.4.1 Kinematic Effects.*** Kinematic effects refer to the impact of sloped terrain on the vertical wind speed. Because the wind follows the terrain near the ground, the wind will have a Cartesian vertical wind speed component near the ground. High above the gound it is likely that the wind is horizontal in Cartestian coordinates. In this section an interpolation procedure between these two limits is provided. The treatment of kinematic effects can lead to unrealistic wind speed profiles and is *not recommended*. The EPA default is *not* to treat kinematic effects.

Kinetic effects can only be invoked in stable conditions, when a Brunt–Väisälä frequencey can be defined:

$$N_{\mathrm{BV}} = \sqrt{\frac{g}{\theta}\frac{d\theta}{dz}} \tag{5.257}$$

where $g$ is the acceleration due to gravity, $\theta$ is the potential temperature; and $N_{\mathrm{BV}}$ is calculated at a user-defined height ZUPT. The Cartesian vertical wind speed is then calculated as follows:

$$w = \left(u\frac{\partial h_{\mathrm{t}}}{\partial x} + v\frac{\partial h_{\mathrm{t}}}{\partial y}\right)\exp(-kz) \tag{15.4}$$

where $h_{\mathrm{t}}$ is the terrain height in an absolute frame of reference; $k$ is given by

$$k = \frac{N_{\mathrm{BV}}}{|V|} \tag{15.5}$$

and $|V|$ is the domain mean wind speed. The wind speed in terrain-following coordinates, $W$, is calculated from $w$ with eq. (15.2).

***15.2.4.2 Slope Flows.*** Slope flow refers to the tendence of cold air to flow down a slope due to its higher density in comparison with the surrounding air. It occurs in stable conditions. An upward slope flow can also exist when a shallow layer of warm air is trapped under an inversion layer. This situation occurs early in the morning. Slope flows were discussed in Section 5.13.4, where a simple model for a downward slope flow was developed.

The treatment of slope flows in CALMET involves a calculational procedure that is very different from the one in Section 5.13.4. It was developed by Scire and Robe (1997) based on a gravity flow model of Mahrt (1982). A slope flow vector with length $S$ is calculated and added to the existing wind vector. The slope flow vector length $S$ is calculated as

$$S=\left[\frac{|q|gx\sin\alpha}{\rho C_p T(C_{\mathrm{D}}+k)}\right]^{1/3}\left[1-\exp\left(-\frac{x}{L_{\mathrm{e}}}\right)\right]^{1/3} \tag{15.6}$$

where $q$ is the sensible heat flux, $x$ is the distance to the crest of the hill, $\alpha$ is the angle of the hill relative to the horizontal, $\rho$ is the density of air, $C_p$ is the heat capacity of air, $T$ the absolute temperature, $C_{\mathrm{D}}$ is a drag coefficient (0.04), $k$ is an entrainment coefficient (0.04); and $L_{\mathrm{e}}$ is an equilibrium length scale given by

$$L_e=\frac{h}{C_{\mathrm{D}}+k} \tag{15.7}$$

where $h$ is the depth of the slope flow given by

$$h=0.05\Delta Z \tag{15.8}$$

where $\Delta Z$ is the elevation drop from the crest of the hill.

The discretization of the terrain can cause large local values of $\alpha$ without physical meaning. To avoid that such artifacts lead to a spurious slope flow, the slope is bounded by the average slope of the terrain:

$$\sin\alpha=\min\left(\sin\alpha_{\mathrm{local}},\frac{\Delta Z}{x}\right) \tag{15.9}$$

For an upslope wind the value of $C_{\mathrm{D}}+k$ is approximately 1. The following equation is used:

$$S=\left(\frac{qg\,\Delta Z}{\rho C_p T}\right)^{1/3} \tag{15.10}$$

In this equation the meaning of $\Delta Z$ is the elevation gain from the valley floor. The vector length $S$ is divided into components in the $x$ and $y$ directions as follows:

$$u_{\mathrm{s}}=\frac{\pm S(\partial h_{\mathrm{t}}/\partial x)}{\sqrt{(\partial h_{\mathrm{t}}/\partial x)^2+(\partial h_{\mathrm{t}}/\partial y)^2}} \tag{15.11}$$

$$u_{\mathrm{s}}=\frac{\pm S(\partial h_{\mathrm{t}}/\partial y)}{\sqrt{(\partial h_{\mathrm{t}}/\partial x)^2+(\partial h_{\mathrm{t}}/\partial y)^2}} \tag{15.12}$$

where + is for upslope flows, and – is for downslope flow; and $u_{\mathrm{s}}$ and $v_{\mathrm{s}}$ are then added to the existing wind speed vectors $u$ and $v$.

***15.2.4.3 Blocking Effects.*** Blocking effects are the CALMET equivalent of the dividing steamline height in AERMOD, which is also discussed in Section 5.13.4. It refers to the interaction between wind flows and obstacles in the terrain in stable atmosphere. The calculation of blocking effects in CALPUFF has the objective of determining whether a wind streamline will move over a hill or around it. It is more sophisticated than the dividing streamline height algorithm in AERMOD in the sense that plumes are able to move sideway around a hill, something that is not possible with a plume in AERMOD. On the other hand, the CALMET algorithm is less sophisticated in the sense that CALMET assumes a constant Brunt–Väisälä frequency $N_{BV}$ (calculated at a user-defined height ZUPT), whereas AERMOD allows for a height-dependent $N_{BV}$.

The calculation of blocking effects is based on the local Froude number [see eq. (5.264)]:

$$\mathrm{Fr} = \frac{V}{N_{BV}\,\Delta h_t} \tag{15.13}$$

where $V$ is the horizontal wind speed [$= (u^2 + v^2)^{1/2}$], and $\Delta h_t$ is given by

$$\Delta h_t = h_{max} - z \tag{15.14}$$

where $h_{max}$ is the highest point in the terrain within a radius TERRAD (a user-defined variable) from the grid point, $z$ is the height of the grid point; and $h_{max}$ and $z$ are measured on the same Cartesian reference system. If Fr is less than a critical Froude number (the user-defined variable CRITFN), then it is assumed that the wind does not have sufficient kinetic energy to move to the top of the hill, so if the uncorrected wind direction is uphill, the direction is modified to flow horizontally in a Cartesian frame of reference (i.e., in the direction where the landscape has zero slope). If Fr is greater than the critical Froude number, the wind direction is unaffected. The default value of CRITFN is 1, consistent with theory in Section 5.13.4.

***15.2.4.4 Divergence Minimization.*** In Chapter 10 we saw that the following equation applies in an incompressible fluid:

$$\frac{\partial u}{\partial x} + \frac{\partial v}{\partial y} + \frac{\partial w}{\partial z} - 0 \tag{10.21}$$

The atmosphere can be considered incompressible at the scale relevant for the calculation discussed here, so eq. (10.21) must be observed. After the preceding three steps, there is no guarantee that the resulting wind field is consistent with eq. (10.21), so a correction is required. If the correction is not made, pollutant puff calculations can include spurious dilution effects due to a spurious creation of air in the mathematical procedure. To that effect the divergence $D$ is defined as

$$D = \frac{\partial u}{\partial x} + \frac{\partial v}{\partial y} + \frac{\partial w}{\partial z} \tag{15.15}$$

Mass conservation requires that $D=0$ with acceptable accuracy. The calculation needed is not strictly a minimization but rather a solution of the divergence problem that is as close as possible to the existing wind field estimation.

The correction to reduce $|D|$ to acceptable values is an iterative procedure as follows:

- Calculate $D$ in each grid point.
- Adjust the four surrounding points (north, south, east, and west of the grid point) of each grid point in such a way that $D$ in the grid point is 0.
- Repeat the procedure until $D < \varepsilon$ in each grid point.

This procedure is iterative because each calculation that renders a $D$ value equal to zero changes the surrounding $D$ values, hence turning zero $D$ values back into non-zero values. However, the $D$ values decrease with each iteration.

### 15.2.5 Step 2: Wind Field Formulation in CALMET

Step 2 formulation for establishing the wind field consists of mainly mathematical adjustments and includes interpolation, smoothing, the O'Brien adjustment of vertical velocities, and an additional divergence minimization.

***15.2.5.1 Interpolation.*** The main purpose of the interpolation step is to introduce data from observational stations into the wind field estimation. Note that in some options the observational data are introduced a second time here. Again, a $1/r^2$ weighing is used, but now it is applied as follows:

$$u_2 = \frac{u_1/R^2 + \sum_k u_{\text{obs},k}/r_k^2}{1/R^2 + \sum_k 1/r_k^2} \tag{15.16}$$

A similar equation applies to the wind speed component in the $y$ direction, $v$. In eq. (15.16), $u_2$ is the corrected wind speed in the $x$ direction in a grid point, $u_1$ is the wind speed in the $x$ direction in the same grid point after completing the step 1 formulation, $u_{\text{obs},k}$ is the wind speed in the $x$ direction observed in meteorological station $k$, $r_k$ is the distance from the grid point to meteorological station $k$, and $R$ is a user-defined weighing parameter.

The meaning of the weighing parameter is as follows: An observation station at distance $R$ has the same weight as the step 1 calculation of the wind speed in the grid point of interest. There are two values of $R$ in the calculation:

- R1 is the value of $R$ applied to the first vertical layer in the grid.
- R2 is the value of $R$ applied to layers 2 and above.

Only observation stations within a certain maximum distance from the grid point are considered in the calculation:

- In the first vertical layer over land the maximum distance is the user-defined variable RMAX1.

- In the second and further vertical layers over land the maximum distance is the user-defined variable RMAX2.
- Over water the maximum distance is the user-defined variable RMAX3.

The RMAX values should reflect the distance the meteorological data are expected to be representative. The width of the valleys in an area is usually an appropriate choice.

Sometimes it is desirable to avoid that the influence of a meteorological station extends beyond the edge of a valley. This can be accomplished by setting a **barrier** at the valley's edge. A barrier is a user-defined line on the wind speed map. A station that has a barrier separating it from a grid point is not used in the calculation of the wind speed in that grid point. With a careful combination of barriers and RMAX values it is possible to base grid points in valleys only on measured data within that valley and not on measured data in other valleys, where the conditions may be different.

So far the issue of vertical extrapolation was not considered in this section. Wind speed data measured at surface stations are usually only available near the surface and need to be extrapolated to the layers aloft in the application of eq. (15.16). CALMET has four options for this extrapolation, the choice of which is controlled by the variable IEXTRP.

When IEXTRP = 1, no extrapolations are made, and the wind speed in the upper layers is assumed to be the same as the wind speed in the surface layer. This choice will systematically underestimate wind speeds and is not recommended.

When IEXTRP = 2, the wind speed is based on the power law:

$$u_2 = u_1 \left( \frac{z_2}{z_1} \right)^p \tag{2.3}$$

where $p = 0.143$ is chosen over land, and $p = 0.286$ is chosen over water. Unfortunately, these are poor choices for $p$, as a simple comparison with Table 2.4 (Chapter 2) will indicate. Hence, this method is not recommended.

When IEXTRP = 3, the wind speed is extrapolated with user-defined scaling factors:

$$u_i = u_1 \cdot \mathrm{FEXTRP}_i \tag{15.17}$$

where $u_i$ is the wind speed in layer $i$, $u_1$ is the wind speed in the first layer, and $\mathrm{FEXTRP}_i$ is a user-defined scaling factor for layer $i$. This method may be acceptable in cases of homogeneous terrain and constant weather but is otherwise not recommended.

When IEXTRP = 4, the wind speed is extrapolated using similarity theory. This is the recommended option among the four.

When this method is used, the wind speed correction is made up to a 200-m height or up to the mixing layer height, whichever is higher. The algorithms are essentially the same as in AERMOD. In some cases, this approach will overestimate the wind speed above the surface layer, but the result is acceptable in most cases. The largest error can be expected in the case of a nocturnal jet.

IEXTRP = 4 is the only option that includes a wind direction extrapolation. The direction extrapolation is based on the observation that the wind direction turns

**TABLE 15.1 Wind Direction Corrections at 200 m, *D*(200 m), at Different Values of Obukhov Length *L***

| *L* (m) | *D*(200 m) (°) |
|---|---|
| -30 | 12 |
| -100 | 10 |
| -370 | 9 |
| 10,000 | 12 |
| 350 | 18 |
| 130 | 28 |
| 60 | 35 |
| 20 | 38 |
| 9 | 39 |

clockwise in the Northern Hemisphere and counterclockwise in the Southern Hemisphere, so that the necessary correction with respect to the ground-level wind angle is additive and subtractive in Northern and Southern Hemispheres, respectively. This raises the questions of what happens at the equator and whether the corrections are latitude dependent. The measurements were made in The Netherlands, which has a latitude of about 52–55°. Hence, the procedure below may overestimate the correction to some extent in most regions.

The correction, made at 0–200 m, is as follows:

$$\frac{D(z)}{D(200\,\mathrm{m})} = d_1\left[1 - \exp\left(-\frac{d_2 z}{200\,\mathrm{m}}\right)\right] \tag{15.18}$$

where $D(z)$ is the turning of the wind at height $z$, $D(200\,\mathrm{m})$ is the turning of the wind at a 200-m height, $d_1 = 1.58$, and $d_2 = 1$. The values of $D(200\,\mathrm{m})$ are tabulated for different values of the Obukhov length $L$ in Table 15.1. When a value of $D(200\,\mathrm{m})$ at a different value of $L$ is needed, then a linear interpolation of the adjacent values versus $1/L$ should be used for the calculation. Above 200 m, the value at 200 m is used.

To calculate a wind direction correction at layer $i$, the value of $D$ at the height of node point $i$ is calculated, as well as the value of $D$ at the height of the first node point (i.e., in the first horizontal layer). The difference between the two $D$ values is the wind direction correction.

***15.2.5.2 Smoothing.*** Some of the corrections discussed above can cause discontinuities in the wind field. To reduce such discontinuities, a smoothing is carried out based on the following equation:

$$u_{i,j,\mathrm{corr}} = 0.5u_{i,j} + 0.125(u_{i-1,j} + u_{i+1,j} + u_{i,j-1} + u_{i,j+1}) \tag{15.19}$$

where $i$ and $j$ are indices referring to the cell number in the $x$ and $y$ directions. A similar equation is applied to the $v$ component of the wind speed. The number of passes of this smoother is a user-defined variable, NSMTH, in each layer.

***15.2.5.3 Vertical Wind Speed Adjustment.*** Most adjustments discussed above do not consider the vertical wind speed. This adjustment is included specifically for vertical wind speeds. There are two options:

- Adjust the vertical wind speed so that the divergence equals 0 in each grid point.
- Adjust the vertical wind speed so that it equals 0 at the top of the domain (O'Brien adjustment).

When the first option is chosen, the equation

$$\frac{\partial u}{\partial x}+\frac{\partial v}{\partial y}+\frac{\partial w}{\partial z}=0 \tag{10.21}$$

is solved for $w$ in each grid point, starting at the bottom layer. When this option is chosen, the next step (divergence minimization) becomes unnecessary.

In some cases, this procedure can lead to unrealistically large values of $w$ at the top of the domain. When this is the case, the second option is recommended. The following correction is applied to all vertical wind speeds:

$$w_{\text{corr}}=w-\frac{z}{z_{\text{top}}}w\Big|_{z_{\text{top}}} \tag{15.20}$$

where $w_{\text{corr}}$ is the corrected vertical wind speed, $z_{\text{top}}$ is the top of the computational domain, and $w\big|_{z_{\text{top}}}$ is the vertical wind speed at the top of the domain before correction.

When this option is chosen, divergence minimization is still needed. In some cases, such as a breeze that extends vertically above the domain, zero vertical wind speed at the top is not realistic, and the first option must be chosen.

***15.2.5.4 Divergence Minimization.*** Step 2 adjustments may undo the divergence minimization that concluded step 1 adjustments. Hence, a second divergence minimization is carried out at this point, unless it was carried out in the previous step, by adjusting the vertical wind speeds. The procedure is the same as in Section 15.2.4.4 and will not be repeated here.

### 15.2.6 Determination of Stability in CALMET

In one of the previous sections the similarity method for determining a wind speed profile was mentioned but not discussed. The stability parameters needed to determine a wind speed profile are also needed to determine the dispersion parameters. Their calculation will be discussed very briefly here.

CALMET uses the energy balance method to calculate values of the sensible heat flux $q$ and the Obukhov length $L$. The heat balance used is similar to the one discussed in Chapter 5, but with the addition of an optional, user-defined anthropogenic heat flux $q_{\text{A}}$:

$$R_N+q_{\text{A}}=q+q_{\text{L}}+q_{\text{G}} \tag{5.49}$$

The anthropogenic heat flux $q_A$ is usually small in comparison with the other terms.

The ground heat flux $q_G$ is calculated with eq. (5.51):

$$q_G = C_G R_N \tag{5.51}$$

Unlike AERMET, CALMET allows the user to choose the value of $C_G$. Values of 0.05–0.25 are recommended in rural areas, whereas urban areas are best calculated with $C_G$ values of 0.25–0.30 (Oke, 1982).

The net radiation $R_N$ is calculated with the radiation balance method of Holtslag and van Ulden (1983), except that no angle dependence of the albedo is used. As is customary in micrometeorology, the Bowen ratio is used to reduce the number of variables in the heat balance, so the sensible heat flux $q$ can be calculated. Details of these calculations can be found in Chapter 5.

In stable conditions the heat balance method is replaced with the similarity approach discussed in Section 5.9. With the exception of some small changes in the coefficients, the procedure is the same as in AERMET. An exception is the case where $u<u_{min}$, the minimum wind speed for the theory to apply. In that case the friction temperature is decreased until the friction velocity predicted by eq. (5.107) is real. For details, see Scire et al. (2000b).

The thermal mixing layer height calculation in *unstable* conditions is similar to the calculation in Section 5.10, but a different set of assumptions is used. Essentially it is assumed that 1.15 times the sensible heat given off by the surface since the onset of unstable conditions is deposited in a layer of constant potential temperature. In addition, allowance is made for a potential temperature jump at the top of the mixing layer (see Section 5.13.2 for a description of this potential temperature jump). The resulting equation for the mixing layer height is as follows:

$$h_{\mathrm{mix},t+\Delta t} = \sqrt{h^2_{\mathrm{mix},t} + \frac{2.3q\Delta t}{\gamma\rho C_p} - \frac{2\Delta\theta_t h_{\mathrm{mix},t}}{\gamma} + \frac{\Delta\theta_{t+\Delta t}}{\gamma}} \tag{15.21}$$

where the potential temperature jump $\Delta\theta_{t+\Delta t}$ is calculated as

$$\Delta\theta_{t+\Delta t} = \sqrt{\frac{2\gamma q\,\Delta t}{\rho C_p}} \tag{15.22}$$

where the subscripts $t$ and $t+\Delta t$ refer to the time, $\gamma$ is the potential temperature gradient above the mixing layer, and $\Delta\theta$ is the temperature jump at the top of the mixing layer. CALMET allows the user to define a minimum sensible heat flux needed to initiate boundary layer growth. In that case, the value is subtracted from $q$ in eq. (15.22). The procedure is rather artificial because the temperature jump at the top of the boundary layer depends on the time step used. Consequently, the mixing layer growth depends more strongly on the time step than can be explained by time discretization. Mixing layer heights predicted by this algorithm are substantially smaller than predictions with the model presented in Section 5.13.2.

When the boundary layer is *neutral*, the mixing layer height is determined by mechanical mixing. CALMET uses the following equation to predict the mixing layer height:

$$h_{\text{mix}} = \frac{Bu_*}{\sqrt{fN_{\text{BV}}}} \qquad (15.23)$$

where $B = 2^{1/2}$, $f$ is the Coriolis parameter, and $N_{\text{BV}}$ is the Brunt–Väisälä frequency of the atmosphere *above* the mixing layer.

A smoothing procedure is applied to reduce fluctuations of $h_{\text{mix}}$ resulting from the calculation scheme. Details are given by Scire et al. (2000b).

In a *stable* boundary layer, mixing height predictions from two different equations are compared, and the lower of the two values is retained. The two equations are

$$h_{\text{mix}} = C\sqrt{\frac{u_* L}{|f|}} \qquad (5.113)$$

$$h_{\text{mix}} = Bu_*^{3/2} \qquad (5.115)$$

where $C = 0.4$ and $B = 2400\,\text{s}^{1.5}\,\text{m}^{0.5}$ (Scire et al., 2000b); and $f$ is the Coriolis parameter.

So far in this section the assumption was made that the terrain was terrestrial. *Above water* the stability properties of the atmosphere are mainly determined by the temperature difference between the air and the water. The following equations can be used:

$$C_{u\text{N}} = (0.75 + 0.067u_{10}) \cdot 10^{-3} \qquad (15.24)$$

$$u_* = u_{10}\sqrt{C_{u\text{N}}} \qquad (15.25)$$

$$L = \frac{\theta_{\text{v}} C_{u\text{N}}^{3/2} u^2}{E_2(\theta_{\text{v}} - \theta_{\text{vs}})} \qquad (15.26)$$

$$z_0 = 2 \times 10^{-6} u_{10}^{2.5} \qquad (15.27)$$

$$h_{\text{mix}} = \frac{c_{\text{w}} u_*}{f} \qquad (15.28)$$

where $C_{u\text{N}}$ is a drag coefficient, $u_{10}$ is the wind speed at a 10-m height, $\theta_{\text{v}}$ is the virtual potential temperature of the air, $\theta_{\text{vs}}$ is the water temperature, $E_2 = 5.096 \times 10^{-3}$, and $c_{\text{w}} = 0.16$. In eq. (15.27) the wind speed is in m s$^{-1}$; the roughness height is in m.

Note that the above equations are overdetermined. The logarithmic wind speed profile links $u$, $u_*$, $L$, and $z_0$ together, so they are not independent. As a consequence, the logarithmic wind speed profile will predict a value of $u_{10}$ that is not necessarily the same as the measured value.

Equations (15.24)–(15.28) are only used above seas and oceans, not above lakes and other small water bodies, where the heat balance approach is used.

### 15.2.7 Precipitation Interpolation

In CALMET, precipitation is calculated by interpolation from hourly measured precipitation values. The calculation can be carried out with a $1/r$ interpolation, a $1/r^2$ interpolation, or a $1/r$-exponential interpolation. The latter is as follows:

$$R = \frac{\sum_k \frac{R_k \exp\left(-r_k^2/\sigma^2\right)}{r_k^2}}{\sum_k \frac{\exp\left(-r_k^2/\sigma^2\right)}{r_k^2}} \tag{15.29}$$

where $R_k$ is the precipitation rate at the $k$th measurement station, $r_k$ is the distance between the grid point and station $k$, and $\sigma$ is a user-defined radius of influence.

## 15.3 DESCRIPTION OF CALPUFF

### 15.3.1 Concentration Calculations in CALPUFF

As indicated in Section 6.10, the direct calculation of air pollution with puff snapshots that are Gaussian in three dimensions is computationally expensive as one puff per second needs to be defined for each source in some cases. To overcome this weakness of direct puff methods, CALPUFF calculates concentrations with either the integrated puff method or the slug method. The slug method is the more accurate of the two methods but requires much more computation time than the integrated puff method. Usually the slug method is not needed unless some receptors are located very close to the source.

In the **integrated puff mode** it is assumed that the puff is circular in the horizontal direction and has a constant size as it moves over a receptor. The size used in the calculation is the size of the puff at its closest distance from the receptor. All other time-dependent variables, such as location and pollutant mass in the puff, are assumed to vary linearly with time during the integration.

A derivation of the integrated puff equation was presented in Section 6.10.2 with the simplifying assumption that the puff conserves mass. CALPUFF does not make this assumption, making the integegrated puff equation more complicated. The following variables are defined: $x_r$ and $y_r$ are the coordinates of the receptor; $x_1$ and $y_1$ are the coordinates of the puff center at the beginning of the time step; $x_2$ and $y_2$ are the the coordinates of the puff center at the end of the time step; $s_1$ and $s_2$ are the distance traveled by the puff at the beginning and the end of the time step; $Q_p(s)$ is the pollutant mass in the puff after having traveled a distance $s$; $h$ is the effective source height; and $h_{mix}$ is the mixing layer height.

In addition, define $\Delta x$ and $\Delta y$ as

$$\Delta x = x_2 - x_1 \quad (15.30)$$

$$\Delta y = y_2 - y_1 \quad (15.31)$$

The average concentration at the receptor location over the time step is as follows:

$$\overline{c} = \frac{g}{2\pi\sigma_y^2}\left\{Q_p(s_1)I_1 + \left[Q_p(s_2) - Q_p(s_1)\right]I_2\right\} \quad (15.32)$$

where the following auxiliary functions are defined:

$$g = \frac{2}{\sqrt{2\pi}\sigma_z}\sum_{n=-\infty}^{+\infty}\exp\left[-\frac{(h+2nh_{\mathrm{mix}})^2}{2\sigma_z^2}\right] \quad (15.33)$$

$$I_1 = \sqrt{\frac{\pi}{2a}}\exp\left(\frac{b^2}{2a}-\frac{c}{2}\right)\left[\mathrm{erf}\left(\frac{a+b}{\sqrt{2a}}\right)-\mathrm{erf}\left(\frac{b}{\sqrt{2a}}\right)\right] \quad (15.34)$$

$$I_2 = \frac{-bI_1}{a}+\frac{1}{a}\exp\left(\frac{b^2}{2a}-\frac{c}{2}\right)\left\{\exp\left(-\frac{b^2}{2a}\right)-\exp\left[-\frac{1}{2}\left(a+2b+\frac{b^2}{a}\right)\right]\right\} \quad (15.35)$$

$$a = \frac{(\Delta x)^2+(\Delta y)^2}{\sigma_y^2} \quad (15.36)$$

$$b = \frac{\Delta x(x_1 - x_r)+\Delta y(y_1 - y_r)}{\sigma_y^2} \quad (15.37)$$

$$c = \frac{(x_1 - x_r)^2+(y_1 - y_r)^2}{\sigma_y^2} \quad (15.38)$$

In the **slug mode**, the concentration profile at time $t$ is given by the following equation [eq. (6.167)]:

$$c(t) = \frac{Qg}{2\sqrt{2\pi}u'\,\sigma_y}\exp\left(-\frac{d_c^2}{2\sigma_y^2}\frac{u^2}{u'^2}\right)\left[\mathrm{erf}\left(\frac{d_{a2}}{\sqrt{2}\sigma_{y2}}\right)-\mathrm{erf}\left(-\frac{d_{a1}}{\sqrt{2}\sigma_{y1}}\right)\right] \quad (15.39)$$

where $g$ is given by eq. (15.33), $Q$ is the pollutant emission rate (in mg s$^{-1}$, not mg puff$^{-1}$), $u$ is the wind speed, and $u'$ is given by

$$u' = \sqrt{u^2+\sigma_u^2} \quad (15.40)$$

In some special cases it is possible to find analytical solutions of eq. (15.39), but in general a numerical integration is needed to calculate an hourly averaged concentration.

### 15.3.2 Dispersion Parameter Calculations in CALPUFF

Plume dispersion parameters in CALPUFF can contain three components, which are added as variances:

$$\sigma_x^2 = \sigma_y^2 = \sigma_{y,t}^2 + \sigma_{y,s}^2 + \sigma_{y,b}^2 \tag{15.41}$$

$$\sigma_z^2 = \sigma_{z,t}^2 + \sigma_{z,b}^2 \tag{15.42}$$

where $\sigma_{y,t}$ and $\sigma_{z,t}$ are the dispersion parameters due to atmospheric turbulence, $\sigma_{y,s}$ is an (optional) initial puff dimension in case an area source is modeled, and $\sigma_{y,b}$ and $\sigma_{z,b}$ are dispersion enhancements resulting from plume rise. As indicated in eq. (15.41), the dispersion parameters are the same in $x$ and $y$ directions in CALPUFF. This is only an approximation because in reality the spread is larger in the $x$ direction than in the $y$ direction, and wind shear causes significant stretching of puffs in the $x$ direction.

Every time the conditions influencing puff growth change, the virtual source concept (see Section 6.8) is used to calculate the appropriate puff growth. In practice, this means that the virtual source concept is invoked with every time step in CALPUFF.

There are five different options for calculating $\sigma_{y,t}$ and $\sigma_{z,t}$ in CALPUFF:

- Calculated from measured values of $\sigma_v$ and $\sigma_w$.
- Calculated based on similarity theory, that is, on $u_*$, $L$, . . . . This is the recommended method.
- Calculated based on stability classes: the Pasquill–Gifford and McElroy–Pooler parameterizations.
- Based on stability classes, using the MESOPUFF II equations.
- Based on measured values of $\sigma_v$ and $\sigma_w$, with the CTDM equations for stable and neutral conditions; based on stability classes: the Pasquill–Gifford and McElroy–Pooler parameterizations in unstable conditions.

The five options will be discussed in more detail.

When the dispersion parameters are calculated based on measured values of $\sigma_v$ and $\sigma_w$, the following equations are used [see Eqs. (6.27) and (6.28)]:

$$\sigma_{x,t} = \sigma_{y,t} = \sigma_v t f_y \tag{15.43}$$

$$\sigma_{z,t} = \sigma_w t f_z \tag{15.44}$$

where the functions $f_y$ and $f_z$ are given by Eqs. (6.65)–(6.67), with $T_{i,L}$ values given by Eqs. (6.47)–(6.49). Hence, the equation for $f_y$ is

$$f_y = \frac{1}{1 + 0.9\left(t/1000\right)^{1/2}} \tag{15.45}$$

In unstable conditions, the equation for $f_z$ is

$$f_z = \frac{1}{1+0.9(t/500)^{1/2}} \tag{15.46}$$

In stable conditions, the equation for $f_z$ is

$$f_z = \frac{1}{1+0.945(t/100)^{0.806}} \tag{15.47}$$

Equations (6.65)–(6.67) tend to overestimate the effect of $T_{i,L}$ on dispersion. On the other hand, Eqs. (6.47)–(6.49) tend to overestimate $T_{i,L}$, especially when the effective source height is low, which compensates. Overall, these equations can be expected to be reasonably accurate.

The second option also uses Eqs. (15.45)–(15.47), but with $\sigma_v$ and $\sigma_w$ values calculated from similarity theory. This is the recommended option if measured turbulence velocities are not available, unless there is a specific reason (e.g., regulations) to choose otherwise. This is likely to become the default option in future versions of CALPUFF. The turbulent velocities are calculated as follows:

When $L<0$ (unstable):

$$\sigma_v = \sqrt{4u_*^2 a_n^2 + 0.35w_*^2} \tag{15.48}$$

When $L<0$ (unstable) and $z<0.1h_{mix}$ (surface layer):

$$\sigma_w = \sqrt{1.6u_*^2 a_n^2 + 2.9u_*^2 (z/-L)^{2/3}} \tag{15.49}$$

with

$$a_n = \exp(-0.9z/h_{mix}) \tag{15.50}$$

When $L<0$ (unstable) and $0.1h_{mix}<z<0.8h_{mix}$ (mixing layer):

$$\sigma_w = \sqrt{1.15u_*^2 a_n^2 + 0.35w_*^2} \tag{15.51}$$

When $L<0$ (unstable) and $0.8h_{mix}<z<h_{mix}$ (entrainment layer):

$$\sigma_w = \sqrt{1.15u_*^2 a_n^2 + a_{c1} 0.35w_*^2} \tag{15.52}$$

with

$$a_{c1} = \frac{1}{2} + \frac{h_{mix} - z}{0.4h_{mix}} \tag{15.53}$$

When $L<0$ (unstable) and $h_{mix}<z<1.2h_{mix}$ (entrainment layer):

$$\sigma_w = \sqrt{1.15u_*^2 a_n^2 + a_{c2} 0.35w_*^2} \tag{15.54}$$

with

$$a_{c2} = \frac{1}{3} + \frac{1.2h_{\text{mix}} - z}{1.2h_{\text{mix}}} \tag{15.55}$$

When $L > 0$ (stable):

$$\sigma_v = u_* \left[ \frac{1.6c_s(z/L) + 1.8a_n}{1 + z/L} \right] \tag{15.56}$$

$$\sigma_w = 1.3u_* \left[ \frac{c_s(z/L) + a_n}{1 + z/L} \right] \tag{15.57}$$

with

$$c_s = \left( 1 + \frac{z}{h_{\text{mix}}} \right)^{3/4} \tag{15.58}$$

When the wind speed is very low, these equations may underestimate $\sigma_y$ and $\sigma_z$. For that reason, the user can define minimum values of $\sigma_y$ and $\sigma_z$ in CALPUFF.

The third option is to calculate dispersion parameters based on stability classes based on the Pasquill–Gifford and McElroy–Pooler parameterizations. This is the same as the ISC parameterization, and the reader is referred to Section 13.2 for details.

The fourth option is to calculate dispersion parameters based on stability classes based on the MESOPUFF II equations. These are simple power law equations:

$$\sigma_y = ax^b \tag{6.12}$$

$$\sigma_z = cx^d \tag{6.13}$$

where the coefficients are given in Table 15.2. Because of the limited applicability of the power law, this formulation is not recommended.

The fifth option is to calculate dispersion parameters based on the CTDM equations in stable and neutral conditions and on the Pasquill–Gifford and McElroy–Pooler parameterizations in unstable conditions. The CTDM equations

**TABLE 15.2 Coefficients of the MESOPUFF II Parameterization for Calculating Dispersion Parameters with eqs. (6.12) and (6.13)**

| Stability Class | $a$ | $b$ | $c$ | $d$ |
|---|---|---|---|---|
| A | 0.36 | 0.9 | 0.00023 | 2.10 |
| B | 0.25 | 0.9 | 0.058 | 1.09 |
| C | 0.19 | 0.9 | 0.11 | 0.91 |
| D | 0.13 | 0.9 | 0.57 | 0.58 |
| E | 0.096 | 0.9 | 0.85 | 0.47 |
| F | 0.063 | 0.9 | 0.77 | 0.42 |

are based on Eqs. (15.43) and (15.44), with measured values of $\sigma_y$ and $\sigma_z$, and the following equations for $f_y$ and $f_z$:

$$f_y = \frac{1}{(1 + ut / 20{,}000\text{m})^{1/2}} \tag{15.59}$$

$$f_z = \frac{1}{\left\{1 + \sigma_w t\left[(1/0.72z) + (N_{\text{BV}} / 0.54\sigma_w)\right]\right\}^{1/2}} \tag{15.60}$$

where $z$ is the height of the puff, and $N_{\text{BV}}$ is the Brunt–Väisälä frequency.

If option 3 or 4 is chosen, the user can choose to make adjustments to the dispersion parameters for **averaging time** and for surface **roughness height**. The adjustment for averaging time is

$$\sigma_{y,2} = \sigma_{y,1}\left(\frac{t_2}{t_1}\right)^p \tag{6.19}$$

with $\sigma_{y,1}$ and $\sigma_{y,2}$ dispersion parameters obtained with averaging times of $t_1$ and $t_2$, respectively, and $p = 0.2$. This equation can be used for averaging times up to 1 h. For longer averaging times, it is better to carry out separate calculations for consecutive hours and to take the average of the results.

For the adjustment for surface roughness, see Scire et al. (2000a). Adjustments for surface roughness are not recommended for sources higher than 100 m.

The calulation of the **plume rise enhancement effect** on dispersion is calculated in a manner similar but not identical to Section 6.5:

$$\sigma_{y,b} = \sigma_{z,b} = \frac{\Delta h}{3.5} \tag{15.61}$$

When the source is an **area source**, the plume is assumed to have an initial lateral dispersion parameter $\sigma_{y,s} > 0$. This initial plume size is included in the dispersion parameters in eq. (15.41).

The calculation is very different when the source is defined as a **volume source**. In that case both lateral and vertical dispersion parameters can have nonzero initial values, $\sigma_{y,0}$ and $\sigma_{z,0}$. In the case of a volume source, a virtual source approach is used to establish an initial plume size. The purpose of a volume source is to simplify calculations when there are several closely spaced sources by treating them as a single source.

In some cases CALPUFF invokes a **puff-splitting algorithm**. The objective of puff splitting is to account for the effect of vertical wind shear on plume dispersion. Plume splitting is invoked when this option is chosen by the user, and the following two additional conditions are met:

- The plume is homogeneous in the vertical direction, that is, the vertical plume spread is large enough to invoke eq. (6.9).
- The current mixing height is much smaller than the maximum mixing height experienced previously by the puff.

In these conditions part of the puff is below the mixing height, and part is above, where different conditions may apply. The number of puffs after splitting is user defined. The topmost puff takes up half of the entire puff height, the second puff takes half of the remaining puff height, and so forth.

CALPUFF has the option of calculating vertical plume dispersion based on a nonsymmetrical **probability density function** (PDF) of pollutant location in a convective boundary layer. This is the result of updrafts and downdrafts in the atmosphere and is one of the main features of AERMOD. The approach in CALPUFF is very similar to the approach in AERMOD and will not bc repeated here.

Note that dispersion enhancement due to plume rise in CALPUFF [eq. (15.61)] is based on plume rise without updraft or downdraft effects.

Sometimes the size of a puff increases due to **vertical puff stretching**, which is not an effect of turbulence but of a vertical divergence of the wind streamlines. This is accounted for in CALPUFF by calculating the average value of the vertical gradient of the vertical wind speed component, and making the following correction:

$$\sigma_{z,\mathrm{corr}} = \sigma_z \left(1 + \Delta t \frac{\partial w}{\partial z}\right) \tag{15.62}$$

where the average is calculated from one standard deviation above the puff center to one standard deviation below the puff center. When the surface or a temperature inversion is less than one standard deviation from the puff center, then they are used as the boundary for averaging.

### 15.3.3 Plume Rise Calculations in CALPUFF

In CALPUFF, **transitional** plume rise is calculated with a modification of eq. (7.25), which is for plumes that are governed by both momentum and buoyancy. The equation is as follows:

$$\Delta z = \left(\frac{3F_{\mathrm{m}}x}{\beta_{\mathrm{j}}^2 u_{\mathrm{s}}^2} + \frac{3F_{\mathrm{b}}x^2}{\beta_1^2 u_{\mathrm{s}}^3}\right)^{1/3} \tag{15.63}$$

where $F_{\mathrm{m}}$ is the momentum flux parameter, $F_{\mathrm{b}}$ is the buoyancy flux parameter, $u_{\mathrm{s}}$ is the wind speed at stack height, $x$ is the downwind distance from the source, $\beta_1$ is the neutral entrainment parameter (= 0.6), and $\beta_{\mathrm{j}}$ is the jet entrainment parameter, calculated as

$$\beta_j = \frac{1}{3} + \frac{u_{\mathrm{s}}}{w} \tag{15.64}$$

where $w$ is the stack gas exit speed.

The **final** plume rise calculation depends on the stability of the boundary layer. In the case of *unstable* or *neutral* conditions the final plume rise is calculated from the distance to final plume rise $x_f$, which is calculated as follows when the puff is positively buoyant ($F_b > 0$):

$$x_f = 3.5x^* \tag{15.65}$$

When the puff is neutrally buoyant ($F_b = 0$) the equation becomes

$$x_f = \frac{4d_s(w + 3u_s)}{u_s w} \tag{15.66}$$

where $d_s$ is the stack diameter, and $x^*$ is given by

$$x^* = 14F_b^{5/8} \ (F_b < 55\,\mathrm{m^4 s^{-3}}) \tag{15.67}$$

$$x^* = 34F_b^{2/5} \ (F_b > 55\,\mathrm{m^4 s^{-3}}) \tag{15.68}$$

The final plume rise is obtained by applying the transitional plume rise equation to $x = x_f$.

In *stable* conditions the final plume rise is calculated directly:

$$\Delta h = \left( \frac{3F_m}{\beta_j^2 u_s^2 s^{1/2}} + \frac{6F_b x^2}{\beta_2^2 u_s^2 s} \right)^{1/3} \tag{15.69}$$

where $\beta_2$ is the stable entrainment parameter (= 0.36), and $s$ is the stability parameter (see eq. (5.48)):

$$s = \frac{g}{T_a} \frac{\partial \theta}{\partial z} \tag{15.70}$$

where $T_a$ is the temperature of the atmosphere.

A problem arises when the wind speed at the stack height is less than 1 m s$^{-1}$. In that case, the above equations overestimate plume rise. In CALPUFF, this problem is solved in unstable and neutral conditions by imposing a minimum wind speed of 1 m s$^{-1}$ at the stack tip for plume rise calculations. In stable conditions, the final plume rise is calculated as

$$\Delta h = \frac{4F_b^{3/8}}{s^{3/8}} \tag{15.71}$$

which is eq. (7.15) with a smaller coefficient.

A special situation occurs if the plume rise as calculated here would cause the plume to move above the mixing layer. This phenomenon is known as **plume penetration**. In the case of plume penetration it is typical that part of the plume leaves the mixing layer, where it is unavailable for immediate mixing, and part of the plume remains in the mixed layer. This is known as partial plume penetration. To calculate plume proportions that stay in the mixed layer and that rise aloft, the penetration parameter is defined:

$$p = \frac{F_b}{u_s b_i (h_i - h_s)} \tag{15.72}$$

where $h_i$ is the inversion layer height and $b_i$ is the strength of the temperature inversion:

$$b_i = g\frac{\Delta T_i}{T_a} \tag{15.73}$$

where $\Delta T_i$ is the temperature jump across the temperature inversion. Based on $p$, the fraction of the plume that remains in the mixing layer, $f$, is calculated:

$$f = 1 \ (p < 0.08) \tag{15.74}$$

$$f = \frac{0.08}{p} - p + 0.08 \ (0.08 < p < 0.3) \tag{15.75}$$

$$f = 0 \ (p > 0.3) \tag{15.76}$$

The puffs are split into two parts, each with their own final plume rise. The final plume rise of the puffs that remain in the mixing layer is as follows:

$$\Delta h_1 - h_s = \left(1 - \frac{f}{3}\right)(h_i - h_s) \tag{15.77}$$

However, if eq. (15.63), calculated at $x{=}x_f$ [eq. (15.65) or (15.66)], leads to a lower value of $\Delta h$ than $\Delta h_1$ calculated with eq. (15.77), then eq. (15.63) is used. The final plume rise of the puff fraction above the mixing layer is calculated as

$$\Delta h_2 - h_s = (2 - f)(h_i - h_s) \tag{15.78}$$

except when $f{=}0$ (full penetration). In that case the following equation is used:

$$\Delta h - h_s = \left[1.8(h_i - h_s)^3 + 18.25\frac{F_b}{u_s s}\right]^{1/3} \tag{15.79}$$

### 15.3.4 Impact of Downwash on Plume Rise and Dispersion Calculations in CALPUFF

The simplest case of downwash is stack tip downwash, which is calculated as a correction on the stack height $h_s$. When the stack gas exit velocity $w$ is less than

1.5 times the wind speed at stack height $u_s$, then the correction is as follows [see eq. (7.51)]:

$$h_{s,corr} = h_s + 2d_s\left(\frac{w}{u_s} - 1.5\right) \tag{15.80}$$

where $d_s$ is the stack tip diameter. When $w > 1.5u_s$, no stack tip downwash is assumed.

For calculating the effect of building downwash on plume rise, we need to calculate the dispersion induced by building downwash first. Two algorithms for that purpose are available in CALPUFF: the Huber–Snyder procedure and the Schulman–Scire procedure.

To determine the proper choice of procedure, a number of variables need to be defined. Apart from $h_s$ (actual source height), these are $H_b$ (actual building height downwind of the source), $H_w$ (projected width of the building perpendicular to the wind direction), $L_b = \min(H_b, H_w)$, $T_{bd}$ (user-defined variable with default value 0.5), and $h_e$ (effective source height at distance $2H_b$ from the source). Three cases are distinguished:

- When $h_e > H_b + 1.5L_b$, no building downwash is assumed. This is in fair agreement with the rule of thumb that a stack should be 2.5 times as high as the tallest nearby building to avoid downwash.
- When building downwash is to be considered, and $h_s > H_b + T_{bd} L_b$, the Huber–Snyder procedure is used.
- When $h_s < H_b + T_{bd} L_b$, the Schulman–Scire procedure is used.

In the **Huber–Snyder procedure**, both $\sigma_y$ and $\sigma_z$ are adjusted when $h_s < 1.2H_b$, whereas only $\sigma_z$ is adjusted when $h_s > 1.2H_b$. Dispersion parameters adjusted for building downwash will be indicated with a prime.

For a **squat building** ($H_w > H_b$) the following dispersion parameter is used in the $z$ direction (for downwind distances $3H_b < x < 10H_b$):

$$\sigma_z' = 0.7H_b + 0.067(x - 3H_b) \tag{15.81}$$

In the $y$ direction, the adjusted dispersion parameter is as follows when $H_b < H_w < 5H_b$, and $3H_b < x < 10H_b$:

$$\sigma_y' = 0.35H_w + 0.067(x - 3H_b) \tag{15.82}$$

If $H_w < 5H_b$, and $3H_b < x < 10H_b$:

$$\sigma_y' = 0.35H_b + 0.067(x - 3H_b) \tag{15.83}$$

For a tall building ($H_w < H_b$) and $3H_b < x < 10H_b$:

$$\sigma_y' = 0.35H_w + 0.067(x - 3H_w) \tag{15.84}$$

$$\sigma_y' = 0.7H_w + 0.067(x - 3H_w) \tag{15.85}$$

In the **Schulman–Scire procedure**, the adjusted dispersion parameters are first calculated with the Huber–Snyder procedure. Then the adjusted $\sigma'_z$ value is decayed as follows:

$$\sigma''_y = A\sigma'_y \tag{15.86}$$

where $A$ is given by

$$A = 1 \quad (h_e < H_b) \tag{15.87}$$

$$A = \frac{H_b + h_e}{2L_b} + 1 \quad (H_b < h_e < H_b + 2L_b) \tag{15.88}$$

$$A = 0 \quad (H_b + 2L_b < hy_e) \tag{15.89}$$

For the impact of plume downwash on plume rise, the variables $\sigma_{y,0}$ and $\sigma_{z,0}$ are defined as the values of $\sigma'_y$ and $\sigma'_z$ at a distance $x=3H_b$ from the source.

When $\sigma_{y,0} < \sigma_{z,0}$ then transitional plume rise in unstable and neutral conditions is calculated by solving the following equation for the plume rise $\Delta z$:

$$(\Delta z)^3 + \frac{3R_0}{\beta_1}(\Delta z)^2 + \frac{3R_0^2}{\beta_1^2}\Delta z = \frac{3F_m x}{\beta_j^2 u_s^2} + \frac{3F_b x^2}{2\beta_1^2 u_s^3} \tag{15.90}$$

where $R_0$ is the dilution radius, given by

$$R_0 = \sqrt{2}\sigma_{z,0} \tag{15.91}$$

The other variables in eq. (15.90) have been defined earlier.

For stable conditions, final plume rise can be calculated as follows:

$$(\Delta h)^3 + \frac{3R_0}{\beta_2}(\Delta h)^2 + \frac{3R_0^2}{\beta_2^2}\Delta h = \frac{3F_m}{\beta_j^2 u_s s^{1/2}} + \frac{6F_b}{\beta_2^2 u_s s} \tag{15.92}$$

Transitional plume rise in stable conditions is calculated with eq. (15.90) until the value predicted by eq. (15.92) is reached.

When $\sigma_{y,0} > \sigma_{z,0}$, then the plume is treated as a line source with the following effective line length:

$$L_e = \sqrt{2\pi}(\sigma_{y,0} - \sigma_{z,0}) \tag{15.93}$$

Transitional plume rise is then calculated as

$$(\Delta z)^3 + \left(\frac{3L_e}{\pi\beta_1} + \frac{3R_0}{\beta_1}\right)(\Delta z)^2 + \left(\frac{6R_0 L_e}{\pi\beta_1^2} + \frac{3R_0^2}{\beta_1^2}\right)\Delta z = \frac{3F_m x}{\beta_j^2 u_s^2} + \frac{3F_b x^2}{2\beta_1^2 u_s^3} \tag{15.94}$$

In stable conditions, final plume rise is calculated as

$$(\Delta z)^3 + \left(\frac{3L_e}{\pi\beta_2} + \frac{3R_0}{\beta_2}\right)(\Delta z)^2 + \left(\frac{6R_0L_e}{\pi\beta_2^2} + \frac{3R_0^2}{\beta_2^2}\right)\Delta z = \frac{3F_m}{\beta_j^2 u_s s^{1/2}} + \frac{6F_b}{\beta_2^2 u_s s} \quad (15.95)$$

### 15.3.5 Area Source Plume Rise Calculations in CALPUFF

Area source plume rise is a special model within CALPUFF that is meant to model large buoyant sources, such as forest fires. It uses the numerical plume rise model also used in PRIME (Schulman et al., 2000) and discussed in Section 7.2.4. The model accounts for buoyancy, drag forces, mixing of air into the plume, and radiative heat loss.

### 15.3.6 Coastal Dispersion Calculations in CALPUFF

Above land the structure of the boundary layer is often different from the structure above sea. As a result, a thermal internal boundary layer (TIBL) develops as the wind enters the land from sea or vice versa. This process is described in more detail in Sections 5.12.2 and 5.13.3.

To account for coastal thermal internal boundary layers, a finer resolution coastline is often needed than is available from terrain data. Hence, CALPUFF can accept a high-resolution definition of the coastline. CALPUFF calculates a high-resolution TIBL whenever the heat flux is at least $5\,\mathrm{W\,m^{-2}}$. The slope of the TIBL height $h$ is calculated as

$$\frac{\partial h}{\partial x} = \frac{q_0(1+2\beta)}{\rho c_p \gamma h u} \quad (5.247)$$

where $q_0$ is a the surface sensible heat flux, which can be dependent on the downwind distance, $\beta$ is the ratio between the downward heat flux at the TIBL height to the upward heat flux at the surface (taken to be 0.2 for a convective boundary layer over land), $\gamma$ is the potential temperature gradient above the TIBL, and $u$ is the mean wind speed within the TIBL.

The actual calculation of $h$ is by a stepwise approximate integration with the following equation:

$$h(x+\Delta x) = \sqrt{\frac{2(1+2\beta)_{q0}}{\gamma\rho C_p u}\Delta x + h^2(x)} \quad (15.96)$$

### 15.3.7 Complex Terrain Calculations in CALPUFF: Introduction

CALPUFF can handle complex terrain in three different ways depending on the size of the complex terrain features and the preference of the user. The available options are:

- Large terrain features, on a scale larger than the CALMET grid, are handled entirely by CALMET, and no adjustments are needed in CALPUFF.
- Small terrain features between the grid points can be specified by the user. CALPUFF has two options for handling such features:

- With the full CTSG (complex terrain algorithm for subgrid features) algorithm that is also used in the air dispersion model CTDM (complex terrain dispersion model). This option is discussed in Section 15.3.8.
- With a simpler set of terrain adjustments. This option is discussed in Section 15.3.9.

### 15.3.8 Complex Terrain Algorithm for Subgrid Scale Features (CTSG)

***15.3.8.1 Introduction.*** The purpose of the CTSG algorithm is to calculate a trajectory of a puff when it encounters a complex terrain feature such as a hill. The movement of such a puff can follow three paths: above the feature, around the feature on the left side, and around the feature on the right side. The path taken by the puff depends on its position relative to the (vertical) **dividing streamline height** $H_{\mathrm{d}}$ and the (horizontal) **stagnation streamline** $y_{\mathrm{d}}$.

The dividing streamline height is the height where the kinetic energy of the wind is just enough to move up the air parcel to the top of the hill against the buoyant forces. It follows that there is only a dividing streamline in stable conditions. In neutral and unstable conditions the entire puff is always above the dividing streamline.

Below the dividing streamline, the air cannot flow above the obstacle, so it must flow around it. It can do so to the left of the obstacle or to the right. The dividing streamline between the two is a streamline that stops (stagnates) against the hill. This streamline is named the stagnation streamline.

Puffs approaching the obstacle can be split into three subpuffs, one for each path. Downwind from the hill the three subpuffs reunite to form a single puff again.

The dividing streamline height $H_{\mathrm{d}}$ is calculated the same way as in AERMOD:

$$\frac{1}{2}u^2\left(H_{\mathrm{d}}\right)=\int_{H_{\mathrm{d}}}^{h_{\mathrm{c}}} N_{\mathrm{BV}}^2\left(h_{\mathrm{c}}-z\right)dz \tag{15.97}$$

where $u(H_{\mathrm{d}})$ is the wind speed at the dividing streamline height, and $h_{\mathrm{c}}$ is the height of the crest of the obstacle.

The dividing streamline height defines three zones in the atmosphere in the crosswind and vertical directions. In addition, three zones are defined in the *down-wind* direction: zone 1 in front of the upwind foot of the hill, zone 2 between the upwind foor and the downwind foot of the hill, and zone 3 behind the downwind foot of the hill. Hence, nine zones are distinguished.

***15.3.8.2 The Obstacle.*** This section discusses how CALPUFF defines the geometry of the obstacle. First, the user describes the obstacle by means of a best-fitting equation of the following form:

$$h_{\mathrm{t}}=\mathrm{relief}\left[\frac{1-\left(|X|/\mathrm{axmax}\right)^{\mathrm{expo}}}{1+\left(|X|/\mathrm{scale}\right)^{\mathrm{expo}}}\right] \tag{15.98}$$

where relief is the height of the hill, axmax is the distance from the foot to the crest of the feature, and scale and expo are variables that define the shape of the feature. However, this is *not* how CALPUFF describes the feature. CALPUFF converts this description into a different set of equations, one above the dividing streamline height and one below.

The part of the feature above $H_d$ is described by a Gaussian equation:

$$h_t - H_d = H \exp\left[-\left(\frac{x}{L_x}\right)^2\right] \exp\left[-\left(\frac{y}{L_y}\right)^2\right] \tag{15.99}$$

where

$$H = \text{relief} - H_d \tag{15.100}$$

The value of $L_x$ is chosen such that the Gaussian equation predicts an $h_t$ value equal to $H_d + H/2$ at the same $x$ value as eq. (15.98).

The part of the feature below $H_d$ is described as an elliptical cylinder:

$$\left(\frac{x}{a}\right)^2 + \left(\frac{y}{b}\right)^2 = 1 \tag{15.101}$$

where $a$ is the distance from the feature crest where the height is min($H_d$, $z_p$), with $z_p$ the puff center height, in the downwind direction, and $b$ has a similar meaning in the crosswind direction.

The reason for this procedure is because there exists an approximate solution for wind flow above a Gaussian hill top, and this solution is used for the part of the feature above $H_d$. Below $H_d$ the wind is assumed to simply flow around the cylinder. The behavior of the puff depends in which of the nine zones defined earlier it is located.

#### 15.3.8.3 Puff Behavior before Reaching the Hill.

Above the dividing streamline height $H_d$, the puff follows the terrain in the vertical direction and follows the undisturbed wind in the horizontal direction.

Below $H_d$ the puff follows the terrain in the horizontal direction, with no deflection in the vertical direction.

#### 15.3.8.4 Puff Behavior above or around the Hill.

In this region the puff is divided into three puffs each moving into a different direction. Hence, CALPUFF starts by calculating the fractions that belong to the three zones, that is, above $H_d$, below $H_d$ left of $y_d$, and below $H_d$ right of $y_d$. Details of the calculation are not given here. They are discussed by Scire et al. (2000a).

*Above* $H_d$, no puff trajectory calculations are made, but rather for each receptor the air parcel trajectory is back-calculated to find the location from which it is coming. Hence, the receptor can be associated with a location within the puff. Streamline distortions and velocity changes are accounted for because they affect the size of the puff. Typically, streamlines will be more closely spaced and flow faster above a hill, so the puffs will be smaller and more widely spaced in the along-wind direction as a result.

*Below* $H_d$, the puff dispersion parameters are calculated as if there was no hill. The puff center is moved around the hill on an elliptical trajectory with the same size and shape as the elliptical cylinder that defines the hill.

**15.3.8.5 Puff behavior behind the Hill.** Behind the hill the three subpuffs are reunited, and the calculation resumes with a single puff.

### 15.3.9 Simple Adjustments for Complex Terrain

CALPUFF offers three simplified adjustments for complex terrain as options. They are the ISC terrain treatment, the plume path coefficient treatment, and a strain-based approach. They are discussed briefly in the following sections.

**15.3.9.1 ISC Terrain Treatment.** This algorithm refers to an older version of ISC, where any terrain was truncated at the stack height and plumes move horizontally, that is, without following the terrain. This eliminates most of the difficulties arising from complex terrain, but it leads to a very crude terrain treatment. ISC3, which is discussed in Chapter 13, has a more sophisticated treatement for complex terrain. The "ISC" treatment in CALPUFF is not recommended.

**15.3.9.2 Plume Path Coefficient Treatment.** In simple terrain, CALPUFF assumes that plumes follow the landscape rather than moving horizontally. The ISC treatment covered in the previous section assumes that plumes move horizontally. Neither approach is correct, and the real plume trajectory will be somewhere in between those two extremes. To achieve a result intermediate between horizontal and terrain following, the plume path coefficient $C$ is defined as the slope of the plume in absolute (Cartesian) coordinates divided by the slope of the terrain in absolute coordinates. In this option, CALPUFF uses $C=0.35$ in stable conditions and $C=0.5$ in neutral and unstable conditions. However, when $h_t$ exceeds the effective source height, a value of $C=1$ is adopted.

**15.3.9.3 Strain-Based Approach.** When the strain-based option is chosen, CALPUFF assumes that a puff does not follow the landscape until the following relationship applies:

$$\frac{H_p}{\sigma_z} = R_c \tag{15.102}$$

where $H_p$ is the height of the plume in terrain-following coordinates, and $R_c$ is a constant chosen equal to 1.8. Once $H_p/\sigma_z$ equals $R_c$ or less, the plume path coefficient is used as discussed in the previous section.

In addition, when the strain-based option is chosen, it is assumed that the plume lateral axis does not follow the landscape. This is a useful characteristic when plumes exist in steep valleys, where the sides of the plume tend to hit the ground due to the valley's shape. Furthermore, assuming that the plume does not follow the landscape can lead to calculated plume locations below the ground level. The plume is effectively confined within the valley. In a real valley the plume will not dilute in these conditions. For that reason, lateral puff growth is suppressed whenever $\sigma_y$ exceeds 0.4 times the width of the valley. To emulate the property of a puff with a horizontal lateral axis in Cartesian coordinates, when the coordinate system used is terrain following, the receptors are put on flagpoles.

When the dividing streamline height $H_d$ equals zero or less, the plume is assumed to follow the landscape, and no flagpole receptors are defined.

In addition, the strain-based approach contains a calculation for the effect of strain on puff growth, where strain means that the distance between streamlines in complex terrain is less than the distance between the same streamlines over flat terrain. In the presence of strain, the puff is temporarily contracted. During this contraction the concentration gradient is greater than in the absence of strain, which leads to enhanced dispersion. For details on this effect, see Scire et al. (2000a).

## REFERENCES


Holtslag A.A.M. and van Ulden A.P. (1983). A simple scheme for daytime estimates of the surface fluxes from routine weather data. *J. Clim. Appl. Meteorol.* **22**, 517–529.

Mahrt L. (1982). Momentum balance of gravity flows. *J. Atmos. Sci.* **39**, 2701–2711.

Oke T.R. (1982). The energetic basis of the urban heat island. *Quart. J. Roy. Meteorol. Soc.* **108**, 1–24.

Schulman L.L., Strimaitis D.G., and Scire J.S. (2000). Development and evaluation of the PRIME plume rise and building downwash model. *J. Air Waste Manage. Assoc.* **50**, 378–390.

Scire J.S. and Robe F.R. (1997). Fine-scale application of the CALMET meteorological model to a complex terrain. Proceedings AWMA 90th Annual Meeting and Exhibition, June 8–13, 1997, Toronto, Ontario, Canada, Paper 97-A1313.

Scire J.S., Strimaitis D.G., and Yamartino R.J. (2000a) *A User's Guide for the CALPUFF Dispersion Model.* Earth Tech, Concord, MA.

Scire J.S., Robe F.R., Fernau M.E., and Yamartino R.J. (2000b). *A User's Guide for the CALMET Meteorological Model.* Earth Tech, Concord, MA.

CHAPTER 16

# CMAQ: A BRIEF DESCRIPTION

## 16.1 INTRODUCTION

CMAQ (Community Multiscale Air Quality) is an air quality model rather than an air dispersion model. This means that the main focus of the model is the transport and chemical transformation of air pollutants, while pollutant dispersion is of secondary importance.

The main purpose of air quality models such as CMAQ is to predict concentrations of $O_3$, $NO_x$, and particulate matter, not to predict concentrations of, for instance, primary organic air pollutants. The air pollutants of interest are mainly secondary or are actively involved in atmospheric reactions.

CMAQ was developed by the EPA in the 1990s to address emerging air quality issues covered by the Clean Air Act Amendments of 1990, taking advantage of the availablity of increasing computational capabilities (Byun and Ching, 1999). EPA's main CMAQ website can be found at http://www.epa.gov/AMD/CMAQ/. The model can be downloaded from the website of the CMAS Center at http://www.cmascenter.org/download.cfm.

Implementing CMAQ is a significant challenge, not only because of the computational requirements, which are still substantial for today's computers, but also because of the complexity of the modeling system, with many interacting modules, and externally generated data files. It is a project not to be undertaken lightly.

The purpose of this chapter is to get a basic understanding of the CMAQ modeling system and a more thorough understanding of the chemistry used in CMAQ.

## 16.2 MAIN FEATURES OF CMAQ

CMAQ is an Eulerian air quality model, that is, it is based on a material balance defined on a grid that is fixed in space. It receives meteorological data from the weather forecast model MM5 and pollution emission data from the model MEPPS.

*Air Dispersion Modeling: Foundations and Applications*, First Edition. Alex De Visscher.

The material balance of an Eulerian dispersion model was discussed in Chapter 11. An equation in terms of concentration was derived [eq. (11.21)] and rewritten in mass fraction form in eq. (11.30). The latter is more appropriate in large domains, where density cannot be considered constant. This is usually the case in applications of CMAQ. However, eq. (11.30) is expressed in rectangular coordinates, which is often not an acceptable approximation in CMAQ calculations because the domain is too large to neglect the curvature of Earth.

Because the curvature of Earth cannot usually be neglected in CMAQ, the material balance, as well as other balances, is expressed in **generalized curvilinear coordinates**. What that means is that the coordinate system is not specified in the material balance but written in tensor form with appropriate factors to describe the local curvature of the coordinate system. Each module that interacts with the material balance has its own coordinate system and translates the generalized equation internally in that system. Hence, it is possible that different modules interacting with each other internally use different coordinate systems.

Mass consistency is one of the main issues in Eulerian air dispersion models. Whereas mass consistency is built in when Gaussian or Lagrangian models are used, the quality of mass consistency in an Eulerian model depends on the quality of the discretization of the material balance. Furthermore, inconsistencies with the mass balance (i.e., the continuity equation) can further undermine mass consistency. Since heat balance, rather than the mass balance, is the main focus in meteorological models, the mass balance condition is usually *not* fulfilled in meteorological data used as input for CMAQ. When the mass inconsistency is the result of the discretization, the problem can be solved with a source term that represents a closure of the material balance. If the mass consistency is the result of a problem with the mass balance, then the problem can be solved with an adjustment of the wind field. In practice, it is usually necessary to include both types of corrections to achieve accurate calculations.

Meteorological data are usually available in 1-h intervals. However, solving the material balance usually requires shorter time steps. For that reason, the meteorological data are interpolated in time to allow for calculations with the required resolution.

An important feature of CMAQ is the cloud model because it drives conditions in many of the other modules. For instance, scavenging and wet deposition of pollutants are driven by clouds, and some atmospheric reactions take place in the cloud water phase. In addition, cloud cover drives photolysis rates in the atmosphere. CMAQ models clouds with both a resolved cloud module and a subgrid cloud module. Details are discussed by Byun and Ching (1999).

Direct simulation of air quality on an Eulerian grid is usually impossible because the range of scales between the smallest and largest levels of turbulence is too large. CMAQ calculations are no exception in this respect. In CMAQ a Reynolds-decomposed material balance is used, and the second-order terms are closed with a closure model like gradient transport theory ($K$ theory). See Section 11.3 for details.

As discussed in Section 11.5, there is not a single numerical method that can adequately solve the material balance in a large Eulerian dispersion model such as CMAQ. This is because air pollutants undergo several processes, like advection, diffusion, and reaction, each with very different numerical properties. To get around that problem, CMAQ uses a technique known as operator splitting, where each process is modeled separately, and their influences on the pollutant concentration are added together.

## 16.3 ADVECTION AND DIFFUSION MODELING IN CMAQ

The advection process is modeled by the following truncated form of eq. (11.21):

$$\frac{\partial c_i}{\partial t} = \frac{\partial c_i u}{\partial x} - \frac{\partial c_i v}{\partial y} - \frac{\partial c_i w}{\partial z} \tag{16.1}$$

This equation was defined in a rectangular coordinate system. However, with some precautions it can also be used in curvilinear coordinates (Byun and Ching, 1999). In CMAQ operation splitting is applied to the three dimensions, which reduces the problem to a one-dimensional problem. CMAQ contains three algorithms for advection calculations as options: the piecewise parabolic method of Colella and Woodward (1984), the method of Bott (1989), and the cubic scheme of Yamartino (1993). These methods are described briefly in Section 11.5.1. Of these three methods, the method of Yamartino (1993) is the most accurate, but it is still prone to some distortion, and it is more time consuming than the other methods. Of these, the Bott scheme is more accurate than the Colella and Woodward scheme, but more likely to generate spurious oscillations.

Modeling vertical advection in CMAQ can be more challenging than horizontal advection because the vertical coordinate is usually an irregular one, like a variable linked to pressure. After the advection calculation, the concentrations are corrected for mass consistency.

**Dispersion** is divided into horizontal and vertical dispersion, which have very different properties, and will be discussed separately. **Vertical dispersion** can be described in CMAQ with a first-order closure technique based on similarity theory, a closure technique based on turbulent kinetic energy, and three nonlocal closure techniques: the Blackadar scheme, the Asymmetrical Convective Model (ACM) and a transilient turbulence model. These techniques are discussed in Sections 11.3.1 and 11.3.2.

Numerically, the local dispersion schemes are integrated numerically using a semi-implicit scheme, the Crank–Nicholson scheme. The principle will be illustrated with vertical dispersion in the simplified case of constant diffusivity and a rectangular coordinate system. The starting point is the following equation:

$$\frac{\partial c_i}{\partial t} = K_z \frac{\partial^2 c_i}{\partial z^2} \tag{16.2}$$

where $i$ is the number of the vertical layer where $c$ is evaluated. The time derivative is discretized as follows:

$$\frac{\partial c_i}{\partial t} = \frac{c_i^{t+\Delta t} - c_i^t}{\Delta t} \tag{16.3}$$

where superscripts are used to denote time. Equation (16.3) represents an average value between time $t$ and time $t + \Delta t$. Hence, the left-hand side of eq. (16.2) should also an average time between time $t$ and time $t + \Delta t$. This is achieved with the following equation:

$$\frac{\partial^2 c_i}{\partial z^2} = \frac{1}{2}\left[\frac{c_{i-1}^t - 2c_i^t + c_{i+1}^t}{(\Delta z)^2} + \frac{c_{i-1}^{t+\Delta t} - 2c_i^{t+\Delta t} + c_{i+1}^{t+\Delta t}}{(\Delta z)^2}\right] \quad (16.4)$$

Substitution of Eqs. (16.3) and (16.4) into eq. (16.2) leads to the following equation:

$$\frac{c_i^{t+\Delta t} - c_i^t}{\Delta t} = \frac{K_z}{2}\left[\frac{c_{i-1}^t - 2c_i^t + c_{i+1}^t}{(\Delta z)^2} + \frac{c_{i-1}^{t+\Delta t} - 2c_i^{t+\Delta t} + c_{i+1}^{t+\Delta t}}{(\Delta z)^2}\right] \quad (16.5)$$

Hence, a linear algebraic equation for each vertical grid point is obtained, with a number of unknowns (the concentrations at time $t + \Delta t$) as equations. This set of equations is solved to obtain the unknowns. Because of the separation of operations, each vertical column of grid points can be solved separately.

The nonlocal closure schemes include a numerical integration routine and do not need to be considered here.

As indicated in Section 11.3.1, horizontal dispersion is more difficult to incorporate accurately than vertical dispersion because a substantial numerical dispersion occurs in the horizontal direction. In CMAQ the assumption is made that a turbulent diffusivity of $4000\,m^2\,s^{-1}$, along with the numerical dispersion at a grid cell size of 4 km, adequately predicts dispersion. A correction is made for other grid cell sizes to account for the difference in numerical dispersion. This is a rather ad hoc approach, especially as the numerical dispersion of the Colella and Woodward (1984) and Bott (1989) routines is much larger than the dispersion of the Yamartino (1993) routine. Furthermore, a value of $4000\,m^2\,s^{-1}$ is a reasonable value above the surface layer in highly convective conditions, but orders of magnitude too high near the surface and in the absence of convective cycles.

Eulerian models inherently have a coarse resolution, so that pollutants from point sources cannot be resolved properly until several hours after their emission. Furthermore, the physics of gradient transport breaks down when the scale is a few kilometers or less. To resolve these issues, a plume-in grid formalism was introduced in CMAQ. The plume-in-grid model is based on continuous parameterization of atmospheric stability and uses the parameterization of Weber et al. (1982) and Hanna (1982) discussed in Section 6.6.7. A more extensive description of the plume-in-grid formalism is given in Section 11.5.5.

## 16.4 ATMOSPHERIC CHEMISTRY MODELING IN CMAQ

### 16.4.1 Photolysis Rates

Atmospheric chemistry is largely initiated by photochemical reactions, so it is crucial to accurately predict photolysis rates in CMAQ. An introduction on photochemistry is given in Section 11.4.3. Photolysis rate constants $j_i$ are calculated with the following equation:

$$j_i = \int_0^\infty \phi(\lambda)\sigma(\lambda)I(\lambda)\,\mathrm{d}\lambda \qquad (11.120)$$

where $\phi$ is quantum yield, $\sigma$ is the absorption cross section, and $I$ is the spectral actinic flux, all functions of the wavelength $\lambda$. Calculating the spectral actinic flux in the atmosphere is computationally demanding because reflection and scattering causes light rays to change direction so solar angle alone does not define the light's path length. Hence, to save computation time, CMAQ calculates photolysis rates in two steps:

- First, a table of photolysis rate constants in clear sky conditions is generated for a set of heights, latitudes, and times of the day.
- Second, actual photolysis rate constants are interpolated from the table and corrected for cloud cover.

Details of the calculation procedure are given by Byun and Ching (1999).

### 16.4.2 Chemical Mechanisms in CMAQ

The main reason for using CMAQ is the level of sophistication of its chemistry. CMAQ can handle complex chemistry, with nonlinear effects, whereas other models can only handle simple chemistry where the effect is linear. Even so, the chemistry in CMAQ is incomplete, as important air pollutants such as $H_2S$ and benzene are not included. The user of CMAQ can choose between two chemical mechanisms, CB4 and RADM2, and select a mumber of extensions for each.

The chemical mechanism CB4 (Carbon Bond version 4) includes 93 reactions, involving 36 chemical species. Eleven of the reactions are photolytic.

Some of the species in CB4, such as ethene, isoprene, and formaldehyde are real molecules, whereas other species represent a chemical bond (e.g., paraffins represent a single carbon–carbon bond; olefins refer to a double carbon–carbon bond), whereas some species represent a group of molecules (e.g., xylene refers to ethylbenzene as well).

The representation of real molecules in terms of species depends on both chemistry and reactivity. For instance, an internal double bond is not described as an olefin but as a higher aldehyde. Olefins are only used to describe a double bond at the end of a hydrocarbon chain.

The chemical model RADM2 follows an approach similar to CB4 but is more extensive. It contains 158 reactions involving 57 species; 21 of the reactions are photolytic.

The models can be used with several **extensions**. The first extension is for **aerosol formation** from organic precursors. To that effect, the model defines a number of virtual species (TOLAER for aerosols formed from toluene, SULAER for aerosols formed from sulfur comounds, etc.) that do not take part in any reactions once produced. Their only function is to keep track of the number of molecules that take part in aerosol formation.

RADM2 also has an **isoprene extension**. It defines four species representing reaction products of isoprene.

Another extension is for **aqueous chemistry**. It keeps track of highly soluble compounds such as formic acid and methyl hydroperoxide, which dissolve readily in clouds whenever present.

Table 16.1 shows the chemical mechanism of CB4 as an example. The species list is found in Table 16.2. The rate constants of the reactions at 25 °C and 1 atm pressure are given in Table 16.3, with the exception of the photolytical rate constants, which depend on the fluence rate. Representative photochemical rate constants for the most important photochemical reactions are given in Sections 11.4.3 and 11.4.4, and in Problem 1 of Chapter 11. Some of the principles underlying the develoment of this mechanism were discussed in Section 11.4.4 and will be further elaborated here.

**TABLE 16.1 Chemistry of the Chemical Mechnism CB4 in CMAQ[a]**

| | |
|---|---|
| 1 | NO2 + $h\nu$ → NO + O |
| 2 | O + [O2] → O3 |
| 3 | O3 + NO → NO2 |
| 4 | O + NO2 → NO |
| 5 | O + NO2 → NO3 |
| 6 | O + NO → NO2 |
| 7 | O3 + NO2 → NO3 |
| 8 | O3 + $h\nu$ → O |
| 9 | O3 + $h\nu$ → O1D |
| 10 | O1D + [N2] → O |
| 11 | O1D + [O2] → O |
| 12 | O1D + [H2O] → 2OH |
| 13 | O3 + OH → HO2 |
| 14 | O3 + HO2 → OH |
| 15 | NO3 + $h\nu$ → 0.89NO2 + 0.89O + 0.11NO |
| 16 | NO3 + NO → 2NO |
| 17 | NO3 + NO2 → NO + NO2 |
| 18 | NO3 + NO2 → N2O5 |
| 19 | N2O5 + [H2O] → 2HNO3 |
| 20 | N2O5 → NO3 + NO2 |
| 21 | NO + NO + [O2] → 2NO2 |
| 22 | NO + NO2 + [H2O] → 2HONO |
| 23 | OH + NO → HONO |
| 24 | HONO + $h\nu$ → OH + NO |
| 25 | HONO + OH → NO2 |
| 26 | HONO + HONO → NO + NO2 |
| 27 | OH + NO2 → HNO3 |
| 28 | OH + HNO3 → NO3 |
| 29 | HO2 + NO → OH + NO2 |
| 30 | HO2 + NO2 → PNA |

*(Continued)*

**TABLE 16.1** *(Continued)*

| | |
|---|---|
| 31 | PNA → HO2 + NO2 |
| 32 | PNA + OH → NO2 |
| 33 | HO2 + HO2 → H2O2 |
| 34 | HO2 + HO2 + [H2O] → H2O2 |
| 35 | H2O2 + $h\nu$ → 2OH |
| 36 | H2O2 + OH → HO2 |
| 37 | CO + OH → HO2 |
| 38 | FORM + OH → HO2 + CO |
| 39 | FORM + $h\nu$ → 2HO2 + CO |
| 40 | FORM + $h\nu$ → CO |
| 41 | FORM + O → OH + HO2 + CO |
| 42 | FORM + NO3 → HNO3 + HO2 + CO |
| 43 | ALD2 + O → C2O3 + OH |
| 44 | ALD2 + OH → C2O3 |
| 45 | ALD2 + NO3 → C2O3 + HNO3 |
| 46 | ALD2 + $h\nu$ → XO2 + 2HO2 + CO + FORM |
| 47 | C2O3 + NO → NO2 + XO2 + FORM + HO2 |
| 48 | C2O3 + NO2 → PAN |
| 49 | PAN → C2O3 + NO2 |
| 50 | C2O3 + C2O3 → 2XO2 + 2FORM + 2HO2 |
| 51 | C2O3 + HO2 → 0.79FORM + 0.79XO2 + 0.79HO2 + 0.79OH + 0.21PACD |
| 52 | OH → XO2 + FORM + HO2 |
| 53 | PAR + OH → 0.87XO2 + 0.13XO2N + 0.11HO2 + 0.11ALD2 + 0.76ROR – 0.11PAR |
| 54 | ROR → 1.1ALD2 + 0.96XOR2 + 0.94HO2 – 2.1PAR + 0.04XO2N + 0.02ROR |
| 55 | ROR → HO2 |
| 56 | ROR + NO2 → NTR |
| 57 | OLE + O → 0.63ALD2 + 0.38HO2 + 0.28XO2 + 0.3CO + 0.2FORM + 0.02XO2N + 0.22PAR + 0.2OH |
| 58 | OLE + OH → FORM + ALD2 + XO2 + HO2 – PAR |
| 59 | OLE + O3 → 0.5ALD2 + 0.74FORM + 0.33CO + 0.44HO2 + 0.22XO2 + 0.1OH + 0.2FACD + 0.2AACD – PAR |
| 60 | OLE + NO3 → 0.91XO2 + 0.09XO2N + FORM + ALD2 – PAR + NO2 |
| 61 | ETH + O → FORM + 0.7XO2 + 1.7HO2 + 0.3OH |
| 62 | ETH + OH → XO2 + 1.56FORM + HO2 + 0.22ALD2 |
| 63 | ETH + O3 → FORM + 0.42CO + 0.12HO2 + 0.4FACD |
| 64 | TOL + OH → 0.08XO2 + 0.36CRES + 0.44HO2 + 0.56TO2 |
| 65 | TO2 + NO → 0.9NO2 + 0.9HO2 + 0.9OPEN + 0.1NTR |
| 66 | TO2 → CRES + HO2 |
| 67 | CRES + OH → 0.4CRO + 0.6XO2 + 0.6HO2 + 0.3OPEN |
| 68 | CRES + NO3 → CRO + HNO3 |
| 69 | CRO + NO2 → NTR |
| 70 | XYL + OH → 0.7HO2 + 0.5XO2 + 0.2CRES + 0.8MGLY + 1.1PAR + 0.3TO2 |
| 71 | OPEN + OH → XO2 + 2CO + 2HO2 + C2O3 + FORM |
| 72 | OPEN + $h\nu$ → C2O3 + HO2 + CO |
| 73 | OPEN + O3 → 0.03ALD2 + 0.62C2O3 + 0.7FORM + 0.03XO2 + 0.69CO + 0.08OH + 0.76HO2 + 0.2MGLY |

*(Continued)*

**TABLE 16.1** *(Continued)*

| | |
|---|---|
| 74 | MGLY + OH → XO2 + C2O3 |
| 75 | MGLY + $h\nu$ → C2O3 + HO2 + CO |
| 76 | ISOP + O → 0.75ISPD + 0.5FORM + 0.25XO2 + 0.25HO2 + 0.25C2O3 + 0.25PAR |
| 77 | ISOP + OH → 0.912ISPD + 0.629FORM + 0.991XO2 + 0.912HO2 + 0.088XO2N |
| 78 | ISOP + O3 → 0.65ISPD + 0.6FORM + 0.2XO2 + 0.066HO2 + 0.266OH + 0.2C2O3 + 0.15ALD2 + 0.35PAR + 0.066CO |
| 79 | ISOP + NO3 → 0.2ISPD + 0.8NTR + XO2 + 0.8HO2 + 0.2NO2 + 0.8ALD2 + 2.4PAR |
| 80 | XO2 + NO → NO2 |
| 81 | XO2 + XO2 → |
| 82 | XO2N + NO → NTR |
| 83 | SO2 + OH → SULF + HO2 |
| 84 | SO2 → SULF |
| 85 | XO2 + HO2 → |
| 86 | XO2N + HO2 → |
| 87 | XO2N + XO2N → |
| 88 | XO2N + XO2 → |
| 89 | ISPD + OH → 1.565PAR + 0.167FORM + 0.713XO2 + 0.503HO2 + 0.334CO + 0.168MGLY + 0.273ALD2 + 0.498C2O3 |
| 90 | ISPD + O3 → 0.114C2O3 + 0.15FORM + 0.85MGLY + 0.154HO2 + 0.268OH + 0.064XO2 + 0.02ALD2 + 0.36PAR + 0.225CO |
| 91 | ISPD + NO3 → 0.357ALD2 + 0.282FORM + 1.282PAR + 0.925HO2 + 0.643CO + 0.85NTR + 0.075C2O3 + 0.075XO2 + 0.075HNO3 |
| 92 | ISPD + $h\nu$ → 0.333CO + 0.067ALD2 + 0.9FORM + 0.832PAR + 1.033HO2 + 0.7XO2 + 0.967C2O3 |
| 93 | ISOP + NO2 → 0.2ISPD + 0.8NTR + XO2 + 0.8HO2 + 0.2NO + 0.8ALD2 + 2.4PAR |

Source: From Byun and Ching (1999).

[a] $h\nu$ referes to a photon; species between square brackets are assumed to have constant concentration and require no component in the mechanism.

**TABLE 16.2 Species List for Chemical Mechanism CB4 in CMAQ**

| Notation | Species |
|---|---|
| NO | Nitric oxide |
| NO2 | Nitrogen dioxide |
| HONO | Nitrous acid |
| N2O5 | Nitrogen pentoxide |
| HNO3 | Nitric acid |
| PNA | Peroxynitric acid |
| O3 | Ozone |
| H2O2 | Hydrogen peroxide |
| SO2 | Sulfur dioxide |
| SULF | Sulfuric acid |
| O | Oxygen atom (triplet) |
| O1D | Oxygen atom (singlet) |
| OH | Hydroxyl radical |
| HO2 | Hydroperoxy radical |

*(Continued)*

**TABLE 16.2** *(Continued)*

| Notation | Species |
|---|---|
| CO | Carbon monoxide |
| PAR | Paraffin carbon bond |
| ETH | Ethene |
| OLE | Olefinic carbon bond |
| TOL | Toluene |
| XYL | Xylene |
| ISOP | Isoprene |
| FORM | Formaldehyde |
| ALD2 | Acetaldehyde and higher aldehydes |
| MGLY | Methyl glyoxal ($CH_3C(O)C(O)H$) |
| CRES | Cresol and higher molecular weight phenols |
| PAN | Peroxyacyl nitrate ($CH_3C(O)OONO_2$) |
| NTR | Organic nitrate |
| C2O3 | Peroxyacyl radical ($CH_3C(O)OO^{\cdot}$) |
| ROR | Secondary organic oxy radical |
| CRO | Methylphenoxy radical |
| XO2 | NO-to-$NO_2$ operation |
| XO2N | NO-to-nitrate operation |
| TO2 | Toluene-hydroxyl radical adduct |
| OPEN | High molecular weight aromatic oxidation ring fragment |
| ISPD | Products of isoprene reactions |

Source: From Byun and Ching (1999).

**TABLE 16.3 Rate Constants of Chemical Reactions in CB4 at 25 °C and 1 atm ($cm^{-3}$ $molecule^{-1}$ $s^{-1}$)**

| Reaction | $k_2$ ($cm^{-3}$ $molecule^{-1}$ $s^{-1}$) |
|---|---|
| 1 | Photochemical |
| 2 | $1.37\times10^{-14}$ |
| 3 | $1.81\times10^{-14}$ |
| 4 | $9.30\times10^{-12}$ |
| 5 | $1.58\times10^{-12}$ |
| 6 | $1.66\times10^{-12}$ |
| 7 | $3.23\times10^{-17}$ |
| 8 | Photochemical |
| 9 | Photochemical |
| 10 | $2.58\times10^{-11}$ |
| 11 | $4.01\times10^{-11}$ |
| 12 | $2.20\times10^{-10}$ |
| 13 | $6.83\times10^{-14}$ |
| 14 | $2.00\times10^{-15}$ |
| 15 | Photochemical |

**TABLE 16.3** *(Continued)*

| Reaction | $k_2$ (cm$^{-3}$ molecule$^{-1}$ s$^{-1}$) |
|---|---|
| 16 | $3.01\times10^{-11}$ |
| 17 | $4.03\times10^{-16}$ |
| 18 | $1.26\times10^{-12}$ |
| 19 | $1.3\times10^{-21}$ |
| 20 | $4.36\times10^{-2a}$ |
| 21 | $1.95\times10^{-38b}$ |
| 22 | $4.40\times10^{-40b}$ |
| 23 | $6.70\times10^{-12}$ |
| 24 | Photochemical |
| 25 | $6.60\times10^{-12}$ |
| 26 | $1.00\times10^{-20}$ |
| 27 | $1.15\times10^{-11}$ |
| 28 | $1.47\times10^{-13}$ |
| 29 | $8.28\times10^{-12}$ |
| 30 | $1.48\times10^{-12}$ |
| 31 | $9.18\times10^{-2a}$ |
| 32 | $4.65\times10^{-12}$ |
| 33 | $2.80\times10^{-12}$ |
| 34 | $6.24\times10^{-30b}$ |
| 35 | Photochemical |
| 36 | $1.66\times10^{-12}$ |
| 37 | $2.40\times10^{-13}$ |
| 38 | $1.00\times10^{-11}$ |
| 39 | Photochemical |
| 40 | Photochemical |
| 41 | $1.65\times10^{-13}$ |
| 42 | $6.30\times10^{-16}$ |
| 43 | $4.39\times10^{-13}$ |
| 44 | $1.62\times10^{-11}$ |
| 45 | $2.50\times10^{-15}$ |
| 46 | Photochemical |
| 47 | $4.39\times10^{-11}$ |
| 48 | $9.41\times10^{-12}$ |
| 49 | $4.23\times10^{-4a}$ |
| 50 | $2.50\times10^{-12}$ |
| 51 | $6.50\times10^{-12}$ |
| 52 | $3.54\times10^{-1a}$ |
| 53 | $8.10\times10^{-13}$ |
| 54 | $2.19\times10^{3a}$ |
| 55 | $1.60\times10^{3a}$ |
| 56 | $1.50\times10^{-11}$ |
| 57 | $4.05\times10^{-12}$ |
| 58 | $2.82\times10^{-11}$ |
| 59 | $1.20\times10^{-17}$ |
| 60 | $7.70\times10^{-15}$ |

*(Continued)*

**TABLE 16.3** *(Continued)*

| Reaction | $k_2$ (cm$^{-3}$ molecule$^{-1}$ s$^{-1}$) |
|---|---|
| 61 | $7.01\times10^{-13}$ |
| 62 | $7.94\times10^{-12}$ |
| 63 | $1.89\times10^{-18}$ |
| 64 | $6.19\times10^{-12}$ |
| 65 | $8.10\times10^{-12}$ |
| 66 | 4.20[a] |
| 67 | $4.10\times10^{-11}$ |
| 68 | $2.20\times10^{-11}$ |
| 69 | $1.40\times10^{-11}$ |
| 70 | $2.51\times10^{-11}$ |
| 71 | $3.00\times10^{-11}$ |
| 72 | Photochemical |
| 73 | $1.01\times10^{-17}$ |
| 74 | $1.70\times10^{-11}$ |
| 75 | Photochemical |
| 76 | $3.60\times10^{-11}$ |
| 77 | $9.97\times10^{-11}$ |
| 78 | $1.29\times10^{-17}$ |
| 79 | $6.74\times10^{-13}$ |
| 80 | $8.10\times10^{-12}$ |
| 81 | $1.33\times10^{-12}$ |
| 82 | $8.10\times10^{-12}$ |
| 83 | $7.51\times10^{-13}$ |
| 84 | $1.36\times10^{-6}$[a] |
| 85 | $6.02\times10^{-12}$ |
| 86 | $6.02\times10^{-12}$ |
| 87 | $1.36\times10^{-12}$ |
| 88 | $2.71\times10^{-12}$ |
| 89 | $3.36\times10^{-11}$ |
| 90 | $7.11\times10^{-18}$ |
| 91 | $1.00\times10^{-15}$ |
| 92 | Photochemical |
| 93 | $1.49\times10^{-19}$ |

Source: From Byun and Ching (1999).
[a]Monomolecular reaction (s$^{-1}$).
[b]Termolecular reaction (cm$^6$ molecule$^{-2}$ s$^{-1}$).

Some reactions in Table 16.1 have species between square brackets. For instance, in reaction 2:

$$O^{\bullet} + [O_2] \rightarrow O_3 \qquad \text{(Reaction 2)}$$

These species, $O_2$, $N_2$, and $H_2O$, are necessary in the rate law but do not require a reaction rate of their own because they are present in a large excess concentration.

For the same reason they are not mentioned when they occur as reaction products. An example is reaction 3, for which the full reaction is

$$O_3 + NO \rightarrow NO_2 + O_2 \qquad \text{(Reaction 3)}$$

In the mechanism there are several cases where a single reaction is in fact a shorthand notation for a series of reactions. An example is reaction 37:

$$CO + OH^{\bullet} \rightarrow HO_2^{\bullet} \qquad \text{(Reaction 37)}$$

The actual reaction is as follows:

$$CO + OH^{\bullet} \rightarrow CO_2 + H^{\bullet} \qquad (16.6)$$

Howoever, the hydrogen radical reacts readily with oxygen:

$$H^{\bullet} + O_2 \rightarrow HO_2^{\bullet} \qquad (16.7)$$

$CO_2$ and $O_2$ are ignored in these reactions. Because the second reaction is intrinsically very fast, it can be assumed instantaneous, and the combination of eqs. (16.6) and (16.7) cannot be distinguished from reaction 37. The overall reaction rate is the reaction rate of eq. (16.6). This reaction is referred to as the **rate limiting** reaction.

Another example is reaction 58:

$$OLE + {}^{\bullet}OH \rightarrow FORM + ALD2 + XO_2 + HO_2^{\bullet} - PAR \qquad \text{(Reaction 58)}$$

Note the minus sign in front of PAR: This species is consumed without affecting the kinetics of the reaction and is therefore included on the product side of the equation, with a negative stoichiometry. One component, $XO_2$, is not a real species, but refers to an ability to convert NO to $NO_2$.

Reaction 58 will be outlined in terms of the actual reactions taking place, using 1-butene ($C_4H_8$) as example. The initial, rate-limiting reaction is the addition of a hydroxyl radical on the double bond:

$$CH_2 = CH - C_2H_5 + {}^{\bullet}OH \rightarrow HO - CH_2 - C^{\bullet}H - C_2H_5 \qquad (16.8)$$

In the next step the unpaired electron reacts with oxygen, forming a peroxyl radical. As oxygen is not tracked in the chemistry, this step is not directly visible in reaction 58:

$$HO - CH_2 - C^{\bullet}H - C_2H_5 + O_2 \rightarrow HO - CH_2 - C(OO^{\bullet})H - C_2H_5 \qquad (16.9)$$

Next, NO is converted into $NO_2$ by reaction with the peroxyl group. This is where the virtual species $XO_2$ is produced:

$$HO - CH_2 - C(OO^{\bullet})H - C_2H_5 + NO \rightarrow HO - CH_2 - C(O^{\bullet})H - C_2H_5 + NO_2 \quad (16.10)$$

An oxy radical usually leads to the production of an aldehyde and breaks a C-C bond in the process. This is the step where ALD2 is produced and PAR is consumed:

$$HO-CH_2-C(O^{\bullet})H-C_2H_5 \rightarrow HO-C^{\bullet}H_2+H(C=O)C_2H_5 \qquad (16.11)$$

In the last step oxygen abstracts a hydrogen atom from the organic radical:

$$HO-C^{\bullet}H_2+O_2 \rightarrow O=CH_2+HO_2^{\bullet} \qquad (16.12)$$

This is where FORM is produced. It follows that reaction 58 is shorthand notation of a series of five reactions, the first one of which is rate limiting. Some reactions in the model are even more complex and describe several competing reactions in parallel, as well as reactions in series. This is apparent from the fractional stoichiometric factors in some of the reactions.

### 16.4.3 Numerical Aspects of Chemistry in CMAQ

Radical chemistry is very difficult to model numerically because of the nature of the process. This is because the reactions producing radicals are slow, whereas the reactions consuming radicals are several orders of magnitude faster. As a result, the dynamics of the radicals is orders of magnitude faster than the overall dynamics of the process, and the differential equations describing the chemical species reflect that dynamics. This is known as a stiff set of differential equations. When the regular differential equation solvers, like the class of Runge–Kutta methods, are used to integrate the differential equations, the time step of the integration must be shorter than the shortest time constant occurring in the system. In the presence of radicals such as $O(^1D)$, billions of time steps would be needed to resolve the chemistry of the atmosphere.

Fortunately, as discussed in Section 11.5.3, there are special solvers for stiff sets of ordinary differential equations. A popular stiff solver that formed the basis for the algorithm used in CMAQ is the Gear (1971) algorithm, which is based on backward differentiation. In itself, the Gear algorithm is still too inefficient for optimal use in CMAQ. For that reason, further improvements were made by Jacobson and Turco (1994) and by Jacobson (1995) and implemented in CMAQ. An alternative option in CMAQ is the use of an iterative quasi-steady-state approximation.

## REFERENCES

Bott A. (1989). A positive definite advection scheme obtained by nonlinear renormalization of the advective fluxes. *Mon. Weather Rev.* **117**, 1006–1015.

Byun D.W. and Ching J.K.S. (1999). *Science Algorithms of the EPA Models-3 Community Multiscale Air Quality (CMAQ) Modeling System*. Report EPA/600/R-99/030. US-EPA, Research Triangle Park, NC.

Colella P. and Woodward P.R. (1984). The piecewise parabolic method (PPM) for gas-dynamical simulations. *J. Comp. Phys.* **54**, 174–201.

Gear C.W. (1971). The automatic integration of ordinary differential equations. *Comm. ACM* **14**, 176–179.

Hanna S.R. (1982). Applications in air pollution modelling. In Nieuwstadt F.T.M. and van Dop H. (eds.) *Atmospheric Turbulence and Air Pollution Modelling*. Reidel, Dordrecht, The Netherlands.

Jacobson M. (1995). Computation of global photochemistry with SMVGEAR II. *Atmos. Environ.* **29**, 2541–2546.

Jacobson M. and Turco R.P. (1994). SMVGEAR: A sparse-matrix, vectorized Gear code for atmospheric models. *Atmos. Environ.* **28**, 273–284.

Weber A.H., Irwin J.S., Petersen W.B., Mathis J.J. Jr., and Kahler J.P. (1982). Spectral scales in the atmospheric boundary layer. *J. Appl. Meteorol.* **21**, 1622–1632.

Yamartino R.J. (1993). Nonnegative, conserved scalar transport using grid-cell-centered, spectrally constrained Blackman cubics for applications on a variable-thickness mesh. *Mon. Weather Rev.* **121**, 753–763.

APPENDIX A

# AUXILIARY CALCULATIONS AND DERIVATIONS

## CHAPTER A5: METEOROLOGY FOR AIR DISPERSION MODELERS

### A.5.1 Pressure and Temperature Dependence of $\Gamma_s$

The purpose of this section is to find an explicit relationship between $\Gamma_s$ and the temperature and pressure, based on eq. (5.25):

$$\frac{dT}{dz} = -\frac{g}{c_{p,\text{air}} + \Delta_{\text{vap}} H \left( {dw_w}/{dT} \right)} \tag{5.25}$$

First, we write the water weight fraction $w_w$ at saturation in terms of total pressure and water vapor pressure:

$$w_w = \frac{m_{\text{water}}}{m_{\text{air}} + m_{\text{water}}} = \frac{m_{\text{water}} / m_{\text{air}}}{1 + m_{\text{water}} / m_{\text{air}}} = \frac{w'_w}{1 + w'_w}$$

where $w'_w$ is defined as

$$w'_w = \frac{m_{\text{water}}}{m_{\text{air}}} = \frac{M_{\text{water}} n_{\text{water}}}{M_{\text{air}} n_{\text{air}}}$$

where $M$ and $n$ stand for molar mass and number of moles, respectively. At constant temperature and pressure the number of moles of an ideal gas is proportional to the partial pressure of the gas. Thus it follows:

*Air Dispersion Modeling: Foundations and Applications*, First Edition. Alex De Visscher.

$$w'_{\mathrm{w}} = \frac{M_{\mathrm{water}}}{M_{\mathrm{air}}} \frac{p_{\mathrm{v}}}{p - p_{\mathrm{v}}} = \frac{18}{29} \frac{p_{\mathrm{v}}}{p - p_{\mathrm{v}}} = 0.62 \frac{p_{\mathrm{v}}}{p - p_{\mathrm{v}}}$$

and $w_{\mathrm{w}}$ is thus calculated as

$$w_{\mathrm{w}} = \frac{0.62 p_{\mathrm{v}} / (p - p_{\mathrm{v}})}{1 + 0.62 p_{\mathrm{v}} / (p - p_{\mathrm{v}})} = \frac{0.62 p_{\mathrm{v}}}{p - p_{\mathrm{v}} + 0.62 p_{\mathrm{v}}} = \frac{0.62 p_{\mathrm{v}}}{p - 0.38 p_{\mathrm{v}}}$$

From this equation it is clear that $w_{\mathrm{w}}$ depends on $p$, so $\Gamma_{\mathrm{s}}$ will depend on $p$ as well. For the $T$ dependence we take the temperature derivative:

$$\frac{dw_{\mathrm{w}}}{dT} = 0.62 \frac{p}{(p - 0.38 p_{\mathrm{v}})^2} \frac{dp_{\mathrm{v}}}{dT}$$

The saturated vapor pressure $p_{\mathrm{v}}$ of liquid water is taken from Wagner and Pruß (2002):

$$p_{\mathrm{v}} = p_{\mathrm{c}} \exp\left[\frac{T_{\mathrm{c}}}{T}(a_1\theta + a_2\theta^{1.5} + a_3\theta^3 + a_4\theta^{3.5} + a_5\theta^4 + a_6\theta^{7.5})\right]$$

where

$p_{\mathrm{c}} = 22.064 \times 10^6$ Pa (critical pressure)
$T_{\mathrm{c}} = 647.096$ K (critical temperature)
$\theta = 1 - T/T_{\mathrm{c}}$ (not to be confused with potential temperature)
$a_1 = -7.85951783$
$a_2 = 1.84408259$
$a_3 = -11.7866497$
$a_4 = 22.6807411$
$a_5 = -15.9618719$
$a_6 = 1.80122502$

As the above equation has a strongly nonlinear temperature relationship, $dp_{\mathrm{v}}/dT$ will be temperature dependent, and so will $\Gamma_{\mathrm{s}}$. In particular:

$$\frac{dp_{\mathrm{v}}}{dT} = -p_{\mathrm{v}} \left(\frac{T_{\mathrm{c}}}{T^2} A + \frac{1}{T_{\mathrm{c}}} B\right)$$

where

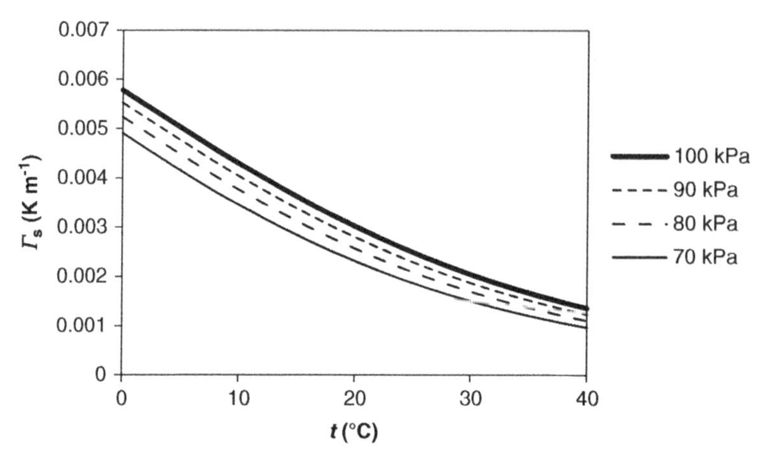


Figure A5.1 Temperature and pressure dependence of $\Gamma_s$.

$$A = a_1\theta + a_2\theta^{1.5} + a_3\theta^3 + a_4\theta^{3.5} + a_5\theta^4 + a_6\theta^{7.5}$$
$$B = a_1 + 1.5a_2\theta^{0.5} + 3a_3\theta^2 + 3.5a_4\theta^{2.5} + 4a_5\theta^3 + 7.5a_6\theta^{6.5}$$

The result of eq. (5.25) with the above equations to supply the required variables is shown in Figure A5.1. An approximate equation that reproduces this relationship closely in the temperature range 0–40 °C is the following:

$$\Gamma_s = \frac{a}{1 + b\exp(ct^d)} \qquad (5.27)$$

where $a = 0.01028$ K m$^{-1}$, $b = 0.75586$, $c = 0.074363$ °C$^{-1}$, and $d = 0.915735$.

## A.5.2 Calculation of Solar Elevation

The solar elevation is given by the following equation:

$$\varphi = \arcsin(\sin\lambda\ \sin\delta + \cos\lambda\ \cos\delta\ \cos h)$$

In this equation $\lambda$ is the latitude, and $\delta$ is the solar declination (the latitude where the sun is in the zenith at the solar noon). It is given by

$$\delta = 23.45° \cdot \sin\left[\frac{360°(d - 284)}{365.25}\right]$$

with $d$ the day of the year. The hour angle $h$ is given by

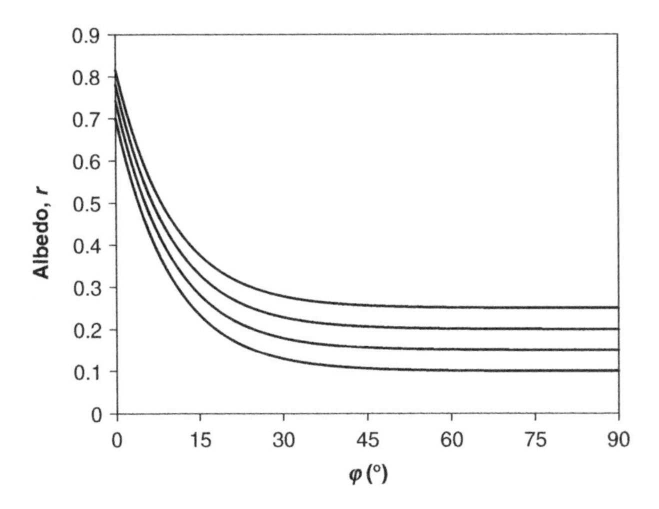


Figure A5.2. Solar elevation dependence of albedo.

$$h = 15°(t - t_0)$$

with $t$ the time (in hours), and $t_0$ the time of the solar noon.

Note that all angles are given in degrees. To convert to radians, multiply by $2\pi/360$.

## A.5.3 Solar Elevation Dependence of Albedo

The albedo, $r$, at a given solar elevation, $\varphi$, can be calculated from the following equation:

$$r = r' + (1 - r')\exp[-0.1\varphi - 0.5(1 - r')^2]$$

with $r'$ the albedo at 90° and $\varphi$ is still given in degrees. Some curves of albedo versus solar elevation are given in Figure A5.2. The albedo increases at low solar elevations.

## A.5.4 Calculation of the Contribution of Viscosity on the Surface Shear Stress

In the derivation of the similarity theory for wind speed profiles (Section 5.7.1), it was assumed that the contribution of viscosity to the shear stress caused by the wind is negligible. The purpose of this section is to test this assumption.

The friction velocity is usually greater than $0.1\ \mathrm{m\,s^{-1}}$. Applying eq. (5.58) and a density of about $1\ \mathrm{kg\,m^{-3}}$, this means that the shear stress at the surface is at least 0.01 Pa.

Assume that the wind speed is zero at the surface, and $1\,\mathrm{m\,s^{-1}}$ at 0.05 m height. The shear stress caused by the air's viscosity under these conditions (i.e., under laminar or viscous flow) is given by Newton's law of viscosity:

$$\tau = -\mu \frac{\partial u}{\partial z}$$

where $\mu$ is the dynamic viscosity of air (about $1.8\times10^{-5}\,\mathrm{Pa\,s}$). Hence, the viscous contribution to the shear stress is $1.8\times10^{-5}\,\mathrm{Pa\,s}\times1\,\mathrm{m\,s^{-1}}/0.05\,\mathrm{m}=3.6\times10^{-4}\,\mathrm{Pa}$, which is only 3.6% of the shear stress. The influence of viscous flow decreases with increasing height because of the decreasing wind speed gradient with increasing height.

### A.5.5 Alternative Equations for $\phi_m(\zeta)$

Function $\phi_m(\zeta)$ is the Monin–Obukhov similarity function for momentum transfer, a dimensionless function that is used in the calculation of the wind speed gradient in similarity theory. The most commonly used function for $\phi_m(\zeta)$ is the one proposed by Dyer (1974):

$\zeta>0$ (stable):

$$\phi_m(\zeta)=1+5\zeta \tag{5.76}$$

$\zeta<0$ (unstable):

$$\phi_m(\zeta)=(1-16\zeta)^{-1/4} \tag{5.77}$$

However, several alternatives have been proposed.

The equations of Businger et al. (1971) are very close to eqs. (5.76) and (5.77), but they needed to assume $k=0.35$ to get a good fit:

$\zeta>0$ (stable):

$$\phi_m(\zeta)=1+4.7\zeta$$

$\zeta<0$ (unstable):

$$\phi_m(\zeta)=(1-15\zeta)^{-1/4}$$

Högström (1988) made careful observations to minimize flow distortion and reviewed literature data in view of this. The study concluded the following:

$\zeta > 0$ (stable):

$$\phi_m(\zeta) = 1 + 6\zeta$$

$\zeta < 0$ (unstable):

$$\phi_m(\zeta) = (1 - 19.3\zeta)^{-1/4}$$

Careful observations showed that the linearity of $\phi_m(\zeta)$ in the stable regime is only valid in a limited $\zeta$ range. Cheng and Brutsaert (2005) found that

$\zeta > 0$ (stable):

$$\phi_m(\zeta) = 1 + 5.8\zeta$$

is valid for $\zeta$ values up to 0.8. Beyond that $\phi_m(\zeta)$ levels off. A general equation was based on:

$\zeta > 0$ (stable):

$$\psi_m(\zeta) = -a \ln\left[\zeta + \left(1 + \zeta^b\right)^{\frac{1}{b}}\right]$$

with $a = 6.1$ and $b = 2.5$. This leads to

$\zeta > 0$ (stable):

$$\phi_m = 1 + a\left[\frac{\zeta + \zeta^b\left(1 + \zeta^b\right)^{\frac{(1-b)}{b}}}{\zeta + \left(1 + \zeta^b\right)^{\frac{1}{b}}}\right]$$

AERMOD (Cimorelli et al., 2004) uses the following integrated Monin–Obukhov similarity function of van Ulden and Holtslag (1985) in stable conditions:

$$\Psi(\zeta) = -17[1 - \exp(-0.29\zeta)]$$

## A.5.6 Derivation of Geostrophic Wind Speed from the Navier–Stokes Equations

The geostrophic wind is the wind in the layers of the troposphere above the planetary boundary layer. It is not affected by the surface. The purpose of this section is to find simple equations to predict geostrophic wind speeds with reasonable accuracy, and to show how this wind differs from winds closer to the surface.

The Navier–Stokes equations are the main building block of computational fluid dynamics. They are discussed in detail in Chapter 10. The equations are as follows:

$$\frac{\partial u}{\partial t}+u\frac{\partial u}{\partial x}+v\frac{\partial u}{\partial y}+w\frac{\partial u}{\partial z}=-\frac{1}{\rho}\frac{\partial p}{\partial x}+fv+\nu\left(\frac{\partial^2 u}{\partial x^2}+\frac{\partial^2 u}{\partial y^2}+\frac{\partial^2 u}{\partial z^2}\right)$$

$$\frac{\partial v}{\partial t}+u\frac{\partial v}{\partial x}+v\frac{\partial v}{\partial y}+w\frac{\partial v}{\partial z}=-\frac{1}{\rho}\frac{\partial p}{\partial y}-fu+\nu\left(\frac{\partial^2 v}{\partial x^2}+\frac{\partial^2 v}{\partial y^2}+\frac{\partial^2 v}{\partial z^2}\right)$$

$$\frac{\partial w}{\partial t}+u\frac{\partial w}{\partial x}+v\frac{\partial w}{\partial y}+w\frac{\partial w}{\partial z}=-\frac{1}{\rho}\frac{\partial p}{\partial z}-g+\nu\left(\frac{\partial^2 w}{\partial x^2}+\frac{\partial^2 w}{\partial y^2}+\frac{\partial^2 w}{\partial z^2}\right)$$

In these equations $u$, $v$, and $w$ are the wind speed components in the $x$, $y$, and $z$ directions; $t$ is time, $g$ is the acceleration due to gravity, and $f$ is the Coriolis parameter, given by

$$f = 2\Omega \sin\phi$$

where $\Omega$ is the rotational speed of Earth ($7.292\times10^{-5}\,\mathrm{s}^{-1}$), and $\phi$ is the latitude.

To simplify, we will assume that the wind moves at constant speed in a straight horizontal line. We also assume that there are no wind speed gradients. This reduces the equations to

$$\frac{1}{\rho}\frac{\partial p}{\partial x}=fv$$

$$\frac{1}{\rho}\frac{\partial p}{\partial y}=-fu$$

$$\frac{1}{\rho}\frac{\partial p}{\partial z}=-g$$

The last equation is simply the hydrostatic equation. The first two indicate that in the Northern Hemisphere ($f>0$) the wind moves perpendicular to the pressure gradient, with the lower pressure to the left when the wind moves forward. Assuming that the $x$ axis points east and the $y$ axis points north, the first equation indicates that the wind moves north ($v>0$) when there is a low pressure in the west and a high pressure in the east ($\partial p/\partial x>0$). The second equation indicates that the wind moves west ($u<0$) when the pressure is low in the south and high in the north ($\partial p/\partial y>0$). This shows in mathematical terms that wind in the Northern Hemisphere always moves counterclockwise around areas of low pressure, and clockwise around areas of high pressure.

As an example, calculate the geostrophic wind speed associated with a pressure gradient of 1 kPa per 1000 km assuming an air density of $1\,\mathrm{kg\,m^{-3}}$ and a latitude of 45°. At this latitude, the Coriolis parameter is $7.292\times10^{-5}/2^{1/2}\,\mathrm{s}^{-1}=5.16\times10^{-5}\,\mathrm{s}^{-1}$.

The velocity is given by the following equation:

$$v = \frac{1}{\rho f}\frac{\partial p}{\partial x}$$

The pressure gradient is 1 kPa/1000 km = $10^3$ Pa/$10^6$ m = $10^{-3}$ Pa m$^{-1}$. Hence, the wind speed is $10^{-3}$ Pa m$^{-1}$/(1 kg m$^{-3}$ × 5.16 × $10^{-5}$ s$^{-1}$) = **19.4 m s$^{-1}$**. Geostrophic winds can be very strong indeed.

Geostrophic winds only occur high above the ground, where the approximations made here are acceptable. Close to the ground, friction (the terms containing the viscosity) are no longer negligible. They tend to drown out the effect of the Coriolis terms, so the above approximate equations no longer apply. Instead, we will assume the following approximation:

$$\frac{1}{\rho}\frac{\partial p}{\partial x} = v\frac{\partial^2 u}{\partial z^2}$$

$$\frac{1}{\rho}\frac{\partial p}{\partial y} = v\frac{\partial^2 v}{\partial z^2}$$

As direct simulations, these are not acceptable approximations. However, one could envision turbulent eddies that are ignored in this approximation as having the effect of increasing the apparent viscosity of the air. This is demonstrated in Chapter 10. Because the wind speed profile is always convex, the second derivative of the wind speed in the $z$ direction is always negative for a positive wind speed, which means that wind speed always flows against the pressure gradient. Now the wind blows west if the pressure is low in the west and high in the east, and the wind blows south when the pressure is low in the south and high in the north. Hence, the wind direction at high altitude can be expected to be 90° clockwise from the wind speed at ground level. In practice, high-altitude winds are not entirely separated from ground-level winds, so the actual angle is smaller than 90°.

## A.5.7 Alternative Equations for $\phi_h(\zeta)$

Function $\phi_h(\zeta)$ is the Monin–Obukhov similarity function for heat transfer, a dimensionless function that is used in the calculation of the wind speed gradient in similarity theory. The most commonly used function for $\phi_h(\zeta)$ is the one proposed by Dyer (1974):

$\zeta > 0$ (stable):

$$\phi_h(\zeta) = \phi_m(\zeta) = 1 + 5\zeta \qquad (5.91)$$

$\zeta < 0$ (unstable):

$$\phi_h(\zeta) = \phi_m^2(\zeta) = (1 - 16\zeta)^{-1/2} \qquad (5.92)$$

However, several alternatives have been proposed.

Businger et al. (1971) found the following equations. As with $\phi_m(\zeta)$, they needed to assume $k=0.35$ to get a good fit:

$\zeta>0$ (stable):

$$\phi_h(\zeta)=0.74+4.7\zeta$$

$\zeta<0$ (unstable):

$$\phi_h(\zeta)=0.74(1-9\zeta)^{-1/2}$$

Högström (1988) made careful observations to minimize flow distortion and reviewed literature data in view of this. The study concluded the following:

$\zeta>0$ (stable):

$$\phi_h(\zeta)=0.95+8\zeta$$

$\zeta<0$ (unstable):

$$\phi_h(\zeta)=0.95(1-11.6\zeta)^{-1/2}$$

Careful observations showed that the linearity of $\phi_h(\zeta)$ in the stable regime is only valid in a limited $\zeta$ range. Cheng and Brutsaert (2005) found that

$\zeta>0$ (stable):

$$\phi_m(\zeta)=1+5.4\zeta$$

is valid for $\zeta$ values up to 0.8. Beyond that $\phi_h(\zeta)$ levels off. A general equation was based on

$\zeta>0$ (stable):

$$\psi_h(\zeta)=-c\ln\left[\zeta+\left(1+\zeta^d\right)^{\frac{1}{d}}\right]$$

with $c=5.3$ and $d=1.1$. This leads to

$\zeta>0$ (stable):

$$\phi_h=1+c\left[\frac{\zeta+\zeta^d\left(1+\zeta^d\right)^{\frac{(1-d)}{d}}}{\zeta+\left(1+\zeta^d\right)^{\frac{1}{d}}}\right]$$

AERMOD (Cimorelli et al., 2004) uses the following equation to calculate the potential temperature gradient above a 100-m height under stable conditions (Stull, 1983; van Ulden and Holtslag, 1985):

$$\frac{\partial\theta}{\partial z}=\frac{\partial\theta(z=100\,\text{m})}{\partial z}\exp\left[-\frac{z-100\,\text{m}}{0.44\max(h_{mix},100\,\text{m})}\right]$$

Park et al. (2009) found the following correlations:

$\zeta < 0$ (stable):

$$\phi_h(\zeta) = (1 - 27.5\zeta)^{1/2}$$

$\zeta < 0$ (unstable):

$$\phi_h(\zeta) = (1 - 13.3\zeta)^{-1/2}$$

Surprisingly, they found different values for the mass transfer of water vapor:

$\zeta < 0$ (stable):

$$\phi_q(\zeta) = 1.21(1 - 60.4\zeta)^{1/3}$$

$\zeta < 0$ (unstable):

$$\phi_q(\zeta) = 1.21(1 - 13.1\zeta)^{-1/2}$$

This would invalidate the notion that turbulent heat transfer coefficients and turbulent mass transfer coefficients are the same. So far there is no independent study confirming this.

## A.5.8 Link between Similarity Theories of Wind Speed and Temperature: The Richardson Number

This section presents some information on how the similarity theory for wind speed is related to the similarity theory for temperature.

From the similarity theory for wind speed, we obtained eq. (5.74):

$$\frac{\partial u}{\partial z} = \frac{u_*}{kz}\phi_m(\zeta)$$

From the similarity theory for temperature, we obtained eq. (5.90):

$$\frac{\partial \theta}{\partial z} = \frac{\theta_*}{kz}\phi_h(\zeta)$$

Equation (5.89) relates the friction temperature to the friction velocity. Solving for the friction temperature leads to

$$\theta_* = \frac{u_*^2 T_0}{kgL}$$

Substituting into eq. (5.90) leads to

$$\frac{\partial \theta}{\partial z} = \frac{u_*^2 T_0}{k^2 zLg}\phi_h(\zeta)$$

Substitute $\zeta = z/L$ to eliminate $L$:

$$\frac{\partial \theta}{\partial z} = \frac{u_*^2 T_0 \zeta}{k^2 z^2 g} \phi_h(\zeta)$$

Take the potential temperature gradient divided by the square of the wind speed gradient:

$$\frac{\partial \theta / \partial z}{\left( \partial u / \partial z \right)^2} = \frac{\left[ u_*^2 T_0 \zeta / (k^2 z^2 g) \right] \phi_h(\zeta)}{\left[ u_*^2 / (k^2 z^2) \right] \phi_m^2(\zeta)}$$

This leads to

$$\frac{g}{T_0} \frac{\partial \theta / \partial z}{\left( \partial u / \partial z \right)^2} = \frac{\zeta \phi_h(\zeta)}{\phi_m^2(\zeta)}$$

This is called the **Richardson number** (Ri), or the gradient Richardson number (Nieuwstadt, 1984). It *only* depends on $\zeta$. Hence, it can be used as a measure of stability.

Based on the Dyer (1974) equations for $\phi_m$ and $\phi_h$, the following equations for Ri are obtained:

$\zeta > 0$ (stable):

$$\mathrm{Ri} = \frac{\zeta}{1 + 5\zeta}$$

$\zeta = 0$ (neutral):

$$\mathrm{Ri} = 0$$

$\zeta < 0$ (unstable):

$$\mathrm{Ri} = \zeta$$

Based on these equations, $\zeta$, and hence $L$, can be calculated immediately from knowledge of a wind speed gradient and a temperature gradient at the same height. This is illustrated in the next section.

The Richardson number should not be confused with the **flux Richardson number**, which is a related measure of stability. It is defined as

$$\mathrm{Ri}_f = \frac{K_h}{K_m} \mathrm{Ri}$$

We saw above that

$$\mathrm{Ri} = \frac{g}{T_0} \frac{\partial \theta / \partial z}{\left( \partial u / \partial z \right)^2}$$

Furthermore, by definition:

$$K_{\mathrm{h}} = -\frac{q / (\rho c_p)}{\partial \theta / \partial z} \quad \text{and} \quad K_{\mathrm{m}} = -\frac{\tau / \rho}{\partial u / \partial z}$$

Hence, we can show that

$$\mathrm{Ri}_f = \frac{\dfrac{g}{T_0} \dfrac{q}{\rho c_p}}{\dfrac{\tau}{\rho} \dfrac{\partial u}{\partial z}} = \frac{\dfrac{g}{T_0} \overline{\theta' w'}}{\overline{u' w'} \dfrac{\partial u}{\partial z}}$$

## A.5.9 Using Potential Temperature to Calculate $u_*$ and $L$

The best way to determine $L$ is by means of a heat balance of the surface, so $q$ is determined fairly directly (Cimorelli et al., 2004). However, such measurements or estimates are not always available. Fortunately, $L$ can also be estimated by combining a wind speed profile with a potential temperature profile. There is usually some loss of accuracy with this approach because potential temperature gradients are always very small and, therefore, difficult to determine accurately.

What follows are two methods to calculate $u_*$ and $L$ without using an energy balance or a sensible heat flux, but instead using the temperature at two different heights. They were both taken from AERMOD (Cimorelli et al., 2004).

The first method assumes that the wind speed at only one height is known but requires knowledge of the roughness length. First, apply eq. (5.81) or (5.82) to the measured wind speed at the height of the wind speed measurement. Solve the equation theoretically for $u_*$. Substitute the equation into eq. (5.89). The result has two unknowns: $L$ and $\theta_*$. Solve for $\theta_*$ and substitute into eq. (5.94) or (5.95), applied to the two measured potential temperatures. The result is an equation with only one unknown: $L$. The equation can be solved for $L$ iteratively.

As an illustration of this approach, it will be applied to the solution of Example 5.8, rounded to the nearest 0.01 K for the potential temperature. The values used are $\theta_1 = 283.57$ K at $z_1 = 5$ m height, and $\theta_2 = 283.71$ K at $z_2 = 10$ m height.

Because the example refers to stable conditions, eq. (5.82), applied to height 2, is solved for $u_*$:

$$u_* = \frac{ku_2}{\ln\left(\frac{z_2}{z_0}\right) + 5\left[\frac{(z_2 - z_0)}{L}\right]}$$

Substitution into eq. (5.89) leads to

$$L = \frac{ku_2^2 T_0}{g\theta_* \left\{\ln\left(\frac{z_2}{z_0}\right) + 5\left[\frac{(z_2 - z_0)}{L}\right]\right\}^2}$$

Solving for $\theta_*$ leads to

$$\theta_* = \frac{ku_2^2 T_0}{gL \left\{\ln\left(\frac{z_2}{z_0}\right) + 5\left[\frac{(z_2 - z_0)}{L}\right]\right\}^2}$$

Substitution into eq. (5.94) leads to

$$\theta_2 = \theta_1 + \frac{u_2^2 T_0}{gL \left\{\ln\left(\frac{z_2}{z_0}\right) + 5\left[\frac{(z_2 - z_0)}{L}\right]\right\}^2} \left(\ln\left(\frac{z_2}{z_1}\right) + 5\frac{z_2 - z_1}{L}\right)$$

To approximate $T_0$, we can simply use one of the potential temperatures. Hence, the equation above has only one variable, $L$. By trial and error, it is found that $L$ is about 180 m. Substitution into an equation for $u_*$ and for $\theta_*$ yields values of about 0.41 m s$^{-1}$ and 0.067 K, respectively. These values are within about 3% of the values actually found in Example 5.8. However, when the temperatures are rounded to the nearest 0.1 K, a value of about 300 m is found for $L$, indicating that this method is extremely sensitive to even the smallest measurement error.

An even simpler approach for determining $L$ has been proposed when the wind speed is known at the two heights where the potential temperatures are known, $z_1$ and $z_2$. The method is based on the gradient Richardson number, Ri, discussed in the previous section:

$$\mathrm{Ri} = \frac{g}{T_0} \frac{\partial\theta / \partial z}{\left(\partial u / \partial z\right)^2}$$

Based on wind speeds $u_1$ and $u_2$, and on potential temperatures $\theta_1$ and $\theta_2$ measured at heights $z_1$ and $z_2$, the Richardson number can be approximated as follows:

$$\mathrm{Ri} = \frac{g}{T_0} \frac{(\theta_2 - \theta_1)/(z_2 - z_1)}{\left[(u_2 - u_1)/(z_2 - z_1)\right]^2} = \frac{g}{T_0} \frac{\theta_2 - \theta_1}{u_2 - u_1}(z_2 - z_1)$$

Ri depends on $\zeta$ and is, therefore, height dependent. What is the appropriate height that corresponds with the above average? It can be proven mathematically that a slope of a logarithmic relationship determined by finite differencing as above applies to the **logarithmic mean height** $z_{\mathrm{lm}}$:

$$z_{\mathrm{lm}} = \frac{z_2 - z_1}{\ln\left(z_2/z_1\right)}$$

Because both wind speed profiles and temperature profiles are approximately logarithmic, and exactly logarithmic in neutral conditions, this is the obvious choice.

$L$ is calculated from

$$\frac{z_{\mathrm{lm}}}{L} = \mathrm{Ri} \quad (Ri < 0)$$

$$\frac{z_{\mathrm{lm}}}{L} = \frac{\mathrm{Ri}}{1 - 5\mathrm{Ri}} \quad (0 < Ri < 0.2)$$

Then $u_*$ and $\theta_*$ are calculated:

$$u_* = \frac{k(u_2 - u_1)}{\phi_{\mathrm{m}}\left(z_{\mathrm{lm}}/L\right)\ln\left(z_2/z_1\right)}$$

$$\theta_* = \frac{k(\theta_2 - \theta_1)}{\phi_{\mathrm{h}}\left(z_{\mathrm{lm}}/L\right)\ln\left(z_2/z_1\right)}$$

As before, we use the data from Example 5.8 to calculate the result. A Ri value of 0.03273 is obtained. Solving the appropriate $\zeta$ relationship leads to

$$\zeta = \frac{\mathrm{Ri}}{1 - 5\mathrm{Ri}}$$

A value of $\zeta = 0.03914$ is obtained. The definition of $z_{\mathrm{lm}}$ yields a value of 7.213 m. Hence, solving the relationship $\zeta = z/L$ for $L$ leads to a value of 184 m, close to the actual value. Again, rounding temperatures to the nearest 0.1 K changes the estimate of $L$ to about 300 m.

The method described by Cimorelli et al. (2004) is slightly different. First, the **geometric mean height** $z_{gm}$ is calculated:

$$z_{gm} = (z_1 z_2)^{1/2}$$

Because the gradient Richardson number is proportional with the height in unstable and near-neutral conditions, it is estimated as follows:

$$\text{Ri} = \frac{g}{T_0} \frac{(\theta_2 - \theta_1) z_{gm}}{(u_2 - u_1)^2} \ln \frac{z_2}{z_1}$$

$L$ is calculated from

$$\frac{z_{gm}}{L} = \text{Ri} \qquad (\text{Ri} < 0)$$

$$\frac{z_{gm}}{L} = \frac{\text{Ri}}{1 - 5\text{Ri}} \qquad (0 < \text{Ri} < 0.2)$$

Then $u_*$ and $\theta_*$ are calculated:

$$u_* = \frac{k(u_2 - u_1)}{\phi_m \left( z_{gm}/L \right) \ln \left( z_2/z_1 \right)}$$

$$\theta_* = \frac{k(\theta_2 - \theta_1)}{\phi_h \left( z_{gm}/L \right) \ln \left( z_2/z_1 \right)}$$

## A.5.10 Intuitive Derivation of Shear Stress on a Horizontal Plane

In turbulent flow, shear stress is generated by slow-moving air parcels mixing into faster moving air, causing a backward tug. The effect of this tug can be calculated based on turbulent covariances. What follows is an intuitive derivation. According to Newton's law,

$$\begin{aligned} \text{Force} &= \text{mass} \times \text{acceleration} = \text{mass} \times d\,\text{velocity} / dt \\ &= d(\text{mass} \times \text{velocity}) / dt = d\,\text{momentum} / dt \end{aligned}$$

It follows that if a net amount of momentum is transferred downward across a horizontal plane of 1 m$^2$ area per unit time, then that amount of momentum is equal in number to the force acting on the air above that 1-m$^2$ plane. In other words, it is equal to the shear stress that the plane acts upon the overlying air.

The momentum of 1 m$^3$ of air above the plane is $\rho u$. The vertical momentum flux is $-\rho\overline{u'w'}$. The minus sign is needed because we are looking at a downward flux. Because this momentum flux is equal to the shear stress, we obtain

$$\tau_{xz} = -\rho\overline{u'w'}$$

## A.5.11 Bi-Gaussian Model for the Convective Boundary Layer: Derivation of Equations

Section 5.13.1 outlines a model for the probability density function of the vertical wind speed in the convective boundary layer (CBL), based on the assumption that the CBL consists of Gaussian updrafts and Gaussian downdrafts. In the derivation of the model, six equations with six unknowns were obtained. The objective of this section is to provide a brief outline of the solution of this set of equations. The equations are as follows:

$$\lambda_1 + \lambda_2 = 1 \tag{5.205}$$

$$\lambda_1\overline{w_1} + \lambda_2\overline{w_2} = 0 \tag{5.206}$$

$$\sigma_w^2 = \lambda_1(\overline{w_1}^2 + \sigma_{w1}^2) + \lambda_2(\overline{w_2}^2 + \sigma_{w2}^2) \tag{5.209}$$

$$S\sigma_w^3 = \lambda_1(\overline{\overline{w_1}^3 + 3\overline{w_1}\sigma_{w1}^2}) + \lambda_2(\overline{\overline{w_2}^3 + 3\overline{w_2}\sigma_{w2}^2}) \tag{5.217}$$

$$\sigma_{w1} = R\overline{w_1} \tag{5.218}$$

$$\sigma_{w2} = -R\overline{w_2} \tag{5.219}$$

See Section 5.13.1 for the meaning of the variables in the equations.

Equation (5.205) can be solved for $\lambda_2$ and used to eliminate this variable from the other equations. Meanwhile, eqs. (5.218) and (5.219) can be used to eliminate $\sigma_{w1}$ and $\sigma_{w2}$:

$$\lambda_2 = 1 - \lambda_1 \tag{A5.1}$$

$$\lambda_1\overline{w_1} + (1-\lambda_1)\overline{w_2} = 0 \tag{A5.2}$$

$$\sigma_w^2 = \lambda_1\overline{w_1}^2(1+R^2) + (1-\lambda_1)\overline{w_2}^2(1+R^2) \tag{A5.3}$$

$$S\sigma_w^3 = \lambda_1\overline{w_1}^3(1+3R^2) + (1-\lambda_1)\overline{w_2}^3(1+3R^2) \tag{A5.2}$$

We still have three equations [eqs. (A5.2)–(A5.4)] with three unknowns, $\lambda_1$, $\overline{w_1}$, and $\overline{w_2}$. Next, eq. (A5.2) is solved for $\overline{w_2}$ and used to eliminate $\overline{w_2}$ from eqs. (A5.3) and (A5.4).

$$\overline{w_2} = -\overline{w_1}\frac{\lambda_1}{1-\lambda_1} \tag{A5.5}$$

$$\sigma_w^2 = \lambda_1 \overline{w_1}^2 (1+R^2) - \frac{\lambda_1^2}{1-\lambda_1}\overline{w_1}^2(1+R^2) = \frac{\lambda_1}{1-\lambda_1}\overline{w_1}^2(1+R^2) \tag{A5.6}$$

$$S\sigma_w^3 = \lambda_1 \overline{w_1}^3 (1+3R^2) + \frac{\lambda_1^3}{(1-\lambda_1)^2}\overline{w_1}^3(1+3R^2) = \frac{\lambda_1(1-2\lambda_1)}{(1-\lambda_1)^2}\overline{w_1}^3(1+3R^2) \tag{A5..7}$$

Equations (A5.6) and (A5.7) have two unknowns, $\lambda_1$ and $\overline{w_1}$. Equation (A5.6) is solved for $\overline{w_1}$ and used to eliminate $\overline{w_1}$ in eq. (A5.7) after squaring:

$$\overline{w_1}^2 = \frac{\sigma_w^2(1-\lambda_1)}{\lambda_1(1+R^2)} \tag{A5.8}$$

$$\begin{aligned} S^2\sigma_w^6 &= \frac{\lambda_1^2(1-2\lambda_1)^2}{(1-\lambda_1)^4}\frac{\sigma_w^6(1-\lambda_1)^3}{\lambda_1^3(1+R^2)^3}(1+3R^2)^2 \\ &= \frac{(1-2\lambda_1)^2}{1-\lambda_1}\frac{\sigma_w^6}{\lambda_1(1+R^2)^3}(1+3R^2)^2 \end{aligned} \tag{A5.9}$$

Equation (A5.9) is written as a polynome equation in $\lambda_1$ and solved:

$$\lambda_1^2\left[4\frac{(1+3R^2)^2}{(1+R^2)^3}+S^2\right]+\lambda_1\left[-4\frac{(1+3R^2)^2}{(1+R^2)^3}-S^2\right]+\frac{(1+3R^2)^2}{(1+R^2)^3}=0 \tag{A5.10}$$

Solving eq. (A5.10) leads to eq. (5.220):

$$\lambda_1 = \frac{1}{2}\left[1-\sqrt{\frac{1}{1+{4}/{(\gamma_1^2\gamma_2 S^2)}}}\right] \tag{5.220}$$

where

$$\gamma_1 = \frac{1+R^2}{1+3R^2} \tag{5.226}$$

$$\gamma_2 = 1+R^2 \tag{5.227}$$

Substitution of eq. (5.220) into eqs. (A5.1) and (A5.8) leads to eq. (5.221) and (5.222):

$$\lambda_2 = \frac{1}{2}\left[1+\sqrt{\frac{1}{1+4/(\gamma_1^2\gamma_2 S^2)}}\right] \tag{5.221}$$

$$\overline{w_1} = \frac{S\sigma_w\gamma_1}{2}\left(1+\sqrt{1+\frac{4}{\gamma_1^2\gamma_2 S^2}}\right) \tag{5.222}$$

Further substitutions lead to eqs. (5.223)–(5.225):

$$\overline{w_2} = \frac{S\sigma_w\gamma_1}{2}\left(1-\sqrt{1+\frac{4}{\gamma_1^2\gamma_2 S^2}}\right) \tag{5.223}$$

$$\sigma_{w1} = \frac{RS\sigma_w\gamma_1}{2}\left(1+\sqrt{1+\frac{4}{\gamma_1^2\gamma_2 S^2}}\right) \tag{5.224}$$

$$\sigma_{w1} = \frac{RS\sigma_w\gamma_1}{2}\left(1-\sqrt{1+\frac{4}{\gamma_1^2\gamma_2 S^2}}\right) \tag{5.225}$$

# CHAPTER A6: GAUSSIAN DISPERSION MODELING: AN IN-DEPTH STUDY

## A.6.1 Parameterizations for $\sigma_y$ and $\sigma_z$ Based on Stability Classes

In Section 6.3 several empirical correlations for the dispersion parameters $\sigma_y$ and $\sigma_z$ based on the Pasquill–Gifford stability classification are presented. This section tabulates the parameters that have been proposed with these correlations.

Klug (1969) proposed eqs. (6.12) and (6.13) for $\sigma_y$ and $\sigma_z$:

$$\sigma_y = ax^b \tag{6.12}$$

$$\sigma_z = cx^d \tag{6.13}$$

with the parameters presented in Table A6.1 (rural). Parameterizations for a 60-min averaging time based on eqs. (6.12) and (6.13) were presented in ASME (1973). The parameters are given in Table A6.2.

Martin (1976) used eq. (6.12) for $\sigma_y$ but extended eq. (6.13) for $\sigma_z$ to eq. (6.14):

$$\sigma_z = cx^d + f \tag{6.14}$$

**TABLE A6.1 Coefficients of Klug (1969) for eqs. (6.12) and (6.13) (Rural)**

| Stability Class | $a$ | $b$ | $c$ | $d$ |
|---|---|---|---|---|
| A | 0.469 | 0.903 | 0.017 | 1.380 |
| B | 0.306 | 0.885 | 0.072 | 1.021 |
| C | 0.230 | 0.855 | 0.076 | 0.879 |
| D | 0.219 | 0.764 | 0.140 | 0.727 |
| E | 0.237 | 0.691 | 0.217 | 0.610 |
| F | 0.273 | 0.594 | 0.262 | 0.500 |

**TABLE A6.2 Coefficients of ASME (1973) for eqs. (6.12) and (6.13) (60-min Averaging Time)**

| Stability Class | $a$ | $b$ | $c$ | $d$ |
|---|---|---|---|---|
| A | 0.40 | 0.91 | 0.40 | 0.91 |
| B | 0.36 | 0.86 | 0.33 | 0.86 |
| D | 0.32 | 0.78 | 0.22 | 0.78 |
| F | 0.31 | 0.71 | 0.06 | 0.71 |

**TABLE A6.3 Coefficients of Martin (1976) for eqs. (6.12) and (6.14) (Rural)**

| | | | $x<1000$ m | | | $x>1000$ m | | |
|---|---|---|---|---|---|---|---|---|
| Stability Class | $a$ | $b$ | $c$ | $d$ | $e$ | $c$ | $d$ | $e$ |
| A | 0.443 | 0.894 | 0.00066 | 1.941 | 9.27 | 0.00024 | 2.094 | –9.6 |
| B | 0.324 | 0.894 | 0.03809 | 1.149 | 3.3 | 0.05498 | 1.098 | 2 |
| C | 0.216 | 0.894 | 0.1128 | 0.911 | 0 | 0.1128 | 0.911 | 0 |
| D | 0.141 | 0.894 | 0.2219 | 0.725 | –1.7 | 1.26 | 0.516 | –13.0 |
| E | 0.105 | 0.894 | 0.2108 | 0.678 | –1.3 | 6.738 | 0.305 | –34.0 |
| F | 0.071 | 0.894 | 0.08647 | 0.74 | –0.35 | 18.05 | 0.18 | –48.6 |

For the paramerization of $\sigma_z$ a distinction between $x<1000$ m and $x>1000$ m was made. The parameters are given in Table A6.3.

Turner (1970) used eqs. (6.15) and (6.16) for $\sigma_y$ and $\sigma_z$:

$$\sigma_y = \exp[a + b\ln x + c(\ln x)^2] \tag{6.15}$$

$$\sigma_z = \exp[d + e\ln x + f(\ln x)^2] \tag{6.16}$$

The coefficients are given in Table A6.4.

ISC uses eqs. (6.17) and (6.18) for $\sigma_y$ in rural terrain (EPA, 1995):

$$\sigma_y = 465.11628x\tan(\theta) \tag{6.17}$$

**TABLE A6.4 Coefficients of Turner (1970) for eqs. (6.15) and (6.16) (Rural)**

| Stability Class | *a* | *b* | *c* | *d* | *e* | *f* |
|---|---|---|---|---|---|---|
| A | –1.104 | 0.9878 | –0.0076 | 4.679 | –1.7172 | 0.2770 |
| B | –1.634 | 1.0350 | –0.0096 | –1.999 | 0.8752 | 0.0136 |
| C | –2.054 | 1.0231 | –0.0076 | –2.341 | 0.9477 | –0.0020 |
| D | –2.555 | 1.0423 | –0.0087 | –3.186 | 1.1737 | –0.0316 |
| E | –2.754 | 1.0106 | –0.0064 | –3.783 | 1.3010 | –0.0450 |
| F | –3.143 | 1.0148 | –0.0070 | –4.490 | 1.4024 | –0.0540 |

**TABLE A6.5 ISC Parameterization for $\sigma_y$ Based on eqs. (6.17) and (6.18) (EPA, 1995)**

| Stability Class | *c* | *d* |
|---|---|---|
| A | 24.1670 | 2.5334 |
| B | 18.3330 | 1.8096 |
| C | 12.5000 | 1.0857 |
| D | 8.3330 | 0.72382 |
| E | 6.2500 | 0.54287 |
| F | 4.1667 | 0.36191 |

$$\theta = 0.017453293[c - d\ln(x)] \tag{6.18}$$

The parameters are shown in Table A6.5. For $\sigma_z$ eq. (6.13) is used, but with several sets of parameters depending on the distance. The parameters are shown in Table A6.6. Note that $x$ is in kilometers in this table. For urban terrain the Briggs (1973) parameterization is used.

CALPUFF also contains the MESOPUFF II parameterization (Scire et al., 1984) as an option. This parameterization is based on a single power law and is not recommended. The coefficients are given in Table A6.7.

## A.6.2 Estimation of Plume Dispersion at Large Distance

In Section 6.6.6, an equation is derived to describe a plume dispersion parameter based on an exponential autocorrelation function. The derivation involves a detailed statistical argument. This section provides a much simpler argument that will show the limiting behavior of plume dispersion at large distances from the source.

At large distance, plume dispersion can be approximated as a random redistribution with standard deviation $\sigma_v T_{i,L}$ every $T_{i,L}$ seconds (i.e., for every $x$ increment of $L_1$ where $L_1 = T_{i,L}\, u$). After $n = x/L_1$ redistributions we get

$$x = nL_{L,y} \quad \sigma_y^2 = n(\sigma_v T_{i,L,y})^2$$

**TABLE A6.6 ISC Parameterization for $\sigma_z$ Based on eq. (6.13)**[1]

| Stability Class | $x$ (km) | $a$ | $b$ |
|---|---|---|---|
| A | < 0.1 | 122.800 | 0.94470 |
| | 0.10–0.15 | 158.080 | 1.05420 |
| | 0.16–0.20 | 170.220 | 1.09320 |
| | 0.21–0.25 | 179.520 | 1.12620 |
| | 0.26–0.30 | 217.410 | 1.26440 |
| | 0.31–0.40 | 258.890 | 1.40940 |
| | 0.41–0.50 | 346.750 | 1.72830 |
| | 0.51–3.11 | 453.850 | 2.11660 |
| | > 3.11 | 5000 | 0 |
| B | < 0.2 | 90.673 | 0.93198 |
| | 0.21–0.40 | 98.483 | 0.98332 |
| | 0.40 | 109.300 | 1.09710 |
| C | all | 61.141 | 0.91465 |
| D | < 0.30 | 34.459 | 0.86974 |
| | 0.31–1.00 | 32.093 | 0.81066 |
| | 1.01–3.00 | 32.093 | 0.64403 |
| | 3.01–10.00 | 33.504 | 0.60486 |
| | 10.01–30.00 | 36.650 | 0.56589 |
| | > 30.00 | 44.053 | 0.51179 |
| E | < 0.10 | 24.260 | 0.83660 |
| | 0.10–0.30 | 23.331 | 0.81956 |
| | 0.31–1.00 | 21.628 | 0.75660 |
| | 1.01–2.00 | 21.628 | 0.63077 |
| | 2.01–4.00 | 22.534 | 0.57154 |
| | 4.01–10.00 | 24.703 | 0.50527 |
| | 10.01–20.00 | 26.970 | 0.46713 |
| | 20.01–40.00 | 35.420 | 0.37615 |
| | > 40.00 | 47.618 | 0.29592 |
| F | < 0.20 | 15.209 | 0.81558 |
| | 0.21–0.70 | 14.457 | 0.78407 |
| | 0.70–1.00 | 13.953 | 0.68465 |
| | 1.01–2.00 | 13.953 | 0.63227 |
| | 2.01–3.00 | 14.823 | 0.54503 |
| | 3.01–7.00 | 16.187 | 0.46490 |
| | 7.01–15.00 | 17.836 | 0.41507 |
| | 15.01–30.00 | 22.651 | 0.32681 |
| | 30.01–60.00 | 27.074 | 0.27436 |
| | > 60.00 | 34.219 | 0.21716 |

[a]Note that $x$ is in kilometers and $\sigma_z$ is in meters.
*Source*: From EPA (1995).

**TABLE A6.7 MESOPUFF II Parameterization Based on eqs. (6.12) and (6.13)**

| Stability Class | *a* | *b* | *c* | *d* |
|---|---|---|---|---|
| A | 0.36 | 0.9 | 0.00023 | 2.10 |
| B | 0.25 | 0.9 | 0.058 | 1.09 |
| C | 0.19 | 0.9 | 0.11 | 0.91 |
| D | 0.13 | 0.9 | 0.57 | 0.58 |
| E | 0.096 | 0.9 | 0.85 | 0.47 |
| F | 0.063 | 0.9 | 0.77 | 0.42 |

because a variance of a sum is the sum of the variances when the terms are uncorrelated. Hence:

$$x = nL_{\mathrm{L,y}} \quad \sigma_y = \sqrt{n}(\sigma_v T_{\mathrm{i,L,y}})$$

In practice, this equation underestimates the dispersion. Although the lateral motion is somewhat less than $\sigma_v T_{\mathrm{i,L,y}}$ after a time $T_{\mathrm{i,L,y}}$, the movements are not entirely uncorrelated. We multiply by a constant $a$ to account for that:

$$x = nL_{\mathrm{L,y}} \quad \sigma_y = a\sqrt{n}(\sigma_v T_{\mathrm{i,L}})$$

Substitute $n = x/L_{\mathrm{L,y}}$ in the above equation:

$$\sigma_y = a\sqrt{\frac{x}{L_{\mathrm{L,y}}}}(\sigma_v T_{\mathrm{i,L,y}})$$

Substitute $T_{\mathrm{i,L,y}} = {}^{L_{\mathrm{L,y}}}\!/_{\bar{u}}$ to obtain

$$\boxed{\sigma_y = a\sqrt{\frac{x}{L_{\mathrm{L,y}}}}\left(\sigma_v \frac{L_{\mathrm{L,y}}}{\bar{u}}\right) = a\frac{\sigma_v}{\bar{u}}\sqrt{L_{\mathrm{L,y}}x}} \quad (x \text{ large})$$

After comparison with

$$\sigma_y = \sigma_v \frac{x}{\bar{u}} f_y$$

this leads to

$$\boxed{f_y = a\sqrt{\frac{L_{\mathrm{L,y}}}{x}}} \quad (x \text{ large})$$

Based on the calculations in Section 6.6.6, we can argue that $a = 2^{1/2}$.

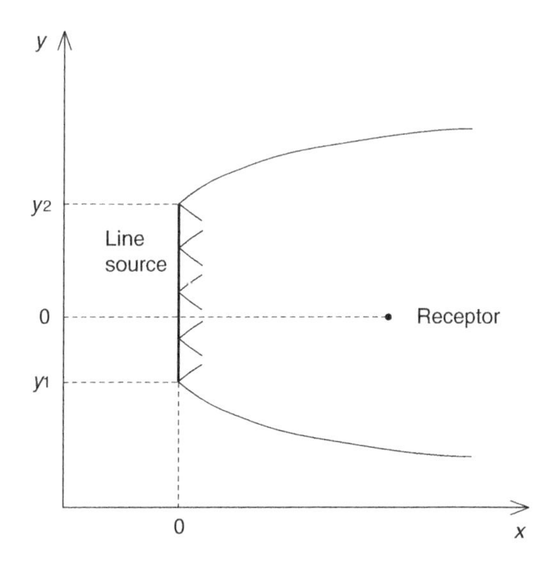


Figure A6.1 Coordinate system for line source calculations.

## A.6.3 Dispersion Calculations for a Finite Line Source Perpendicular to the Wind Direction

Assume a line source with length $L$ and total emission mass flow rate $Q$. Choose the origin of the $y$ axis on the line source, directly upwind of the receptor (i.e., at the receptor, $y=0$, and at the source, $x=0$). Define $y_1$ and $y_2$ as the $y$ coordinates of the ends of the source, so that $y_1<y_2$, and $y_2-y_1=L$ (Fig. A6.1).

When a point source is defined as a section of the line source from $y$ to $y+dy$, then the strength (emission mass flow rate) of the point source is $Q\ dy/L$. The concentration at the receptor resulting from the point source is

$$d\overline{C_y} = C_x \varphi_y \varphi_z \frac{dy}{L}$$

where each of the variables have the same definition as in eq. (6.1). To calculate the concentration resulting from the entire line source, we need to integrate the above equation:

$$\bar{C} = \int_{y_1}^{y_2} d\overline{C_y} = \int_{y_1}^{y_2} C_x \varphi_y \varphi_z \frac{dy}{L}$$

Because all variables except $\varphi_y$ are independent of $y$, the equation can be simplified to

$$\bar{C} = \frac{C_x \varphi_z}{L} \int_{y_1}^{y_2} \varphi_y\ dy = \frac{C_x \varphi_z}{L} \frac{1}{\sqrt{2\pi}\sigma_y} \int_{y_1}^{y_2} \exp\left(-\frac{1}{2}\frac{y^2}{\sigma_y^2}\right) dy$$

The error function erf($x$) is defined as follows:

$$\text{erf}(x) = \frac{2}{\sqrt{\pi}} \int_0^x \exp(-t^2)\, dt$$

Applying the substitution $t = y/(\sqrt{2}\sigma_y)$ and $x = y_i/(\sqrt{2}\sigma_y)$, the error function can be rewritten as

$$\text{erf}(x) = \frac{2}{\sqrt{\pi}} \int_0^{y_i} \exp\left(-\frac{1}{2}\frac{y^2}{\sigma_y^2}\right) d\frac{y}{\sqrt{2}\sigma_y} = \frac{2}{\sqrt{2\pi}\sigma_y} \int_0^{y_i} \exp\left(-\frac{1}{2}\frac{y^2}{\sigma_y^2}\right) dy = \text{erf}\left(\frac{y_i}{\sqrt{2}\sigma_y}\right)$$

Hence, we obtain

$$\bar{c} = \frac{C_x \varphi_z}{2L}\left[\text{erf}\left(\frac{y_2}{\sqrt{2}\sigma_y}\right) - \text{erf}\left(\frac{y_1}{\sqrt{2}\sigma_y}\right)\right]$$

In the absence of an overlying temperature inversion, this equation can be written in full as

$$\bar{c} = \frac{Q}{2\bar{u}L\sqrt{2\pi}\sigma_z}\left[\text{erf}\left(\frac{y_2}{\sqrt{2}\sigma_y}\right) - \text{erf}\left(\frac{y_1}{\sqrt{2}\sigma_y}\right)\right]\left\{\exp\left[-\frac{1}{2}\frac{(z-h)^2}{\sigma_z^2}\right] + \exp\left[-\frac{1}{2}\frac{(z+h)^2}{\sigma_z^2}\right]\right\}$$

# CHAPTER A7: PLUME–ATMOSPHERE INTERACTIONS

## A.7.1 Derivation of the Momentum Balance along the Wind Direction for Denser-than-Air Plumes

Description of denser-than-air plumes require at least solving the material balance, the momentum balance in the $x$ direction, and the species balance. Here an approximate momentum balance in the $x$ direction is derived. The reasoning is similar to the reasoning applied by Zeman (1982).

In this derivation density changes are only accounted for when they affect buoyancy. Pressures and densities in this model are height averages. Sometimes it is important to keep track of the domain to which the average refers.

The momentum balance is based on the differential plume element shown in Figure 7.7. It is given by the following equation:

$$\text{Momentum produced} = \text{force} \qquad (7.103)$$

This equation can be rewritten as

$$\text{Momentum out} - \text{momentum in} + \text{momentum accumulated} = \text{force}$$

where

$$\begin{aligned}\text{Momentum out} &= \text{mass flow rate out} \times \text{velocity out}\\ &= \rho\left(u+\frac{\partial u}{\partial x}dx\right)\left(h+\frac{\partial h}{\partial x}dx\right)\cdot\left(u+\frac{\partial u}{\partial x}dx\right)\\ &\approx \rho\left(u^2h+2uh\frac{\partial u}{\partial x}dx+u^2\frac{\partial h}{\partial x}dx\right)\end{aligned}$$

Here we neglect terms with two differentials as well as the $x$ dependence of the density.

The momentum in is given by:

$$\begin{aligned}\text{Momentum in} &= \text{momentum in left} + \text{momentum in by entrainment (top)}\\ &= \rho u^2 h + \rho_\mathrm{a} w_\mathrm{e} u_\mathrm{a} dx\end{aligned}$$

The momentum accumulated is:

$$\begin{aligned}\text{Momentum accumulated} &= \text{mass} \times \text{acceleration} + \text{velocity} \times \text{mass accumulation}\\ &= \rho h\frac{\partial u}{\partial t}dx + u\rho\frac{\partial h}{\partial t}dx\end{aligned}$$

The net force (positive from left to right) is calculated as

$$\begin{aligned}\text{Force} &= \text{force from left} + \text{force from top (horizontal component)}\\ &\quad - \text{force from right} - \text{friction}\\ &= ph + p_\mathrm{a}\frac{\partial h}{\partial x}dx - \left(p+\frac{\partial p}{\partial x}dx\right)\left(h+\frac{\partial h}{\partial x}dx\right) - \tau\, dx\\ &\approx ph + p_\mathrm{a}\frac{\partial h}{\partial x}dx - ph - p\frac{\partial h}{\partial x}dx - h\frac{\partial p}{\partial x}dx - \tau\, dx\end{aligned}$$

It is obvious that the first and third terms cancel each other. Less obvious is that the second and fourth terms also cancel. At the top of the plume $p=p_\mathrm{a}$. Because these terms refer to the height range $h$ to $h+dh$, the difference between $p$ and $p_\mathrm{a}$ in these terms is proportional to $dh$. This means that the difference between these two terms has two differentials and can be ignored. Hence, the equation simplifies to

$$\text{Force} \approx -h\frac{\partial p}{\partial x}dx - \tau\ dx = -h\frac{\partial p}{\partial x}dx - \rho u_*^2\ dx$$

where the friction velocity $u_*$ is introduced to describe the shear stress at the surface.

Putting all terms together, the following equation is obtained:

$$\begin{aligned}&\rho\left(u^2h+2uh\frac{\partial u}{\partial x}dx+u^2\frac{\partial h}{\partial x}dx\right) - \rho u^2 h - \rho_\mathrm{a} w_\mathrm{e} u_\mathrm{a} dx\\ &+ \rho h\frac{\partial u}{\partial t}dx + u\rho\frac{\partial h}{\partial t}dx = -h\frac{\partial p}{\partial x}dx - \rho u_*^2\, dx\end{aligned}$$

Eliminating terms that cancel each other and dividing by $\rho h dx$ yields

$$2u\frac{\partial u}{\partial x}+\frac{u^2}{h}\frac{\partial h}{\partial x}-\frac{\rho_a w_e u_a}{\rho h}+\frac{\partial u}{\partial t}+\frac{u}{h}\frac{\partial h}{\partial t}=-\frac{1}{\rho}\frac{\partial p}{\partial x}-\frac{u_*^2}{h}$$

Substitute the following equation, which is a rearrangement of eq. (7.101):

$$\frac{\partial h}{\partial t}=-u\frac{\partial h}{\partial x}-h\frac{\partial u}{\partial x}+w_b+w_e$$

This leads to

$$2u\frac{\partial u}{\partial x}+\frac{u^2}{h}\frac{\partial h}{\partial x}-\frac{\rho_a w_e u_a}{\rho h}+\frac{\partial u}{\partial t}+\frac{u}{h}\left(-u\frac{\partial h}{\partial x}-h\frac{\partial u}{\partial x}+w_b+w_e\right)=-\frac{1}{\rho}\frac{\partial p}{\partial x}-\frac{u_*^2}{h}$$

Elimination of terms that cancel each other, and rearranging, leads to

$$\frac{\partial u}{\partial t}+u\frac{\partial u}{\partial x}=\frac{w_e}{h}\left(\frac{\rho_a u_a}{\rho}-u\right)-\frac{uw_b}{h}-\frac{1}{\rho}\frac{\partial p}{\partial x}-\frac{u_*^2}{h}$$

We will assume that the pressure gradient is the result of changing hydrostatic pressures. Furthermore, we will assume that the atmospheric pressure is constant, except for the hydrostatic effect. Hence, we can write

$$\frac{\partial p}{\partial x}=\frac{\partial(p-p_a)}{\partial x}$$

At the top of the plume ($z=h$), $p=p_a$. Assuming that the height dependence of the density is negligible, we can use the hydrostatic equation to calculate the pressure in the plume, and the pressure of the surrounding atmosphere at the same height:

$$p(x)=p_a(h)+\rho g\ \Delta h$$

$$p_a(x)=p_a(h)+\rho_a g\Delta h$$

Hence:

$$p(x)-p_a(x)=(\rho-\rho_a)g\Delta h$$

The average pressure difference across the plume depth is

$$\overline{p(x)-p_a(x)}=\frac{1}{\Delta h}\int_0^{\Delta h}(\rho-\rho_a)g\Delta h\,dh=\frac{g\Delta h}{2}(\rho-\rho_a)$$

In practice, densities in the plume are not constant with height. To account for that, a correction factor $\beta$ is introduced:

$$\overline{p(x)-p_a(x)}=\frac{\beta g\Delta h}{2}(\rho-\rho_a)$$

and $\beta = 1$ when the density difference is constant and is $\frac{2}{3}$ when the density difference is proportional to $(h - z)$. Hence, the momentum equation becomes

$$\frac{\partial u}{\partial t} + u\frac{\partial u}{\partial x} = \frac{w_e}{h}\left(\frac{\rho_a u_a}{\rho} - u\right) - \frac{uw_b}{h} - \frac{\beta g}{2\rho}\frac{\partial}{\partial x}[h(\rho - \rho_a)] - \frac{u_*^2}{h}$$

### A.7.2 Derivation of the Species Balance for Denser-than-Air Plumes

Description of denser-than-air plumes require at least solving the material balance, the momentum balance in the $x$ direction, and the species balance. Here an approximate species balance is derived. The reasoning is similar to the reasoning applied by Zeman (1982).

The species balance is based on the differential plume element shown in Figure 7.7. It is given by the following equation:

$$\text{Pollutant accumulation} = \text{pollutant in} - \text{pollutant out} \qquad (7.106)$$

The accumulation consists of two terms: one corresponding with the change of the concentration in the element, and one corresponding with the change of the height of the volume element:

$$\text{Pollutant accumulation} = \rho V\frac{\partial c}{\partial t} + \rho c\frac{\partial V}{\partial t} = \rho \cdot h\,dx \cdot \frac{\partial c}{\partial t} + \rho c \cdot \frac{\partial h}{\partial t}\,dx$$

where $c$ is the mass fraction of the pollutant in the plume. The term pollutant in consists of two flows: one carried into the element with the wind and one by emission from the bottom. It is assumed that there is no pollutant entering the volume element by entrainment. The equation for pollutant in is

$$\text{Pollutant in} = \rho cuh + w_b \rho c_b\,dx$$

The pollutant out is the pollution carried out of the element by the wind:

$$\begin{aligned}\text{Pollutant out} &= \rho\left(c + \frac{\partial c}{\partial x}dx\right)\left(u + \frac{\partial u}{\partial x}dx\right)\left(h + \frac{\partial h}{\partial x}dx\right)\\ &= \rho cuh + \rho uh\frac{\partial c}{\partial x}dx + \rho ch\frac{\partial u}{\partial x}dx + \rho cu\frac{\partial h}{\partial x}dx\end{aligned}$$

Completing all terms in the species balance leads to

$$\begin{aligned}\rho \cdot h\,dx \cdot \frac{\partial c}{\partial t} + \rho c \cdot \frac{\partial h}{\partial t}dx &= \rho cuh + \rho c_b w_b\,dx - \rho cuh\\ &- \rho uh\frac{\partial c}{\partial x}dx - \rho ch\frac{\partial u}{\partial x}dx - \rho cu\frac{\partial h}{\partial x}dx\end{aligned}$$

Eliminating terms that cancel each other and dividing by $\rho h\,dx$ leads to

$$\frac{\partial c}{\partial t}+\frac{c}{h}\frac{\partial h}{\partial t}=\frac{c_{\mathrm{b}}w_{\mathrm{b}}}{h}-u\frac{\partial c}{\partial x}-c\frac{\partial u}{\partial x}-\frac{cu}{h}\frac{\partial h}{\partial x}$$

Substitute the material balance, eq. (7.101), in the following form:

$$\frac{\partial h}{\partial t}=-u\frac{\partial h}{\partial x}-h\frac{\partial u}{\partial x}+w_{\mathrm{b}}+w_{\mathrm{e}}$$

The following equation is obtained:

$$\frac{\partial c}{\partial t}-\frac{cu}{h}\frac{\partial h}{\partial x}-\frac{ch}{h}\frac{\partial u}{\partial x}+\frac{cw_{\mathrm{b}}}{h}+\frac{cw_{\mathrm{e}}}{h}=\frac{c_{\mathrm{b}}w_{\mathrm{b}}}{h}-u\frac{\partial c}{\partial x}-c\frac{\partial u}{\partial x}-\frac{cu}{h}\frac{\partial h}{\partial x}$$

Eliminating terms that cancel each other, and rearranging, leads to

$$\frac{\partial c}{\partial t}+u\frac{\partial c}{\partial x}=\frac{w_{\mathrm{b}}}{h}(c_{\mathrm{b}}-c)-\frac{cw_{\mathrm{e}}}{h}$$

## A.7.3 Derivation of an Alternative Deposition Equation for Particles

In the literature, eq. (7.137) is often put forward to calculate the deposition velocity of particles.

The equation is as follows:

$$v_{\mathrm{d}}=\frac{1}{r_{\mathrm{t}}}=\frac{1}{r_{\mathrm{a}}+r_{\mathrm{b}}+r_{\mathrm{a}}r_{\mathrm{b}}v_{\mathrm{s}}}+v_{\mathrm{s}} \tag{7.137}$$

It was argued in Section 7.4.1.1 that eq. (7.137) is based on an analogue that is not entirely applicable. Here an alternative equation will be derived based on diffusion and advection theory. First, the necessary equations will be derived in the absence of particle settling. It is assumed that particles settle by diffusion through two layers. The diffusion equation is Fick's law:

$$F=-D\frac{dc}{dz}$$

where $D$ is the diffusion constant ($\mathrm{m^2\,s^{-1}}$). Because the flux is constant, this equation is integrated directly to

$$F=-D\frac{\Delta c}{\Delta z}$$

Assume that the concentration of the pollutant is $c_{\mathrm{a}}$ at the top of the first layer (i.e., in the free atmosphere), $c_{\mathrm{s}}$ between the two layers (i.e., at the top of the laminar film), and 0 at the bottom of the second layer (i.e., at the surface). Assume further that the thickness of the first layer is $\Delta z_a$, and the thickness of the second layer is $\Delta z_b$. The diffusivity in the first layer is $D_a$; the diffusivity in the second layer is $D_b$. The flux $F$ (negative in this case) is constant, and equal in both layers:

$$-F = D_a \frac{c_a - c_s}{\Delta z_a} = D_b \frac{c_s}{\Delta z_b}$$

Compare this with the two-resistance model, where the flux is described as

$$-F = \frac{c_a - c_s}{r_a} = \frac{c_s}{r_b}$$

We can link the two-resistance model with the diffusion model as follows:

$$r_a = \frac{\Delta z_a}{D_a}$$

$$r_b = \frac{\Delta z_b}{D_b}$$

Settling can be included in the diffusion model by describing it as an advective flow. Hence, the flux model becomes

$$F = -D\frac{dc}{dz} - v_s c$$

The minus sign indicates that settling leads to a negative flux (down). To integrate this equation, it is rearranged as follows:

$$\frac{dc}{c + F/v_s} = -\frac{v_s}{D} dz$$

Integrating from $c_a$ to $c_s$ with diffusivity $D_a$ and layer thickness $\Delta z_a$ leads to the following equation:

$$\ln\left(\frac{c_s + F/v_s}{c_a + F/v_s}\right) = \frac{v_s}{D_a}\Delta z_a = v_s r_a$$

Similarly, integrating from $c_s$ to 0 with diffusivity $D_b$ and layer thickness $\Delta z_b$ leads to the following equation:

$$\ln\left(\frac{\frac{F}{v_s}}{c_s + F/v_s}\right) = \frac{v_s}{D_b}\Delta z_b = v_s r_b$$

Adding these two equations together, one obtains

$$\ln\left(\frac{F/v_{\mathrm{s}}}{c_{\mathrm{a}} + F/v_{\mathrm{s}}}\right) = v_{\mathrm{s}}(r_{\mathrm{a}} + r_{\mathrm{b}})$$

Taking the exponential, and rearranging, leads to

$$c_{\mathrm{a}} = -\frac{F}{v_{\mathrm{s}}}\{1 - \exp[-v_{\mathrm{s}}(r_{\mathrm{a}} + r_{\mathrm{b}})]\}$$

Or the flux can be expressed as

$$-F = \frac{v_{\mathrm{s}}}{1 - \exp[-v_{\mathrm{s}}(r_{\mathrm{a}} + r_{\mathrm{b}})]} c_{\mathrm{a}}$$

Comparing with the definition of the deposition velocity [eq. (7.129)], the following equation for $v_{\mathrm{d}}$ is obtained:

$$v_{\mathrm{d}} = \frac{v_{\mathrm{s}}}{1 - \exp[-v_{\mathrm{s}}(r_{\mathrm{a}} + r_{\mathrm{b}})]}$$

## A.7.4 Diffusion of a Compound into a Sphere

To fully understand wet deposition, we need to find out if the mass transfer rate of pollutants into droplets is mainly governed by air-side mass transfer limitation or water-side mass transfer limitation. In this section, we will outline a numerical model to calculate the time required to saturate a water droplet with a solute by diffusion from a constant interface concentration. Molecular diffusion follows Fick's law:

$$F = -D_{\mathrm{w}} \frac{dc_{\mathrm{w}}}{dr}$$

where $F$ is the flux outward ($\mathrm{mg\,m^{-2}\,s^{-1}}$), $D_{\mathrm{w}}$ is the diffusivity of the pollutant in the water phase ($\mathrm{m^2\,s^{-1}}$), $c_{\mathrm{w}}$ is the water phase concentration of the pollutant ($\mathrm{mg\,m^{-3}}$), and $r$ is the distance from the center of the droplet (m).

The material balance is as follows:

$$\text{Accumulation} = \text{in} - \text{out}$$

This material balance is applied to pollutant in a differential shell of the droplet, from $r$ to $r+dr$. The accumulation is

$$\text{Accumulation} = 4\pi r^2 dr \frac{\partial c_{\mathrm{w}}}{\partial t}$$

The material flow into the differential shell is

$$\text{In} = 4\pi r^2 \cdot \left(-D_w \frac{\partial c_w}{\partial r}\right)$$

The material flow out of the differential shell is

$$\text{Out} = 4\pi(r+dr)^2 \cdot \left[-D_w \frac{\partial}{\partial r}\left(c_w + \frac{\partial c_w}{\partial r} dr\right)\right]$$

Hence, the material balance can be written as

$$4\pi r^2\, dr \frac{\partial c_w}{\partial t} = -4\pi r^2 \cdot D_w \frac{\partial c_w}{\partial r} + 4\pi(r+dr)^2 \cdot D_w \frac{\partial}{\partial r}\left(c_w + \frac{\partial c_w}{\partial r} dr\right)$$

The last term can be expanded into six terms, one of which cancels the first term in the right-hand side, and three of which have one or two extra differentials. Hence, the above equation can be rewritten as

$$4\pi r^2\, dr \frac{\partial c_w}{\partial t} = 4\pi \cdot 2r\, dr \cdot D_w \frac{\partial c_w}{\partial r} + 4\pi r^2 \cdot D_w \frac{\partial^2 c_w}{\partial r^2} dr$$

Dividing by $4\pi r^2 dr$ leads to

$$\frac{\partial c_w}{\partial t} = \frac{2D_w}{r}\frac{\partial c_w}{\partial r} + D_w \frac{\partial^2 c_w}{\partial r^2} \tag{7.239}$$

This partial differential equation can be integrated numerically in the form of eq. (7.239), with the following initial and boundary conditions:

$$c_w = 0 \qquad (t = 0)$$

$$\frac{\partial c_w}{\partial z} = 0 \qquad (r = 0)$$

$$c_w = c_{w,i} \qquad (r = r_d)$$

However, the number of combinations of variables can be reduced greatly by nondimensionalizing the problem. To that effect, the following substitutions are carried out: $c_w = c'_w\, c_{i,w}$, $t = t' r_d^2 / D_w$, and $r = r' r_d$. The result is

$$\frac{\partial c'_w c_{i,w}}{\partial t' r_d^2 / D_w} = \frac{2D_w}{r' r_d}\frac{\partial c'_w c_{i,w}}{\partial r' r_d} + D_w \frac{\partial^2 c'_w c_{i,w}}{\partial r'^2 r_d^2}$$

Multiplying by $r_d^2 c_{i,w}^{-1} D_w^{-1}$ leads to

$$\frac{\partial c'_w}{\partial t'} = \frac{2}{r'}\frac{\partial c'_w}{\partial r'} + \frac{\partial^2 c'_w}{\partial r'^2} \tag{7.241}$$

This equation is integrated numerically with the following boundary conditions:

$$c'_{\mathrm{w}} = 0 \qquad (t' = 0) \tag{7.242}$$

$$\frac{\partial c'_{\mathrm{w}}}{\partial r'} = 0 \qquad (r' = 0) \tag{7.243}$$

$$c'_{\mathrm{w}} = 1 \qquad (r' = 1) \tag{7.244}$$

In the rest of this section the primes will be left out for convenience of the notation.

The simplest way to integrate eq. (7.241) numerically is by converting it to a set of ordinary differential equations using finite differences. First, the continuous function for the concentration $c_{\mathrm{w}}$ is subdivided into a set of values $c_1, c_2, c_3, \ldots, c_{\mathrm{N}}$ on a grid within the droplet with spacing $\Delta r$. The coordinates are $r_0=0$, $r_1 = \Delta r$, and $r_2=2\Delta r, \ldots, r_{\mathrm{N}}=N\,\Delta r=1$. This leads to the following relationship:

$$\Delta r = \frac{1}{N}$$

The following substitutions are made in the differential equation:

$$\frac{\partial c_i}{\partial r} = \frac{c_{i+1} - c_{i-1}}{2\Delta r}$$

$$\frac{\partial^2 c_i}{\partial r^2} = \frac{c_{i+1} - 2c_i + c_{i-1}}{(\Delta r)^2}$$

for $i=1, \ldots, N-1$. The initial value is $c_{\mathrm{i}}=0$ at $t=0$ for $i=0, \ldots, N$. The boundary conditions are

$$c_0 = \frac{4}{3}c_1 - \frac{1}{3}c_2$$

$$c_N = 1$$

Hence, there are $N-1$ ordinary differential equations to be solved simultaneously. This can be done with a differential equation solver.

When the concentrations are known at all points of the grid, the average concentration can be calculated by weighted averaging using the volume of each differential shell of the sphere as the weight. This leads to eq. (7.245):

$$S_{\mathrm{w}} = \frac{\int_0^1 4\pi r'^2 c'\, dr}{\int_0^1 4\pi r'^2\, dr} = 3\int_0^1 r'^2 c'\, dr \tag{7.245}$$

Because the concentrations are normalized using the interface concentration, what is calculated here is the saturation $S_{\mathrm{w}}$. A convenient way to carry out the integration is with one of Simpson's rules. For instance, with Simpson's one-third rule, one obtains

$$S_w = 3 \times \frac{\Delta r}{3}(I_0 + 4I_1 + 2I_2 + 4I_3 + 2I_4 + \cdots + 2I_{N-2} + 4I_{N-1} + I_N)$$

where $I_i = c_i \, r_i^2$.

## MATERIALS ONLINE

- Folder "Diffusion in droplet," files "data.m," "f.m," and "main.m": Calculation of liquid-side mass transfer of a pollutant in a droplet.

# CHAPTER A8: GAUSSIAN MODEL APPROACHES IN URBAN OR INDUSTRIAL TERRAIN

## A.8.1 Calculation of Average Plume Height of a Ground-Level Plume

When a pollutant is emitted at ground level ($h=0$), then there is no defined wind speed that can be used in the Gaussian equation. In this case, it is best to define an average plume height $h_a$ and use the wind speed at $h_a$ in the Gaussian equation. In Chapter 8, the average plume height is defined as

$$h_a = \frac{\int_0^\infty zuc\,dz}{\int_0^\infty uc\,dz} \tag{8.48}$$

Substituting eq. (8.49) in eq. (8.48) and canceling common factors leads to

$$h_a = \frac{\int_0^\infty zu \exp\left(-\frac{1}{2}\frac{z^2}{\sigma_z^2}\right) dz}{\int_0^\infty u \exp\left(-\frac{1}{2}\frac{z^2}{\sigma_z^2}\right) dz}$$

Two cases will be considered here: no height dependence of the wind speed (i.e., an unweighted average plume height) and a wind speed proportional to height. The first case allows us to eliminate $u$ directly in both integrals:

$$h_a = \frac{\int_0^\infty z \exp\left(-\frac{1}{2}\frac{z^2}{\sigma_z^2}\right) dz}{\int_0^\infty \exp\left(-\frac{1}{2}\frac{z^2}{\sigma_z^2}\right) dz}$$

The solution to this equation is based on the following two integrals from Appendix B:

$$\int_0^x \exp(-at^2)\,dt = \frac{\sqrt{\pi}}{2\sqrt{a}}\operatorname{erf}(\sqrt{a}x)$$

$$\int x\exp(-ax^2)\,dx = -\frac{1}{2a}\exp(-ax^2)+C$$

The following result is obtained:

$$h_a = \frac{\sigma_z^2}{\sqrt{\pi/2}\,\sigma_z} = \sqrt{\frac{2}{\pi}}\sigma_z$$

When the wind speed is proportional to height, the average plume height equation is

$$h_a = \frac{\int_0^\infty z^2 \exp\left(-\frac{1}{2}\frac{z^2}{\sigma_z^2}\right)dz}{\int_0^\infty z\exp\left(-\frac{1}{2}\frac{z^2}{\sigma_z^2}\right)dz}$$

An additional integral is needed from Appendix B:

$$\int x^2\exp(-ax^2)\,dx = -\frac{1}{4}\sqrt{\frac{\pi}{a^3}}\operatorname{erf}(\sqrt{a}x) - \frac{x}{2a}\exp(-ax^2)+C$$

The following result is obtained:

$$h_a = \frac{\sqrt{\pi/2}\,\sigma_z^3}{\sigma_z^2} = \sqrt{\frac{\pi}{2}}\sigma_z$$

Approximate values are $0.8\sigma_z$ and $1.25\sigma_z$, respectively. A reasonable average is $h_a = \sigma_z$.

# CHAPTER A9: STOCHASTIC MODELING APPROACHES

## A.9.1 Derivation of the Langevin Equation for Homogeneous Turbulence from the Fokker–Planck Equation

The derivation of Langevin equations for stochastic Lagrangian particle models from the Fokker–Planck equation is based on the work of Thomson (1987). The Fokker–Planck equation is

$$\frac{\partial p}{\partial t} + v\frac{\partial p}{\partial y} = -\frac{\partial}{\partial v}(ap) + \frac{1}{2}\frac{\partial^2}{\partial v^2}(b^2 p) \tag{9.42}$$

where $p$ is the time-dependent probability density function of the location and velocity, that is, $p(y,v,t)\,dy\,dv$ is the probability that an air parcel at time $t$ is located between $y$ and $y+dy$, and has a velocity between $v$ and $v+dv$.

We apply the **well-mixed criterion** to the Fokker–Planck equation. This criterion states that if the distribution of air parcels is initially homogeneous (i.e., $p$ is independent of $y$), then it will remain homogeneous (i.e., its time derivative is zero). It follows that $p$ is only dependent on $v$, and eq. (9.42) can be simplified to

$$0 = -\frac{\partial}{\partial v}(ap) + \frac{1}{2}\frac{\partial^2}{\partial v^2}(b^2 p)$$

and $p$ can be expressed as the probability of $v$ given a certain value of $y$, multiplied by the probability of $y$:

$$p(y,v,t) = p(y,v) = p(v|y)\cdot p(y)$$

Because $p(y)$ is a constant (homogeneous pollutant concentration), it can be taken out of the Fokker–Planck equation. Hence, from now on it will be assumed in this section that $p$ refers to $p(v|y)$ or simply $p(v)$ (homogeneous turbulence).

One derivative with respect to $v$ can be put outside brackets:

$$\frac{\partial}{\partial v}\left[-ap + \frac{1}{2}\frac{\partial}{\partial v}(b^2 p)\right] = 0$$

It follows that the sum of the terms between brackets is a constant, $\phi$:

$$-ap + \frac{1}{2}\frac{\partial}{\partial v}(b^2 p) = \phi$$

If $\phi \neq 0$, then $p$ will equal $-\phi/a$ for $v$ tending toward infinity if it tends toward a constant value, or it will tend to infinity. Neither of these cases are physically realistic. It follows that $\phi$ must be zero in this case. Assume that $p$ follows a Gaussian equation, which is a realistic assumption when the turbulence is homogeneous, and there are no convective cycles. Hence:

$$p = \frac{1}{\sqrt{2\pi}\sigma_v}\exp\left(-\frac{1}{2}\frac{v^2}{\sigma_v^2}\right)$$

In this equation the mean velocity is zero, so $v$ is actually the velocity fluctuation, $v'$. Hence the derivative with respect to $v$ is

$$\frac{\partial p}{\partial v} = -\frac{1}{2}\frac{2v}{\sigma_v^2}\frac{1}{\sqrt{2\pi}\sigma_v}\exp\left(-\frac{1}{2}\frac{v^2}{\sigma_v^2}\right) = -\frac{v}{\sigma_v^2}p$$

Substitution into the Fokker–Planck equation leads to

$$-ap - \frac{1}{2}b^2\frac{v}{\sigma_v^2}p = 0$$

Hence, $a$ is given by

$$a = -\frac{1}{2} b^2 \frac{v}{\sigma_v^2} = -\frac{v}{T_{\text{i,L}}}$$

which is also the solution found in Section 9.2.2.

## A.9.2 Rescaling Stochastic Lagrangian Calculations for Variable $T_{\text{i,L}}$

In Section 9.2.2 and the previous section, the Langevin equation was derived for homogeneous turbulence. In principle, the equations are still valid when $T_{\text{i,L}}$ is variable, provided that sufficiently short time steps are taken to ensure that $T_{\text{i,L}}$ does not change markedly during a single time step. In practice, this is difficult to achieve near the surface, where $T_{\text{i,L}}$ approaches zero and, therefore, experiences large relative changes over short distances. Hence, to resolve stochastic Lagrangian calculations near the surface, it is recommended to apply a rescaling approach that converts the variable $T_{\text{i,L}}$ calculation into a mathematically equivalent constant $T_{\text{i,L}}$ calculation. The rescaling approach discussed here was developed by Wilson et al. (1981). In this section it will be assumed that $\sigma_w$ is constant. Starting point is the Langevin equation for homogeneous turbulence:

$$dw' = -\frac{w'}{T_{\text{i,L}}} dt + \sqrt{\frac{2\sigma_w^2}{T_{\text{i,L}}}} dW$$

and $T_{\text{i,L}}$ is assumed to depend on height alone and can be written as $T_{\text{i,L}}(z)$. The objective of the rescaling is to obtain a Langevin equation with height-independent coefficients.

Define a reference height $H$, characterized by a time scale $T_{\text{i,L}}(H)$. Now define a scaled time $t_H$ as

$$dt = \frac{T_{\text{i,L}}(z)}{T_{\text{i,L}}(H)} dt_H$$

Noting that the Wiener process $W$ has a standard deviation proportional to $t^{1/2}$, we can define the Wiener process in the rescaled time, $W_H$, as

$$dW = \sqrt{\frac{T_{\text{i,L}}(z)}{T_{\text{i,L}}(H)}} dW_H$$

Substitution into the Langevin equation leads to

$$dw' = -\frac{w'}{T_{\text{i,L}}(H)} dt_H + \sqrt{\frac{2\sigma_w^2}{T_{\text{i,L}}(H)}} dW_H$$

This equation appears to fulfill the requirements, but there is one problem: $w'$ and $\sigma_w$ are defined with respect to $t$, not $t_H$, which leads to an inconsistency. To maintain consistency, we must rescale $z$. To that effect, we consider $w'$ in terms of $t_H$:

$$w' = \frac{dz'}{dt} = \frac{T_{\mathrm{i,L}}(H)}{T_{\mathrm{i,L}}(z)} \frac{dz'}{dt_H}$$

Now define the scaled length coordinate $z_H$ as

$$dz' = \frac{T_{\mathrm{i,L}}(z)}{T_{\mathrm{i,L}}(H)} dz'_H$$

Substitution into the definition of $w'$ leads to

$$w' = \frac{dz'_H}{dt_H}$$

Hence, the rescaled Langevin equation should be used in rescaled spatial coordinates $z_H$.

In *neutral* conditions, $T_{\mathrm{i,L}}$ is proportional to $z$, at least in the surface layer. Hence:

$$\frac{T_{\mathrm{i,L}}(z)}{T_{\mathrm{i,L}}(H)} = \frac{z}{H}$$

Hence, the scaled vertical coordinate is given by:

$$dz = \frac{z}{H} dz_H$$

Choose a domain boundary $z_2$ close enough to the surface that no important processes occur below $z_2$. Typically, $z_0$ is a good choice; $z = z_2$ corresponds with $z_H = 0$ so that the above equation can be integrated as

$$\ln\left(\frac{z}{z_2}\right) = \frac{z_H}{H}$$

Hence, the coordinate transformations are

$$z_H = H \ln\left(\frac{z}{z_2}\right)$$

$$z = z_2 \exp\left(\frac{z_H}{H}\right)$$

When the domain is divided into layers of constant thickness in $z_H$, then high resolution is achieved near the surface, whereas the resolution is lower far from the surface.

The main disadvantage of this scaling technique is that the movement of the particles is not synchronized in time, so that some planning is required to ensure that concentration calculations are based on synchronous particle locations.

# CHAPTER A10: COMPUTATIONAL FLUID DYNAMICS AND METEOROLOGICAL MODELING

## A.10.1 Coriolis Force of an Object Moving North

From Figure 10.1 (Section 10.2.1) it can be seen that an object moving north on a rotating Earth will experience an acceleration toward the east. This acceleration is felt as an apparent force known as the Coriolis force. To calculate the effect, we will consider the movements on an inertial frame. It will be assumed that time and space increments considered are small enough to allow the use of a first-order Taylor series in the calculations.

Assume that the trajectory starts north of the equator, at latitude $\varphi$ (in radians). At this location Earth moves east on the inertial frame with velocity $u_e$:

$$u_e = R\cos\varphi \cdot \Omega$$

where $R$ is the radius of Earth. Because the object moves north at the beginning of the trajectory, its eastward velocity $u_o$ is the same as the local velocity of Earth at $t=0$:

$$u_o(t=0) = R\cos\varphi \cdot \Omega$$

The object moves north at a constant velocity $v$ for a time $t$. The distance traveled is

$$\Delta y = vt$$

The latitude change associated with this movement is

$$\Delta\varphi = \frac{\Delta y}{R}$$

At this new location the velocity of Earth on the inertial frame is

$$\begin{aligned} u_e(\Delta y) &= u_e(y=0) + \frac{\partial u_e}{\partial \varphi}\frac{\partial \varphi}{\partial y}\Delta y = R\cos\varphi \cdot \Omega - \frac{R\sin\varphi \cdot \Omega \Delta y}{R} \\ &= R\cos\varphi \cdot \Omega - \sin\varphi \cdot \Omega \cdot vt \end{aligned}$$

After a time $t$, Earth has moved the following distance in the eastern direction:

$$x_e(\Delta y, t) = u_e(\Delta y)t = R\cos\varphi \cdot \Omega t - \sin\varphi \cdot \Omega \cdot vt^2$$

In the absence of force or acceleration in the east direction, the thrown object travels the following distance east in a time $t$:

$$x_0(t) = u_0(t=0)t = R\cos\varphi \cdot \Omega t$$

The apparent eastward movement of the object is the movement in the inertial plane, less the movement of Earth in the inertial plane:

$$x(t) = x_0(t) - x_e(\Delta y, t) = R\cos\varphi \cdot \Omega t - (R\cos\varphi \cdot \Omega t - \sin\varphi \cdot \Omega \cdot vt^2) = \sin\varphi \cdot \Omega \cdot vt^2$$

The apparent acceleration of the object is the second derivative of the location:

$$a(t) = \frac{\partial^2 x}{\partial t^2} = 2\sin\varphi \cdot \Omega \cdot v$$

An apparent force can be associated with this acceleration by using Newton's law:

$$F_{c,x} = \rho\, dx\, dy\, dz \cdot 2\Omega \sin\varphi \cdot v = \rho\, dx\, dy\, dz \cdot fv \tag{10.6}$$

This is the Coriolis force.

## REFERENCES


ASME (1973). *Recommended Guide for the Prediction of the Dispersion of Airborne Effluents*, 2nd ed. ASME, New York.

Briggs G.A. (1973). *Diffusion Estimation of Small Emissions*. Contribution No. 79, Atmospheric Turbulence and Diffusion Laboratory, Oak Ridge, TN.

Businger J.A., Wyngaard J.C., Izumi Y., and Bradley E.F. (1971). Flux-profile relationships in the atmospheric surface layer. *J. Atmos. Sci.* **28**, 181–189.

Cheng Y. and Brutsaert W. (2005). Flux-profile relationships for wind speed and temperature in the stable atmospheric boundary layer. *Bound. Layer Meteorol.* **114**, 519–538.

Cimorelli A.J., Perry S.G., Venkatram A., Weil J.C., Paine R.J., Wilson R.B., Lee R.F., Peters W.D., Brode R.W., and Paumier J.O. (2004). *AERMOD: Description of Model Formulation*. Report EPA-454/R-03-004, US-EPA, Research Triangle Park, NC.

Dyer A.J. (1974). A review of flux-profile relationships. *Bound. Layer Meteorol.* **7**, 363–372.

EPA (1995). *User's Guide for the Industrial Source Complex (ISC3) Dispersion Models. Volume II – Description of Model Algorithms*. Report EPA-454/B-95-003b, US-EPA, Research Triangle Park, NC.

Högström U. (1988). Non-dimensional wind and temperature profiles in the atmospheric surface layer: A re-evaluation. *Bound. Layer Meteorol.* **42**, 55–78.

Klug W. (1969). A method for determining diffusion conditions from synoptic observations. *Staub-Reinhalt. Luft* **29**, 14–20.

Martin D.O. (1976). Comment on the change of concentration standard deviations with distance. *J. Air Pollut. Control Assoc.* **26**, 145–146.

Nieuwstadt F.T.M. (1984). Some aspects of the turbulent stable boundary layer. *Bound. Layer Meteorol.* **30**, 31–55.

Park S.J., Park S.U., Ho C.H., and Mahrt L. (2009). Flux-gradient relationship of water vapor in the surface layer obtained from CASES-99 experiment. *J. Geophys. Res.* **114**, D08115.

Scire J.S., Lurmann F.W., Bass A., and Hanna S.R. (1984) *User's Guide to the MESOPUFF II Model and Related Processor Programs*. Report EPA-600/8-84-013, US-EPA, Research Triangle Park, NC.

Stull R.B. (1983). A heat flux history length scale for the nocturnal boundary layer. *Tellus* **35A**, 219–230.

Thomson D.J. (1987). Criteria for the selection of stochastic models of particle trajectories in turbulent flows. *J. Fluid Mech.* **180**, 529–556.

Turner D.B. (1970). *Workbook of Atmospheric Dispersion Estimates*. US EPA, Washington DC.

van Ulden A.P. and Holtslag A.A.M. (1985). Estimation of atmospheric boundary layer parameters for diffusion applications. *J. Climate Appl. Meteorol.* **24**, 1196–1207.

Wagner W. and Pruß A. (2002). The IAPWS formulation 1995 for the thermodynamic properties of ordinary water substance for general and scientific use. *J. Phys. Chem. Ref. Data* **31**, 387–535.

Wilson J.D., Thurtell G.W., and Kidd G.E. (1981). Numerical simulation of particle trajectories in inhomogeneous turbulence, I: Systems with constant turbulent velocity scale. *Bound. Layer Meteorol.* **21**, 295–313.

Zeman O. (1982). The dynamics and modeling of heavier-than-air, cold gas releases. *Atmos. Environ.* **16**, 741–751.

APPENDIX B

# AUXILIARY DATA AND METHODS

**B1 Values of the Gaussian Function with Zero Mean and Unit Standard Deviation, $N(x,1,0)$ Note that $N(-x,1,0)=N(x,1,0)$.**

| $x$ | N($x$,1,0) | $x$ | N($x$,1,0) | $x$ | N($x$,1,0) |
|---|---|---|---|---|---|
| 0 | 0.398942 | | | | |
| 0.1 | 0.396953 | 1.1 | 0.217852 | 2.1 | 0.043984 |
| 0.2 | 0.391043 | 1.2 | 0.194186 | 2.2 | 0.035475 |
| 0.3 | 0.381388 | 1.3 | 0.171369 | 2.3 | 0.028327 |
| 0.4 | 0.368270 | 1.4 | 0.149727 | 2.4 | 0.022395 |
| 0.5 | 0.352065 | 1.5 | 0.129518 | 2.5 | 0.017528 |
| 0.6 | 0.333225 | 1.6 | 0.110921 | 2.6 | 0.013583 |
| 0.7 | 0.312254 | 1.7 | 0.094049 | 2.7 | 0.010421 |
| 0.8 | 0.289692 | 1.8 | 0.078950 | 2.8 | 0.007915 |
| 0.9 | 0.266085 | 1.9 | 0.065616 | 2.9 | 0.005953 |
| 1.0 | 0.241971 | 2.0 | 0.053991 | 3.0 | 0.004432 |

**B2 Values of Arctan $n$**

| $n$ | arctan $n$ | $n$ | arctan $n$ | $n$ | arctan $n$ |
|---|---|---|---|---|---|
| 1 | 0.785398 | 1.5 | 0.982794 | 3.2 | 1.267911 |
| 1.01 | 0.790373 | 1.55 | 0.99783 | 3.4 | 1.284745 |
| 1.02 | 0.795299 | 1.6 | 1.012197 | 3.6 | 1.299849 |
| 1.03 | 0.800175 | 1.65 | 1.025932 | 3.8 | 1.313473 |
| 1.04 | 0.805003 | 1.7 | 1.039072 | 4 | 1.325818 |
| 1.05 | 0.809784 | 1.75 | 1.051650 | 4.2 | 1.337053 |
| 1.06 | 0.814516 | 1.8 | 1.063698 | 4.4 | 1.347320 |
| 1.07 | 0.819202 | 1.85 | 1.075245 | 4.6 | 1.356736 |
| 1.08 | 0.823841 | 1.9 | 1.086318 | 4.8 | 1.365401 |
| 1.09 | 0.828434 | 1.95 | 1.096945 | 5 | 1.373401 |

(*Continued*)

*Air Dispersion Modeling: Foundations and Applications*, First Edition. Alex De Visscher.

**B2** *(Continued)*

| $n$ | arctan $n$ | $n$ | arctan $n$ | $n$ | arctan $n$ |
|---|---|---|---|---|---|
| 1.1 | 0.832981 | 2 | 1.107149 | 5.5 | 1.390943 |
| 1.12 | 0.841942 | 2.1 | 1.126377 | 6 | 1.405648 |
| 1.14 | 0.850726 | 2.2 | 1.144169 | 6.5 | 1.418147 |
| 1.16 | 0.859337 | 2.3 | 1.160669 | 7 | 1.428899 |
| 1.18 | 0.867780 | 2.4 | 1.176005 | 7.5 | 1.438245 |
| 1.2 | 0.876058 | 2.5 | 1.190290 | 8 | 1.446441 |
| 1.25 | 0.896055 | 2.6 | 1.203622 | 8.5 | 1.453688 |
| 1.3 | 0.915101 | 2.7 | 1.216091 | 9 | 1.460139 |
| 1.35 | 0.933248 | 2.8 | 1.227772 | 9.5 | 1.465919 |
| 1.4 | 0.950547 | 2.9 | 1.238737 | 10 | 1.471128 |
| 1.45 | 0.967047 | 3 | 1.249046 | $\infty$ | 1.570796 |

## B3 INTEGRALS

Exponential functions:

$$\int \exp(ax)dx = \frac{1}{a}\exp(ax) + C$$

$$\int x\exp(ax)dx = \left(\frac{x}{a} - \frac{1}{a^2}\right)\exp(ax) + C$$

$$\int x^2 \exp(ax)\,dx = \left(\frac{x^2}{a} - \frac{2x}{a^2} + \frac{2}{a^3}\right)\exp(ax) + C$$

$$\int x^3 \exp(ax)dx = \left(\frac{x^3}{a} - \frac{3x^2}{a^2} + \frac{6x}{a^3} - \frac{6}{a^4}\right)\exp(ax) + C$$

Gaussian-type functions:

$$\int_0^x \exp(-t^2)dt = \frac{\sqrt{\pi}}{2}\operatorname{erf}(x)$$

$$\int_0^x \exp(-at^2)dt = \frac{\sqrt{\pi}}{2\sqrt{a}}\operatorname{erf}\left(\sqrt{a}x\right)$$

$$\int x\exp(-ax^2)dx = -\frac{1}{2a}\exp(-ax^2) + C$$

$$\int x^2 \exp(-ax^2)dx = \frac{1}{4}\sqrt{\frac{\pi}{a^3}}\operatorname{erf}\left(\sqrt{a}x\right) - \frac{x}{2a}\exp(-ax^2) + C$$

Fractions containing polynomes:

$$\int \frac{dx}{x} = \ln(x) + C$$

$$\int \frac{dx}{x+a} = \ln(x+a) + C$$

$$\int \frac{dx}{ax^2+bx+c} = \frac{2}{\sqrt{4ac-b^2}} \arctan\left(\frac{2ax+b}{\sqrt{4ac-b^2}}\right) + C \text{ if } b^2 < 4ac$$

$$\int \frac{dx}{ax^2+bx+c} = \frac{-2}{ax+b} + C \text{ if } b^2 = 4ac$$

$$\int \frac{dx}{ax^2+bx+c} = \frac{1}{\sqrt{b^2-4ac}} \ln\left|\frac{2ax+b-\sqrt{b^2-4ac}}{2ax+b+\sqrt{b^2-4ac}}\right| + C \quad \text{if } b^2 > 4ac$$

$$\int \frac{(mx+n)\,dx}{ax^2+bx+c} = \frac{m}{2a} \ln\left|ax^2+bx+c\right| + \frac{2an-bm}{a\sqrt{4ac-b^2}} \arctan\left(\frac{2ax+b}{\sqrt{4ac-b^2}}\right) + C \text{ if } b^2 < 4ac$$

$$\int_0^d \frac{dz}{z^3+E} = \frac{1}{3E^{2/3}}\left[\ln\left(\frac{d+E^{1/3}}{E^{1/3}}\right) - \frac{1}{2}\ln\left|\frac{d^2-E^{1/3}d+E^{2/3}}{E^{2/3}}\right| + \frac{3}{\sqrt{3}}\arctan\left(\frac{2d}{\sqrt{3}E^{1/3}}\right)\right]$$

## B4 GRAPHICAL INTEGRATION RULES

This section provides equations for the evaluation of the integral

$$y = \int_{x_0}^{x_n} f(x)\,dx$$

if $f(x)$ is given in tabular form, that is, $f(x_0) = f_0$, $f(x_1) = f_1$, and so forth. The methods are based on the notion that the integral $y$ is the area under the curve describing $f$ in a diagram from $x_0$ to $x_n$.

The simplest, but least accurate, calculation method is the **trapezoid rule** (see Fig. B1)

$$y = \int_{x_0}^{x_1} f(x)\,dx = (x_1 - x_0)\frac{f_0+f_1}{2} = \frac{\Delta x}{2}(f_0 + f_1) \tag{B4.1}$$

The integral of the function between two adjacent data points is the area of the trapezoid connecting the two points to the $x$ axis, $\Delta x$. When more than two data points are given, then the area under the curve is obtained by dividing the area up into trapezoids. Hence

$$y = (x_1 - x_0)\frac{f_0+f_1}{2} + (x_2 - x_1)\frac{f_1+f_2}{2} + \cdots + (x_n - x_{n-1})\frac{f_{n-1}+f_n}{2} \tag{B4.2}$$

The main advantage of the trapezoid rule is that the increments $x_{i+1} - x_i$ do not need to be the same. When they are, the equation simplifies to

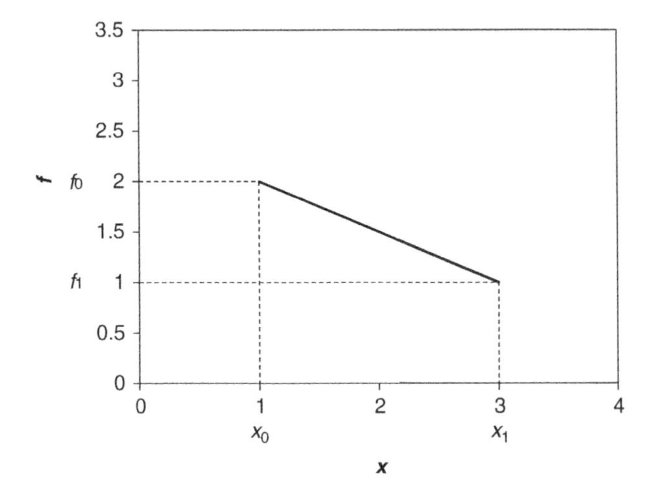


Figure B1 Trapezoid rule.

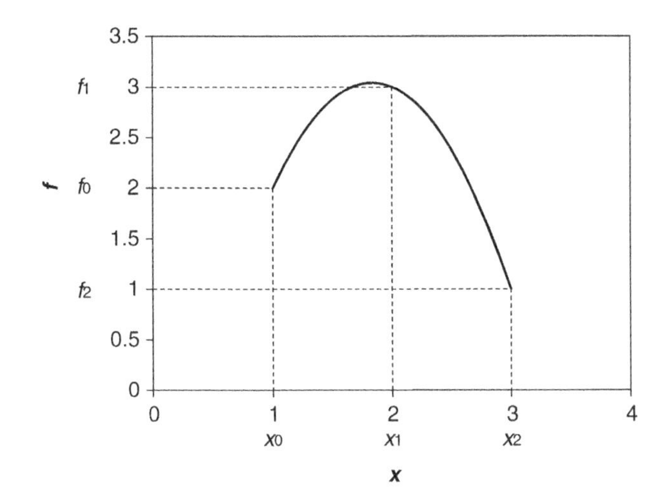


Figure B2 Simpson's 1/3 rule.

$$y = \frac{\Delta x}{2}\left(f_0 + 2f_1 + 2f_2 + \cdots + 2f_{n-1} + f_n\right) \tag{B4.3}$$

The error of this method is proportional to $(\Delta x)^2$.

The second rule is **Simpson's 1/3 rule**:

$$y = \int_{x_0}^{x_2} f(x)dx = \frac{\Delta x}{3}(f_0 + 4f_1 + f_2) \tag{B4.4}$$

where $\Delta x$ is the difference between two consecutive $x$ values. This equation is based on the area under the parabola connecting the three data points (Fig. B2). When there are more than three data points, the equation can be used consecutively, obtaining

$$y=\int_{x_0}^{x_n} f(x)dx=\frac{\Delta x}{3}(f_0+4f_1+2f_2+4f_3+2f_4+\cdots+2f_{n-2}+4f_{n-1}+f_n) \quad \text{(B4.5)}$$

For this method to be valid, the number of data points must be odd, and the spacing between the data points must be constant. The error of Simpson's 1/3 rule is proportional to $(\Delta x)^3$.

When Simpson's approach is applied to a third-order polynome connecting four equally spaced points, then Simpson's 3/8 rule is obtained:

$$y=\int_{x_0}^{x_3} f(x)dx=\frac{3\Delta x}{8}(f_0+3f_1+3f_2+f_3) \quad \text{(B4.6)}$$

Again, the method can be applied with a larger number of data points:

$$y=\int_{x_0}^{x_n} f(x)dx=\frac{\Delta x}{3}(f_0+3f_1+3f_2+2f_3+3f_4+\cdots+2f_{n-3}+3f_{n-2}+3f_{n-1}+f_n) \quad \text{(B 4.7)}$$

The integration is different if the data are not discrete points but averages across intervals. In that case the integral is simply the summation of the $f$ values multiplied by the width of the interval to which the $f$ value applies, taking care that the entire range of independent variable is covered without overlap.

# B5 NUMERICAL INTEGRATION OF ORDINARY DIFFERENTIAL EQUATIONS

## B5.1 In Excel: Euler's Method

Euler's method is the simplest numerical solution method for ordinary differential equations, but it is also one of the least accurate methods. Consider the following ordinary differential equation:

$$\frac{dy}{dt}=f(y,t) \quad \text{(B5.1)}$$

Euler's method simply replaces the differentials by finite differences and evaluates the function $f$ at the initial conditions:

$$\frac{\Delta y}{\Delta t}=\frac{y_{n+1}-y_n}{t_{n+1}-t_n}=f(y_n,t_n) \quad \text{(B5.2)}$$

Solving for $y_{n+1}$ leads to

$$y_{n+1}=y_n+f(y_n,t_n)(t_{n+1}-t_n) \quad \text{(B5.3)}$$

Hence, with knowledge of an initial condition $(t_0, y_0)$, the value of $y$ at any value $t$ can be calculated by repeated application of eq. (B5.3), for $n=0$, $n=1$, ... .

This is illustrated in the Excel file "Appendix B5. Eulers method.xlsx." In the file a numerical solution of the ordinary differential equation

$$\frac{dy}{dt} = -ay \tag{B5.4}$$

is given and compared with the analytical solution. By experimenting with the variables, it is possible to see that the accuracy of the method depends strongly on the step size in relation to the rate of dynamics of the differential equation. Fairly small steps are needed to get accurate results.

## B5.2 In Matlab: Runge–Kutta and Other Methods

Matlab has a programming language to run scripts and contains functions for many purposes, including the numerical integration of ordinary differential equations. In the example below this is illustrated with the simultaneous integration of three ordinary differential equations of the form of eq. (B5.4). The program consists of three files: the main file "main.m," which is called by the user and which calls the other files as needed to run the calculations and display the results, the data file "data.m," which contains all data needed for the calculation, and the function "f.m," which defines the differential equations. The three files are reproduced below, and included on the CD in the subfolder "Demo numerical integration."

File "main.m":

```
    % Main file: Numerical integration of a set of
differential equations
    clear all  % Clear memory
    tic        % Set timer
    data       % Call data file
    % Run numerical integration
    options=odeset('RelTol', 1e-6, 'AbsTol', 1e-8,
'InitialStep', 0.01);
    [T,Y]=ode45(@f,tspan,y0,options,a,b,c);
    % Output
    plot(T,Y)
    T
    Y
    [s1,s2]=size(T);
    jcomp=3; % Component number
    figure
    plot(T,Y(1:s1,jcomp))
    Y(1:s1,jcomp)
    toc % Read timer
```

File "data.m"

```
    % Data File
    tspan=0:10 % Span of the independent variable
    y0(1)=1; % Initial conditions
    y0(2)=2;
    y0(3)=3;
    a=0.1; % Model Parameters
```

```
b=0.2;
c=0.3;
```

File "f.m"

```
% Function defining the differential equations
function dydt=f(t,y,a,b,c)
dydt=zeros(3,1);
dydt(1)=-a*y(1);
dydt(2)=-b*y(2);
dydt(3)=-c*y(3);
```

The program is run by running the "main.m" file. This file first runs the file "data.m," which first generates a matrix tspan=[0 1 2 ... 10] containing the values of $t$ where output is desired. It then generates a matrix $y0$ containing the initial conditions, [1 2 3]. Then the model parameters are defined.

In the main file the options of the numerical integration are set with a relative error tolerance of $10^{-6}$ and an absolute error tolerance of $10^{-8}$. The initial step size (0.01) is the third option.

The external function ode45 is called to execute the numerical integration. ode45 is a 4+5-th order Runge–Kutta–Fehlberg algorithm with variable step size (Press et al., 1992). It is an efficient, versatile algorithm provided that the set of differential equations is not stiff. When a stiff set of differential equations is to be solved (e.g., atmospherical chemical kinetics), then the function ode15s is recommended. ode15s has Gear's algorithm available. The right-hand sides of the differential equations are defined in the function "f.m." In the function, the time (independent variable) is given by the variable $t$, which is a number, whereas in the main file a matrix $T$ is constructed containing all the times of output. This is the transposition of tspan defined in the data file. The dependent variable (e.g., concentration) is stored in the $3\times1$ matrix $y$ in the function, whereas an array of dependent variables at different times is stored in the $3\times11$ matrix $Y$ in the main file. Each row in $Y$ is a set of $y$ values at a different time.

After the numerical integration a number of output statements are included in the main file.

## B5.3 Higher Order Ordinary Differential Equations

Higher order ordinary differential equations are equations involving a second or higher order derivative. In the case of a second-order equation, the general form is

$$\frac{d^2y}{dt^2} = f\left(y, t, \frac{dy}{dt}\right) \tag{B5.5}$$

To solve this equation, define a new variable:

$$u = \frac{dy}{dt} \tag{B5.6}$$

Hence, eq. (B5.5) can be converted to a set of two differential equations:

$$\frac{du}{dt} = f(y,t,u) \tag{B5.7}$$

$$\frac{dy}{dt} = u \tag{B5.8}$$

These equations can be solved with any of the methods discussed earlier for differential equations.

## B6 NUMERICAL INTEGRATION OF PARTIAL DIFFERENTIAL EQUATIONS

Partial differential equations are equations of the form

$$\frac{\partial y}{\partial t} = f\left(y,t,x,\frac{\partial y}{\partial x},\frac{\partial^2 y}{\partial x^2}\right) \tag{B6.1}$$

where $t$ is usually time, and $x$ is a spatial coordinate. There can be multiple spatial coordinates in the equation. There are several ways to solve partial differential equations numerically. The methods used in this book involve defining a grid and developing an ordinary differential equation in each grid point representing a local approximation of the partial differential equation. Hence, the partial differential equation is converted into a set of ordinary differential equations, which can be solved using the methods discussed in Section B5.

A grid of points $x_1$, $x_2$, … is defined spanning the relevant domain of variable $x$, typically, but not necessarily, with equal spacing $\Delta x$. The values of $y$ in these points are $y_1$, $y_2$, … . The simplest ways to define partial derivatives in $x$ are the following:

$$\frac{\partial y_n}{\partial x} = \frac{y_{n+1} - y_{n-1}}{2\Delta x} \tag{B6.2}$$

$$\frac{\partial^2 y_n}{\partial t^2} = \frac{y_{n-1} - 2y_n + y_{n+1}}{(\Delta x)^2} \tag{B6.3}$$

Hence, eq. (6.1) is converted to a set of equations of the form:

$$\frac{\partial y_n}{\partial t} = f\left(y,t,x,\frac{y_{n+1} - y_{n-1}}{2\Delta x},\frac{y_{n+1} - 2y_n + y_{n-1}}{(\Delta x)^2}\right) \tag{B6.4}$$

Equation (B6.4) is an ordinary differential equation. This technique is known as the finite differences technique. To calculate the first and last grid points, the boundary conditions are used.

Equation (B6.2) can cause numerical instability when it dominates in the differential equation, especially when inaccurate integration routines such as Euler's method are used. Numerical instability means that the solution found by the algorithm has alternating high and low values, with the exact solution in between. This is illustrated in Chapter 11. The stability of the algorithm can be improved by replacing eq. (B6.2) by

$$\frac{\partial y_n}{\partial x} = \frac{y_{n+1} - y_n}{\Delta x} \tag{B6.5}$$

Equation (B6.5) is inherently less accurate than eq. (B6.2) so we are gaining stability at the expense of accuracy. When the second-order derivatives dominate the differential equation, then eq. (B6.3) will have a stabilizing effect on the algorithm. A way to increase both the accuracy and the stability of the solution is by replacing eqs. (B6.2) and (B6.3) by higher order equations based on a larger number of grid points.

## B7 Values of $\mathrm{erf}(x) = (2/\pi^{1/2})\int_0^x \exp(-t^2)\,dt$.

Note that $\mathrm{erf}(-x) = -\mathrm{erf}(x)$.

| $x$ | erf($x$) | $x$ | erf($x$) | $x$ | erf($x$) |
|---|---|---|---|---|---|
| 0.00 | 0.0000 | | | | |
| 0.05 | 0.05637 | 0.55 | 0.56332 | 1.10 | 0.88021 |
| 0.10 | 0.11246 | 0.60 | 0.60386 | 1.20 | 0.91031 |
| 0.15 | 0.16800 | 0.65 | 0.64203 | 1.30 | 0.93401 |
| 0.20 | 0.22270 | 0.70 | 0.67780 | 1.40 | 0.95229 |
| 0.25 | 0.27633 | 0.75 | 0.71116 | 1.50 | 0.96611 |
| 0.30 | 0.32863 | 0.80 | 0.74210 | 1.60 | 0.97635 |
| 0.35 | 0.37938 | 0.85 | 0.77067 | 1.70 | 0.98379 |
| 0.40 | 0.42839 | 0.90 | 0.79691 | 1.80 | 0.98909 |
| 0.45 | 0.47548 | 0.95 | 0.82089 | 1.90 | 0.99279 |
| 0.50 | 0.52050 | 1.00 | 0.84270 | 2.00 | 0.99532 |

## B8 Properties of Dry Air

| $t$ (°C) | $c_p$ (J kg$^{-1}$ K$^{-1}$) | $\mu$ ($10^{-5}$ Pa s) |
|---|---|---|
| −30 | 1002.86 | 1.5650 |
| −25 | 1002.96 | 1.5911 |
| −20 | 1003.08 | 1.6169 |
| −15 | 1003.22 | 1.6426 |
| −10 | 1003.37 | 1.6680 |
| −5 | 1003.54 | 1.6933 |
| 0 | 1003.72 | 1.7184 |
| 5 | 1003.91 | 1.7432 |
| 10 | 1004.12 | 1.7680 |
| 15 | 1004.34 | 1.7925 |
| 20 | 1004.58 | 1.8169 |
| 25 | 1004.83 | 1.8411 |
| 30 | 1005.09 | 1.8651 |
| 35 | 1005.38 | 1.8890 |
| 40 | 1005.67 | 1.9127 |
| 45 | 1005.99 | 1.9362 |
| 50 | 1006.32 | 1.9596 |

*Source*: NIST Chemistry Webbook; http://webbook.nist.gov/chemistry.

## B9 Properties of Water

| $t$ (°C) | $\mu$ ($10^{-3}$ Pa s) | $\rho$ (kg m$^{-3}$) | $p_v$ ($10^3$ Pa) | $\Delta_{vap}H°$ (kJ kg$^{-1}$) |
|---|---|---|---|---|
| 0 | 1.7909 | 999.84 | 0.61165 | 2500.9 |
| 1 | 1.7309 | 999.90 | 0.65709 | 2498.6 |
| 2 | 1.6734 | 999.94 | 0.70599 | 2496.2 |
| 3 | 1.6189 | 999.97 | 0.75808 | 2493.8 |
| 4 | 1.5672 | 999.97 | 0.81355 | 2491.4 |
| 5 | 1.5181 | 999.97 | 0.87258 | 2489.0 |
| 6 | 1.4714 | 999.94 | 0.93536 | 2486.7 |
| 7 | 1.4270 | 999.90 | 1.0021 | 2484.3 |
| 8 | 1.3847 | 999.85 | 1.0730 | 2481.9 |
| 9 | 1.3444 | 999.78 | 1.1483 | 2479.6 |
| 10 | 1.3059 | 999.70 | 1.2282 | 2477.2 |
| 11 | 1.2691 | 999.61 | 1.3130 | 2474.8 |
| 12 | 1.2340 | 999.50 | 1.4028 | 2472.5 |
| 13 | 1.2004 | 999.38 | 1.4981 | 2470.1 |
| 14 | 1.1683 | 999.25 | 1.5990 | 2467.7 |
| 15 | 1.1375 | 999.10 | 1.7058 | 2465.4 |
| 16 | 1.1081 | 998.95 | 1.8188 | 2463.0 |
| 17 | 1.0798 | 998.78 | 1.9384 | 2460.6 |
| 18 | 1.0527 | 998.60 | 2.0647 | 2458.3 |
| 19 | 1.0266 | 998.41 | 2.1983 | 2455.9 |
| 20 | 1.0016 | 998.21 | 2.3393 | 2453.5 |
| 21 | 0.97755 | 997.99 | 2.4882 | 2451.2 |
| 22 | 0.95442 | 997.77 | 2.6453 | 2448.8 |
| 23 | 0.93216 | 997.54 | 2.8111 | 2446.4 |
| 24 | 0.91073 | 997.30 | 2.9858 | 2444.0 |
| 25 | 0.89008 | 997.05 | 3.1699 | 2441.7 |
| 26 | 0.87018 | 996.79 | 3.3639 | 2439.3 |
| 27 | 0.85099 | 996.52 | 3.5681 | 2436.9 |
| 28 | 0.83248 | 996.24 | 3.7831 | 2434.6 |
| 29 | 0.81460 | 995.95 | 4.0092 | 2432.2 |
| 30 | 0.79735 | 995.65 | 4.2470 | 2429.8 |
| 31 | 0.78067 | 995.34 | 4.4969 | 2427.4 |
| 32 | 0.76456 | 995.03 | 4.7596 | 2425.1 |
| 33 | 0.74898 | 994.70 | 5.0354 | 2422.7 |
| 34 | 0.73390 | 994.37 | 5.3251 | 2420.3 |
| 35 | 0.71932 | 994.03 | 5.6290 | 2417.9 |
| 36 | 0.70519 | 993.68 | 5.9479 | 2415.5 |
| 37 | 0.69152 | 993.33 | 6.2823 | 2413.1 |
| 38 | 0.67827 | 992.97 | 6.6328 | 2410.8 |
| 39 | 0.66543 | 992.59 | 7.0002 | 2408.4 |
| 40 | 0.65298 | 992.22 | 7.3849 | 2406.0 |

*Source*: NIST Chemistry Webbook; http://webbook.nist.gov/chemistry.

**B10 Estimated Properties of Some Pollutants at 25 °C**

| Pollutant | Henry constant (–) | Diffusivity in Air ($10^{-5}\,m^2\,s^{-1}$) | Diffusivity in Water ($10^{-9}\,m^2\,s^{-1}$) | Vapor Pressure ($10^3$ Pa) |
|---|---|---|---|---|
| Benzene | 0.225 | 0.91 | 1.14 | 12.7 |
| Toluene | 0.255 | 0.82 | 1.03 | 3.80 |
| CO | 45 | 2.02 | 2.01 | — |
| $CO_2$ | 1.186 | 1.59 | 2.00 | 6,430 |
| $NH_3$ | 0.00068 | 2.23 | 2.46 | 1,000 |
| NO | 22 | 2.00 | 2.55 | — |
| $NO_2$ | 1–3.4 | 1.60 | 1.40 | 86.1 |
| $H_2S$ | 0.4 | 1.67 | 2.03 | 2,030 |
| $SO_2$ | 0.034 | 1.27 | 1.75 | 401 |

*Sources*: Poling et al. (2001); NIST Chemistry Webbook; http://webbook.nist.gov/chemistry.

## MATERIALS ONLINE

- "Appendix B5. Eulers method.xlsx": Demonstration of a numerical integration of a differential equation with Euler's method.
- Subfolder "Demo numerical integration", files "main.m", "data.m", and "f.m": Demonstration of a numerical integration of a set of differential equations with the Runge–Kutta–Fehlberg method.

## REFERENCES

Poling B.E., Prausnitz J.M., and O'Connell J.P. (2001). *The Properties of Gases and Liquids*. McGraw-Hill, Boston.

Press W.H., Flannery B.P., Teukolsky S.A., and Vetterling W.T. (1992). *Numerical Recipes in C. The Art of Scientific Computing*. 2nd ed. Cambridge University Press, Cambridge, UK.

APPENDIX C

# THEORY OF NEAR SURFACE TURBULENCE APPLIED TO WIND SPEED PROFILES, DRY DEPOSITION, AIR–WATER EXCHANGE, AND CANOPY EFFECTS

## C1 INTRODUCTION

Atmospheric turbulence near the surface determines wind speed profiles near the surface, and it also determines the rate of dry deposition of air pollutants and the rate of volatilization of pollutants from surface water to the air. The theories describing these processes are very related, and it is possible to develop a theory that describes all these processes. Such a theory would draw from concepts discussed in Chapters 5, 7, and 8 and does not fit in any of the chapters in the book. Nevertheless, such a theory would be very instructive, as it would enhance the understanding of all these interrelated processes. The purpose of this appendix is to develop such a theory. We will limit ourselves to the neutral atmosphere. The key assumption underlying the theory is that the turbulent momentum diffusivity (i.e., the turbulent kinematic viscosity) equals the turbulent mass diffusivity. This is probably a justified assumption in the case of a neutral atmosphere, and it is the one of the foundations of dry deposition theory. There are some indications that the turbulent mass diffusivity is actually greater than the turbulent momentum diffusivity (Wilson et al., 1981; Stull, 1988). This is an important issue for future clarification, but the possibility will not be explored here.

In this appendix, we will start from the "law of the wall," which describes the wind speed profile along an aerodynamically smooth surface, and derive some properties of the turbulence that give rise to this velocity profile. Next, we will apply this

*Air Dispersion Modeling: Foundations and Applications*, First Edition. Alex De Visscher.

information to the transfer of pollutants to a smooth wall and compare the result with standard dry deposition theory.

Next, we will consider the case of a water surface and assume that it is an aerodynamically smooth surface, but with enhanced turbulence due to wave formation. Next, we will derive a dry deposition model for a water surface and consider the repercussions for the process of volatilization. We will consider the water-side mass transfer by assuming that the water phase is a scaled mirror image of the atmosphere.

Next, we will move to aerodynamically rough surfaces. When the roughness elements are smaller than the viscous sublayer covering the surface, a simple adjustment of the theory for smooth surfaces is needed. This case will not be considered here. When the roughness elements are larger than the viscous sublayer, a canopy model is needed. A new canopy model will be developed in this appendix.

## C2 ANALYSIS OF THE LAW OF THE WALL FOR AERODYNAMICALLY SMOOTH SURFACES

von Kármán (1930) studied the velocity profile near an aerodynamically smooth wall and found the following relationship:

$$\frac{u}{u_*} = 5.75\log\left(\frac{zu_*}{\nu}\right) + C \quad \text{(C1)}$$

where $z$ (m) is distance above the surface, $u_*$ (m s$^{-1}$) is the friction velocity, and $\nu$ (m$^{-2}$ s$^{-1}$) is the kinematic viscosity. The coefficient $C$ was estimated at 5.5 by von Kármán (1930). Later studies found values ranging from 4 to 6 (Zanoun et al., 2003). This law of the wall translates into the following contemporary formulation:

$$u = \frac{u_*}{k}\ln\frac{z}{z_0} \quad \text{(5.64)}$$

where $z_0$ is the roughness length, given by

$$z_0 = \frac{\alpha_0 \nu}{u_*} \quad \text{(C2)}$$

where $\alpha_0$=0.111 when $C$=5.5, 0.202 when $C$=4, and 0.091 when $C$=6. For this appendix, we will assume a value of $\alpha_0$=0.111. This leads to $z_0$ values on the order of $10^{-5}$ m for aerodynamically smooth surfaces.

The law of the wall is not valid at distances closer than 30 $\nu/u_*$ (Garratt, 1994; Absi, 2009). Based on this information we will screen potential descriptions of turbulence near a smooth wall. We know that at a distance from the wall, where the law of the wall is valid, the turbulent diffusivity (mass or momentum) is as follows:

$$K_m = ku_*z \quad \text{(5.187)}$$

At these distances the kinematic viscosity is negligible in comparison with the turbulent viscosity. At the wall, where there is no turbulence, the momentum transfer is governed by the kinematic viscosity:

$$K_{\mathrm{m}} = \nu \tag{C3}$$

The simplest way to obtain a model that reduces to eqs. (5.187) and (C3) is the following:

$$K_{\mathrm{m}} = k u_* z + \nu \tag{C4}$$

To derive a wind speed profile from the momentum diffusivity, a turbulent version of Newton's viscosity law is used:

$$\frac{\tau_{xz}}{\rho} = K_{\mathrm{m}} \frac{\partial u}{\partial z} \tag{5.174}$$

If $\tau_{xz}$ is the shear stress experienced by the surface, no minus sign is needed in the equation. The shear stress $\tau$ can be assumed constant over short distances and related to the friction velocity:

$$u_* = \sqrt{\frac{\tau_{xz}}{\rho}} \tag{5.167}$$

This means that we are defining friction velocity in a way that includes viscous friction as well as turbulent friction. This definition will reduce to the regular definition of friction velocity as relating to turbulent friction only at larger distances from the wall.

Combining eqs. (5.174) and (5.167) leads to

$$\frac{\partial u}{\partial z} = \frac{u_*^2}{K_{\mathrm{m}}} \tag{C5}$$

Combining eqs. (C5) and (C4), the following equation is obtained:

$$\frac{\partial u}{\partial z} = \frac{u_*^2}{k u_* z + \nu} \tag{C6}$$

Solving this differential equation with boundary condition $u=0$ at $z=0$ (no-slip condition) leads to

$$u = \frac{u_*}{k} \ln\left(\frac{k u_* z}{\nu} + 1\right) \tag{C7}$$

At large distances from the wall, the term 1 in the logarithm is negligible. Hence, the equation reduces to

$$u = \frac{u_*}{k} \ln\left(\frac{k u_* z}{\nu}\right) \tag{C8}$$

which is an equation of the form of eqs. (5.64) and (C2), but with $\alpha_0 = 1/k = 2.5$. The model predicts a roughness length that is much too large. This is because the model seriously underestimates the thickness of the viscous sublayer.

We need a correlation for the thickness of the viscous sublayer. Because this is the layer too close to the wall to develop turbulence, we can expect that its thickness, $d$, is proportional to the Kolmogorov length microscale:

$$d = a\eta \tag{C9}$$

where $\eta$ is defined as

$$\eta = \nu^{3/4} \varepsilon^{-1/4} \tag{6.184}$$

where $\varepsilon$ ($m^2 s^{-3}$) is the energy dissipation rate. In neutral conditions, $\varepsilon$ can be estimated with the following equation:

$$\varepsilon = \frac{u_*^3}{kz} \tag{6.186}$$

Setting $z = d$, and combining eqs. (C9), (6.184), and (6.186), we obtain the following equation:

$$d = a\nu^{3/4} \frac{k^{1/4} d^{1/4}}{u_*^{3/4}} \tag{C10}$$

Solving for $d$ leads to

$$d = k^{1/3} a^{4/3} \frac{\nu}{u_*} = \alpha \frac{\nu}{u_*} \tag{C11}$$

where $\alpha$ is a coefficient that needs to be determined empirically.

For $z > d$, eq. (C4) is assumed to be valid in all model variants. Three different equations will be considered for $z < d$. The first one is eq. (C3). Substitution into eq. (C5) leads to

$$\frac{\partial u}{\partial z} = \frac{u_*^2}{\nu} \tag{C12a}$$

The solution to this equation with boundary condition $u=0$ for $z=0$ is

$$u(d) = \frac{u_*^2}{\nu} d \tag{C12b}$$

Solving eq. (C6) with eq. (C12b) as the new boundary condition leads to

$$u - u(d) = \frac{u_*}{k} \ln\left(\frac{ku_*z + \nu}{ku_*d + \nu}\right) \tag{C13}$$

Substituting eq. (C12b) and rearranging leads to

$$u = \frac{u_*^2}{\nu} d + \frac{u_*}{k} \ln\left(\frac{ku_*z + \nu}{ku_*d + \nu}\right) \tag{C14}$$

where $d$ is calculated with eq. (C11). At large distances from the wall, eq. (C14) reduces to eq. (5.64) with the following roughness length:

$$z_0 = d \exp\left(-\frac{u_*kd}{\nu}\right) \tag{C15}$$

To obtain the law of the wall with $C=5.5$, a value of $\alpha=11.6$ is needed in the calculation of $d$. For $C$ values of 4 and 6, $\alpha$ values of 9.7 and 12.3 are needed, respectively.

An $\alpha$ value of about 10 means that the law of the wall is approached at distances as close to the wall as $10\nu/u_*$. This is closer than actually observed. The law of the wall is generally observed starting at a distance of about $30\nu/u_*$ (Absi, 2009). Furthermore, as we will see in the next section, this model does not predict the correct particle size dependence of deposition rates. Hence, we need to modify the model.

The next model that was developed involves the following momentum diffusivity close to the wall ($z<d$):

$$K_{\mathrm{m}} = az^2 + \nu \tag{C16}$$

where $a$ is chosen to predict the same value of $K_{\mathrm{m}}$ as eq. (C4) at $z=d$. Hence, equating eq. (C4) to eq. (C16) at $z=d$ leads to the following equation between $z=0$ and $z=d$:

$$a = \frac{ku_*}{d} \tag{C17}$$

Substituting eq. (C16) into eq. (C5) leads to

$$\frac{\partial u}{\partial z} = \frac{u_*^2}{az^2 + \nu} \tag{C18}$$

Rearranging this equation leads to

$$du = \frac{u_*^2}{az^2 + \nu} dz \tag{C19}$$

This equation is integrated with the no-slip boundary condition. To that effect, the following equation is used from Appendix B:

$$\int \frac{dx}{ax^2 + bx + c} = \frac{2}{\sqrt{4ac - b^2}} \arctan\left( \frac{2ax + b}{\sqrt{4ac - b^2}} \right) + C \text{ if } b^2 < 4ac \tag{C20}$$

Hence:

$$u(d) = \frac{u_*^2}{\sqrt{a\nu}} \arctan\left( \sqrt{\frac{a}{\nu}} d \right) \tag{C21}$$

The $u_*^2$ dependence of $u(d)$ in eq. (C21) is deceptive because $a$ also depends on $u_*$. Hence, substituting eqs. (C17) and (C11) into eq. (C21) leads to

$$u(d) = u_* \sqrt{\frac{\alpha}{k}} \arctan\left( \sqrt{k\alpha} \right) \tag{C22}$$

The wind speed profile for $z > d$ is given by eq. (C13). Hence:

$$u = u_* \sqrt{\frac{\alpha}{k}} \arctan\left( \sqrt{k\alpha} \right) + \frac{u_*}{k} \ln\left( \frac{ku_* z + \nu}{ku_* d + \nu} \right) \tag{C23}$$

At a sufficiently large distance from the wall, this reduces to the law of the wall with $C = 5.5$ when $\alpha = 52.4$. For $C = 4$ and 6, $\alpha$ values of 40.8 and 56.4 are needed, respectively. These values are somewhat higher than the expected value of about 30, based on velocity profiles.

The last model that was developed for aerodynamically smooth surfaces is based on the following equation for $K_m$ at $z < d$:

$$K_m = az^3 + \nu \tag{C24}$$

The requirement of continuity of $K_m$ at $z = d$ leads to the following equation for $a$:

$$a = \frac{ku_*}{d^2} \tag{C25}$$

Substituting eq. (C24) into eq. (C5) leads to

$$\frac{\partial u}{\partial z} = \frac{u_*^2}{az^3 + \nu} \tag{C26}$$

Rearranging this equation leads to

$$du = \frac{u_*^2}{a}\frac{1}{z^3 + \nu / a}dz \tag{C27}$$

This equation can be solved with the no-slip boundary condition, based on the following integral:

$$\int_0^d \frac{dz}{z^3 + E} = \frac{1}{3E^{2/3}}\left[\ln\left(\frac{d+E^{1/3}}{E^{1/3}}\right) - \frac{1}{2}\ln\left|\frac{d^2 - E^{1/3}d + E^{2/3}}{E^{2/3}}\right| + \frac{3}{\sqrt{3}}\arctan\left(\frac{2d}{\sqrt{3}E^{1/3}}\right)\right] \tag{C28}$$

Hence, the velocity at $z=d$ is

$$u(d) = \frac{u_*^2}{3aE^{2/3}}\left[\ln\left(\frac{d+E^{1/3}}{E^{1/3}}\right) - \frac{1}{2}\ln\left|\frac{d^2 - E^{1/3}d + E^{2/3}}{E^{2/3}}\right| + \frac{3}{\sqrt{3}}\arctan\left(\frac{2d}{\sqrt{3}E^{1/3}}\right)\right] \tag{C29}$$

Substitution of $a$ and $E$ $(=\nu/a)$ leads to

$$u(d) = \frac{\alpha^{2/3}u_*}{3k^{1/3}}\left[\ln\left(\frac{d+E^{1/3}}{E^{1/3}}\right) - \frac{1}{2}\ln\left|\frac{d^2 - E^{1/3}d + E^{2/3}}{E^{2/3}}\right| + \frac{3}{\sqrt{3}}\arctan\left(\frac{2d}{\sqrt{3}E^{1/3}}\right)\right] \tag{C30}$$

where $E$ is given by

$$E = \frac{\alpha^2 \nu^3}{k u_*^3} \tag{C31}$$

Again, the wind speed at $z>d$ is given by eq. (C13). Hence, the following equation is obtained:

$$u = \frac{\alpha^{2/3}u_*}{3k^{1/3}}\left[\ln\left(\frac{d+E^{1/3}}{E^{1/3}}\right) - \frac{1}{2}\ln\left|\frac{d^2 - E^{1/3}d + E^{2/3}}{E^{2/3}}\right| + \frac{3}{\sqrt{3}}\arctan\left(\frac{2d}{\sqrt{3}E^{1/3}}\right)\right] + \frac{u_*}{k}\ln\left(\frac{ku_*z+\nu}{ku_*d+\nu}\right) \tag{C32}$$

At large distances from the wall, the wind speed profile reduces to the law of the wall, with $C=5.5$ when $\alpha=45.6$, $C=4$ when $\alpha=37.4$, and $C=6$ when $\alpha=48.5$. This range of values is still somewhat above the expected 30, but this model is the one most consistent with our understanding of dry deposition, as will be demonstrated in the next section.

## C3 DRY DEPOSITION TO AERODYNAMICALLY SMOOTH SURFACES

In the previous section we have covered four different models describing the wind speed profile at an aerodynamically smooth surface, three of which are consistent with the law of the wall. In this section we will derive equations for the dry deposition rate, and compare them with what is generally known about these rates.

In Section 7.5 our current understanding of dry deposition is discussed. Dry deposition is governed by three resistances in series, the aerodynamic resistance $r_a$, the quasi-laminar layer resistance $r_b$, and the surface resistance $r_c$. The first two of these resistances are governed by atmospheric turbulence. For that reason, we will not consider the surface resistance here, or assume that it is zero. An exception is a water surface, where the surface resistance can be related to the condition of the atmosphere.

We will consider the dry deposition of both gases and particulate matter (aerosols). For the latter, the quasi-laminar resistance is usually combined with the surface resistance and incorporates transport by Brownian motion, impaction, and interception. We will only consider Brownian motion here, which is the dominant removal mechanism for particulate matter with a diameter of less than 0.1 μm.

The aerodynamic resistance and the quasilaminar resistance are somewhat overlapping concepts because the former is usually taken as the resistance associated with the atmosphere above a height $z_0$. Based on the models introduced in the previous section, it is clear that a height of $z=d$ would have been a better choice, except for the fact that this is usually an unknown quantity. For that reason, we will simply consider $r_a+r_b$, and compare it with predictions from the models developed here.

The deposition velocity is defined through a relationship between the deposition flux and the pollutant concentration at a reference height:

$$F = -v_d c \tag{7.129}$$

The deposition velocity is inversely related to the resistance to deposition:

$$v_d = \frac{1}{r_d} \tag{C33}$$

It will be assumed that the surface resistance and the settling velocity (in the case of particles) are negligible. Hence, the resistance to deposition is the sum of the aerodynamic resistance and the quasi-laminar layer resistance:

$$r_d = r_a + r_b \tag{C34}$$

As discussed in Chapter 7, the following are commonly used equations for $r_a$ and $r_b$:

$$r_a = \frac{1}{ku_*} \ln\left(\frac{z}{z_0}\right) \tag{7.146}$$

$$r_b = \frac{\alpha_b \mathrm{Sc}^{2/3}}{ku_*} \tag{7.158}$$

where estimates for $\alpha_b$ range from 0.133 to 10, with indications that higher values of $\alpha_b$ are associated with smooth surfaces and low values with rough surfaces (Shepherd, 1974; Slinn et al., 1978; Slinn and Slinn, 1980; Zhang et al., 2001). For aerodynamically smooth surfaces, values in the 2–10 range can be expected. Hence, for a friction velocity of 0.1 m s$^{-1}$ and a reference height of 1 m, values of $r_a$+$r_b$ in the 300–500 s m$^{-1}$ range can be expected for gases with a Sc number of about 1, and values of $r_a + r_b$ in the 5000–25,000 s m$^{-1}$ range can be expected for particulate matter with a Sc = 1000.

On the other hand, the vertical flux of a pollutant is related to its concentration gradient as follows:

$$-F = K_z \frac{dc}{dz} \tag{7.141}$$

After integration one obtains

$$c - c_0 = -F \int_0^{z_{ref}} \frac{dz}{K_z} \tag{C35}$$

where $c_0$ is the pollutant concentration at $z$=0. Because the surface resistance is assumed to be negligible, we can assume $c_0$=0. Equation (C35) reduces to

$$c = -F \int_0^{z_{ref}} \frac{dz}{K_z} \tag{C36}$$

Combining eqs. (C33) and (7.128) leads to

$$r_d = -\frac{c}{F} \tag{C37}$$

Substituting eq. (C36) leads to

$$r_d = \int_0^{z_{ref}} \frac{dz}{K_z} \tag{C38}$$

The values of $r_d$ obtained with the different models will be compared with predictions of conventional deposition theory.

We will assume that $K_z = K_m$ for the mechanical (turbulent) part of the transfer, whereas the kinematic viscosity is replaced by the molecular diffusivity, $D$ ($m^2 s^{-1}$). Hence, eqs. (C3), (C4), (C16), and (C24) can be replaced by

$$K_z = D \tag{C39}$$

$$K_z = ku_* z + D \tag{C40}$$

$$K_z = az^2 + D \tag{C41}$$

$$K_z = az^3 + D \tag{C42}$$

We can calculate deposition resistances by integration of the reciprocal of the $K_z$ equations. For instance, the first model considered in Section C2 corresponds with eq. (C40). Hence:

$$r_d = \int_0^{z_{ref}} \frac{dz}{ku_* z + D} \tag{C43}$$

Integration leads to

$$r_d = \frac{1}{ku_*} \ln\left(\frac{ku_* z_{ref}}{D} + 1\right) \tag{C44}$$

For a friction velocity of 0.1 $m s^{-1}$ and a reference height of 1 m, a value of $r_d = 197 s m^{-1}$ is obtained for Sc = 1 and a value of 370 $s m^{-1}$ for Sc = 1000. These values are too low for an aerodynamically smooth surface. This confirms the finding of the previous section that eq. (C40) [and hence eq. (C4)] does not describe the viscous sublayer.

The second model consists of eq. (C39) for $z<d$ and eq. (C40) for $z>d$. Substituting in eq. (C38) leads to

$$r_d = \int_0^d \frac{dz}{D} + \int_d^{z_{ref}} \frac{dz}{ku_* z + D} \tag{C45}$$

Solving the integrals leads to

$$r_d = \frac{d}{D} + \frac{1}{ku_*} \ln\left(\frac{ku_* z_{ref} + D}{ku_* d + D}\right) \tag{C46}$$

Assuming $u_* = 0.1 m s^{-1}$, $z_{ref} = 1$ m, and $\nu = 1.5\times10^5 m^2 s^{-1}$, a value of $r_d = 270 s m^{-1}$ is found for a gas with Sc = 1, and a value of $r_d = 116{,}000 s m^{-1}$ is found for particles with Sc = 1000. While the gas value is realistic, the particle value is too large to be realistic. In this model $r_d$ is far too dependent on Sc. We conclude that this model does not provide a realistic description of the turbulence near a wall. Note that the value for Sc = 1 should exactly match $r_a$ for similarity reasons. Any deviation is due to rounding errors of $\alpha$.

The third model consists of eq. (C41) for $z<d$ and eq. (C40) for $z>d$. Substituting in eq. (C38) leads to

$$r_{\mathrm{d}} = \int_0^d \frac{dz}{az^2 + D} + \int_d^{z_{\mathrm{ref}}} \frac{dz}{ku_* z + D} \tag{C47}$$

Solving the integrals leads to

$$r_{\mathrm{d}} = \frac{1}{\sqrt{aD}} \arctan\left(\sqrt{\frac{a}{D}} d\right) + \frac{1}{ku_*} \ln\left(\frac{ku_* z_{\mathrm{ref}} + D}{ku_* d + D}\right) \tag{C48}$$

Assuming $u_* = 0.1\,\mathrm{m\,s^{-1}}$, $z_{\mathrm{ref}} = 1\,\mathrm{m}$, and $\nu = 1.510^5\,\mathrm{m^2\,s^{-1}}$, a value of $r_{\mathrm{d}} = 275\,\mathrm{s\,m^{-1}}$ is found for a gas with Sc = 1, and a value of $r_{\mathrm{d}} = 5780\,\mathrm{s\,m^{-1}}$ is found for particles with Sc = 1000. These are realistic values, although the dependence on Sc, which is approximately a square root relationship at large Sc values, is less strong than expected, and the value of $r_{\mathrm{d}}$ is at the lower end of the expected range.

The third model consists of eq. (C42) for $z<d$ and eq. (C40) for $z>d$. Substituting in eq. (C38) leads to

$$r_{\mathrm{d}} = \int_0^d \frac{dz}{az^3 + D} + \int_d^{z_{\mathrm{ref}}} \frac{dz}{ku_* z + D} \tag{C49}$$

Solving the integrals leads to

$$r_{\mathrm{d}} = \frac{1}{3aE_D^{2/3}} \left[ \ln\left(\frac{d + E_D^{1/3}}{E_D^{1/3}}\right) - \frac{1}{2} \ln\left|\frac{d^2 - E_D^{1/3} d + E_D^{2/3}}{E_D^{2/3}}\right| + \frac{3}{\sqrt{3}} \arctan\left(\frac{2d}{\sqrt{3} E_D^{1/3}}\right) \right] + \frac{1}{ku_*} \ln\left(\frac{ku_* z_{\mathrm{ref}} + D}{ku_* d + D}\right) \tag{C50}$$

where $E_D = D/a$. Assuming $u_* = 0.1\,\mathrm{m\,s^{-1}}$, $z_{\mathrm{ref}} = 1\,\mathrm{m}$, and $\nu = 1.5 \times 10^{-5}\,\mathrm{m^2\,s^{-1}}$, a value of $r_{\mathrm{d}} = 275\,\mathrm{s\,m^{-1}}$ is found for a gas with Sc = 1, and a value of $r_{\mathrm{d}} = 15{,}800\,\mathrm{s\,m^{-1}}$ is found for particles with Sc = 1000. These are realistic values.

Of the models discussed so far, we can discard the first two and continue with the last two.

## C4 ROUGHNESS OF A SMOOTH WATER SURFACE

As discussed in Chapter 5, the surface roughness of a water surface can be described by the equation of Charnock (1955):

$$z_0 = \frac{\alpha_{\mathrm{c}} u_*^2}{g} \tag{5.66}$$

where $\alpha_c$=0.012 (Peña and Gryning, 2008). This coefficient was determined based on data at open sea, for friction velocities exceeding 0.15 m s$^{-1}$. Under these conditions, the water surface cannot be considered aerodynamically smooth, and this case will not be considered. However, wind tunnel experiments have been conducted with water surfaces that show roughness lengths lower than the value for a smooth surface, at least at low wind speeds. One such study is by Mller and Schumann (1970), who found a surface roughness ranging from $9\times10^{-9}$ m at $u_*$=0.1 m s$^{-1}$ to $1.8\times10^{-5}$ m at $u_*$=0.4 m s$^{-1}$.

It is remarkable that the trend is increasing, whereas on an immobile, aerodynamically smooth surface the surface roughness decreases with increasing friction velocity. In this section, we will only consider low wind speeds, where we will assume that the water surface is still aerodynamically smooth but affects the thickness of the viscous sublayer. As gravity has an effect on wave formation, it is assumed that the acceleration due to gravity, $g$, is an additional variable that affects the value of $d$. With four variables ($d$, $u_*$, $\nu$, and $g$) we can create two dimensionless numbers, $\pi_1$ and $\pi_2$. The following were chosen:

$$\pi_1 = \frac{du_*}{\nu} \tag{C51}$$

$$\pi_2 = \frac{u_*^2}{dg} \tag{C52}$$

where $\pi_1$ has the form of the Reynolds number, based on the friction velocity and the thickness of the viscous sublayer. We assume the following relationship between $\pi_1$ and $\pi_2$:

$$\pi_1 = \alpha' \pi_2^n \tag{C53}$$

where $\alpha'$ is a constant, to be determined empirically. Substituting eqs. (C51) and (C52) into eq. (C53) and solving for $d$ leads to

$$d = \alpha'^{1/(n+1)} \frac{u_*^{(2n-1)/(n+1)} \nu^{1/(n+1)}}{g^{n/(n+1)}} = \alpha_w \frac{u_*^{(2n-1)/(n+1)} \nu^{1/(n+1)}}{g^{n/(n+1)}} \tag{C54}$$

Some interesting features are found when $n=\frac{1}{7}$ is chosen. Equation (C54) then becomes

$$d = \alpha_w \frac{\nu^{7/6} g^{1/6}}{u_*^{3/2}} \tag{C55}$$

Based on this equation, it is not possible to obtain a surface roughness that increases with increasing wind speed when eq. (C16) is used to model the viscous sublayer. However, when eq. (C24) is used to model the viscous sublayer, then the roughness does increase with wind speed. With eq. (C16) a stronger inverse dependence on $u_*$ is needed to achieve an increasing value of $z_0$ with increasing $u_*$.

Surface roughness data of Möller and Schumann (1970) were analyzed with eqs. (C23) and (C32). The result was an almost perfect power law relationship in the entire friction velocity range in both cases ($r^2$>0.998):

$$d = \alpha_0 u_*^{n_0} \tag{C56}$$

In the case of eq. (C23) the coefficients are $\alpha_0$=1.214×10$^{-4}$, and $n_0$=2.332; in the case of eq. (C32) the coefficients are $\alpha_0$=3.143×10$^{-5}$ and $n_0$=2.969. The power law indices are so close to $\frac{7}{3}$ and 3 that we will use these values instead. In that case the optimal values of $\alpha_0$ are 1.213×10$^{-4}$ and 3.003×10$^{-5}$, respectively.

Comparing the power law of eq. (C56) and its coefficients with eq. (C54), the coefficients for eq. (C54) can be obtained. For the wind speed profile of eq. (C23), the following equation is obtained:

$$d = \alpha_w \frac{\nu^{5/3} g^{2/3}}{u_*^3} \tag{C57}$$

where $\alpha_w$=718.4. For the wind speed profile of eq. (C32), the following equation is obtained:

$$d = \alpha_w \frac{\nu^{13/9} g^{4/9}}{u_*^{7/3}} \tag{C58}$$

where $\alpha_w$=408.4.

## C5 AIR-SIDE MASS TRANSFER TO A SMOOTH WATER SURFACE

Mass transfer coefficients, or dry deposition resistances, can be calculated for a water surface in the same way as for a smooth solid surface, but with the equations obtained in the previous section for the calculation of $d$. These values are compared with literature values.

Möller and Schumann (1970) also conducted dry deposition measurements at $u_*$=0.4 m s$^{-1}$ with gases and particles with a wide range of Sc numbers and found $r_d$ values ranging from 25 to 30 s m$^{-1}$ for gases with Sc values near 1, to 5000–10,000 s m$^{-1}$ for particles with Sc values above 10,000. Equation (C48), with $d$ values calculated with eq. (C57), leads to $r_d$ values ranging from 47.0 s m$^{-1}$ at Sc=1 to 2210 s m$^{-1}$ at Sc=10,000. Equation (C50), when $d$ values are calculated with eq. (C58), leads to $r_d$ values ranging from 53.9 s m$^{-1}$ at Sc=1 to 12,500 s m$^{-1}$ at Sc=10,000. At low Sc values both model variants overestimate the measured resistances, whereas at high Sc numbers the experimental data is between the predictions of the two models, but somewhat closer to the third-order model. These data provide experimental verification that both model variants provide reasonable predictions.

Similar wind tunnel experiments were conducted by Liss (1973). In this study values of $k_L$ and $k_G$ from the two-film model (Section 7.5.1.3) were measured at different wind speeds. Liss found an average $z_0$ value of $5\times10^{-5}$ m, substantially larger than the $z_0$ values of Möller and Schumann (1970). Furthermore, Liss only vaguely observed an increasing trend of $z_0$ with increasing wind speed. For that reason, adjusted values of $\alpha_w$ and $n$ are explored. Based on the values, predictions are made of $k_G$, which is equivalent to $(r_a+r_b)^{-1}$. Good agreement between the $k_G$ values observed by Liss and the values predicted by eq. (C50) are obtained when values of $\alpha_w=60$, and $n=\frac{1}{4}$ are used. With these values, an average $z_0$ value of about $7\times10^{-5}$ m is obtained. With increasing wind speed, $z_0$ passes through a maximum at intermediate wind speed, explaining the lack of clear wind speed dependence of the measured $z_0$ values. The equation for $d$ is as follows:

$$d = \alpha_w \frac{\nu^{4/3} g^{1/3}}{u_*^2} \tag{C59}$$

## C6 WATER-SIDE MASS TRANSFER TO A SMOOTH WATER SURFACE

In the preceding sections we have derived equations describing the mass transfer of pollutants from a bulk air phase to a solid or liquid surface. In this section these equations will be modified to describe the transfer of pollutants from a water surface to the water bulk.

The literature indicates that the resistance to mass transfer from a water surface to the water bulk is proportional to $\mathrm{Sc}^{1/2}$ (Schwarzenbach et al., 1993). In the previous section it was shown that $r_b$ is proportional to $\mathrm{Sc}^{1/2}$ when the turbulent diffusivity in the quasi-laminar layer is proportional to $z^2$, where $z$ is the height above the surface. For that reason, we will assume that the same assumption applies to the quasi-laminar layer on the liquid side; $z$ will now denote depth below the water surface.

It is assumed that the shear stress on the water surface is transferred to the water phase. Hence:

$$\tau_{0,\mathrm{air}} = \rho_{\mathrm{air}} u_{*,\mathrm{air}}^2 = \tau_{0,\mathrm{water}} = \rho_{\mathrm{water}} u_{*,\mathrm{water}}^2 \tag{C60}$$

Hence:

$$u_{*,\mathrm{water}} = \sqrt{\frac{\rho_{\mathrm{air}}}{\rho_{\mathrm{water}}}} u_{*,\mathrm{air}} = \delta \cdot u_{*,\mathrm{air}} \tag{C61}$$

where $\delta$ is defined as $\delta=(\rho_{\mathrm{air}}/\rho_{\mathrm{water}})^{1/2}$. Hence eq. (C48) can be rewritten to calculate $r_d$ for the liquid phase, and hence $k_L$:

$$r_d = \frac{1}{\sqrt{a_w D_w}} \arctan\left(\sqrt{\frac{a_w}{D_w}} d_w\right) + \frac{1}{k u_{*w}} \ln\left(\frac{k u_{*w} z_{\mathrm{ref,w}} + D_w}{k u_{*w} d_w + D_w}\right) = k_L^{-1} \tag{C62}$$

where subscript w refers to the water phase; and $a_w$ is given by an equation equivalent to eq. (C17):

$$a_w = \frac{ku_{*w}}{d_w} \tag{C63}$$

and $d_w$ is given by an equation equivalent to eq. (C54):

$$d_w = \alpha_w \frac{u_{*w}^{\frac{(2n-1)}{(n+1)}} \nu_w^{1/(n+1)}}{g^{n/(n+1)}} \tag{C64}$$

Remarkably, using eqs. (C62)–(C64) with the same parameters found for the air phase (i.e., $\alpha_w=60$; $n=\frac{1}{4}$) show good agreement with the values of $k_L$ determined experimentally by Liss (1973), indicating that the equations derived here have indeed a basis in physical reality. The following equation for $d_w$ is thus obtained:

$$d_w = \alpha_w \frac{\nu_w^{4/3} g^{1/3}}{u_{*w}^2} \tag{C65}$$

In eq. (C62), the first term usually dominates the resistance to mass transfer. Furthermore, the arctangent in the equation is usually very close to $\pi/2$. Hence, the equation can be simplified to the following, with only slight loss of accuracy:

$$r_d \approx \frac{\pi}{2\sqrt{a_w D_w}} = \frac{\pi}{2}\sqrt{\frac{d_w}{ku_{*w} D_w}} = \frac{\pi}{2}\sqrt{\frac{\alpha_w \nu_w^{4/3} g^{1/3}}{ku_{*w}^3 D_w}} = k_L^{-1} \tag{C66}$$

We find that the liquid-side mass transfer coefficient is proportional to $u_{*w}^{3/2}$ and to $D_w^{1/2}$.

## C7 WIND SPEEDS IN AN URBAN CANOPY

Several investigators developed analytical models to describe the wind speed profiles between the obstacles in a canopy. Most applications are based on exponential wind speed profiles, which result from the models of Inoue (1963) and Cionco (1965, 1978). A useful parameterization of this model was developed by Macdonald (2000), who correlated the attenuation coefficient of the model to the frontal area parameter $\lambda_f$.

In this section a new canopy model will be developed in order to provide a better understanding of the concepts of roughness length, displacement height, and wind speed attenuation coefficient. The parameterization of Macdonald (2000), and the data it is based on, will be a useful testing ground for the new canopy model. The model is related to a recent model of Wang (2012), with a number of important modifications.

As in the model of Wang (2012) it will be assumed that the momentum of the wind is absorbed partly on the canopy elements and partly on the surface. A force

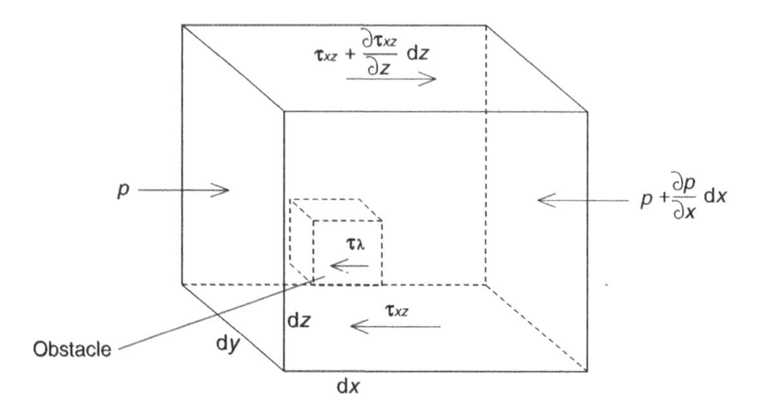


Figure C1 Force balance for in-canopy wind speed calculations.

balance is made based on Figure C1. In the figure, there are two pressures (one left, one right) and three shear stresses (one at the bottom, one at the top, and one on vertical surfaces) that must be balanced. The shear stress on vertical surfaces must be multiplied by the effective vertical surface area, where "effective" expresses the fact that the area felt by the airstream is larger than the solid surface area because building wakes also remove momentum from the airstream. Hence, the effective surface area projected on a vertical plane parallel to the wind speed, per unit plot area, $\lambda$, is expected to be proportional to the frontal surface area parameter, $\lambda_f$, but larger. The force balance is

$$p\,dy\,dz - \left(p + \frac{\partial p}{\partial x}dx\right)dy\,dz - \tau_{xz}\,dx\,dy + \left(\tau_{xz} + \frac{\partial \tau_{xz}}{\partial z}dz\right)dx\,dy - \tau_{\lambda}\frac{\lambda dz}{H_r}dx\,dy = 0 \quad \text{(C67)}$$

where $\tau$ is the shear stress on vertical surfaces, and $H_r$ is the average obstacle height. Dividing by $dx\,dy\,dz$, and removing canceling terms, leads to

$$-\frac{\partial p}{\partial x} + \frac{\partial \tau_{xz}}{\partial z} - \tau_{\lambda}\frac{\lambda}{H_r} = 0 \quad \text{(C68)}$$

Above the obstacles, there is no vertical area to remove momentum from the wind flow, and the equation reduces to

$$-\frac{\partial p}{\partial x} + \frac{\partial \tau_{xz}}{\partial z} = 0 \quad \text{(C69)}$$

We know from the mechanical considerations in Section 5.11.5 that the shear stress $\tau_{xz}$ is approximately constant in the surface layer, above the roughness elements. It follows that the along-wind pressure gradient can be considered negligible for the purpose of this model. Hence, between the roughness elements, eq. (C68) can be approximated as

$$\frac{\partial \tau_{xz}}{\partial z} - \tau_{\lambda} \frac{\lambda}{H_{\mathrm{r}}} = 0 \tag{C70}$$

Assuming that the shear stress $\tau_{\lambda}$ is proportional to the shear stress $\tau_{xz}$, and, as indicated above, the area parameter $\lambda$ is proportional to the area parameter $\lambda_{\mathrm{f}}$, the following expression applies:

$$\tau_{\lambda}\lambda = \gamma \tau_{xz} \lambda_{\mathrm{f}} \tag{C71}$$

where $\gamma$ is an as yet unknown constant. Substitution into eq. (C70) leads to

$$\frac{\partial \tau_{xz}}{\partial z} - \tau_{xz} \frac{\gamma \lambda_{\mathrm{f}}}{H_{\mathrm{r}}} = 0 \tag{C72}$$

The equation is solved with the following boundary condition at $z=H_{\mathrm{r}}$:

$$\tau_{xz} = \rho u_*^2 \tag{C73}$$

The solution is

$$\tau_{xz} = \rho u_*^2 \exp\left[ -\gamma \lambda_{\mathrm{f}} \left( 1 - \frac{z}{H_{\mathrm{r}}} \right) \right] \tag{C74}$$

We define a local friction velocity $u_{\tau}$ as

$$\tau_{xz} = \rho u_{\tau}^2 \tag{C75}$$

Hence the profile of the local friction velocity is

$$u_{\tau} = u_* \exp\left[ -\frac{\gamma \lambda_{\mathrm{f}}}{2} \left( 1 - \frac{z}{H_{\mathrm{r}}} \right) \right] \tag{C76}$$

Above the roughness defining obstacles, the momentum diffusivity can be described as

$$K_{\mathrm{m}} = k u_* \left( z - d \right) + \nu \tag{C77}$$

where $\nu$ is the kinematic viscosity of the air. It will be assumed here that this equation is valid only when $u_*$ is constant. This is the case if the derivative of eq. (C77) is valid. Hence:

$$\frac{dK_{\mathrm{m}}}{dz} = k u_* \tag{C78}$$

In the canopy, the friction velocity in eq. (C78) should be replaced by the local friction velocity:

$$\frac{dK_{\mathrm{m}}}{dz} = ku_{\mathrm{r}} = ku_{*}\exp\left[-\frac{\gamma\lambda_{\mathrm{f}}}{2}\left(1-\frac{z}{H_{\mathrm{r}}}\right)\right] \tag{C79}$$

Equation (C79) is integrated with the following initial condition at $z=0$:

$$K_{\mathrm{m}} = \nu \tag{C80}$$

The solution is

$$K_{\mathrm{m}} = ku_{*}\frac{2H_{\mathrm{r}}}{\gamma\lambda_{\mathrm{f}}}\exp\left(-\frac{\gamma\lambda_{\mathrm{f}}}{2}\right)\left[\exp\left(\frac{\gamma\lambda_{\mathrm{f}}}{2}\frac{z}{H_{\mathrm{r}}}\right)-1\right]+\nu \tag{C81}$$

The advantage of this calculation scheme is that the displacement height $d$ springs naturally from eq. (C81), as it must yield the same result as eq. (C77) at a height $z=H_{\mathrm{r}}$. Hence:

$$ku_{*}\frac{2H_{\mathrm{r}}}{\gamma\lambda_{\mathrm{f}}}\exp\left(-\frac{\gamma\lambda_{\mathrm{f}}}{2}\right)\left[\exp\left(\frac{\gamma\lambda_{\mathrm{f}}}{2}\right)-1\right]+\nu = ku_{*}(H_{\mathrm{r}}-d)+\nu \tag{C82}$$

Solving for $d/H_{\mathrm{r}}$ leads to

$$\frac{d}{H_{\mathrm{r}}} = 1-\frac{2}{\gamma\lambda_{\mathrm{f}}}\left[1-\exp\left(-\frac{\gamma\lambda_{\mathrm{f}}}{2}\right)\right] \tag{C83}$$

In Figure C2 the results of eq. (C83) with $\gamma=20$ is compared with data of Macdonald (2000) and the average relationship recommended by Grimmond and Oke (1999) and

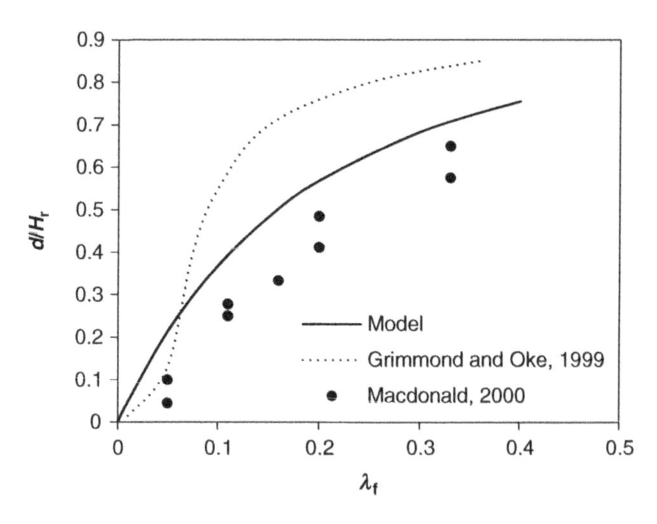


Figure C2 Displacement height versus frontal area parameter. Model predictions ($\gamma=20$) compared with recommendations of Grimmond and Oke (1999) and data of Macdonald (2000).

Hanna and Britter (2002). The data compiled by Grimmond and Oke (1999) support values of $\gamma$ ranging from 12 to 50, where the data of Macdonald (2000) is at the lower limit. A value of 20 provides a reasonable average.

The wind speed profile is calculated with a turbulent version of Newton's viscosity law:

$$K_{\mathrm{m}} \frac{\partial u}{\partial z} = \frac{\tau_{xz}}{\rho} = u_{\tau}^{2} \tag{C84}$$

Substitution of eqs. (C76) and (C81) into eq. (C84) leads to

$$\left\{ k u_{*} \frac{2H_{\mathrm{r}}}{\gamma\lambda_{\mathrm{f}}} \exp\left(-\frac{\gamma\lambda_{\mathrm{f}}}{2}\right)\left[\exp\left(\frac{\gamma\lambda_{\mathrm{f}}}{2}\frac{z}{H_{\mathrm{r}}}\right)-1\right]+\nu \right\} \frac{\partial u}{\partial z} = u_{*}^{2} \exp\left[-\gamma\lambda_{\mathrm{f}}\left(1-\frac{z}{H_{\mathrm{r}}}\right)\right] \tag{C85}$$

Equation (C85) can be rewritten as

$$\frac{\partial u}{\partial z} = \frac{u_{*}^{2} \exp(-\gamma\lambda_{\mathrm{f}}) \exp\left(\gamma\lambda_{\mathrm{f}} \frac{z}{H_{\mathrm{r}}}\right)}{k u_{*} \frac{2H_{\mathrm{r}}}{\gamma\lambda_{\mathrm{f}}} \exp\left(-\frac{\gamma\lambda_{\mathrm{f}}}{2}\right)\left(\exp\left(\frac{\gamma\lambda_{\mathrm{f}}}{2}\frac{z}{H_{\mathrm{r}}}\right)-1\right)+\nu} \tag{C86}$$

or as

$$\frac{\partial u}{\partial z} = \frac{u_{*} \exp\left(-\frac{\gamma\lambda_{\mathrm{f}}}{2}\right)}{k \frac{2H_{\mathrm{r}}}{\gamma\lambda_{\mathrm{f}}}} \cdot \frac{\exp\left(\gamma\lambda_{\mathrm{f}} \frac{z}{H_{\mathrm{r}}}\right)}{\exp\left(\frac{\gamma\lambda_{\mathrm{f}}}{2}\frac{z}{H_{\mathrm{r}}}\right)-(1-\delta)} \tag{C87}$$

where $\delta$ is given by

$$\delta = \frac{\nu\gamma\lambda_{\mathrm{f}} \exp(\gamma\lambda_{\mathrm{f}}/2)}{2H_{\mathrm{r}} k u_{*}} \tag{C88}$$

Integration with initial condition $u=0$ at $z=0$ leads to

$$u = \frac{u_{*} \exp\left(-\frac{\gamma\lambda_{\mathrm{f}}}{2}\right)}{k \frac{2H_{\mathrm{r}}}{\gamma\lambda_{\mathrm{f}}}} \cdot \int_{0}^{z} \frac{\exp\left(\gamma\lambda_{\mathrm{f}} \frac{z}{H_{\mathrm{r}}}\right)}{\exp\left(\frac{\gamma\lambda_{\mathrm{f}}}{2}\frac{z}{H_{\mathrm{r}}}\right)-(1-\delta)} dz \tag{C89}$$

In choosing the initial condition like this, it is assumed that eq. (C81) is valid in the quasi-laminar layer. Equation (C89) can be written as

$$u=\frac{u_* \exp\left(-\frac{\gamma\lambda_f}{2}\right)}{k\frac{2H_r}{\gamma\lambda_f}}\cdot\frac{2H_r}{\gamma\lambda_f}\int_0^z \frac{\exp\left(\frac{\gamma\lambda_f}{2}\frac{z}{H_r}\right)}{\exp\left(\frac{\gamma\lambda_f}{2}\frac{z}{H_r}\right)-(1-\delta)}d\exp\left(\frac{\gamma\lambda_f}{2}\frac{z}{H_r}\right) \tag{C90}$$

Solving the integral leads to

$$u=\frac{u_*}{k}\exp\left(-\frac{\gamma\lambda_f}{2}\right)\left(\exp\left(\frac{\gamma\lambda_f}{2}\frac{z}{H_r}\right)-1+(1-\delta)\ln\left\{\frac{\exp[(\gamma\lambda_f/2)(z/H_r)]-(1-\delta)}{\delta}\right\}\right) \tag{C91}$$

This equation can be used for heights up to $z=H_r$. The wind speed at $z=H_r$ is

$$u_{H_r}=\frac{u_*}{k}\exp\left(-\frac{\gamma\lambda_f}{2}\right)\left\{\exp\left(\frac{\gamma\lambda_f}{2}\right)-1+(1-\delta)\ln\left[\frac{\exp(\gamma\lambda_f/2)-(1-\delta)}{\delta}\right]\right\} \tag{C92}$$

For wind speeds at $z>H_r$, the logarithmic equation is used:

$$u=\frac{u_*}{k}\ln\left(\frac{z-d}{z_0}\right) \tag{C93}$$

where $d$ is based on eq. (C83), and $z_0$ is based on the requirement that eq. (C93) predicts the same wind speed as eq. (C92) at $z=H_r$:

$$\frac{z_0}{H_r}=\left(1-\frac{d}{H_r}\right)\exp\left(-\frac{ku_{H_r}}{u_*}\right) \tag{C94}$$

To test the model results, wind speeds at $z/H_r=0.1$–1 at intervals of 0.1 are calculated with the model, and the model of Inoue (1963) is fitted to the wind speeds. The model of Inoue (1963) is as follows:

$$u=u_{H_r}\exp\left[-a\left(1-\frac{z}{H_r}\right)\right] \tag{8.17}$$

Values of $a$ obtained this way are compared with values obtained experimentally by Macdonald (2000). In the calculations, a height $H_r=0.1$ m was chosen and a friction velocity of 0.24 m s$^{-1}$, which is approximately the average value in Macdonald's experiments. The result is shown in Figure C3. With a value of $\gamma=20$, the agreement is very good, which justifies our earlier choice of this value. The model predicts that

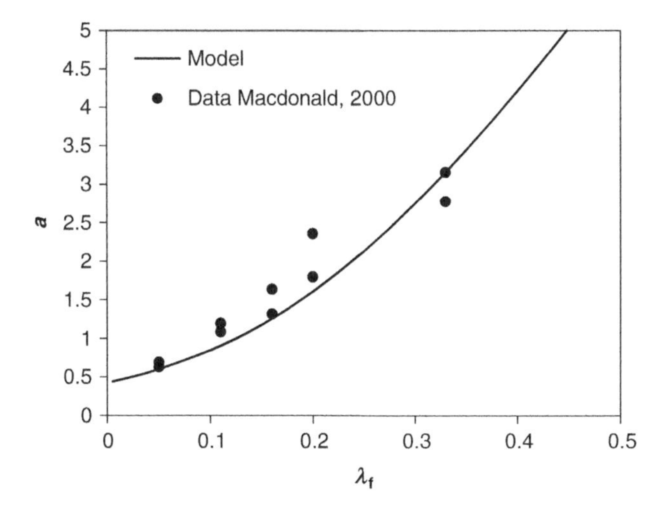


Figure C3 Canopy wind speed attenuation factor measured by Macdonald (2000) and obtained with the new model ($\gamma=20$).

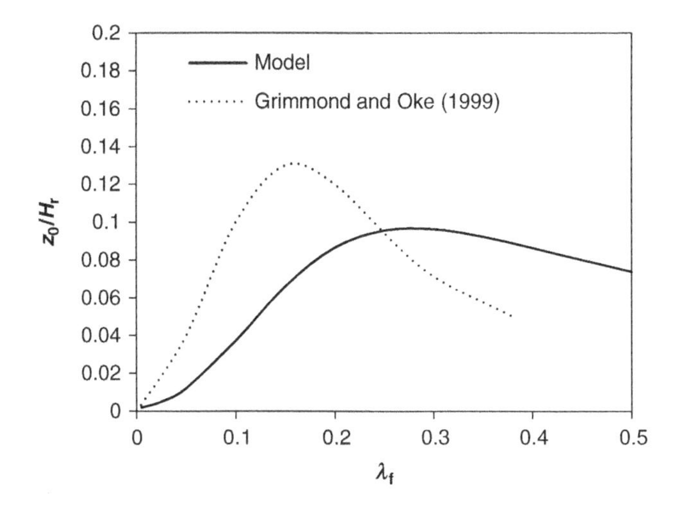


Figure C4 Relative roughness length versus frontal area parameter, obtained with the new model ($\gamma=20$) and recommended by Grimmond and Oke (2000).

$a$ values for obstacle heights of 10 m are about 40% lower than for obstacle heights of 0.1 m. This may indicate that wind tunnel studies are not entirely representative for real cities.

A second way to test the model is by comparing $z_0/H_r$ values with the recommended trend of Grimmond and Oke (1999). This is shown in Figure C4. The agreement is good qualitatively, as both relationships predict that $z_0$ goes through a maximum. Considering the large scatter of the data on which the recommended trend of Grimmond and Oke (1999) is based, which is about as large as the trend itself, this is a good agreement. It is concluded that the model presented here provides a reasonable picture of the actual urban canopy.

## REFERENCES

Absi R. (2009). A simple eddy viscosity formulation for turbulent boundary layers near smooth walls. *C. R. Mecanique* **337**, 158–165.

Charnock H. (1955). Wind stress on a water surface. *Quart. J. Roy. Meteorol. Soc.* **81**, 639–640.

Cionco R.M. (1965). A mathematical model for air flow in a vegetative canopy. *J. Appl. Meteorol.* **4**, 517–522.

Cionco R.M. (1978). Analysis of canopy index values for various canopy densities. *Bound. Layer Meteorol.* **15**, 81–93.

Garratt J.R. (1994). *The Atmospheric Boundary Layer*. Cambridge University Press, Cambridge, UK.

Grimmond C.S.B. and Oke T.R. (1999). Aerodynamic properties of urban areas derived from analysis of surface form. *J. Appl. Meteorol.* **38**, 1262–1292.

Hanna S.R. and Britter R.E. (2002). *Wind Flow and Vapor Cloud Dispersion at Industrial and Urban Sites*. American Institute of Chemical Engineers, Center for Chemical Process Safety, New York.

Inoue E. (1963). On the turbulent structure of air flow within crop canopies. *J. Meteorol. Soc. Japan* **41**, 317–326.

Liss P.S. (1973). Processes of gas exchange across an air water interface. *Deep Sea Res.* **20**, 221–238.

Macdonald R.W. (2000). Modelling the mean velocity profile in the urban canopy layer. *Bound. Layer Meteorol.* **97**, 25–45.

Möller U. and Schumann G. (1970). Mechanisms of transport from the atmosphere to the earth's surface. *J. Geophys. Res.* **75**, 3013–3019.

Peña A. and Gryning S.E. (2008). Charnock's roughness length model and nondimensional wind profiles over the sea. *Bound. Layer Meteorol.* **128**, 191–203.

Schwarzenbach R.P., Gschwend P.M., and Imboden D.M. (1993). *Environmental Organic Chemistry*. Wiley, New York.

Shepherd J.G. (1974). Measurements of the direct deposition of sulphur dioxide onto grass and water by the profile method. *Atmos. Environ.* **8**, 69–74.

Slinn S.A. and Slinn W.G.N. (1980). Predictions for particle deposition on natural waters. *Atmos. Environ.* **14**, 1013–1016.

Slinn W.G.N., Hasse L., Hicks B.B., Hogan A.W., Lai D., Liss P.S., Munnich K.O., Sehmel G.A., and Vittori O. (1978). Some aspects of the transfer of atmospheric trace constituents past the air-sea interface. *Atmos. Environ.* **12**, 2055–2087.

Stull R.B. (1988). *An Introduction to Boundary Layer Meteorology*. Kluwer Academic, Dordrecht, The Netherlands.

von Kármán T. (1930). Mechanische Änlichkeit and Turbulenz. *Nachr. Ges. Wiss. Goettingen Fachgr. I Math.* **5**, 58–76.

Wang W. (2012). An analytical model for mean wind profiles in sparse canopies. *Bound. Layer Meteorol.* **142**, 383–399.

Wilson J.D., Thurtell G.W., and Kidd G.E. (1981). Numerical simulation of particle trajectories in inhomogeneous turbulence, III: Comparison of predictions with experimental data for the atmospheric surface layer. *Bound. Layer Meteorol.* **21**, 443–463.

Zanoun E.S., Durst F., and Nagib H. (2003). Evaluating the law of the wall in two dimensional fully developed turbulent channel flows. *Phys. Fluids* **15**, 3079–3089.

Zhang L., Gao S, Padro J., and Barrie L. (2001). A size aggregated dry deposition scheme for an atmospheric aerosol module. *Atmos. Environ.* **35**, 549–560.

# INDEX

The page numbers in **bold** refer to the section of the book that provides key information about the subject.



*Air Dispersion Modeling: Foundations and Applications*, First Edition. Alex De Visscher.

Printed and bound by CPI Group (UK) Ltd, Croydon, CR0 4YY

16/04/2025

14658530-0005